Springer Series in Computational Mathematics

29

Springer
Berlin
Heidelberg
New York
Barcelona
Hong Kong
London
Milan
Paris
Singapore
Tokyo

Pieter Wesseling

Principles
of Computational
Fluid Dynamics

With 150 Figures

 Springer

Pieter Wesseling
Faculty of Information Technology and Systems
Delft University of Technology
Mekelweg 4
2628 Delft, The Netherlands
e-mail: p.wesseling@its.tudelft.nl
WWW: ta.twi.tudelft.nl/nw/users/wesseling

Library of Congress Cataloging-in-Publication Data

Wesseling, Pieter, Dr.
Principles of computational fluid dynamics / Pieter Wesseling.
p. cm. – (Springer series in computational mathematics, ISSN 0179-3632; 29)
Includes bibliographical references and index.
ISBN 3540678530 (alk. paper)
1. Fluid dynamics – Data processing. I. Title. II. Series.
QA911 .W35 2000   532'.05'0285 – dc21
00-046339

Mathematics Subject Classification (1991): 76M, 65M

ISSN 0179-3632
ISBN 3-540-67853-0 Springer-Verlag Berlin Heidelberg New York

Springer-Verlag Berlin Heidelberg New York
a member of Springer Science+Business Media

© Springer-Verlag Berlin Heidelberg 2001
Printed in Germany

Typesetting: By the author using a Springer TEX macro package
Cover design: *design & production* GmbH, Heidelberg
Printed on acid-free paper   SPIN 10987797   46/3111XT – 5 4 3 2

Preface

The technological value of computational fluid dynamics has become undisputed. A capability has been established to compute flows that can be investigated experimentally only at reduced Reynolds numbers, or at greater cost, or not at all, such as the flow around a space vehicle at re-entry, or a loss-of-coolant accident in a nuclear reactor. Furthermore, modern computational fluid dynamics has become indispensable for design optimization, because many different configurations can be investigated at acceptable cost and in short time. A distinguishing feature of the present state of computational fluid dynamics is, that large commercial computational fluid dynamics computer codes have arisen, and found widespread use in industry. The days that a great majority of code users were also code developers are gone. This attests to the importance and a certain degree of maturity of computational fluid dynamics as an engineering tool. At the same time, this creates a need to go back to basics, and to disseminate the basic principles to a wider audience. It has been observed on numerous occasions, that even simple flows are not correctly predicted by advanced computational fluid dynamics codes, if used without sufficient insight in both the numerics and the physics involved. The present book aims to elucidate the principles of computational fluid dynamics. With a variation on Lamb's preface to his classic Hydrodynamics, owing to the elaborate nature of some of the methods of computational fluid dynamics, it has not always been possible to fit an adequate account of them into the frame of this book.

When technology progresses from the pre-competitive to the competitive stage, unavoidably, something like an information stop sets in. To protect investments, and because of the relatively long learning curve to be traversed in order to become familiar with a large computer code, a certain sluggishness of change makes itself felt. These consequences of the widespread distribution of large computational fluid dynamics codes needs to be counteracted by the dynamics of unencumbered scientific enquiry, not to pursue change for change's sake, but because much improvement seems feasible. Therefore I hope the book will be helpful not only to users of computational dynamics codes, but also to researchers in the field.

The book has grown out of graduate courses for doctoral students and practicing engineers, held under the auspices of the J.M. Burgers Center, the national inter-university graduate school for fluid dynamics in The Netherlands. I expect teachers of advanced courses of computational fluid dynamics courses will find this a useful book. For an introductory course the book seems too advanced, but I have found selected material from the manuscript useful in teaching an introductory undergraduate course.

Other relatively recent introductions to the subject of computational fluid dynamics that the reader will find useful are Ferziger and Perić (1996), Fletcher (1988), Hirsch (1988), Hirsch (1990), Peyret and Taylor (1985), Roache (1998a), Shyy (1994), Sod (1985), Tannehill, Anderson, and Pletcher (1997), Versteeg and Malalasekera (1995), Wendt (1996). The two volumes by Hirsch give an especially wide coverage. The present book differs from these works in the following respects. More mathematical and numerical analysis is given, but the mathematical background of the reader is assumed not to go beyond what physicists and engineers are generally familiar with. The maximum principle for differential equations and numerical schemes gets generous attention, in order to put discussions of spurious 'wiggles', accuracy of schemes on nonuniform grids, and accuracy of numerical boundary conditions on a firm footing. Singular perturbation theory is introduced to predict qualitative features of the flow, to which numerical methods can be adapted for better accuracy and efficiency. In particular, singular perturbation theory is used with a fair amount of rigor to demonstrate convincingly how it is possible to achieve accuracy and computing cost uniform in the Reynolds number, showing that a 'numerical windtunnel' that operates at arbitrarily high Reynolds number is feasible, notwithstanding the effect of 'numerical viscosity'. Much attention is given to the principles and the application of von Neumann stability analysis, giving useful stability conditions, some of them new, for many schemes used in practice. Godunov's order barrier and how to overcome it by slope-limited schemes is discussed extensively. The theory of scalar conservation laws including the nonconvex case is treated. Distributive iteration is used as a unifying framework for describing iterative methods for the incompressible Navier-Stokes equations. The principles of Krylov subspace and multigrid methods for efficient solution of the large sparse algebraic systems that arise are introduced. Much attention is given to the complications brought about by geometric complexity of the flow domain, including an introduction to tensor analysis. A chapter on unified methods to compute incompressible and compressible flows is included. In order to help the reader along who wants to delve deeper and to quickly reach the current research frontier, references to more advanced literature are provided.

Errata and MATLAB software related to a number of examples discussed in the book my be obtained via the author's website, to be found at
`ta.twi.tudelft.nl/nw/users/wesseling`

Combining the writing of a textbook of this size with the daily tasks of a university professor was not always easy, and would have been impossible without the support of the numerical team, and in particular our secretary Tatiana Tijanova. Her dedication, love of perfection and capability to cope with repeated stress were of vital importance for keeping the manuscript organized, and finally bringing it into publishable form. I am indebted to dr. C. Vuik for advice on Chap. 7, to professor G.S. Stelling for checking up on Chap. 8, and to professor F.T.M. Nieuwstadt for casting a critical eye on what I wrote about turbulence. The enthusiasm of the students in the graduate courses on computational fluid dynamics of the J.M. Burgers Center, and the cooperation with my fellow teacher professor A.E.P. Veldman, were inspiring and stimulating. The moral support of my wife Tineke was and remains invaluable.

Delft, June 2000   P. Wesseling

Table of Contents

1. The basic equations of fluid dynamics

1.1 Introduction

Fluid dynamics is a classic discipline. The physical principles governing the
flow of simple fluids and gases, such as water and air, have been understood
since the times of Newton. Sect. IX of the second book of Newton's Principia
starts with what came to be known as the Newtonian stress hypothesis :
"The resistance arising from the want of lubricity in the parts of a fluid, is,
other things being equal, proportional to the velocity with which the parts of
the fluid are separated from another". This hypothesis is followed by Propo-
sition LI, in which the flow generated by a rotating cylinder in an unbounded
medium is considered. The period of the orbit of a fluid particle is found to
be proportional to the distance r from the cylinder axis. This is not correct.
The source of the error is that the master balances force instead of torque;
this may be of some consolation to beginning students who find mechanics difficult. The closing remark "All of this can be tested in deep stagnant
water" must be taken with a grain of salt. Newton was more interested in
celestial mechanics than in fluid dynamics. His aim was to test Descartes's
vortex theory of planetary motion, which would gain credibility if the period of the orbit of a particle in this flow would be proportional to $r^{3/2}$; in
fact, it is proportional to r^2. The mathematical formulation of the laws that
govern the dynamics of fluids has been complete for a century and a half.
In the nineteenth century and the beginning of the twentieth, eminent scientists and engineers were drawn to the subject, and gave it clarity, unification
and elegance, as exemplified in the classic work of Lamb (1945), that first
appeared in 1879. In the preface to the 1932 edition Lamb writes, that the
subject has in recent years received considerable developments, classic fluid
dynamics having a widening field of practical applications. This has remained
true ever since, especially because in the last forty years or so classic fluid
dynamics finds itself in the company of computational fluid dynamics. This
new discipline still lacks the elegance and unification of its classic counter-
part, and is in a state of rapid development, so that we can do no more than
give a glimpse of its current status. But first, we take a look at classic fluid
dynamics.

Continuum hypothesis

The dynamics of fluids is governed by the conservation laws of classical physics, namely conservation of mass, momentum and energy. From these laws partial differential equations are derived and, under appropriate circumstances, simplified. It is customary to formulate the conservation laws under the assumption that the fluid is a continuous medium (*continuum hypothesis*). Physical properties of the flow, such as density and velocity can then be described as time-dependent scalar or vector fields on $\mathbb{R}^3$, for example $\rho(t, \boldsymbol{x})$ and $\boldsymbol{u}(t, \boldsymbol{x})$.

For the flowing medium we restrict ourselves here to gases and liquids. More general media, such as mixtures of gases and liquids (*multiphase flows*), will not be considered. For a liquid the continuum hypothesis is always satisfied in practice. A gas satisfies the continuum hypothesis (to a sufficient degree) if K $\ll$ 1, with K the *Knudsen number*, defined as K $= \lambda/L$, with λ the mean free path and L the length scale of the flow phenomenon under study. Consider for example flow over a flat plate with free-stream velocity V (Fig. 1.1). It is known that at the plate a boundary layer is generated, in

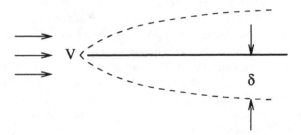

Fig. 1.1. Flow over a flat plate

which the velocity changes from zero to V. The thickness δ of this boundary layer is the relevant length scale. If the fluid is air at room temperature then $\lambda = 0.4$ μm. With $V = 1$ m/s, experiment and boundary layer theory tell us that $\delta \cong 2.5$ cm at 0.5 m downstream of the leading edge, so that here K $= 0.16 * 10^{-4}$. Hence, at this location momentum exchange due to friction takes place over a length scale of about 60,000 mean free paths, and the continuum hypothesis is very well satisfied. Perhaps this would not be so quite near the leading edge of an extremely sharp flat plate, but that need not concern us here. We will throughout assume the continuum hypothesis, and the flowing medium, be it gas or liquid, will often be called the fluid. The most common situation in technology where the continuum hypothesis has to be abandoned is the flow of very rarefied gases. Such a flow regime occurs at a certain stage of atmospheric re-entry of space vehicles. Unexpectedly perhaps, for flows of the interstellar medium often K $\ll$ 1, because of the size of the galactic length scale, so that the continuum hypothesis can be safely applied there.

Lagrangean and Eulerian formulation

The continuum hypothesis enables us to speak of the properties of a flow at a point in space, and of the physical properties of an infinitesimally small volume element of the fluid, to which we shall refer for brevity as a *material particle* . A flow can be described exhaustively by specification of the physical properties of each material particle as a function of time. This kind of specification of a flow is called the *Lagrangean formulation*. Alternatively, a flow may be described by specification of the time history of the flow properties at every fixed point of the domain. This is called the *Eulerian formulation*. The second formulation is usually more accessible for analysis and computation than the first, but sometimes the Lagrangean formulation may be preferable, for example when fluid interfaces have to be tracked. In most cases we ask for flow properties at fixed locations, such as the pressure at a wall, and this information is provided directly by the Eulerian formulation, to which we will adhere throughout this book. The Eulerian and Lagrangean points of view meet in the transport theorem, to be discussed in Sect. 1.3.

Selection of topics

The reader is assumed to be familiar with the principles of fluid dynamics, of vector analysis, of numerical linear algebra and of the numerical analysis of partial differential equations.

Fluid dynamics is a vast discipline, utilizing many different mathematical models. As the Mach number M (to be introduced later) varies, we encounter incompressible ($M \ll 1$), subsonic ($0 < M < 1$), transonic ($M \cong 1$), supersonic ($M > 1$) and hypersonic ($M \gg 1$) flow. In hypersonic flow, chemical processes taking place in the fluid have to be accounted for, giving rise to the discipline of aerothermochemistry. Multiphase flows play a large role in chemical engineering and reservoir engineering. Flows in porous media are governed by the Darcy equations. In hydraulic engineering the shallow-water equations are predominant. In ship hydrodynamics the free surface (water-air interface) often has to be accounted for. Capillary forces may be important. As the Reynolds number (to be introduced shortly) increases, transition from laminar to turbulent flow occurs, giving rise to a plethora of more or less semi-empirical turbulence models. Rotation causes special effects, important in oceanography, and in atmospheric and planetary fluid dynamics. We have made a selection of topics. The book is focussed on the incompressible and compressible Navier-Stokes equations, restricting ourselves to $M \lesssim 2$, thus catering mainly to the needs of industrial and environmental fluid dynamics and aeronautics. We will also pay attention to the shallow-water equations. Fortunately, many of the underlying principles carry over to cases not treated. In particular, a thorough understanding of the analytical and computational aspects of the comparatively simple convection-diffusion equation gives valuable insight in more complex models. Therefore this equation will receive much attention.

Although most practical flows are turbulent, we restrict ourselves here to laminar flow, because this book is on numerics only. Turbulence modeling is a vast subject in itself, that is briefly discussed in Sect. 1.13, where pointers to the literature are given for further study. The numerical principles uncovered for the laminar case carry over to the turbulent case. To facilitate this, viscosity is usually assumed variable.

Fluid dynamics is governed by partial differential equations. These may be solved numerically by finite difference, finite volume, finite element and spectral methods. In engineering applications, finite difference and finite volume methods are predominant. In order to limit the scope of this work, we will confine ourselves to finite difference and finite volume methods.

Since in computational fluid dynamics mathematical modeling aspects invariably play an important role, we devote the remainder of this chapter to a thorough derivation of the basic equations of fluid dynamics and their main simplifications. Of course, the subject cannot be adequately reviewed in a single chapter. For a more extensive treatment, see Batchelor (1967), Chorin and Marsden (1979), Kreiss and Lorenz (1989), Lamb (1945), Landau and Lifshitz (1959), Sedov (1971), Zucrow and Hoffman (1976), Zucrow and Hoffman (1977). A brief introduction to the history of the subject, with references to further literature, is given by Eberle, Rizzi, and Hirschel (1992).

Good starting points for exploration of the Internet for material related to computational fluid dynamics are the following websites:

    `www.cfd-online.com/`
    `www.princeton.edu/~gasdyn/fluids.html`

and the ERCOFTAC (European Research Community on Flow, Turbulence and Combustion) site:

    `imhefwww.epfl.ch/ERCOFTAC/`

Readers well-versed in fluid dynamics may skip the remainder of this chapter, perhaps after taking note of the notation introduced in the next section. But those less familiar with this discipline will find it useful to continue with the present chapter.

1.2 Vector analysis

Cartesian tensor notation

The basic equations will be derived in a right-handed Cartesian coordinate system $(x_1, x_2, ..., x_d)$ with d the number of space dimensions. Boldfaced lower case Latin letters denote vectors, for example, $\boldsymbol{x} = (x_1, x_2, ..., x_d)$. Greek letters denote scalars. In *Cartesian tensor notation*, which we shall often use, differentiation is denoted as follows:

$$\phi_{,\alpha} = \partial\phi/\partial x_\alpha \, .$$

Greek subscripts refer to coordinate directions, and the *summation convention* is used: summation takes place over Greek indices that occur twice in a term or product, for example:

$$u_\alpha v_\alpha = \sum_{\alpha=1}^{d} u_\alpha v_\alpha \, , \quad \phi_{,\alpha\alpha} = \sum_{\alpha=1}^{d} \partial^2\phi/\partial x_\alpha^2 \, .$$

We will also use *vector notation*, instead of the *subscript notation* just explained, and may write div$\boldsymbol{u}$, if this is more elegant or convenient than the tensor equivalent $u_{\alpha,\alpha}$; and sometimes we write grad ϕ for the vector $(\phi_{,1}, \phi_{,2}, \phi_{,3})$.

Divergence theorem

We need the following fundamental theorem:

Theorem 1.2.1. *For any volume $V \subset \mathbb{R}^d$ with piecewise smooth closed surface S and any differentiable scalar field ϕ we have*

$$\int_V \phi_{,\alpha} dV = \int_S \phi n_\alpha dS \, ,$$

where $\boldsymbol{n}$ is the outward unit normal on S.

For a proof, see for example Aris (1962).

A direct consequence of this theorem is:

Theorem 1.2.2. *(Divergence theorem).*
For any volume $V \subset \mathbb{R}^d$ with piecewise smooth closed surface S and any differentiable vector field $\boldsymbol{u}$ we have

$$\int_V \mathrm{div}\boldsymbol{u} dV = \int_S \boldsymbol{u} \cdot \boldsymbol{n} dS \, ,$$

where $\boldsymbol{n}$ is the outward unit normal on S.

Proof. Apply Theorem 1.2.1 with $\phi_{,\alpha} = u_\alpha$, $\alpha = 1, 2, ..., d$ successively and add.   □

A vector field satisfying $\text{div}\,\boldsymbol{u} = 0$ is called *solenoidal*.

Stokes's theorem

The *curl* of a vector field is defined by

$$\text{curl}\,\boldsymbol{u} = \begin{pmatrix} u_{3,2} - u_{2,3} \\ u_{1,3} - u_{3,1} \\ u_{2,1} - u_{1,2} \end{pmatrix} .$$

That is, the x_1-component of the vector $\text{curl}\,\boldsymbol{u}$ is $u_{3,2} - u_{2,3}$, etc. Often, the curl is called rotation, and a vector field satisfying $\text{curl}\,\boldsymbol{u} = 0$ is called *irrotational*.

Theorem 1.2.3. *(Stokes's theorem)*
Let C be a closed curve and let S be any piecewise smooth surface bounded by C. Then for every differentiable vector field $\boldsymbol{u}$ we have

$$\oint_C \boldsymbol{u} \cdot d\boldsymbol{x} = \iint_S (\text{curl}\,\boldsymbol{u}) \cdot \boldsymbol{n}\,dS \, ,$$

with $d\boldsymbol{x}$ the curve element along C and $\boldsymbol{n}$ the unit normal to S, right-handed with respect to the direction of traversing C.

For a proof, see for example Aris (1962).

Potential flow

Using Stokes's theorem it can be shown (cf. Aris (1962)) that if a vector field $\boldsymbol{u}$ satisfies $\text{curl}\,\boldsymbol{u} = 0$ there exists a scalar field φ such that

$$\boldsymbol{u} = \text{grad}\,\varphi \tag{1.1}$$

(or $u_\alpha = \varphi_{,\alpha}$). The scalar φ is called the *potential*, and flows with velocity field $\boldsymbol{u}$ satisfying (1.1) are called potential flows or *irrotational* flows (since $\text{curl}\,\text{grad}\,\varphi = 0$).

Helmholtz and Clebsch representations

Useful representations of vector fields are given by the following theorems.

Theorem 1.2.4. *(Helmholtz representation)*
For any bounded continuous vector field $\boldsymbol{u}$ *which vanishes at infinity there exist a scalar field* φ *and a solenoidal vector field* $\boldsymbol{b}$ *such that*

$$\boldsymbol{u} = \operatorname{grad}\varphi + \operatorname{curl}\boldsymbol{b}\,.$$

Theorem 1.2.5. *(Clebsch representation)*
For any bounded continuous vector field $\boldsymbol{u}$ *which vanishes at infinity there exist scalar fields* ϕ, ψ *and* θ *such that*

$$\boldsymbol{u} = \operatorname{grad}\varphi + \theta\operatorname{grad}\psi\,. \tag{1.2}$$

For proofs, see Aris (1962). The scalars in (1.2) are often called Clebsch or Monge potentials.

The volume element

The *outer product* of two vectors $\boldsymbol{u}$ and $\boldsymbol{v}$ is defined by

$$\boldsymbol{u} \times \boldsymbol{v} = (u_2 v_3 - u_3 v_2)\boldsymbol{e}_{(1)} + (u_3 v_1 - u_1 v_3)\boldsymbol{e}_{(2)} + (u_1 v_2 - u_2 v_1)\boldsymbol{e}_{(3)}\,,$$

with $\boldsymbol{e}_{(\alpha)}$ the unit vector in the x_α-direction. Note that $\boldsymbol{u} \times \boldsymbol{v} = -\boldsymbol{v} \times \boldsymbol{u}$.

The *permutation symbol* $\varepsilon_{\alpha\beta\gamma}$ is defined as follows:

$$\varepsilon_{\alpha\beta\gamma} = \begin{cases} 0, & \text{if any two of } \alpha, \beta, \gamma \text{ are the same}\,, \\ 1, & \text{if } \alpha\beta\gamma \text{ is an even permutation of } 123\,, \\ -1, & \text{if } \alpha\beta\gamma \text{ is an odd permutation of } 123\,. \end{cases}$$

An even permutation is in order of increasing magnitude, with 1 coming after 3, for example 231; the odd case is the reverse, such as 213. Using the permutation symbol, the outer product can be written in subscript notation as

$$\boldsymbol{u} \times \boldsymbol{v} = \varepsilon_{\alpha\beta\gamma} u_\alpha v_\beta \boldsymbol{e}_{(\gamma)}\,.$$

The volume V of the parallelepiped spanned by the vectors $\boldsymbol{u}, \boldsymbol{v}$ and $\boldsymbol{w}$ (taken in right-handed order) is given by (the proof is left as Exercise 1.2.3)

$$V = \boldsymbol{u} \cdot (\boldsymbol{v} \times \boldsymbol{w}) = \varepsilon_{\alpha\beta\gamma} u_\alpha v_\beta w_\gamma\,. \tag{1.3}$$

Let (y_1, y_2, y_3) be an arbitrary right-handed curvilinear coordinate system, related to the Cartesian coordinates (x_1, x_2, x_3) by

$$x_\alpha = x_\alpha(y_1, y_2, y_3)\,.$$

The incremental change in position vector associated with an incremental change dy_α in y_α is

$$d\boldsymbol{x}^{(\alpha)} = \frac{\partial \boldsymbol{x}}{\partial y_\alpha} dy_\alpha \quad \text{(no summation)} . \tag{1.4}$$

From (1.3) and (1.4) it follows that the volume element is given by

$$dV = d\boldsymbol{x}^{(1)} \cdot (d\boldsymbol{x}^{(2)} \times d\boldsymbol{x}^{(3)}) =$$

$$= \varepsilon_{\alpha\beta\gamma} \frac{\partial x_\alpha}{\partial y_1} dy_1 \frac{\partial x_\beta}{\partial y_2} dy_2 \frac{\partial x_\gamma}{\partial y_3} dy_3 = J dy_1 dy_2 dy_3 , \tag{1.5}$$

where

$$J = \varepsilon_{\alpha\beta\gamma} \frac{\partial x_\alpha}{\partial y_1} \frac{\partial x_\beta}{\partial y_2} \frac{\partial x_\gamma}{\partial y_3} . \tag{1.6}$$

By inspection it is easily verified that the determinant of a 3×3 matrix A with elements $a_{\alpha\beta}$ is given by

$$\det(A) = \varepsilon_{\alpha\beta\gamma} a_{\alpha 1} a_{\beta 2} a_{\gamma 3} .$$

Hence, J in (1.6) is the determinant of a matrix A with elements

$$a_{\alpha\beta} = \frac{\partial x_\alpha}{\partial y_\beta} ,$$

so that A is recognized as the *Jacobian matrix* of the mapping (1.4); the quantity J is called the *Jacobian*.

Two dimensions

Two-dimensional versions of the above results are easily obtained by putting the third component and/or $\partial/\partial x_3$ equal to zero. For example, in two dimensions,

$$\text{curl}\, \boldsymbol{u} = u_{2,1} - u_{1,2} .$$

Often, $\omega = u_{2,1} - u_{1,2}$ can be safely handled as a scalar in a two-dimensional context, but in fact curl$\boldsymbol{u}$ is a vector with only the x_3-component not zero.

Exercise 1.2.1. Prove Theorem 1.2.1 for the special case that V is the unit cube.

Exercise 1.2.2. Show that curl$\boldsymbol{u}$ is solenoidal.

Exercise 1.2.3. Show that $c = \boldsymbol{a} \times \boldsymbol{b}$ is perpendicular to $\boldsymbol{a}$ and $\boldsymbol{b}$ in right-handed direction, and that the magnitude $|c| = |\boldsymbol{a}| \cdot |\boldsymbol{b}| \cos \varpi$, with ϖ the angle between $\boldsymbol{a}$ and $\boldsymbol{b}$. Prove (1.3).

1.3 The total derivative and the transport theorem

Streamlines, particle paths and streaklines

A *streamline* is a curve that is everywhere tangent to the velocity vector $u(t, x)$ at a given time t. Hence, a streamline may be parametrized with a parameter s such that a streamline is a curve defined by

$$dx/ds = u(t, x) .$$

Let $x(t, y)$ be the position of a material particle at time $t > 0$, that at time $t = 0$ had initial position y. The *particle path* that has been traced out by the particle is the curve $x(s, y)$, $0 \leq s \leq t$. The particle path is related to the velocity field by

$$\partial x(s, y)/\partial s = u(s, x) , \quad 0 \leq s \leq t .$$

A *streakline* is defined as the curve on which lie all those particles that at some earlier point of time passed through a certain point of space, say y. This is the curve traced out by releasing a continuous stream of dye in the flow at location y.

In general, streamlines, particle paths and streaklines are all different. But when the flow is stationary, i.e. the velocity field is time-independent, they coincide.

The total derivative

Obviously, the velocity field $u(t, x)$ of the flow satisfies

$$u(t, x) = \frac{\partial x(t, y)}{\partial t} . \tag{1.7}$$

The time-derivative of a property ϕ of a material particle, called a *material property* (for example its temperature), is denoted by $D\phi/Dt$. This is called the *total derivative*. All material particles have some ϕ, so ϕ is defined everywhere in the flow, and is a scalar field $\phi(t, x)$. We have

$$\frac{D\phi}{Dt} = \frac{\partial}{\partial t} \phi(t, x(t, y)) , \tag{1.8}$$

where the partial derivative has to be taken with y constant, since the total derivative tracks variation for a particular material particle. We obtain

$$\frac{D\phi}{Dt} = \frac{\partial \phi}{\partial t} + \frac{\partial x_\alpha(t, y)}{\partial t} \frac{\partial \phi}{\partial x_\alpha} .$$

By using (1.7) we get

$$\frac{D\phi}{Dt} = \frac{\partial \phi}{\partial t} + u_\alpha \phi_{,\alpha} .$$

The transport theorem

A *material volume* $V(t)$ is a volume of fluid that moves with the flow and consists permanently of the same material particles.

Theorem 1.3.1. *(Reynolds's transport theorem)*
For any material volume $V(t)$ and differentiable scalar field ϕ we have

$$\frac{d}{dt} \int_{V(t)} \phi dV = \int_{V(t)} (\frac{\partial \phi}{\partial t} + \text{div } \phi \boldsymbol{u}) dV . \qquad (1.9)$$

For the proof we need the following lemma.

Lemma 1.3.1. *Let $\boldsymbol{x}(t, \boldsymbol{y})$ describe the trajectory of a material particle, and let $J = \det(a_{\alpha\beta})$, $a_{\alpha\beta} = \partial x_\alpha / \partial y_\beta$. Then*

$$\frac{\partial J}{\partial t} = J \text{div } \boldsymbol{u} . \qquad (1.10)$$

Proof. By expanding the determinant it is easy to see that

$$\frac{\partial J}{\partial t} = J_1 + J_2 + J_3 ,$$

with J_α obtained from J by replacing the elements in row number α by their time derivatives. We have

$$\frac{\partial a_{\alpha\beta}}{\partial t} = \frac{\partial}{\partial t} \frac{\partial x_\alpha}{\partial y_\beta} = \frac{\partial u_\alpha}{\partial y_\beta} = \frac{\partial u_\alpha}{\partial x_\gamma} \frac{\partial x_\gamma}{\partial y_\beta} = u_{\alpha,\gamma} a_{\gamma\beta} .$$

Hence

$$J_1 = \begin{vmatrix} u_{1,\gamma} a_{\gamma 1} & u_{1,\gamma} a_{\gamma 2} & u_{1,\gamma} a_{\gamma 3} \\ a_{21} & a_{22} & a_{23} \\ a_{31} & a_{32} & a_{33} \end{vmatrix} =$$

$$\begin{vmatrix} u_{1,1} & u_{1,2} & u_{1,3} \\ 0 & 1 & 0 \\ 0 & 0 & 1 \end{vmatrix} \begin{vmatrix} a_{11} & a_{12} & a_{13} \\ a_{21} & a_{22} & a_{23} \\ a_{31} & a_{32} & a_{33} \end{vmatrix} = u_{1,1} J .$$

Similarly, $J_2 = u_{2,2} J$, $J_3 = u_{3,3} J$, which completes the proof. □

Proof of Theorem 1.3.1:
The initial position $\boldsymbol{y}$ and its position $\boldsymbol{x}$ at time t of a material particle are related by $\boldsymbol{x}(t, \boldsymbol{y})$. This provides a mapping between $V(t)$ and $V(0)$, that we can use to transform the integral over $V(t)$ to an integral over $V(0)$. In y_α-coordinates, the volume element is given by (1.5), so that, using (1.8) and (1.10),

$$\frac{d}{dt}\int_{V(t)}\phi dV = \frac{d}{dt}\int_{V(0)}\phi J dy_1 dy_2 dy_3 = \int_{V(0)}(J\frac{\partial\phi}{\partial t} + \phi J \text{div} \boldsymbol{u})dy_1 dy_2 dy_3 .$$

Transforming back to x_α-coordinates, we obtain

$$\frac{d}{dt}\int_{V(t)}\phi dV = \int_{V(t)}(\frac{D\phi}{Dt} + \phi\text{div}\boldsymbol{u})dV = \int_{V(t)}(\frac{\partial\phi}{\partial t} + \text{div}\phi\boldsymbol{u})dV .$$

□

A somewhat similar theorem applies to the rate of change of *circulation* around a *material contour*. The circulation around a closed curve C is defined as

$$\Gamma(C) = \oint_C \boldsymbol{u} \cdot d\boldsymbol{x} ,$$

with $d\boldsymbol{x}$ a curve element. A material contour $C(t)$ is a simple (*i.e.* non-intersecting) closed curve that consists of material particles. We will prove

Theorem 1.3.2. *(Rate of change of circulation).*
For any material contour $C(t)$ and differentiable vector field $\boldsymbol{u}(t, \boldsymbol{x})$ we have

$$\frac{d}{dt}\oint_{C(t)} \boldsymbol{u} \cdot d\boldsymbol{x} = \oint_{C(t)} \frac{D\boldsymbol{u}}{Dt} \cdot d\boldsymbol{x} . \tag{1.11}$$

Proof. As in the proof of the preceding theorem, we transform the integral to an integral over the initial curve $C(0)$:

$$\frac{d}{dt}\oint_{C(t)} \boldsymbol{u} \cdot d\boldsymbol{x} = \frac{d}{dt}\oint_{C(0)} \boldsymbol{u}(t, \boldsymbol{x}(t, \boldsymbol{y})) \cdot \frac{\partial\boldsymbol{x}(t, \boldsymbol{y})}{\partial y_\alpha} dy_\alpha .$$

Since $\partial\boldsymbol{u}(t, \boldsymbol{x}(t, \boldsymbol{y}))/\partial t = D\boldsymbol{u}/Dt$ and $\partial^2\boldsymbol{x}(t, \boldsymbol{y})/\partial t\partial y_\alpha = \partial\boldsymbol{u}/\partial y_\alpha$, this can be written as

$$\frac{d}{dt}\oint_{C(t)} \boldsymbol{u} \cdot d\boldsymbol{x} = \oint_{C(0)} \frac{\partial\boldsymbol{u}}{\partial t} \cdot \frac{\partial\boldsymbol{x}}{\partial y_\alpha} dy_\alpha + \oint_{C(0)} \boldsymbol{u} \cdot \frac{\partial\boldsymbol{u}}{\partial y_\alpha} dy_\alpha = \oint_{C(t)} \frac{D\boldsymbol{u}}{Dt} \cdot d\boldsymbol{x} ,$$

because $\oint_{C(0)} \boldsymbol{u} \cdot \frac{\partial\boldsymbol{u}}{\partial y_\alpha} dy_\alpha = \frac{1}{2}\oint_{C(0)} \frac{\partial\boldsymbol{u}\cdot\boldsymbol{u}}{\partial y_\alpha} dy_\alpha = 0.$

□

We are now ready to formulate the conservation laws for mass, momentum and energy.

1.4 Conservation of mass

The continuity equation

The mass conservation law says that the rate of change of mass in an arbitrary material volume $V(t)$ equals the rate of mass production in $V(t)$. This can be expressed as

$$\frac{d}{dt} \int_{V(t)} \rho dV = \int_{V(t)} \sigma dV \;, \tag{1.12}$$

where $\rho(t, \boldsymbol{x})$ is the density of the material particle at time t and position $\boldsymbol{x}$, and $\sigma(t, \boldsymbol{x})$ is the rate of mass production per volume. In practice, $\sigma \neq 0$ only in multiphase flows, in which case (1.12) holds for each phase separately. We take $\sigma = 0$, and use the transport theorem to obtain

$$\int_{V(t)} (\frac{\partial \rho}{\partial t} + \mathrm{div} \rho \boldsymbol{u}) dV = 0 \;.$$

Since this holds for every $V(t)$ the integrand must be zero:

$$\frac{\partial \rho}{\partial t} + \mathrm{div} \rho \boldsymbol{u} = 0 \;. \tag{1.13}$$

This is the *mass conservation law*, also called the *continuity equation*.

Incompressible flow

An incompressible flow is a flow in which the density of each material particle remains the same during the motion:

$$\rho(t, \boldsymbol{x}(t, \boldsymbol{y})) = \rho(0, \boldsymbol{y}) \;. \tag{1.14}$$

Hence

$$\frac{D\rho}{Dt} = 0 \;.$$

Because

$$\mathrm{div} \rho \boldsymbol{u} = \rho \mathrm{div} \boldsymbol{u} + u_\alpha \rho_{,\alpha} \;,$$

it follows from the mass conservation law (1.13) that

$$\mathrm{div} \boldsymbol{u} = 0 \;. \tag{1.15}$$

This is the form that the mass conservation law takes for incompressible flow.

Sometimes incompressibility is erroneously taken to be a property of the fluid rather than of the flow. But as we shall see later, compressibility depends only on the speed of the flow. If the magnitude of the velocity of the flow is of the order of the speed of sound in the fluid ($\sim$ 340 m/s in air at sea level at 15°C, $\sim$ 1.4 km/s in water at 15°C, depending on the amount of dissolved air) the flow is compressible; if the velocity is much smaller than the speed of sound, incompressibility is a good approximation. It is true that in liquids flow velocities anywhere near the speed of sound cannot normally be reached, due to the enormous pressures involved and the phenomenon of *cavitation*. But we will now discuss other effects that cause density variations.

Incompressible flows with variable density

Incompressibility does not imply $\rho =$ constant. An example of incompressible flow with $\rho \neq$ constant is the flow of water with salt dissolved in it with concentration $c(\boldsymbol{x}, t)$. Neglecting diffusion, the amount of solute in a material particle is constant:

$$c(t, \boldsymbol{x}(t, \boldsymbol{y})) = c(0, \boldsymbol{y}) . \tag{1.16}$$

The density is a function of c only. Therefore (1.14) is satisfied, and the flow is incompressible according to our definition. If diffusion is taken into account, the concentration is governed by the law of conservation of solute, given by

$$\frac{Dc}{Dt} = \frac{1}{\rho}(kc_{,\alpha})_{,\alpha} + \frac{1}{\rho}q . \tag{1.17}$$

This equation will be derived in Sect. 1.11. Now (1.16), hence (1.14) is not satisfied. From equations (1.13) and (1.17) it follows that the mass conservation law becomes

$$\operatorname{div}\boldsymbol{u} = -\frac{1}{\rho}\frac{D\rho}{Dt} = -\frac{1}{\rho}\frac{d\rho}{dc}\frac{Dc}{Dt} = -\frac{1}{\rho}\frac{d\ln\rho}{dc}\{(kc_{,\alpha})_{,\alpha} + q\} . \tag{1.18}$$

Since $\ln\rho$ is a known function of c we have effectively eliminated ρ, just as in the incompressible case.

1.5 Conservation of momentum

Body forces and surface forces

Newton's law of conservation of momentum implies that the rate of change of momentum of a material volume equals the total force on the volume. There are *body forces* and *surface forces*. A body force acts on a material particle, and is proportional to its mass. Let the volume of the material particle be $dV(t)$ and let its density be ρ. Then we can write

$$\text{body force} = \boldsymbol{f}^b \rho dV(t) \ . \tag{1.19}$$

A surface force works on the surface of $V(t)$ and is proportional to area. The surface force working on a surface element $dS(t)$ of $V(t)$ can be written as

$$\text{surface force} = \boldsymbol{f}^s dS(t) \ . \tag{1.20}$$

Conservation of momentum

The law of conservation of momentum applied to a material volume gives

$$\frac{d}{dt} \int_{V(t)} \rho u_\alpha dV = \int_{V(t)} f_\alpha^b dV + \int_{S(t)} f_\alpha^s dS \ . \tag{1.21}$$

By substituting $\phi = \rho u_\alpha$ in the transport theorem (1.9), this can be written as

$$\int_{V(t)} \{\frac{\partial \rho u_\alpha}{\partial t} + (\rho u_\alpha u_\beta)_{,\beta}\} dV = \int_{V(t)} \rho f_\alpha^b dV + \int_{S(t)} f_\alpha^s dS \ . \tag{1.22}$$

It may be shown (see Aris (1962)) there exist nine quantities $\tau_{\alpha\beta}$ such that

$$f_\alpha^s = \tau_{\alpha\beta} n_\beta \ , \tag{1.23}$$

where $\tau_{\alpha\beta}$ is the *stress tensor* (more on tensors in Chap. 11) and $\boldsymbol{n}$ is the outward unit normal on dS. By applying Theorem 1.2.1 with ϕ replaced by $\tau_{\alpha\beta}$ and n_α by n_β, equation (1.22) can be rewritten as

$$\int_{V(t)} \{\frac{\partial \rho u_\alpha}{\partial t} + (\rho u_\alpha u_\beta)_{,\beta}\} dV = \int_{V(t)} (\rho f_\alpha^b + \tau_{\alpha\beta,\beta}) dV \ .$$

Since this holds for every $V(t)$, we must have

$$\frac{\partial \rho u_\alpha}{\partial t} + (\rho u_\alpha u_\beta)_{,\beta} = \tau_{\alpha\beta,\beta} + \rho f_\alpha^b \ , \tag{1.24}$$

which is the momentum conservation law. The left-hand side is called the *inertia term*, because it comes from the inertia of the mass of fluid contained in $V(t)$ in equation (1.21).

An example where $\boldsymbol{f}^b \neq 0$ is stratified flow under the influence of gravity. This will be discussed in Sect. 1.14.

Constitutive relation

In order to complete the system of equations it is necessary to relate the stress tensor to the motion of the fluid. Such a relation is called a *constitutive relation*. A discussion of constitutive relations would lead us too far. The simplest constitutive relation is (see Batchelor (1967))

$$\tau_{\alpha\beta} = -p\delta_{\alpha\beta} + 2\mu(e_{\alpha\beta} - \frac{1}{3}\Delta\delta_{\alpha\beta}) , \tag{1.25}$$

where p is the pressure, $\delta_{\alpha\beta}$ is the Kronecker delta, μ is the dynamic viscosity coefficient, $e_{\alpha\beta}$ is the *rate of strain tensor*, defined by

$$e_{\alpha\beta} = \frac{1}{2}(u_{\alpha,\beta} + u_{\beta,\alpha}) ,$$

and

$$\Delta = e_{\alpha\alpha} = \text{div}\boldsymbol{u} .$$

In many fluids μ depends on temperature, but not on pressure. Fluids satisfying (1.25) are called *Newtonian fluids*. Examples are gases and liquids such as water and mercury. Examples of non-Newtonian fluids are polymers and blood. *Rheology* is the discipline that studies the motion of non-Newtonian fluids. We will not discuss rheological flows.

The Navier-Stokes equations

Substitution of (1.25) in (1.24) gives

$$\frac{\partial \rho u_\alpha}{\partial t} + (\rho u_\alpha u_\beta)_{,\beta} = -p_{,\alpha} + 2\{\mu(e_{\alpha\beta} - \frac{1}{3}\Delta\delta_{\alpha\beta})\}_{,\beta} + \rho f_\alpha^b . \tag{1.26}$$

These are the *Navier-Stokes equations*. Because of the continuity equation (1.13), one may also write

$$\rho\frac{Du_\alpha}{Dt} = -p_{,\alpha} + 2\{\mu(e_{\alpha\beta} - \frac{1}{3}\Delta\delta_{\alpha\beta})\}_{,\beta} + \rho f_\alpha^b . \tag{1.27}$$

Still another form may be obtained by using the identity (which may be verified by inspection)

$$u_\beta u_{\alpha,\beta} = \frac{1}{2}(u_\beta u_\beta)_{,\alpha} - (\boldsymbol{u} \times \boldsymbol{\omega})_\alpha , \tag{1.28}$$

where the *vorticity* $\boldsymbol{\omega}$ is defined by

$$\boldsymbol{\omega} = \text{curl}\boldsymbol{u} .$$

Substitution of (1.28) in (1.27) gives

$$\rho\frac{\partial u_\alpha}{\partial t} + \frac{1}{2}\rho(u_\beta u_\beta)_{,\alpha} - \rho(\boldsymbol{u} \times \boldsymbol{\omega})_\alpha = -p_{,\alpha} + 2\{\mu(e_{\alpha\beta} - \frac{1}{3}\Delta\delta_{\alpha\beta})\}_{,\beta} + \rho f_\alpha^b \,.$$

$$(1.29)$$

In incompressible flows $\Delta = 0$. If, furthermore, $\mu = $ constant then we can use $u_{\beta,\alpha\beta} = u_{\beta,\beta\alpha} = 0$ to obtain

$$\rho\frac{Du_\alpha}{Dt} = -p_{,\alpha} + \mu u_{\alpha,\beta\beta} + \rho f_\alpha^b \,.$$

$$(1.30)$$

This equation was first derived by Navier (1823), Poisson (1831), de Saint-Venant (1843) and Stokes (1845).

Making the equations dimensionless

In fluid dynamics there are exactly four independent physical units: those of length, velocity, mass and temperature, to be denoted by L, U, M and T_r, respectively. From these all other units can be and should be derived in order to avoid the introduction of superfluous coefficients in the equations. For instance, the appropriate unit of time is L/U; the unit of force F follows from Newton's law as MU^2/L. Often it is useful not to choose these units arbitrarily, but to derive them from the problem at hand, and to make the equations dimensionless. This leads to the identification of the dimensionless parameters that govern a flow problem. An example follows.

The Reynolds number

Let L and U be typical length and velocity scales for a given flow problem, and take these as units of length and velocity. The unit of mass is chosen as $M = \rho_r L^3$ with ρ_r a suitable value for the density, for example the density in the flow at upstream infinity, or the density of the fluid at rest. Dimensionless variables are denoted by a prime:

$$\boldsymbol{x}' = \boldsymbol{x}/L, \quad \boldsymbol{u}' = \boldsymbol{u}/U, \quad \rho' = \rho/\rho_r \,.$$

$$(1.31)$$

In dimensionless variables, equation (1.26) takes the following form:

$$\frac{L}{U}\frac{\partial \rho' u_\alpha'}{\partial t} + (\rho' u_\alpha' u_\beta')_{,\beta} = -\frac{1}{\rho_r U^2}p_{,\alpha} + \frac{2}{\rho_r U L}\{\mu(e_{\alpha\beta}' - \frac{1}{3}\Delta'\delta_{\alpha\beta})\}_{,\beta} + \frac{L}{U^2}\rho' f_\alpha^b \,,$$

$$(1.32)$$

where now the subscript $,\alpha$ stands for $\partial/\partial x_\alpha'$, and $e_{\alpha\beta}' = \frac{1}{2}(u_{\alpha,\beta}' + u_{\beta,\alpha}')$, $\Delta' = e_{\alpha\alpha}'$. We introduce further dimensionless quantities as follows:

$$t' = Ut/L, \quad p' = p/\rho_r U^2, \quad (f^b)' = \frac{L}{U^2} f^b. \tag{1.33}$$

By substitution in (1.32) we obtain the following *dimensionless form* of the Navier-Stokes equations, deleting the primes:

$$\frac{\partial \rho u_\alpha}{\partial t} + (\rho u_\alpha u_\beta)_{,\beta} = -p_{,\alpha} + 2\{Re^{-1}(e_{\alpha\beta} - \frac{1}{3}\Delta\delta_{\alpha\beta})\}_{,\beta} + f_\alpha^b,$$

where the *Reynolds number* Re is defined by

$$Re = \frac{\rho_r U L}{\mu}.$$

The dimensionless form of (1.30) is

$$\frac{Du_\alpha}{Dt} = -p_{,\alpha} + Re^{-1} u_{\alpha,\beta\beta} + f_\alpha^b. \tag{1.34}$$

The transformation (1.31) shows that the inertia term is of order $\rho_r U^2/L$ and the viscous term is of order $\mu U/L^2$. Hence, Re is measure of the ratio of inertial and viscous forces in the flow. This can also be seen immediately from equation (1.34). For Re $\gg$ 1 inertia dominates, for Re $\ll$ 1 friction (the viscous term) dominates. Both are balanced by the pressure gradient.

In the case of constant density, equations (1.34) and (1.15) form a complete system of four equations with four unknowns. The solution depends on the single dimensionless parameter Re only. What values does Re have in practice? At a temperature of 15°C and atmospheric pressure, for air we have $\mu/\rho = 1.5 * 10^{-5}$ m^2/s, whereas for water $\mu/\rho = 1.1 * 10^{-6}$ m^2/s. In the International Civil Aviation Organization Standard Atmosphere, $\mu/\rho = 4.9 * 10^{-5}$ m^2/s at an altitude of 12.5 km. This gives for the flow over an aircraft wing in cruise condition at 12.5 km with wingcord $L = 3$ m and $U = 900$ km/h: Re $= 1.5 * 10^7$. In a windtunnel experiment at sea level with $L = 0.5$ m and $U = 25$ m/s we obtain Re $= 8.3 * 10^5$. For landing aircraft at sea-level with $L = 3$ m and $U = 220$ km/h we obtain Re $= 1.2 * 10^7$. For a house in a light wind with $L = 10$ m and $U = 0.5$ m/s we have Re $= 3.3*10^5$. Air circulation in a room with $L = 4$ m and $U = 0.1$ m/s gives Re $= 2.7*10^4$. A large ship with $L = 200$ m and $U = 7$ m/s gives Re $= 1.3 * 10^8$, whereas a yacht with $L = 7$ m and $U = 3$ m/s has Re $= 1.9 * 10^7$. A small fish with $L = 0.1$ m and $U = 0.2$ m/s has Re $= 1.8 * 10^4$.

All these examples have in common that Re $\gg$ 1, which is indeed almost the rule in flows of industrial and environmental interest. One might think that flows around a given shape will be quite similar for different values of Re, as long as Re $\gg$ 1, but nothing is farther from the truth. At Re $= 10^7$ a flow may be significantly different from the flow at Re $= 10^5$, in the same geometry. This strong dependence on Re complicates predictions based on scaled down experiments. Therefore computational fluid dynamics plays an important role in extrapolation to full scale. The rich variety of solutions of (1.34) that evolves as Re → ∞ is one of the most surprising and intesting features of fluid dynamics, with important consequences for technological applications. A 'route to chaos' develops as Re → ∞, resulting in *turbulence*. Intricate and intriguing flow patterns occur, accurately rendered in masterful drawings by Leonardo da Vinci, and photographically recorded in Hinze (1975), Nakayama and Woods (1988), Van Dyke (1982) and Hirsch (1988). We return briefly to the subject of turbulence in Sect. 1.13.

The complexity of flows used to be thought surprising, since the physics underlying the governing equations is simply conservation of mass and momentum. Since about 1960, however, it is known that the sweeping generalizations about determinism of Newtonian mechanics made by many scientists (notably Laplace) in the nineteenth century were wrong. Even simple classic nonlinear dynamical systems often exibit a complicated seemingly random behavior, with such a sensitivity to initial conditions, that their long-term behavior cannot be predicted in detail. For a discussion of the modern view on (un-)predictability in Newtonian mechanics, see Lighthill (1986).

The Stokes equations

For Re ↓ 0 the system (1.34) simplifies to the *Stokes equations*. If we multiply (1.34) by Re and let Re ↓ 0, the pressure drops out, which cannot be correct, since we would have four equations (Stokes and mass conservation) for three unknowns u_α. It follows that $p = \mathcal{O}(\mathrm{Re}^{-1})$. We therefore substitute

$$p = \mathrm{Re}^{-1}p' . \tag{1.35}$$

From (1.35) and (1.33) it follows that the dimensional (physical) pressure is $\mu U p'/L$. Substitution of (1.35) in (1.34), multiplying by Re and letting Re ↓ 0 gives the Stokes equations:

$$u_{\alpha,\beta\beta} - p_{,\alpha} = 0 . \tag{1.36}$$

These linear equations together with (1.15) were solved by Stokes (1851) for flow around a sphere.

The Stokes paradox

One might think that (1.36) governs flows with Re ≪ 1 (very slow or very viscous flows), but Stokes (1851) shows that (1.36) does not have a solution for unbounded flow around an infinite cylinder, with uniform flow at infinity. Since such a flow seems physically acceptable, there must be something lacking in the mathematical model (1.36). In fact, *in two dimensions* the Stokes equations do not in general provide a good low Re approximation to the Navier-Stokes equations, that is uniformly valid everywhere in the flow. This surprising fact later came to be called the *Stokes paradox*. An important contribution to the understanding of the situation was made by Oseen (1910).

The Oseen equations

Suppose that the flow deviates only little from a uniform flow, as for example far from a body placed in an unbounded uniform flow. Let the flow be governed by (1.34). We write

$$u_1 = 1 + \varepsilon v_1, \quad u_\alpha = \varepsilon v_\alpha, \quad \alpha > 1, \quad \varepsilon \ll 1 .$$

This we substitute in (1.34) with $f^b = 0$. We introduce $p = \varepsilon p'$, take $\varepsilon \downarrow 0$, delete the prime and obtain the *Oseen equations*:

$$v_{\alpha,1} = -p_{,\alpha} + \mathrm{Re}^{-1} v_{\alpha,\beta\beta} ,$$

proposed by Oseen (1910).

The Oseen equations not only describe almost uniform flows, but also slow (Re $\ll 1$) flows (if one replaces v again by u), in any number of dimensions.

The study of the limit Re $\downarrow 0$ of the Navier-Stokes equation is far from trivial, and has given an important impetus to *singular perturbation theory*. See Van Dyke (1975) for more information about the asymptotic behavior of Navier-Stokes solutions as Re $\downarrow 0$. Singular perturbation theory will be introduced in Chap. 2, and will be found to provide useful a priori information about the nature of solutions to flow problems, also for Re $\gg 1$.

1.6 Conservation of energy

The first law of thermodynamics

The first law of thermodynamics says that work done on a closed system plus heat added to it equals the increase of the sum of *kinetic* and *internal energy* of the system. The kinetic energy per unit mass of a material particle is $\frac{1}{2} u_\alpha u_\alpha$. Its internal energy per unit mass is denoted by e. This is a *state variable*, like p and ρ. According to thermodynamics, for simple systems there are at most two independent state variables, on which the others depend algebraically. We will restrict ourselves to simple systems. In principle one may select any two of the state variables as independent variables.

The energy equation

By applying the first law of thermodynamics to a material volume $V(t)$ we find

$$\frac{d}{dt} \int_{V(t)} \rho E dV = W + Q \, , \qquad (1.37)$$

with E the *total energy* per unit mass, given by

$$E = e + \frac{1}{2} u_\alpha u_\alpha \, .$$

Furthermore, W is the rate of work expended by the surroundings on the fluid in $V(t)$, and Q is the rate of heat addition. The body force (1.19) does work at a rate $\boldsymbol{u} \cdot \boldsymbol{f}^b \rho dV(t)$, and the surface force (1.20) does work at a rate $\boldsymbol{u} \cdot \boldsymbol{f}^s dS(t)$, so that

$$W = \int_{V(t)} u_\alpha f_\alpha^b \rho dV + \int_{S(t)} u_\alpha f_\alpha^s dS \, .$$

Using (1.23) and Theorem 1.2.1 this can be written as

$$W = \int_{V(t)} \{\rho u_\alpha f_\alpha^b + (u_\alpha \tau_{\alpha\beta})_{,\beta}\} dV \, . \qquad (1.38)$$

Assuming that heat is added to each material particle at a rate q per unit of mass, and that there is a heat flux σ per unit of area through $S(t)$, we find

$$Q = \int_{V(t)} \rho q dV + \int_{S(t)} \sigma dS \, .$$

Let heat diffusion be governed by Fourier's law:

$$\sigma = k\boldsymbol{n} \cdot \text{grad} T \, ,$$

with k the *thermal conductivity*, and T the temperature, which is another state variable. Using the divergence theorem we find

$$Q = \int_{V(t)} \{\rho q + (kT_{,\alpha})_{,\alpha}\} dV \, . \qquad (1.39)$$

By application of the transport theorem 1.3.1 to (1.37) and substitution of (1.38) and (1.39) we obtain

$$\int\limits_{V(t)} \{\frac{\partial \rho E}{\partial t} + (\rho u_\alpha E),_\alpha\}dV = \int\limits_{V(t)} \{(u_\alpha \tau_{\alpha\beta}),_\beta + (kT,_\alpha),_\alpha + \rho u_\alpha f_\alpha^b + \rho q\}dV .$$

Since this holds for every $V(t)$, we have

$$\frac{\partial \rho E}{\partial t} + (\rho u_\alpha E),_\alpha = (u_\alpha \tau_{\alpha\beta}),_\beta + (kT,_\alpha),_\alpha + \rho u_\alpha f_\alpha^b + \rho q . \qquad (1.40)$$

This is the *energy equation*.

It is useful to put the energy equation in another form. By using the mass conservation equation, the energy equation can be written as follows:

$$\rho \frac{DE}{Dt} = (u_\alpha \tau_{\alpha\beta}),_\beta + (kT,_\alpha),_\alpha + \rho u_\alpha f_\alpha^b + \rho q . \qquad (1.41)$$

Similarly, the momentum equation (1.24) can be written as

$$\rho \frac{Du_\alpha}{Dt} = \tau_{\alpha\beta,\beta} + \rho f_\alpha^b .$$

Multiplication by u_α and summation gives

$$\rho \frac{D}{Dt}(\tfrac{1}{2} u_\alpha u_\alpha) = u_\alpha \tau_{\alpha\beta,\beta} + \rho u_\alpha f_\alpha^b . \qquad (1.42)$$

Subtraction of (1.42) from (1.41) and substitution of (1.25) gives

$$\rho \frac{De}{Dt} = -p\Delta - \tfrac{2}{3}\mu\Delta^2 + 2\mu u_{\alpha,\beta} e_{\alpha\beta} + (kT,_\alpha),_\alpha + \rho q .$$

Because $e_{\alpha\beta} = e_{\beta\alpha}$ we have $u_{\alpha,\beta}e_{\alpha\beta} = e_{\alpha\beta}e_{\alpha\beta}$. Furthermore,

$$e_{\alpha\beta}e_{\alpha\beta} - \frac{1}{3}\Delta^2 = (e_{\alpha\beta} - \frac{1}{3}\Delta\delta_{\alpha\beta})(e_{\alpha\beta} - \frac{1}{3}\Delta\delta_{\alpha\beta}) ,$$

since $\delta_{\alpha\beta}\delta_{\alpha\beta} = 3$. Thus we obtain the following form of the energy equation :

$$\rho \frac{De}{Dt} = -p\Delta + \psi + (kT,_\alpha),_\alpha + \rho q , \qquad (1.43)$$

where

$$\psi = 2\mu(e_{\alpha\beta} - \tfrac{1}{3}\Delta\delta_{\alpha\beta})(e_{\alpha\beta} - \tfrac{1}{3}\Delta\delta_{\alpha\beta}) .$$

The terms in this equation have the following physical interpretation. The left-hand side represents transport of internal energy by convection. The first term at the right is work done by pressure in a (if departures from thermodynamic equilibrium are neglected) reversible process (the thermodynamic concepts of reversibility and irreversibility will be discussed in the next section). The second term ψ is non-negative, and represents irreversible generation of heat by *dissipation* of mechanical energy by frictional heating. The third term represents transfer of heat by diffusion. The fourth term represents heat addition or withdrawal by external sources.

1.7 Thermodynamic aspects

Equation of state

We have obtained a total of five conservation equations, namely one for mass, one for momentum in each coordinate direction, and one for energy. There are seven unknowns: ρ, u_α ($\alpha = 1, 2, 3$), p, T and e. The system of equations is completed by *equations of state*. As said before, for simple thermodynamic systems there are at most two independent state variables, on which the other state variables depend. Which state variables can best be selected to be independent depends on circumstances. We present two examples: the compressible flow of a *perfect gas* and incompressible flow with heat transfer.

Perfect gas

A gas is called perfect if the following equation of state holds:

$$p = \rho RT , \qquad (1.44)$$

with T the temperature measured in K (Kelvin). The Kelvin scale is related to the centigrade scale by a shift: $0°C = 273.16$ K. When expressed in Kelvin, T is called the *absolute temperature*. The parameter R is a characteristic constant for a particular gas.

According to thermodynamics, for a perfect gas the internal energy e depends on T only, which gives us a second equation of state:

$$e = e(T) . \qquad (1.45)$$

We now have seven equations: five conservation laws and two equations of state.

Because the conservation laws contain time-evolution equations for ρ and e, it is best to take these as our two independent state variables. Because of the explicit occurence of T in (1.44) and the simplicity of relation (1.45) it may be a little more convenient to use T instead of e. Given ρ and T, the other state variables p and e follow from the two equations of state (1.44) and (1.45).

Enthalpy and entropy

For later use we introduce here two more state variables, namely *enthalpy* and *entropy*. The enthalpy h is defined by

$$h = e + p/\rho . \qquad (1.46)$$

For a perfect gas we have

$$h = e(T) + RT = h(T) .$$ (1.47)

A law governing h in flows will be presented in the next section.

We define

$$c_v = \frac{de}{dT} , \quad c_p = \frac{dh}{dT} ,$$ (1.48)

where c_v is called the *specific heat at constant volume* and c_p is called the *specific heat at constant pressure*. From (1.47) it follows that

$$R = c_p - c_v .$$ (1.49)

A gas is called *calorically perfect* if $c_v =$ constant. From (1.49) it follows that then also $c_p =$ constant.

The entropy per unit mass s is defined by

$$T\delta s = \delta e + p\delta(1/\rho) ,$$ (1.50)

where δ indicates small changes in the state variables. This can be rewritten as follows. From (1.46) it follows that

$$\delta e = \delta h - \delta(p/\rho) .$$

Substitution in (1.50) gives

$$T\delta s = \delta h - \frac{1}{\rho}\delta p .$$ (1.51)

A *reversible* thermodynamic process is a process in which heat is transformed into work or vice-versa in such a way that the process can be reversed without loss. An *adiabatic* thermodynamic process is a process in which no heat is transferred to or from the system. Therefore, in an adiabatic flow $q = 0$ in the energy equation, and solid walls (if any) are thermally insulated. The *second law of thermodynamics* says that in a closed system, i.e. an adiabatic system that exchanges no work with its environment, the entropy of the system does not decrease. It may be shown that only if the process is also reversible the entropy mains constant.

Law for entropy

A law governing the variation of s in the flow of a perfect gas can be obtained as follows. Application of (1.50) to a material particle gives the *Gibbs equation*:

$$T\frac{Ds}{Dt} = \frac{De}{Dt} - \frac{p}{\rho^2}\frac{D\rho}{Dt} \ .$$

Substitution of (1.43) and (1.13) gives

$$T\frac{Ds}{Dt} = \frac{1}{\rho}\psi + \frac{1}{\rho}(kT_{,\alpha})_{,\alpha} + q \ . \tag{1.52}$$

This is another form of the energy equation. We see that frictional heating increases the entropy, since $\psi \geq 0$. The second and third term of the right-hand side present interaction of the material particle with the environment, so that the material particle is not a closed system; accordingly, these terms may have either sign. In adiabatic inviscid and non-heat-conducting flow ($q = \mu = k = 0$) we find $Ds/Dt = 0$, in which case the flow is called *isentropic*. Note that although s is constant for each material particle, different particles may have different entropy. If all material particles have the same constant entropy, the flow is called *homentropic*.

Homentropic flow

In homentropic flow, $s = s_0 = $ constant, so that there is only one independent state variable left, and the equations of state simplify. Take for example s and ρ as independent state variables. From equation (1.50) it follows that for a perfect gas infinitesimal changes of state variables satisfy

$$0 = \delta s = c_v\frac{\delta T}{T} - \frac{R}{\rho}\delta\rho \ . \tag{1.53}$$

For a perfect gas equation (1.44) gives

$$\delta p = RT\delta\rho + \rho R\delta T \ .$$

By elimination of δT with (1.53) we obtain

$$\delta p = RT(1 + R/c_v)\delta\rho \ .$$

By using (1.44) to eliminate T and (1.49) this can be written as

$$\frac{\delta p}{p} = \gamma\frac{\delta\rho}{\rho} \ .$$

with $\gamma = c_p/c_v$ the ratio of specific heats ($\gamma = 7/5$ for air). From this we obtain the following equation of state for homentropic flow of a perfect gas:

$$p/\rho^\gamma = p_0/\rho_0^\gamma = \text{constant} \ . \tag{1.54}$$

In isentropic flow (1.54) holds along particle paths, with the constant differing per particle path.

In Chap. 10 we will go deeper into the role of the entropy. For more information on the thermodynamics of flows, see Liepmann and Roshko (1957).

Low subsonic flow with heat transfer

Consider a flow with a non-uniform temperature distribution. The density of a fluid particle may change due to heat conduction. The speed of the flow is assumed to be small with respect to the speed of sound; such a flow is called low subsonic. As will be seen in Sect. 1.12, this implies that the pressure variations in the flow are so small, that their effect on the other thermodynamic quantities is negligible. Therefore we use p as one independent variable; T will be the second. Equation (1.51) gives the variation of s with p and T. For $\partial s/\partial T$ we get

$$T\frac{\partial s}{\partial T} = \frac{\partial h}{\partial T} = c_p \ .$$

Substitution in the energy equation (1.52) gives, assuming $\partial s/\partial p = 0$ for the reason given above,

$$c_p \frac{DT}{Dt} = \frac{1}{\rho}\psi + \frac{1}{\rho}(kT_{,\alpha})_{,\alpha} + q \ . \tag{1.55}$$

This is the equation of *heat transfer* for low subsonic flow. If the viscosity is small, i.e. the Reynolds number is large, the frictional heating term ψ may be neglected.

Barotropic flow

A flow is called *barotropic* if there exists a function w such that for small variations the following relation holds:

$$\frac{1}{\rho}\delta p = \delta w \ . \tag{1.56}$$

This occurs, for example, if $\rho = \rho_0 = $ constant, in which case

$$w = p/\rho_0 \ ,$$

or if pressure is a function of density only, *i.e.* we have an equation of state

$$p = p(\rho) \ .$$

This is the case if a state variable θ is constant. Taking θ and ρ as our two independent state variables, we have an equation of state $p(\theta, \rho) = $ constant, so that $w = \int_{\rho_0}^{\rho} \frac{1}{\rho}\frac{\partial p}{\partial \rho}d\rho$. An example is *isothermal* flow, i.e. flow in which $T = T_0 = $ constant, of a perfect gas. Taking $\theta = T$, the equation of state (1.44) becomes

$$p = \rho R T_0$$

and we obtain

$$w = R T_0 \ln(\rho/\rho_0) \, ,$$

from which it follows that indeed

$$w_{,\alpha} = \frac{1}{\rho} R T_0 \rho_{,\alpha} = \frac{1}{\rho} p_{,\alpha} \, .$$

A more important practical example is homentropic flow of perfect gas with $s = s_0 = $ constant. Then we have the equation of state (1.54), from which it follows that

$$w = \frac{\gamma}{\gamma - 1} \frac{p}{\rho} \, . \tag{1.57}$$

If the gas is calorically perfect we have for the enthalpy

$$h = c_p T = \frac{\gamma}{\gamma - 1} \frac{p}{\rho} = w \, ,$$

but also if the gas is not calorically perfect we have that homentropic flow of a perfect gas is barotropic, with $w = h$:

$$\frac{1}{\rho} p_{,\alpha} = h_{,\alpha} \, . \tag{1.58}$$

This can be seen as follows. Writing $p_0/\rho_0^\gamma = c$ and using (1.54), we obtain

$$h_{,\alpha} = c_p T_{,\alpha} = \frac{c_p}{R}\left(\frac{p}{\rho}\right)_{,\alpha} = \frac{\gamma}{\gamma - 1} c^{1/\gamma} (p^{1 - 1/\gamma})_{,\alpha} = (c/p)^{1/\gamma} p_{,\alpha} = \frac{1}{\rho} p_{,\alpha} \, . \tag{1.59}$$

Exercise 1.7.1. Show that (1.57) implies (1.56) for homentropic flow.

1.8 Bernoulli's theorem

The energy equation for an ideal fluid

An *ideal fluid* is a fluid in which there is no friction and no heat conduction, i.e.

$$\mu = k = 0 \, . \tag{1.60}$$

For $Re \gg 1$ this often is a good approximation. With assumption (1.60) the energy equation (1.40) simplifies as follows. We have

$$(u_\alpha \tau_{\alpha\beta})_{,\beta} = -(pu_\alpha)_{,\alpha} \ .$$

Substitution in (1.40) with $k = 0$ gives

$$\frac{\partial \rho E}{\partial t} + \{u_\alpha(\rho E + p)\}_{,\alpha} = \rho u_\alpha f_\alpha^b + \rho q \ . \tag{1.61}$$

By using the definition (1.46) of enthalpy this can be rewritten as

$$\frac{\partial \rho E}{\partial t} + (\rho u_\alpha H)_{,\alpha} = \rho u_\alpha f_\alpha^b + \rho q \ , \tag{1.62}$$

where $H = h + \frac{1}{2}u_\alpha u_\alpha$ is the total enthalpy. This is the energy equation for an ideal fluid.

Another useful form of the energy equation for an ideal fluid is obtained by rewriting (1.61) as

$$\rho \frac{DE}{Dt} + u_\alpha p_{,\alpha} + pu_{\alpha,\alpha} = \rho u_\alpha f_\alpha^b + \rho q \ . \tag{1.63}$$

With (1.13) we find

$$u_{\alpha,\alpha} = -\frac{1}{\rho}\frac{D\rho}{Dt} \ , \tag{1.64}$$

so that (1.63) can be rewritten as

$$\rho \frac{DE}{Dt} + \frac{Dp}{Dt} - \frac{\partial p}{\partial t} - \frac{p}{\rho}\frac{D\rho}{Dt} = \rho u_\alpha f_\alpha^b + \rho q \ ,$$

or, recalling (1.46),

$$\frac{DH}{Dt} = \frac{1}{\rho}\frac{\partial p}{\partial t} + u_\alpha f_\alpha^b + q \ , \tag{1.65}$$

where $H = h + \frac{1}{2}u_\alpha u_\alpha$ is the total enthalpy.

Bernoulli's theorem

Let us assume that the external force field f^b is *conservative*. That is, there exists a potential Φ such that

$$f^b = -\mathrm{grad}\Phi \ .$$

Such a Φ exists if $\mathrm{curl}\, f^b = 0$. Assume furthermore that the flow is *adiabatic*, i.e. there are no heat sources: $q = 0$, and that the flow is stationary, i.e. $\partial p/\partial t = 0$. Then from (1.65) the following theorem follows, derived by Daniel Bernoulli in 1738:

Theorem 1.8.1. *(Bernoulli's theorem)*
For stationary flow of an ideal fluid without heat sources in the presence of a conservative force field we have

$$\frac{D}{Dt}(H + \Phi) = 0, \quad H = E + \frac{p}{\rho} .$$

According to Bernoulli's theorem, $H + \Phi$ constant along particle paths in stationary flows, and hence also along streamlines. It often happens that all streamlines originate in a region where $H + \Phi = $ constant. Then every streamline carries the same value of $H + \Phi$, so that $H + \Phi$ is constant throughout the flow. This may lead to considerable simplification of the governing equations.

1.9 Kelvin's circulation theorem and potential flow

Rate of change of circulation

Consider the circulation $\Gamma(C(t))$ arround a closed material curve $C(t)$ as discussed in Sect. 1.3:

$$\Gamma(C(t)) = \oint_{C(t)} \boldsymbol{u} \cdot d\boldsymbol{x} .$$

According to (1.11),

$$\frac{d\Gamma(C(t))}{dt} = \oint_{C(t)} \frac{D\boldsymbol{u}}{Dt} \cdot d\boldsymbol{x} .$$

We substitute (1.27) and obtain

$$\frac{d\Gamma(C(t))}{dt} = \oint_{C(t)} (\frac{1}{\rho}\tau_{\alpha\beta,\beta} + f_\alpha^b) dx_\alpha . \tag{1.66}$$

Inviscid barotropic flow

For the circulation in an inviscid barotropic flow in a conservative force field we have the following conservation law, derived by Lord Kelvin in 1869:

Theorem 1.9.1. *(Kelvin's circulation theorem)*
In inviscid barotropic flow in a conservative force field the circulation around closed material curves is constant.

Proof. The theorem follows immediately from (1.66):

$$\frac{d\Gamma(C(t))}{dt} = \oint_{C(t)} (-\frac{1}{\rho}p_{,\alpha} + f_{\alpha}^{b})dx_{\alpha} = - \oint_{C(t)} (w + \Phi)_{,\alpha}dx_{\alpha} = 0 \; .$$

$\square$

Irrotational flow

Let $S(t)$ be an open *material surface* (i.e. a surface consisting of material particles) bounded by a reducible closed material curve $C(t)$ (a reducible curve is a curve which can be shrunk to a point by continuous deformation without leaving the flow domain). Then according to Stokes's theorem 1.2.3 we have

$$\Gamma(C(t)) = \oint_{C(t)} \boldsymbol{u} \cdot d\boldsymbol{x} = \iint_{S(t)} \boldsymbol{\omega} \cdot \boldsymbol{n}dS \; ,$$

with $\boldsymbol{\omega}$ the *vorticity*, defined as

$$\boldsymbol{\omega} = \text{curl}\boldsymbol{u} \; .$$

If $\boldsymbol{\omega} \equiv 0$ the flow is called *irrotational*. If the flow is irrotational at $t = 0$ and if Kelvin's circulation theorem applies, then

$$0 = \Gamma(C(t)) = \iint_{S(t)} \boldsymbol{\omega} \cdot \boldsymbol{n}dS$$

for every $C(t)$ and $S(t)$, so that $\boldsymbol{\omega}(t, \boldsymbol{x}) \equiv 0$. Hence we may assert that an inviscid barotropic flow in a conservative force field that is initially irrotational remains irrotational. Similarly, if $C(t)$ was in an irrotational part of the flow domain at some previous time t_0 then $\Gamma(C(t)) = 0$ (if Kelvin's circulation theorem applies). Hence we may assert that along particle lines originating in an irrotational part of the flow we have $\boldsymbol{\omega} = 0$.

Incompressible potential flow

If $\boldsymbol{\omega} \equiv 0$ then we have (1.1):

$$\boldsymbol{u} = \text{grad}\varphi \; ,$$

with φ the velocity potential. If the flow is incompressible then (1.15) gives

$$\varphi_{,\alpha\alpha} = 0 \, .$$

This is the potential equation for incompressible flow. It is simply Laplace's equation.

Compressible potential flow

Let the flow be compressible and irrotational. Assume that the flow is inviscid and barotropic and that the force field is conservative field (otherwise the assumption of irrotationality would be highly unlikely). The momentum equation in the form (1.29) gives

$$\frac{\partial \varphi_{,\alpha}}{\partial t} + (\tfrac{1}{2}\varphi_{,\beta}\varphi_{,\beta} + w + \varPhi)_{,\alpha} = 0 \, .$$

It follows that

$$\frac{\partial \varphi}{\partial t} + \frac{1}{2}\boldsymbol{u} \cdot \boldsymbol{u} + w + \varPhi = H_0 \, , \qquad (1.67)$$

with H_0 a function of time, that follows from the boundary conditions. For example, for potential flow about an oscillating wing without body forces with uniform velocity of magnitude U at infinity, $H_0 = \tfrac{1}{2}U^2 + w_\infty$.

The mass conservation equation for potential flow is

$$\frac{\partial \rho}{\partial t} + (\rho\varphi_{,\alpha})_{,\alpha} = 0 \, . \qquad (1.68)$$

Equations (1.67) and (1.68) contain two unknowns φ and ρ. The most important example of inviscid barotropic flow is homentropic flow. For a perfect gas we have (1.59), so that $w = h$. Assuming the gas to be calorically perfect,

$$h = \frac{\gamma}{\gamma - 1}\frac{p}{\rho} = \frac{\gamma}{\gamma - 1}\frac{p_r}{\rho_r^\gamma}\rho^{\gamma-1} = h_r\left(\frac{\rho}{\rho_r}\right)^{\gamma-1} \, ,$$

with subscript r referring to some reference state. We obtain with (1.67):

$$\frac{\rho}{\rho_r} = \left(\frac{h}{h_r}\right)^{1/(\gamma-1)} = \{(H_0 - \varPhi - \frac{1}{2}\boldsymbol{u} \cdot \boldsymbol{u} - \frac{\partial \varphi}{\partial t})/h_r\}^{1/(\gamma-1)} \, . \qquad (1.69)$$

The system now consists of (1.68) and (1.69).

Specializing further to steady flow without body forces, equation (1.67) becomes

$$\frac{1}{2}\boldsymbol{u} \cdot \boldsymbol{u} + h = H_0 \, , \qquad (1.70)$$

so that H_0 equals the value of h in points where $\boldsymbol{u} = 0$, hence H_0 is the stagnation enthalpy. We have recovered Bernoulli's theorem. With stagnation as reference state, indicated by subscript zero, equation (1.69) becomes

$$\frac{\rho}{\rho_0} = (1 - \frac{1}{2}\boldsymbol{u} \cdot \boldsymbol{u}/H_0)^{1/(\gamma-1)} . \qquad (1.71)$$

In the stationary case, equation (1.68) becomes

$$(\rho\varphi_{,\alpha})_{,\alpha} = 0 . \qquad (1.72)$$

The system of equations now is (1.72) and (1.71). The reference density is found as follows. We have

$$\rho_0 = \frac{\gamma p_0}{(\gamma - 1)H_0} . \qquad (1.73)$$

From (1.54) it follows that

$$p_0 = p_\infty (\rho_0/\rho_\infty)^\gamma . \qquad (1.74)$$

Let conditions at infinity be given. Combination of (1.73) and (1.74) results in

$$\rho_0 = \{\frac{(\gamma - 1)H_0\rho_\infty^\gamma}{\gamma p_\infty}\}^{1/(\gamma-1)} . \qquad (1.75)$$

This fully specifies the equations for potential flow. Equation (1.72) with ρ given by (1.71) and (1.75) is called the *full potential equation*. The adjective full refers to the fact that, apart from the assumptions required for potential flow, no further simplifications have been made.

It is instructive to rewrite the equation for ρ in the following manner. As will be shown in Sect. 1.12, the local speed of sound a is given by

$$a^2 = \frac{dp}{d\rho} = \gamma\frac{p}{\rho} = (\gamma - 1)h ,$$

where we have used (1.54). Hence

$$H_0 = \frac{1}{2}\boldsymbol{u} \cdot \boldsymbol{u} + \frac{\gamma}{\gamma - 1}\frac{p}{\rho} = \frac{1}{2}\boldsymbol{u} \cdot \boldsymbol{u} + a^2/(\gamma - 1) .$$

The *Mach number* M is defined as

$$M = |\boldsymbol{u}|/a .$$

Hence, equation (1.71) can be written as

$$\frac{\rho}{\rho_0} = (1 + \frac{\gamma - 1}{2}M^2)^{-1/(\gamma-1)} .$$

In Sect. 13.3 of Hirsch (1990) it is shown that (1.72) is elliptic for M < 1 and hyperbolic for M > 1.

The field of computing potential flows is mature; see Hirsch (1990) Chap. 13–15 for a survey. In this book we will not pay further attention to the potential flow model. Potential flow codes are in daily use for quickly computing flows around aircraft and in flow machinery in the predesign phase. When coupled with boundary layer computations (also not discussed here), accurate predictions of lift and drag are possible when flow separation does not occur. But in the presence of separation or strong shock waves the more complete Euler or Navier-Stokes equations must be used. These form the main subject matter of this book.

1.10 The Euler equations

Flow of ideal fluids is governed by the Euler equations. We have the mass conservation equation (1.13), the momentum equations (1.26) with $\mu = 0$ and the energy equation for an ideal fluid (1.62):

$$\frac{\partial \rho}{\partial t} + (\rho u_\alpha)_{,\alpha} = 0 \,, \tag{1.76}$$

$$\frac{\partial \rho u_\alpha}{\partial t} + (\rho u_\alpha u_\beta)_{,\beta} + p_{,\alpha} = \rho f_\alpha^b \,, \tag{1.77}$$

$$\frac{\partial \rho E}{\partial t} + (\rho u_\alpha H)_{,\alpha} = \rho u_\alpha f_\alpha^b + \rho q \,. \tag{1.78}$$

We have five equations and seven unknowns: ρ, u_α, p, e and h. The system is completed by the equations of state, discussed in Sect. 1.7. For a perfect gas the equations of state are (1.44), (1.45). Elimination of T gives a relation

$$p = p(\rho, e) \,,$$

and the definition (1.46) of the enthalpy h completes the system of equations:

$$h = e + p/\rho \,.$$

Homentropic and barotropic flow of ideal fluids

There are various cases in which the Euler equations simplify, because the energy equation (1.78) can be replaced by an algebraic relation. An example is homentropic flow. As shown in Sect. 1.7, we have

$$\frac{Dp/\rho^\gamma}{Dt} = 0 \,. \tag{1.79}$$

We now have an equation too many. The point is that the energy equation is satisfied automatically. This may be seen by using the alternative form of the energy equation (1.43) with $\mu = k = 0$:

$$\frac{De}{Dt} + \frac{p}{\rho} u_{\alpha,\alpha} = q \ . \tag{1.80}$$

By using (1.64) and the equation of state for a perfect gas (1.44) we get

$$\frac{De}{Dt} - RT\frac{D\ln\rho}{Dt} = q \ . \tag{1.81}$$

Substitution of (1.48) gives

$$c_v \frac{DT}{Dt} - RT\frac{D\ln\rho}{Dt} = q \ , \tag{1.82}$$

and with (1.49) we obtain

$$\frac{D\ln(T\rho^{1-\gamma})}{Dt} = \frac{q}{c_v T} \ . \tag{1.83}$$

Along a particle path $p/\rho^\gamma = $ constant, hence $RT\rho^{1-\gamma} = $ constant, and the left-hand side of (1.83) vanishes. Hence we have a paradox, unless q vanishes. This is to be expected, since the assumption of homentropic flow implies that no heat is to be added or subtracted. Assuming $q = 0$, the energy equation is satisfied. It is replaced by (1.79). If all particle paths emanate from a reservoir with constant entropy, the flow is isentropic, and equation (1.79) reduces to an algebraic relation:

$$p/\rho^\gamma = p_0/\rho_0^\gamma = C \ , \tag{1.84}$$

where the subscript zero refers to the reservoir conditions. By using (1.84), equation (1.77) can be rewritten as

$$\frac{\partial \rho u_\alpha}{\partial t} + (\rho u_\alpha u_\beta)_{,\beta} + \gamma C \rho^{\gamma-1} \rho_{,\alpha} = \rho f_\alpha^b \ , \tag{1.85}$$

from which we will later draw an analogy with the shallow-water equations.

1.11 The convection-diffusion equation

Conservation law for material properties

Let φ be a material property, i.e. a scalar that corresponds to a physical property of material particles, such as heat or concentration of a solute. Assume that φ is conserved and can change only through exchange between material particles or through external sources. Let φ be defined per unit of mass. Then the conservation law for φ is:

$$\frac{d}{dt} \int_{V(t)} \rho\varphi dV = \int_{S(t)} \boldsymbol{f} \cdot \boldsymbol{n} dS + \int_{V(t)} q dV \,.$$

Here $\boldsymbol{f}$ is the *flux vector*, governing the rate of transfer through the surface, and q is the source term. For $\boldsymbol{f}$ we assume *Fick's law* (called Fourier's law if φ is temperature):

$$\boldsymbol{f} = k \operatorname{grad} \varphi \,,$$

with k the diffusion coefficient. By arguments that are now familiar it follows that

$$\frac{\partial \rho\varphi}{\partial t} + \operatorname{div}(\rho\varphi\boldsymbol{u}) = (k\varphi_{,\alpha})_{,\alpha} + q \,. \tag{1.86}$$

This is the *convection-diffusion equation*. The left-hand side represents transport of φ by convection with the flow, the first term at the right represents transport by diffusion. The balance between convection and diffusion is characterized by the *Péclet number* Pe, defined as

$$\text{Pe} = UL/(\rho_0 k_0) \,,$$

where L is a typical length scale of the flow, and U, ρ_0 and k_0 are typical magnitudes of $|\boldsymbol{u}|$, ρ and k. For Pe $\gg 1$ we have dominating convection, for Pe $\ll 1$ diffusion dominates.

By using the mass conservation law, equation (1.86) can be written as

$$\rho \frac{D\varphi}{Dt} = (k\varphi_{,\alpha})_{,\alpha} + q \,. \tag{1.87}$$

The convection-diffusion equation governs the concentration distribution of solutes. Note that we have already encountered several times equations that closely resemble the convection-diffusion equation, such as the energy equation. Since $De/Dt = c_v DT/Dt$, equation (1.43) is a convection-diffusion equation for T with special source terms. Also the momentum equation (1.30) comes close to being a convection-diffusion equation. Many aspects of numerical approximation in computational fluid dynamics already show up in the numerical analysis of the relatively simple convection-diffusion equation, which is why we will devote two special chapters to this equation.

1.12 Conditions for incompressible flow

Equations of motion for isentropic flow

Suppose the motion of the fluid consists of small isentropic oscillations around a state of static equilibrium in which $\boldsymbol{u} = 0$, in the absence of body forces.

Taking ρ and s as our independent state variables, there is an equation of state $p = p(\rho, s)$. For isentropic flow the mass conservation equation (1.13) can be written as

$$\frac{1}{a^2}\frac{Dp}{Dt} + \rho \mathrm{div}\boldsymbol{u} = 0 , \tag{1.88}$$

defining

$$a^2 = (\frac{\partial p}{\partial \rho})_s ,$$

where the subscript s indicates that the entropy s is kept constant. Note that $(\partial p/\partial \rho)_s > 0$, because increasing p (compression) gives increasing ρ. For a perfect gas we have the equations of state (1.44) and (1.54), from which it follows that

$$a^2 = \gamma\frac{p}{\rho} = \gamma RT .$$

Because $\mu = 0$ in isentropic flow, the momentum equation (1.27) becomes

$$\rho\frac{D\boldsymbol{u}}{Dt} = -\mathrm{grad}p . \tag{1.89}$$

Equations (1.88) and (1.89) are the equations of motion for isentropic flow.

The speed of sound

Let the equilibrium state be denoted by subscript zero. From (1.89) it follows that $p_0 = $ constant and from the equation of state (1.54) it follows that $\rho_0 = $ constant. Linearizing equations (1.88) and (1.89) around the equilibrium state we obtain, writing $p = p_0 + p_1$,

$$\frac{1}{a_0^2}\frac{\partial p_1}{\partial t} + \rho_0\mathrm{div}\boldsymbol{u} = 0 ,$$

$$\rho_0\frac{\partial \boldsymbol{u}}{\partial t} = -\mathrm{grad}p_1 . \tag{1.90}$$

By taking the divergence of (1.90) and eliminating $\boldsymbol{u}$ we obtain

$$\frac{1}{a_0^2}\frac{\partial^2 p_1}{\partial t^2} = (p_1)_{,\alpha\alpha} .$$

This is the *wave equation* governing acoustics. In one space dimension the general solution is, as can be seen by inspection,

$$p_1 = f(a_0t + x_1) + g(a_0t - x_1) ,$$

with f and g arbitrary; f and g present left- and right-traveling waves with phase velocity a_0, so that we see that a_0 is the speed of sound.

Conditions for incompressibility

Let L and U be characteristic length and velocity scales of the flow. This means that flow properties vary only slightly over distances small compared to L, and that variations of u have a typical magnitude U. For example, for flow around a sphere U can be taken the velocity at infinity. For low Reynolds numbers a suitable choice for L is the diameter D of the sphere, but at large Reynolds numbers a correct choice for L is a typical boundary layer thickness, leading to $L = D\mathrm{Re}^{-1/2}$.

The flow may be regarded as incompressible if the density of a particle does not vary much:

$$\left|\frac{1}{\rho}\frac{D\rho}{Dt}\right| \ll \frac{U}{L}.$$

Using the equation of state $p = p(\rho, s)$, we can write

$$\frac{1}{\rho}\frac{D\rho}{Dt} = \frac{1}{\rho a^2}\frac{Dp}{Dt} - \frac{1}{\rho a^2}\left(\frac{\partial p}{\partial s}\right)_\rho \frac{Ds}{Dt}.$$

It can be shown (cf. Batchelor (1967)), using (1.52), that the second term of the right-hand side contributes significantly only under unlikely circumstances, such as gas flow with $LU = 10^{-1}$ cm^2/s and $T = 100°$C. Retaining only the first term, the condition for incompressibility becomes

$$\left|\frac{1}{\rho a^2}\frac{Dp}{Dt}\right| \ll \frac{U}{L}.$$

We now assume isentropic flow, because normally the influence of viscosity and heat conductivity is not such as to have a decisive effect on the order of magnitude of pressure variations. Hence we can use (1.89) to obtain

$$\frac{Dp}{Dt} = \frac{\partial p}{\partial t} + u \cdot \mathrm{grad}p = \frac{\partial p}{\partial t} - \frac{1}{2}\rho\frac{Du \cdot u}{Dt}.$$

Since in general there is no reason why these two terms should balance, we obtain the following two conditions for incompressibility:

$$\left|\frac{1}{\rho a^2}\frac{\partial p}{\partial t}\right| \ll \frac{U}{L}, \tag{1.91}$$

$$\left|\frac{1}{a^2}\frac{Du \cdot u}{Dt}\right| \ll \frac{U}{L}. \tag{1.92}$$

Let τ ba a typical time scale of the flow. The magnitude of pressure variation is $\rho UL/\tau$, as can be seen from (1.89), hence $\frac{1}{\rho a^2}\partial p/\partial t$ is of order $UL/(a\tau)^2$, so that (1.91) becomes

$$\frac{L^2}{a^2\tau^2} \ll 1 \, . \tag{1.93}$$

In acoustics L is the wavelength of a sound wave, i.e. $L = a\tau$, and (1.93) is not satisfied. We are not surprised to see that in acoustics we cannot assume incompressibility.

Next, consider (1.92). We have $D(\boldsymbol{u} \cdot \boldsymbol{u})/Dt = \partial(\boldsymbol{u} \cdot \boldsymbol{u})/\partial t + u_\alpha \partial(\boldsymbol{u} \cdot \boldsymbol{u})/\partial x_\alpha$. The order of magnitude of these term is U^2/τ and U^3/L respectively. Hence, (1.92) gives the following two conditions:

$$\frac{UL}{a^2\tau} \ll 1 \, , \tag{1.94}$$

$$\frac{U^2}{a^2} \ll 1 \, . \tag{1.95}$$

Condition (1.94) is not independent, but follows from (1.93) and (1.95) (which give $(UL/a^2\tau)^2 \ll 1$).

In conclusion, the conditions for incompressibility can be written as

$$1/\tau \ll a/L \, , \tag{1.96}$$

and

$$\mathrm{M} \ll 1 \, , \tag{1.97}$$

where $\mathrm{M} = U/a$ is the Mach number. The first condition limits the frequency of unsteady phenomena in the flow; the second limits the magnitude of the velocity of the flow. In practice, for $\mathrm{M} \lesssim 0.3$ incompressibility is a good approximation.

1.13 Turbulence

As already mentioned in Sect. 1.5, it is a fact of nature that for sufficiently large Reynolds numbers flows show rapid apparently random fluctuations, even when the controlling factors of the flow, such as body geometry and upstream conditions, are stationary. Such flows are called *turbulent*. Turbulent flow is a complicated example of a *chaotic dynamical system*. For an introduction to turbulence, see Hinze (1975), Tennekes and Lumley (1982) or Libby (1996). Turbulence remains one of the great unsolved problems of physics. Accurate prediction of turbulent flows starting from first principles is out of the question for realistic applications in engineering, as will be shown below. Other fundamentally sound prediction methods have not (yet) been found. The difficulty is that turbulence is both nonlinear and stochastic.

Direct numerical simulation of turbulence

Since turbulence is governed by the Navier-Stokes equations, turbulent flow can be computed from first principles by solving the Navier-Stokes equations. We now show that this is not feasible for general engineering computations.

Let $\mathcal{U}$ and $\mathcal{L}$ be the velocity and length scales of the large eddies in a turbulent flow, and let us define the macroscale Reynolds number by

$$\mathrm{Re}_m = \mathcal{U}\mathcal{L}/\nu \,,$$

with ν the dynamic viscosity coefficient. The length scale of the smallest eddies is called the Kolmogorov scale, and is denoted by η. It has been shown by Kolmogorov (see Tennekes and Lumley (1982)) that

$$\eta/\mathcal{L} = \mathcal{O}(\mathrm{Re}_m^{-3/4}) \,.$$

As an example, let us consider the turbulent flow in a pipe with diameter L and mean velocity U. The macroscale velocity $\mathcal{U}$ and length $\mathcal{L}$ are found to be approximately

$$\mathcal{L} = L/10 \,, \quad \mathcal{U} = \sqrt{\tau_w/\rho} \,,$$

with τ_w the wall friction. The Blasius resistance formula (Goldstein (1965), Sect. 155) gives:

$$\mathcal{U} = 0.2U\mathrm{Re}^{-1/8} \,, \quad \mathrm{Re} = UL/\nu \,.$$

Hence the macroscale Reynolds number is

$$\mathrm{Re}_m = \frac{1}{50}\mathrm{Re}^{7/8} \,.$$

This gives us away from the pipe wall

$$\eta/L \cong 1.9\mathrm{Re}^{-21/32} \,.$$

To resolve the Kolmogorov scale, the mesh size h of the computational grid must satisfy $h < \eta$. With $h = \eta/2$, we obtain for the number of grid cells in one direction $N = L/h \cong \mathrm{Re}^{21/32}$, so that the total number of cells is

$$N^3 \cong \mathrm{Re}^{63/32} \cong \mathrm{Re}^2 \,.$$

For $\mathrm{Re} = 10^5$, for example, this gives us $N^3 \cong 10^{10}$, which is clearly not sustainable on the computer infrastructures available at present and in the foreseeable future. In more complicated geometries and higher Reynolds numbers the required number of grid cells is even much larger. But at moderate Reynolds numbers and in simple geometries direct numerical simulation is a valuable tool for studying the fundamental properties of turbulence. Databases generated by direct numerical simulation can be used in the development and checking of approximate models. Such databases can be found at the following website:

ercoftac.mech.surrey.ac.uk

Some recent publications on direct numerical simulation are: Eggels, Unger, Weiss, Westerweel, Adrian, Friedrich, and Nieuwstadt (1994), Le, Moin, and Kim (1997), Moin and Madesh (1998), Na and Moin (1998), Reynolds (1990), Skote, Henningson, and Henkes (1998), Verstappen and Veldman (1997), Voke, Kleiser, and Chollet (1994). Earlier publications are quoted in Libby (1996). The numerical principles discussed in the present book apply to direct numerical simulation of turbulence.

Large-eddy simulation of turbulence

With large-eddy simulation, the large turbulent eddies are resolved numerically, but, in order to diminish the demands put on computer resources, the small eddies are modeled heuristically. This is called subgrid-scale modeling. The structure of the small eddies is to a large extent independent of the particular geometry at hand, and largely universal. Therefore there is some hope of obtaining a universal model for the small eddies, but this matter is still not completely settled. It would lead to far here to discuss subgrid-scale modeling. The numerical principles discussed in the present book apply to large-eddy simulation of turbulence. Application of the numerical method presented in Sect. 13.5 is described in Manhart *et al.* (1998).

For large-eddy simulation of the flow around large aircraft at $Re = 10^7$, Chapman (1979) has estimated a requirement of $8 \cdot 10^8$ grid cells and 10^4 Mwords storage, assuming that large eddies are resolved only where they occur, namely in the boundary layer at the surface of the aircraft, and in the wake. A crude estimate of computing time may be obtained as follows. Taking as a rough guess for the number of floating-point operations per grid cell and per time step 500 flop, and assuming 500 time steps are required to gather sufficient flow statistics, we arrive at an estimate of a total of 200 teraflop (1 teraflop $= 10^{12}$ flop). So this type of computation is becoming feasible with the teraflop machines that are making their appearance. But at present, large-eddy simulation is still largely in the research stage, and has not yet developed in an engineering tool. Some recent publications on large-eddy simulation are: Boersma *et al.* (1997), Boersma and Nieuwstadt (1996), Ferziger (1996), Galperin and Orszag (1993), Reynolds (1990), Voke, Kleiser, and Chollet (1994). Large-eddy simulation databases can be found in the ERCOFTAC website mentioned before.

Reynolds decomposition

In engineering practice, turbulent flows are generally modeled by the Reynolds averaged Navier-Stokes (RANS) equations, in order to achieve computer time and memory requirements that are feasible for industrial applications. Before presenting the RANS equations, we discuss the Reynolds decomposition. A quantity $A(\boldsymbol{x}, t)$ can be decomposed into a mean value $\bar{A}$ and a fluctuation A' as follows:

$$A = \bar{A} + A' , \tag{1.98}$$

defining $\bar{A}$ as

$$\bar{A}(t, \boldsymbol{x}) = \frac{1}{T} \int_{-T/2}^{T/2} A(t + \tau, \boldsymbol{x}) d\tau , \tag{1.99}$$

where T is a timespan that is large compared with the time scale of turbulent fluctuations but small compared to the time scale of other time-dependent features of the flow. Such a choice of T is often possible, because in fully developed turbulence (Re sufficiently large) the time scale of the turbulent fluctuations is often very small compared to the time scale of other unsteady features of the flow, such as the period of an oscillating body; often, the mean flow is stationary. Equations (1.98) and (1.99) define the *Reynolds decomposition*; $\bar{A}$ is called the *Reynolds average* (Reynolds (1895)).

In order to avoid products of fluctuation of density and fluctuations of other quantities one also uses the *density-weighted mean* or Favre (1965) average $\tilde{A}$, defined by

$$\tilde{A} = \overline{\rho A}/\bar{\rho} ,$$

with the corresponding fluctuation

$$A'' = A - \tilde{A} .$$

Note that

$$\bar{\bar{A}} = \bar{A}, \quad \tilde{\bar{A}} = \tilde{A}, \quad \bar{A'} = 0, \quad \tilde{A''} = 0 , \quad \overline{A''} = \bar{A} - \tilde{A}, \quad \overline{\rho A''} = 0 .$$

It follows that

$$\begin{aligned} \overline{AB} &= \bar{A}\bar{B} + \overline{A'B'} , \\ \overline{\rho AB} &= \bar{\rho}(\tilde{A}\tilde{B} + \widetilde{A''B''}) = \bar{\rho}\tilde{A}\tilde{B} + \overline{\rho A''B''} . \end{aligned} \tag{1.100}$$

Reynolds averaged Navier-Stokes equations

Taking the mean of the mass conservation equation (1.13) gives

$$\frac{\partial \bar{\rho}}{\partial t} + \operatorname{div} \bar{\rho}\tilde{\boldsymbol{u}} = 0 \ .$$

Next, consider the momentum equation (1.26). Applying (1.100), we have

$$\overline{\rho u_\alpha u_\beta} = \bar{\rho}\tilde{u}_\alpha \tilde{u}_\beta + \overline{\rho u''_\alpha u''_\beta} \ .$$

Taking the mean of (1.26) gives

$$\frac{\partial \bar{\rho}\tilde{u}_\alpha}{\partial t} + (\bar{\rho}\tilde{u}_\alpha \tilde{u}_\beta)_{,\beta} = -\bar{p}_{,\alpha} + \sigma^v_{\alpha\beta} + \sigma^r_{\alpha\beta} + \bar{\rho}\tilde{f}^b_\alpha \ , \qquad (1.101)$$

where the viscous stress $\sigma^v_{\alpha\beta}$ is defined by

$$\sigma^v_{\alpha\beta} = 2(\overline{\mu e_{\alpha\beta}} - \frac{1}{3}\overline{\mu \Delta}\delta_{\alpha\beta}) \ ,$$

and $\sigma^r_{\alpha\beta}$ is the *Reynolds stress*, defined by

$$\sigma^r_{\alpha\beta} = -\overline{\rho u''_\alpha u''_\beta} \ .$$

Equations (1.101) are called the *Reynolds averaged Navier-Stokes equations*. Note that the turbulent fluctuations cause additional momentum transfer through the Reynolds stress. The Reynolds stress is quite often much larger than the viscous stress. Frequently, the Reynolds stress is related to the mean velocity field by

$$\sigma^r_{\alpha\beta} = -\mu_t(\tilde{u}_{\alpha,\beta} + \tilde{u}_{\beta,\alpha}) + \frac{1}{3}\overline{\rho u_\gamma u_\gamma}\delta_{\alpha\beta} \ ,$$

where μ_t is called the eddy viscosity, still undetermined. This is called Boussinesq's closure hypothesis. The eddy viscosity needs to be modeled, and $\frac{1}{3}\overline{\rho u_\gamma u_\gamma}$ is subsumed under the pressure. Unfortunately, significant deviations from Boussinesq's closure hypothesis are common.

Closure problem

In order to complete the system of equations, the Reynolds stress $\sigma^r_{\alpha\beta}$ has to be expressed in terms of the mean quantities $\bar{\rho}$, $\tilde{\boldsymbol{u}}$ and $\bar{p}$. However, no general law for this is known. This is known as the *closure problem*. In practice semi-empirical relations are introduced, leading to so-called turbulence models.

Many turbulence models have been proposed. The quest for adequate turbulence models has been vexed by a lack of sufficiently general and sound simplifying principles. In algebraic models, Boussinesq's closure hypothesis is used, with an algebraic equation to relate the eddy viscosity to the primry unknowns. The computing cost of algebraic turbulence models is not much larger than that of the laminar Navier-Stokes equations, and the numerical aspects do not go beyond what is covered in the present book. See Bradshaw (1997), Libby (1996) and Wilcox (1993) for a survey of turbulence modeling.

In one- and two-equation models, the eddy viscosity depends on certain quantities that obey partial differential equations. These quantities are, typically, the turbulence kinetic energy k and the dissipation per unit turbulence kinetic energy ω or the turbulence dissipation ε. This leads to the $k - \varepsilon$ and $k - \omega$ models, which are, generally speaking, more dependable than algebraic models, but still suffer from the weakness of Boussinesq's closure hypothesis. The equations governing k, ω and ε are modeled heuristically, and involve large source and sink terms, that require delicacy in numerical approximations, because k, ω and ε must remain strictly non-negative. This involves numerical considerations not covered in the present book; see Mohammadi and Pironneau (1994), Wilcox (1993), Zijlema, Segal, and Wesseling (1995), Zijlema (1996), Wesseling, Zijlema, Segal, and Kassels (1997) for these aspects. Computing cost is higher than for algebraic models, but significantly less than for the class of turbulence models discussed next.

In Reynolds stress models, Boussinesq's closure hypothesis is abandoned, and differential equations are formulated for the individual components of the Reynolds stress tensor. The mathematical properties of the resulting system of differential equations are not yet well understood, and research into numerical approximations continues. Computing cost approaches that of large-eddy simulation. Some publications on Reynolds stress models are Dol, Hanjalić, and Kenjereš (1997), Hanjalić (1994), Launder (1990), Launder (1996), Shih (1997). Boussinesq's clossure hypothesis can also be replaced by something better in the context of two-equation models, as discussed in Launder (1996) and in Chap. 6 of Wilcox (1993); this keeps computing cost lower than for Reynolds stress models.

Because of the inadequacies of the existing turbulence models as well as the very large computational complexity of advanced models, the prediction of turbulent flows is generally considered one of the 'grand challenges' of computational science. In view of our remarks on predictability of nonlinear dynamical systems in Sect. 1.5, we can only hope to obtain accurate predictions of statistical averages of flow quantities. And even for these the time interval over which these predictions can be accurate is often severely limited. For example, the predictability horizon for the atmospheric flows that determine the daily weather at middle latitudes is thought to be at most a few weeks, in the hypothetical circumstance that the initial and boundary conditions are specified exactly. The butterfly on a windowsill in Rio de Janeiro that by the flutter of its wings causes a tornado in Texas (Lorenz (1993)) has become proverbial. Aristotle's (384 BC) admonition should be kept in mind: "It is the mark of an educated mind to rest satisfied with the degree of precision that the nature of the subject admits, and not to seek exactness when only an approximation is possible".[1]

1.14 Stratified flow and free convection

Stratified flow

Flows in which heavier and lighter layers of fluid occur are called *stratified*. If the fluid is initially at rest, but the density distribution becomes unstable (heavy fluid on top of light fluid), for example by heating from below, flow ensues. *Free convection* is flow driven by gravity through non-uniformity of the density distribution.

Consider flow in the presence of gravity. The momentum equation (1.27) becomes

$$\frac{Du_\alpha}{Dt} = -\frac{1}{\rho}p_{,\alpha} + \frac{1}{\rho}\sigma_{\alpha\beta,\beta} + g_\alpha , \tag{1.102}$$

with g the gravitational acceleration vector, and

$$\sigma_{\alpha\beta} = 2\mu\left(e_{\alpha\beta} - \frac{1}{3}\Delta\delta_{\alpha\beta}\right) .$$

Let there be density variations in the fluid due to non-uniformity of concentration c of a solute, or due to temperature variations. Let the density satisfy

$$\rho = \rho(c) .$$

The transport equation for the solute is (1.87):

$$\frac{Dc}{Dt} = \frac{1}{\rho}(kc_{,\alpha})_{,\alpha} + \frac{1}{\rho}q .$$

The mass conservation equation is given by (1.18):

$$\text{div}\boldsymbol{u} = -\frac{1}{\rho}\frac{d\ln\rho}{dc}\{(kc_{,\alpha})_{,\alpha} + q\} . \tag{1.103}$$

These are the equations for stratified flow. If $d\ln\rho/dc$ and variations of c are small, then (1.103) may be approximated by $\text{div}\boldsymbol{u} = 0$.

---

[1] The author is indebted to Kenjereš (1999) for this quotation.

Hydrostatic equilibrium

In *hydrostatic equilibrium* $u = 0$. Equation (1.102) reduces to

$$\frac{1}{\rho}p_{,\alpha} = g_\alpha .\qquad(1.104)$$

This equation can be solved as follows. Consider a fluid with an equation of state $\rho = \rho(p, s)$. Assume $s = $ constant. This means that there is thermodynamic equilibrium and that no heat transfer takes place. Choose one axis in the direction opposite to g, and call the coordinate along this axis ξ; we may think of ξ as height above the surface of the earth, and take $\xi = 0$ on the surface. Then (1.104) becomes

$$\frac{1}{\rho}\frac{dp}{d\xi} = -g , \quad g = |g| .\qquad(1.105)$$

This equation can be solved as follows. Denoting conditions at $\xi = 0$ by a subscript zero, the equation of state for a perfect gas is

$$\frac{p}{\rho^\gamma} = \frac{p_0}{\rho_0^\gamma} = C .$$

By elimination of ρ, equation (1.105) becomes

$$\frac{dp^{(\gamma-1)/\gamma}}{d\xi} = -\frac{\gamma-1}{\gamma}C^{-1/\gamma}g .$$

The solution is

$$p = p_0[1 - (\gamma - 1)g\xi/a_0^2]^{\gamma/(\gamma-1)} ,\qquad(1.106)$$

where $a_0 = \sqrt{\gamma p_0/\rho_0}$ is the speed of sound at $\xi = 0$. If

$$g\xi/a_0^2 \ll 1 ,\qquad(1.107)$$

the solution (1.106) can be approximated by the following linear pressure distribution:

$$p = p_0 - \rho_0 g\xi .\qquad(1.108)$$

This is also obtained directly if we assume in (1.105) $\rho = \rho_0$. Under what circumstances does (1.107) hold? Suppose we study the atmosphere. We have $g = 9.8$ m/s^2, and under standard circumstances at sea level $a_0 = 314$ m/s. Then (1.107) gives $\xi \ll 10$ km.

If the equation of state is arbitrary, we can proceed as follows. We have

$$\left(\frac{\partial p}{\partial \rho}\right)_s = a^2 ,$$

so that (1.105) can be rewritten as

$$\frac{\partial \ln \rho}{\partial \xi} = -g/a^2 \; ,$$

with formal solution

$$\ln \rho / \rho_0 = -g \int_0^{\xi} \frac{1}{a^2} d\xi \; .$$

If (1.107) is satisfied, we have approximately $\rho = \rho_0$, and (1.108) is recovered.

In general Cartesian coordinates, equation (1.108) becomes

$$p = p_0 + \rho_0 g_\alpha x_\alpha \; . \tag{1.109}$$

The Boussinesq equations

Consider a hydrostatic equilibrium state with $p = p_h$ given by (1.109) and $\rho = \rho_0$. Let the hydrostatic equilibrium state be perturbed by variations of a variable that influences the density, for example temperature or concentration of a solute; in any case, denote this variable by T. The fluid may or may not be set in motion as a result. We have an equation of state $\rho = \rho(p, T)$. If the velocity is small compared to the speed of sound, the dependence of ρ on p may be neglected, and we may write $\rho = \rho(T)$. Let $T = T_0$ be the uniform distribution of T for which we have $\rho(T_0) = \rho_0$. Let

$$|\rho'|/\rho_0 \ll 1, \quad |p'|/p_h \ll 1 \; , \tag{1.110}$$

with $\rho' = \rho - \rho_0$, $p' = p - p_h$. We approximate the terms in (1.102) using (1.110), assuming the flow is driven by gravity. Because the pressure gradient and gravity may balance, as in hydrostatic equilibrium, it is the difference in these two terms that drives the flow. These terms will therefore be approximated together. We may write

$$-\frac{1}{\rho}p_{,\alpha} + g_\alpha = g_\alpha - \frac{1}{\rho_0}p_{h,\alpha}\left(1 - \frac{\rho'}{\rho_0}\right) - \frac{1}{\rho_0}p'_{,\alpha} + \tag{1.111}$$

$$\mathcal{O}\left(\frac{p'\rho'}{\rho_0^2}, \left(\frac{\rho'}{\rho_0}\right)^2 p_h\right) \cong g_\alpha \frac{\rho'}{\rho_0} - \frac{1}{\rho_0}p'_{,\alpha} \; .$$

Using (1.110) we see that the neglected terms are negligible compared to the terms kept. Furthermore,

$$\frac{1}{\rho}\sigma_{\alpha\beta} = \frac{1}{\rho_0}\left(1 + \mathcal{O}\left(\frac{\rho'}{\rho_0}\right)\right)\sigma_{\alpha\beta} \cong \frac{1}{\rho_0}\sigma_{\alpha\beta} \; . \tag{1.112}$$

At first sight this approximation seems unjustified, because we neglect a term of $\mathcal{O}(\rho'/\rho_0)$, while terms of this magnitude are kept in (1.111). However, because (1.111) is assumed to be the driving term, $\sigma_{\alpha\beta,\beta} = \mathcal{O}(\frac{\rho'}{\rho_0}, \frac{1}{\rho_0}p'_{,\alpha})$, so that the neglected term in (1.112) is in fact quadratic in the perturbations. Using the equation of state we may write

$$\rho' \cong -b\rho_0 T' , \qquad (1.113)$$

with $T' = T - T_0$, and

$$b = -\frac{1}{\rho_0}\frac{d\rho(T_0)}{dT} ,$$

with b the thermal expansion coefficient of the fluid. Substitution of (1.111)–(1.113) in (1.102) gives

$$\frac{Du_\alpha}{Dt} = -\frac{1}{\rho_0}p'_{,\alpha} + \frac{1}{\rho_0}\sigma_{\alpha\beta,\beta} - g_\alpha b T' . \qquad (1.114)$$

These are the *Boussinesq equations*. The flow that results if the hydrostatic state is unstable is called *free convection*.

Dimensionless Boussinesq equations

The following units are chosen:

density: ρ_0
length: L (a typical length scale of the flow domain)
velocity: $U = \frac{\mu_0}{\rho_0 L}$ (with μ_0 a reference viscosity value)
time: $\tau = L/U$
temperature: ΔT (a typical magnitude of the variations in T that are imposed by the boundary conditions)
pressure: $P = \rho_0 U^2$

The dimensionless form of (1.114) is found to be, keeping the same symbols (i.e. $\boldsymbol{u} := \boldsymbol{u}/U$, $\mu := \mu/\mu_0$, $g_\alpha := g_\alpha \tau^2/L$ etc.):

$$\frac{Du_\alpha}{Dt} = -p'_{,\alpha} + \sigma_{\alpha\beta,\beta} - G g_\alpha T' , \qquad (1.115)$$

where the dimensionless number G, called the *Grashoff number*, is defined by

$$G = b\Delta T g L^3 \rho_0^2/\mu_0^2 .$$

The system of equations is completed by the mass conservation equation and the energy equation for T'. The mass conservation equation is given by (1.18). Using (1.110) this is approximated by

$$\mathrm{div}\boldsymbol{u} = 0 . \tag{1.116}$$

The energy equation is given by (1.55) with $\psi = 0$. In dimensionless variables we have $\rho = 1$, and with $q = 0$ we obtain the following dimensionless form of the energy equation:

$$\frac{DT}{Dt} = \mathrm{Pr}^{-1}(kT_{,\alpha})_{,\alpha} , \tag{1.117}$$

where the dimensionless number Pr is the *Prandtl number*, defined by

$$\mathrm{Pr} = k_0/(c_p\mu_0) .$$

The equations governing free convection (1.115)–(1.117) are called the Boussinesq equations.

1.15 Moving frame of reference

In applications with a rotating geometry, such as helicopter blades, turbomachinery or the atmosphere or seas and oceans, it is necessary to use a rotating coordinate system. Therefore we will present the laws that govern fluid motion in a moving Cartesian coordinate system.

Assume that we have a Cartesian reference frame that is not inertial but that is moving. Let the instantaneous linear acceleration of the origin be $\boldsymbol{a}$ and let the instantaneous rotation around the origin be $\boldsymbol{\Omega}$. Batchelor (1967) Sect. 3.2 shows that the mass, momentum and energy conservation equations are identical in form to those in an inertial frame, provided we add the fictitious body force

$$\tilde{\boldsymbol{f}}^b = -\boldsymbol{a} - 2\boldsymbol{\Omega} \times \boldsymbol{u} - \frac{d\boldsymbol{\Omega}}{dt} \times \boldsymbol{x} - \boldsymbol{\Omega} \times (\boldsymbol{\Omega} \times \boldsymbol{x})$$

per unit mass to the real body force $\boldsymbol{f}^b$. This modification is to take place both in the momentum equation (1.26) and in the energy equation (1.40).

The term $-2\boldsymbol{\Omega} \times \boldsymbol{u}$ is the Coriolis acceleration. Note that $\boldsymbol{u}$ is the velocity relative to the moving frame. The last term can be rewritten as follows. Let $\boldsymbol{x} = \boldsymbol{y} + \boldsymbol{r}$ with $\boldsymbol{y}$ the orthogonal projection of $\boldsymbol{x}$ on $\boldsymbol{\Omega}$:

$$\boldsymbol{y} = (\boldsymbol{x} \cdot \boldsymbol{\Omega}/\boldsymbol{\Omega} \cdot \boldsymbol{\Omega})\boldsymbol{\Omega} .$$

Then $\boldsymbol{\Omega} \times \boldsymbol{x} = \boldsymbol{\Omega} \times \boldsymbol{r}$, and $\boldsymbol{r} \cdot \boldsymbol{\Omega} = 0$, so that

$$\boldsymbol{\Omega} \times (\boldsymbol{\Omega} \times \boldsymbol{x}) = \boldsymbol{\Omega} \times (\boldsymbol{\Omega} \times \boldsymbol{r}) = (\boldsymbol{\Omega} \cdot \boldsymbol{r})\boldsymbol{\Omega} - (\boldsymbol{\Omega} \cdot \boldsymbol{\Omega})\boldsymbol{r}$$
$$= -\Omega^2 \boldsymbol{r} , \quad \Omega^2 = \boldsymbol{\Omega} \cdot \boldsymbol{\Omega} .$$

This is recognized as centrifugal acceleration.

The Cartesian components of $\tilde{f}^b$ are:

$$\tilde{f}_\alpha^b = -a_\alpha - \varepsilon_{\alpha\beta\gamma}\{2\Omega_\beta u_\gamma + \frac{d\Omega_\beta}{dt}x_\gamma\} + \Omega^2 r_\alpha \,, \quad \boldsymbol{r} = \boldsymbol{x} - \boldsymbol{y} \,.$$

1.16 The shallow-water equations

Depth-averaged continuity equation

Consider a three-dimensional domain in which water flows with a free surface under the influence of gravity. The situation is illustrated in Fig. 1.2. The

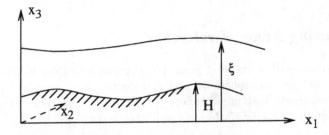

Fig. 1.2. A free surface flow problem

x_3-axis is in the direction opposite to that of gravity. The water elevation is $\xi(t, x_1, x_2)$ and the bottom topography is given by $H(t, x_1, x_2)$, so that the water depth is $d = \xi - H$. The free surface may also be defined by $F(t, \boldsymbol{x}) \equiv x_3 - \xi(t, x_1, x_2) = 0$. Since the free surface moves with the flow, a material particle that is at the free surface stays at the free surface (except in singular points), so that $DF/Dt = 0$, or

$$\frac{\partial\xi}{\partial t} + u_1\frac{\partial\xi}{\partial x_1} + u_2\frac{\partial\xi}{\partial x_2} - u_3 = 0 \,. \tag{1.118}$$

Similarly,

$$\frac{\partial H}{\partial t} + u_1\frac{\partial H}{\partial x_1} + u_2\frac{\partial H}{\partial x_2} - u_3 = 0 \,. \tag{1.119}$$

In the *shallow-water approximation* the depth-averaged horizontal velocity components U_1 and U_2 are used as unknowns instead of u_1, u_2:

$$U_\alpha = \frac{1}{d}\int_H^\xi u_\alpha dx_3, \quad \alpha = 1, 2 \,.$$

The density ρ is assumed to be constant. Integration of the continuity equation (1.15) over x_3 gives

$$\int_H^\xi (\frac{\partial u_1}{\partial x_1} + \frac{\partial u_2}{\partial x_2})dx_3 + u_3(t, x_1, x_2, \xi) - u_3(t, x_1, x_2, H) = 0 . \qquad (1.120)$$

We have, summing over $\alpha = 1, 2$:

$$\int_H^\xi \frac{\partial u_\alpha}{\partial x_\alpha}dx_3 = \frac{\partial}{\partial x_\alpha}(dU_\alpha) - u_\alpha(t, x_1, x_2, \xi)\frac{\partial \xi}{\partial x_\alpha} + u_\alpha(t, x_1, x_2, H)\frac{\partial H}{\partial x_\alpha} .$$

By using (1.118) and (1.119) and substitution in (1.120) we obtain the *depth-averaged continuity equation*:

$$\frac{\partial d}{\partial t} + \frac{\partial}{\partial x_1}(dU_1) + \frac{\partial}{\partial x_2}(dU_2) = 0 , \qquad (1.121)$$

where we have used $\partial(\xi - H)/\partial t = \partial d/\partial t$.

Depth-averaged momentum equations

Since many interesting applications of the (to be derived) shallow-water equations arise in geophysics, it is of interest to take the rotation of the earth into account by choosing a rotating frame of reference. But the domain is assumed to be small enough to be assumed flat, and we will continue the use of Cartesian coordinates. Including the effect of Coriolis and gravitational acceleration, according to the preceding section we have a body force

$$\boldsymbol{f}^b = \boldsymbol{g} - 2\boldsymbol{\Omega} \times \boldsymbol{u}$$

in the momentum equations (1.27), where $\boldsymbol{\Omega}$ is the rotation vector of the earth. The centrifugal acceleration is negligible in geophysical applications.

In the shallow-water approximation it is assumed that $|u_3| \ll |u_1|, |u_2|$ and u_3 is approximated by zero. This gives, neglecting the vertical component of the Coriolis acceleration because it is very small compared to gravity:

$$2\boldsymbol{\Omega} \times \boldsymbol{u} = (-fu_2, fu_1, 0) , \quad f = 2|\boldsymbol{\Omega}| \sin \phi , \qquad (1.122)$$

with ϕ the geographic latitude. With $u_3 = 0$, the x_3-component of (1.27) becomes:

$$0 = -\frac{\partial p}{\partial x_3} - \rho g ,$$

so that

$$p = F(x_1, x_2) - \rho g x_3 \, ,$$

with F still to be determined. At the free surface p equals the atmospheric pressure p_a. This determines F, and we obtain

$$p = p_a + \rho g(\xi - x_3). \tag{1.123}$$

The remaining momentum equations become

$$\rho \frac{Du_1}{Dt} = -p_{,1} + \mu u_{1,\beta\beta} + \rho f u_2 \, ,$$

$$\rho \frac{Du_2}{Dt} = -p_{,2} + \mu u_{2,\beta\beta} - \rho f u_1 \, .$$

To exploit the shallow-water approximation, we integrate over the depth. We stipulate that the range of Latin indices is $\{1, 2\}$. The summation convection is temporarily extended to Latin indices. Because $\operatorname{div} \boldsymbol{u} = 0$ we have $u_\beta \partial u_j / \partial x_\beta = \partial u_j u_\beta / \partial x_\beta$. Integration of the inertia terms gives:

$$\int_H^\xi \frac{\partial u_j u_\beta}{\partial x_\beta} dx_3 = \frac{\partial}{\partial x_k} \int_H^\xi u_j u_k dx_3$$

$$-\{u_j(u_k \frac{\partial \xi}{\partial x_k} - u_3)\}|_{x_3=\xi} + \{u_j(u_k \frac{\partial H}{\partial x_k} - u_3)\}u_j|_{x_3=H}$$

$$= \frac{\partial}{\partial x_k} \int_H^\xi u_j u_k dx_3 + \frac{\partial \xi}{\partial t} u_j|_{x_3=\xi} - \frac{\partial H}{\partial t} u_j|_{x_3=H} \, .$$

Furthermore,

$$\int_H^\psi \frac{\partial u_j}{\partial t} dx_3 = \frac{\partial}{\partial t}(dU_j) - \frac{\partial \xi}{\partial t} u_j|_{x_3=\xi} + \frac{\partial H}{\partial t} u_j|_{x_3=H} \, .$$

Hence

$$\int_H^\xi \frac{Du_j}{Dt} dx_3 = \frac{\partial}{\partial t}(dU_j) + \frac{\partial}{\partial x_k} \int_H^\xi u_j u_k dx_3 \, .$$

Let us write

$$u_j = U_j + \tilde{u}_j \, .$$

Then

$$\int_H^\xi u_j u_k dx = dU_j U_k + \sigma_{jk,k} \, ,$$

with

$$\sigma_{jk} = \int_H^{\xi} \tilde{u}_j \tilde{u}_k dx_3 \ .$$

Next, σ_{jk} is neglected, which is justified only if $|\tilde{u}_j| \ll |U_j|$, in other words, if u_α depends only weakly on x_3. Due to the no-slip condition, this will not be the case near the bottom. For more accurate modeling, depth-averaging would have to be done over more than one layer, or abandoned altogether, but we will not do this here. Proceeding with these approximations, we now have:

$$\int_H^{\xi} \frac{Du_j}{Dt} dx_3 \cong \frac{\partial}{\partial t}(dU_j) + \frac{\partial}{\partial x_k}(dU_j U_k) \ .$$

Because of (1.121) this may also be written as

$$\int_H^{\xi} \frac{Du_j}{Dt} dx_3 \cong d\frac{\partial U_j}{\partial t} + dU_k \frac{\partial U_j}{\partial x_k} \ .$$

The next term to be depth-averaged is the pressure gradient. By using (1.123) we obtain (assuming p_a = constant)

$$\frac{1}{\rho} \int_H^{\xi} p_{,j} dx_3 = gd\xi_{,j} \ .$$

The viscous stress term is not depth-averaged. The influence of friction makes itself felt mainly at the bottom, but since depth-averaging implies that the no-slip condition is not applied, the influence of bottom friction is accounted for in a simplified semi-empirical way by means of a force term F_j, given by

$$F_j = -gU_j |U|/C^2 d \ ,$$

with C the so-called *Chézy coefficient*, and $|U| = (U_\alpha U_\alpha)^{1/2}$. The empirical parameter C depends on the bottom roughness.

Depth-averaging of the Coriolis acceleration gives

$$\int_H^{\xi} fu_j d\xi = fdU_j \ .$$

The depth-averaged momentum equations can now be summarized as follows:

$$\frac{\partial U_1}{\partial t} + U_1 \frac{\partial U_1}{\partial x_1} + U_2 \frac{\partial U_1}{\partial x_2} + g \frac{\partial \xi}{\partial x_1} - f U_2 + \frac{g U_1 |U|}{C^2 d} = 0 \,,$$
$$\frac{\partial U_2}{\partial t} + U_1 \frac{\partial U_2}{\partial x_1} + U_2 \frac{\partial U_2}{\partial x_2} + g \frac{\partial \xi}{\partial x_2} + f U_1 + \frac{g U_2 |U|}{C^2 d} = 0 \,. \tag{1.124}$$

Together with (1.121) these form the shallow-water equations.

For a more extended discussion of the mathematical modeling of flows of shallow water, see Vreugdenhil (1994). The main assumption for the validity of the shallow-water equations is that the water depth is much smaller than the horizontal length scales. This implies that the wavelength of surface waves should be much langer than the depth (more than 20 times as large, say, cf. Vreugdenhil (1994) p.20). A systematic derivation of the shallow-water equations using asymptotic expansions is given in Stoker (1957), Sect. 2.4.

Gasdynamics analogy

If we neglect Coriolis acceleration and bottom friction, take the bottom height $H(x_1, x_2) = \text{constant}$ and give d the new name ρ, the shallow-water equations (1.121), (1.124) take on the following appearance, with $\alpha \in \{1, 2\}$:

$$\frac{\partial \rho}{\partial t} + (\rho U_k)_{,k} = 0 \,, \tag{1.125}$$

$$\frac{D U_j}{Dt} + g \rho_{,j} = 0 \,. \tag{1.126}$$

By multiplying (1.125) by U_j and (1.126) by ρ and adding we obtain

$$\frac{\partial \rho U_j}{\partial t} + (\rho U_j U_k)_{,k} + \rho g \rho_{,j} = 0 \,. \tag{1.127}$$

Equations (1.125) and (1.127) are equivalent to the isentropic Euler equations (1.76), (1.85), for ideal fluids with ratio of specific heats $\gamma = 2$. This is called *the gasdynamics analogy*.

2. Partial differential equations: analytic aspects

2.1 Introduction

As seen in the preceding chapter, fluid dynamics is governed by partial differential equations. Therefore knowledge of the numerical analysis of partial differential equations is indispensable in computational fluid dynamics. Introductions to this subject of varying degree of difficulty are Fletcher (1988), Hackbusch (1986), Hall and Porsching (1990), Hirsch (1988), Lapidus and Pinder (1982), Mitchell and Griffiths (1994), Morton and Mayers (1994), Grossmann and Roos (1994), Strikwerda (1989), Quarteroni and Valli (1994), Richtmyer and Morton (1967); the last book in particular is useful for practitioners of computational fluid dynamics.

Of course, in the study of the numerical aspects of partial differential equations, their analytic aspects play an important role. Although the reader is assumed to be familiar with the basics of the analysis of partial differential equations, we will devote this chapter to this subject, in order to highlight a few important topics that receive less attention elsewhere, and that will play a role later. In particular, we will discuss *maximum principles*, in order to put discussions of numerical 'wiggles' (see Chap. 4) on a firm footing. Furthermore, we will treat those aspects of *singular perturbation theory* that are required for a thorough understanding of the convection-diffusion equation. Also, we will pay more attention than usual to the treatment of mixed derivatives. These have often been neglected in the past, presumably because mixed derivatives seldom arise in mathematical physics. However, at present the predominant approach to handle arbitrarily shaped flow domains in the context of finite volume discretization is by means of *boundary fitted coordinates* (see Chap. 11). Because these coordinates are generally non-orthogonal, the governing equations frequently have mixed derivatives in the transformed coordinates.

More advanced information on the analytic aspects of partial differential equations can be found in Courant and Hilbert (1989), Garabedian (1964), Hackbusch (1986), Kreiss and Lorenz (1989), Protter and Weinberger (1967).

We will discuss in this chapter (special cases of) the following linear partial differential equation:

$$L\varphi \equiv \varphi_t - (a_{\alpha\beta}\varphi_{,\beta})_{,\alpha} + b_\alpha\varphi_{,\alpha} + c\varphi = s \,,$$
$$\varphi, s : \Omega \times (0, T] \to \mathbb{R}, \quad \Omega \subset \mathbb{R}^d, \quad \alpha, \beta = 1, 2, ..., d \,. \tag{2.1}$$

where t stands for time, φ_t for $\partial\varphi/\partial t$, and d is the number of space dimensions. Boundary conditions will be specified later. The coefficients $a_{\alpha\beta}, b_\alpha$ and c are given functions of t and $\boldsymbol{x} \in \Omega$.

2.2 Classification of partial differential equations

Stationary case

We start with assuming that φ does not depend on time, so that the time-derivative in (2.1) is deleted. Furthermore, for the time being it is convenient to rewrite the equation as

$$L\varphi \equiv -a_{\alpha\beta}\varphi_{,\alpha\beta} + b_\alpha\varphi_{,\alpha} + c\varphi = s \,, \tag{2.2}$$

which implies the following replacement;

$$b_\alpha := b_\alpha - a_{\beta\alpha,\beta} \,.$$

Since $\varphi_{,\alpha\beta} = \varphi_{,\beta\alpha}$ we can assume without loss of generality that

$$a_{\alpha\beta} = a_{\beta\alpha} \tag{2.3}$$

(if (2.3) does not hold, simply redefine $a_{\alpha\beta} = a_{\beta\alpha} := \frac{1}{2}(a_{\alpha\beta} + a_{\beta\alpha})$). Consequently, the matrix A with elements $a_{\alpha\beta}$ is symmetric and has real eigenvalues.

Three special types of partial differential equations are distinguished, according to the following definition.

Definition 2.2.1. *Classification of partial differential equations.*
Equation (2.2) is called

(i) elliptic in $\boldsymbol{x}$ if the eigenvalues of A are nonzero and have the same sign;
(ii) hyperbolic in $\boldsymbol{x}$ if the eigenvalues of A are nonzero and precisely one eigenvalue has sign different from all others;
(iii) parabolic in $\boldsymbol{x}$ if precisely one eigenvalue is zero, while the others have the same sign and $\mathrm{Rank}(A, \boldsymbol{b}) = d$, with $\boldsymbol{b}$ the vector with elements b_α.

In the elliptic case the symmetric matrix A is definite. Without loss of generality we assume in this case A to be positive definite, because if this is not the case we may reverse sign in (2.2).

It can be shown that the type of a partial differential equation is invariant under coordinate transformation (see Hackbusch (1986) Sect. 1.2). Note that there are undefined cases.

Nonstationary case

The nonstationary case $\partial \varphi / \partial t + L\varphi = s$ can be subsumed under the classification for the stationary case $L\varphi = s$ by introducing time t as an additional coordinate x_{d+1}, and extending the range of α and β to $1, 2, ..., d + 1$. The result is the following change in A and b:

$$A = \begin{pmatrix} A_s & 0 \\ 0 & 0 \end{pmatrix}, \quad b = \begin{pmatrix} b_s \\ 1 \end{pmatrix},$$

where the subscript s refers to the stationary case. Note that if the stationary case is elliptic, then the nonstationary case is parabolic.

Two-dimensional stationary case

In this case we find that equation (2.2) is
(i) elliptic in x if

$$a_{11}a_{22} - a_{12}^2 > 0, \tag{2.4}$$

(ii) hyperbolic in x if

$$a_{11}a_{22} - a_{12}^2 < 0, \tag{2.5}$$

(iii) parabolic in x if

$$a_{11}a_{22} - a_{12}^2 = 0 \quad \text{and} \quad \text{Rank} \begin{pmatrix} a_{11} & a_{12} & b_1 \\ a_{12} & a_{22} & b_2 \end{pmatrix} = 2. \tag{2.6}$$

The archetypes of the three cases are

$$-\varphi_{,\alpha\alpha} = 0, \tag{2.7}$$

(*Laplace's equation*, elliptic),

$$-\varphi_{,11} + \varphi_{,22} = 0, \tag{2.8}$$

(the *wave equation*, hyperbolic),

$$\varphi_{,1} - \varphi_{,22} = 0, \tag{2.9}$$

(the *diffusion equation*, parabolic).
The three types require different boundary conditions (more on this later) and different numerical methods.

The coefficients and hence the type depend on x, so that the type may change in Ω. A classic example is the *Tricomi equation*:

$$-\varphi_{,11} - x_2\varphi_{,22} = 0 .$$

This equation is elliptic for $x_2 > 0$ and hyperbolic for $x_2 < 0$; for $x_2 = 0$ the type is not defined. This situation arises in transonic flow, where we have ellipticity where the flow is subsonic and hyperbolicity where it is supersonic.

Example 2.2.1. Equation of undefined type
Consider

$$\frac{\partial \varphi}{\partial t} - 2\varphi_{,12} = 0 .$$

What is the type of this equation? It is left to the reader to verify, that the matrix to consider is

$$A = \begin{pmatrix} 0 & 1 & 0 \\ 1 & 0 & 0 \\ 0 & 0 & 0 \end{pmatrix} .$$

The eigenvalues are $0, \pm 1$, so that the above equation is unclassified.   □

Physical significance of classification

The classification corresponds to different types of qualitative behavior of solutions, and hence to differences in the underlying physics. This can be shown as follows. A plane wave solution is given by

$$\varphi = e^{iw(\boldsymbol{x})} , \quad w(\boldsymbol{x}) = \boldsymbol{n} \cdot \boldsymbol{x} - b ,$$

where the function $w(\boldsymbol{x})$ describes the wave fronts or *characteristic surfaces*, given by $w(\boldsymbol{x}) = $ constant. Substitution in (2.2) gives (in order to bring out the physical differences between the three cases it suffices to restrict ourselves to $b_\alpha = c = s = 0$, $a_{\alpha\beta} = $ constant)

$$a_{\alpha\beta}n_\alpha n_\beta = 0 , \tag{2.10}$$

with $\boldsymbol{n} = \mathrm{grad} w$ the normal to the wave front. It is convenient to rewrite (2.10) as

$$\boldsymbol{n}^T A \boldsymbol{n} = 0 . \tag{2.11}$$

Let $\boldsymbol{v}^{(\alpha)}$, $\alpha = 1, 2, ..., d$ be the set of orthonormal eigenvectors of the matrix A, with corresponding eigenvalues λ_α. Because A is symmetric, $\boldsymbol{v}^{(\alpha)}$ and λ_α are real. Let us write

$$\boldsymbol{n} = c_\alpha \boldsymbol{v}^{(\alpha)} .$$

Substitution in (2.11) gives

$$\lambda_\alpha c_\alpha^2 = 0 \ . \tag{2.12}$$

If (2.2) is elliptic then $\lambda_\alpha > 0$, which contradicts (2.12), so that there is no solution for $\boldsymbol{n}$; hence wave-like behavior of the solution is not to be expected. If $\lambda_\alpha > 0$ then A is positive definite and vice-versa, so that

$$a_{\alpha\beta}\xi_\alpha\xi_\beta > 0 \ , \quad \forall \, \xi_\alpha \neq 0 \ . \tag{2.13}$$

This is often taken as the definition of ellipticity of (2.2).

In the hyperbolic case one eigenvalue, say λ_1, is negative, and solutions are obtained with

$$c_1^2 = \big(\sum_{\alpha > 1} \lambda_\alpha c_\alpha^2 \big)/\lambda_1 \tag{2.14}$$

so that wave-like solutions exist. Furthermore, $\boldsymbol{n}$ can point in any direction, since there are $d-1$ free parameters in (2.14); hence, wave propagation may take place in any direction. In the parabolic case, let λ_1 be the zero eigenvalue. The only solutions of (2.12) are $c_1 \neq 0$, $c_\alpha = 0$, $\alpha > 1$, so that wave propagation can take place in one specific direction only, corresponding to a *time-like* variable.

Another way to illuminate the physical significance of the classification is to look at the role of the boundary conditions. As we will see, in the elliptic case boundary conditions have to be prescribed all along the boundary of the domain, and a local change in boundary data influences the solution everywhere. One might therefore say that elliptic equations model *equilibrium* phenomena. In the hyperbolic case a local change in boundary data propagates its influence only in part of the domain, showing a wave-like behavior, so that hyperbolic equations model *propagation* phenomena. In the parabolic case, as we just saw, invariably a *time-like* independent variable can be identified. Changes in boundary data at a certain time make themselves felt everywhere, but only at later times. The time-like variable cannot be reversed without jeopardizing the well-posedness of the problem, as will be seen. Note that equations (2.7) and (2.8) are invariant under the transformation $x_\alpha := -x_\alpha$, but that (2.9) is not invariant under the transformation $x_1 := -x_1$. This *irreversibility* is why the variable concerned is called time-like. In short, parabolic equations model *diffusion* processes.

First order systems

Extension of the classification to equations more general than (2.2) can be done on the basis of the criterion whether or not wave-like solutions exist. It is always possible to write a system of partial differential equations as a first order system of m equations with m unknowns:

$$F_0 U_t + F_\alpha U_{,\alpha} = Q \,, \qquad (2.15)$$

with $U = (U_1, ..., U_m)$ and where $F_0(t, \boldsymbol{x}, U)$, $F_\alpha(t, \boldsymbol{x}, U)$ are $m \times m$ matrices.

We consider the case where F_0 and F_α are made constant by 'freezing' them to their local value at $t = t_p$, $\boldsymbol{x} = \boldsymbol{x}_p$, $U = U_p$. We now ask whether plane waves can be solutions of the homogeneous version of (2.15). Plane waves are solutions of the following type:

$$U = \hat{U} e^{iw(t, \boldsymbol{x})}, \quad \hat{U} = \text{constant} \,, \quad w(t, \boldsymbol{x}) = \boldsymbol{n} \cdot \boldsymbol{x} - \lambda t \,.$$

This wave travels in the direction $\boldsymbol{n}$ with velocity $\lambda / |\boldsymbol{n}|$. Substitution shows that plane waves are solutions if

$$(n_\alpha F_\alpha - \lambda F_0)\hat{U} = 0 \,. \qquad (2.16)$$

Here $\boldsymbol{n} \in \mathbb{R}^d$ is real. This is called a generalized eigenvalue problem. Eigensolutions $\hat{U}$ may exist if λ and $\boldsymbol{n}$ satisfy some constraint:

$$\lambda = \lambda(\boldsymbol{n}) \,.$$

The classification of (2.15) in $\{t_p, \boldsymbol{x}_p, U_p\}$ depends on the number of plane waves that exist. The classification is as follows.

Definition 2.2.2. *The system (2.15) is called hyperbolic in $(t_p, \boldsymbol{x}_p, U_p)$ if all eigenvalues $\lambda = \lambda(\boldsymbol{n})$ of (2.16) are real and if there exist m linearly independent eigenvectors.*

If, on the other hand, no real eigenvalues exist at all, then wave-like solutions are not possible, and the system is called elliptic:

Definition 2.2.3. *The system (2.15) is called elliptic in $(t_p, \boldsymbol{x}_p, U_p)$ if the eigenvalue problem (2.16) has no real eigenvalues $\lambda = \lambda(\boldsymbol{n})$.*

A generalization of parabolicity to the case of first order systems is:

Definition 2.2.4. *The system (2.15) is called parabolic in $(t_p, \boldsymbol{x}_p, U_p)$ if all eigenvalues $\lambda = \lambda(\boldsymbol{n})$ of (2.16) are real and if there exist only $m - 1$ linearly independent eigenvectors.*

These definitions can be applied to the stationary case by deleting λ and F_0, and by checking whether $\boldsymbol{n}$ is real or not, and by counting the eigenvectors.

If the system (2.15) does not fall into one of these three categories it is called *hybrid*. Because higher-order systems can be reformulated as first order systems, this classification scheme can be generally applied.

Example 2.2.2. Second order equation
Consider the following special case of (2.2):

$$-a_{\alpha\beta}\varphi_{,\alpha\beta} = 0 \,. \qquad (2.17)$$

This is equivalent to the following first order system:

$$\varphi_{,\alpha} = U_\alpha, \quad -a_{\alpha\beta}U_{\alpha,\beta} = 0 . \tag{2.18}$$

In order to obtain a system of type (2.15) we eliminate φ by differentiation and combination:

$$\varphi_{,\alpha d} - \varphi_{,d\alpha} = 0, \quad \alpha = 1, 2, ..., d-1 ,$$

and we obtain the following system of d equations for d unknowns:

$$\begin{aligned} U_{\alpha,d} - U_{d,\alpha} &= 0, \quad \alpha = 1, 2, ..., d-1 , \\ -a_{\alpha\beta}U_{\alpha,\beta} &= 0 . \end{aligned} \tag{2.19}$$

The reader may verify that in the case of Laplace's equation and $d = 2$ equations (2.19) are the *Cauchy-Riemann* equations:

$$U_{1,2} - U_{2,1} = 0, \quad U_{\alpha,\alpha} = 0 .$$

These equations are equivalent to Laplace's equation, but for $d > 2$ equations (2.19) and (2.17) are not equivalent. Solutions of (2.17) are solutions of (2.19), but not vice-versa. For example, for $d = 3$ a solution of (2.19) is obtained with $U_3 = 0$, $U_\alpha = U_\alpha(x_1, x_2)$, $\alpha = 1, 2$ and $-a_{\alpha\beta}U_{\alpha,\beta} = 0$. However, in general this will not satisfy $U_{1,2} = U_{2,1}$, so that we cannot solve for φ from (2.18). In other words, the transformation (2.18) introduces *spurious solutions*.

By comparing (2.19) and (2.15) and specializing to $d = 3$ we find $F_0 = 0$ and

$$F_1 = - \begin{pmatrix} 0 & 0 & 1 \\ 0 & 0 & 0 \\ a_{11} & a_{21} & a_{31} \end{pmatrix}, \quad F_2 = - \begin{pmatrix} 0 & 0 & 0 \\ 0 & 0 & 1 \\ a_{12} & a_{22} & a_{32} \end{pmatrix},$$

$$F_3 = - \begin{pmatrix} -1 & 0 & 0 \\ 0 & -1 & 0 \\ a_{13} & a_{23} & a_{33} \end{pmatrix} .$$

For (2.16) to have a non-trivial solution of the type $e^{iw(x)}$, $w(x) = n \cdot x$ we must have $\det(\sum n_\alpha F_\alpha) = 0$, or

$$0 = \begin{vmatrix} n_3 & 0 & -n_1 \\ 0 & n_3 & -n_2 \\ -n_\alpha a_{1\alpha} & -n_\alpha a_{2\alpha} & -n_\alpha a_{3\alpha} \end{vmatrix} = -n_3 n_\alpha n_\beta a_{\alpha\beta} . \tag{2.20}$$

One real solution is obtained for $n_3 = 0$, giving $w = w(x_1, x_2)$. From (2.16) it follows that $\hat{U}_3 = 0$, $\sum_{\alpha=1}^{2} \sum_{\beta=1}^{2} w_{,\alpha} a_{\beta\alpha} \hat{U}_\beta = 0$, which has non-zero real solutions for every $w(x_1, x_2)$, giving solutions of (2.17) of the form $U_\alpha = \hat{U}_\alpha e^{iw}$, $\alpha = 1, 2$, $U_3 = 0$. However, in general these will not satisfy $U_{1,2} - U_{2,1} = 0$, or

$w_{,2}\hat{U}_1 - w_{,1}\hat{U}_2 = 0$. Hence, these are spurious solutions. The other possibility following from (2.20) is

$$n_\alpha n_\beta a_{\alpha\beta} = 0 ,$$

which is just (2.11), bringing us back to the classification of second order systems. We see, that apart from spurious solutions, the classification of first order systems agrees with the classification of second order systems.   □

Example 2.2.3. Parabolic equation
Consider equation (2.9). It can be written as a first order system by defining

$$U_1 = \varphi , \quad U_2 = \varphi_{,2} .$$

We elemine φ by differentiation, as in Example 2.2.2, and obtain the following first order system:

$$U_{1,2} = U_2 ,$$
$$U_{1,1} - U_{2,2} = 0 .$$

This gives us the following first order system:

$$F_1 U_{,1} + F_2 U_{,2} = Q , \quad F_1 = \begin{pmatrix} 0 & 0 \\ 1 & 0 \end{pmatrix} , \quad F_2 = \begin{pmatrix} 1 & 0 \\ 0 & -1 \end{pmatrix} , \quad Q = \begin{pmatrix} U_2 \\ 0 \end{pmatrix} .$$

The eigenproblem becomes

$$0 = \begin{vmatrix} n_2 & 0 \\ n_1 & -n_2 \end{vmatrix} = -n_2^2 ,$$

giving $n = (n_1, 0)$ with n_1 arbitrary. The eigenvector $\hat{U}$ has to satisfy $\hat{U}_1 = 0$, so that there is only one eigenvector:

$$\hat{U} = \begin{pmatrix} 0 \\ \hat{U}_2 \end{pmatrix}$$

Hence, the system is parabolic according to Definition 2.2.4, and we find this classification to be in agreement with Definition 2.2.1.   □

Exercise 2.2.1. Derive the form taken by the wave equation (2.8) after the following change of coordinates:

$$y_1 = x_1 + x_2, \quad y_2 = x_1 - x_2 ,$$

and determine the type of the resulting equation. Show that every solution of the wave equation can be written as

$$\varphi(\boldsymbol{x}) = \psi(x_1 + x_2) + \eta(x_1 - x_2) . \tag{2.21}$$

Exercise 2.2.2. Show that

$$\varphi(t, x) = \frac{1}{\sqrt{4\pi t}} \int\limits_{-\infty}^{\infty} \varphi_0(\xi) \exp(-(x - \xi)^2/4t)d\xi$$

satisfies the diffusion equation (2.9). More difficult (see Hackbusch (1986)): show that

$$\lim_{t \downarrow 0} \varphi(t, x) = \varphi_0(x) .$$

Exercise 2.2.3. Prove (2.4)–(2.6).

Exercise 2.2.4. Determine, depending on the parameter c, the type of

$$-\varphi_{,11} + 2c\varphi_{,12} - \varphi_{,22} = 0 .$$

Exercise 2.2.5. Let a coordinate transformation $x \to y$ be given by

$$x = By ,$$

with B a constant non-singular $d \times d$ matrix. Show that this transformation leaves the type of equation (2.2) invariant. Hint: use Sylvester's lemma: If $C = B^T AB$ and $A = A^T$, then the number of positive, zero and negative eigenvalues of A and C is the same.

Exercise 2.2.6. Repeat Exercise 2.2.5 for equation (2.1). Next, consider the transformation

$$\begin{pmatrix} t \\ x \end{pmatrix} = B \begin{pmatrix} \tau \\ y \end{pmatrix} ,$$

with B a constant non-singular $(d+1) \times (d+1)$ matrix. Show that the type of equation (2.1) is invariant.

2.3 Boundary conditions

We start with the stationary case of (2.1):

$$L\varphi \equiv -(a_{\alpha\beta}\varphi_{,\beta})_{,\alpha} + b_\alpha\varphi_{,\alpha} + c\varphi = s , \tag{2.22}$$

and assume this equation to be elliptic for all $x \in \Omega$.

Well-posed problems

To ensure the existence of a unique solution, suitable boundary conditions have to be imposed at the boundary $\partial\Omega$ of Ω. In order to be amenable to computation in the presence of rounding and truncation errors, and in general also in order to correctly model physical phenomena, the solution φ should depend continuously on the data, i.e. on the right-hand-side s and on the prescribed boundary values. In other words, small perturbations in the data should not cause large changes in the solution. This leads us to the concept of *well-posed problems*. Let the boundary condition be given by

$$B\varphi = f \quad \text{on} \quad \partial\Omega \tag{2.23}$$

with B some operator, typical examples of which are given in (2.26)–(2.28), or B could be the zero operator on part of $\partial\Omega$, in which case no boundary condition is prescribed there. Let $\Phi \ni \varphi$, $S \ni s$ and $F \ni f$ be suitable function spaces. Then the concept of well-posedness can be defined as follows:

Definition 2.3.1. *Well-posedness*
The problem specified by (2.22) and (2.23) is called well-posed if for all $f \in F$, $s \in S$ the following two conditions are satisfied:

(i) There exists a unique solution $\varphi \in \Phi$;
(ii) For every two sets of data f_1, s_1 and f_2, s_2 in F and S the corresponding solutions φ_1 and φ_2 satisfy

$$\|\varphi_1 - \varphi_2\|_\Phi \leq C\{\|f_1 - f_2\|_F + \|s_1 - s_2\|_S\} \tag{2.24}$$

with C some fixed constant.

For mathematical precision, the function spaces Φ, F and S have to be specified (their choice depends on the problem, notably on the smoothness of $\partial\Omega$), but the general idea is clear.

The following three types of boundary conditions lead to well-posed elliptic boundary value problems, assuming

$$c \geq 0 \tag{2.25}$$

(we will not prove this; there is a huge amount of literature on this subject):

$$\varphi = f \quad \text{on} \quad \partial\Omega \qquad \qquad \text{(Dirichlet)}, \tag{2.26}$$
$$n_\alpha a_{\alpha\beta}\varphi_{,\beta} = f \quad \text{on} \quad \partial\Omega \qquad \text{(Neumann)}, \tag{2.27}$$
$$n_\alpha a_{\alpha\beta}\varphi_{,\beta} + a\varphi = f, \quad a > 0 \quad \text{on} \quad \partial\Omega \quad \text{(Robin)}, \tag{2.28}$$

with n the outward unit normal on $\partial\Omega$. Instead of (2.27) often the condition $n_\alpha\varphi_{,\alpha} = f$ is given as Neumann condition, but physics always leads to (2.27);

if $a_{\alpha\beta} = 0$ for $\beta \neq \alpha$ the two versions are equivalent.

When Ω is *simply connected* and $\partial\Omega$ and the data f, s satisfy suitable smoothness conditions, then (2.26) and (2.28) are known to result in a well-posed problem. This is also the case if $\partial\Omega$ is divided in segments, in each of which we have precisely one of the conditions (2.26)–(2.28), with (2.26) or (2.28) in at least one of the seqments.

Compatibility condition

In the case of a pure Neumann boundary value problem the situation is more complicated if $c = 0$ in (2.22). Obviously, if a solution exists an arbitrary constant can be added to it. Usually $b = 0$ in this situation. Assuming this, integration of (2.22) over Ω and application of the divergence theorem gives

$$\int_{\partial\Omega} f dS = -\int_{\Omega} s d\Omega . \tag{2.29}$$

If and only if this so-called *compatibility condition* is satisfied, solutions exist. They can be stably computed, and the problem is still considered to be well-posed, although we have no uniqueness in the strict sense. Equation (2.29) expresses a physical conservation principle: transport through the boundary balances production in the interior.

We illustrate these theoretical considerations with some examples.

Example 2.3.1. Hadamard's problem
Let $\Omega = (0, 1) \times (0, 1)$, and consider the following problem:

$$-\varphi_{,\alpha\alpha} = 0 \quad \text{in} \quad \Omega ,$$

with boundary conditions

$$\varphi(x_1, 0) = 0, \quad -\varphi_{,2}(x_1, 0) = f(x_1) , \tag{2.30}$$
$$\varphi(0, x_2) = \varphi(1, x_2) = 0 . \tag{2.31}$$

Notice that these conditions are not of the type discussed before, since we have two conditions at the seqment $x_2 = 0$ of $\partial\Omega$, and none at $x_2 = 1$. With $f(x_1) = -\frac{1}{m} \sinh m\pi x_1$, separation of variables gives the following solution:

$$\varphi(x) = \frac{1}{\pi m^2} f(x_1) \sinh m\pi x_2.$$

With the maximum principle we can establish uniqueness. Nevertheless the problem is *ill-posed*, which can be seen as follows. If $f(x_1) = 0$ the solution is $\varphi(x) = \varphi_1(x) \equiv 0$. With f as above, we have $\varphi(x_1, 1) = \varphi_2(x_1, 1) \equiv$

$\frac{1}{m\pi^2}f(x_1)\sinh m\pi$. Given an ε there is an M such that $|f(x_1)| < \varepsilon$ for all $m > M$. However, due to the exponential growth of sinh we can for arbitrarily large K always find an $m > M$ such that $|\varphi_1(x_1, 1) - \varphi_2(x_1, 1)| > K$, so that (2.24) does not hold.   □

Example 2.3.2. Two-dimensional potential flow
Let $\Omega = \{\boldsymbol{x} \in \mathbb{R}^2 : x_\alpha x_\alpha \geq 1\}$, i.e. the exterior of the unit circle. Consider the following problem:

$$-\varphi_{,\alpha\alpha} = 0 \quad \text{in} \quad \Omega\,,$$
$$n_\alpha \varphi_{,\alpha} = 0 \quad \text{at} \quad x_\alpha x_\alpha = 1\,,$$
$$\lim_{|\boldsymbol{x}|\to\infty} \varphi/x_1 = U\,.$$

This describes potential flow around a cylinder with free-stream velocity U in the x_1-direction at infinity. At first sight it seems that a difficulty is the prescription of an infinite value for φ at infinity, but this difficulty is easily surmounted by a change of the dependent variable: $\psi = \varphi - Ux_1$. Using a conformal mapping method, well-known in classical fluid dynamics, the following one-parameter family of solutions is found:

$$\varphi = U(r + 1/r)\cos\theta + \gamma\theta/2\pi\,,$$

with (r, θ) polar coordinates, and γ an arbitrary constant. The physical meaning of γ is that it governs the amount of circulation, and in fluid dynamics it is determined by an additional condition, called the *Kutta condition*. The reason that uniqueness is lacking here is that Ω is not simply connected.   □

Example 2.3.3. Helmholtz equation
Let $\Omega = (0, \pi) \times (0, \pi)$, and consider

$$-\varphi_{,\alpha\alpha} - \lambda^2\varphi = 0 \quad \text{in} \quad \Omega\,, \quad \varphi = 0 \quad \text{on} \quad \partial\Omega\,.$$

Here (2.25) is violated. A solution is obviously $\varphi \equiv 0$. For $\lambda = 1, 2, 3, \ldots$ (*eigenvalues*) we also have non-zero solutions (*eigenfunctions*) given by

$$\varphi = \sin \lambda x_1 \sin \lambda x_2\,,$$

so that there is no uniqueness. Although this problem does not satisfy our definition of well-posedness, it is meaningful and can be stably computed as an eigenvalue problem.   □

The parabolic case

Let equation (2.1) be parabolic. Then a well-posed problem is obtained if an *initial condition* is given:

$$\varphi(0, \boldsymbol{x}) = g(\boldsymbol{x}), \quad x \in \Omega$$

and, furthermore, boundary conditions on $\partial\Omega$ that would fit an elliptic problem, as given by (2.26)–(2.28), with in this case $f = f(t, \boldsymbol{x})$. If singularities are to be avoided, initial and boundary conditions should agree at $t = 0$ and $\boldsymbol{x} \in \partial\Omega$. Note that at $t = T$ no condition is to be given.

Consider the following pure initial value problem for the diffusion equation (2.9):

$$\varphi_t - \varphi_{,11} = 0 , \quad x_1 \in \mathbb{R} , \quad \varphi(0, x_1) = f(x_1) , \quad 0 < t \leq T . \qquad (2.32)$$

The solution is given by:

$$\varphi(t, x_1) = \frac{1}{2\sqrt{\pi|t|}} \int_{-\infty}^{\infty} f(y) \exp\{-(x_1 - y)^2/4t\} dy ,$$

which is easily verified. Clearly, for $t \ll -1$ the solution is very sensitive to perturbations in f, so that the problem is not well-posed. This means that the condition $t > 0$ in (2.32) is essential: time cannot be reversed. This corresponds to the intuitive notion that from a smooth temperature distribution the corresponding, perhaps non-smooth, temperature distribution at a sufficiently removed earlier instant of time cannot be stably determined. Note that for all f,

$$\varphi(\infty, x_1) = \text{constant} = \int_{-\infty}^{\infty} f(y) dy .$$

This irreversibility of time is a hallmark of parabolic problems.

Example 2.3.4. Backward solution of the diffusion equation
Consider the diffusion equation

$$\varphi_t - \varphi_{,11} = 0, \quad 0 < t < T, \quad x_1 \in (0, 1) ,$$

with boundary conditions

$$\varphi(t, 0) = \varphi(t, 1) = 0$$

and initial (or rather 'final') condition

$$\varphi(T, x_1) = \frac{1}{m} \sin m\pi x_1 .$$

By separation of variables the following solution is obtained:

$$\varphi = \frac{1}{m} \exp(m^2\pi^2(T - t)) \sin m\pi x_1 .$$

so that

$$\varphi(0, x_1) = \frac{1}{m} \exp(m^2 \pi^2 T) \sin m\pi x_1 ,$$

which can be made to differ from zero by an arbitrary amount by choosing m large enough. By an argument similar to that employed in Example 2.3.1, we see that the problem is ill-posed.   □

The hyperbolic case will not be discussed here, but in Chap. 8. For the wave equation (2.8) one can determine which boundary conditions lead to a well-posed problem by requiring that the two functions ψ and η in the general representation (2.21) are determined uniquely. Illustrations are given in the following exercises.

Exercise 2.3.1. Using (2.21), show that boundary conditions (2.30), (2.31) lead to a well-posed problem for the wave equation (2.8).

Exercise 2.3.2. Let $\Omega = (0, 1) \times (0, 1)$, and consider the wave equation (2.8) with a Dirichlet condition prescribed at all of $\partial\Omega$. Using (2.21), show that in general a solution does not exist. Note that according to the theoretical results presented earlier, this boundary condition leads to a well-posed problem for the Laplace equation.

2.4 Maximum principles

Physical interpretation

In this and the following section we discuss qualitative properties of the solution of (2.1), giving a-priori information that can be used advantageously in the development of numerical approximations. It is assumed that (2.13) holds, so that (2.1) is parabolic. An intuitive idea about the behavior of solutions of (2.1) is obtained by associating with this equation a physical interpretation. For example, (2.1) models the temperature distribution in a fluid with temperature φ, heat source distribution q, velocity field $\boldsymbol{u}$ and heat conduction tensor $a_{\alpha\beta}$. If $q \leq 0$, no heating takes place and it is intuitively clear that if φ has a local maximum φ_m at $(t_m, \boldsymbol{x}_m)$, then at $t < t_m$ a value $\varphi \geq \varphi_m$ is to be found somewhere. Furthermore, large values of φ may be imposed on $\partial\Omega$ by a Dirichlet boundary condition. Hence, we arrive at the following hypothesis: local maxima can occur only for $t = 0$ and/or $\boldsymbol{x} \in \partial\Omega$. Such a *maximum principle* can be very useful. We now give it a mathematical basis. More background may be found in Protter and Weinberger (1967) and Sperb (1981).

The one-dimensional stationary case

In this case equation (2.1) may be written as, taking $c = 0$, writing D instead of a_{11} and writing u instead of b,

$$u\frac{d\varphi}{dx} - \frac{d}{dx}(D\frac{d\varphi}{dx}) = s, \quad x \in (a, b) .$$

Suppose φ has a local maximum in $x = x_m \in (a, b)$. Then $d\varphi(x_m)/dx = 0$, and $d^2\varphi(x_m)/dx^2 \leq 0$. Assuming D to be differentiable, we find

$$s = u\frac{d\varphi}{dx} - D\frac{d^2\varphi}{dx^2} - \frac{dD}{dx}\frac{d\varphi}{dx} ,$$

so that $s(x_m) \geq 0$. It follows that if $s < 0$, then φ cannot have a local maximum in (a, b). This result is strengthened in the following theorem.

Theorem 2.4.1. *One-dimensional maximum principle*
Let φ satisfy

$$u\frac{d\varphi}{dx} - \frac{d}{dx}(D\frac{d\varphi}{dx}) \leq 0, \quad x \in (0, 1) , \tag{2.33}$$

with $D > 0$ differentiable and bounded and φ not constant. Then φ has a local maximum only in $x = 0$ and/or $x = 1$, and $d\varphi/dn > 0$ in a boundary point with a local maximum, with $d\varphi/dn$ the outward derivative in $x = 0$ or $x = 1$.

Proof. (cf. Sperb (1981)). We have

$$L\varphi \equiv v\frac{d\varphi}{dx} - \frac{d^2\varphi}{dx^2} \leq 0, \quad x \in (0, 1) ,$$

with $v = u/D - d\ln D/dx$. Assume $\varphi(c) = M$, $0 < c < 1$, is a local maximum. Let $(a, b) \subset (0, 1)$ be a neighbourhood of c in which $\varphi(x) \leq M$. Because φ is not constant it is possible to choose a and b such that $\varphi(a) < M$ and/or $\varphi(b) < M$. Assume $\varphi(b) < M$; if not, the following argument is easily repeated using $\varphi(a) < M$. Then there is a point $x = d$, $c < d < b$ with $\varphi(d) < M$. Let

$$\psi(x) = e^{\alpha(x-c)} - 1 ,$$

with $\alpha > 0$ a constant still to be determined. We have

$$L\psi = (\alpha v - \alpha^2)e^{\alpha(x-c)} .$$

Since v is bounded we can choose α such that $L\psi < 0$ in (a, d). Let

$$\theta(x) = \varphi(x) + \varepsilon\psi(x) ,$$

with ε a constant such that

$$\varepsilon < (M - \varphi(d))/\psi(d).$$

It is easily seen that $\theta(x) < M$ in (a, c), $\theta(c) = M$, $\theta(d) < M$. This means that $\theta(x)$ has a maximum $\bar{M} \geq M$ in (a, d). Furthermore,

$$L\theta = L\varphi + \varepsilon L\psi < 0 \quad \text{in } (a, d).$$

But then θ cannot have a maximum in (a, d), according to the result which preceeds the theorem, and we have a contradiction. Hence, φ cannot have a local maximum in $(0, 1)$. It remains to prove the property of $d\varphi/dn$. Suppose that $\varphi(0) = M$ and $\varphi(x) \leq M$, $x \in (0, 1)$, with $\varphi(d) < M$ for some $d \in (a, b)$. Now define

$$\psi(x) = e^{\alpha x} - 1,$$

and choose $\alpha > 0$ such that $L\psi < 0$ in $(0, d)$. Defining $\theta(x)$ and ε as before we have $L\theta < 0$ in $(0, d)$ and the maximum of θ in $[0, d]$ must therefore occur at $x = 0$ or $x = d$. But $\theta(0) = \varphi(0) = M > \theta(d)$ because of our choice of ε. Therefore we have

$$\frac{d\theta(0)}{dx} = \frac{d\varphi(0)}{dx} + \varepsilon \frac{d\psi(0)}{dx} = \frac{d\varphi(0)}{dx} + \varepsilon\alpha \leq 0,$$

which implies

$$\frac{d\varphi(0)}{dn} = -\frac{d\varphi(0)}{dx} > 0.$$

The procedure is similar if $\varphi(1) = M$.   □

In a similar way we can prove that if φ satisfies

$$u\frac{d\varphi}{dx} - \frac{d}{dx}(D\frac{d\varphi}{dx}) \geq 0, \quad x \in (0, 1),$$

and if the other conditions are satisfied, then φ has a local minimum only in $x = 0$ and/or $x = 1$, and $d\varphi/dn < 0$ in boundary points with a local minimum. Hence, in the frequently occurring case that the equality sign holds in (2.33), there can be no local extrema in $(0, 1)$. This means that interior extrema occurring in numerical solutions, often called numerical wiggles, are numerical artifacts. Furthermore, if we have a homogeneous Neumann boundary condition

$$\frac{d\varphi(0)}{dn} = 0$$

then there can be no local extremum at $x = 0$, and similarly in the case of such a boundary condition at $x = 1$.

The general stationary case

We rewrite (2.2) with $c = 0$ as

$$u_\alpha \varphi_{,\alpha} - a_{\alpha\beta}\varphi_{,\alpha\beta} = s, \quad \boldsymbol{x} \in \Omega \subset \mathbb{R}^d, \tag{2.34}$$

with in every point of $\partial\Omega$ precisely one of the boundary conditions (2.26)–(2.28). By saying that $\partial\Omega$ is *smooth* we mean that for every $\boldsymbol{x} \in \partial\Omega$ there exists an open sphere in Ω that is tangent to $\partial\Omega$ in $\boldsymbol{x}$. The following theorem holds:

Theorem 2.4.2. *(Maximum principle)*
Let (2.34) hold with $s \leq 0$, let any outward derivative $\partial\varphi/\partial\nu \leq 0$ on $\partial\Omega' \subset \partial\Omega$, and let $\partial\Omega$ be smooth. Then local maxima occur only on $\partial\Omega \setminus \partial\Omega'$, or $\varphi = constant$ in Ω.

Proof. See Protter and Weinberger (1967) Sect. 2.3 Theorems 5 and 7. □

By reversing sign we obtain a similar *minimum principle*. Note that $n_\alpha a_{\alpha\beta}\varphi_{,\beta}$ (cf. (2.27)) is an example of an outward derivative. In the frequently occurring case that $s = 0$ and a homogeneous Neumann condition $\partial\varphi/\partial n = 0$ on $\partial\Omega'$, both maxima and minima, i.e. *extrema*, can only occur on $\partial\Omega \setminus \partial\Omega'$. Should extrema occur in Ω in a numerical solution, then we know that such extrema are (undesirable) numerical wiggles. Discretizations that satisfy a maximum principle similar to Theorem 2.4.2, and hence exclude numerical wiggles, are called *monotone schemes*. Monotone schemes will be developed later. The maximum principle can be used in the derivation of *global error estimates*, that we also will present later.

The general nonstationary case

Consider the following version of (2.1):

$$\varphi_t + u_\alpha \varphi_{,\alpha} - a_{\alpha\beta}\varphi_{,\alpha\beta} = a, \quad \boldsymbol{x} \in \Omega, \quad 0 < t \leq T. \tag{2.35}$$

We have

Theorem 2.4.3. *(Maximum principle)*
Let $s \leq 0$ in (2.35), let any outward derivative $\partial\varphi/\partial\nu \leq 0$ on $\partial\Omega' \subset \partial\Omega$, and let $\partial\Omega$ be smooth. Then if a local maximum occurs at $t = t^* > 0$ in the interior of Ω or on $\partial\Omega'$, then $\varphi = constant$, $0 \leq t \leq t^*$.

Proof. See Protter and Weinberger (1967) Sect. 3.3, Theorems 5 and 6. □

Hence if $\varphi \neq$ constant, local maxima can occur only at $t = 0$ and portions of the boundary where $\partial\varphi/\partial\nu > 0$.

We have formulated maximum principles for elliptic and parabolic partial differential equations. For hyperbolic equations there are maximum principles of a different type that will not be discussed; see Protter and Weinberger (1967).

2.5 Boundary layer theory

Consider the following version of (2.1):

$$\varphi_t + u_\alpha \varphi_{,\alpha} - \varepsilon\varphi_{,\alpha\alpha} = s\,, \quad x \in \Omega\,, \quad 0 < t \leq T\,, \quad \varepsilon > 0\,. \tag{2.36}$$

Boundary layer theory or singular perturbation theory studies what happens when $\varepsilon \downarrow 0$. Equation (2.36) is parabolic, and we prescribe initial and boundary conditions in accordance with the rules of Sect. 2.3.

Let the following initial condition be given:

$$\varphi(0, x) = \varphi_0(x)\,.$$

At every point of $\partial\Omega$ precisely one of the three types of boundary condition (2.26)–(2.28) is precribed.

The convection equation

When $\varepsilon \ll 1$ it is natural to approximate (2.36) by putting $\varepsilon = 0$. For simplicity we also take $s = 0$, and obtain the *convection equation*:

$$\varphi_t + u_\alpha \varphi_{,\alpha} = 0\,. \tag{2.37}$$

This equation is hyperbolic. Solutions cannot satisfy all of the initial and boundary conditions, as will be seen shortly.

Characteristics

Let us define curves in (t, x) space by relations $t = t(s)$, $x = x(s)$, satisfying

$$\frac{dt}{ds} = 1, \quad \frac{dx_\alpha}{ds} = u_\alpha\,. \tag{2.38}$$

Then equation (2.37) reduces to

$$\frac{d\varphi}{ds} = 0 \ . \tag{2.39}$$

Since the curves defined by (2.38) are particle paths, equation (2.39) expresses the fact that φ is constant along particle paths. These curves are called *characteristics* in the theory of hyperbolic systems.

One-dimensional case

Although the one-dimensional case is very simple, we will dwell a bit upon it, in order to elucidate the principles of singular perturbation theory. These principles will be applied subsequently in a more general setting.

Let $\Omega = (0,1)$. Equation (2.36) reduces to, taking $s = 0$,

$$\varphi_t + u_1\varphi_{,1} - \varepsilon\varphi_{,11} = 0, \quad 0 < x_1 < 1, \quad 0 < t \leq T . \tag{2.40}$$

We assume the following initial and boundary conditions:

$$\varphi(0, x_1) = \varphi_0(x_1), \quad 0 < x_1 < 1 , \tag{2.41}$$
$$\varphi(t, 0) = f_0(t), \quad \varphi(t, 1) = f_1(t), \quad 0 < t \leq T . \tag{2.42}$$

For $\varepsilon \ll 1$ one would expect that equation (2.40) can be approximated by

$$\varphi_t + u_1\varphi_{,1} = 0 \ . \tag{2.43}$$

Solutions of (2.43) are constant along the characteristics. Figure 2.1 gives a sketch of the characteristics of (2.43) (as defined by (2.38)), assuming $u_1 < 0$. Consider the characteristic C_1. On C_1, $\varphi = \varphi(C_1) = $ constant. In P_1 we have

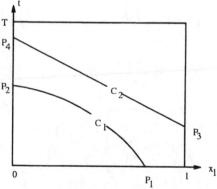

Fig. 2.1. Characteristics for equation (2.43).

the initial condition (2.41), in P_2 the first of the boundary conditions (2.42).

Only one of them can be satisfied in general. What value to take for $\varphi(C_1)$ such that the solution of (2.40)–(2.42) is approximated? The difficulty has to do with the change of type that (2.36) undergoes when ε is replaced by zero. For $\varepsilon = 0$, (2.36) is hyperbolic; for $\varepsilon > 0$, it is parabolic.

Singular perturbation theory

The answer to the foregoing question is provided by *singular perturbation theory*. Introductions to this subject are given in Eckhaus (1973), Kevorkian and Cole (1981) and Van Dyke (1975). In the present case, $\varphi = \varphi(C_1) = \varphi(P_1)$ is a good approximation for $\varepsilon \ll 1$ to the solution of (2.40)–(2.42) in $1 \geq x_1 > \delta = \mathcal{O}(\varepsilon)$, whereas (2.43) has to be replaced by a so-called *boundary layer equation* to obtain an approximation in $\delta > x_1 \geq 0$. This can be seen as follows. First, assume that we indeed have $\varphi(C_1) = \varphi(P_1)$ in $1 \geq x_1 > \delta$ with $\delta \ll 1$. In $\delta > x_1 \geq 0$ we expect a rapid change of φ from $\varphi(P_1)$ to $\varphi(P_2)$. For derivatives of φ we expect

$$\frac{\partial^m \varphi}{\partial x_1^m} = \mathcal{O}(\delta^{-m}) , \tag{2.44}$$

so that perhaps the diffusion term in (2.40) cannot be neglected in the boundary layer; this will depend on the size of δ. Assume

$$\delta = \mathcal{O}(\varepsilon^\alpha) , \tag{2.45}$$

with α to be determined. In order to exhibit the dependence of the magnitude of derivatives on ε we introduce a *stretched coordinate* $\tilde{x}_1$:

$$\tilde{x}_1 = x_1 \varepsilon^{-\alpha} , \tag{2.46}$$

which is chosen such that $\tilde{x}_1 = \mathcal{O}(1)$ in the boundary layer. It follows from (2.44)–(2.46) that

$$\frac{\partial^m \varphi}{\partial \tilde{x}_1^m} = \mathcal{O}(1) \tag{2.47}$$

in the boundary layer. In the stretched coordinate equation (2.40) becomes:

$$\varphi_t + \varepsilon^{-\alpha} u_1 \frac{\partial \varphi}{\partial \tilde{x}_1} - \varepsilon^{1-2\alpha} \frac{\partial^2 \varphi}{\partial \tilde{x}_1^2} = 0 . \tag{2.48}$$

Letting $\varepsilon \downarrow 0$ and using (2.47), equation (2.48) takes various forms, depending on α. The correct value of α follows from the requirement, that the solution of the $\varepsilon \downarrow 0$ limit of equation (2.48) satisfies the boundary condition at $x_1 = 0$, and the so-called matching principle.

Matching principle

As $\tilde{x}_1$ increases, the solution of (the $\varepsilon \downarrow 0$ limit of) equation (2.48) has to somehow join up with the solution of (2.43), i.e. approach the value $\varphi(C_1)$. In singular perturbation theory this condition is formulated precisely, and is known as the *matching principle*:

$$\lim_{\tilde{x}_1 \to \infty} \varphi_{\text{inner}}(t, \tilde{x}_1) = \lim_{x_1 \downarrow 0} \varphi_{\text{outer}}(t, x) .$$

Here φ_{inner}, also called the *inner solution*, is the solution of the *inner equation* or *boundary layer equation*, which is the limit as $\varepsilon \downarrow 0$ of equation (2.48) for the correct value of α, which we are trying to determine. Furthermore, φ_{outer}, also called the *outer solution*, is the solution of the *outer equation*, which is the limit as $\varepsilon \downarrow 0$ of the original equation, i.e. equation (2.43). In our case the matching principle becomes

$$\lim_{\tilde{x}_1 \to \infty} \varphi_{\text{inner}}(t, \tilde{x}_1) = \varphi(C_1) . \tag{2.49}$$

As already mentioned, the other condition to be satisfied is the boundary condition at $\tilde{x}_1 = 0$:

$$\varphi(t, 0) = f_0(t) . \tag{2.50}$$

For $\alpha < 0$ (corresponding to compression rather than stretching) the limit as $\varepsilon \downarrow 0$ of (2.48) is

$$\varphi_t = 0 . \tag{2.51}$$

Obviously, the solution of (2.51) cannot satisfy (2.50), so that the case $\alpha < 0$ has to be rejected. With $\alpha = 0$ equation (2.43) is obtained, which cannot satisfy both conditions at $t = 0$ and $x_1 = \tilde{x}_1 = 0$, as we saw.

For $0 < \alpha < 1$ the limit of (2.48) is

$$u_1 \frac{\partial \varphi}{\partial \tilde{x}_1} = 0 ,$$

so that the inner solution is given by

$$\varphi(t, \tilde{x}_1) = g(t) ,$$

and in general equations (2.49) and (2.50) cannot be satisfied, so that we cannot have $0 < \alpha < 1$.

For $\alpha = 1$ equation (2.48) becomes as $\varepsilon \downarrow 0$:

$$u_1(t, 0) \frac{\partial \varphi}{\partial \tilde{x}_1} - \frac{\partial^2 \varphi}{\partial \tilde{x}_1^2} = 0 , \tag{2.52}$$

where we have used that $u_1(t, x) = u_1(t, \varepsilon \tilde{x}_1) \to u_1(t, 0)$ as $\varepsilon \downarrow 0$. The general solution of (2.52) is

$$\varphi = A(t) + B(t)e^{u\tilde{x}_1}, \qquad (2.53)$$

with $u = u_1(t, 0)$. We can satisfy both (2.49) and (2.50), remembering that we had assumed $u_1 < 0$, with

$$A(t) = \varphi(C_1), \quad B(t) = f_0(t) - \varphi(C_1).$$

This gives us the inner solution. In terms of the unstretched variable x_1 the inner solution is given by

$$\varphi = \varphi(C_1) + \{f_1(t) - \varphi(C1)\}e^{ux_1/\varepsilon}.$$

We see a rapid exponential variation from $f_1(t)$ to $\varphi(C_1)$ in a thin layer of thickness $\delta = \mathcal{O}(\varepsilon)$, confirming our earlier statement about the behavior of the solution. Fig. 2.2 gives a sketch of the inner and outer solutions as a function of x_1. An asymptotic approximation for $\varepsilon \downarrow 0$ that is valid everywhere is given by $\varphi_{\text{inner}} + \varphi_{\text{outer}} - \varphi(C_1)$ (not shown in the figure).

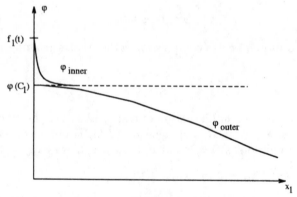

Fig. 2.2. Sketch of inner and outer solutions

The distinguished limit

The limit as $\varepsilon \downarrow 0$ of the stretched equation (2.48) for the special value of $\alpha = 1$ for which the solution of the resulting inner equation can satisfy both the boundary condition and the matching principle is called the *distinguished limit*. In order to show that this limit is unique we will also investigate the remaining values of α that we did not yet consider, namely $\alpha > 1$. Now equation (2.48) gives the following inner equation:

$$\frac{\partial^2 \varphi}{\partial \tilde{x}_1^2} = 0 \ ,$$

with the general solution

$$\varphi = A(t) + B(t)\tilde{x}_1 \ .$$

The limit of φ as $\tilde{x}_1 \to \infty$ does not exist, so that the matching principle cannot be satisfied. Hence, $\alpha = 1$ is the only value that gives a distinguished limit.

The only element of arbitrariness that remains in this analysis is the assumption that we have a boundary layer at $x_1 = 0$. Why no boundary layer at $t = 0$, and $\varphi(C_1) = \varphi(P_2)$ (cf. Fig. 2.1)? This can be investigated by assuming a boundary layer at $t = 0$, and determining whether a distinguished limit exists or not. This is left as an exercise. It turns out there is no boundary layer at $t = 0$. Hence, singular perturbation theory determines uniquely the asymptotic behavior of the solution along the characteristic C_1.

It is left to the reader to verify in a similar way that along characteristic C_2 (cf. Fig. 2.1), which does not originate in $t = 0$ but at the boundary $x_1 = 1$, we have $\varphi(C_2) = \varphi(P_3)$, except in a boundary layer with thickness $\mathcal{O}(\varepsilon)$ at $x_1 = 0$. When $u_1 > 0$ there is a boundary layer at $x_1 = 1$, and no boundary layer at $x_1 = 0$. We see that *boundary layers will not be present at inflow boundaries* (i.e. parts of the domain boundary $\partial\Omega$ where the flow enters the domain, that is where $\boldsymbol{u} \cdot \boldsymbol{n} < 0$ with $\boldsymbol{n}$ the outward normal on $\partial\Omega$). This holds in any number of space dimensions.

The role of boundary conditions

The occurrence of boundary layers is strongly influenced by the type of boundary condition. Let (2.42) be replaced by

$$\frac{\partial \varphi(t, 0)}{\partial x_1} = f_0(t), \quad \varphi(t, 1) = f_1(t), \quad 0 < t \leq T \ . \tag{2.54}$$

As before, a boundary layer of thickness $\mathcal{O}(\varepsilon)$ is found at $x_1 = 0$, and the boundary layer equation is given by (2.52), with solution (2.53). Taking boundary condition (2.54) into account we find

$$B(t) = \varepsilon f_0(t)/u \ ,$$

so that $B(t) \to 0$ as $\varepsilon \downarrow 0$. Hence, to first order, there is no boundary layer, and the outer solution (solution of (2.43)) is uniformly valid in $0 < x_1 < 1$.

Two-dimensional case

Let $\Omega = (0, 1 \times (0, 1)$. If we take $s = 0$, and consider the time-independent case for brevity, equation (2.36) reduces to

$$u_\alpha \varphi_{,\alpha} - \varepsilon \varphi_{,\alpha\alpha} = 0, \quad \boldsymbol{x} \in (0, 1) \times (0, 1) . \tag{2.55}$$

Assume Dirichlet boundary conditions:

$$\begin{aligned}
\varphi(x_1, 0) &= f_1(x_1), \quad \varphi(1, x_2) = f_2(x_2), \quad \varphi(x_1, 1) = f_3(x_1), \\
\varphi(0, x_2) &= f_4(x_2) .
\end{aligned} \tag{2.56}$$

For $\varepsilon \downarrow 0$ we obtain the following outer equation:

$$u_\alpha \varphi_{,\alpha} = 0 . \tag{2.57}$$

This is a hyperbolic equation, with characteristics defined by

$$\frac{dx_\alpha}{ds} = u_\alpha, \quad \alpha = 1, 2 . \tag{2.58}$$

For a sketch of possible characteristics we can reuse Fig. 2.1 by replacing t with x_2. Along C_1, $\varphi = \varphi(C_1) = $ constant. Again the question arises, whether $\varphi(C_1)$ will take on the prescribed value $\varphi(P_1)$ or $\varphi(P_2)$. First, let us postulate a boundary layer at $x_1 = 0$. Reasoning as before, we find the following boundary layer equation:

$$u\frac{\partial \varphi}{\partial \tilde{x}_1} - \frac{\partial^2 \varphi}{\partial \tilde{x}_1^2} = 0, \quad \tilde{x}_1 = x_1/\varepsilon, \quad u = u_1(0, x_2) . \tag{2.59}$$

The following solution satisfies both the boundary condition at $x_1 = 0$ and the matching principle, *provided $u < 0$*:

$$\varphi(\tilde{x}_1, x_2) = \varphi(P_1) + \{f_4(x_2) - \varphi(P_1)\}e^{u\tilde{x}_1} .$$

In case we have a Neumann boundary condition at $x_1 = 0$ there is no boundary layer at $x_1 = 0$. As in the one-dimensional case, when $x_1 = 0$ is an inflow boundary ($u > 0$) there is also no boundary layer at $x_1 = 0$.

Parabolic and ordinary boundary layers

Next, consider the boundary $x_2 = 0$. If it is an outflow boundary, i.e. $v \equiv u_2(x_1, 0) < 0$, then there is again a boundary layer of thickness $\mathcal{O}(\varepsilon)$ in case of a Dirichlet boundary condition. The boundary layer solution is found to be

$$\varphi(x_1, \tilde{x}_2) = \varphi(P_2) + \{f_1(x_1) - \varphi(P_2)\}e^{v\tilde{x}_2}, \quad \tilde{x}_2 = x_2/\varepsilon . \tag{2.60}$$

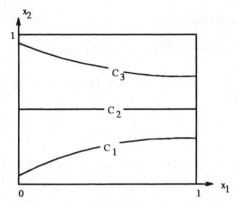

Fig. 2.3. Characteristics of equation (2.57) in a channel flow

Now consider the case that $x_2 = 0$ is a solid wall, so that $v = 0$. The shape of the characteristics of the outer equation (2.57) might be as in Fig. 2.3, where also $x_2 = 1$ is assumed to be a solid wall, so that we have a channel flow. Since $u_2(x_1, 0) = 0$, the curve $x_2 = 0$ is a characteristic of the outer equation (2.57) according to (2.58), so that the solution along this characteristic is given by

$$\varphi(x_1, 0) = f_4(0) \tag{2.61}$$

if $x_1 = 0$ is a inflow boundary, or by

$$\varphi(x_1, 0) = f_2(0) \tag{2.62}$$

if $x_1 = 1$ is an inflow boundary. To verify this using singular perturbation theory is left as an exercise. If both $x_1 = 0$ and $x_1 = 1$ are inflow boundaries the pattern of characteristics has to be qualitatively different from Fig. 2.3, assuming that $u_{\alpha,\alpha} = 0$ (incompressible flow field). This situation will not be considered here. Whether we have (2.61) or (2.62), in both cases the outer solution cannot satisfy boundary condition (2.56) at $x_2 = 0$. Hence, we expect a boundary layer at $x_2 = 0$. Obviously, this boundary layer will be of different type than obtained until now, because the boundary layer solution cannot be given by (2.60), since now we have $v = 0$. In order to derive the boundary layer equation, the same procedure is followed as before. We transform (2.55) to the stretched coordinate $\tilde{x}_2 = x_2 \varepsilon^{-\alpha}$, with α to be determined. Equation (2.55) becomes, with $u_2 = 0$:

$$u_1 \frac{\partial \varphi}{\partial x_1} - \varepsilon \frac{\partial^2 \varphi}{\partial x_1^2} - \varepsilon^{1-2\alpha} \frac{\partial^2 \varphi}{\partial \tilde{x}_2^2} = 0 \,. \tag{2.63}$$

The boundary condition is

$$\varphi(x_1, 0) = f_1(x_1) \,, \tag{2.64}$$

and the matching principle gives

$$\lim_{\tilde{x}_2 \to \infty} \varphi(x_1, \tilde{x}_2) = \lim_{x_2 \downarrow 0} \varphi_0(x_1, x_2) . \tag{2.65}$$

Now we take the limit of (2.63) as $\varepsilon \downarrow 0$. For $\alpha < 1/2$ the outer equation at $x_2 = 0$ is recovered:

$$u_1 \frac{\partial \varphi}{\partial x_1} = 0 ,$$

of which the solution obviously cannot satisfy (2.64) and (2.65). For $\alpha = 1/2$ the limit of (2.63) is

$$u \frac{\partial \varphi}{\partial x_1} - \frac{\partial^2 \varphi}{\partial \tilde{x}_2^2} = 0 , \tag{2.66}$$

with $u = u(x_1) = \lim_{\varepsilon \to 0} u_1(x_1, \tilde{x}_2 \sqrt{\varepsilon}) = u_1(x_1, 0)$. This is a parabolic partial differential equation, which in general cannot be solved explicitly, but for which it is known that boundary conditions at $\tilde{x}_2 = 0$ and $\tilde{x}_2 = \infty$ give a well-posed problem. An analytic expression for the solution of the boundary layer equation (2.66) with $u = 1$ will be given in Sect. 4.7. Hence, $\alpha = 1/2$ gives the distinguished limit, and (2.66) is the boundary layer equation. The thickness of this type of boundary layer is $\mathcal{O}(\sqrt{\varepsilon})$, which is much larger than for the preceding type.

In order to specify a unique solution, in addition an 'initial' condition has to be specified. Assuming $u > 0$, this has to be done at $x_1 = 0$. From (2.56) we obtain the following initial condition for the boundary layer solution:

$$\varphi(0, \tilde{x}_2) = f_4(\tilde{x}_2 \sqrt{\varepsilon}) ,$$

which to the present asymptotic order of approximation (we will not go into higher order boundary layer theory) may be replaced by

$$\varphi(0, \tilde{x}_2) = f_4(0) .$$

It is left to the reader to verify that $\alpha > 1/2$ does not give a distinguished limit.

The cause of the difference between the two boundary layer equations (2.59) (an ordinary differential equation) and (2.66) (a partial differential equation) is the angle which the characteristics of the outer equation (2.57) make with the boundary layer. In the first case this angle is non-zero (cf. Fig. 2.1), in the second case the characteristics do not intersect the boundary layer. The first type is called an *ordinary boundary layer* (the boundary layer equation is an ordinary differential equation), whereas the second type is called a *parabolic boundary layer* (parabolic boundary layer equation).

Summarizing, in the case of the channel flow depicted in Fig. 2.3, for $\varepsilon \ll 1$ there are parabolic boundary layers of thickness $\mathcal{O}(\sqrt{\varepsilon})$ at $x_2 = 0$ and $x_2 = 1$, and an ordinary boundary layer of thickness $\mathcal{O}(\varepsilon)$ at the outflow boundary, unless a Neumann boundary condition is prescribed there.

On outflow boundary conditions

It frequently happens that physically no outflow boundary condition is known, but that this is required mathematically. If $\varepsilon = \mathcal{O}(1)$ such a physical model is incomplete, but for $\varepsilon \ll 1$ an artificial (invented) outflow condition may safely be used to complete the mathematical model, because this does not affect the solution to any significant extent. Furthermore, an artificial condition of Neumann type is to be preferred above one of Dirichlet type. This may be seen as follows.

Consider the following physical situation: an incompressible flow with given velocity field $\boldsymbol{u}$ through a channel, the walls of which are kept at a known temperature. We want to know the temperature of the fluid, especially at the outlet. This leads to the following mathematical model. The governing equation is (2.55), with φ the temperature. Assume $\varepsilon \ll 1$, and $u_1 > 0$. We have φ prescribed at $x_1 = 0$ and at $x_2 = 0, 1$, but at $x_1 = 1$ we know nothing. Hence, we cannot proceed with solving (2.55), either analytically or numerically. Now let us just postulate some temperature profile at $x_1 = 1$:

$$\varphi(1, x_2) = f_2(x_2).$$

An ordinary boundary layer will develop at $x_1 = 1$, with solution, derived in the way discussed earlier, given by

$$\varphi(x_1, x_2) = \varphi_0(1, x_2) + \{f_2(x_2) - \varphi_0(1, x_2)\}e^{u(x_1-1)\varepsilon} . \tag{2.67}$$

This shows that the invented temperature profile $f_2(x_2)$ influences the solution only in the thin (artificially generated) boundary layer at $x_1 = 1$. This means that the computed temperature outside this boundary layer will be correct, regardless what we take for $f_2(x_2)$. When $\varepsilon = \mathcal{O}(1)$ this is no longer true, and more information from physics is required. In physical reality there will not be a boundary layer at all at $x_1 = 1$, of course. Therefore a more satisfactory artificial outflow boundary condition is

$$\frac{\partial \varphi(1, x_2)}{\partial x_1} = 0 ,$$

since with this Neumann boundary condition there will be no boundary layer at $x_1 = 1$ in the mathematical model.

Exercise 2.5.1. Consider equations (2.40)–(2.42) with $u_1 > 0$. Show that for $\varepsilon \ll 1$ there is a boundary layer of thickness $\mathcal{O}(\varepsilon)$ at and only at $x_1 = 1$.

Exercise 2.5.2. Consider equations (2.55) and (2.56), with the Dirichlet boundary condition at $x_1 = 0$ replaced by a homogeneous Neumann boundary condition. Show that when $u_1 < 0$ and $\varepsilon \ll 1$ there is no boundary layer at $x_1 = 0$.

Exercise 2.5.3. Derive equation (2.67).

3. Finite volume and finite difference discretization on nonuniform grids

3.1 Introduction

Nonuniform grids are often used to obtain accuracy in regious where the solution varies rapidly. We will see that on nonuniform grids, finite volume and finite difference discretization are not equivalent. On arbitrary nonuniform grids the local truncation error is usually larger than on uniform grids, or grids on which the mesh size varies smoothly. This has sometimes led to confusion. Cell-centered finite volume discretization is sometimes advised against, because the local truncation error is larger than for vertex-centered finite volumes, and is in fact of the same order even as the term that is approximated. Nevertheless, this is a good discretization method that is popular in reservoir engineering and porous media flow computation, and in computational fluid dynamics in general. The source of the confusion is that the relation between the local and global truncation error is complicated. Surprisingly, the global truncation error is small, as we will see. Of course, it is the global truncation error that counts.

In order to make the matter clear we will present a detailed study of global truncation errors for simple cases. This will shed light not only on the accuracy of discretizations on nonuniform grids, but also on the accuracy required in the approximation of boundary conditions. The fact that often the local truncation error is larger at the boundary than in the interior has been another source of confusion in the past. We will see that even if the local truncation error is of low order at the boundary and in the interior, the global truncation error is still second order for cell-centered finite volume discretization.

Furthermore, we will study the discretization of the diffusion equation with discontinuous diffusion coefficient, because there seem to be no texts giving a comprehensive account of discretization methods in this situation, which is common in reservoir engineering and porous media flow.

Elementary introductions to finite difference and finite volume discretization are given in Forsythe and Wasow (1960), Mitchell and Griffiths (1994), Morton and Mayers (1994) and Strikwerda (1989).

3.2 An elliptic equation

This chapter is devoted to the discretization of (special cases of) the general single second-order elliptic equation, that can be written as

$$L\varphi \equiv -(a_{\alpha\beta}\varphi_{,\alpha})_{,\beta} + (b_\alpha\varphi)_{,\alpha} + c\varphi = q \quad \text{in} \quad \Omega \subset \mathbb{R}^d. \qquad (3.1)$$

The diffusion tensor $a_{\alpha\beta}$ is assumed to be symmetric: $a_{\alpha\beta} = a_{\beta\alpha}$. Boundary conditions will be discussed later. We assume:

Definition 3.2.1. *Uniform ellipticity.*
Equation (3.1) is called uniformly elliptic if there exists a constant $C > 0$ such that

$$a_{\alpha\beta}(\boldsymbol{x})v_\alpha v_\beta \geq C v_\alpha v_\alpha, \quad \forall\, \boldsymbol{v} \in \mathbb{R}^d, \quad \boldsymbol{x} \in \Omega. \qquad (3.2)$$

For $d = 2$ this is equivalent to equation (2.4). Property (3.2) is invariant under coordinate transformations.

The domain Ω

The domain Ω is taken to be the d-dimensional unit cube. This is not a serious limitation, because the current main trend in grid generation consists of the decomposition of the physical domain in subdomains, each of which is mapped onto a cubic computational domain. In general, such mappings change the coefficients in (3.1). As a result, special properties, such as separability or the coefficients being constant, may be lost. This is the price to pay for geometric generality.

The weak formulation

Assume that $a_{\alpha\beta}$ is discontinuous along some manifold $\Gamma \subset \Omega$, which we will call an *interface*; then equation (3.1) is called an *interface problem*. Equation (3.1) now has to be interpreted in the *weak sense*, as follows. From (3.1) it follows that

$$(L\varphi, \psi) = (q, \psi), \quad \forall\, \psi \in H, \quad (\varphi, \psi) \equiv \int_\Omega \varphi\psi d\Omega,$$

where H is a suitable Sobolev space, that we do not need to specify here. Define

$$a(\varphi, \psi) \equiv \int_\Omega a_{\alpha\beta}\varphi_{,\alpha}\psi_{,\beta}d\Omega - \int_{\partial\Omega} a_{\alpha\beta}\varphi_{,\alpha}n_\beta\psi d\Gamma,$$
$$b(\varphi, \psi) \equiv \int_\Omega (b_\alpha\varphi)_{,\alpha}\psi d\Omega,$$

with n_β the x_β component of the outward unit normal on the boundary $\partial\Omega$ of Ω. Application of the divergence theorem gives

$$(L\varphi, \psi) = a(\varphi, \psi) + b(\varphi, \psi) + (c\varphi, \psi) \ . \tag{3.3}$$

The weak formulation of (3.1) is

Find $\varphi \in \tilde{H}$ such that $a(\varphi, \psi) + b(\varphi, \psi) + (c\varphi, \psi) = (q, \psi), \quad \forall\, \psi \in H$,

$$\tag{3.4}$$

with $\tilde{H}$ another suitable Sobolev space. For suitable choices of $H, \tilde{H}$ and boundary conditions, existence and uniqueness of the solution of (3.4) has been established. For more details on the weak formulation (not needed here), see for example Ciarlet (1978) and Hackbusch (1986).

The jump condition

Consider the case with one interface Γ, which divides Ω in two parts Ω_1 and Ω_2, in each of which $a_{\alpha\beta}$ is continuous. At Γ, $a_{\alpha\beta}(\boldsymbol{x})$ is discontinuous. Let superscripts 1 and 2 denote quantities on Γ at the side of Ω_1 and Ω_2, respectively. Application of the divergence theorem to (3.3) gives

$$a(\varphi, \psi) = - \int_{\Omega\backslash\Gamma} (a_{\alpha\beta}\varphi_{,\alpha})_{,\beta}\psi d\Omega + \int_\Gamma (a^1_{\alpha\beta}\varphi^1_{,\alpha} - a^2_{\alpha\beta}\varphi^2_{,\alpha})n^1_\beta\psi d\Gamma \ . \tag{3.5}$$

Hence, the solution of (3.4) satisfies (3.1) in Ω/Γ, together with the following *jump condition* on the interface Γ:

$$a^1_{\alpha\beta}\varphi^1_{,\alpha}n^1_\beta = a^2_{\alpha\beta}\varphi^2_{,\alpha}n^1_\beta \quad \text{on} \quad \Gamma \ . \tag{3.6}$$

This means that where $a_{\alpha\beta}$ is discontinuous, so is $\varphi_{,\alpha}$. This has to be taken into account in constructing discrete approximations.

The pressure equation in reservoir engineering

With $\boldsymbol{b} = c = 0$ and a scalar diffusion coefficient (i.e. $a_{\alpha\beta} = 0$, $\alpha \neq \beta$ and $a_{11} = a_{22} = a_{33} = a$) equation (3.1) reduces to

$$-(a\varphi_{,\alpha})_{,\alpha} = q \ .$$

This equation plays an important role in the theory of flow in porous media, and closely resembles the pressure equation in the IMPES (implicit pressure, explicit saturation) model in reservoir engineering. Usually the domain contains interfaces across which $a(\boldsymbol{x})$ has large jumps. For an introduction to the IMPES model, see Aziz and Settari (1979).

3.3 A one-dimensional example

The basic ideas of finite difference and finite volume discretization taking discontinuities in $a_{\alpha\beta}$ into account will be explained for the following example:

$$-(a\varphi_{,1})_{,1} = q, \quad x \in \Omega \equiv (0,1). \tag{3.7}$$

Boundary conditions will be given later.

Finite difference discretization

A computational grid $G \subset \bar{\Omega}$ is defined by

$$G = \{x \in \mathbb{R}: x = x_j = jh, \quad j = 0, 1, ..., n, \quad h = 1/n\}.$$

Forward and backward difference operators are defined by

$$\Delta\varphi_j \equiv (\varphi_{j+1} - \varphi_j)/h, \quad \nabla\varphi_j = (\varphi_j - \varphi_{j-1})/h.$$

A finite difference approximation of (3.7) is obtained by replacing d/dx_1 by Δ or ∇. A nice symmetric formula is

$$-\frac{1}{2}\{\nabla(a\Delta) + \Delta(a\nabla)\}\varphi_j = q_j, \quad j = 1, ..., n-1, \tag{3.8}$$

where $q_j = q(x_j)$ and φ_j is the numerical approximation of $\varphi(x_j)$. Written out in full, equation (3.8) gives

$$\{-(a_{j-1} + a_j)\varphi_{j-1} + (a_{j-1} + 2a_j + a_{j+1})\varphi_j - (a_j + a_{j+1})\varphi_{j+1}\}/2h^2 = q_j,$$
$$j = 1, ..., n-1. \tag{3.9}$$

If the boundary condition at $x = 0$ is $\varphi(0) = f$ (Dirichlet), we eliminate φ_0 from (3.9) with $\varphi_0 = f$. If the boundary condition is $a(0)\varphi_{,1}(0) = f$ (Neumann), we write down (3.9) for $j = 0$ and replace the quantity $-(a_{-1} + a_0)\varphi_{-1} + (a_{-1} + a_0)\varphi_0$ by $2hf$. If the boundary condition is $c_1\varphi_{,1}(0) + c_2\varphi(0) = f$ (Robin), we again write down (3.9) for $j = 0$, and replace the quantity just mentioned by $2h(f - c_2\varphi_0)a(0)/c_1$. The boundary condition at $x = 1$ is handled in a similar way.

An interface problem

In order to show that (3.8) can be inaccurate for interface problems, we consider the following example: $q = 0$, and

$$a(x) = \varepsilon, \quad 0 < x \le x^*, \quad a(x) = 1, \quad x^* < x < 1.$$

The boundary conditions are: $\varphi(0) = 0$, $\varphi(1) = 1$. The jump condition (3.6) becomes

$$\varepsilon \lim_{x \uparrow x^*} \varphi_{,1} = \lim_{x \downarrow x^*} \varphi_{,1} \,. \tag{3.10}$$

By postulating a piecewise linear solution, the solution of (3.7) and (3.10) is found to be

$$\varphi = \alpha x \,, \quad 0 \le x < x^* \,, \quad \varphi = \varepsilon \alpha x + 1 - \varepsilon \alpha \,, \quad x^* \le x \le 1 \,,$$
$$\alpha = 1/(x^* - \varepsilon x^* + \varepsilon) \,. \tag{3.11}$$

Assume $x_k < x^* \le x_{k+1}$. By postulating a piecewise linear solution

$$\varphi_j = \alpha j \,, \quad 0 \le j \le k \,, \quad \varphi_j = \beta j - \beta n + 1 \,, \quad k+1 \le j \le n \,, \tag{3.12}$$

one finds that the solution of (3.9), with the boundary conditions given above, is given by (3.12) with

$$\beta = \varepsilon \alpha \,, \quad \alpha = \left(\varepsilon \frac{1-\varepsilon}{1+\varepsilon} + \varepsilon(n-k) + k \right)^{-1} \,.$$

Hence

$$\varphi_k = \frac{x_k}{\varepsilon h(1-\varepsilon)/(1+\varepsilon) + (1-\varepsilon)x_k + \varepsilon} \,. \tag{3.13}$$

Let $x^* = x_{k+1}$. The exact solution in x_k is according to (3.11)

$$\varphi(x_k) = \frac{x_k}{(1-\varepsilon)x_{k+1} + \varepsilon} \,.$$

Hence, the error satisfies

$$\varphi_k - \varphi(x_k) = \mathcal{O}\left(\varepsilon \frac{1-\varepsilon}{1+\varepsilon} h \right) \,.$$

As another example, let $x^* = x_k + h/2$. The numerical solution in x_k is still given by (3.13). The exact solution in x_k is according to (3.11)

$$\varphi(x_k) = \frac{x_k}{(1-\varepsilon)x_k + \varepsilon + h(1-\varepsilon)/2} \,.$$

The error in x_k satisfies

$$\varphi_k - \varphi(x_k) = \mathcal{O}\left(\frac{(1-\varepsilon)^2}{\varepsilon(1+\varepsilon)} h \right) \,.$$

When $a(x)$ is continuous ($\varepsilon = 1$) the error is zero. For general continuous $a(x)$ the error is $\mathcal{O}(h^2)$. When $a(x)$ is discontinuous, the error of the scheme (3.8) increases to $\mathcal{O}(h)$.

Finite volume discretization

By starting from the weak formulation (3.4) and using *finite volume discretization* one may obtain $\mathcal{O}(h^2)$ accuracy for discontinuous $a(x)$. The domain Ω is covered by cells or finite volumes Ω_j:

$$\Omega_j = (x_j - h/2,\, x_j + h/2)\,, \quad j = 1, .., n-1\,.$$

Let $\psi(x)$ be the characteristic function of Ω_j:

$$\psi(x) = 0\,, \quad x \notin \Omega_j\,; \quad \psi(x) = 1\,, \quad x \in \Omega_j\,.$$

A convenient unified treatment of both cases: $a(x)$ continuous and $a(x)$ discontinuous, is as follows. We approximate $a(x)$ by a piecewise constant function that has a constant value a_j in each Ω_j. Of course, this works best if discontinuities of $a(x)$ lie at boundaries of finite volumes Ω_j. One may take $a_j = a(x_j)$, or

$$a_j = h^{-1} \int\limits_{\Omega_j} a\,d\Omega\,.$$

With this approximation of $a(x)$ one obtains from (3.5)

$$a(\varphi, \psi) = - \int\limits_{\Omega_j} (a\varphi_{,1})_{,1}\,d\Omega$$

$$= -a\varphi_{,1}\big|_{x_j - h/2}^{x_j + h/2} \quad \text{if } 1 \le j \le n-1\,. \tag{3.14}$$

By taking successively $j = 1, 2, ..., n-1$, equation (3.4) leads to $n-1$ equations for the $n-1$ unknowns φ_j ($\varphi_0 = 0$ and $\varphi_n = 1$ are given), after making further approximations in (3.14).

In order to approximate $a\varphi_{,1}(x_j + h/2)$ we proceed as follows. Since $au_{,1}$ is smooth, $\varphi_{,1}(x_j + h/2)$ does not exist if $a(x)$ jumps at $x = x_j + h/2$, because of the jump condition. Hence, it is a bad idea to discretize $\varphi_{,1}(x_j + h/2)$. Instead, we write

$$\varphi\big|_j^{j+1} = \int\limits_{x_j}^{x_{j+1}} \varphi_{,1}\,dx = \int\limits_{x_j}^{x_{j+1}} \frac{1}{a}a\varphi_{,1}\,dx \cong (a\varphi_{,1})_{j+1/2} \int\limits_{x_j}^{x_{j+1}} \frac{1}{a}\,dx\,, \tag{3.15}$$

where we have exploited the smoothness of $a\varphi_{,1}$. The piecewise constant approximation of a gives

$$\int\limits_{x_j}^{x_{j+1}} \frac{1}{a}\,dx = h/w_j\,,$$

with w_j the *harmonic average* of a_j and a_{j+1}:

$$w_j \equiv 2a_j a_{j+1}/(a_j + a_{j+1}) \ .$$

We obtain the following approximation:

$$(a\varphi,_1)_{j+1/2} \cong w_j(\varphi_{j+1} - \varphi_j)/h \ . \tag{3.16}$$

With equations (3.14) and (3.16), the weak formulation (3.4) leads to the following discretization:

$$w_{j-1}(\varphi_j - \varphi_{j-1})/h - w_j(\varphi_{j+1} - \varphi_j)/h = hq_j \ , \quad j = 1, 2, ..., n - 1 \ , \tag{3.17}$$

with

$$q_j \equiv h^{-1} \int_{\Omega_j} q dx \ .$$

When $a(x)$ is smooth, $w_j \approx (a_j + a_{j+1})/2$, and we recover the finite difference approximation (3.9).

Equation (3.17) can be solved in a similar way as (3.9) for the interface problem under consideration. Assume $x^* = x_k + h/2$. Hence

$$w_j = \varepsilon \ , \quad 1 \leq j < k \ ; \quad w_k = 2\varepsilon/(1 + \varepsilon) \ ; \quad w_j = 1 \ , \quad k < j \leq n - 1 \ .$$

By postulating a solution as in (3.12) one finds

$$\beta = \alpha\varepsilon \ , \quad \alpha = h/(x^* - \varepsilon x^* + \varepsilon) \ .$$

Comparison with (3.11) shows that $\varphi_j = \varphi(x_j)$: the numerical error is zero. In more general circumstances the error will be $\mathcal{O}(h^2)$. Hence, finite volume discretization is more accurate than finite difference discretization for interface problems.

Discontinuity inside a finite volume

What happens when $a(x)$ is discontinuous *inside* a finite volume Ωj, at $x^* = x_j$, say? One has, with ψ as before, according to (3.5):

$$a(\varphi, \psi) = -a\varphi,_1 \big|_{x_j - h/2}^{x_j + h/2} + \lim_{x \uparrow x_j} a\varphi,_1 - \lim_{x \downarrow x_j} a\varphi,_1 \ .$$

The exact solution u satisfies the jump condition (3.6); thus the last two terms cancel. Approximating $\varphi,_1$ by finite differences one obtains

$$a(\varphi, \psi) \cong -a_{j+1/2}(\varphi_{j+1} - \varphi_j)/h + a_{j-1/2}(\varphi_j - \varphi_{j-1})/h \ .$$

This leads to the following discretization:

$$[-a_{j-1/2}\varphi_{j-1} + (a_{j-1/2} + a_{j+1/2})\varphi_j - a_{j+1/2}\varphi_{j+1}]/h = hq_j . \qquad (3.18)$$

For smooth $a(x)$ this is very close to the finite difference discretization (3.9), but for discontinuous $a(x)$ there is an appreciable difference: (3.18) remains accurate to $\mathcal{O}(h^2)$, like (3.17); the proof of this left as an exercise.

We conclude that for interface problems finite volume discretization is more suitable than finite difference discretization.

Exercise 3.3.1. The discrete maximum and l_2 norms are defined by, respectively,

$$|\varphi|_\infty = \max\{|\varphi_j| : 0 \le j \le n\} , \quad |\varphi|_0 = h\{\sum_{j=0}^{n} \varphi_j^2\}^{1/2} . \qquad (3.19)$$

Estimate the error in the numerical solution given by (3.12) in these norms.

Exercise 3.3.2. Show that the solution of (3.18) is exact for the model problem treated in this section.

3.4 Vertex-centered discretization

Vertex-centered grid

We now turn to the discretization of (3.1) in more dimensions. It suffices to study the two-dimensional case. The computational grid G is defined by

$$G \equiv \{\boldsymbol{x} \in \bar{\Omega} : \boldsymbol{x} = jh , \; j = (j_1, j_2) , \; j_\alpha = 0, 1, ..., n_\alpha , \; h = (h_1, h_2) ,$$
$$h_\alpha = 1/n_\alpha\} .$$

G is the union of a set of cells, the vertices of which are the grid points $\boldsymbol{x} \in G$. This is called a *vertex-centered grid*. Figure 3.1 gives a sketch. The solution of (3.1) or (3.4) is approximated in $\boldsymbol{x} \in G$, resulting in a *vertex-centered discretization*.

Finite difference discretization

Forward and backward difference operators Δ_α and ∇_α are defined by

$$\Delta_\alpha \varphi_j \equiv (\varphi_{j+e_\alpha} - \varphi_j)/h_\alpha , \quad \nabla_\alpha \varphi_j \equiv (\varphi_j - \varphi_{j-e_\alpha})/h_\alpha ,$$

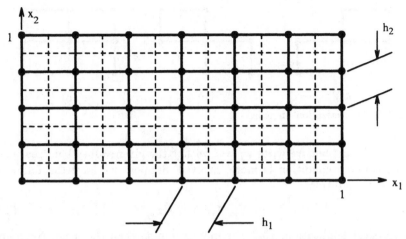

Fig. 3.1. Vertex-centered grid. (• grid points; - - - finite volume boundaries.)

where $e_1 \equiv (1,0)$, $e_2 \equiv (0,1)$. Of course, the summation convention does not apply here. *Finite difference approximations* of (3.1) are obtained by replacing $\partial/\partial x_\alpha$ by Δ_α or ∇_α or a linear combination of the two.

We mention a few possibilities. A nice symmetric formula is

$$-\frac{1}{2}\{\nabla_\beta(a_{\alpha\beta}\Delta_\alpha) + \Delta_\beta(a_{\alpha\beta}\nabla_\alpha)\}\varphi + \frac{1}{2}(\nabla_\alpha + \Delta_\alpha)(b_\alpha\varphi) + c\varphi = q \ . \quad (3.20)$$

The finite difference scheme (3.20) relates φ_j to φ in the neighbouring grid points $\boldsymbol{x}_{j\pm e_\alpha}$, $\boldsymbol{x}_{j\pm e_1 \mp e_2}$. This set of grid points together with $\boldsymbol{x}_j$ is called the *stencil* of (3.20). It is depicted in Fig. 3.2(a). This stencil is not symmetric. The points $\boldsymbol{x}_{j\pm e_1 \mp e_2}$ enter only in the stencil when $a_{12} \neq 0$. The local discretization error is $\mathcal{O}(h_1^2, h_2^2)$, and so is the global discretization error, if the right-hand side of (3.1) is sufficiently smooth, if the boundary conditions are suitably implemented, and if $a_{\alpha\beta}$ is continuous. It is left to the reader to write down a finite difference approximation with stencil as in Fig. 3.2(b). The average of Figures 3.2(a) and 3.2(b) gives 3.2(c), which has the advantage of being symmetric. This means that when the solution has a certain symmetry, the discrete approximation will also have this symmetry. With Fig. 3.2(a) or 3.2(b) this will in general be only approximately the case. A disadvantage of Fig. 3.2(c) is that the corresponding matrix is less sparse.

Boundary conditions

Although elementary, a brief discussion of the implementation of boundary conditions is given, because a full discussion with $a_{12}(\boldsymbol{x}) \neq 0$ is hard to find in the literature. If $\boldsymbol{x}_j \in \partial\Omega$ and a Dirichlet condition is given, then (3.20) is

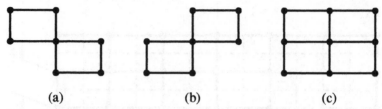

(a)   (b)   (c)

Fig. 3.2. Discretization stencils.

not used in $\boldsymbol{x}_j$, but we write $\varphi_j = f$ with f the given value. The treatment of a Neumann condition is more involved. Suppose we have the following Neumann condition:

$$a_{1\alpha}\varphi_{,\alpha}(1, x_2) = f(x_2) \tag{3.21}$$

(in physical applications (3.21) is more common than the usual Neumann condition $\varphi_{,1} = f$, and (3.21) is somewhat easier to implement numerically).

Let $\boldsymbol{x}_j$ lie on $x_1 = 1$. Equation (3.20) is written down in $\boldsymbol{x}_j$. This involves φ_j values in points outside G (*virtual points*). By means of (3.21) the virtual values are eliminated, as follows. First the virtual values arising from the second-order term are discussed. Let us write

$$-\frac{1}{2}\{\nabla_\beta(a_{\alpha\beta}\Delta_\alpha) + \Delta_\beta(a_{\alpha\beta}\nabla_\alpha)\}\varphi_j = q_j^{-3}\varphi_{j-e_2} + q_j^{-2}\varphi_{j+e_1-e_2} + q_j^{-1}\varphi_{j-e_1}$$
$$+q_j^0\varphi_j + q_j^1\varphi_{j+e_1} + q_j^2\varphi_{j-e_1+e_2} + q_j^3\varphi_{j+e_2}\,, \tag{3.22}$$

with

$$
\begin{aligned}
q_j^{-3} &= -(a_{22,j-e_2} + a_{22,j})/2h_2^2 - (a_{12,j-e_2} + a_{12,j})/2h_1h_2\,, \\
q_j^{-2} &= (a_{12,j-e_2} + a_{12,j+e_1})/2h_1h_2\,, \\
q_j^{-1} &= -(a_{11,j-e_1} + a_{11,j})/2h_1^2 - (a_{12,j-e_1} + a_{12,j})/2h_1h_2\,, \quad (3.23) \\
q_j^1 &= q_{j+e_1}^{-1}\,, \quad q_j^2 = q_{j-e_1+e_2}^{-2}\,, \quad q_j^3 = q_{j+e_2}^{-3}\,, \\
q_j^0 &= -\sum_{m\neq 0} q_j^m\,.
\end{aligned}
$$

By Taylor expansion one finds that approximately

$$-q_j^{-1}(\varphi_j - \varphi_{j-e_1}) + q_j^1(\varphi_{j+e_1} - \varphi_j) - q_j^2(\varphi_j - \varphi_{j-e_1+e_2})$$
$$+q_j^{-2}(\varphi_{j+e_1-e_2} - \varphi_j) \simeq \tfrac{2}{h_1}a_{1\alpha}\varphi_{,\alpha}(x_j) = \tfrac{2}{h_1}f(x_2)\,.$$

This equation is used to eliminate the virtual values from (3.22). The first order term $(b_1\varphi)_{,1}$ is discretized as follows at $\boldsymbol{x} = \boldsymbol{x}_j$:

$$(b_1\varphi)_{,1} = b_{1,1}\varphi + b_1\varphi_{,1} = b_{1,1}\varphi + b_1(f - a_{12}\varphi_{,2})/a_{11}\,,$$

and $\varphi_{,2}$ is replaced by $\frac{1}{2}(\Delta_2 + \nabla_2)\varphi_j$.

Finite volume discretization

For smooth $a_{\alpha\beta}(\boldsymbol{x})$ there is little difference between finite difference and finite volume discretization, but for discontinuous $a_{\alpha\beta}$ it is more natural to use finite volume discretization, because this uses the weak formulation (3.4), and because it is more accurate, as we saw in the preceding section.

The domain Ω is covered by finite volumes or cells Ω_j, satisfying

$$\Omega = \bigcup_j \Omega_j , \quad \Omega_i \cap \Omega_j = \emptyset, \quad i \neq j .$$

The boundaries of the finite volumes are the broken lines in Fig. 3.1. Except at the boundaries, the grid points $\boldsymbol{x}_j$ are at the centre of Ω_j.

The point of departure is the weak formulation (3.4), with $a(\varphi, \psi)$ given by (3.5). Let ψ be the characteristic function of Ω_j:

$$\psi(\boldsymbol{x}) = 0 , \quad \boldsymbol{x} \notin \Omega_j , \quad \psi(\boldsymbol{x}) = 1 , \quad \boldsymbol{x} \in \Omega_j .$$

The exact solution satisfies the jump condition; thus the integral along Γ in (3.5) can be neglected. One obtains

$$a(\varphi, \psi) + b(\varphi, \psi) + c(\varphi, \psi) = -\int_{\Omega_j} (a_{\alpha\beta}\varphi_{,\alpha})_{,\beta} d\Omega + \int_{\Omega_j} (b_\alpha\varphi)_{,\alpha} d\Omega + \int_{\Omega_j} c\varphi d\Omega$$

$$= \int_{\Gamma_j} a_{\alpha\beta}\varphi_{,\alpha} n_\beta d\Gamma + \int_{\Gamma_j} b_\alpha\varphi n_\alpha d\Gamma + \int_{\Omega_j} c\varphi d\Omega = \int_{\Omega_j} q d\Omega ,$$

$$\text{(3.24)}$$

where we have used the divergence theorem, assuming that $a_{\alpha\beta}(\boldsymbol{x})$ is continuous in Ω_j, and where Γ_j is the boundary of Ω_j. We approximate the terms in (3.24) separately, as follows:

$$\int_{\Omega_j} q d\Omega \simeq |\Omega_j| q_j , \quad \int_{\Omega_j} c\varphi d\Omega \simeq |\Omega_j| c_j \varphi_j ,$$

where $|\Omega_j|$ is the area of Ω_j. For the integrals over Γ_j we first discuss the integral over the part AB of Γ_j, with $A = \boldsymbol{x}_j + (h_1/2, -h_2/2)$, $B = \boldsymbol{x}_j + (h_1/2, h_2/2)$; Ω_j is assumed not to be adjacent to $\partial\Omega$. The following approximations are made:

$$\int_A^B b_1\varphi dx_2 \cong h_2(b_1\varphi)_C , \quad \int_A^B a_{\alpha 1}\varphi_{,\alpha} dx_2 \cong h_2(a_{\alpha 1}\varphi_{,\alpha})_C , \quad \text{(3.25)}$$

where C is the center of AB : $C = \boldsymbol{x}_j + (h_1/2, 0)$. The right-hand sides of (3.25) have to be approximated further.

Continuous coefficients

First, assume that $a_{\alpha\beta}(\boldsymbol{x})$ is continuous. We write

$$\int\limits_A^B b_1\varphi dx_2 \cong h_2 b_1(C)(\varphi_j + \varphi_{j+e_1})/2 \;, \quad \int\limits_A^B a_{11}\varphi_{,1} dx_2 \cong h_2 a_{11}(C)\Delta_1\varphi_j$$

$$(3.26)$$

and

$$\int\limits_A^B a_{12}\varphi_{,2} dx_2 \cong h_2 a_{12}(C)(\nabla_2\varphi_{j+e_1} + \Delta_2\varphi_j)/2 \tag{3.27}$$

or

$$\int\limits_A^B a_{12}\varphi_{,2} dx_2 \cong h_2 a_{12}(C)(\Delta_2 + \nabla_2)(\varphi_j + \varphi_{j+e_1})/4 \tag{3.28}$$

or

$$\int\limits_A^B a_{12}\varphi_{,2} dx_2 \cong h_2 a_{12}(C)(\Delta_2\varphi_{j+e_1} + \nabla_2\varphi_j)/2 \;. \tag{3.29}$$

Discontinuous coefficients

Assume that $a_{\alpha\beta}(\boldsymbol{x})$ is continuous in Ω_j, but may be discontinuous at the boundaries of Ω_j. In the approximation of the right-hand sides of (3.25), the jump condition (3.6) has to be taken into account.

For simplicity we assume $a_{12} = 0$. Proceeding in the same way as in the one-dimensional case in the preceding section we obtain

$$\int\limits_A^B a_{11}\varphi_{,1} dx_2 \cong h_2 w_j \Delta_1\varphi_j \;, \tag{3.30}$$

with

$$w_j \equiv 2a_{11,j}a_{11,j+e_1}/(a_{11,j} + a_{11,j+e_1}) \;. \tag{3.31}$$

The convective term in (3.25) may be approximated as follows:

$$h_2(b_1\varphi)_C \cong h_2 b_1(C)(\varphi_j + \varphi_{j+e_1})/2 \;. \tag{3.32}$$

If a_{11} is strongly discontinous it is more accurate to approximate $\varphi(C)$ as follows. According to the jump conditon (3.6), with $a_{12} = 0$, we have approximately

$$a_{11}(\boldsymbol{x}_j)(\varphi(C) - \varphi_j)/h_1 = a_{11}(\boldsymbol{x}_{j+e_1})(\varphi_{j+e_1} - \varphi(C))/h_1 .$$

Solving for $\varphi(C)$ gives

$$\varphi(C) = \{a_{11}(\boldsymbol{x}_j)\varphi_j + a_{11}(\boldsymbol{x}_{j+e_1})\varphi_{j+e_1}\}/(a_{11}(\boldsymbol{x}_j) + a_{11}(\boldsymbol{x}_{j+e_1})) .$$

Substitution in (3.25) results in

$$h_2(b_1\varphi)_C \cong \frac{1}{2}h_2 b_1(C)w_j\{\varphi_j/a_{11}(\boldsymbol{x}_{j+e_1}) + \varphi_{j+e_1}/a_{11}(\boldsymbol{x}_j)\} . \tag{3.33}$$

The integrals along the other faces of Ω_j are approximated in a similar fashion.

Just as in the one-dimensional example discussed earlier, one may also assume that $a_{\alpha\beta}(\boldsymbol{x})$ is continuous across the boundaries of the finite volumes, but may be discontinuous at the solid lines in Fig. 3.1. Then we approximate $a_{\alpha\beta}(\boldsymbol{x})$ by a constant in each cell bounded by solid lines. The integral over AB is split into two parts: over AC and over CB. One obtains, for example,

$$\int_A^C a_{\alpha 1}\varphi_{,\alpha}dx_2 \cong h_2\{a_{11}(A)\Delta\varphi_j + a_{12}(A)\varphi_{,2}\} , \tag{3.34}$$

where $\varphi_{,2}$ has to be approximated further. Now the case $a_{12}(x) \neq 0$ is easily handled, because the jump conditions do not interfere with the approximation of $\varphi_{,2}$. For example,

$$\varphi_{,2} \cong \nabla_2(\varphi_j + \varphi_{j+e_1})/2 .$$

For the convective term the approximation (3.32) may be used.

Boundary conditions

The boundary conditions are treated as follows. If we have a Dirichlet condition at $\boldsymbol{x}_j$ we simply substitute the given value for φ_j. Suppose we have a Neumann condition, for example at $x_1 = 1$:

$$a_{\alpha 1}\varphi_{,\alpha}(1, x_2) = f(x_2) . \tag{3.35}$$

Let AB lie on $x_1 = 1$. Then we have

$$\int_A^B a_{\alpha 1}\varphi_{,\alpha}dx_2 \cong h_2 f(x_{2,j}) \tag{3.36}$$

and

$$\int_A^B b_1 \varphi \, dx_2 \cong h_2 b_{1,j} \varphi_j \ .$$

Exercise 3.4.1. Derive a discretization using the stencil of Fig. 3.2(b).
(Hint: only the discretization of the mixed derivative needs to be changed.)

3.5 Cell-centered discretization

Cell-centered grid

The domain Ω is divided in cells as before (solid lines in Fig. 3.1), but now
the grid points are the centers of the cells, see Fig. 3.3. The computational
grid G is defined by

$$G = \{ x \in \Omega : \ x = x_j = (j - p)h, j = (j_1, j_2), p = (\tfrac{1}{2}, \tfrac{1}{2}),$$
$$h = (h_1, h_2), j_\alpha = 1, 2, ..., n_\alpha, h_\alpha = 1/n_\alpha \} \ .$$

The cell with center x_j is called Ω_j. Note that in a cell-centered grid there
are no grid points on the boundary $\partial \Omega$.

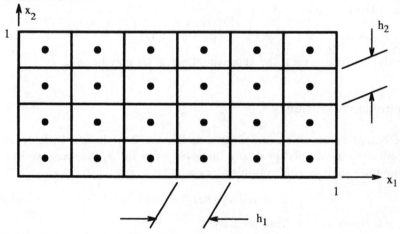

Fig. 3.3. Cell-centered grid. ($\bullet$ grid points; — finite volume boundaries.)

Finite difference discretization

Finite difference discretizations are obtained in the same way as in Sect. 3.4. Equation (3.20) can be used.

Boundary conditions

Suppose Ω_j is adjacent to $x_1 = 1$. Let a Dirichlet condition be given at $x_1 = 1$: $\phi(1, x_2) = f(x_2)$. Then (3.20) is written down at $\pmb{x}_j$, and ϕ values outside G are eliminated with the Dirichlet condition:

$$\varphi_{j+e_1} = 2f(x_{2,j}) - \varphi_j .$$

When we have a Neumann condition at $x_1 = 1$ as given in (3.21) then the procedure is similar to that in Sect. 3.4. Equation (3.20) is written down at $\pmb{x}_j$. Quantities involving values outside G (virtual values) are eliminated with the Neumann condition. Using the notation of equation (3.23), we have approximately

$$q_j^1(\varphi_{j+e_1} - \varphi_j) + q_j^{-2}(\varphi_{j+e_1-e_2} - \varphi_j) \simeq -a_{1\alpha}\varphi_{,\alpha}(1, x_2)/h_1 = -f(x_2)/h_1 .$$

This equation is used to eliminate the virtual values.

Finite volume discretization

In the interior, cell-centered finite volume discretization is identical to vertex-centered finite volume discretization. When $a_{\alpha\beta}(\pmb{x})$ is continuous in Ω then one obtains equations (3.26)–(3.29). When $a_{\alpha\beta}(\pmb{x})$ is continuous in Ω_j but is allowed to be discontinuous at the boundaries of Ω_j then one obtains equations (3.30)–(3.33). We require $a_{12}(\pmb{x}) = 0$ in this case. When $a_{\alpha\beta}(\pmb{x})$ is allowed to be discontinuous only at line segments connecting cell centers in Fig. 3.3, then one obtains equation (3.34).

Boundary conditions

Because now there are no grid points on the boundary, the treatment of boundary conditions is different from the vertex-centered case.

Let the face AB of the finite volume Ω_j lie at $x_1 = 1$. If we have a Dirichlet condition $\varphi(1, x_2) = f(x_2)$ then we put

$$\int_A^B a_{1\alpha}\varphi_{,\alpha}dx_2 \simeq 2h_2 a_{11,j}(f(C) - \varphi_j)/h_1 + h_2 a_{12}df(C)/dx_2$$

and

$$\int_A^B b_1 \varphi dx_2 \cong h_2 b_1(C) f(C) \,,$$

where C is the midpoint of AB.

If a Neumann condition (3.35) is given at $x_1 = 1$ then we use (3.36) and

$$\int_A^B b_1 \varphi dx_2 \cong h_2 b_1(C) \varphi(C) \,, \tag{3.37}$$

where $\varphi(C)$ has to be approximated further. With upwind differencing, to be discussed shortly, this is easy. Higher order accuracy can be obtained with

$$\varphi(C) \cong \varphi_j + \frac{1}{2} h_1 \varphi_{,1}(C) \,,$$

where the trick is to find a simple approximation to $\varphi_{,1}(C)$. If $a_{\alpha\beta}(\boldsymbol{x})$ is continuous in Ω_j, then we can put, using (3.35),

$$\varphi_{,1}(C) = \{ f(C) - a_{12,j} \varphi_{,2}(C) \} / a_{11,j}$$
$$\simeq \{ f(C) - \frac{1}{2} a_{12,j} (\nabla_2 + \Delta_2) \varphi_j \} / a_{11,j} \,.$$

If $a_{\alpha\beta}(\boldsymbol{x})$ is discontinuous at lines connecting finite volume centers (grid points) then (3.35) gives

$$a_{11,j}^2 \varphi_{,1}(C) + a_{12,j}^2 \varphi_{,2}^2(C) = f(C) \,,$$
$$a_{11,j}^1 \varphi_{,1}(C) + a_{12,j}^1 \varphi_{,2}^1(C) = f(C) \,,$$

where the superscripts 1 and 2 indicate $\lim_{x_2 \uparrow x_2(C)}$ and $\lim_{x_2 \downarrow x_2(C)}$, respectively. Taking the average of the preceding two equations and approximating $\varphi_{,2}$ one obtains

$$\varphi_{,1}(C) = \frac{1}{2} \{ f(C) - a_{12}^2 \Delta_2 \varphi_j \} / a_{11,j}^2 + \frac{1}{2} \{ f(C) - a_{12}^1 \nabla_2 \varphi_j \} / a_{11,j}^1 \,.$$

3.6 Upwind discretization

The mesh Péclet number condition

Assume $a_{12} = 0$. Write the discretization obtained in the interior of Ω with one of the methods just discussed as

$$q_j^{-3} \varphi_{j-e_2} + q_j^{-1} \varphi_{j-e_1} + q_j^0 \varphi_j + q_j^1 \varphi_{j+e_1} + q_j^3 \varphi_{j+e_2} = r_j \,.$$

As will be discussed later, for the matrix A of the resulting linear system to have the desirable property of being an M-*matrix*, it is necessary that

$$q_j^\nu \leq 0 , \quad \nu = \pm 1, \pm 3 . \tag{3.38}$$

Let us see whether this is the case. First, take $a_{\alpha\beta}(\boldsymbol{x})$ and $b_\alpha(\boldsymbol{x})$ constant. Then, apart from a scaling factor, all discretization methods discussed lead in the interior of Ω to

$$q_j^{-3} = -\frac{1}{2}h_1 b_2 - h_1 a_{22}/h_2 , \quad q_j^{-1} = -\frac{1}{2}h_2 b_1 - h_2 a_{11}/h_1 ,$$
$$q_j^1 = \frac{1}{2}h_2 b_1 - h_2 a_{11} h_1 , \quad q_j^3 = \frac{1}{2}h_1 b_2 - h_1 a_{22}/h_2 , \quad q_j^0 = -\sum_{\nu \neq 0} q_j^\nu .$$

From (3.38) it follows that the mesh Péclet numbers P_α, defined as

$$\mathrm{P}_\alpha = |b_\alpha| h_\alpha / a_{\alpha\alpha} \quad \text{(no summation)}$$

must satisfy

$$\mathrm{P}_\alpha \leq 2 . \tag{3.39}$$

With variable $a_{\alpha\beta}(\boldsymbol{x})$ and $b_\alpha(\boldsymbol{x})$ the expressions for q_j^ν become more complicated. Let us take, for example, cell-centered finite volume discretization, with $a_{\alpha\beta}(\boldsymbol{x})$ continuous inside the finite volumes, but possibly discontinuous at their boundaries. Then one obtains

$$q_j^{-3} = -h_1 b_{2,j-e_2/2} v_{j-e_2}/2a_{22,j-e_2} - h_1 v_{j-e_2}/h_2 ,$$
$$q_j^{-1} = -h_2 b_{1,j-e_1/2} w_{j-e_1}/2a_{11,j-e_1} - h_2 w_{j-e_1}/h_1 ,$$
$$q_j^1 = h_2 b_{1,j+e_1/2} w_j/2a_{11,j} - h_2 w_j/h_1 ,$$
$$q_j^3 = h_1 b_{2,j+e_2/2} v_j/2a_{22,j} - h_1 v_j/h_2 ,$$
$$q_j^0 = h_1 v_{j-e_2}/h_2 + h_2 w_{j-e_1}/h_1 + h_1 v_j/h_2$$
$$\quad + h_2 b_{1,j+e_1/2} w_j/2a_{11,j+e_1} - h_2 b_{1,j-e_1/2} w_{j-e_1}/(2a_{11,j-e_1})$$
$$\quad + h_1 b_{2,j+e_2/2} v_j/2a_{22,j+e_2} - h_1 b_{2,j-e_2/2} v_{j-e_2}/(2a_{22,j-e_2}) ,$$

where w_j is defined by (3.31), and $v_j = 2a_{22,j} a_{22,j+e_2}/(a_{22,j} + a_{22,j+e_2})$. Again, for A to be an M-matrix, equation (3.39) must satisfied, with P_α replaced by $\mathrm{P}_{\alpha,j}$, defined by

$$\mathrm{P}_{\alpha,j} = |b_{\alpha,j+e_\alpha/2}| h_\alpha / a_{\alpha\alpha,j} \quad \text{(no summation)} .$$

Upwind discretization

In computational fluid dynamics applications, often (3.39) is not satisfied. In order to have an M-matrix, the first derivatives in the equation may be discretized differently, namely by *upwind discretization*. This generates only non-positive contributions to q_{j}^{ν}, $\nu \neq 0$.

First we describe the concept of *flux splitting*. The convective fluxes $b_\alpha \phi$ are split according to

$$b_\alpha \varphi = f_\alpha^+ + f_\alpha^- \ .$$

First-order upwind discretization is obtained by the following splitting:

$$f_\alpha^\pm = \frac{1}{2}(b_\alpha \varphi \pm |b_\alpha|\varphi) \ ,$$

and by the following finite difference approximation:

$$(b_\alpha \varphi)_{,\alpha} \cong \nabla_\alpha f_\alpha^+ + \Delta_\alpha f_\alpha^- \ . \tag{3.40}$$

In the finite volume context, upwind discretization is obtained with (cf. (3.26), (3.32) and (3.37))

$$\int_A^B b_1 \varphi dx_2 \cong h_2(f_{1,j}^+ + f_{1,j+e_1}^-) \ . \tag{3.41}$$

Upwind discretization reduces the truncation error to $\mathcal{O}(h_\alpha)$. Much has been written in the computational fluid dynamics litarature about the pros and cons of upwind discretization. We will go into this later. The interested reader may consult Roache (1972) or Gresho and Lee (1981).

The mixed derivative

When $a_{12}(\boldsymbol{x}) \neq 0$, condition (3.38) may be violated, even when $P_\alpha = 0$. In practice, however, usually $a_{12}(\boldsymbol{x}) \neq 0$ does not cause the matrix A to deviate much from the M-matrix property, so that the behavior of the numerical solution methods applied is not seriously affected. See Mitchell and Griffiths (1994) and Exercise 3.6.1 for a discretization of the mixed derivative that leaves (3.38) intact.

Boundary conditions

Upwind discretization makes the application of boundary conditions easier than before, provided we have the physically common situation of a Dirichlet condition at an inflow boundary ($b_\alpha n_\alpha < 0$ with $\boldsymbol{n}$ the outward normal on $\partial\Omega$).

In the vertex-centered case, if $x_1 = 1$ is an inflow boundary, the Dirichlet condition is applied directly, and (3.41) is not required. If $x_1 = 1$ is an outflow boundary ($b_1 > 0$), (3.40) gives

$$(b_1\varphi)_{,1} \simeq \nabla_1 f_{1,j}^+ , \tag{3.42}$$

whereas (3.41) becomes

$$\int_A^B b_1\varphi dx_2 \simeq h_2 f_{1,j}^+ , \tag{3.43}$$

so that no virtual values need to be evaluated. In the cell-centered case with finite differences, if $x_1 = 1$ is an inflow boundary, a suitable approximation at this boundary is

$$(b_1\varphi)_{,1} \simeq 2(b_1(1, x_2)g(x_2) - b_{1,j}\varphi_j)/h_1 ,$$

with $g(x_2)$ the prescribed Dirichlet value, whereas in the outflow case we have (3.42). With finite volumes we have in the case of inflow

$$\int_A^B b_1\varphi dx_2 \simeq h_2 b_1(1, x_2)g(x_2)$$

and equation (3.43) in the case of outflow.

Exercise 3.6.1. Show that in order to satisfy (3.38) in the case that $a_{12} \neq 0$ one should use the seven-point stencil of Fig. 3.2(a) if $a_{12} < 0$ and the stencil of Fig. 3.2(b) if $a_{12} > 0$ (cf. Exercise 3.4.1). Assume $a_{\alpha\beta} = $ constant and $b_\alpha = c = 0$, and determine conditions that should be satisfied by a_{12} for (3.38) to hold; compare these with (2.4).

3.7 Nonuniform grids in one dimension

We now turn to nonuniform grids. We consider only the one-dimensional case, because much can be learned from it with little effort. Two-dimensional applications will be studied in the next chapter.

Difference between finite volume and finite difference discretization

Let us consider the convection-diffusion equation with constant coefficients:

$$L\varphi \equiv -D\frac{d^2\varphi}{dx^2} + u\frac{d\varphi}{dx} = q . \tag{3.44}$$

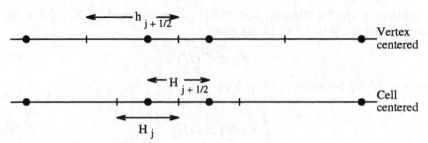

Fig. 3.4. A nonuniform grid.

Figure 3.4 shows a nonuniform grid. There is no need to consider the boundaries. Central finite difference discretization is given by

$$
\begin{aligned}
L_f \varphi_j &\equiv -D\{(\varphi_{j+1} - \varphi_j)/h_{j+1} - (\varphi_j - \varphi_{j-1})/h_j\}/h_{j+1/2} \\
&\quad + u(\varphi_{j+1} - \varphi_{j-1})/(h_j + h_{j+1}) \\
&= \varphi_{j-1}(-D/h_j - u/2)/h_{j+1/2} + 2\varphi_j D/(h_j h_{j+1}) \\
&\quad + \varphi_{j+1}(-D/h_{j+1} + u/2)/h_{j+1/2} = q_j \; ,
\end{aligned}
$$

where $h_{j+1/2} \equiv (h_j + h_{j+1})/2$, and $h_j = H_{j-1/2}$, cf. Fig. 3.4.

With vertex-centered finite volume discretization the volume boundaries are placed midway between the nodes. We write

$$
\int_{\Omega_j} L\varphi dx = F|_{j-1/2}^{j+1/2} = h_{j+1/2} q_j
$$

with

$$
F = F^v + F^c \; , \quad F^v = -Dd\varphi/dx \; , \quad F^c = u\varphi \; .
$$

We will call F the flux, F^v the viscous flux and F^c the convective flux. Central approximation gives

$$
F_{j+1/2}^v \cong -D(\varphi_{j+1} - \varphi_j)/h_{j+1} \; , \quad F_{j+1/2}^c \cong \frac{1}{2} u(\varphi_j + \varphi_{j+1}) \; .
$$

The resulting discrete approximation is

$$
L_v \varphi_j = h_{j+1/2} q_j \; , \quad L_v \equiv h_{j+1/2} L_f \; . \tag{3.45}
$$

Hence, finite difference discretization and vertex-centered finite volume discretization are equivalent.

With cell-centered finite volume discretization the domain is first divided into cells, and the nodes are placed in the cell centers, as shown in Fig. 3.4. Let H_j be the size of the cell with center at x_j. The cell-centered flux approximations are

$$F_{j+1/2}^v \cong -D(\varphi_{j+1} - \varphi_j)/H_{j+1/2} \, , \quad F_{j+1/2}^c \cong \frac{1}{2}u(\varphi_j + \varphi_{j+1}) \, , \quad (3.46)$$

with $H_{j+1/2} \equiv (H_j + H_{j+1})/2$. The following discrete approximation results:

$$L_c\varphi_j \equiv \varphi_{j-1}(-D/H_{j-1/2} - u/2) + \hspace{3cm} (3.47)$$
$$\varphi_j D(1/H_{j-1/2} + 1/H_{j+1/2}) + \varphi_{j+1}(-D/H_{j+1/2} + u/2) = H_j q_j \, .$$

If the grid is uniform the three discretizations L_f, L_v and L_c are identical, but on non-uniform grids L_c is different from L_v and L_f, since $H_{j-1/2} = h_j$, but $H_j \neq h_{j+1/2}$, cf. Fig. 3.4. This results in a striking difference in local truncation error, as we shall now see.

Definition 3.7.1. The *local truncation error* of the discrete operator L_d is

$$\tau_j \equiv L_d\{\varphi(x_j) - \varphi_j\} \, , \quad j = 1, ..., J \, , \hspace{2cm} (3.48)$$

with $\varphi(x)$ the exact solution.

Definition 3.7.2. *Consistency*
If L is the differential operator to be approximated, and $L_d\varphi_j$ approximates $h_j^p L\varphi(x_j)$, then L_d is called *consistent* if

$$\lim_{\Delta \downarrow 0} \tau_j/h_j^p = 0 \, , \quad j = 1, ..., J \, , \quad \Delta = \max\{h_j : j = 1, ..., J\} \, .$$

This implies that the size of τ is measured in the maximum norm. If we use a weaker norm, such as proposed for example in Spijker (1971), then an inconsistent (by our definition) scheme may get reclassified as consistent. In fact, this is the case for the inconsistent schemes that we will encounter. But we will stick for simplicity to the above definition. As a consequence, an inconsistent scheme may be convergent, as will be seen.

By Taylor expansion we find for the vertex-centered scheme (3.45), denoting the local truncation error by τ^v:

$$\tau_j^v = (-\frac{1}{3}D\varphi^{(3)} + \frac{1}{2}u\varphi^{(2)})h_{j+1/2}(h_{j+1} - h_j) + \mathcal{O}(\Delta_v^3) \, , \hspace{1cm} (3.49)$$

where $\varphi^{(\alpha)} = d^\alpha\varphi(x_j)/dx^\alpha \, , \quad \alpha = 2, 3$, and

$$\Delta_v \equiv \max\{h_j : j = 1, ..., J\} \, ,$$

with J number of nodes. The local truncation error τ^c of the cell-centered scheme (3.47) is found to be

$$\tau_j^c = (-\frac{1}{4}D\varphi^{(2)} + \frac{1}{4}u\varphi^{(1)})(H_{j+1} - 2H_j + H_{j+1}) \hspace{1cm} (3.50)$$
$$+ (-\frac{1}{6}D\varphi^{(3)} + \frac{1}{4}u\varphi^{(2)})(H_{j+1/2}^2 - H_{j-1/2}^2) + \mathcal{O}(\Delta_c^3) \, ,$$

where

$$\Delta_c = \max\{H_j \, : \, j = 1, .., J - 1\} \, ,$$

with $J - 1$ the number of cells.

The grids are called *smooth* if

$$|h_{j+1} - h_j| = \mathcal{O}(\Delta_v^2)$$

in the vertex-centered case, and

$$|H_{j+1/2} - H_{j-1/2}| = \mathcal{O}(\Delta_c^2) \quad \text{and} \quad |H_{j+1} - 2H_j + H_{j-1}| = \mathcal{O}(\Delta_c^3)$$

in the cell-centered case. On smooth grids,

$$\tau = \mathcal{O}(\Delta^3) \tag{3.51}$$

in both cases. But on arbitrary grids

$$\tau_j^v = \mathcal{O}(\Delta_v^2) \, , \quad \tau_j^c = \mathcal{O}(\Delta_c) \, . \tag{3.52}$$

We see that the cell-centered scheme is not even consistent. However, it would be wrong to think on the basis of (3.51) and (3.52) that the accuracy on smooth grids is necessarily better than on arbitrary grids, or that vertex-centered discretization is more accurate than cell-centered discretization. This follows from a study of the error, to the presented below. As noted earlier, the seemingly inaccurate cell-centered finite volume discretization is in widespread use, also on nonsmooth grids, and is known as block-centered discretization in reservoir engineering (Aziz and Settari (1979)). Fortunately, it is not necessary to advise against its use because of its large local truncation error, even though it is not consistent.

Global truncation error

Definition 3.7.3. *Global truncation error*
The *global truncation error* is defined as

$$e_j \equiv \varphi(x_j) - \varphi_j \, ,$$

with $\varphi(x)$ the exact solution.

According to (3.48) the local and global truncation errors are related by

$$L_d e_j = \tau_j \, . \tag{3.53}$$

Estimates of e_j are given in Tikhonov and Samarskii (1963), Manteuffel and White, Jr. (1986), Forsyth, Jr. and Sammon (1988) and Weiser and Wheeler

(1988). Both in the vertex-centered case (3.45) and in the cell-centered case (3.47) we have

$$e_j = \mathcal{O}(\Delta^2) . \tag{3.54}$$

Because this satisfactory accuracy result is the most surprising for the in-consistent cell-centered scheme, we will partially sketch a proof of (3.54) for this case only, following the theory developed in Forsyth, Jr. and Sammon (1988) for the two-dimensional diffusion equation. We will include a convection term, but restrict ourselves for simplicity to the one-dimensional constant coefficient case.

Alternative discretizations of the convection term

Before sketching a proof of (3.54) we introduce two additional discretizations of the convection term. One may try to improve accuracy on nonsmooth grids by replacing (3.46) by linear interpolation for the convective flux:

$$F^c_{j+1/2} \cong u(H_j\varphi_{j+1} + H_{j+1}\varphi_j)/(H_j + H_{j+1}) . \tag{3.55}$$

A third possibility is the following finite difference approximation (studied in Veldman and Rinzema (1992) for the vertex-centered case), obtained by fitting a parabola through $\varphi_{j-1}, \varphi_j, \varphi_{j+1}$:

$$H_j u d\varphi(x_j)/dx \cong \frac{uH_j}{H_{j-1/2} + H_{j+1/2}} \left\{ \frac{H_{j-1/2}}{H_{j+1/2}}(\varphi_{j+1} - \varphi_j) \right. \tag{3.56}$$
$$\left. + \frac{H_{j+1/2}}{H_{j-1/2}}(\varphi_j - \varphi_{j-1}) \right\} .$$

The contributions of these approximations of the convection term to the local turncation error τ^c are

$$\frac{1}{8}H_j(H_{j+1} - H_{j-1})ud^2\varphi(x_j)/dx^2 + \mathcal{O}(\Delta_c^3) . \tag{3.57}$$

for (3.55) and $\mathcal{O}(\Delta_c^3)$ for (3.56). As expected, these discretizations have a smaller local truncation error than (3.46). On uniform grids the three dis-cretizations are identical.

Estimate of global truncation error for cell-centered scheme

The validity of (3.54) can be made plaussible as follows. Define the following grid functions:

$$\mu_j^1 \equiv H_j^2 , \quad \mu_j^2 \equiv \sum_{k=1}^{j} H_{k-1/2}^3 , \quad \mu_j^3 \equiv \sum_{k=1}^{j}(H_k^2 + H_{k-1}^2)H_{k-1/2} , \tag{3.58}$$

where $H_0 \equiv 0$. We find with L_c defined by (3.47) and $\psi_k(x)$, $k = 1, 2, 3$ smooth functions to be chosen later:

$$L_c(\psi_1(x_j)\mu_j^1) = D\psi_1(x_j)(-2H_{j+1} + 4H_j - 2H_{j-1}) + \qquad (3.59)$$

$$+ \{D\frac{d\psi_1(x_j)}{dx} - \frac{1}{2}u\psi_1(x_j)\}(H_{j-1}^2 - H_{j+1}^2) + \mathcal{O}(\Delta_c^3) ,$$

$$L_c(\psi_2(x_j)\mu_j^2) = D\psi_2(x_j)(H_{j-1/2}^2 - H_{j+1/2}^2) + \mathcal{O}(\Delta_c^3) , \qquad (3.60)$$

$$L_c(\psi_3(x_j)\mu_j^3) = D\psi_3(x_j)(H_{j-1}^2 - H_{j+1}^2) + \mathcal{O}(\Delta_c^3) . \qquad (3.61)$$

We choose

$$\psi_1 = \frac{1}{8}(\varphi^{(2)} - \frac{u}{D}\varphi^{(1)}) , \psi_2 = \frac{1}{6}\varphi^{(3)} - \frac{u}{4D}\varphi^{(2)} ,$$
$$\psi_3 = -\frac{d\psi_1}{dx} + \frac{u}{2D}\psi_1$$

and define

$$e_j^k \equiv \psi_k(x_j)\mu_j^k , \quad k = 1, 2, 3 . \qquad (3.62)$$

Remembering (3.53), comparison of (3.59)–(3.61) with (3.50) shows that

$$L_c(e_j - e_j^1 - e_j^2 - e_j^3) = \mathcal{O}(\Delta_c^3) . \qquad (3.63)$$

The right-hand side is of the same order as the local truncation error in the uniform grid case, which makes it plausible that

$$e_j - e_j^1 - e_j^2 - e_j^3 = \mathcal{O}(\Delta_c^2) . \qquad (3.64)$$

Since $e_j^k = \mathcal{O}(\Delta_c^2)$, $k = 1, 2, 3$ we find

$$e_j = \mathcal{O}(\Delta_c^2) . \qquad (3.65)$$

which is what we wanted to show.

Of course, for a complete proof the boundary conditions have to be taken into account. Furthermore, it has to be shown that (3.63) gives (3.64). This can be done using the maximum principle (to be discussed later) but we will not do this. The maximum principle holds if the discretization matrix is an M-matrix. If it is not, (3.64) becomes doubtful. In the present case we do have an M-matrix, provided the mesh Péclet number is small enough. This will be more fully discussed in the next chapter.

Complete proofs of (3.65) are given in Manteuffel and White, Jr. (1986) (for vertex-centered discretizations, and for a slightly different cell-centered discretization) and in Forsyth, Jr. and Sammon (1988) (for D variable, but $u = 0$); in the same situation second order convergence in the discrete l_2-norm is proved in Weiser and Wheeler (1988). Hence, the inconsistent scheme defined by (3.47) has second order convergence on arbitrary grids, so that its widespread application is justified.

A similar method can be applied to the other two discretizations of the convection term. For (3.55), with associated contribution to the local truncation error τ^c given by (3.57), we choose μ^1, μ^2 and μ^3 as in (3.58), and

$$\psi_1 = \frac{1}{8}\varphi^{(2)}, \quad \psi_2 = \frac{1}{6}\varphi^{(3)}, \quad \psi_3 = -d\psi_1/dx.$$

With e^k acording to (3.62) we again have (3.64). This holds also for (3.56).

Spectro-consistent discretization scheme

In the continuous case convection corresponds to a skew-symmetric operator, whereas diffusion corresponds to a symmetric semi-positive definite operator. In Verstappen and Veldman (1998) it is shown that it is beneficial for accuracy if these properties are shared by the discretization scheme. Furthermore, in the nonstationary case the scheme is automatically stable in the energy norm if the symmetric part of the spatial discretization operator is semi-positive definite. In Verstappen and Veldman (1998) this is called *spectro-consistent discretization*. We will check these properties.

The discrete scheme leads to a linear system that we may denote as

$$(B + C)\varphi_h = b,$$

with B the diffusion matrix and C the convection matrix. The cell-centered scheme (3.47) gives

$$B = D \begin{pmatrix} \bullet & & & \\ & -\dfrac{1}{H_{j-1/2}} & \dfrac{1}{H_{j-1/2}} + \dfrac{1}{H_{j+1/2}} & -\dfrac{1}{H_{j+1/2}} \\ & & \bullet & & \bullet \end{pmatrix},$$

$$C = \frac{u}{2}\begin{pmatrix} \bullet & \bullet & \bullet & \\ & -1 & 0 & 1 \\ & \bullet & \bullet & \bullet \end{pmatrix}.$$

The conditions for a spectro-consistent scheme are obviously satisfied. However, for the seemingly more accurate discretization of the convection term (3.55) we obtain

$$C = \frac{u}{2}\begin{pmatrix} \bullet & & & & \\ -\dfrac{H_{j-1}}{H_{j-3/2}} & \dfrac{H_j}{H_{j-1/2}} - \dfrac{H_{j-2}}{H_{j-3/2}} & \dfrac{H_{j-1}}{H_{j-1/2}} & \\ & -\dfrac{H_j}{H_{j-1/2}} & \dfrac{H_{j+1}}{H_{j+1/2}} - \dfrac{H_{j-1}}{H_{j-1/2}} & \dfrac{H_j}{H_{j+1/2}} \\ & & \bullet & & \bullet \end{pmatrix}$$

which is clearly not skew-symmetric. Furthermore, the generic row of $(C + C^T)/2$ is given by

$$\frac{u}{2}\left(\bullet \quad \frac{H_{j-1} - H_j}{2H_{j-1/2}} \quad \frac{H_{j+1}}{H_{j+1/2}} - \frac{H_{j-1}}{H_{j-1/2}} \quad \frac{H_j - H_{j+1}}{2H_{j+1/2}} \quad \bullet \right),$$

so that the symmetric part of $B + C$ is not necessarily semi-positive definite. Hence, with (3.55) we do not have a spectro-consistent scheme, which spells trouble, as shown by examples presented in Verstappen and Veldman (1998).

Discussion

Despite the differences in local truncation error on nonsmooth grids for the three discretizations of the convection term (3.46), (3.55) and (3.56), we have made it plausible that in all three cases the global truncation error $e_j = \mathcal{O}(\Delta_c^2)$, regardless of grid smoothness. Hence, there is no reason to think that (3.55) is more accurate than (3.46), or that (3.56) is more accurate than (3.55). In fact, the opposite may be true on nonsmooth grids. If $H_{j-1/2}/H_{j+1/2} \gg 1$ then (3.56) approximates backward differencing, which is likely to cause numerical wiggles when $u > 0$. This is confirmed by numerical experiments with (3.56) on a vertex-centered grid reported in Veldman and Rinzema (1992). When $H_j/H_{j+1} \gg 1$ and $H_{j-1}/H_j \gg 1$ then (3.55) also approximates backward differencing.

Another indication of possible trouble is the fact that (3.55) and (3.56) may make a negative contribution to the main diagonal of the resulting discretization matrix, i.e. to the coefficient of φ_j. As a result, the discretization matrix is not an M-matrix, and (3.63) does not give (3.64). Our conclusion is, that the inconsistent scheme (3.47) is perfectly all right, and that the derivation of more accurate schemes should be approached with caution on nonsmooth grids, because making the local truncation error smaller may make the global truncation error larger. That this may happen is easy to see. According to (3.53)

$$e_j = L_d^{-1} \tau_j .$$

A change in L_d that makes $|\tau_j|$ smaller but that makes $\|L_d^{-1}\|$ larger may make $|e_j|$ larger. Of course, the choice of norm is important here.

Finally, we have shown that (3.55) does not lead to a spectro-consistent scheme. This can also be shown for (3.56).

Accuracy at boundaries

Let a Dirichlet condition be given at $x = 0$:

$$\varphi(0) = f .$$

For simplicity we assume $u = 0$ and $D = 1$. The vertex-centered scheme (3.45) becomes at the boundary

$$L_v \varphi_1 \equiv -\frac{1}{h_2} \varphi_2 + \varphi_1 \left(\frac{1}{h_1} + \frac{1}{h_2} \right) = h_{3/2} q_1 + f/h_1 ,$$

with local truncation error τ_1^v given by (3.49), as in the interior.

For the cell-centered scheme (3.47) this boundary condition is implemented as described in Sect. 3.5, resulting in

$$L_c \varphi_1 = -\frac{\varphi_2}{H_{3/2}} + \varphi_1 \left(\frac{1}{H_{3/2}} + \frac{2}{H_1} \right) = H_1 q_1 + 2f/H_1 , \qquad (3.66)$$

with local truncation error

$$\tau_1^c = \frac{1}{4}(-H_2 + 2H_1) \frac{d^2 \varphi(x_1)}{dx^2} + \frac{1}{6} \left(\frac{1}{4} H_1^2 - H_{3/2}^2 \right) \frac{d^3 \varphi(x_1)}{dx^3} + \mathcal{O}(\Delta_c^3) . \quad (3.67)$$

In the case of a Neumann condition

$$-d\varphi(0)/dx = f ,$$

the two schemes become at the boundary, following Sections 3.4 and 3.5:

$$L_v \varphi_0 = -(\varphi_1 - \varphi_0)/h_1 = \frac{1}{2} h_1 q_0 + f ,$$
$$L_c \varphi_1 = -(\varphi_2 - \varphi_1)/H_{3/2} = H_1 q_1 + f ,$$

with local truncation errors in the grid-point x_1, respectively,

$$\tau_1^v = -\frac{1}{6} h_1^2 \frac{d^3 \varphi(0)}{dx^3} + \mathcal{O}(\Delta_v^3) ,$$

$$\tau_1^c = \frac{1}{4}(H_1 - H_2) \frac{d^2 \varphi(x_1)}{dx^2} - \frac{1}{24}(3H_1^2 + 4H_{3/2}^2) \frac{d^3 \varphi(x_1)}{dx^3} \qquad (3.68)$$
$$+ \mathcal{O}(\Delta_c^3) . \qquad (3.69)$$

We make the following observations. On smooth grids with the Dirichlet boundary condition we see that

$$\tau_1^v = \mathcal{O}(\Delta_v^3) , \quad \tau_1^c = \mathcal{O}(\Delta_c) ,$$

so that even on uniform grids the cell-centered scheme is not consistent at the boundary! However, as before, this is no reason to prefer vertex-centered over cell-centered discretization. Nor is it necessary to improve the accuracy of the cell-centered (or, on non-smooth grids, also the vertex-centered) implementation of the Dirichlet condition, because the global truncation error is $\mathcal{O}(\Delta^2)$ in all cases (see Manteuffel and White, Jr. (1986)) and Forsyth, Jr. and Sammon (1988)).

We can make this plausible by a kind of backward error analysis, as follows. Suppose we use a perturbed boundary value $\tilde{f}$ in the cell-centered scheme (3.66). Then the local truncation error is found to be

$$\tilde{\tau}_1^c = \tau_1^c - \frac{2}{H_1}(\tilde{f} - f) ,$$

with τ_1^c given by (3.67). Hence, we approximate the exact solution of the differential equation with a perturbed boundary value

$$\tilde{f} = f + \frac{H_1}{2}\tau_1^c$$

with zero local truncation error at the boundary, so that we expect the numerical solution to be close to the perturbed exact solution. Since $\tilde{f} = f + \mathcal{O}(\Delta_c^2)$ we expect the perturbation of the exact solution to be $\mathcal{O}(\Delta_c^2)$, and the global truncation error also to be $\mathcal{O}(\Delta_c^2)$.

On smooth grids with the Neumann condition,

$$\tau_1^v = \mathcal{O}(\Delta_v^2) , \quad \tau_1^c = \mathcal{O}(\Delta_c^2) ,$$

which is an order lower than in the interior. Again, this is no reason to improve the accuracy of the discretization near the boundary, because even on nonsmooth grids we have (3.54), as shown in Tikhonov and Samarskii (1963), Manteuffel and White, Jr. (1986), Forsyth, Jr. and Sammon (1988) and Weiser and Wheeler (1988).

Final remarks

We conclude that both the vertex- and the cell-centered schemes discussed with straightforward implementation of boundary conditions are second order accurate. This is true not only for the one-dimensional convection-diffusion equation (3.44), (to which we have restricted ourselves here to diminish technical details), but more generally. Second order accuracy is shown for the one-dimensional convection-diffusion equation with variable coefficients in Manteuffel and White, Jr. (1986). The two-dimensional diffusion equation

$$-(a\varphi_{,\alpha})_{,\alpha} = q$$

with smooth diffusion coefficient $a(\pmb{x})$ is fully discussed in Forsyth, Jr. and Sammon (1988) and Weiser and Wheeler (1988). The case of discontinuous

$a(x)$ (important for flow in porous media and reservoir engineering) is included in Tikhonov and Samarskii (1963).

In the next chapter we treat the two-dimensional convection-diffusion equation, including singular perturbation aspects. Our main purpose will be to make it plausible, that with suitable local mesh refinement it is possible to compute flows at high Reynolds number accurately, despite the fact that the local truncation error (causing numerical viscosity) seems large compared to the viscous forces.

Exercise 3.7.1. (cf. Veldman and Rinzema (1992)). Perform numerical experiments for the following special case of (3.44):

$$-D\frac{d^2\varphi}{dx^2} + \frac{d\varphi}{dx} = 0 , \quad 0 \le x \le 1 , \quad \varphi(0) = 0 , \quad \varphi(1) = 1 ,$$

with exact solution

$$\varphi(x) = (e^{x/D} - 1)/(e^{1/D} - 1) .$$

Use the cell-centered finite volume method. Choose $D = 10^{-2}$ and 10^{-3}. Choose m cells of equal size in $(1-5D, 1)$ and vary m. Compare the accuracy of the three convection term discretizations (3.47), (3.55) and (3.56).

Exercise 3.7.2. Show that (3.56) does not lead to a spectro-consistent scheme.

4. The stationary convection-diffusion equation

4.1 Introduction

The convection-diffusion equation has been derived in Sect. 1.11. We take
$\rho = 1$, and obtain

$$\frac{\partial \varphi}{\partial t} + u_\alpha \varphi_{,\alpha} - (D\varphi_{,\alpha})_{,\alpha} = q, \quad \boldsymbol{x} \in \Omega \subset \mathbb{R}^d, \quad 0 < t \leq T . \tag{4.1}$$

For the physical significance of the terms in this equation, see Sect. 1.11.
The equation is assumed to be linear, with φ the only unknown. A number of important aspects of the numerical analysis of the equations of fluid
dynamics show up in this simple equation. Its simplicity allows a thorough
analysis of these aspects, which will be given in this chapter. Readers who
have some experience in computational fluid dynamics may think at first sight
that our treatment of such a simple linear equation is too detailed, but it is
a fact that sometimes controversial and not always well understood important issues, notably the occurrence of numerical 'wiggles', the specification of
outflow boundary conditions, singular perturbation aspects (the occurrence
of boundary layers when $D \ll 1$, in a sense to be made precise shortly, a
common situation in fluid dynamics) and the role of false (numerical) viscosity, can be brought out and clarified completely in the context of this simple
equation.

Dimensionless form

Let L and U be typical length and velocity scales, and let $\varPhi$ be a typical scale
for φ. Define dimensionless variables according to

$$\begin{aligned} \varphi' &= \varphi/\varPhi, \quad u'_\alpha = u_\alpha/U, \quad t' = tU/L, \quad x'_\alpha = x_\alpha/L, \\ q' &= qL/(U\varPhi) . \end{aligned}$$

Then (4.1) takes the following dimensionless form, upon deleting the primes:

$$\frac{\partial \varphi}{\partial t} + u_\alpha \varphi_{,\alpha} - (\mathrm{Pe}^{-1}\varphi_{,\alpha})_{,\alpha} = q ,$$

where

$$Pe \equiv UL/D \qquad (4.2)$$

is the Péclet number, already encountered in Sect. 1.11.

Initial and boundary conditions

At time $t = 0$ the following initial condition is given:

$$\varphi(0, \boldsymbol{x}) = \varphi_0(\boldsymbol{x}), \quad \boldsymbol{x} \in \Omega .$$

At every point of the boundary $\partial\Omega$ of Ω we have precisely one of the following types of boundary condition:

Dirichlet: $\varphi(t, \boldsymbol{x}) = f(t, \boldsymbol{x}), \quad \boldsymbol{x} \in \partial\Omega, \quad 0 < t \leq T ,$

Neumann: $k\dfrac{\partial\varphi(t, \boldsymbol{x})}{\partial n} = f(t, \boldsymbol{x}), \quad \boldsymbol{x} \in \partial\Omega, \quad 0 < t \leq T ,$

Robin: $k\dfrac{\partial\varphi(t, \boldsymbol{x}}{\partial n} + a\varphi = f(t, \boldsymbol{x}), \quad \boldsymbol{x} \in \Omega, \quad 0 < t \leq T ,$

with $a > 0$ and n the outward unit normal on $\partial\Omega$.

Classification

According to Sect. 2.2, equation (4.1) is parabolic; if $D = 0$ it is hyperbolic. If $0 < D \ll 1$, or more correctly, if $Pe \gg 1$, hyperbolic aspects emerge, although according to the classification scheme the equation is parabolic. A sort of mixture of parabolic and hyperbolic behavior is typical of the convection-diffusion equation, and more generally, of the equations of fluid dynamics. This is why typical numerical methods for parabolic methods are less suitable for the convection-diffusion equation with $Pe \gg 1$. Exploration of the 'no-man's land' between the parabolic and hyperbolic realms is one of the aims of the present chapter.

The *stationary case* is obtained by deleting the time-derivative. In the stationary case the foregoing remarks hold with the word parabolic replaced by the word elliptic.

The consequences of $Pe \gg 1$ are so far-reaching that this chapter is not just a completion of the preceding one.

Exercise 4.1.1. Show that (4.1) is parabolic.

4.2 Finite volume discretization of the stationary convection-diffusion equation in one dimension

The conservation form

The stationary (time-independent) version of the convection-diffusion equation (4.1) is

$$L\varphi \equiv (u_\alpha\varphi)_{,\alpha} - (D\varphi_{,\alpha})_{,\alpha} = q, \quad \boldsymbol{x} \in \Omega . \tag{4.3}$$

Here we have replaced $u_\alpha\varphi_{,\alpha}$ by $(u_\alpha\varphi)_{,\alpha}$, assuming $u_{\alpha,\alpha} = 0$ (divergence-free velocity field), so that (4.3) is in *conservation form*. An equation $L\varphi = q$ is said to be in conservation form if $\int_\Omega L\varphi d\Omega$ reduces to an integral over $\partial\Omega$. This is the case here, since with Theorem 1.2.1 we find

$$\int_\Omega Ld\varphi = \int_{\partial\Omega} (u_\alpha\varphi - D\varphi_{,\alpha})n_\alpha dS . \tag{4.4}$$

The terminology 'conservation form' derives from the fact that in the non-stationary case, given by $\partial\varphi/\partial t + L\varphi = q$, we have that $\int_\Omega \varphi d\Omega$ is constant (conserved) if there is no production ($q = 0$) and no transport through the boundary $\partial\Omega$ (right-hand side of (4.4) equal to zero).

One-dimensional case, central approximation

We now restrict ourselves to one dimension. As a result everything becomes very simple. This should not tempt the reader, however, to skip this part. Because the one-dimensional case is so simple we can be perfectly clear about important issues about which one often encounters somewhat foggy notions, and which are harder to settle in more dimensions because of more intricate technicalities involved. In general we will not repeat the arguments to be presented when we go to higher dimensions later. The issues we have in mind are treatment and accuracy of outflow boundary conditions, how to handle numerical wiggles, upwind discretization and artificial viscosity, and how to balance accuracy and low order wiggle-free discretization. Our exhaustive treatment of the one-dimensional case will allow us to be brief about the higher-dimensional case later.

We rewrite (4.3) as, replacing u_1 by u,

$$L\varphi \equiv \frac{du\varphi}{dx} - \frac{d}{dx}(D\frac{d\varphi}{dx}) = q, \quad x \in (0,1) . \tag{4.5}$$

Let the following boundary conditions be given:

$$\varphi(0) = a, \quad d\varphi(1)/dx = b .$$

Finite volume discretization of (4.5) was discussed in the preceding chapter, mainly for constant u and D. Extension to variable coefficients is easy. We discuss only cell-centered discretization; the vertex-centered case can be handled in similar way. An arbitrary cell-centered grid is given in Fig. 4.1. As in the preceding chapter, the cell with center x_j is called Ω_j, its size H_j and the interface between Ω_j and Ω_{j+1} is called $x_{j+1/2}$. With finite volume

Fig. 4.1. Non-uniform cell-centered grid.

discretization, equation (4.5) is integrated over Ω_j, giving

$$\int_{\Omega_j} L\varphi d\Omega = F|_{j-1/2}^{j+1/2} = \int_{\Omega_j} qd\Omega \cong H_j q_j ,$$

with $F|_{j-1/2}^{j+1/2} \equiv F_{j+1/2} - F_{j-1/2}$, $F_{j+1/2} = F(x_{j+1/2})$, $F(x) \equiv u\varphi - Dd\varphi/dx$. Often, $F(x)$ is called the *flux*. The following scheme is obtained:

$$F_{j+1/2} - F_{j-1/2} = H_j q_j , \quad j = 1, 2, ..., J . \tag{4.6}$$

Conservative discretization

Summation of equation (4.6) over all cells results in

$$F_{J+1/2} - F_{1/2} = \sum_{j=1}^{J} H_j q_j . \tag{4.7}$$

We see that only boundary fluxes remain, to that equation (4.7) mimics the conservation property (4.4) of the differential equation. Therefore the scheme (4.6) is called *conservative*. This property is generally beneficial for accuracy and physical realism.

Discretization of the flux

Central discretization of the convection term is done by approximating the convective flux in the same way as in Sect. 3.7:

$$(u\varphi)_{j+1/2} \cong \frac{1}{2} u_{j+1/2}(\varphi_j + \varphi_{j+1}) . \tag{4.8}$$

Upwind discretization is given, as in Sect. 3.6, by

$$(u\varphi)_{j+1/2} \cong \frac{1}{2}(u_{j+1/2} + |u_{j+1/2}|)\varphi_j + \frac{1}{2}(u_{j+1/2} - |u_{j+1/2}|)\varphi_{j+1} . \quad (4.9)$$

This means that $u\varphi$ is biased in upstream direction, which is why this is called upwind discretization.

We allow $D(x)$ to be discontinuous at cell interfaces, as in Sect. 3.3. The discretization of the diffusion term in Sect. 3.3 is easily generalized to non-uniform grids as follows (cf. (3.15)):

$$\varphi|_j^{j+1} \cong (D\frac{d\varphi}{dx})_{j+1/2} \int_{x_j}^{x_{j+1}} \frac{1}{D}dx .$$

Assuming D to be constant in the cells, we obtain

$$\int_{x_j}^{x_{j+1}} \frac{1}{D}dx = H_{j+1/2}/w_j , \quad H_{j+1/2} \equiv \frac{1}{2}(H_j + H_{j+1}) ,$$

with

$$w_j \equiv \frac{D_j D_{j+1}(H_j + H_{j+1})}{D_j H_{j+1} + D_{j+1}H_j} .$$

It follows that the diffusive part of the flux is approximated by

$$(D\frac{d\varphi}{dx})_{j+1/2} \cong w_j(\varphi_{j+1} - \varphi_j)/H_{j+1/2} . \quad (4.10)$$

The boundary conditions are implemented as described in Sect. 3.5.

An example

We will now discuss the numerical consequences of Pe $\gg 1$, with Pe defined by (4.2). Let u and D be constant and let $u = U > 0$, $q = 0$. Equation (4.5) can be rewritten as

$$\frac{d\varphi}{dx} - \text{Pe}^{-1}\frac{d^2\varphi}{dx^2} = 0, \quad x \in (0, 1) . \quad (4.11)$$

The general solution is

$$\varphi(x) = A + Be^{x\text{Pe}} , \quad (4.12)$$

with A and B constants.

For the boundary conditions we consider the following two cases:

$$\varphi(0) = a, \quad \varphi(1) = b \tag{4.13}$$

and

$$\varphi(0) = a, \quad d\varphi(1)/dx = b . \tag{4.14}$$

Equation (4.12) gives

$$\varphi(x) = a + (b - a) \frac{e^{(x-1)\mathrm{Pe}} - e^{-\mathrm{Pe}}}{1 - e^{-\mathrm{Pe}}} \tag{4.15}$$

with boundary condition (4.13) and

$$\varphi(x) = a + \frac{b}{\mathrm{Pe}} (e^{(x-1)\mathrm{Pe}} - e^{-\mathrm{Pe}}) \tag{4.16}$$

with boundary condition (4.14). Of course, these solutions satisfy the maximum principle of Theorem 2.4.1, since they are monotone. Furthermore, for Pe ≫ 1 they exhibit the boundary layer behavior predicted in Sect. 2.5.

Equation (4.16) confirms the physical insight that it is wrong to prescribe a Neumann boundary condition at an inflow boundary. This can be seen as follows. If $x = 1$ is an inflow boundary, then Pe < 0. For Pe ≪ −1 and $x < 1$ equation (4.16) gives

$$\varphi(x) \cong \frac{b}{\mathrm{Pe}} e^{-\mathrm{Pe}} ,$$

so that $\varphi(x)$ is very sensitive to perturbations in b, which means that the problem is ill-posed in a practical sense.

Next, we study the numerical solution of (4.11). We choose the cell-centered grid depicted in Fig. 4.1 uniform with $H_j = h$. We start with boundary conditions (4.14). The boundary conditions are implemented as in Sect. 3.5, resulting in, using the central discretization (4.8):

$$\begin{aligned} F_{1/2} &= a - 2\mathrm{Pe}^{-1}(\varphi_1 - a)/h , \\ F_{J+1/2} &= \varphi_J + \tfrac{1}{2}hb - \mathrm{Pe}^{-1}b . \end{aligned}$$

In the interior,

$$F_{j+1/2} = \frac{1}{2}(\varphi_j + \varphi_{j+1}) - \mathrm{Pe}^{-1}(\varphi_{j+1} - \varphi_j)/h .$$

Substitution in (4.6) results in

$$\varphi_1(1 + \frac{6}{p}) + \varphi_2(1 - \frac{2}{p}) = a(\frac{4}{p} + 2),$$

$$-\varphi_{j-1}(1 + \frac{2}{p}) + \frac{u}{p}\varphi_j + \varphi_{j+1}(1 - \frac{2}{p}) = 0, \quad j = 2, 3, ..., J - 1,$$

(4.17)

$$-\varphi_{J-1}(1 + \frac{2}{p}) + \varphi_J(1 + \frac{2}{p}) = bh(\frac{2}{p} - 1),$$

where $p = h\mathrm{Pe}$ is called the *mesh Péclet number*. The general solution of (4.17) is found in the standard way by postulating solutions of the form $\varphi_j = z^j$, resulting in, if $p \neq 0, \pm2$:

$$\varphi_j = A + Bz^j, \quad z = (2 + p)/(2 - p). \tag{4.18}$$

The constants A and B follow from the first and third equation of (4.17). For $p = 0$, i.e. $\mathrm{Pe} = 0$, only the diffusion term is left in (4.11), and the discrete solution is given by

$$\varphi_j = A + Bj.$$

For $p = 2$ the solution of (4.17) is given by

$$\varphi_j = a, \quad j = 1, 2, ..., J, \tag{4.19}$$

and for $p = -2$ it is not determined. This is another indication that one should not apply a Neumann condition at an inflow boundary. Assume $\mathrm{Pe} \geq 0$, and consider only (4.18). From the first and third equation of (4.17) we find

$$A + \frac{2}{2 - p}B = a, \quad Bz^J = bh\frac{2 - p}{2p},$$

so that

$$\varphi_j = a + \frac{bh}{p}z^{-J}\{-1 + (1 - \frac{p}{2})z^j\}. \tag{4.20}$$

Next, consider boundary condition (4.13). it is left to the reader to show that in this case the discrete equation in grid point $j = J$ becomes

$$-\varphi_{J-1}(1 + \frac{2}{p}) + \varphi_J(-1 + \frac{6}{p}) = b(\frac{4}{p} - 2)$$

With this boundary condition we find for $p = 0$ or $p = 2$ that the numerical solution is given by, respectively, equation (4.18) or (4.19), whereas otherwise we find

$$\varphi_j = a + \frac{b - a}{z^J - 1}(-1 + \frac{2 - p}{2}z^j). \tag{4.21}$$

We now discuss the quality of the numerical solutions (4.20) and (4.21). Our most important observation is, that if $z < 0$ then z^j alternates in sign, so that the numerical solution shows local extrema called 'wiggles'. However, if $z \geq 0$ the numerical solution is monotone, like the exact solution. We have $z \geq 0$ if and only if

$$|p| \leq 2 . \tag{4.22}$$

Exact and numerical solutions for boundary conditions (4.13) are compared in Fig. 4.2. Clearly, for $p > 2$ the numerical solution shows wiggles and is not acceptable. With boundary conditions (4.14) (Neumann at outflow) and

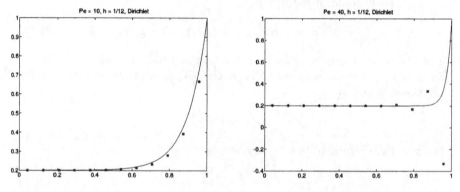

Fig. 4.2. Exact solution ($-$, equation(4.15)) and numerical solution (*, equation(4.21)).

$b = 0$ the exact and numerical solutions, given by $\varphi(x) = a$ and (4.20), are identical. With $b = 1$ the results of Fig. 4.3 are obtained. Although there is

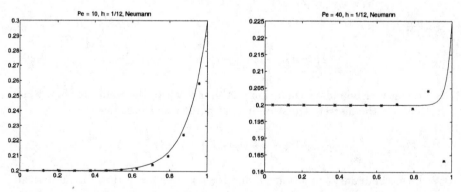

Fig. 4.3. Exact solution ($-$, equation(4.16)) and numerical solution (*, equation (4.20)).

less variation in the boundary layer (note the difference in scale) the relative size of the numerical wiggles for Pe = 40 is just as bad as in Fig. 4.2.

In practical situations often Pe is so large, that it is not feasible to make h small enough to satisfy (4.22). However, this is not always a problem, since by one of three measures wiggles may be largely or completely avoided. These measures are:

(i) Apply a homogeneous ($b = 0$) Neumann boundary condition at outflow;
(ii) Change the discretization such that the discrete system satisfies a similar maximum principle as the differential equation;
(iii) Apply local mesh refinement in the boundary layer.

Measure (i) has just been illustrated: the global discretization error is zero in the present simple case. Although this is not true in more complicated multidimensional cases, with this outflow boundary condition wiggles are usually absent. Measure (ii) will be discussed later. Measure (iii) is numerically illustrated in the next section.

Wiggles and stability

Schemes that allow wiggles are sometimes called *instable*. But this is not correct, as we will now show; an extensive discussion of stability will be given later. Stability always refers to the influence of perturbations of the data (boundary and initial conditions and right-hand side). Consider the numerical solution (4.21). Let the boundary value a be perturbed by an amount δa. The ensuing perturbation of the numerical solution $\delta \varphi$ is given by

$$\delta \varphi_j = \delta a(z^J - z^j + z^j p/2)/(z^J - 1) .$$

The numerical solution is stable in the norm $\|.\|$ with respect to perturbation of a if there exists a constant C independent of the mesh size $h = 1/J$ such that

$$\|\delta \varphi\| \leq C\delta a . \tag{4.23}$$

Let $z < 0$, so that we have wiggles, and let $z < -1$. Then

$$\|\delta \varphi\|_\infty \leq \delta a(2 + |p|/2)|z^J|/|z^J - 1| ,$$

so that (4.23) holds with $C = 2(2 + |p|/2)$, since

$$2 \geq 1 + 1/|z^J - 1| = \frac{|z^J - 1| + 1}{|z^J - 1|} \geq \frac{|z|^J}{|z^J - 1|} .$$

The case $0 > z \geq -1$ and the case of perturbation of b is left to the reader. This shows that the numerical scheme, although it generates wiggles, is nevertheless stable.

Exercise 4.2.1. Devise alternative ways to implement the boundary condition $d\varphi(1)/dx = b$ for

$$\frac{du\varphi}{dx} - \frac{d}{dx}(D\frac{d\varphi}{dx}) = q \tag{4.24}$$

on a cell-centered grid, and also on a vertex-centered grid, with global truncation error $\mathcal{O}(h^2)$.

Exercise 4.2.2. We study the accurate implementation of a Neumann boundary condition for variable coefficient u. Discretize (4.24) with

$$\frac{du\varphi}{dx} \cong \frac{1}{2h}(u\varphi)|_{i-1}^{i+1}$$

on the grid of Fig. 4.1, with boundary conditions

$$\varphi(0) = a, \quad d\varphi(1)/dx = b.$$

Can you make at the Neumann boundary

$$\tau_J = \mathcal{O}(h^2)$$

by a $\mathcal{O}(h^2)$ perturbation of the boundary data? Investigate the accuracy by numerical experiments. Take $u = 1 + c\sin n\pi x$, $c = 0.7$, $n = 1$, $a = 1$, $b = 2$, $q = 0$, $D = 10^{-2}$ and $D = 1$.

Exercise 4.2.3. Prove that the scheme (4.17) is stable by using its exact solution.

4.3 Numerical experiments on locally refined one-dimensional grid

Choice of grid

We want to apply local grid refinement in the boundary layer, such that the mesh Péclet number p satisfies (4.22) locally in the boundary layer. We will solve (4.11) numerically. Boundary layer theory tells us there is a boundary layer at $x = 1$ with thickness $\delta = \mathcal{O}(\mathrm{Pe}^{-1})$. After some trial and error we put $\delta = 4/\mathrm{Pe}$. The grid is again cell-centered. In the boundary layer the cell-size H_j = constant, and we put $1/3$ of the total number (J) of cells in the boundary layer, hence $H_j = 3\delta/J$. The remaining cells also have the same size $(H_j = 3(1-\delta)/2J)$. Figure 4.4 gives a qualitative impression of the grid. Note that as Pe changes, the grid points move. There is a large jump in mesh size at the edge of the boundary layer. As discussed in Sect. 3.7, the cell-centered scheme is locally inconsistent in the maximum norm, but is predicted to be nevertheless convergent. We will see what happens in practice.

Fig. 4.4. Local grid refinement in boundary layer.

Numerical experiments

Equation (4.11) is discretized on the grid just defined in the way described in the preceding section. Let the boundary conditions be of Dirichlet type (4.13) with $a = 0.2$, $b = 1$. The exact solution is given by (4.15). Numerical results are shown in Fig. 4.5.

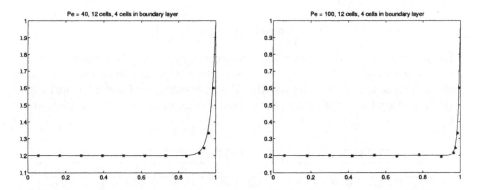

Fig. 4.5. Numerical solution with local grid refinement in boundary layer (*: numerical solution: –: exact solution.)

The total number of cells is 12, as in Fig. 4.2. But much better accuracy is obtained. Inside the refinement region the mesh Péclet number equals $p = H_j \text{Pe} = 12/J = 1$, satisfying (4.22) for all Pe , so that there can be no local extrema in the numerical solution in the refinement region, according to principles to be discussed in the next section. Outside the refinement region, $p = (\frac{3}{2}\text{Pe} - 6)/J$, so that $p = 4.5$ for Pe $= 40$ and $p = 12$ for Pe $= 100$, violating (4.22), so that local extrema can occur. But Fig. 4.5 shows that the numerical solution is quite acceptable, even for Pe $= 100$. Wiggles are almost absent. In order to eliminate unacceptably large wiggles as those in Fig. 4.2, local refinement or changing a boundary condition is preferable to switching to an upwind scheme, which results in a numerical boundary layer that is much too thick. This illustrates why the advice "look at the wiggles, they tell you something" is given in Gresho and Lee (1981). Wiggles are a signal that there are boundary layers that are not resolved, and that local grid refinement is called for.

It may happen that the small wiggles in Fig. 4.5 are unacceptable. This may be the case when the unknown must remain strictly non-negative for physical or numerical reasons. This occurs in the modeling of turbulent or chemically reacting flows. Then one may apply a monotone scheme (to be discussed in the next section), for example an upwind scheme, on the locally refined grid of Fig. 4.4. An accurate numerical solution is obtained, with correct resolution of the boundary layer.

Accuracy

The maximum norm of the global truncation error $e_j \equiv \varphi(x_j) - \varphi_j$ has been measured:

$$\|e\|_\infty = \max\{|e_j|, \quad j = 1, ..., J\}$$

Results are presented in Table 4.1. Apparently, $e_j = \mathcal{O}(J^{-2})$, confirming the error analysis of Sect. 3.7. Secondly, we see that $\|e\|_\infty$ is almost independent of Pe. This is due to the adaptive (i.e. Pe-dependent) local grid refinement in the boundary layer. Of course, for a given J computing work and storage are

	Pe = 40			Pe = 100			Pe = 200		
J	12	24	48	12	24	48	12	24	48
$\|e\|_\infty$	0.0852	0.0230	0.0060	0.0853	0.0230	0.0060	0.0854	0.0230	0.0060

Table 4.1. Error norm on locally refined grid.

also independent of Pe. We may conclude the *computing cost and accuracy are uniform in* Pe.

This is an important observation. A not uncommon misunderstanding is that numerical predictions of high Reynolds (or Péclet in the present case) number flows are inherently untrustworthy, because numerical discretization errors ('numerical viscosity') dominate the small viscous forces. This is not true, provided appropriate measures are taken, as was just shown. Local grid refinement in boundary layers enables us to obtain accuracy independent of the Reynolds number. Because in practice Re (or its equivalent such as the Péclet number) is often very large (see Sect. 1.5) it is an important (but not impossible) challenge to realize this also in more difficult multi-dimensional situations. A thorough analysis of Pe-uniform accuracy for a two-dimensional singular perturbation problem will be given in Sect. 4.7.

4.4 Schemes of positive type

We now discuss measure (ii) of the end of Sect. 4.2.

A discrete maximum principle

Lemma 4.4.1. *Let c_n and x_n, $n = 1, 2, ..., N$ be reals satisfying*

$$\sum_{n=1}^{N} c_n = 0, \quad c_n < 0 \text{ for } n > 1 \qquad (4.25)$$

and

$$\sum_{n=1}^{N} c_n x_n \leq 0.$$

Then

$$x_n = x_1, \quad n = 1, ..., N \quad or \quad x_1 < \max\{x_n : n > 1\}.$$

Proof. From (4.25) it follows that $c_1 > 0$. Without loss of generality we may take $c_1 = 1$, so that

$$x_1 \leq -\sum_{n=2}^{N} c_n x_n. \qquad (4.26)$$

This shows that x_1 is not larger than a weighted average of x_n, $n > 1$, so that the lemma is obvious, but the proof is completed anyway. Let $x_m = \max\{x_n : n = 2, 3, ..., N\}$, and let there exist some $p > 1 : x_p < x_m$. Then (4.26) gives

$$x_1 < -x_m \sum_{n=2}^{N} c_n = x_m.$$

If there is no such p then we have $x_1 = x_m$, $m = 1, ..., N$. $\qquad \square$

Let L_h be a linear discrete operator defined by

$$L_h \psi_j = \sum_{k \in K} \alpha(j, k) \psi_{j+k}, \quad j = 1, ..., J,$$

where K is some index set, and $\alpha(j, k)$ are coefficients. For example, in the L_h of equation (4.17) we have $K = \{-1, 0, 1\}$.

Definition 4.4.1. The operator L_h is *of positive type* if

$$\sum_{k \in K} \alpha(j, k) = 0, \quad j = 2, ..., J - 1 \qquad (4.27)$$

and

$$\alpha(j, k) < 0, \quad k \neq 0, \quad j = 2, ..., J - 1. \qquad (4.28)$$

Theorem 4.4.1. *Discrete maximum principle.*
If L_h is of positive type and

$$L_h \psi_j \leq 0, \quad j = 2, ..., J - 2$$

then local maxima of ψ can occur only if $j = 1$ or $j = J$, or $\psi_j = constant$.

Proof. For arbitrary $j \in \{2, ..., J - 1\}$ let $c_0 = \alpha(j, 0)$, $c_k = \alpha(j, k)$, $k \neq 0$. Then Lemma 4.4.1 applies after suitable re-indexing of c_k. Considering each interior grid point in turn, the theorem follows.   □

Corrollary Let (4.27), (4.28) also hold for $j = J$. Then a local maximum of ψ can occur only for $j = 1$.
Proof Trivial extension of the proof of Theorem 4.4.1.   □

If $L_h \psi_j \geq 0$ for $j = 2, ..., J - 1$ and L_h is of positive type then it can be shown in a similar way that local *minima* can occur only for $j = 1$ and $j = J$. Hence, if $L_h \psi_j = 0$ then local *extrema* can only occur for $j = 1$ or $j = J$. We see that if the discretization is of positive type, then it obeys a similar maximum principle as the differential equation. Hence, spurious wiggles cannot occur. Therefore it seems attractive to work only with schemes of positive type. Unfortunately, these have the following drawback.

Order barrier

A prime concern is to have accuracy and computing cost uniform in Pe. Hence, we would like to have at least an accurate and positive scheme for the case Pe $= \infty$. However, we have the following theorem.

Theorem 4.4.2. *Order barrier.*
Linear discretization schemes of positive type for the following differential equation:

$$\frac{d\varphi}{dx} = q(x)$$

are at most first order accurate.

Remark A linear discretization scheme is a scheme that gives a linear system of equations if the differential equation that is approximated is linear. By 'first order accurate' we mean that $\tau_j = \mathcal{O}(\Delta)$, $\Delta = \max\{H_1, ..., H_J\}$, if the discretization is scaled such that the differential equation is approximated. The order barrier theorem holds on both cell-centered and vertex-centered grids.

Proof. We restrict ourselves to a uniform grid with cell-size h. Hence, cell-centered and vertex-centered discretizations are identical. In general, in grid points sufficiently far from the boundaries the discretization is given by

$$L_h\psi_j = \sum_{k\in K} \alpha(k)\psi_{j+k} \,.$$

The local discretization error satisfies

$$\begin{aligned}\tau_j &= L_h\varphi(x_j) - q_k = \sum_{k\in K} \alpha(k)\{\varphi(x_j) + kh\frac{d\varphi(x_j)}{dx} \\ &\quad + \tfrac{1}{2}k^2 h^2 \frac{d^2\varphi(x_j)}{dx^2} + \mathcal{O}(h^3)\} - q_j \,.\end{aligned}$$

For τ_j to be $\mathcal{O}(h)$ it is necessary that

$$\sum_{k\in K}\alpha(j) = 0, \quad \sum_{k\in K} k\alpha(k) = 1/h \,.$$

Hence, $\alpha(k) = \mathcal{O}(h^{-1})$. For the scheme to be second order accurate it is necessary that in addition

$$\sum_{k\in K} k^2\alpha(k) = 0 \,. \tag{4.29}$$

However, since L_h is of positive type, $\alpha(k) < 0$ for $k \neq 0$, so that (4.29) cannot be satisfied. □

A similar order barrier theorem for the time dependent case is discussed in Sect. 9.2. A way to get around the order barrier is to use nonlinear schemes. We will come back to this in Sect. 4.8. In practice, usually a better balance between accuracy and efficiency is obtained if $\tau_j = \mathcal{O}(\Delta^2)$ instead of $\tau_j = \mathcal{O}(\Delta)$ on smooth grids.

Monotone schemes

A scheme for which $L_h\psi \geq 0$ implies $\psi \geq 0$ is called *monotone*. For a monotone scheme the solution decreases (or increases) with the right-hand side, cf. Exercise 4.4.1. Schemes of positive type are monotone (cf. Exercise 4.4.2), but the reverse does not hold.

Mesh Péclet condition

Consider equation (4.5), discretized according to (4.8) and (4.10). This results in

$$\alpha(j,-1) = -\frac{1}{2}u_{j-1/2} - w_{j-1}/H_{j-1/2} \, ,$$

$$\alpha(j,1) = \frac{1}{2}u_{j+1/2} - w_j/H_{j+1/2} \, , \qquad (4.30)$$

$$\alpha(j,0) = -\alpha(j,-1) - \alpha(j,1) \, .$$

Do we have a scheme of positive type? Application of conditions (4.28) for all interior $j = 2, ..., J-1$ results in

$$-2 < u_{j+1/2}H_{j+1/2}/w_j < 2 \, . \qquad (4.31)$$

Define the mesh Péclet number in the present variable coefficient case as

$$p_{j+1/2} \equiv |u_{j+1/2}|H_{j+1/2}/w_j \, . \qquad (4.32)$$

Equation (4.31) results in

$$p_{j+1/2} < 2 \, . \qquad (4.33)$$

If $D(x)$ is smooth one may just as well replace w_j by $D_{j+1/2}$, in which case the customary definition of the mesh Péclet number is recovered from (4.32). Equation (4.33) is the generalization of (4.22) to variable coefficients and non-uniform grids. As is to be expected from the order barrier theorem, because (4.30) corresponds to a second order scheme, (4.33) is violated as $D(x) \downarrow 0$.

Exercise 4.4.1. Let L_h be of positive type, and let $L_h\varphi^{(\alpha)} = f^{(\alpha)}$, $\alpha = 1, 2$. Show that if $f_j^{(2)} \leq f_j^{(1)}$, $j = 1, ..., J$ then $\varphi_j^{(2)} \leq \varphi_j^{(1)}$.

Exercise 4.4.2. Show that a scheme of positive type is monotone.

4.5 Upwind discretization

A scheme that gives us a scheme of positive type for all Pe is obtained if we approximate the convective flux $u\varphi$ such that, in the terminology of Definition 4.4.1, a positive contribution is made to $\alpha(j,0)$ and a non-positive contribution to $\alpha(j,k)$, $k \neq 0$. This may be achieved by using the upwind discretization (4.9). It is left to the reader to show that on a uniform grid the local truncation error satisfies (assuming the discretization is scaled such that the differential equation is approximated)

$$\tau_j = \mathcal{O}(\Delta) \, , \qquad (4.34)$$

in accordance with the order barrier theorem 4.4.2.

Artificial viscosity

Another, at first sight more crude, way to make L_h of positive type is to maintain central discretization but to artificially increase the diffusion coefficient D by an amount D_a such that (4.33) is satisfied. In other words, we add an *artificial viscosity* term to the original differential equation (4.1), and solve instead

$$\frac{du\varphi}{dx} - \frac{d}{dx}(D\frac{d\varphi}{dx}) - \frac{d}{dx}(D_a\frac{d\varphi}{dx}) = q \ .$$

From (4.33) it follows that in order to obtain a scheme of positive type we must have (assuming D smooth, and putting $w_j = D_{j+1/2}$)

$$D_{a,j+1/2} \geq |u_{j+1/2}|H_{j+1/2}/2 - D_{j+1/2} \ .$$

This means that $D_{a,j+1/2} = \mathcal{O}(H_{j+1/2})$ suffices, so that the local truncation error of the artificial viscosity scheme will be $\mathcal{O}(H)$ on a uniform grid, as is to be expected from the order barrier Theorem 4.4.2.

At first sight, modification of the differential equation to be solved seems a crude stratagem to obtain a scheme of positive type, and one might think that upwind discretization is more sophisticated. However, the two are closely related, because for the following choice of the artificial viscosity:

$$D_{a,j+1/2} = |u_{j+1/2}|H_{j+1/2}/2 \tag{4.35}$$

the artificial viscosity scheme is identical to the upwind scheme. Turning the argument around we see, that compared to the central scheme the upwind scheme introduces an additional local truncation error representing an increase of the effective viscosity by an amount given by (4.35). This is why it is sometimes thought that high Reynolds number (or rather, in the present situation, high Péclet number, i.e. $D \ll 1$) flows cannot be computed with linear schemes of positive type, because the numerical Péclet number is governed by $D + D_a$, and is therefore $\mathcal{O}(H_{j+1/2}^{-1})$. However, it has already been seen in Sect. 4.3 that to achieve accury and computing cost uniform in Pe it is necessary to apply local grid refinement in the boundary layer, where $H_j = \mathcal{O}(Pe^{-1})$, bringing the numerical Péclet number close to the physical Péclet number, and making upwind discretization unnecessary even. Outside the boundary layer we are still stuck with a numerical Péclet number which is too low. But in the numerical examples of Sect. 4.3, $\varphi =$ constant outside the boundary layer, so that the artificial viscosity term is negligible, and uniform (in Pe) accuracy and computing cost seem to come into reach with linear schemes of positive type. The situation that $\varphi =$ constant outside the boundary layer is not so atypical as one might think, for in real high Reynolds number flows velocity gradients are often small outside boundary layers, so that the artificial viscosity terms are small. We will shortly further explore these issues numerically.

Hybrid scheme

Because of the loss of accuracy incurred it is a good idea to use the upwind scheme only in regions where (4.33) is violated, and use the central scheme elsewhere. This can be done as follows. In the finite volume discretization of the convection term $(u\varphi)_{i+1/2}$ has to be approximated. Central discretization according to (4.8) is defined by

$$(u\varphi)_{j+1/2} \cong (u\varphi)_{c,j+1/2} \equiv \frac{1}{2} u_{j+1/2}(\varphi_j + \varphi_{j+1}) \,,$$

and upwind discretization according to (4.9) is defined by

$$(u\varphi)_{j+1/2} \cong (u\varphi)_{u,j+1/2} \equiv \frac{1}{2}(u_{j+1/2} + |u|_{j+1/2})\varphi_j \\ + \frac{1}{2}(u_{j+1/2} - |u_{j+1/2}|)\varphi_{j+1} \,.$$

Let us define a switch function $s(p_{j+1/2})$ with the mesh Péclet number for a non-uniform mesh $p_{j+1/2}$ defined by (4.32), and define *hybrid discretization* by

$$(u\varphi)_{j+1/2} \cong s(p_{j+1/2})(u\varphi)_{u,j+1/2} + (1 - s(p_{j+1/2}))(u\varphi)_{c,j+1/2} \,, \qquad (4.36)$$

which gives

$$(u\varphi)_{j+1/2} \cong \frac{1}{2}(u_{j+1/2} + s(p_{j+1/2})|u_{j+1/2}|)\varphi_j \\ + \frac{1}{2}(u_{j+1/2} - s(p_{j+1/2})|u_{j+1/2}|)\varphi_{j+1} \,.$$

For iterative convergence in nonlinear cases it is beneficial if $s(p)$ switches smoothly between 0 and 1, for example

$$\begin{aligned} s(p) &= 0, & p &< 1.9\,, \\ s(p) &= 10(p - 1.9), & 1.9 &\leq p < 2\,, & (4.37) \\ s(p) &= 1, & p &\geq 2\,. \end{aligned}$$

The hybrid scheme has been introduced in Spalding (1972), and is further discussed in Patankar (1980).

Exercise 4.5.1. Prove (4.34), assuming constant coefficients $u(x)$ and $k(x)$.

Exercise 4.5.2. Consider the following problem:

$$\frac{du\varphi}{dx} - \mathrm{Pe}^{-1}\frac{d^2\varphi}{dx^2} = q(x), \quad 0 < x < 1, \quad \varphi(0) = 0, \quad \varphi(1) = 1 \qquad (4.38)$$

with

$$u(x) = 1 - b \sin \pi x , \qquad (4.39)$$

and Pe constant. Choose the following exact solution:

$$\varphi(x) = a(\sin \alpha \pi x - \sin \alpha \pi) + \frac{e^{(x-1)Pe} - e^{-Pe}}{1 - e^{-Pe}} , \qquad (4.40)$$

and derive the corresponding right-hand side $q(x)$.

Note that with $a = b = 0$ we have $q(x) = 0$, and we recover the test case of Sections 4.2 and 4.3. Let the grid be non-uniform, and choose it as in Sect. 4.3, or uniform. Implement the hybrid scheme, and include also the upwind and central schemes by freezing $s(p)$ at 0 or 1. Implement defect correction (see Sect. 4.6).

Define and perform numerical tests that illustrate, confirm or refute the points made in Sections 4.2–4.5 about occurrence of spurious wiggles and accuracy. Plot exact and numerical solutions as in Sections 4.2 and 4.3. Compute the maximum norm of the error, and find out how it depends on the number of cells J.

(a) Solve the same problem as in Fig. 4.2, but with the upwind scheme. Discuss the results.

(b) Solve the same problem as in Fig. 4.5, but with the upwind scheme. Discuss the results.

(c) Solve the same problem as in Fig. 4.5, but with the hybrid scheme. Discuss the results.

(d) Choose $a = 1$, $b = 0.95$, $\alpha = 3$. Use the grid of Fig. 4.5. Compare the upwind, hybrid and central schemes.

4.6 Defect correction

The algorithm

Defect correction is a method to improve the accuracy of a lower order discretization, without having to solve for a higher order discretization. Denote the system of equations corresponding to a lower and a higher order discretization by, respectively,

$$\bar{L}_h \varphi_h = \bar{q}_h, \quad L_h \psi_h = q_h .$$

Defect correction is an iterative method defined by

$$\bar{L}_h \varphi_h^{(0)} = \bar{q}_h ,$$
$$\bar{L}_h \varphi_h^{(n)} = q_h - L_h \varphi_h^{(n-1)} + \bar{L}_h \varphi_h^{(n-1)}, \quad n = 1, 2, \dots .$$

If $\bar{L}_h$ is a first order scheme and L_h is a second order scheme, then already $\varphi_h^{(1)}$ is of second order accuracy. This we will now show.

Accuracy of defect correction

Denote the boundary value problem to be solved by

$$L\varphi = q, \quad \varphi \in \Phi, \quad q \in Q, \tag{4.41}$$

with Φ and Q suitable function spaces. Let

$$\varphi_h^{(n)} \in \Phi_h, \quad n = 0, 1, \dots,$$

and

$$q_h \in Q_h, \quad \bar{q}_h \in Q_h .$$

Define the following operators:

$$R_h : Q \to Q_h, \quad \bar{R}_h : Q \to Q_h, \quad \hat{R}_h : \Phi \to \Phi_h$$

such that

$$q_h = R_h q, \quad \bar{q}_h = \bar{R}_h q$$

and, on a two-dimensional grid for example,

$$(\hat{R}_h \varphi)_{jk} = \varphi(\boldsymbol{x}_{jk}) .$$

The operators R_h and $\bar{R}_h$ may be different, depending on how the boundary conditions are approximated. Here we need not be specific about the choice of these operators. The local truncation error (defined by (3.48)) of scheme $\bar{L}_h$ can be written as

$$\tau_h = \bar{L}_h \hat{R}_h \varphi - q_h = (\bar{L}_h \hat{R}_h - R_h L)\varphi .$$

Assuming $\bar{L}_h$ to be first order, we have

$$\|\bar{L}_h \hat{R}_h - R_h L\|_{Q_h \leftarrow \Phi} \leq C_1 \Delta , \tag{4.42}$$

with Δ the maximum mesh size. Here operator norms are defined in the usual way, for example

$$\|\bar{L}_h\|_{Q_h \leftarrow \Phi_h} = \sup\{\|\bar{L}_h \varphi_h\|_{Q_h}/\|\varphi_h\|_{\Phi_h} : 0 \neq \varphi_h \in \Phi_h\} . \tag{4.43}$$

We will come back to the choice of norms later. Assuming L_h to be second order accurate, we have

$$\|L_h \hat{R}_h - R_h L\|_{Q_h \leftarrow \Phi} \leq C_2 \Delta^2 . \tag{4.44}$$

The two schemes are assumed to be relatively consistent, in the following sense:

$$\|L_h - \bar{L}_h\|_{Q_h \leftarrow \Phi_h} \leq C_3 \Delta . \tag{4.45}$$

$\bar{L}_h$ is assumed to be regular:

$$\|\bar{L}_h^{-1}\|_{\phi_h \leftarrow Q_h} \leq C_4 . \tag{4.46}$$

The constants $C_1, C_2, \ldots$ are assumed to be independent of Δ. Equations (4.42)–(4.46) are our set of assumptions.

We will show, following the general framework presented in Hackbusch (1985) Sect. 14.2.2, that

$$\|\varphi_h^{(1)} - \varphi_h^*\|_{\Phi_h} = \mathcal{O}(h^2) .$$

where $\varphi_h^* = \hat{R}_h \varphi$. We have

$$\bar{L}_h(\varphi_h^{(0)} - \varphi_h^*) = (\bar{R}_h L - \bar{L}_h \hat{R}_h)\varphi .$$

From (4.42) and (4.46) it follows that

$$\|\varphi_h^{(0)} - \varphi_h^*\|_{\Phi_h} \leq C_1 C_4 \Delta \|\varphi\|_\Phi . \tag{4.47}$$

Furthermore,

$$\begin{aligned}
\bar{L}_h(\varphi_h^{(1)} - \varphi_h^*) &= q_h + (\bar{L}_h - L_h)(\varphi_h^{(0)} - \varphi_h^*) - L_h \varphi_h^* \\
&= (R_h L - L_h \hat{R}_h)\varphi + (\bar{L}_h - L_h)(\varphi_h^{(0)} - \varphi_h^*) ,
\end{aligned}$$

so that, using (4.44), (4.45), (4.46) and (4.47)

$$\|\varphi_h^{(1)} - \varphi_h^*\|_{\Phi_h} \leq \Delta^2(C_2 + C_1 C_3 C_4)\|\varphi\|_\Phi ,$$

which is what we wanted to show.

Whether the assumptions (4.42)–(4.46) are satisfied depends on the choice of the norms. In the Taylor expansions for the local truncation errors of $\bar{L}_h$ and L_h derivatives of φ occur, so for (4.42) and (4.44) it is necessary that $\|\varphi\|_\Phi$ measures φ and its derivatives up to the order of those that occur in the local truncation error Taylor expansions. To elucidate (4.45), suppose that (4.41) is the convection-diffusion equation, and that $\bar{L}_h$ is an upwind version of L_h obtained by addition of artificial viscosity with coefficient $D_a = \mathcal{O}(h)$, as in the preceding section, so that on a uniform grid

$$\bar{L}_h = L_h + D_a A_h ,$$

with A_h the discretization of the Laplacian. Choosing

$$\|\varphi_h\|_{\Phi_h} = \|A_h\varphi_h\|_{Q_h} ,$$

equation (4.45) is trivially satisfied with $C_3 = D_a/h$. It remains to study (4.46). We have

$$\|\bar{L}_h^{-1}q_h\|_{\Phi_h} = \|A_h\bar{L}_h^{-1}q_h\|_{Q_h} .$$

Assume that $\bar{L}_h$ is scaled such that

$$A_h = A/h^2, \quad \bar{L}_h = (D + D_a)A/h^2 + B/h ,$$

with the elements of the matrices A and B independent of h. Then

$$A_h\bar{L}_h^{-1} = A\{(D + D_a)A + hB\}^{-1} = (D + D_a)^{-1}(I + \mathcal{O}(h)) ,$$

where the $\mathcal{O}(h)$ term represents the norm of a matrix of rank $\mathcal{O}(h^{-1})$ and elements of size $\mathcal{O}(h)$. To make the remainder of our line of argument rigorous the size of this $\mathcal{O}(h)$ term would have to be investigated further, which we will not do. Proceeding, we get

$$\|A_hL_h^{-1}q_h\|_Q = (D + D_a)^{-1}\|q_h\|_{Q_h}(1 + \mathcal{O}(h)) ,$$

so that (4.46) holds with

$$C_4 = (D + D_a)^{-1}(1 + \mathcal{O}(h)) .$$

Numerical illustration

We solve the boundary value problem (4.38)–(4.40). We choose $a = 0.2$, $b = -0.95$, $\alpha = 1$. The grid is cell-centered, similar to the one shown in Fig. 4.4. The thickness of the refinement region is $\delta = 6/\text{Pe}$. The number of cells in the refinement region is m_f and in the remainder of the grid it is m_c. Table 4.2 presents results. One defect correction has been applied for method 3. In many cases the central scheme shows small wiggles, see for example Fig. 4.6. Figure 4.7 shows results for upwind discretization and defect correction. In all cases, the defect correction solution is visually wiggle-free and has error almost as small as with the central scheme. The accuracy of the upwind scheme is almost independent of Pe; this will be shown theoretically in a situation with a parabolic boundary layer in Sect. 4.7. For this the maximum principle is required. The central scheme does not satisfy the maximum principle, and its wiggles get worse when Pe increases, which explains why the central scheme is not second order for Pe = 400.

Exercise 4.6.1. Repeat these numerical experiments with the software you developed for Exercise 4.5.2. Compute also results for the hybrid scheme. Does the hybrid scheme produce better results than the upwind scheme? Apply defect correction to the hybrid scheme.

m_c	m_f	method	$\|e\|_\infty$	method	$\|e\|_\infty$	method	$\|e\|_\infty$
				Pe = 40			
8	6	1	0.1087	2	0.1033	3	0.1037
16	12	1	0.0412	2	0.0284	3	0.0284
32	24	1	0.0298	2	0.0074	3	0.0074
				Pe = 400			
8	6	1	0.1070	2	0.1068	3	0.1063
16	12	1	0.0471	2	0.0289	3	0.0288
32	24	1	0.0328	2	0.0086	3	0.0092

Table 4.2. Maximum norm of the error for problem (4.38)–(4.40). Method 1: upwind discretization; method 2: central discretization; method 3: defect correction.

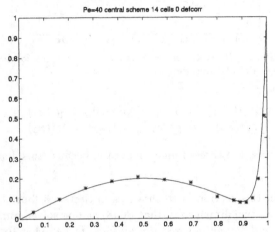

Fig. 4.6. Solution for Pe = 40, $m_c = 8$, $m_f = 6$. — : exact solution; * : numerical solution with central discretization.

4.7 Péclet-independent accuracy in two dimensions

Problem statement

In this section we study the following version of the two-dimensional stationary convection-diffusion equation:

$$L\varphi \equiv (u_\alpha \varphi)_{,\alpha} - (D\varphi_{,\alpha})_{,\alpha} = 0 \ , \quad \alpha = 1, 2 \ , \quad (x_1, x_2) \in (0, 1) \times (0, 2) \ . \tag{4.48}$$

We assume that we have solid walls at $x_2 = 0, 2$ so that $u_2(x_1, 0) = u_2(x_1, 2) = 0$. Let $u_1 < 0$, so that $x_1 = 0$ is an outflow boundary. In view of what we learned in Sect. 2.5 we choose a homogeneous Neumann boundary condition at $x_1 = 0$, and Dirichlet boundary conditions at the other parts of the boundary:

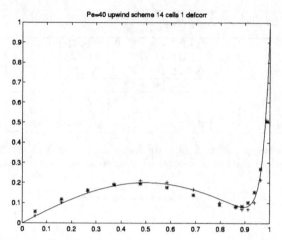

Fig. 4.7. Solution for Pe $= 40$, $m_c = 8, m_f = 6$. — : exact solution; * : upwind discretization; + : defect correction.

$$\varphi_{,1}(0, x_2) = 0, \qquad \varphi(x_1, 0) = g^1(x_1),$$
$$\varphi(x_1, 2) = g^2(x_1), \quad \varphi(1, x_2) = g^3(x_2). \tag{4.49}$$

We will not discuss the three-dimensional case, because this does not provide new insights.

Our purpose in this section is to show, as in Sect. 4.3, but this time more rigorously and in two dimensions, that (4.48) can be solved numerically such that *accuracy and computing cost are uniform in* Pe. Therefore fear that it is impossible to compute high Péclet (Reynolds) number flow accurately is unfounded, as argued before. We now make our case even more convincing. Our treatment will be detailed and technical, but elementary.

In order to illuminate the main point with as little technical detail as possible, we specialize to $u_1 = -1$, $u_2 \equiv 0$ and D constant. Furthermore, the problem is assumed to be symmetric around $x^2 = 1$, so that the solution needs to be computed only in half the domain, namely $(0, 1) \times (0, 1)$. Equation (4.48) is rewritten as

$$L\varphi \equiv -\varphi_{,1} - \varepsilon\varphi_{,\alpha\alpha} = 0,$$

with

$$\varepsilon = 1/\text{Pe} = D/u$$

where Pe is the Péclet number. The purpose is to find a discretization with accuracy and number of grid cells uniform in ε as $\varepsilon \downarrow 0$.

Choice of grid

In view of Sect. 2.5 we expect a parabolic (because $u^2 = 0$) boundary layer at $x_2 = 0$ with thickness $\mathcal{O}(\varepsilon^{1/2})$. Because a homogeneous Neumann condition is applied at the outflow boundary, there is no boundary layer at $x_1 = 0$. In order to make accuracy and computing work uniform in ε we choose a grid with local refinement in the boundary layer, as sketched in Fig. 4.8. The region of refinement has thickness σ. The choice of σ will be discussed later, and is such that the boundary layer falls inside the refinement region. The refined part of the grid is called G_f, the interface between the refined and

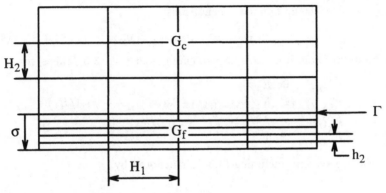

Fig. 4.8. Computational grid

unrefined parts is called Γ and the remainder of the grid is called G_c. The mesh sizes in G_f and G_c are uniform, as indicated in Fig. 4.8. Note that the location of the horizontal grid lines depends on ε.

Finite volume discretization

The cell centers are labeled by integer two-tuples (i, j) in the usual way: Ω_{ij} is the cell with center at (x_i, y_j). Hence, for example, $(i + 1/2, j)$ refers to the center of a vertical cell edge. Cell-centered discretization is used as described in Sect. 3.5 (two-dimensional uniform grid) and Sect. 3.7 (one-dimensional nonuniform grid). For ease of reference the discretization is summarized below. The finite volume method gives

$$\int_{\Omega_{ij}} L\varphi d\Omega = F^1\big|_{i-1/2,j}^{i+1/2,j} + F^2\big|_{i,j-1/2}^{i,j+1/2} \, ,$$

with the fluxes F^1 and F^2 given by

$$F^1 \equiv -\varphi - \varepsilon\varphi_{,1} \, , \quad F^2 = -\varepsilon\varphi_{,2} \, .$$

Using upwind discretization for the first derivative, the fluxes are approximated as follows:

$$F^1_{i+1/2,j} \cong -K_j\{\varphi_{i+1,j} + \varepsilon(\varphi_{i+1,j} - \varphi_{ij})/H_1\}\,,$$
$$F^2_{i,j+1/2} \cong -2H_1\varepsilon(\varphi_{i,j+1} - \varphi_{ij})/(K_j + K_{j+1})\,,$$

where K_j is the vertical dimension of Ω_{ij} : $K_j = h_2$ in G_f and $K_j = H_2$ in G_c. The cells are numbered $i = 1,...,I$ and $j = 1,...,J$ in the x_1- and x_2-directions, respectively.

In order to facilitate the error analysis that is to follow we place additional unknowns $\varphi_{i,1/2}$, $\varphi_{I+1/2,j}$ on the Dirichlet boundaries and extend the system of equations with the boundary conditions:

$$\varphi_{i,1/2} = g^1_i\,,\quad \varphi_{I+1/2,j} = g^2_j\,. \tag{4.50}$$

The boundary conditions are implemented as in Sect. 3.5. This gives

$$\begin{aligned}
F^1_{1/2,j} &\cong K_j\varphi_{1j}\,,\\
F^1_{I+1/2,j} &\cong K_j\{-\varphi_{I+1/2,j} - 2\varepsilon(\varphi_{I+1/2,j} - \varphi_{Ij})/H_1\}\,,\\
F^2_{i,1/2} &= -2H_1\varepsilon(\varphi_{i,1} - \varphi_{i,1/2})/h_2\,,\\
F^2_{i,J+1/2} &= 0\,.
\end{aligned}$$

The discrete system, including (4.50), is denoted by

$$L_c\varphi = q\,. \tag{4.51}$$

Local truncation error

The local discretization error is easily found from (3.50), (3.67) and (3.69). However, the $\mathcal{O}(\Delta^3_c)$ terms in these equations depend on Pe $\equiv 1/\varepsilon$. Because we want to study the accuracy as $\varepsilon \downarrow 0$, the dependence of the $\mathcal{O}(\Delta^3_c)$ terms on ε has to be exhibited.

Let us write the local truncation error as $\tau = \tau^1 + \tau^2$, where τ^α is associated with the x_α-derivatives. By Taylor expansion one finds, to leading order in the mesh sizes, writing $x = x_1$, $y = x_2$ in the remainder of this section,

$$|\tau^1_{1j}| \le H^2_1 K_j C_1,\quad j = 1,...,J\,, \tag{4.52}$$
$$|\tau^1_{ij}| \le H^2_1 K_j C_2,\quad i = 2,...,I-1,\quad j = 1,...,J\,, \tag{4.53}$$
$$|\tau^1_{Ij}| \le H_1 K_j C_3,\quad j = 1,...,J\,, \tag{4.54}$$
$$|\tau^2_{i1}| \le H_1 h_2 C_4,\quad i = 1,...,I\,, \tag{4.55}$$
$$|\tau^2_{ij}| \le H_1 h^3_2 C_5,\quad i = 1,...,I,\quad j = 2,...,j_\Gamma - 1\,, \tag{4.56}$$
$$|\tau^2_{ij}| \le H_1 H^3_2 C_6,\quad i = 1,...,I,\quad j = j_\Gamma + 2,...,J-1\,, \tag{4.57}$$
$$|\tau^2_{iJ}| \le H_1 H^2_2 C_7,\quad i = 1,...,I\,, \tag{4.58}$$

with j_Γ such that the cells Ω_{ij_Γ} are just below the interface Γ (cf. Fig. 4.8), and, replacing x_1 by x and x_2 by y,

$$C_1 = \sup\{|\frac{1}{2}\frac{\partial^2\varphi}{\partial x^2}| : (x,y) \in \Omega\} + \sup\{|\frac{\varepsilon}{24}\frac{\partial^3\varphi}{\partial x^3}| : (x,y) \in \Omega\}, \quad (4.59)$$

$$C_2 = \sup\{|\frac{1}{2}\frac{\partial^2\varphi}{\partial x^2}| : (x,y) \in \Omega\} + \sup\{|\frac{\varepsilon}{12}\frac{\partial^4\varphi}{\partial x^4}| : (x,y) \in \Omega\}, \quad (4.60)$$

$$C_3 = \sup\{|\frac{1}{2}\frac{\partial\varphi}{\partial x}| : (x,y) \in \Omega\} + \sup\{|\frac{\varepsilon}{4}\frac{\partial^2\varphi}{\partial x^2}| : (x,y) \in \Omega\}, \quad (4.61)$$

$$C_4 = \sup\{|\frac{\varepsilon}{4}\frac{\partial^2\varphi}{\partial y^2}| : (x,y) \in \Omega_f\}, \quad (4.62)$$

$$C_5 = \sup\{|\frac{\varepsilon}{12}\frac{\partial^4\varphi}{\partial y^4}| : (x,y) \in \Omega_f\}, \quad (4.63)$$

$$C_6 = \sup\{|\frac{\varepsilon}{12}\frac{\partial^4\varphi}{\partial y^4}| : (x,y) \in \Omega_c\}, \quad (4.64)$$

$$C_7 = \sup\{|\frac{\varepsilon}{24}\frac{\partial^3\varphi}{\partial y^3}| : 0 < x < 1,\ 1 - H_2 < y < 1\}. \quad (4.65)$$

Here Ω_f is the part of the domain Ω that is covered by the refined grid G_f, and $\Omega_c = \Omega \setminus \Omega_f$.

For the cells adjacent to Γ we need expansions similar to (3.50). We find, for $i = 1, ..., I$,

$$\tau^2_{ij_\Gamma} = -\frac{1}{4}\varepsilon\varphi_{,22}H_1(H_2 - h_2) - \frac{1}{24}\varepsilon\varphi_{,222}H_1(H_2 + 3h_2)(H_2 - h_2)$$
$$-\frac{1}{24}\varepsilon\varphi(\boldsymbol{x}^*_\Gamma)_{,2222}H_1\{h_2^3 + \frac{1}{8}(h_2 + H_2)^3\}, \quad (4.66)$$

$$\tau^2_{i,j_\Gamma+1} = \frac{1}{4}\varepsilon\varphi_{,22}H_1(H_2 - h_2) - \frac{1}{24}\varepsilon\varphi_{,222}H_1(h_2 + 3H_2)(H_2 - h_2)$$
$$-\frac{1}{24}\varepsilon\varphi(\boldsymbol{x}^*_\Gamma)_{,2222}H_1\{H_2^3 + \frac{1}{8}(h_2 + H_2)^3\}. \quad (4.67)$$

Where no argument is given the derivatives of the exact solution $\varphi(\boldsymbol{x})$ are evaluated in the same point as τ^α ; $\boldsymbol{x}^*_\Gamma$ stands for various points in the vicinity of Γ. More precisely, we have

$$\boldsymbol{x}^*_\Gamma \in X,$$

with X the union of the two rows of cells adjacent to Γ. On the Dirichlet boundaries, because (4.50) is exact,

$$\tau_{i,1/2} = \tau_{I+1/2,j} = 0.$$

Because of finite volume integration, a factor H_1K_j has been gained. Hence, a second order scheme has $\tau = \mathcal{O}(\Delta^4)$, $\Delta = \max\{H_1, K_j\}$. We see that (due to upwind discretization) the scheme is first order in the interior and at Neumann boundaries, but inconsistent in cells adjacent to Γ and to Dirichlet boundaries, in the maximum norm. This at first sight unfavorable property of cell-centered discretization on nonuniform grids is why we have chosen cell-centered discretization; vertex-centered discretization turns out to be consistent. It will be seen that nevertheless the global accuracy is $\mathcal{O}(H_1 + H_2^2)$.

The discrete maximum principle

The concept of operator of positive type of Sect. 4.4 is easily generalized to two dimensions. Define temporarily $j = (j_1, j_2)$, $k = (k_1, k_2)$ with j_α and k_α integers. Then the discrete system (4.51) can be written as

$$L_c \varphi_j = \sum_{k \in K} \alpha(j, k) \varphi_{j+k} = q_j \;,$$

with $K \equiv \{(0,0),\ (0, \pm 1),\ (\pm 1, 0)\}$. Define

$$G \equiv \{1, ..., I\} \times \{1, ..., J\} \;.$$

We say that L_c is an operator of positive type if

$$\sum_{k \in K} \alpha(j, k) = 0 \;, \quad \alpha(j, k) < 0 \quad \text{for} \quad k \neq (0,0) \quad \forall j \in G \;. \tag{4.68}$$

It is easily seen that condition (4.68) is satisfied. Hence (cf. Sect. 4.4) if

$$L_c \varphi_j \leq 0 \;, \quad j \in G \;,$$

then φ cannot have a local maximum for $j \in G$. Therefore the maximum of φ is to be found on the Dirichlet boundary G_D, with

$$G_D \equiv \{1, ..., I\} \times \{1/2\} \cup \{I + 1/2\} \times \{1, ..., J\} \;.$$

Error estimation with the maximum principle

By $\varphi < \psi$ we mean

$$\varphi_j < \psi_j \;, \quad j \in G \cup G_D \;,$$

and by $|\varphi|$ we mean the grid function with value $|\varphi_j|$ in $\boldsymbol{x}_j$. The global discretization error satisfies

$$L_c e = \tau \;.$$

Suppose we have a grid function $\boldsymbol{E}$, which will be called a barrier function, such that

$$L_c E \geq |\tau| \,. \tag{4.69}$$

Then

$$L_c(e - E) \leq 0 \,.$$

According to the maximum principle,

$$e - E \leq \max\{-E_j : j \in G_D\} \,,$$

where we have used $e_j = 0$ for $j \in G_D$. Furthermore,

$$L_c(-e - E) \leq 0 \,,$$

so that

$$-e - E \leq \max\{-E_j : j \in G_D\} \,.$$

If follows that

$$|e| \leq |E| + \max\{|E_j| : j \in G_D\} \,, \tag{4.70}$$

which is the desired estimate.

Expansion for global truncation error

From now on, i, j and k are again integers. Similar to (3.63) we write for the global truncation error $e_{ij} = \varphi(\boldsymbol{x}_{ij}) - \varphi_{ij}$:

$$e = \sum_{k=1}^{4} e^k \,. \tag{4.71}$$

We choose e^1, e^2 and e^3 such that $L_h(e^1 + e^2 + e^3)$ equals the first two terms in (4.66) and (4.67) and is of higher order for $j \neq j_\Gamma, j_\Gamma + 1$; e^4 will be estimated with the maximum principle. Let

$$e_{ij}^k = \psi_k(\boldsymbol{x}_{ij})\mu_j^k \,, \quad k = 1, 2, 3 \,,$$

with

$$\mu_j^1 = 0 \,, \quad j \leq j_\Gamma \,, \quad \mu_j^1 = H_2^2 - h_2^2 \,, \quad j \geq j_\Gamma + 1 \,,$$

$$\mu_j^2 = 0 \,, \quad j \leq j_\Gamma \,, \quad \mu_{j_\Gamma+1}^2 = \frac{1}{48}(H_2 + 3h_2)(H_2^2 - h_2^2) \,,$$

$$\mu_j^2 = \mu_{j-1}^2 + \frac{1}{6}H_2(H_2^2 - h_2^2) \,, \quad j \geq j_\Gamma + 2 \,,$$

$$\mu_j^3 = 0 \,, \quad j \leq j_\Gamma \,,$$

$$\mu_j^3 = -\frac{1}{2}(H_2 + h_2)(H_2^2 - H_2^2) \,, \quad j \geq j_\Gamma + 1 \,,$$

$$\psi_1 = \frac{1}{8}\varphi_{,22} \,, \quad \psi_2 = \varphi_{,222} \,, \quad \psi_3 = \psi_{1,2} \,.$$

We have

$$
\begin{aligned}
L_c\{\mu_j\psi(\boldsymbol{x}_{ij})\} = {} & \mu_j L_c\psi(\boldsymbol{x}_{ij}) \\
& -\varepsilon\psi(\boldsymbol{x}_{ij})\{\mu_{j+1} - \mu_j)/K_{j+1/2} - (\mu_j - \mu_{j-1})/K_{j-1/2}\} \\
& -\varepsilon\psi_{,2}(\boldsymbol{x}_{ij})(\mu_{i+1} - \mu_{j-1}) + \mathcal{O}(\frac{1}{2}\mu_j\varepsilon H_2\psi_{,22}(\boldsymbol{x}^*)) ,
\end{aligned}
\tag{4.72}
$$

where $\boldsymbol{x}^*$ is some point in the cells (i, j) or $(i, j \pm 1)$. $L_c\psi$ can be estimated by (4.52)–(4.65), replacing φ by ψ. After some cumbersome manipulations we find, to leading order in the mesh sizes:

$$
\begin{aligned}
L_c e^4 &= \tau^1 + \tilde{\tau}^2 , \\
\tilde{\tau}^2_{ij} &= \tau^2_{ij} , \quad i = 1, ..., I , \quad j \le j_\Gamma - 1 , \\
|\tilde{\tau}^2_{ij}| &\le H_1 H_2^3 C_8 , \quad i = 1, ..., I , \quad j_\Gamma \le j \le J - 1 , \\
C_8 &= \sup\{|\frac{1}{3}\varepsilon\varphi_{,2222}| : \boldsymbol{x} \in X \cup \Omega_c\} , \\
|\tilde{\tau}^2_{iJ}| &\le 9 H_1 H_2^2 C_7 , \quad i = 1, ..., I .
\end{aligned}
$$

Note that e^1, e^2, e^3 are $\mathcal{O}(H_2^2)$. It remains to estimate e^4. In fact, we have taken out the $\mathcal{O}(1)$ local errors near the boundary Γ of the refinement zone, and shown that these give only a second order contribution to the global error.

Error estimate

We will estimate e^4 with the maximum principle. Without any risk of confusion we will not distinguish between grid functions and functions of $\boldsymbol{x}$. A barrier function is chosen as follows, in a dimension-by-dimension approach that is extendable to more dimensions:

$$
\boldsymbol{E}(\boldsymbol{x}) = B_1 \boldsymbol{E}^1(x) + B_2 \boldsymbol{E}^2(y) ,
$$

with

$$
\begin{aligned}
\boldsymbol{E}^1(x) &= 1 + (1 - x)(2 + x + \frac{1}{2}H_1) , \quad 0 \le x \le 1 - H_1/4 , \\
\boldsymbol{E}^1(x) &= 0 , \quad x > 1 - H_1/4 , \\
\boldsymbol{E}^2(y) &= 0 , \quad 0 \le y \le h_2/4 , \\
\boldsymbol{E}^2(y) &= 1 + y(2\sigma + 3\sigma^2 - y)/\sigma^2 , \quad h_2/4 < y \le \sigma , \\
\boldsymbol{E}^2(y) &= \boldsymbol{E}^2(\sigma) + (y - \sigma)(3 - y) , \quad \sigma < y \le 1 ,
\end{aligned}
$$

where $\sigma = (j_\Gamma - \frac{1}{2})h_2$. We find, for $j = 1, ..., J$,

$$L_c E_{ij}^1 > K_j H_1 , \quad i = 2, ..., I-1 ,$$
$$L_c E_{Ij}^1 > K_j .$$

Furthermore, for $i = 1, ..., I$,

$$L_c E_{i1}^2 > 2\varepsilon H_1/h_2 ,$$
$$L_c E_{ij}^2 > 2\varepsilon H_1 h_2/\sigma^2 , \quad j = 2, ..., j_\Gamma - 1 ,$$
$$L_c E_{ij}^2 > \frac{3}{2}\varepsilon H_1 H_2 , \quad j = j_\Gamma, ..., J-1 ,$$
$$L_c E_{ij}^2 > \varepsilon H_1 , \quad j = J .$$

With

$$B_1 = H_1 \max\{C_1 , C_2 , C_3\}$$

we find

$$L_c B_1 E^1 > |\tau^1| .$$

Furthermore, let

$$B_2 = \varepsilon^{-1} \max\{h_2^2 C_4/2 , h_2^2 C_5 \sigma^2/2 , 9H_2^2 C_7 , 2H_2^2 C_8/3\} .$$

Then

$$L_c B_2 E^2 > |\tilde{\tau}^2| .$$

Therefore, E satisfies (4.69). Because $0 \le E^1 < 4$ and $0 \le E^2 \le 5$ we have

$$|E| \le 4B_1 + 5B_2 .$$

Hence equation (4.70) results in the following estimate:

$$|e^4| \le 8B_1 + 10B_2 .$$

The other terms in the error expansion (4.71) are easily estimated. We have, to leading order in the mesh sizes,

$$|e^1| \le \frac{1}{8}H_2^2 \sup\{|\varphi_{,22}| : x \in \Omega_c\} ,$$
$$|e^2| \le \frac{1}{6}H_2^2 \sup\{|\varphi_{,222}| : x \in \Omega_c\} , \qquad (4.73)$$
$$|e^3| \le \frac{1}{16}H_2^2 \sup\{|\varphi_{,222}|; x \in \Omega_c\} .$$

Note that for ε fixed

$$e = \mathcal{O}(H_1 + H_2^2) ,$$

so that the fact that the scheme is locally inconsistent in the maximum norm near Γ and the Dirichlet boundaries does not affect the global truncation error, as was made plausible in Sect. 3.7 for the one-dimensional case.

Dependence of the error on ε

We now turn to the main purpose of this section, which is to show how the accuracy and the computing work can be made independent of ε as $\varepsilon \downarrow 0$. For the analysis of the dependence of the bound obtained on ε we need to estimate derivatives of φ. For this the first term of a matched asymptotic expansion for the solution will be used. In order to show rigorously that higher order terms in this expansion may be neglected a laborious analysis of higher order terms would be needed, from which we will refrain. Numerical experiments will give further validation of the results obtained.

For the terminology and the method followed, see Sect. 2.5 and the literature on singular perturbation theory, for example Eckhaus (1973), Kevorkian and Cole (1981) Van Dyke (1975). The outer equation is

$$-\partial \varphi_0/\partial x = 0, \quad \varphi_0(1,y) = g^3(y) .$$

Its solution, called the outer solution, is given by

$$\varphi_0 = g^3(y) .$$

The inner or boundary layer solution φ_b that approximates $\varphi - \varphi_0$ is the solution of the following boundary layer problem:

$$-\frac{\partial \varphi_b}{\partial x} - \varepsilon \frac{\partial^2 \varphi_b}{\partial y^2} = 0, \quad 0 < x < 1, \quad 0 < y < \infty ,$$

$$\tag{4.74}$$

$$\varphi_b(1,y) = 0, \quad \varphi_b(x,\infty) = 0, \quad \varphi_b(x,0) = g(x) ,$$

with

$$g(x) = g_1(x) - g_3(0) .$$

In order to avoid a corner singularity it is assumed that $g_1(1) = g_3(0)$. The solution of (4.74) is given by

$$\varphi_b = \sqrt{\frac{2}{\pi}} \int_{y/\sqrt{2\varepsilon(1-x)}}^{\infty} e^{-\frac{1}{2}t^2} g(x + \frac{y^2}{2\varepsilon t^2}) dt .$$

A uniformly valid asymptotic approximation to φ is given by $\varphi_b + \varphi_0$. We find, using $g(1) = 0$:

$$\frac{\partial^2 \varphi_b}{\partial y^2} = -\frac{1}{\varepsilon}\sqrt{\frac{2}{\pi}} \int_{y/\sqrt{2\varepsilon(1-x)}}^{\infty} e^{-\frac{1}{2}t^2} g'(x + \frac{y^2}{2\varepsilon t^2}) dt .$$

Assuming $|g'(x)| \leq M_1$, $0 < x < 1$ we obtain

$$|\frac{\partial^2 \varphi_b}{\partial y^2}| \leq \frac{1}{\varepsilon} M_1 \mathrm{erfc}(y/2\sqrt{\varepsilon}) \,. \tag{4.75}$$

Using the inequality

$$\frac{2/\sqrt{\pi}}{z + \sqrt{z^2 + 2}} e^{-z^2} < \mathrm{erfc}(z) < \frac{2/\sqrt{\pi}}{z + \sqrt{z^2 + 4/\pi}} e^{-z^2} \quad (z \geq 0) \tag{4.76}$$

we may without losing much sharpness replace (4.75) by

$$\left|\frac{\partial^2 \varphi_b}{\partial y^2}\right| \leq \frac{4}{\sqrt{\pi \varepsilon}} M_1 \frac{1}{y + \sqrt{y^2 + 16\varepsilon/\pi}} e^{-y^2/4\varepsilon} \,. \tag{4.77}$$

It also necessary to estimate $\partial^4 \varphi_b / \partial y^4$. In order to avoid a corner singularity at $(1,0)$ it is assumed that

$$g'(1) = 0 \,.$$

Then

$$\frac{\partial^4 \varphi_b}{\partial y^4} = -\frac{1}{\varepsilon^2} \sqrt{\frac{2}{\pi}} \int\limits_{y/\sqrt{2\varepsilon(1-x)}}^{\infty} e^{-\frac{1}{2}t^2} g''(x + \frac{y^2}{2\varepsilon t^2}) dt \,.$$

Assuming $|g''(x)| \leq M_2$, $0 < x < 1$ we obtain

$$|\frac{\partial^4 \varphi_b}{\partial y^4}| \leq \frac{1}{\varepsilon^2} M_2 \mathrm{erfc}(y/2\sqrt{\varepsilon}) \,.$$

Again using (4.76) one obtains

$$|\frac{\partial^4 \varphi_b}{\partial y^4}| \leq \frac{4}{\sqrt{\pi \varepsilon^3}} M_2 \frac{1}{y + \sqrt{y^2 + 16\varepsilon/\pi}} e^{-y^2/4\varepsilon} \,.$$

Finally, $\partial^3 \varphi_b / \partial y^3$ has to be estimated. We obtain

$$\frac{\partial^3 \varphi_b}{\partial y^3} = -\frac{1}{\varepsilon} \sqrt{\frac{2}{\pi}} \int\limits_{y/\sqrt{2\varepsilon(1-x)}}^{\infty} e^{-\frac{1}{2}t^2} \frac{y}{\varepsilon t^2} g''(x + \frac{y^2}{2\varepsilon t^2}) dt$$

$$= -\frac{1}{\varepsilon y} \sqrt{\frac{2}{\pi}} \int\limits_{y/\sqrt{2\varepsilon(1-x)}}^{\infty} g'(x + \frac{y^2}{2\varepsilon t^2})(t^2 - 1) e^{-\frac{1}{2}t^2} dt \,.$$

Hence

$$|\frac{\partial^3 \varphi_b}{\partial y^3}| \leq \frac{1}{\varepsilon y} M_1 \sqrt{\frac{2}{\pi}} \int_{y/\sqrt{2\varepsilon}}^{\infty} |t^2 - 1| e^{-\frac{1}{2}t^2} dt \ . \tag{4.78}$$

Let us consider the estimate for $e^{(1)}$ given in (4.73). Using $\varphi \cong \varphi_b + \varphi_0$ we obtain, using (4.77),

$$\sup\{\left|\frac{\partial^2 \varphi}{\partial y^2}\right| : (x, y) \in \Omega_c\} \leq M_3 + \frac{2}{\sqrt{\pi \varepsilon}} M_1 \frac{1}{\sigma} e^{-\sigma^2/4\varepsilon} \ ,$$

with

$$M_3 = \sup\{|g_3''(y)| : 0 < y < 1\} \ .$$

For $e^{(2)}$ and $e^{(3)}$ we need, using (4.78) and assuming $\sigma/\sqrt{2\varepsilon} > 1$,

$$\sup\{\left|\frac{\partial^3 \varphi}{\partial y^3}\right| : (x, y) \in \Omega_c\} \leq M_4 + \frac{1}{\varepsilon\sigma} M_1 \sqrt{\frac{2}{\pi}} \int_{\sigma/\sqrt{2\varepsilon}}^{\infty} (t^2 - 1) e^{-\frac{1}{2}t^2} dt$$

$$= M_4 + \varepsilon^{-3/2} M_1 \frac{1}{\sqrt{\pi}} e^{-\sigma^2/4\varepsilon} \ ,$$

where

$$M_4 = \sup\{|g^{3'''}(y)| : 0 < y < 1\} \ .$$

For B_2 we have to estimate C_4, C_5, C_7 and C_8. For this is needed:

$$\sup\{\left|\frac{\partial^2 \varphi}{\partial y^2}\right| : (x, y) \in \Omega_f\} \leq M_1/\varepsilon + M_3 \ ,$$

$$\sup\{\left|\frac{\partial^4 \varphi}{\partial y^4}\right| : (x, y) \in \Omega_f\} \leq M_2/\varepsilon^2 + M_5 \ ,$$

$$\sup\{\left|\frac{\partial^4 \varphi}{\partial y^4}\right| : (x, y) \in X \cup \Omega_c\} \leq M_5 + \varepsilon^{-3/2} M_2 \frac{2}{\sqrt{\pi}} \tilde{\sigma} e^{-\tilde{\sigma}^2/4\varepsilon} \ ,$$

$$\sup\{\left|\frac{\partial^3 \varphi}{\partial y^3}\right| : 0 < x < 1, \ 1 - H_2 < y < 1\} \lesssim$$

$$M_4 + \frac{1}{\varepsilon} M_1 \sqrt{\frac{2}{\pi}} \int_{1/\sqrt{2\varepsilon}}^{\infty} (t^2 - 1) e^{-\frac{1}{2}t^2} dt = M_4 + \varepsilon^{-3/2} M_1 \frac{1}{\sqrt{\pi}} e^{-1/4\varepsilon} \ ,$$

where

$$\tilde{\sigma} = \sigma - h_2, \quad M_5 = \sup\{|g^{3''''}(y)| : 0 < y < 1\} \ ,$$

and where we have assumed

$$\tilde{\sigma}/\sqrt{2\varepsilon} > 1 \ .$$

Finally, the x-derivatives are $\mathcal{O}(1)$.

Thickness of the refinement region

We now have to choose σ and h_2 such that the error $e^1 + e^2 + e^3 + e^4$ is uniformly bounded in ε. Choosing

$$\sigma = \sqrt{-6\varepsilon \ln \varepsilon} \qquad (4.79)$$

we see that

$$e_1 + e_2 + e_3 = \mathcal{O}(H_2^2)$$

uniformly in ε. Furthermore, to have B_2 uniformly bounded it is necessary that

$$h_2 = H_2 \sqrt{-c\varepsilon / \ln \varepsilon} , \qquad (4.80)$$

with c some positive constant. Then we have

$$\frac{1}{\varepsilon} h_2^2 C_4 / 2 \;\leq\; \frac{1}{8} c H_2^2 (M_1 + \varepsilon M_3) / \ln \varepsilon^{-1} , \qquad (4.81)$$

$$\frac{1}{2\varepsilon} h_2^2 C_5 \sigma^2 \;\leq\; \frac{5}{24} c H_2^2 (M_2 + \varepsilon^2 M_5) , \qquad (4.82)$$

$$\frac{9}{\varepsilon} H_2^2 C_7 \;\leq\; \frac{3}{8} H_2^2 (M_4 + \varepsilon^{-3/2} M_1 \frac{1}{\sqrt{\pi}} e^{-1/4\varepsilon}) , \qquad (4.83)$$

$$\frac{2}{3\varepsilon} H_2^2 C_8 \;\leq\; \frac{2}{9} H_2^2 (\frac{1}{\sqrt{\pi}} \varepsilon^{-1/4} \sqrt{-5 \ln \varepsilon} M_1 + M_5) . \qquad (4.84)$$

In (4.83), the difference between σ and $\tilde{\sigma}$ has been neglected, which is asymptotically correct. Uniform boundedness of B_2 follows from (4.81)–(4.84). The following error estimate results in the maximum norm:

$$\|e\| = \mathcal{O}(H_1 + H_2^2) \qquad (4.85)$$

uniformly in ε.

Equations (4.79), (4.80) and (4.85) are our main results. Although the goal of uniform accuracy has been achieved, the computing work increases (slowly) as $\varepsilon \downarrow 0$. If there are n_c horizontal grid-lines in G_c, then the number of horizontal grid lines n_f in G_f satisfies

$$n_f = \mathcal{O}(n_c \ln \varepsilon^{-1}) , \qquad (4.86)$$

which follows from (4.79) and (4.80). However, numerical experience shows that the logarithmic factors in (4.79) and (4.80) come into play only for unrealistically small values of H_2, so that in practice one may work with

$$\sigma = \mathcal{O}(\sqrt{\varepsilon}), \quad h_2 = \mathcal{O}(\sqrt{\varepsilon} H_2) . \qquad (4.87)$$

Now the computing cost is uniform in ε, whereas numerical experiments still show uniform accuracy.

The analysis of vertex-centered discretization, where the nodes are the vertices of the cells in Fig. 4.8 rather than the centers, is easier, because an error expansion like (4.71) is not required. By using the maximum principle in a similar way, (4.79), (4.80) and (4.85) may be derived.

It is also possiblle to allow in addition to the parabolic boundary layer at $y = 0$ an ordinary boundary layer at $x = 0$. An additional refinement region of thickness $\mathcal{O}(\varepsilon \ln \varepsilon^{-1})$ needs to be introduced at $x = 0$ with mesh size $h_1 = \mathcal{O}(H_1 \varepsilon \ln \varepsilon^{-1})$. The same method of analysis can be followed.

In Farrell, Hemker, and Shishkin (1996), Hemker and Shishkin (1994), Shishkin (1990) it is proposed to choose

$$\sigma = \mathcal{O}\sqrt{\varepsilon} \ln n_f , \quad n_f = n , \tag{4.88}$$

so that the computing cost is uniform in ε. The corresponding grid is called a Shishkin grid. By a less elementary proof than above it is shown that the accuracy is uniform in ε, but a result less sharp than (4.85) is obtained. Numerical evidence points in the direction of (4.85).

Because the analysis uses the maximum principle, upwind discretization is necessary. But in practice, as will be shown, central discretization also gives good results, with $\mathcal{O}(H_1^2 + H_2^2)$ accuracy.

Numerical experiments

The following exact solution as assumed:

$$\varphi = \frac{1}{\sqrt{2 - x}} \left\{ \exp(-\frac{y^2}{4\varepsilon(2 - x)}) + \exp(-\frac{(2 - y)^2}{4\varepsilon(2 - x)}) \right\} .$$

The right-hand side and boundary conditions in (4.48) and (4.49) are chosen accordingly. Because the solution is extremely smooth in Ω_c it turns out that in G_c the number of cells in the vertical direction can be fixed at 4; the maximum of the error is found to always occur in G_f. We take

$$\sigma = 8\sqrt{\varepsilon} ,$$

neglecting the logarithmic factor in (4.79).

Table 4.3 gives results for the cell-centered upwind case. Equation (4.85) is confirmed. Exactly the same results are obtained for $\varepsilon = 10^{-5}$ and $\varepsilon = 10^{-7}$, showing ε-uniform accuracy, despite the fact that the logarithmic factors in

(4.79) and (4.80) have been neglected, to make work and storage independent of ε. The maximum error occurs in the interior of the boundary layer. Table 4.4 gives results for central discretization. Visual inspection of the re-

nx	ny	error $* 10^4$
8	32	54
32	64	14
128	128	3.6

Table 4.3. Maximum error as function of number of grid-cells for $\varepsilon = 10^{-3}$; cell-centered upwind discretization. nx: horizontal number of cells; ny : vertical number of cells in G_f.

nx	ny	error $* 10^4$
8	16	92
16	32	28
32	64	7.8
64	128	2.1

Table 4.4. Cell-centered central discretization; $\varepsilon = 10^{-3}$.

sults shows no visible wiggles. The rate of convergence is somewhat worse than the hoped for $\mathcal{O}(H_1^2 + H_2^2)$, but here again the same results are obtained for $\varepsilon = 10^{-7}$, showing uniformity in ε.

Table 4.5 gives results for the vertex-centered upwind case. A rate of conver-

nx	ny	error $* 10^4$
8	32	69
32	64	15
128	128	3.8

Table 4.5. Vertex-centered upwind discretization; $\varepsilon = 10^{-3}$.

gence of $\mathcal{O}(H_1 + H_2^2)$ is clearly demonstrated. The accuracy is about the same as for cell-centered discretization. Exactly the same results are obtained for $\varepsilon = 10^{-5}$ and 10^{-7}, again showing ε-uniform accuracy.

Finally, table 4.6 gives results for vertex-centered central discretization. The rate of convergence is $\mathcal{O}(H_1^2 + H_2^2)$. Further refinement shows that the finest grid of table 4.6 is the finest that rounding errors allow. No wiggles are observed. Results for $\varepsilon = 10^{-5}$ and $\varepsilon = 10^{-7}$ are virtually identical.

nx	ny	error $* 10^4$
8	16	30
16	32	6.76
32	64	1.88

Table 4.6. Vertex-centered central discretization; $\varepsilon = 10^{-3}$.

Further numerical experiments, including grids choosen according to (4.86) and (4.88), may be found in Wesseling (1996b).

We may conclude that in practice work and accuracy can be made to be uniform in ε, both for cell- and vertex-centered discretization, by suitable local mesh refinement according to (4.87) or (4.88). Hence, in principle, high Reynolds number flows are amenable to computation.

4.8 More accurate discretization of the convection term

Introduction

In many applications in fluid dynamics the Reynolds number satisfies $\mathrm{Re} \gg 1$, and as a consequence the diffusion terms are small in large parts of the domain. It is therefore especially important to approximate the convection term accurately, the more so because in inviscid flow models, such as the Euler equations, diffusion terms are not present. When improving the accuracy of the discretization of the convection terms, the order barrier on schemes of positive type (Theorem 4.4.2) makes itself felt. In order to obtain higher order accuracy without spurious oscillations , a way around this barrier has to be found. This leads to the introduction of *flux limiting*.

The κ-scheme

Let us reconsider equation (4.5), discretized with the finite volume method on the grid of Fig. 4.1 according to (4.6). Let us split the flux function F in its convective and diffusion parts:

$$F = F^c + F^v, \quad F^c = u\varphi, \quad F^v = -Dd\varphi/dx.$$

As compared to the second order central scheme (4.8) or the first order upwind scheme (4.9), with the κ-scheme one tries to enhance the accuracy by using an additional cell-center value, namely φ_{j-1} for $u_{j+1/2} \geq 0$ or φ_{j+2} for $u_{j+1/2} < 0$. In order to maintain second order accuracy, to the convective flux function of the central second order scheme a higher order term is added, as follows:

$$F^c_{j+1/2} \cong u_{j+1/2}\{\frac{1}{2}(\varphi_j + \varphi_{j+1}) + \frac{1-\kappa}{4}(-\varphi_{j-1} + 2\varphi_j - \varphi_{j+1})\}, \quad u_{j+1/2} \geq 0$$

$$F^c_{j+1/2} \cong u_{j+1/2}\{\frac{1}{2}(\varphi_j + \varphi_{j+1}) + \frac{1-\kappa}{4}(-\varphi_j + 2\varphi_{j+1} - \varphi_{j+2})\}, \quad u_{j+1/2} < 0$$

$$(4.89)$$

The approximation to the convection term becomes

$$\left(\frac{du\varphi}{dx}\right)_j \cong \frac{1}{h_j}(F^c_{j+1/2} - F^c_{j-1/2}) . \tag{4.90}$$

This one-parameter family of schemes, introduced by van Leer (1977), will be called the κ-scheme. If $u = $ constant > 0 the stencil is given by

$$\frac{u}{h_j}\left[\frac{1-\kappa}{4} \quad \frac{3\kappa-5}{4} \quad \frac{3-3\kappa}{4} \quad \frac{1+\kappa}{4} \quad 0\right] .$$

For $\kappa = -1$ we obtain the one-sided second order upwind scheme (Warming and Beam (1976)). With $\kappa = 0$ one obtains Fromm's zero average phase error scheme (Fromm (1968)), if in this scheme the terms quadratic in the Courant number are neglected, which are meant to improve time accuracy, of no concern here; this scheme results from optimizing, among 5-point schemes, for the propagation of a step function over one time step in the absence of diffusion (Wesseling (1973)). For $\kappa = 1/3$ the third order upwind biased scheme of Anderson, Thomas, and van Leer (1986) results. For $\kappa = 1/2$ we have the QUICK (quadratic upstream interpolation for convective kinematics) scheme proposed by Leonard (1979). This corresponds to approximation of $\varphi_{j+1/2}$ by interpolation of highest possible order (i.e. quadratic) between three surrounding grid points, but does not lead to a local truncation error of maximum possible order (cf. Exercise 4.8.1). Finally, $\kappa = 1$ gives the central scheme (4.32).

The κ-scheme provides a convenient framework for formulating upwind biased schemes. For $\kappa = 1/3$, the κ-scheme is third order accurate; for all other values of κ, it is second order.

A remark on the local truncation error

The global and local truncation errors in grid point j have been defined in the preceding chapter as

$$e_j \equiv \varphi(x_j) - \varphi_j, \quad \tau_j \equiv L_d e_j ,$$

with $\varphi(x)$ the exact solution and L_d the discrete operator. It is assumed that L_d is scaled such that it approximates the differential operator. Let $D = 0$, and assume $\tau_j = \mathcal{O}(h^p)$ and $e_j = \mathcal{O}(h^q)$. Since (4.90) is an exact consequence of the differential equation and contains no error, it may come as a surprise that q is not maximized by making $F^c_{j\pm1/2}$ as accurate as possible, which is the case for $\kappa = 1/2$, resulting in $p = q = 2$. Instead, q is maximized for $\kappa = 1/3$, resulting in $p = q = 3$. The underlying cause of this apparent discrepancy is, that (4.90) is not *equivalent* to the differential equation. It follows from the differential equation, but the differential equation does not follow from (4.90). The situation is illustrated in Exercise 4.8.1.

The fourth order central scheme

Again using a 5-point stencil, but no upwind bias, $\tau_j = \mathcal{O}(h^4)$ on a uniform grid is obtained by

$$F^c_{j+1/2} \cong u_{j+1/2}\{\frac{7}{12}(\varphi_j + \varphi_{j+1}) - \frac{1}{12}(\varphi_{j-1} + \varphi_{j+2})\}\,.$$

If $u = $ constant the stencil for the resulting approximation of $(du\varphi/dx)_j$ is given by

$$\frac{u}{12h}[1 \quad -8 \quad 0 \quad 8 \quad -1]\,. \tag{4.91}$$

Approximation of $\varphi_{j+1/2}$ by interpolation to highest order gives a different stencil, with $\tau_j = \mathcal{O}(h^3)$, cf. Exercise 4.8.1.

Nonlinear schemes

Because the order of accuracy of the preceding schemes is higher than one, the order barrier Theorem 4.4.2 predicts that these schemes are not of positive type. This is immediately confirmed by a look at their stencils. Hence, the maximum principle of Theorem 4.4.1 does not apply, and spurious wiggles may occur. Of course it is desirable to combine higher order accuracy with the positive type property, in order to avoid spurious wiggles. To achieve this, a way around the order barrier theorem has to be found. This may be done by using *nonlinear schemes*, i.e. schemes that are nonlinear, ever though the differential equation that is approximated is linear. A great variety of nonlinear additions to linear schemes has been proposed in the pursuit of positivity. Many of these can be formulated in a unifying framework called the normalized variable diagram, which we will first discuss. A different but equivalent way to formulate nonlinear schemes of positive type is provided by the frequently used concept of flux limiting. This will also be discussed.

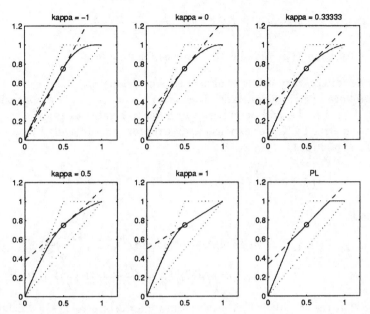

Fig. 4.9. Normalized variable diagram: $f = f(\tilde{\varphi})$. o : the point $P = (1/2, 3/4)$; - - - : κ-scheme; —— : $f(\tilde{\varphi})$.

Normalized variable diagram

Point of departure is an upwind biased linear scheme, for which we take the κ-scheme (4.89). For brevity we assume $u_j \geq 0$; the contrary case may be handled by symmetry. In Gaskell and Lau (1988), Leonard (1988) the following *normalized variable* is introduced:

$$\tilde{\varphi} = \frac{\varphi - \varphi_{j-1}}{\varphi_{j+1} - \varphi_{j-1}} \,. \tag{4.92}$$

From (4.89) it follows that for the κ-scheme we have

$$\tilde{\varphi}_{j+1/2} = \frac{2-\kappa}{2}\tilde{\varphi}_j + \frac{1+\kappa}{4} \,. \tag{4.93}$$

The graph of $\tilde{\varphi}_{j+1/2}$ versus $\tilde{\varphi}_j$ is called in Leonard (1988) the *characteristic* of the scheme and can be plotted in the *normalized variable diagram*, see Fig. 4.9 for example. A general nonlinear scheme is defined by

$$\tilde{\varphi}_{j+1/2} = f(\tilde{\varphi}_j) \,. \tag{4.94}$$

For the κ-scheme, f is linear.

If $u_j < 0$ we get by symmetry:

$$\tilde{\varphi}_{j+1/2} = f(\tilde{\varphi}_{j+1}) , \quad \tilde{\varphi} = \frac{\varphi - \varphi_{j+2}}{\varphi_j - \varphi_{j+2}} , \tag{4.95}$$

and the function $f(\tilde{\varphi})$ remains the same as for $u_j \geq 0$.

The order of accuracy of the nonlinear scheme defined by (4.94) can be studied as follows. Let $u = $ constant. Then $\varphi_{j+1/2} - \varphi_{j-1/2}$ has to approximate $h_j(d\varphi/dx)_j$. Let the grid be uniform. In order to be able to analyze the local truncation error by Taylor expansion, f is approximated locally by, assuming that f'' is bounded,

$$f(\tilde{\varphi}_j) \cong f(\frac{1}{2}) + (\tilde{\varphi}_j - \frac{1}{2})f'(\frac{1}{2}) + \mathcal{O}\left(\tilde{\varphi}_j - \frac{1}{2}\right)^2 . \tag{4.96}$$

One finds

$$\varphi_{j+1/2} - \varphi_{j-1/2} = h(d\varphi/dx)_j + h^2(d^2\varphi/dx^2)_j\{2f(\frac{1}{2}) - \frac{3}{2}\} \tag{4.97}$$
$$+ h^3(d^3\varphi/dx^3)_j\{\frac{7}{6} - f(\frac{1}{2}) - \frac{1}{2}f'(\frac{1}{2})\} + \mathcal{O}(h^4) .$$

We see that the scheme is first order accurate for arbitrary $f(\frac{1}{2})$, which justifies the choice made for the normalized variables, and the decision to develop $f(\tilde{\varphi}_j)$ arround $\tilde{\varphi}_j = 1/2$ in (4.96). If

$$f(1/2) = 3/4 \tag{4.98}$$

the scheme is second order. As expected, the κ-scheme satisfies (4.98) for all κ. Finally, if

$$f'(1/2) = 5/6 \tag{4.99}$$

the scheme is third order. As expected, the $\kappa = 1/3$ scheme satisfies (4.99). In other words, the characteristics of second order schemes pass through the point $P = (1/2, 3/4)$ in the normalized variable diagram (Fig. 4.9), and for third order schemes are tangent in P to the characteristic of the $\kappa = 1/3$ scheme.

Next, we derive further conditions to be satisfied by f. In order for the scheme to exclude spurious wiggles and to be more than first order accurate, f has to be nonlinear. We consider a number of possibilities for the values of φ_{j-1}, φ_j and φ_{j+1}, see Fig. 4.10. If $\varphi_j \in [\varphi_{j-1}, \varphi_{j+1}]$ (i.e. $0 \leq \tilde{\varphi}_j \leq 1$) the distribution of values is monotone, and monotonicity is preserved if we choose $\varphi_{j+1/2} \in [\varphi_j, \varphi_{j+1}]$. For $f(\tilde{\varphi}_j) = \tilde{\varphi}_{j+1/2}$ this leads to

$$\tilde{\varphi}_j \leq f(\tilde{\varphi}_j) \leq 1, \quad 0 \leq \tilde{\varphi}_j \leq 1 . \tag{4.100}$$

If $\varphi_j \neq [\varphi_{j-1}, \varphi_{j+1}]$ the distribution of values is not monotone, and in order to steer the solution to monotonicity the first order upwind scheme $\varphi_{j+1/2} = \varphi_j$ is applied, resulting in

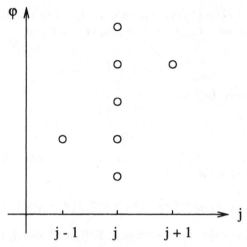

Fig. 4.10. Various values of φ_j in comparison to φ_{j-1} and φ_{j+1}.

$$f(\tilde{\varphi}_j) = \bar{\varphi}_j, \quad \bar{\varphi}_j \notin [0, 1] . \tag{4.101}$$

Finally, we remark that in the neglected $\mathcal{O}(h^4)$ term in (4.97) there are terms proportional to f'', so that it seems advisable to keep f'' bounded. Examples of nonlinear characteristics satisfying these criteria are given in Fig. 4.9. In the origin and in (1,1) f'' is not bounded (since here the characteristic is continued with slope 1 according to (4.101)), but this is allowed, since here the scheme switches to first order upwind, so that lower order terms dominate in the expansion (4.97). Shortly, we will show that it is desirable to sharpen (4.100).

The flux limiter

A different way to formulate nonlinear schemes is as follows. In order to choose the flux $F^c_{j+1/2} = u_{j+1/2}\varphi_{j+1/2}$ we may also write, instead of (4.94),

$$\varphi_{j+1/2} = \varphi_j + \frac{1}{2}\psi(r_j)(\varphi_{j+1} - \varphi_j) . \tag{4.102}$$

Van Leer (1974)has proposed to make ψ a function of the ratio of consecutive gradients:

$$r_j = \frac{\varphi_j - \varphi_{j-1}}{\varphi_{j+1} - \varphi_j} .$$

The function $\psi(r)$ determines a correction on the upwind flux $u_{j+1/2}\varphi_j$, and is called the *flux limiter*. For reasons to become clear in Sect. 9.4, ψ will later also be called *slope limiter*. A survey of flux limiters (in the nonstationary case) is given by Sweby (1984). The relation between the flux limiter ψ and the normalized variable function is as follows. We have

$$r_j = \frac{\tilde{\varphi}_j}{1 - \tilde{\varphi}_j}, \quad \tilde{\varphi}_j = \frac{r_j}{1 + r_j} \ . \tag{4.103}$$

Equation (4.102) can be rewritten as

$$\tilde{\varphi}_{j+1/2} = \tilde{\varphi}_j + \frac{1}{2}\psi(r_j)(1 - \tilde{\varphi}_j) \ .$$

Comparison with (4.94) and substitution of (4.103) gives the following relation, suppressing indices:

$$f(\tilde{\varphi}) = \{r + \frac{1}{2}\psi(r)\}/(1 + r) \ , \quad \psi(r) = 2\{f(\tilde{\varphi}) - \tilde{\varphi}\}/(1 - \tilde{\varphi}) \ . \tag{4.104}$$

From (4.104) follows that if f satisfies (4.100) and (4.101), then ψ satisfies

$$0 \le \psi(r) \le 2 \ , \quad 0 \le r \le \infty \ ; \quad \psi(r) = 0 \ , \quad r < 0 \ .$$

Flux limited schemes of positive type

We will derive conditions for flux limited schemes to be of positive type. Let us consider the general case where $u(x)$ may change sign. The scheme for the convection term is written as

$$L_h\varphi_j = \frac{1}{h_j}F^c|_{j-1/2}^{j+1/2}, \quad F^c_{j+1/2} = (u\varphi)_{j+1/2} \ . \tag{4.105}$$

If $u_{j+1/2} \ge 0$, $\varphi_{j+1/2}$ is given by (4.102). If $u_{j+1/2} < 0$, upwind bias is applied in the other direction. By symmetry we obtain:

$$\varphi_{j+1/2} = \varphi_{j+1} + \frac{1}{2}\psi(\frac{1}{r_{j+1}})(\varphi_j - \varphi_{j+1}) \ , \quad u_{j+1/2} < 0 \ . \tag{4.106}$$

The discretization in the case where $u_{j+1/2}$ is arbitrary can now be written as

$$F^c_{j+1/2} = F^+_{j+1/2} + F^-_{j+1/2} \ , \quad F^\pm = (u\varphi^\pm)_{j+1/2} \ , \tag{4.107}$$

$$\varphi^+_{j+1/2} = \varphi_j + \frac{1}{2}\psi(r_j)(\varphi_{j+1} - \varphi_j) \ , \quad u_{j+1/2} \ge 0 \ ,$$

$$F^+_{j+1/2} = 0 \ , \quad u_{j+1/2} < 0 \ ,$$

$$\varphi^-_{j+1/2} = \varphi_{j+1} + \frac{1}{2}\psi(\frac{1}{r_{j+1}})(\varphi_j - \varphi_{j+1}) \ , \quad u_{j+1/2} < 0 \ ,$$

$$F^-_{j+1/2} = 0 \ , \quad u_{j+1/2} \ge 0 \ ,$$

$$L_h\varphi_j = \frac{1}{h_j}(F^+ + F^-)|_{j-1/2}^{j+1/2} \ .$$

In order to derive conditions on the limiter $\psi(r)$ for scheme (4.107) to be of positive type we write it in the following form:

$$L_h \varphi_j = \{a_j(\varphi_{j+1} - \varphi_j) + b_j(\varphi_{j-1} - \varphi_j)\}/h_j , \qquad (4.108)$$

where a_j and b_j are functions of $\varphi_j, \varphi_{j\pm1}, \varphi_{j\pm2}, \dots$. We can write

$$a_j = \frac{F^-_{j+1/2} - F^-_{j-1/2}}{\varphi^-_{j+1/2} - \varphi^-_{j-1/2}} \cdot \frac{\varphi^-_{j+1/2} - \varphi^-_{j-1/2}}{\varphi_{j+1} - \varphi_j} ,$$

$$b_j = \frac{F^+_{j+1/2} - F^+_{j-1/2}}{\varphi^+_{j+1/2} - \varphi^+_{j-1/2}} \cdot \frac{\varphi^+_{j+1/2} - \varphi^+_{j-1/2}}{\varphi_{j-1} - \varphi_j} .$$

Note that $dF^-/d\varphi^- \leq 0$ and $dF^+/d\varphi^+ \geq 0$, so that by the mean value theorem we have

$$\frac{F^-_{j+1/2} - F^-_{j-1/2}}{\varphi^-_{j+1/2} - \varphi^-_{j-1/2}} \leq 0 , \qquad \frac{F^+_{j+1/2} - F^+_{j-1/2}}{\varphi^+_{j+1/2} - \varphi^+_{j-1/2}} \geq 0 .$$

Comparison of (4.108) with Definition 4.4.1 shows that for the scheme to be of positive type it is necessary and sufficient that

$$a_j \leq 0 , \quad b_j \leq 0 .$$

It follows that we must have

$$\frac{\varphi^-_{j+1/2} - \varphi^-_{j-1/2}}{\varphi_{j+1} - \varphi_j} \geq 0 , \qquad \frac{\varphi^+_{j+1/2} - \varphi^+_{j-1/2}}{\varphi_{j-1} - \varphi_j} \leq 0 . \qquad (4.109)$$

Using (4.102) and (4.106) we see that we must have

$$2 - \psi(r) + \frac{\psi(s)}{s} \geq 0 , \quad \forall r, s \in \mathbb{R} . \qquad (4.110)$$

This is necessary and sufficient. To have $a_{j+1/2}$ and $b_{j+1/2}$ bounded we must also have

$$\frac{\varphi^-_{j+1/2} - \varphi^-_{j-1/2}}{\varphi_{j+1} - \varphi_j} \leq M , \qquad \frac{\varphi^+_{j+1/2} - \varphi^+_{j-1/2}}{\varphi_{j-1} - \varphi_j} \geq -M ,$$

for some $M > 0$. This gives

$$2 - \psi(r) + \frac{\psi(s)}{s} \leq 2M , \quad \forall r, s \in \mathbb{R} . \qquad (4.111)$$

Combination of (4.110) and (4.111) gives

$$-2 \leq -\psi(r) + \frac{\psi(s)}{s} \leq 2M - 2 , \quad \forall r, s \in \mathbb{R} . \qquad (4.112)$$

If we require

$$1 - M \leq \psi(r) \leq m \leq 2 \,, \qquad (4.113)$$

then equation (4.112) gives

$$m - 2 \leq \frac{\psi(s)}{s} \leq M - 1 \,. \qquad (4.114)$$

Equations (4.113) and (4.114) imply that the graph of $\psi(r)$ must lie in the shaded region of Fig. 4.11. For $m = 2, M = 3$ we obtain the first order TVD

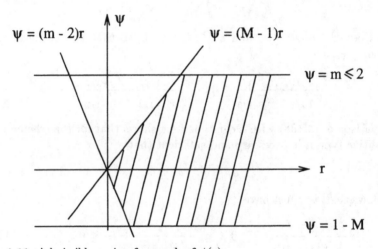

Fig. 4.11. Admissible region for graph of $\psi(r)$.

region derived by Sweby (1984) for the nonstationary case.

The aim of flux-limiting is, of course, to obtain schemes of positive type of second (or higher) order of accuracy. For second order accuracy we must satisfy equation (4.98). Using (4.103) and (4.104) this gives

$$\psi(1) = 1 \,.$$

To have this point in the admissible region we must take $M \geq 2$. To make sure that the graph of $\psi(r)$ passes through $\psi(1) = 1$ we may formulate the following additional requirement:

$$1 \geq \psi(r) \geq r \,, \quad 1 \leq \psi(r) \leq r \,. \qquad (4.115)$$

The results in the admissible region (shaded) of Fig. 4.12, which corresponds to the second order TVD region of Sweby (1984) if $M = 3, m = 2$.

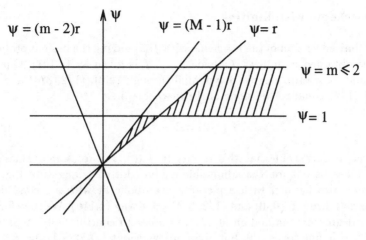

Fig. 4.12. Admissible region for graph of $\psi(r)$ for second order schemes.

Normalized variable diagram for schemes of positive type

Conditions to impose on flux limiters to obtain schemes of positive type are easily transported to the normalized variable framework by means of the correspondence relations (4.103), (4.104). The normalized variable diagram corresponding to the flux limiter diagram of Fig. 4.12 is obtained as follows. The interval $0 \leq r < \infty$ corresponds to $0 \leq \tilde{\varphi} < 1$. The condition $\psi \leq m \leq 2$ corresponds to

$$f(\tilde{\varphi}) \leq \frac{m}{2} + (1 - \frac{m}{2})\tilde{\varphi} \,.$$

The condition $\psi \leq (M - 1)r$ corresponds to

$$f(\tilde{\varphi}) \leq \frac{M + 1}{2}\tilde{\varphi} \,.$$

Condition (4.115) corresponds to

$$\frac{1}{2} + \frac{1}{2}\tilde{\varphi} \geq f(\tilde{\varphi}) \geq \frac{3}{2}\tilde{\varphi}, \quad 0 \leq \tilde{\varphi} \leq \frac{1}{2} \,;$$

$$\frac{1}{2} + \frac{1}{2}\tilde{\varphi} \leq f(\tilde{\varphi}) \leq \frac{3}{2}\tilde{\varphi}, \quad \frac{1}{2}\tilde{\varphi} \leq 1 \,.$$

The resulting admissible region in the normalized variable diagram in the case $m = 2$, $M = 3$ is the interior of the dotted triangles in Fig. 4.9. As noted before (cf. equation (4.97)), characteristics of second order schemes have to pass through $P = (1/2, 3/4)$. For third order accuracy, the slope in P has to be equal to that of the $\kappa = 1/3$ scheme, i.e. $f'(1/2) = 5/6$. Finally, we want f to be continuous and f'' bounded on $(0, 1)$.

The κ-scheme with limiting

A flux limited version of the κ-scheme with $f(\tilde{\varphi})$ having the desired properties may be obtained as follows. Outside $(0,1)$, f is given by (4.101). One $[0,1]$ one could fit a third degree polynomial through $(0,0)$, $(1/2, 3/4)$ and $(1,1)$ with $f'(1/2)$ equal to the slope of the κ-scheme, i.e.

$$f'(1/2) = 1 - \kappa/2 . \tag{4.116}$$

However, it turns out that this results in $f'(0) > 2$ for some values of κ, which takes us outside the admissible region (dotted triangles of Fig. 4.9). This may also happen with a piecewise parabolic fit with a second degree polynomial through $(0, 0)$ and $(1/2, 3/4)$ satisfying (4.116). We therefore use a third degree polynomial on $[0, 1/2)$, in order to control $f'(0)$. A parabola on $[1/2, 1]$ is fine for $\kappa \geq 0$, but gives an overshoot $(f(\tilde{\varphi}) > 1)$ for $\kappa < 0$. In that case we use a third degree polynomial with $f'(1) = 0$. This results in:

$$
\begin{aligned}
f(\tilde{\varphi}) &= \tilde{\varphi}, \qquad \tilde{\varphi} \notin (0,1), \\
f(\tilde{\varphi}) &= 2\tilde{\varphi} + (\kappa - 1)\tilde{\varphi}^2 - 2\kappa\tilde{\varphi}^3, \quad 0 \leq \tilde{\varphi} < 1/2, \\
f(\tilde{\varphi}) &= 1 + \frac{1}{2}\kappa(\tilde{\varphi} - 1) + (\kappa - 1)(\tilde{\varphi} - 1)^2, \quad 0 \leq \kappa \leq 1, \quad 1/2 \leq \tilde{\varphi} \leq 1, \\
f(\tilde{\varphi}) &= 1 - (1 + \kappa)(\tilde{\varphi} - 1)^2 - 2\kappa(\tilde{\varphi} - 1)^3, \quad -1 \leq \kappa < 0, \quad 1/2 \leq \tilde{\varphi} \leq 1.
\end{aligned}
$$

The resulting characteristics are given in Fig. 4.9 for $\kappa = -1$ (second order upwind scheme), $\kappa = 0$ (Fromm scheme), $\kappa = 1/3$ (third order upwind biased scheme), $\kappa = 1/2$ (QUICK scheme) and $\kappa = 1$ (second order central scheme).

The corresponding limiter function is found to be, using (4.103) and (4.104),

$$
\begin{aligned}
\psi(r) &= 0, \qquad r < 0, \\
\psi(r) &= \frac{2r}{(1+r)^2}\{1 + (3 - \kappa)r + (2 - 3\kappa)r^2\}, \quad 0 \leq r \leq 1, \\
\psi(r) &= \frac{1}{1+r}\{\kappa + (2 - \kappa)r\}, \quad r > 1, \quad 0 \leq \kappa \leq 1, \\
\psi(r) &= \frac{2}{(1+r)^2}\{\kappa + (1 - \kappa)r + r^2\}, \quad r > 1, \quad -1 \leq \kappa < 0.
\end{aligned}
$$

According to experience, somewhat better accuracy is obtained if the κ-scheme (i.e. the dashed lines in Fig. 4.9) is used as much as possible. We therefore choose the characteristic function $f(\tilde{\varphi})$ piecewise linear, similar to the SMART scheme proposed in Gaskell and Lau (1988), and we will call it PL. The PL characteristic function is shown in Fig. 4.9. This type of characteristic function can be implemented very efficiently, as shown in Leonard and Drummond (1995), in the following way. The median function $\text{med}(a, b, c)$ takes three real numbers and returns the number that lies in between the other two or coincides with one of the other two. The median function is easily implemented as

$$\text{med}(a, b, c) = \max[\min(a, b), \min\{\max(a, b), c\}] \, .$$

From figure 4.9 it is clear that the graph of

$$f_1(\tilde{\varphi}) \equiv \text{med}\{2\tilde{\varphi}, 1, \tilde{\varphi}\}$$

is given by $f_1(\tilde{\varphi}) = \tilde{\varphi}$ for $\tilde{\varphi} \notin (0, 1)$ whereas otherwise the two upper sides of the triangle are followed. It follows that

$$f(\tilde{\varphi}) = \text{med}(f, (\tilde{\varphi}), \frac{2 - \kappa}{2}\tilde{\varphi} + \frac{1 + \kappa}{4}, \tilde{\varphi})$$

gives the PL characteristic depicted in figure 4.9. Noting that $d+\text{med}(a, b, c) = \text{med}(a + d, b + d, c + d)$ and $d\text{med}(a, b, c) = \text{med}(da, db, dc)$, we can switch from normalized to original variables for efficiency:

$$f_2(\varphi_j) \equiv \text{med}(2\varphi_j - \varphi_{j-1}, \varphi_j, \varphi_{j+1}) \, ,$$
$$\varphi_{j+1/2} = \text{med}(f_2(\varphi_j), \varphi^{\kappa}_{j+1/2}, \varphi_j) \, ,$$
$$\varphi^{\kappa}_{j+1/2} = \frac{1}{2}(\varphi_j + \varphi_{j+1}) + \frac{1 - \kappa}{4}(-\varphi_{j-1} + 2\varphi_j - \varphi_{j+1}) \, .$$

The corresponding PL limiter function is found to be, using (4.103) and (4.104):

$$\begin{aligned}
\psi(r) &= 0 \, , & r &< 0 \, , \\
\psi(r) &= 2r \, , & 0 &\leq r < 2/5 \, , \\
\psi(r) &= (2 + r)/3 \, , & 2/5 &\leq r < 4 \, , \\
\psi(r) &= 2 \, , & r &\geq 4 \, .
\end{aligned}$$

Nonlinear characteristics for the QUICK scheme ($\kappa = 1/2$) have been proposed in Leonard (1988), Gaskell and Lau (1988). Some well-known flux limiter functions are:

van Albada, van Leer, and Roberts (1982):

$$\psi(r) = \frac{r^2 + r}{r^2 + 1} \, ,$$

Van Leer (1977b):

$$\psi(r) = \frac{r + |r|}{1 + |r|} \, ,$$

Roe (1986) ('superbee')

$$\psi(r) = \max\{0, \min(2r, 1), \min(r, 2)\} \, .$$

With (4.103) and (4.104) the corresponding characteristic functions are found to be, respectively:

van Albada:

$$f(\tilde{\varphi}) = \tilde{\varphi} + \frac{\tilde{\varphi} - \tilde{\varphi}^2}{1 + (1 - 2\varphi)^2} \, ,$$

van Leer:

$$
\begin{aligned}
f(\tilde{\varphi}) &= 2\tilde{\varphi} - \tilde{\varphi}^2, \quad \tilde{\varphi} \in [0, 1] \, , \\
f(\tilde{\varphi}) &= \tilde{\varphi}, \quad \tilde{\varphi} \notin [0, 1] \, ,
\end{aligned}
$$

Roe (superbee):

$$
\begin{aligned}
f(\tilde{\varphi}) &= 2\tilde{\varphi}, && 0 \le \tilde{\varphi} < \tfrac{1}{3} \, , \\
f(\tilde{\varphi}) &= \tfrac{1}{2}(1 + \tilde{\varphi}), && \tfrac{1}{3} \le \tilde{\varphi} < \tfrac{1}{2} \, , \\
f(\tilde{\varphi}) &= \tfrac{3}{2}\tilde{\varphi}, && \tfrac{1}{2} \le \tilde{\varphi} < \tfrac{2}{3} \, , \\
f(\tilde{\varphi}) &= 1, && \tfrac{2}{3} \le \tilde{\varphi} \le 1 \, , \\
f(\tilde{\varphi}) &= \tilde{\varphi}, && \tilde{\varphi} \notin [0, 1] \, .
\end{aligned}
$$

All of these characteristics are in the triangle of Fig. 4.9, and hence are wiggle-free and TVD. Also, they pass through (1/2, 3/4), but the second order accuracy of the 'superbee' flux limiter is in doubt, because $f''(1/2)$ is unbounded; cf. the assumption preceding (4.96). The 'superbee' scheme is known to suffer from 'clipping', i.e. near a maximum the numerical solution shows a plateau slightly below the exact maximum, and analogously for minima. Because $f'(1/2)$ does not exist, the 'superbee' scheme cannot be regarded as a nonlinear version of the κ-scheme. For the van Albada and van Leer flux limiters we have $f'(1/2) = 1$, corresponding to $\kappa = 0$ (cf. (4.116), so that these may be regarded as flux limited versions of the Fromm scheme.

Exercise 4.8.1. Let the grid be uniform. Determine $F^c_{j+1/2}$ by interpolating $\varphi_{j+1/2}$ to the highest possible order of accuracy between φ_{j-1}, φ_j and φ_{j+1}, and verify that this gives the $\kappa = 1/2$ scheme. Determine the order of the local truncation error in the approximation of $d\varphi/dx$ for $\kappa = 1/2$ and $\kappa = 1/3$. Next, determine $F^c_{j+1/2}$ by interpolation of $\varphi_{j+1/2}$ to highest order between $\varphi_{j-1}, ..., \varphi_{j+2}$. Determine the order of the resulting local truncation error in the approximation of $d\varphi/dx$. Do this also for (4.91), and compare.

Exercise 4.8.2. The following flux limiter is mentioned in Roe (1986):

$$\psi(r) = \max\{0, \min(r, 1)\} \, .$$

Determine the corresponding characteristic $f(\tilde{\varphi})$. Is the scheme of positive type? What is its order of accuracy? (Note that $f'(1/2)$ does not exist). The minmod function is defined as,

$$\text{minmod}(x, y) = \begin{cases} x & \text{if } |x| < |y| \text{ and } xy > 0, \\ y & \text{if } |x| > |y| \text{ and } xy > 0, \\ 0 & \text{if } xy < 0. \end{cases}$$

Show that $\psi(r) = \text{minmod}(1, r)$. This is why this is called the minmod limiter.

Exercise 4.8.3. Verify that the PL limiter is inside the admissible region of Fig. 4.12.

5. The nonstationary convection-diffusion equation

5.1 Introduction

The equation to be studied in this chapter is the nonstationary convection-diffusion equation:

$$\frac{\partial \varphi}{\partial t} + u_\alpha \varphi_{,\alpha} - (D\varphi_{,\alpha})_{,\alpha} = q, \quad \boldsymbol{x} \in \Omega \subset \mathbb{R}^m, \quad 0 < t \leq T \qquad (5.1)$$

See Sect. 2.3 for initial and boundary conditions and further comments, and Theorem 2.4.3 for the maximum principle satisfied by (5.1). Discretization of the spatial derivatives has been discussed in the preceding chapters. In the present chapter, attention will be focussed on temporal discretization. As a consequence, numerical stability will loom large. As in the preceding chapter, the aim is not merely to discuss the numerical solution of the relatively simple equation (5.1), but to prepare the ground for the numerical solution of the equations of fluid dynamics. This will guide our selection of numerical methods to be discussed, from the plethora of schemes available. Further introduction to the subject may be found in Morton (1996), Strikwerda (1989).

Accuracy requirements

The choice of the temporal and spatial step sizes is governed by accuracy and stability considerations. In the ideal case accuracy and stability restrictions on the time step τ should be about the same. If τ is much more restricted by stability than by accuracy, efficiency may be gained by switching to a possibly more expensive but more stable time discretization method, and vice-versa. If $\mathcal{T}$ and $\mathcal{L}_\alpha$ are the time and length scales in the x_α-direction of the exact solution, then for accuracy it is required that the time step τ and the spatial mesh sizes h_α satisfy

$$\tau \lesssim \mathcal{T}/c_0, \quad h_\alpha \lesssim \mathcal{L}_\alpha/c_\alpha \quad \text{(no summation)} . \qquad (5.2)$$

How large $c_0, ..., c_m$ should be depends on the accuracy required and the order of the discretization. If the global truncation error is $\mathcal{O}(\tau^2)$ then c_0 can be smaller than when it is $\mathcal{O}(\tau)$. It seems safe to assume that in practice one often needs at least $c_0, ..., c_m \gtrsim 10$. Later, on various occasions estimates of $\mathcal{T}$ and $\mathcal{L}_\alpha$ will be presented.

5.2 Example of instability

Let $m = 1$, $u_1 = q = 0$, $D = 1$ and $\Omega = \mathbb{R}$. Let the spatial grid be uniform with mesh size h, and let τ be the time step, kept constant. Let φ_j^n be the numerical approximation to $\varphi(n\tau, jh)$, with $\varphi(t, x)$ the exact solution. Equation (5.1) is discretized as follows:

$$(\varphi_j^{n+1} - \varphi_j^n)/\tau - (\varphi_{j+1}^n - 2\varphi_j^n + \varphi_{j-1}^n)/h^2 = 0 , \tag{5.3}$$

which may be rewritten as

$$\varphi_j^{n+1} = a\varphi_{j+1}^n + (1 - 2a)\varphi_j^n + a\varphi_{j-1}^n, \quad a = \tau/h^2 . \tag{5.4}$$

Let the initial condition be given by

$$\varphi(0, x) = \varepsilon \cos \pi x/h , \tag{5.5}$$

with $0 < \varepsilon \ll 1$. The exact solution is

$$\varphi(t, x) = \varepsilon e^{-\pi^2 t/h^2} \cos \pi x/h .$$

The initial condition for (5.4) is

$$\varphi_j^0 = \varepsilon(-1)^j ,$$

and the solution of (5.4) is found to be

$$\varphi_j^n = \varepsilon(1 - 4a)^n(-1)^j .$$

Hence, if $a > 1/2$, $|\varphi_j^n| \to \infty$ as $n \to \infty$, quite in contrast with the exact solution. The error $e_j^n = \varphi(n\tau, jh) - \varphi_j^n$ is also unbounded for finite fixed time $T = n\tau$, since

$$e_j^{T/\tau} = \varepsilon\{e^{-\pi^2 T/h^2} - (1 - 4a)^{T/\tau}\}(-1)^j , \tag{5.6}$$

which is unbounded as $\tau \downarrow 0$, $h \downarrow 0$ with $a > 1/2$. One might think that with more smooth initial data scheme (5.3) may perhaps provide a good approximation of the exact solution with $a > 1/2$, but because in practice rounding errors always give perturbations of a rough sort, $a > 1/2$ invariably gives bad results.

For $0 < a \leq 1/2$ we have

$$(1 - 4a)^{T/\tau} = \exp\{\frac{T}{\Delta t}\ln(1 - 4a)\} = \exp\{\frac{T}{\tau}(-4a - 8a^2 - ...)\}$$
$$= \exp\{-\frac{4T}{h^2}(1 + 2a + ...)\}.$$

Substitution in (5.6) shows that the numerical solution has a wrong decay rate, so that the error $e^{T/\tau}$, though now bounded, does not vanish upon mesh refinement. This does not mean that scheme (5.3) is not good. The cause of non-convergence is the extreme roughness of the initial condition (5.5). To show this we now present a proof of convergence for the present simple case. Let the initial condition be general but independent of h. Substitution of the exact solution in (5.4) gives

$$\varphi\{(n + 1)\tau, jh\} = a\varphi\{\tau, (j + 1)h\} + (1 - 2a)\varphi(\tau, jh) \qquad (5.7)$$
$$+ a\varphi\{\tau, (j - 1)h\} + \tau_j^n ,$$

with the local truncation error τ_j^n (not to be confused with the time step τ) given by

$$\tau_j^n = \frac{1}{2}\tau^2\frac{\partial^2\varphi(t^*, x)}{\partial t^2} - \frac{1}{12}ah^4\frac{\partial^4(t, x^*)}{\partial x^4} , \qquad (5.8)$$

where $n\tau \le t^* \le (n + 1)\tau$ and $(j - 1)h \le x^* \le (j + 1)h$. Subtraction of (5.7) and (5.4) gives

$$e_j^{n+1} = ae_{j+1}^n + (1 - 2a)e_j^n + ae_{j-1}^n + \tau_j^n .$$

If $a \le 1/2$ we have

$$\|e^{n+1}\| \le \|e^n\| + \|\tau^n\| , \qquad (5.9)$$

with $\|\cdot\|$ the maximum norm. According to (5.8),

$$\|\tau^n\| \le \tau(C_1\tau + C_2h^2) ,$$

with

$$C_1 = \sup\{\frac{1}{2}|\partial^2\varphi/\partial t^2| : 0 \le t \le T, \quad x \in \mathbb{R}\}$$

$$C_2 = \sup\{\frac{1}{12}|\partial^4\varphi/\partial x^4| : 0 \le t \le T, \quad x \in \mathbb{R}\}$$

Repeated application of (5.9) gives, assuming $e_j^0 = 0$,

$$\|e^{T/\tau}\| \le T(C_1\tau + C_2h^2)$$

so that we have convergence: the error can be made arbitrary small by mesh refinement, provided $a \le 1/2$. We also see what goes wrong with initial condition (5.5): $C_2 = \mathcal{O}(1/h^4)$.

5.3 Stability definitions

For a more extensive treatment of stability theory than given here, see Richt-myer and Morton (1967).

Notation

Let the interval $(0, T)$ be divided in N equal subintervals of length τ. By φ_j^n we mean the numerical approximation to $\varphi(t, \boldsymbol{x})$ at $t = n\tau$ in the grid point $\boldsymbol{x}_j$, with $j = (j_1, ..., j_m)$ grid point indices in the spatial grid $G_h \subset \Omega$, as before. The subscript h indicates, that we will consider a family of grids depending on the parameter $h \in \mathbb{R}$. Usually, no confusion will arise from using the symbol φ both for the exact and numerical solution, but if necessary we will be more specific. Let $\boldsymbol{\varphi}$ stand for the algebraic vector with elements φ_j, ordered in some way. We will consider numerical schemes that can be written as

$$(B_1 \varphi^{n+1})_j = (B_0 \varphi^n)_j + (B_2 q^n)_j + (B_3 q^{n+1})_j , \tag{5.10}$$

with $B_0, B_1, ... B_3$ linear operators (that if we wish may be represented by ma-trices). Schemes that use more than two time levels will be considered later. Of course, B_1 is assumed to be nonsingular. Among other things, $B_0, \cdots, B_3$ depend on the spatial mesh sizes h_α and the time step τ. Let these be de-creasing functions of a parameter $h \geq 0$, equal to zero for $h = 0$. We rewrite equation (5.10) as

$$\varphi_j^{n+1} = (C_0(h)\varphi^n)_j + (C_2(h)\varphi^n)_j + (C_3(h)\varphi^{n+1})_j , \tag{5.11}$$

with $C_k = B_1^{-1} B_k$.

Local truncation error

As before (Sect. 3.7) the local truncation error is defined as the result ob-tained by applying the discrete scheme to the global truncation error (exact minus numerical solution). Since the numerical solution satisfies the scheme (disregarding rounding error) we obtain:

Definition 5.3.1. *Local truncation error*
The *local truncation error* of the discrete scheme (5.11) is

$$\tau_j^n \equiv \varphi((n+1)\tau, \boldsymbol{x}_j) - C_0(h)\varphi(n\tau, \boldsymbol{x}_j) - C_2(h)q_j^n - C_3(h)q_j^{n+1} ,$$

with φ the exact solution of (5.1).

Consistency and convergence

Let (5.11) be an approximation of (5.1) times $\tau^{\alpha}|\partial\Omega_j|^{\beta}$ with $|\partial\Omega_j|$ the volume of some cell around $\boldsymbol{x}_j$. Then consistency is defined similar to definition 3.7.2:

Definition 5.3.2. *Consistency*
The scheme (5.11) is called *consistent* if

$$\lim_{h\downarrow 0}\tau_j^n/(\tau^{\alpha}|\partial\Omega_j|^{\beta}) = 0, \quad j \in G_h, \quad 1 \le n \le T/\tau .$$

This implies that the maximum norm is used to measure $\boldsymbol{\tau}^n$, and the same difficulty crops up as in Sect. 3.7, namely that if a weaker norm is used an inconsistent scheme may get reclassified as consistent, and may still be convergent. However, we will study stability only on uniform grids, where this difficulty does not occur.

Similar to Definiton 3.7.3, we define

Definition 5.3.3. *Global truncation error*
The *global truncation error* is defined as

$$e_j^n = \varphi(n\tau, \boldsymbol{x}_j) - \varphi_j^n, \quad j \in G_h, \quad 1 \le n \le T/\tau .$$

Furthermore, we define

Definition 5.3.4. *Convergence*
A scheme is called *convergent* if the global truncation error satisfies

$$\lim_{h\downarrow 0}e_{j_h}^{T/\tau} = 0, \quad \boldsymbol{x}_{j_h} \text{ fixed.}$$

Here it is assumed that τ and the spatial mesh sizes belong to a decreasing sequence such that T/τ is an integer and $\boldsymbol{x}_{j_h}$ is a grid point in the spatial grid. Convergence means that the exact solution can be approximated arbitrarily closely at a fixed time and position by decreasing τ and refining the grid.

Stability definitions

As we saw in the preceding section, consistency does not imply convergence. In addition, stability is required. Let $\boldsymbol{\delta}^0$ be a perturbation of $\boldsymbol{\varphi}^0$. The resulting perturbation of $\boldsymbol{\varphi}^n$ is called $\boldsymbol{\delta}^n$. From (5.11) it follows that

$$\boldsymbol{\delta}^{n+1} = C_0(h)\boldsymbol{\delta}^n . \tag{5.12}$$

Let $\| \cdot \|_h$ be some norm for functions $G_h \to \mathbb{R}$. Stability means that $\boldsymbol{\delta}^n$ remains bounded as $n \to \infty$, for all $\boldsymbol{\delta}^0$. Two useful definitions are:

Definition 5.3.5. *Zero-stability*
A scheme is called *zero-stable* if there exists a function $C(T)$ and a function $\tau_0(h)$ such that

$$\|\boldsymbol{\delta}^{T/\tau}\|_h \leq C(T)\|\boldsymbol{\delta}^0\|_h \tag{5.13}$$

for all $\tau \leq \tau_0(h)$ and all $h \leq h_0$ for some fixed h_0.

The appellation "zero-stability" refers to the fact that the limit $h \downarrow 0$ is considered.

Definition 5.3.6. *Absolute stability*
A scheme is called *absolutely stable* if there exists a constant C and a function $\tau_0(h)$ such that

$$\|\boldsymbol{\delta}^n\|_h \leq C\|\boldsymbol{\delta}^0\|_h \tag{5.14}$$

for h fixed, all $n > 0$ and all $\tau \leq \tau_0(h)$.

The difference with zero-stability is that here h is fixed.

Lax's equivalence theorem

Definition 5.3.5 considers the perturbation at a fixed time T as $h \downarrow 0$, which is the same limit as in the definition of convergence. It can be shown that convergence implies zero-stability, and zero-stability plus consistency imply convergence. This is known as *Lax's equivalence theorem* (cf. Richtmyer and Morton (1967)). But absolute stability is also good to have, because it allows n to grow indefinitely, making it possible to continue time stepping until a steady state is reached. Absolute and zero-stability are not completely equivalent.

A remark on stability analysis

The purpose of stability analysis is to find a suitable function $\tau_0(h)$ such that (5.13) and (5.14) hold. This is the case if the linear operator in (5.12) satisfies

$$\|C_0^n(h)\|_h \leq C.$$

In the case of absolute stability, h and hence the dimensions of the matrix C_0 are fixed, and linear algebra can be used to find conditions under which $\|C_0^n(h)\|$ is bounded as $n \to \infty$. But in the case of zero-stability, $n \to \infty$ and $h \downarrow 0$ simultaneously, and we have to study not just the behavior of a linear operator, but of a family of linear operators. The dimensions of the matrix increase as $n \to \infty$. This is not a familiar situation in linear algebra. We will see that Fourier analysis is well-suited to the study of both kinds of stability, if the boundary conditions are periodic. For stability analysis for other kinds of boundary conditions, see Hindmarsh, Gresho, and Griffiths (1984) and Richtmyer and Morton (1967).

Choice of norms

The stability of a scheme depends on the norm. The l_p norm is defined as

$$\|\boldsymbol{\delta}\|_p = \{\frac{1}{N}\sum_{j\in G_h}|\delta_j|^p\}^{1/p}, \quad 1 \le p < \infty,$$

$$\|\boldsymbol{\delta}\|_\infty = \max\{|\delta_j| : j \in G_h\},$$

where N is the number of grid-points in G_h. In the hyperbolic case $(D = 0)$ there are schemes that are stable for $p = 2$ but instable for every other $p \in [1, \infty)$, cf. Richtmyer and Morton (1967). It can be shown that l^2-stability is always necessary (Geveci (1982)), otherwise the calculation blows up due to catastrophic growth of rounding errors. We will see that precisely for $p = 2$ stability is easily studied with Fourier analysis.

Exercise 5.3.1. Consider (5.1). Let $m = 1$, $u_1 = 0$, $D = $ constant. Choose the following time discretization:

$$\varphi^{n+1} - \varphi^n - \tau D(\omega\varphi_{,11}^{n+1} + (1 - \omega)\varphi_{,11}^n) = \omega\tau q^{n+1} + (1 - \omega)\tau q^n .$$

Let the spatial mesh size $h = $ constant. Show that space discretization with the finite volume method results in

$$h(\varphi_j^{n+1} - \varphi_j^n) + L_h(\omega\varphi_j^{n+1} + (1 - \omega)\varphi_j^n) = \omega h\tau q_j^{n+1} + (1 - \omega)h\tau q_j^n ,$$

where

$$L_h\varphi_j = (-\varphi_{j-1} + 2\varphi_j - \varphi_{j-1})\tau D/h .$$

Show that the local truncation error satisfies

$$\tau_j^n = \frac{1}{2}h\tau^2\{(1 - \omega)\frac{\partial^2\varphi}{\partial t^2}(t^n, x_j) - \omega\frac{\partial^2\varphi}{\partial t^2}(t^{n+1}, x_j)\}$$

$$+ \frac{1}{6}h\tau^3\frac{\partial^3\varphi}{\partial t^3}(t^*, x_j) - \frac{1}{24}\tau h^3 D\varphi_{,1111}(t^{**}, x^*)$$

with $t^n \le t^*, t^{**} \le t^{n+1}$ and $x_{j-1} \le x^* \le x_{j+1}$. Note that with $\omega = 1/2 + \mathcal{O}(\tau)$ we have improved temporal accuracy. Determine α and β according to Definition 5.3.2.

5.4 The discrete maximum principle

As we shall see, the discrete maximum principle is sufficient (but not necessary) for stability. Define

$$\tilde{L}_h(\varphi^{n+1}, \varphi^n) = B_1\varphi^{n+1} - B_0\varphi^n \,, \tag{5.15}$$

with B_0 and B_1 as in (5.10). Let I denote the set of grid point indices, i.e.

$$I = \{j : j = (j_1, ..., j_m), \quad j_\alpha = 0, 1, ..., n_\alpha - 1, \quad \alpha = 1, ..., m\} \,.$$

Let in grid point $j \in I$ equation (5.15) be equivalent with

$$\tilde{L}_h(\varphi_j^{n+1}, \varphi_j^n) = \sum_{k \in K}(\alpha_{jk}\varphi_{j+k}^{n+1} - \beta_{jk}\varphi_{j+k}^n) \,, \tag{5.16}$$

with $k = (k_1, ..., k_m)$ and K some set that follows from the stencil of the discretization. Let $I_D \subset I$ be the set of boundary points where a Dirichlet condition is applied. In the cell-centered case, where there are no points on the boundary, we extend I with the (fractional) Dirichlet boundary point indices. Quite similar to Definition 4.4.1, we define

Definition 5.4.1. The operator $\tilde{L}_h$ is a *of positive type* if

$$\begin{aligned} \sum_{k \in K}(\alpha_{jk} - \beta_{jk}) = 0, \quad j \in I\backslash I_D \,, \\ \alpha_{jk} < 0, \quad k \neq 0 \quad \text{and} \quad \beta_{jk} > 0, \quad k \in K, \quad j \in I\backslash I_D \,. \end{aligned} \tag{5.17}$$

Note that it is the coefficient of φ_j^{n+1} that has to be positive.

Theorem 5.4.1. *Discrete maximum principle*
If $\tilde{L}_h$ is of positive type and

$$\tilde{L}_h(\varphi_j^{n+1}, \varphi_j^n) \leq 0, \quad j \in I\backslash I_D, \quad n = 1, 2, ..., \tag{5.18}$$

then local maxima can occur only for $j \in I_D$ and $n = 0$, or $\varphi^n = \varphi^{n-1} = ... = \varphi^0 = constant$.

This is quite analogous to the maximum principle for the differential equation (Theorem 2.4.3). The proof is similar to that of Theorem 4.4.1.

For the perturbation δ^n in the stability definitions of the preceding section we have

$$\tilde{L}_h(\delta_j^{n+1}, \delta_j^n) = 0, \quad j \in I \,. \tag{5.19}$$

Note that $\delta_j^n = 0$, $j \in I_D$. If $\tilde{L}_h$ is of positive type we have both a maximum and a minimum principle, because the right-hand-side of (5.19) is zero, and by changing the sign, Theorem 5.4.1 gives a minimum principle. Hence

$$|\delta_j^n| \leq \max\{|\delta_j^0| : j \in I\} \,,$$

so that

$$\|\boldsymbol{\delta}^n\|_\infty \leq \|\boldsymbol{\delta}^0\|_\infty \,.$$

The following theorem follows:

Theorem 5.4.2. *If $\tilde{L}_h$ is of positive type, then the scheme (5.10) is both zero-stable and absolutely stable in the maximum norm.*

It will be seen later, that the positive type property is not necessary for stability. Multistep schemes (using more than two time levels for discretization in time) never have the positive type property, if their temporal accuracy is $\mathcal{O}(\tau^2)$ or better.

5.5 Fourier stability analysis

In general it is difficult to derive estimates like (5.13) and (5.14). But if the coefficients in the scheme are constant, the mesh uniform, the grid a rectangular block and the boundary conditions periodic, then Fourier analysis applies and the required estimates are not difficult to obtain.

In practice, of course, the coefficients are usually not constant. The scheme is called *locally stable* if we have stability for the constant coefficients scheme that results from taking local values of the coefficients, and to assign these values to the coefficients in the whole domain (frozen coefficients method). In Richtmyer and Morton (1967), Chapt. 5 stability theory for schemes with variable coefficients is discussed. Local stability in the whole domain is necessary for stability in the variable coefficient case. Sufficient conditions are discussed in Richtmyer and Morton (1967). Quite often, it is sufficient that the scheme is *dissipative*, see Definition 9.3.1. We will discuss stability only for constant coefficients.

Stability theory for non-periodic boundary conditions is complicated. But for explicit time stepping schemes it takes a while before the influence of the boundary conditions makes itself felt in the interior, so that Fourier stability theory applies during a certain initial time span. As a consequence, stability with periodic boundary conditions is desirable, even if the boundary conditions are of different type. For stability analysis in the presence of non-periodic boundary conditions, see Richtmyer and Morton (1967).

In the remainder of this section we present the basic principles of Fourier stability analysis. In the next section, application is made to the convection-diffusion equation.

Discrete Fourier transform

Let I denote the set of grid point indices, i.e.

$$I = \{j : j = (j_1, ..., j_m), \quad j_\alpha = 0, 1, ..., n_\alpha - 1, \quad \alpha = 1, ..., m\} \, .$$

Furthermore, define the set Θ of wavenumbers as

$$\begin{aligned} \Theta = \{\theta : \theta = (\theta_1, ..., \theta_\alpha), \quad \theta_\alpha = 2\pi k_\alpha/n_\alpha \, , \\ k_\alpha = -m_\alpha, -m_\alpha + 1, ..., m_\alpha + p_\alpha, \quad \alpha = 1, ..., m\} \, , \end{aligned}$$

where $p_\alpha = 0$, $m_\alpha = (n_\alpha - 1)/2$ for n_α odd and $p_\alpha = 1$, $m_\alpha = n_\alpha/2 - 1$ for n_α even. Define

$$j\theta = \sum_{\alpha=1}^{m} j_\alpha \theta_\alpha \, .$$

We have

Theorem 5.5.1. *Discrete Fourier transform in m dimensions.*
Every grid function $\varphi : I \to \mathbb{R}$ can be written as

$$\varphi_j = \sum_{\theta \in \Theta} c_\theta e^{ij\theta} \, , \tag{5.20}$$

with

$$c_\theta = N^{-1} \sum_{j \in I} \varphi_j e^{-ij\theta}, \quad N = \prod_{\alpha=1}^{m} n_\alpha \, .$$

The proof is elementary, and can be found, for example, in Wesseling (1992).

A consequence of the preceding theorem is

$$\sum_{\theta \in \Theta} c_\theta e^{ij\theta} = 0, \quad \forall j \in I \quad \Rightarrow \quad c_\theta = 0, \quad \forall \theta \in \Theta \, . \tag{5.21}$$

The l_2-norm is defined by

$$\|\varphi\| = N^{-\frac{1}{2}} \{\sum_{j \in I} (\varphi_j)^2\}^{1/2}, \quad \|c\| = N^{-\frac{1}{2}} \{\sum_{\theta \in \Theta} |c_\theta|^2\}^{1/2}$$

We will need

Theorem 5.5.2. *Parseval.*
If φ and c are related by (5.20), then in the l_2-norm

$$\|\varphi\| = N^{1/2} \|c\| \, .$$

The periodic constant coefficients initial value problem

If the grid is uniform, the coefficients in (5.1) are constant and the boundary conditions are periodic, then a typical equation of the system (5.12) can be written as

$$\sum_{k \in K} \alpha_k \delta_{j+k}^{n+1} = \sum_{k \in K} \beta_k \delta_{j+k}^n, \quad j \in I . \tag{5.22}$$

Because of the periodic boundary conditions, δ^n and equation (5.22) can be extended across the boundary of I. Write

$$\delta_j^n = \sum_{\theta \in \Theta} c_\theta^n e^{ij\theta} .$$

Substitution in (5.22) gives

$$\sum_{\theta \in \Theta} e^{ij\theta} \left[c_\theta^{n+1} \sum_{k \in K} \alpha_k e^{ik\theta} - c_\theta^n \sum_{k \in K} \beta_k e^{ik\theta} \right] = 0 .$$

Because of (5.21) the term between brackets is zero, so that

$$c_\theta^{n+1} = g(\theta) c_\theta^n ,$$

with the *amplification factor* $g(\theta)$ defined as

$$g(\theta) = (\sum_{k \in K} \beta_k e^{ik\theta}) / (\sum_{k \in K} \alpha_k e^{ik\theta}) . \tag{5.23}$$

The von Neumann condition

Since $c_\theta^n = g^n(\theta) c_\theta^0$,

$$\|c^n\| \le \bar{g}^n \|c^0\|, \quad \bar{g} = \max\{|g(\theta)| : \forall \theta\} ,$$

with the l_2-norm. Using Theorem 5.5.2, we see that a sufficient condition for zero-stability is that there exists a constant C such that

$$\bar{g}^{(T/\tau)} \le C \tag{5.24}$$

for $0 \le \tau \le \tau_0(h)$. Since

$$C^{\tau/T} = \exp(\frac{\tau}{T} \ln C) = 1 + \mathcal{O}(\tau) , \tag{5.25}$$

we may write

$$\bar{g} \le 1 + \mathcal{O}(\tau) . \tag{5.26}$$

This is the *von Neumann condition* for zero-stability. This condition is also necessary, because if $\bar{g} \geq 1 + \mu$, $\mu > 0$, then there is a θ with $|g(\theta)| = 1 + \mu$. Choosing $\delta_j^0 = e^{ij\theta}$ gives $\|\delta^n\|/\|\delta\| = (1+\mu)^n$, $n = T/\tau$, which is unbounded as $\tau \downarrow 0$. Note that the $\mathcal{O}(\tau)$ term (5.26) is not allowed to depend on h; this follows from (5.24)–(5.26).

For absolute stability,

$$\bar{g}^n \leq C, \quad \forall n,$$

so that

$$\bar{g} \leq 1$$

is sufficient, and, by the same argument as before, also necessary.

We will study a number of particular cases that are representative of frequently used schemes in computational fuid dynamics.

5.6 Principles of von Neumann stability analysis

Introduction

The derivation of conditions that are sufficient or necessary for the von Neumann condition to hold for a particular numerical discretization scheme requires detailed and often tedious analysis of each particular scheme. It turns out that for many schemes used in practice only partial (for example, restricted to one spatial dimension) stability analysis results, or none at all, are available. Therefore we present below an approach that gives general results for the full convection-diffusion equation in an arbitrary number of space dimensions. The reader who wants to derive a stability condition will benefit from mastering this section. The outline is as follows. First, the symbol or Fourier transform of the spatial discretization scheme is introduced. Stability of an associated time stepping scheme is shown to be equivalent to a certain geometric condition on the symbol, namely that it must lie in a certain region of the complex plane, called the stability domain of the time stepping scheme. In Sect. 5.7 we derive a number of universal properties of the symbol. Complexity due to multi dimensions is mastered by using the Schwartz inequality. These universal properties make it easy to obtain in Sect. 5.8 stability conditions for particular cases that are important in computational fluid dynamics.

Spatial discretization

The convection-diffusion equation in m dimensions is given by

$$\frac{\partial \varphi}{\partial t} + L\varphi = 0, \quad L\varphi = \sum_{\alpha=1}^{m} (u_\alpha \frac{\partial}{\partial x_\alpha} - D \frac{\partial^2}{\partial x_\alpha^2})\varphi \ . \tag{5.27}$$

For Fourier stability analysis, u_α and D are taken constant, the domain is the unit cube and the boundary conditions are periodic. With the κ-scheme, the spatial discretization of L on a uniform grid with mesh sizes $h_1, ..., h_m$ is given by $\tau L_h = C_h + D_h$, with

$$\begin{aligned} C_h &= \frac{1}{4} \sum_\alpha c_\alpha \{(1 - \kappa)\varphi_{j-2e_\alpha} - (5 - 3\kappa)\varphi_{j-e_\alpha} \\ &\quad + (3 - 3\kappa)\varphi_j + (1 + \kappa)\varphi_{j+e_\alpha}\} \ , \end{aligned} \tag{5.28}$$

and

$$D_h = \frac{1}{2} \sum_\alpha d_\alpha(-\varphi_{j-e_\alpha} + 2\varphi_j - \varphi_{j+e_\alpha}) \ , \tag{5.29}$$

where $j = (j_1, ..., j_m)$, $e_1 = (1, 0, ..., 0)$, $e_2 = (0, 1, 0, ..., 0)$ etc., and

$$c_\alpha = |u_\alpha|\tau/h_\alpha, \quad d_\alpha = 2D\tau/h_\alpha^2 \ , \tag{5.30}$$

with τ the time step to be used later in the discretization of $\partial\varphi/\partial t$. In (5.28), $u_\alpha \geq 0$ has been assumed; the contrary case may be treated by symmetry. The dimensionless numbers c_α are called the *CFL* (Courant-Friedrichs-Lewy) *numbers*, after the authors of a pioneering paper published in 1928, and reprinted in English in Courant, Friedrichs, and Lewy (1967). The dimensionless numbers d_α will be called *diffusion numbers*. The first order upwind and fourth order central schemes are given by, respectively,

$$C_h = \frac{1}{\tau} \sum_\alpha c_\alpha(\varphi_j - \varphi_{j-e_\alpha})$$

and

$$C_h = \frac{1}{12\tau} \sum_\alpha c_\alpha(\varphi_{j-2e_\alpha} - 8\varphi_{j-e_\alpha} + 8\varphi_{j+e_\alpha} - \varphi_{j+2e_\alpha}) \ , \tag{5.31}$$

while D_h is given by (5.29).

The symbol

After spatial discretization we are left with the following system of ordinary differential equations:

$$d\varphi_j/dt = -L_h\varphi_j \ . \tag{5.32}$$

The *symbol* or Fourier transform $\hat{L}_h(\theta)$ of L_h is defined by

$$\hat{L}_h(\theta) = e^{-ij\theta} L_h e^{ij\theta} . \tag{5.33}$$

One finds

$$\begin{aligned}
\tau \hat{L}_h(\theta) &= \hat{C}_h(\theta) + \hat{D}_h(\theta) , \\
\hat{C}_h(\theta) &= \gamma_1(\theta) + i\gamma_2(\theta), \quad \hat{D}_h(\theta) = \delta(\theta) .
\end{aligned} \tag{5.34}$$

For (5.28) we have, also if $u_\alpha < 0$,

$$\gamma_1(\theta) = 2(1 - \kappa) \sum_\alpha c_\alpha s_\alpha^2, \quad \gamma_2(\theta) = \sum_\alpha c_\alpha \{(1 - \kappa)s_\alpha + 1\} \sin \theta_\alpha , \tag{5.35}$$

where $s_\alpha = \sin^2 \frac{1}{2}\theta_\alpha$, whereas for the fourth order central (or FOC for brevity) scheme (5.31)

$$\gamma_1(\theta) = 0, \quad \gamma_2(\theta) = \frac{1}{6} \sum_\alpha c_\alpha (8 \sin \theta_\alpha - \sin 2\theta_\alpha) = \sum_\alpha c_\alpha (1 + \frac{2}{3} s_\alpha) \sin \theta_\alpha .$$

Defining

$$\tilde{c}_\alpha = (1 - \kappa) c_\alpha \quad (\kappa\text{-scheme}), \quad \tilde{c}_\alpha = 0 \quad (\text{FOC scheme})$$

we see that for both schemes

$$\gamma_1(\theta) = 2 \sum_\alpha \tilde{c}_\alpha s_\alpha^2, \quad \gamma_2(\theta) = \sum_\alpha c_\alpha \{(1 - \kappa)s_\alpha + 1\} \sin \theta_\alpha , \tag{5.36}$$

provided we define $\kappa = 1/3$ for the FOC scheme. For all schemes,

$$\delta(\theta) = 2 \sum_\alpha d_\alpha s_\alpha . \tag{5.37}$$

The symbol of the first order upwind scheme is obtained by taking $\kappa = 1$ and replacing d_α by $d_\alpha + c_\alpha$. Therefore this scheme does not need to be considered explicitly in the sequel.

Stability

Following the definition of stability in Sect. 5.3, a perturbation $\delta_j(0)$ is added to the initial condition of (5.32). Because of linearity, the equation for the perturbed solution is the same as for the unperturbed solution, and it is convenient not to change notation, but to denote the perturbation by $\varphi_j(t)$. According to Sect. 5.5 we may write

$$\varphi_j(t) = \sum_{\theta \in \Theta} \hat{\varphi}_\theta(t) e^{ij\theta} .$$

Substitution in (5.32) gives

$$\sum_{\theta \in \Theta} \{ d\hat{\varphi}_\theta / dt + \hat{L}_h(\theta) \hat{\varphi}_\theta \} e^{ij\theta} = 0 .$$

From (5.21) it follows that

$$d\hat{\varphi}_\theta / dt = -\hat{L}_h(\theta) \hat{\varphi}_\theta . \tag{5.38}$$

According to Parseval's theorem 5.5.2, $\|\boldsymbol{\varphi}(t)\| = N^{1/2} \|\boldsymbol{\varphi}_\theta(t)\|$. Hence (5.13) implies

$$\|\hat{\varphi}(T)\| \leq C(T) \|\hat{\varphi}(0)\| .$$

For this it is sufficient that

$$|\hat{\varphi}_\theta(T)| \leq C(T) |\hat{\varphi}_\theta(0)|, \quad \forall \theta \in \Theta . \tag{5.39}$$

This also necessary, because the initial perturbation $\varphi_j(0)$ is arbitrary, and one may take $\varphi_j(\theta) = \hat{\varphi}_\theta e^{ij\theta}$, giving equality in (5.39). Since the exact solution of (5.27) does not grow in time (cf. Exercise (5.6.1)) we take $C(T) = 1$ (allowing $C(T) > 1$ relaxes stability restrictions on τ in a negligible way). Hence we require that after discretizing (5.32) in time the resulting solution satisfies

$$|\hat{\varphi}_\theta^n| \leq |\hat{\varphi}_\theta^0|, \quad \forall \theta \in \Theta .$$

The *stability domain* of numerical methods for the differential equation $dy/dt = \lambda y, \quad \lambda \in \mathbb{C}$ (with $\mathbb{C}$ the complex plane) is defined as

$$S = \{ \lambda \tau \in \mathbb{C} : |y^{n+1}| \leq |y^n| \} . \tag{5.40}$$

Comparison with (5.38) shows that for von Neumann stability it is necessary and sufficient that $-\tau \hat{L}_h(\theta) \in S, \quad \forall \theta \in \Theta$. In order to facilitate the analysis we sharpen this condition just a little by requiring it to hold for all θ. This leads to the stability condition that we will use:

$$S_L \subseteq S, \quad S_L \equiv \{ -\tau \hat{L}_h(\theta) : \quad \forall \theta \} . \tag{5.41}$$

This completes our discussion of the basics of von Neumann stability analysis. In order to deduce practical stability conditions in two and three dimensions from the basic condition (5.41), we will have to go into some technical details in the next two sections.

Exercise 5.6.1. Determine the symbol of the operator L in equation (5.27). Show that solutions of (5.27) on an unbounded domain do not grow in time.

5.7 Useful properties of the symbol

Preliminaries

Since S_L does not depend on the time discretization to be used, it is useful to study some of its properties before investigating stability of time discretizations by means of (5.41).

Let $\tilde{d}_\alpha \equiv d_\alpha + \tilde{c}_\alpha$ and $\tilde{d} \equiv \sum_\alpha \tilde{d}_\alpha$. Then for both the κ- and the FOC scheme we may write, since $0 \leq s_\alpha \leq 1$,

$$(\delta + \gamma_1)^2 = 4\{\sum_\alpha d_\alpha s_\alpha + \tilde{c}_\alpha s_\alpha^2\}^2 \leq 4\{\sum \tilde{d}_\alpha s_\alpha\}^2 .$$

Noting that $\tilde{d}_\alpha \geq 0$, writing $\tilde{d}_\alpha s_\alpha = \sqrt{\tilde{d}_\alpha}(s_\alpha\sqrt{\tilde{d}_\alpha})$ and using the *Schwartz inequality* (which we will apply frequently) : $|\boldsymbol{a} \cdot \boldsymbol{b}| \leq \|\boldsymbol{a}\|\|\boldsymbol{b}\|$ for $\boldsymbol{a}, \boldsymbol{b} \in \mathbb{R}^m$, we obtain:

$$(\delta + \gamma_1)^2 \leq 4\tilde{d}\sum_\alpha \tilde{d}_\alpha s_\alpha^2 . \tag{5.42}$$

Furthermore, for both schemes, writing $\bar{\kappa} \equiv (1 - \kappa)$, we have

$$\gamma_2^2 = [\sum_\alpha (c_\alpha/\sqrt{d_\alpha})\{\sqrt{d_\alpha}(\bar{\kappa}s_\alpha + 1)\sin\theta_\alpha\}]^2 .$$

Applying Schwarz we obtain

$$\gamma_2^2 \leq 4\sum_\alpha c_\alpha^2/d_\alpha \sum_\alpha d_\alpha s_\alpha(1 - s_\alpha)(\bar{\kappa}s_\alpha + 1)^2 . \tag{5.43}$$

Noting that $|s_\alpha| \leq 1$ it follows that

$$\gamma_2^2 \leq 4(2 - \kappa)^2 \sum_\alpha c_\alpha^2/d_\alpha \sum_\alpha d_\alpha s_\alpha(1 - s_\alpha) . \tag{5.44}$$

Furthermore,

$$\gamma_2^4 = [\sum_\alpha (c_\alpha^{2/3}d_\alpha^{-1/6})\{c_\alpha^{1/3}d_\alpha^{1/6}(\bar{\kappa}s_\alpha + 1)\sin\theta_\alpha\}]^4 .$$

Applying Schwarz gives

$$\gamma_2^4 \leq \{\sum_\alpha (c^4/d_\alpha)^{1/3}\}^2[\sum_\alpha (c_\alpha^2/d_\alpha^{1/2})^{1/3}\{d_\alpha^{1/2}(\bar{\kappa}s_\alpha + 1)^2\sin^2\theta_\alpha\}]^2 .$$

Another application of Schwartz results in

$$\gamma_2^4 \leq 16\{\sum_\alpha (c_\alpha^4/d_\alpha)^{1/3}\}^3 \sum_\alpha d_\alpha(\bar{\kappa}s_\alpha + 1)^4 s_\alpha^2(1 - s_\alpha)^2 . \tag{5.45}$$

Furthermore, for arbitrary $a > 0$ we have for both schemes, using (5.42):

$$\left(\frac{\delta + \gamma_1}{a} - 1\right)^2 = 1 + \frac{4\tilde{d}}{a^2}\sum_\alpha \tilde{d}_\alpha s_\alpha^2 - \frac{4}{a}\sum_\alpha (d_\alpha s_\alpha + \tilde{c}_\alpha s_\alpha^2) \ .$$

If $\tilde{d} \le a$ this gives

$$\left(\frac{\delta + \gamma_1}{a} - 1\right)^2 \le 1 + \frac{4}{a}\sum_\alpha d_\alpha(s_\alpha^2 - s_\alpha) \ . \tag{5.46}$$

Theorems on S_L

Next, we present some theorems on the location of S_L in the complex plane $\mathbb{C}$. The usefulness of these theorems will become apparent later.

Theorem 5.7.1. *(Rectangle)*
If

$$\tilde{d} \le a/2 \quad and \quad \sum_\alpha c_\alpha \le b/(2-\kappa) \tag{5.47}$$

then S_L is contained in the rectangle defined by

$$-a \le v \le 0, \quad |w| \le b, \quad v + iw = z \ .$$

The first condition is necessary.

Proof. We have to show that $-\tau \hat{L}_h(\theta) \in$ rectangle, $\forall \theta$, or

$$0 \le \delta + \gamma_1 \le a \quad and \quad |\gamma_2| \le b, \quad \forall \theta \ . \tag{5.48}$$

We have

$$\delta + \gamma_1 = 2\sum_\alpha (d_\alpha s_\alpha + \tilde{c}_\alpha s_\alpha^2) \ . \tag{5.49}$$

Hence $0 \le \delta + \gamma_1 \le 2\tilde{d}$ and the first part of (5.48) follows. Necessity of the first condition of (5.47) follows by taking $s_\alpha = 1$. Furthermore,

$$|\gamma_2| \le \sum_\alpha c_\alpha\{(1-\kappa)s_\alpha + 1\}|\sin\theta_\alpha| \le (2-\kappa)\sum_\alpha c_\alpha \le b \ .$$

$\square$

The conditions (5.47) are equivalent to the following time step restriction for the κ-scheme:

$$\tau \le \min\left\{\frac{a}{2D}/\sum_\alpha h_\alpha^{-2}(2 + (1-\kappa)p_\alpha), \ \frac{b}{2-\kappa}/\sum_\alpha (|u_\alpha|/h_\alpha)\right\} \ , \tag{5.50}$$

where the mesh Péclet number p_α is defined as

$$p_\alpha = |u_\alpha| h_\alpha / D. \tag{5.51}$$

For the FOC scheme we obtain

$$\tau \le \min\Big\{ \frac{a}{4D} / \sum_\alpha h_\alpha^{-2}, \ \frac{3}{5} b / \sum_\alpha (|u_\alpha|/h) \Big\}, \tag{5.52}$$

whereas the condition for the first order upwind scheme is

$$\tau \le \min\Big\{ \frac{a}{2D} / \sum_\alpha h_\alpha^{-2}(2 + p_\alpha), \ b / \sum_\alpha (|u_\alpha|/h_\alpha) \Big\}. \tag{5.53}$$

Theorem 5.7.2. *(Ellipse)*
If

$$\tilde{d} \le a \quad and \quad \sum_\alpha c_\alpha^2 / d_\alpha \le (2 - \kappa)^{-2} b^2 / a \tag{5.54}$$

then S_L is contained in the ellipse

$$(v/a + 1)^2 + (w/b)^2 = 1, \quad v + iw = z, \quad a > 0.$$

The first condition is necessary.

Proof. We have to show that $\left(\frac{\delta + \gamma_1}{a} - 1\right)^2 + \left(\frac{\gamma_2}{b}\right)^2 \le 1$. Necessity of the first condition follows by using (5.49) and taking $s_\alpha = 1$. Continuing with sufficiency, using (5.44), (5.46) and (5.54) we obtain

$$\left(\frac{\delta + \gamma_1}{a} - 1\right)^2 + \left(\frac{\gamma_2}{b}\right)^2 \le 1 + \frac{4}{a} \sum_\alpha d_\alpha (s_\alpha^2 - s_\alpha + s_\alpha - s_\alpha^2) \le 1.$$

$\square$

For $\kappa = 1$, necessity of *both* conditions is shown in Wesseling (1995). The conditions (5.54) are equivalent to the following time step restrictions: for the κ-scheme:

$$\tau \le \min\Big\{ \frac{a}{D} / \sum_\alpha h_\alpha^{-2}(2 + (1 - \kappa)p_\alpha), \ 2D(2 - \kappa)^{-2} \frac{b^2}{a} / \sum_\alpha u_\alpha^2 \Big\}. \tag{5.55}$$

For the FOC scheme:

$$\tau \le \min\Big\{ \frac{a}{2D} / \sum_\alpha h_\alpha^{-2}, \ \frac{18 b^2}{25 a} D / \sum_\alpha u_\alpha^2 \Big\}. \tag{5.56}$$

For the first order upwind scheme:

$$\tau \le \min\Big\{ \frac{a}{D} / \sum_\alpha h_\alpha^{-2}(2 + p_\alpha), \ 2D \frac{b^2}{a} / \sum_\alpha u_\alpha^2 \Big\}. \tag{5.57}$$

Theorem 5.7.3. *(Oval)*

If

$$\tilde{d} \leq a \tag{5.58}$$

and one or both of the two following conditions hold:

$$\sum_\alpha (c_\alpha^4/d_\alpha)^{1/3} \leq q_1 (b^4/a)^{1/3}, \quad \sum_\alpha c_\alpha \leq q_2 b^2/a, \tag{5.59}$$

where

$$q_1 = \frac{1}{4}(8 - 4\kappa)^{-5/3}(15 - 5\kappa - r)^{4/3}(5\kappa - 3 + r)^{1/3}(9 - 7\kappa + r)^{1/3},$$

$$r = (25\kappa^2 - 54\kappa + 33)^{1/2},$$

$$q_2 = (1 - \kappa)^{3/2}(8/5 - 4\kappa/5)^{-5/2}, \quad -1 \leq \kappa < 1/2,$$

$$q_2 = \frac{1}{2}(1 - \kappa), \quad 1/2 \leq \kappa \leq 1,$$

then S_L is contained in the oval given by

$$(v/a + 1)^2 + (w/b)^4 = 1, \quad v + iw = z, \quad a > 0.$$

Condition (5.58) is necessary.

Proof. Necessity of (5.58) follows by taking $s_\alpha = 1$. Continuing with sufficiency, assume that (5.58) and the first condition of (5.59) hold. Using (5.45) and (5.46) we obtain

$$\left(\frac{\delta + \gamma_1}{a} - 1\right)^2 + \left(\frac{\gamma_2}{b}\right)^4 \leq$$
$$1 + \frac{4}{a}\sum_\alpha d_\alpha (s_\alpha - s_\alpha^2)\{-1 + 4q_1^3(\bar{\kappa}s_\alpha + 1)^4 s_\alpha(1 - s_\alpha)\}. \tag{5.60}$$

Since

$$\max\{(\bar{\kappa}s + 1)^4 s(1 - s) : 0 \leq s \leq 1\} = 1/4q_1^3,$$

no term in the sum in equation (5.60) is positive, so that $S_L \subseteq$ oval. Next, assume that (5.58) and the second condition of (5.59) hold. Because of (5.58), $0 \leq \delta + \gamma_1 \leq 2a$, hence, with $\tilde{c} = \sum_\alpha \tilde{c}_\alpha$ and using Schwartz,

$$\left(\frac{\delta + \gamma_1}{a} - 1\right)^2 = \frac{\delta}{a}\left(\frac{\delta + \gamma_1}{a} + \frac{\gamma_1}{a} - 2\right) + \left(\frac{\gamma_1}{a} - 1\right)^2$$

$$\leq \left(\frac{\delta}{a} - 2\right)\frac{\gamma_1}{a} + \left(\frac{\gamma_1}{a}\right)^2 + 1$$

$$\leq 2\left(\frac{\tilde{d} - \tilde{c}}{a} - 1\right)\frac{2}{a}\sum_\alpha \tilde{c}_\alpha s_\alpha^2 + \frac{4}{a^2}\tilde{c}\sum_\alpha \tilde{c}_\alpha s_\alpha^4 + 1$$

$$\leq 1 - \frac{4}{a^2}\tilde{c}\sum_\alpha \tilde{c}_\alpha (s_\alpha^2 - s_\alpha^4).$$

Furthermore, similar to (5.45),

$$\gamma_2^4 \leq 16\{\sum_\alpha (c_\alpha^4/\tilde{c}_\alpha)^{1/3}\}^3 \sum_\alpha \tilde{c}_\alpha (\bar{\kappa}s_\alpha + 1)^4 s_\alpha^2 (1 - s_\alpha)^2$$

$$= \frac{16}{\bar{\kappa}}\{\sum_\alpha c_\alpha\}^3 \sum_\alpha \tilde{c}_\alpha (\bar{\kappa}s_\alpha + 1)^4 s_\alpha^2 (1 - s_\alpha)^2 .$$

Hence, using the second condition of (5.59),

$$\left(\frac{\delta + \gamma_1}{a} - 1\right)^2 + \left(\frac{\gamma_2}{b}\right)^4 \leq$$

$$1 + \frac{4}{a^2}\tilde{c}\sum_\alpha [\tilde{c}_\alpha s_\alpha^2 (1 - s_\alpha)\{\frac{4q_2^2}{\bar{\kappa}^2}(\bar{\kappa}s_\alpha + 1)^4 (1 - s_\alpha) - 1 - s_\alpha\}] .$$

Observing that

$$\max\{(\bar{\kappa}s + 1)^4 (1 - s)/(1 + s) : 0 \leq s \leq 1, \ 0 \leq \bar{\kappa} \leq 1/2\} = 1$$

and that

$$\max\{(\bar{\kappa}s + 1)^4 (1 - s) : \ 0 \leq s \leq 1, \ 1/2 \leq \bar{\kappa} \leq 2\} = \bar{\kappa}^2/4q_2^2$$

we see that each term in the preceding sum is non-positive. $\qquad\square$

Note that the first condition of (5.59) is not useful in the hyperbolic case ($D = 0$, i.e. $d_\alpha = 0$), whereas the second condition is not useful for $\kappa = 1$; the two conditions complement each other.

The time step restrictions equivalent to (5.58) and (5.59) are: for the κ-scheme:

$$\tau \leq \frac{a}{D}/\sum_\alpha h_\alpha^{-2}(2 + (1 - \kappa)p_\alpha) \quad \text{and}$$

$$\tau \leq \max\{q_1 \left(\frac{2Db^4}{a}\right)^{1/3} \left(\sum_\alpha u_\alpha^4/h_\alpha^2\right)^{-1/3} , \ q_2\frac{b^2}{a}/\sum_\alpha (|u_\alpha|/h_\alpha)\} . \tag{5.61}$$

For the FOC scheme:

$$\tau \leq \frac{a}{2D}/\sum_\alpha h_\alpha^{-2} \quad \text{and}$$

$$\tau \leq \max\{\tilde{q}_1 \left(\frac{2Db^4}{a}\right)^{1/3} \left(\sum_\alpha u_\alpha^4/h_\alpha^2\right)^{-1/3} , \ \frac{3b^2}{16a^2}/\sum_\alpha (|u_\alpha|/h_\alpha)\} , \tag{5.62}$$

where $\tilde{q}_1 = \frac{1}{5}\left(\frac{5+\sqrt{10}}{12}\right)(\sqrt{10} - 1)^{5/3} \cong 0.6360$.
For the first order upwind scheme:

$$\tau \leq \max\{\frac{a}{D}/\sum_\alpha h_\alpha^{-2}(2 + p_\alpha), \ \left(\frac{2Db^4}{a}\right)^{1/3} (\sum_\alpha u_\alpha^4/h_\alpha)^{-1/3}\} . \tag{5.63}$$

Theorem 5.7.4. *(Half ellipse)*
If

$$\tilde{d} \le \frac{a}{2} \quad and \quad \frac{2\tilde{d}}{b^2} \sum_\alpha c_\alpha^2/\tilde{d}_\alpha \le (2-\kappa)^{-2}(1 + \sqrt{1 - 4\tilde{d}^2/a^2}) , \qquad (5.64)$$

then S_L is contained in the left half of the ellipse given by

$$(v/a)^2 + (w/b)^2 = 1, \quad v + iw = z .$$

The first condition is necessary.

Proof. That $-\tau \hat{L}_h(\theta)$ is not in the right half of the ellipse follows from $\delta + \gamma_1 \ge 0$. Necessity of the first condition follows as before. Continuing with sufficiency, using (5.42), (5.44) (which also holds with d_α replaced by $\tilde{d}_\alpha$) and (5.64) we have

$$\left(\frac{\delta + \gamma_1}{a}\right)^2 + \left(\frac{\gamma_2}{b}\right)^2 \le 4\frac{\tilde{d}}{a^2} \sum_\alpha \tilde{d}_\alpha s_\alpha \{s_\alpha + p(1 - s_\alpha)\} ,$$

where $p = (a^2/2\tilde{d}^2)(1 + \sqrt{1 - 4\tilde{d}^2/a^2})$. Since $p > 2$, $\max\{s(s + p(1 - s)) : 0 \le s \le 1\} = p^2/\{4(p-1)\} = a^2/(4\tilde{d}^2)$, so that $(\delta+\gamma_1)^2/a^2 + (\gamma_2/b)^2 \le 1$. $\square$

The conditions (5.64) are equivalent to the following time step restrictions: for the κ-scheme:

$$\tau \le \min\left[\frac{a}{2D}/\sum_\alpha (h_\alpha^{-2}(2 + (1 - \kappa)p_\alpha), \quad \frac{b}{2 - \kappa}\left(1 + \sqrt{1 - 4\tilde{d}^2/a^2}\right)^{1/2}\right.$$

$$\left. \cdot \{2\sum_\alpha h_\alpha^{-2}(2 + (1 - \kappa)p_\alpha) \sum_\alpha u_\alpha^2(2 + (1 - \kappa)p_\alpha)^{-1}\}^{-1/2}\right] . \qquad (5.65)$$

For the FOC scheme:

$$\tau \le \min\left[\frac{a}{4D}\sum_\alpha h_\alpha^{-2}, \quad \frac{3}{5}b(1 + \sqrt{1 - 4\tilde{d}^2/a^2})^{1/2}\{2\sum_\alpha h_\alpha^{-2} \sum u_\alpha^2\}^{-1/2}\right] .$$

$$(5.66)$$

For the first order upwind scheme, with $\tilde{d} = \tau D \sum h_\alpha^{-2}(2 + p_\alpha)$,

$$\tau \le \min\left[\frac{a}{2D}/\sum_\alpha h_\alpha^{-2}(2 + p_\alpha), \quad b\left(1 + \sqrt{1 - 4\tilde{d}^2/a^2}\right)^{1/2}\right.$$

$$\left. \cdot \{2\sum_\alpha h_\alpha^{-2}(2 + p_\alpha) \sum_\alpha u_\alpha^2(2 + p_\alpha)^{-1}\}^{-1/2}\right] . \qquad (5.67)$$

These inequalities are implicit, because $\tilde{d}$ depends on τ. However, checking the admissibility of a given τ is straighforward, whereas generation of a suitable τ is easily done by some iterative process. Furthermore, without losing much sharpness, the term $\sqrt{1 - 4\tilde{d}^2/a^2}$ may be replaced by 0.

Theorem 5.7.5. *(Parabola)*
If

$$\sum_\alpha c_\alpha^2/d_\alpha \le q_3 b^2 , \qquad (5.68)$$

where

$$q_3 = \tfrac{1}{2}, \qquad \qquad \tfrac{1}{2} \le \kappa \le 1 ,$$
$$q_3 = \tfrac{27}{8}(1 - \kappa)(2 - \kappa)^{-3}, \quad -1 \le \kappa < 1/2 ,$$

then S_L is contained in the parabola given by

$$v + (w/b)^2 = 0, \quad v + iw = z .$$

Proof. We have to show that $-\delta - \gamma_1 - (\gamma_2/b)^2 \le 0$. Using (5.43) and (5.68) we have

$$-\delta - \gamma_1 + (\gamma_2/b)^2 \le 2 \sum_\alpha d_\alpha s_\alpha \{-1 + 2q_3(1 - s_\alpha)(\bar\kappa s_\alpha + 1)^2\} .$$

Bij observing that

$$\max\{(1 - s)(\bar\kappa s + 1)^2 : \ 0 \le s \le 1\} = 1/2q_3$$

the proof is completed.   □

Condition (5.68) is equivalent to the following time step restriction: for the κ-scheme:

$$\tau \le 2Dq_3 b^2 / \sum_\alpha u_\alpha^2 . \qquad (5.69)$$

For the FOC scheme (5.69) holds with $q_3 = \tfrac{1}{4}3^5/5^3 \cong 0.4860$. For the first order upwind scheme we obtain

$$\tau \le \frac{1}{2}Db^2 / \sum_\alpha u_\alpha^2(2 + p_\alpha)^{-1} . \qquad (5.70)$$

The preceding theorems (exept Theorem 5.7.1) have been put forward in Wesseling (1995), Wesseling (1996c), as has the method that will be followed in the next section to derive stability conditions.

5.8 Derivation of von Neumann stability conditions

We will now use the properties of S_L derived in the preceding section to obtain von Neumann stability conditions for time discretizations that are frequently used in computational fluid dynamics, or that are of historical interest. After space discretization according to (5.28), (5.29) we are left with the system of ordinary differential equations (5.32). A good source of information on numerical methods for ordinary differential equations is Hairer, Nørsett, and Wanner (1987), Hairer and Wanner (1991), where an introduction to stability domains may be found.

All that remains to be done here is to choose suitable unions or intersections of the sets considered in the theorems of the preceding section (rectangle, ellipses, oval, parabola) inside the stability domain S of the time discretization method used to discretize (5.32), and useful (i.e. not too conservative and easy to evaluate) sufficient conditions for von Neumann stability tumble like ripe apples. Most of the following results are not generally known or new. The techniques used to derive the preceding theorems can be also used to obtain stability conditions for cases where the rectangle, ellipses, oval and parabola do not fit nicely in S; an example (leapfrog-Euler) will be given.

For understanding the principles of our approach to the derivation of von Neumann stability conditions it is not necessary to study every case presented below in detail. For understanding, it suffices to study the cases of Adams-Bashforth and leapfrog-Euler, for example. A number of other schemes are included because they are frequently used in practice, and the stability conditions presented are not widely known or new. This section may be regarded as a catalog of time stepping schemes and stability conditions. Many references will be given to publications where these schemes are applied to the Navier-Stokes equations.

Explicit Euler

This time discretization method when applied to (5.32) is given by

$$\varphi^{n+1} - \varphi^n = -\tau L_h \varphi^n .$$

The temporal local discretization error is $\mathcal{O}(\tau)$. After Fourier transformation this becomes

$$\varphi_\theta^{n+1} - \varphi_\theta^n = z\varphi_\theta^n , \quad z = -\tau \hat{L}_h(\theta) . \tag{5.71}$$

Substitution of $\varphi_\theta^n = \xi^n$ leads to the following *characteristic equation*:

$$\xi - (1 + z) = 0 .$$

The left-hand side is called the *stability polynomial*. Non-growth of $|\varphi_\theta^n|$ is equivalent to the root(s) of the stability polynomial to be inside the unit disk. Hence z must satisfy $|z + 1| \leq 1$. From this follows the stability domain S of this method:

$$S = \{z \in \mathbb{C} : \ |z+1| \le 1\} . \tag{5.72}$$

Now we apply stability condition (5.41). The ellipse of Theorem 5.7.2 with $a = b = 1$ fits perfectly in the disk (5.72), so that this theorem gives

Theorem 5.8.1. *(Stability of explicit Euler scheme)*
A sufficient condition for von Neumann stability of the explicit Euler scheme is

$$\tilde{d} \le 1 \quad and \quad \sum_\alpha c_\alpha^2/d_\alpha \le (2 - \kappa)^{-2} .$$

Corresponding time step conditions follow inmediately from (5.55)–(5.57). For $\kappa = 1$ (second order central scheme) these conditions are also shown to be necessary in Wesseling (1995), and identical to the necessary and sufficient conditions obtained for this scheme in Hindmarsh, Gresho, and Griffiths (1984), Hirt (1968), Morton (1971).

The contribution to the local discretization error due to temporal discretization is called the temporal discretization error. No temporal discretization error is made if $d\varphi/dt = -L_h\varphi$, so that $d\varphi_\theta/dt = \varphi_\theta e^{tz/\tau}$, hence

$$\varphi_\theta(t + \tau) = \varphi_\theta(t)e^z . \tag{5.73}$$

Substitution of $\varphi_\theta(t)$ in the numerical scheme (5.71) gives the temporal local discretization error $\hat{\tau}_h$:

$$\hat{\tau}_h = \frac{1}{\tau}\varphi_\theta(t)(e^z - 1 - z) = \mathcal{O}(z^2/\tau) = \mathcal{O}(\tau) ,$$

where we have scaled by a factor $1/\tau$, because (5.71) approximates the Fourier transform of the differential equation times τ.

Adams-Bashforth

In order to obtain better temporal accuracy, one may include more time levels. This leads to *multistep schemes*. An example is the second order Adams-Bashforth (or AB2, for short) scheme, frequently used to solve the Navier-Stokes equations, for example in Andersson, Billdal, Eliasson, and Rizzi (1990), Balaras and Benocci (1994), Benocci and Pinelli (1990), Biringen and Cook (1988), Breuer and Rodi (1994), Gao (1994), Gavrilakis, Tsai, Voke, and Leslie (1986), Gerz, Schumann, and Elgobashi (1989), Hoffmann and Benocci (1994), Kajishima, Miyake, and Nishimoto (1990), Kobayashi and Kano (1986), Kobayashi, Morinishi, and Oh (1992), Kost, Bai, Mitra, and Fiebig (1992), Kost, Mitra, and Fiebig (1992), Krettenauer and Schumann (1992), Kristoffersen and Andersson (1993), Lê, Ryan, and Dang Tran

(1992), Moeng (1984), Schmidt and Schumann (1989), Shimomura (1991), Thomas and Williams (1994), Voke and Gao (1995). The scheme is given by

$$\varphi^{n+2} - \varphi^{n+1} = -\frac{1}{2}\tau L_h(3\varphi^{n+1} - \varphi^n) \,. \tag{5.74}$$

This being a three-time-level scheme, it is not self-starting. A start can be made by one explicit Euler step. Including one step with an $\mathcal{O}(\tau)$ scheme does not affect the global $\mathcal{O}(\tau^2)$ accuracy. The first step may even be unstable.

The temporal accuracy is determined in a similar way as for Euler. Taking the Fourier transform of (5.74) gives

$$\varphi_\theta^{n+2} - \varphi_\theta^{n+1} = \frac{1}{2}z(3\varphi_\theta^{n+1} - \varphi_\theta^n) \,.$$

Substitution of the exact solution (5.73) and scaling by $1/\tau$ gives

$$\hat{\tau}_h = \frac{1}{\tau}\varphi_\theta(t)(e^{2z} - e^z - \frac{3}{2}ze^z + \frac{1}{2}z) = \mathcal{O}(z^3/\tau) = \mathcal{O}(\tau^2) \,.$$

In the same way as before, the stability polynomial is found to be:

$$\xi^2 - \xi - \frac{z}{2}(3\xi - 1) \,.$$

For $z \in S$ the roots are inside the unit disk. On the boundary ∂S, one or more roots have modulus 1, and can be written as $\xi = e^{i\mu}$ for some $\mu \in [0, 2\pi)$. Hence ∂S can be found by solving

$$e^{2i\mu} - e^{i\mu} - \frac{z}{2}(3e^{i\mu} - 1) = 0 \,,$$

resulting in

$$z = 2e^{i\mu}(e^{i\mu} - 1)/(3e^{i\mu} - 1) \tag{5.75}$$

When μ is varied, z traverses ∂S. The result is shown in Fig. 5.1. All stability domains to be encountered are symmetric with respect to the real axis, therefore only the upper part is shown.

We will now fit an oval and an ellipse in S. We have $|z| \ll 1$ for $|\mu| \ll 1$. Expansion of (5.75) for small $|\mu|$ gives

$$z \cong (4i\mu - \mu^4)/(10 - 6\cos\mu) = v + iw \,.$$

Hence on ∂S we have near the origin $w \cong \pm(-4v)^{1/4}$, $v \le 0$. Near $z = 0$ we have on the oval of Theorem 5.7.3: $w \cong \pm b(-2v/a)^{1/4}$. We choose $a = 1/2$ and with $b = 1$ the oval and ∂S osculate in $z = 0$. In order to bring the oval partly inside ∂S we choose $b < 1$. With $a = 1/2$, $b = 2^{-1/4}$ the oval in Fig. 5.1 is obtained. Near $z = -1$ this oval is outside ∂S. We do not wish to make the

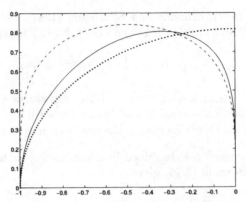

Fig. 5.1. Stability domain S of second order Adams-Bashforth scheme (—) with oval (- - -) and ellipse ($\cdots$)

oval smaller, however, because this would make the oval deviate more from ∂S near $z = 0$, resulting in an unnecessarily strict stability restriction. Staying close to ∂S near $z = 0$ is of importance, since in convection dominated flows $|\text{Im}\{\tau\hat{L}_h(\theta)\}| \gg |\text{Re}\{\tau\hat{L}_h(\theta)\}|$. Therefore we also select the half ellipse of Theorem 5.7.4. Near $z = -1$ we have $\mu = \pi + \varepsilon$, $|\varepsilon| \ll 1$ in (5.75), resulting in the following local behavior of ∂S:

$$z \cong -1 + (5\varepsilon^2 + 12i\varepsilon)/16 = -1 + v + iw, \quad w = \pm 3\sqrt{v/5}, \quad v \geq 0 \ .$$

On the ellipse of Theorem 5.7.4 we have near $z = -1$, choosing $a = 1$: $z = -1 + v + iw$ with $w \cong \pm b\sqrt{2v}$. The ellipse and ∂S osculate in $z = -1$ for $b = 3/\sqrt{10}$. In order to bring the ellipse partly inside ∂S we choose b slightly smaller; $a = 1$, $b = \sqrt{2/3}$ results in the ellipse partly shown in Fig. 5.1. The ellipse and the oval intersect in $z = -1/4 \pm i\sqrt{5/8}$. Numerical solution of the roots $\xi_{1,2}$ of the stability polynomial shows that in the point of intersection we have $\max\{|\xi_1|, |\xi_2|\} = 0.9934 < 1$, so that {ellipse $\cap$ oval} $\subset S$. Hence, using Theorems 5.7.3 and 5.7.4 we obtain

Theorem 5.8.2. *(Stability of second order Adams-Bashforth scheme)*
A sufficient condition for von Neumann stability of the second order Adams-Bashforth scheme is

$$\tilde{d} \leq 1/2 \quad and \quad \tilde{d}\sum_\alpha c_\alpha^2/\tilde{d}_\alpha \leq (8 - 4\kappa)^{-2}(1 + \sqrt{1 - 4\tilde{d}^2/a^2})$$
$$and \quad \{\sum_\alpha c_\alpha^4/d_\alpha)^{1/3} \leq q_1 \quad or \quad \sum_\alpha \leq q_2\sqrt{2}\} \ .$$

Corresponding time step restrictions follow easily from (5.61)–(5.63) and (5.65)–(5.67).

Comparing the AB2 scheme with the explicit Euler scheme, we remark that the accuracy of AB2 is $\mathcal{O}(\tau^2)$, and that the computing work is almost the same. Straightforward implementation of AB2 requires storage of the solution vector at an additional time level (although this may be economized upon, cf. Exercise 5.8.3). One might think that explicit Euler is less stability restricted than AB2, because S (the unit circle at $z = -1$) is rather larger than S for AB2 (see Fig. 5.1). However, near $z = 0$, $|\text{Im}\{\tau \hat{L}_h(\theta)\}|$ is less restricted (∂S stays closer to the imaginary axis, which we try to exploit expressly by the introduction of the oval) for AB2 than for Euler, which may for convection dominated cases result in weaker stability restrictions for AB2 than for Euler (cf. Exercise 5.8.1). All things considered AB2 seems more attractive than explicit Euler. The AB2 scheme is in widespread use for computing nonstationary viscous flows, especially in the context of large-eddy and direct simulation of turbulence. The lack of rigorous stability conditions before 1995 does not seem to have hampered this development.

The third order Adams-Bashforth (or AB3 for brevity) scheme is given by

$$\varphi^{n+3} - \varphi^{n+2} = -\frac{1}{12}\tau L_h(23\varphi^{n+2} - 16\varphi^{n+1} + 5) \, .$$

The stability polynomial is found to be:

$$\xi^3 - \xi^2 - \frac{1}{12}z(23\xi^2 - 16\xi + 5) \, .$$

The boundary ∂S of the stability domain S is computed in the same way as for the AB2 scheme, and is displayed in Fig. 5.2. The stability domain is covered

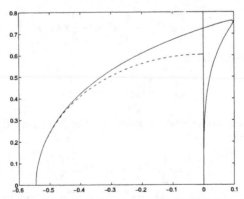

Fig. 5.2. Stability domain S of third order Adams-Bashforth scheme (—) with osculating ellipse (- - -)

to a satisfactory extent by the half ellipse of Theorem 5.7.4 that osculates in $z = -6/11$; its parameters are found to be $a = 6/11$, $b = \frac{72}{11}\sqrt{2/235} \cong 0.6038$ (cf. Exercise 5.8.4). Application of Theorem 5.7.4 gives

Theorem 5.8.3. *(Stability of third order Adams-Bashforth scheme)*
A sufficient condition for von Neumann stability of the third order Adams-Bashforth scheme is

$$\tilde{d} \le \frac{3}{11} \quad and \quad \tilde{d} \sum_{\alpha} c_{\alpha}^2 / \tilde{d}_{\alpha} \le 0.1823(2 - \kappa)^{-2}(1 + \sqrt{1 - (11\tilde{d}/3)^2}) \,.$$

The temporal discretization error is $\mathcal{O}(\tau^3)$. A look at the stability domains of Figures 5.1 and 5.2 leads one to expect that for convection dominated flows ($|\text{Im}(\tau\tilde{L}_h(\theta))|$ large) the stability restrictions will not be much more restrictive than for the AB2 scheme. This expectation is borne out in the example of Exercise 5.8.5. Compared to AB2, the solution at an additional time level has to be stored. The scheme can be started at $t = 0$ by an explicit Euler step, followed by an AB2 step. This scheme seems to be as attractive as AB2, and is also used in practice.

The BDF scheme

This is an example of an implicit three-level scheme. BDF stands for backward differentiation scheme; the method is also called the second order Gear scheme. This method has been applied to the Navier-Stokes equations in, for example Belov, Martinelli, and Jameson (1994), Breuer and Hänel (1990), Dwyer, Soliman, and Hafez (1986), Ho and Lakshminarayana (1993), Rogers and Kwak (1990), Rogers, Kwak, and Kiris (1991). The BDF scheme is given by

$$\varphi^{n+2} - \frac{4}{3}\varphi^{n+1} + \frac{1}{3}\varphi^n = -\frac{2}{3}\tau L_h \varphi^{n+2} \,.$$

The stability polynomial is

$$(1 - \frac{2}{3}z)\xi^2 - \frac{4}{3}\xi + \frac{1}{3} \,.$$

Similar to the Adams-Bashforth scheme, on the boundary ∂S of the stability domain S we have

$$z = \frac{1}{2}e^{-2i\mu} - 2e^{-i\mu} + \frac{3}{2} \,.$$

When μ is varied, z traverses ∂S. The result is shown in Fig. 5.3. Note that the contour is in the right half of the complex plane. Since the roots of the stability polynomial satisfy $|\xi_{1,2}| > 1$ inside the contour (as can be seen by computing $\xi_{1,2}$ in $z = 2$, for example), the stability domain S is in this case the *exterior* of ∂S. We see that in this case making the scheme implicit has the intended effect of making S large. Since $\text{Re}(-\tau\hat{L}_h(\theta)) \le 0$, the BDF scheme is unconditionally stable.

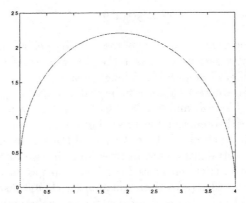

Fig. 5.3. Stability domain of BDF scheme

Leapfrog-Euler

This method has been widely used for large-eddy simulation of turbulence, for example in Arnal and Friedrich (1992), Boersma, Eggels, Pourquié, and Nieuwstadt (1994), Deardorff (1970), Eggels, Unger, Weiss, Westerweel, Adrian, Friedrich, and Nieuwstadt (1994), Friedrich and Unger (1991), Grötzbach (1982), Grötzbach and Wörner (1994), Manhart and Wengle (1994), Mason (1989), Schmitt and Friedrich (1988), Schumann (1975), Wagner and Friedrich (1994), Werner and Wengle (1989). The idea is to improve upon the explicit Euler method by improving the temporal accuracy of the convection term and allowing higher CFL numbers. This is done by time-discretizing the convection term with the leapfrog method, which has no numerical diffusion: both the symbol of the differential operator and the discrete approximation are purely imaginary for central schemes. Leapfrog cannot be applied to the diffusion term as well, because this results in instability; this is left to the reader to show in Exercise 5.8.6. Therefore explicit Euler is applied to the diffusion term. The resulting scheme is given by

$$\varphi^{n+2} - \varphi^n = -2(C_h\varphi^{n+1} + D_h\varphi^n) \,.$$

This is an example of a *mixed method*: the time discretizations of the convection and diffusion terms are different. After Fourier transformation this becomes

$$\varphi_\theta^{n+2} - \varphi_\theta^n = -2(\hat{C}_h(\theta)\varphi_\theta^{n+1} + \hat{D}_h(\theta)\varphi_\theta^n) \,,$$

or, using (5.34),

$$\varphi_\theta^{n+2} - \varphi_\theta^n = -2(\gamma(\theta)\varphi_\theta^{n+1} + \delta(\theta)\varphi_\theta^n) \,,$$

with $\gamma = \gamma_1 + i\gamma_2$, and δ (given by (5.37)) real. The stability polynomial is given by

$$\xi^2 + 2\gamma\xi + 2\delta - 1 \ . \tag{5.76}$$

For mixed methods, the coefficients of the stability polynomial do not depend on a single complex parameter z, as in the preceding cases (where $z = \gamma + \delta$), so that the stability domain S is not longer a subset of the complex plane. In the case of (5.76), where δ happens to be real, the roots depend on the three parameters γ_1, γ_2 and δ, and S is a subset of $\mathbb{R}^3$. It would not be difficult to redevelop the theorems of the preceding section in a three-dimensional setting, but a visual check whether the sets of these theorems are contained in S is much less straightforward in three than in two dimensions. We will not do this, and restrict ourselves for mixed schemes to the case $\gamma_1 = 0$, i.e. central second order ($\kappa = 1$) or fourth order (FOC) discretization of the convection term. Hence, the roots of the stability polynomial depend on two parameters only, δ and γ_2, and we can continue to work in the complex plane, with $z = -\delta - i\gamma_2$.

As before, on ∂S we have, with $z = v + iw$

$$e^{2i\mu} - 2iwe^{i\mu} - 2v - 1 = 0, \quad 0 \le \mu < 2\pi \ .$$

By solving for v and w one obtains

$$\sin 2\mu - 2w\cos\mu = 0 \ ,$$
$$\cos 2\mu + 2w\sin\mu - 2v = 1 \ .$$

For $\mu \ne \pi/2, \ 3\pi/2$ this gives $v = 0$, $w = \sin\mu$, whereas for $\mu = \pi/2, \ 3\pi/2$ one obtains $v \pm w = -1$. The resulting stability domain S is presented in Fig. 5.4. The ellipses, oval and parabola of the theorems of the preceding

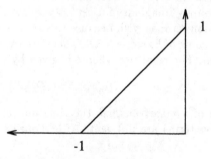

Fig. 5.4. Stability domain S of leapfrog-Euler scheme.

section fit badly in S, and we proceed directly.

Theorem 5.8.4. *(Stability of leapfrog-Euler scheme)*

a) *For the $\kappa = 1$ scheme, the leapfrog-Euler scheme is von Neumann stable if and only if*

$$\sum_\alpha (d_\alpha + \sqrt{d_\alpha^2 + c_\alpha^2}) \leq 1 . \tag{5.77}$$

b) For the fourth order central scheme a sufficient condition for von Neumann stability of the leapfrog-Euler method is

$$\sum_\alpha (d_\alpha + \sqrt{d_\alpha^2 + \frac{25}{9}c_\alpha^2}) \leq 1 .$$

Proof. It is necessary and sufficient that $\delta + |\gamma_2| \leq 1$. For $\kappa = 1$ this means

$$\sum_\alpha \{d_\alpha(1 - \cos\theta_\alpha) + c_\alpha|\sin\theta_\alpha|\} \leq 1 , \quad \forall \theta_\alpha .$$

Since $\theta_1, ..., \theta_m$ are independent, the maximum of the sum is obtained by maximizing each term. Defining $f(\theta) = d(1 - \cos\theta) + c\sin\theta$, $0 \leq \theta \leq \pi$, we have $f'(\theta) = 0$ for $\theta = \pi - \gamma$, $\tan\gamma = c/d$, $0 \leq \gamma < \pi/2$, resulting in $\max\{f(\theta) : 0 \leq \theta < \pi\} = d + \sqrt{c^2 + d^2}$, which concludes the proof of part a).

Continuing with part b), for the FOC scheme a necessary and sufficient condition is

$$\sum_\alpha \{d_\alpha(1 - \cos\theta_\alpha) + \frac{1}{6}c_\alpha|8\sin\theta_\alpha - \sin 2\theta_\alpha|\} \leq 1, \quad \forall \theta_\alpha .$$

Since $|8\sin\theta_\alpha - \sin 2\theta_\alpha| \leq 10|\sin\theta_\alpha|$ it is sufficient if

$$\sum_\alpha \{d_\alpha(1 - \cos\theta_\alpha) + \frac{5}{3}c_\alpha|\sin\theta_\alpha|\} \leq 1, \quad \forall \theta_\alpha .$$

Determining the maxima of the individual terms as before, the proof of part b) is easily completed.   □

The corresponding time step restriction is found to be

$$\tau \leq \{D\sum_\alpha h_\alpha^{-2}(2 + \sqrt{4 + qp_\alpha^2})\}^{-1} ,$$

with $q = 1$ for the second order central ($\kappa = 1$) scheme and $q = 25/9$ for the FOC scheme.

The one-dimensional version of part a) of this theorem has appeared in Chan (1984), with a less elementary proof. The following stability condition for the $\kappa = 1$ scheme has been put forward on heuristic grounds in Schumann (1975) and has been generally used in practice since:

$$\sum_\alpha (2d_\alpha + c_\alpha) \leq 1 . \tag{5.78}$$

A proof (less elementary than above) has been given in Pourquié (1994). Condition (5.78) is stronger than (5.77), cf. Exercise 5.8.8.

One of the roots of the stability polynomial (5.76), say ξ_2, is spurious; ξ_1 is called essential and approximates the exact amplification factor $\exp\{-\tau \dot{L}(\theta)\}$. Ideally, one should have $|\xi_2| < |\xi_1|$ (this is the case for the multistep schemes discussed before), but for $D = 0$, i.e. $\delta = 0$, $|\xi_2| = |\xi_1|$. The trouble this causes may be seen from the fact that for $D = 0$ the numerical scheme contains a spurious undamped eigenmode $\varphi_j^n = (-1)^{n+j_1+\cdots+j_m}$ if we disregard the boundary conditions. Furthermore, on the part of ∂S where $v + |w| = 1$ we have $|\xi_2| = 1 \geq |\xi_1|$. In order to damp spurious modes, users of the leapfrog-Euler scheme stay well within S and usually apply the so-called Asselin filter (Asselin (1972)). After φ^{n+2} has been computed, φ^{n+1} is modified according to

$$\varphi_j^{n+1} = (1 - c)\varphi_j^{n+1} + \frac{1}{2}c(\varphi_j^n + \varphi_j^{n+2}) ,$$

with typically $c \in [0.05, 0.1]$.

Because of these difficulties, and because the temporal accuracy is only $\mathcal{O}(\tau)$ (although if convection dominates practical results come close to $\mathcal{O}(\tau^2)$) users now tend to favor the Adams-Bashforth scheme.

Adams-Bashforth-Euler

Instead of leapfrog one can also combine second order Adams-Bashforth with explicit Euler, avoiding the trouble with the spurious root and the need for artificial damping. This approach has been used for the Navier-Stokes equations in Le Thanh (1992), Le Thanh, Troff, and Ta Phuoc Loc (1991), Verstappen and Veldman (1994). The method is given by

$$\varphi^{n+2} - \varphi^{n+1} = -C_h\left(\frac{3}{2}\varphi^{n+1} - \frac{1}{2}\varphi^n\right) - D_h\varphi^{n+1} .$$

The stability polynomial is given by

$$\xi^2 + \xi\left(\frac{3}{2}\gamma + \delta - 1\right) - \frac{1}{2}\gamma .$$

As before, this being a mixed scheme we restrict ourselves to the case that $\gamma = i\gamma_2$. The boundary ∂S of the stability domain S is given by

$$e^{2i\mu} - e^{i\mu}\left(\frac{3}{2}iw + v + 1\right) + \frac{1}{2}iw = 0, \quad 0 \leq \mu \leq 2\pi, \quad z = v + iw . \quad (5.79)$$

The stability domain is plotted in Fig. 5.5. For comparison, the stability domain of the second order Adams-Bashforth scheme is also shown. Sufficient

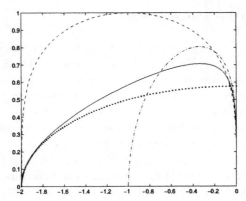

Fig. 5.5. Stability domain S of Adams-Bashforth-Euler scheme (—) with oval (- - -) and ellipse (· · ·), and S for second order Adams-Bashforth scheme (- · − · ·).

stability conditions are obtained in a similar way as for the Adams-Bashforth method. On ∂S, near $z = 0$, expansion of (5.79) for $|\mu| \ll 1$ shows that we have $v \cong -\mu^4/4$, $w \cong \mu - \frac{5}{3}\mu^3$, hence $w \cong \pm(-4v)^{1/4}$. Choosing $a = 1$, on the oval of Theorem 5.7.3 we have near $z = 0$: $w \cong \pm b(-2v)^{1/4}$. Hence, the osculating oval has $b = 2^{1/4}$. In order to bring the oval partly inside ∂S we take $b = 1$. Near $z = -2$ we find in a similar way that on ∂S we have $w = \pm\sqrt{(v+2)/3}$, whereas with $a = 2$ on the ellipse of Theorem 5.7.4 we have $w = \pm b\sqrt{v+2}$. We choose $b = 1/\sqrt{3}$. The resulting oval and ellipse are shown in Fig. 5.5. Application of Theorems 5.7.3 and 5.7.4 to the second order central ($\kappa = 1$) and FOC schemes gives, defining $d = \sum_\alpha d_\alpha$:

Theorem 5.8.5. *(Stability of Adams-Bashforth-Euler scheme)*

a) *For the second order central scheme a sufficient von Neumann stability condition is*

$$d \leq 1 \quad and \quad \sum_\alpha (c_\alpha^4/d_\alpha)^{1/3} \leq 1 \quad and \quad d\sum_\alpha c_\alpha^2/d_\alpha \leq \frac{1}{6}(1 + \sqrt{1 - d^2}) \, .$$

b) *For the fourth order central scheme a sufficient von Neumann stability condition is*

$$d \leq 1 \quad and \quad \sum_\alpha (c_\alpha^2/d_\alpha)^{1/3} \leq q \quad and \quad d\sum_\alpha c_\alpha^2/d_\alpha \leq \frac{3}{50}(1 + \sqrt{1 - d^2}) \, ,$$

with

$$q = \frac{1}{5}\left(\frac{5 + \sqrt{10}}{12}\right)^{1/3}(\sqrt{10} - 1)^{5/3} \cong 0.6360 \, .$$

Corresponding stability restrictions on the time step τ follow from (5.61), (5.62), (5.65) and (5.66).

Adams-Bashforth-Crank-Nicolson

This is an example of a mixed scheme of IMEX (implicit-explicit) type. The idea is to enlarge the stability domain by treating the diffusion terms implicitly. In practical applications, the convection term is often nonlinear, which is an incentive to treat this term explicitly. Second order Adams-Bashforth is applied to convection and Crank-Nicolson to diffusion. The method has been applied to the Navier-Stokes equations in, for example, Braun, Fiebig, and Mitra (1994), Cantaloube and Lê (1992), Esposito (1992), Gavrilakis (1993), Grötzbach and Wörner (1994), Huser and Biringer (1992), Kim and Moin (1985), Moin and Kim (1982), Takemoto and Nakamura (1986), Takemoto and Nakamura (1989), Wörner and Grötzbach (1992), Zang, Street, and Koseff (1994). The method is given by

$$\varphi^{n+2} - \varphi^{n+1} = -C_h\left(\frac{3}{2}\varphi^{n+1} - \frac{1}{2}\varphi^n\right) - \frac{1}{2}D_h\left(\varphi^{n+2} + \varphi^{n+1}\right) .$$

The stability polynomial is found to be

$$\xi^2\left(1 + \frac{1}{2}\delta\right) + \xi\left(-1 + \frac{3}{2}\gamma + \frac{1}{2}\delta\right) - \frac{1}{2}\gamma .$$

For the same reasons as for the two preceding schemes we restrict ourselves to the case that $\gamma = i\gamma_2$. The boundary ∂S of the stability domain S is given by

$$e^{2i\mu}\left(1 - \frac{1}{2}v\right) - e^{i\mu}\left(1 + \frac{3}{2}iw + \frac{1}{2}v\right) + \frac{i}{2}w = 0, \quad z = v + iw ,$$

and is plotted in Fig. 5.6, together with the parabola of Theorem 5.7.5 with $b = 2\sqrt{3}$ and the oval of Theorem 5.7.3 with $a = 1/2$, $b = (3/4)^{1/4}$. We have

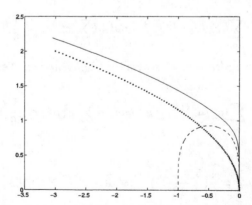

Fig. 5.6. Stability domain of Adams-Bashforth-Crank-Nicolson (—) with oval (- - -) and parabola (···)

{parabola $\cup$ oval} $\subset S$. If we now allow $\gamma_1 \neq 0$, $-\tau \hat{L}_h(\theta)$ shifts to the left in the complex plane, because according to (5.35) we have $\gamma_1 \geq 0$. A look at Fig. 5.6 shows that when $-\tau \hat{L}_h(\theta)$ moves left it stays inside S. Hence the restriction $\gamma_1 = 0$ can be dropped, so that the following theorem holds for the κ- and FOC schemes. Application of Theorems 5.7.3 and 5.7.5 results in

Theorem 5.8.6. *(Stability of Adams-Bashforth-Crank-Nicolson scheme) A sufficient condition for von Neumann stability of the Adams-Bashforth-Crank-Nicolson scheme is*

$$\tilde{d} \leq 1/2 \ and \ \left\{ \sum_{\alpha} (c_\alpha^4/d_\alpha)^{1/3} \leq q_1 \left(\frac{3}{2}\right)^{1/3} \ or \ \sum_{\alpha} c_\alpha \leq q_2\sqrt{3} \right\}$$

or

$$\sum_{\alpha} c_\alpha^2/d_\alpha \leq q_3 \frac{4}{3}$$

with q_1, q_2, q_3 as in Theorems 5.7.3 and 5.7.5.

Corresponding time step restrictions follow from (5.61)–(5.63) and (5.69), (5.70).

Extrapolated BDF

This is an IMEX version of BDF, in which the discretization of convection is made explicit by estimating φ^{n+2} by extrapolation from φ^{n+1} and φ^n. The method is given by

$$\varphi^{n+2} - \frac{4}{3}\varphi^{n+1} + \frac{1}{3}\varphi^n = -\frac{2}{3}D_h\varphi^{n+2} - C_h\left(\frac{4}{3}\varphi^{n+1} - \frac{2}{3}\varphi^n\right).$$

This method has been studied in Varah (1980) for the one-dimensional case. The stability polynomial is

$$\xi^2\left(1 + \frac{2}{3}\delta\right) + \frac{4}{3}\xi(\gamma - 1) + \frac{1}{3} - \frac{2}{3}\gamma.$$

Again, this being a mixed scheme, we restrict ourselves to the case that $\gamma = i\gamma_2$. The boundary of the stability domain ∂S is given by

$$e^{2i\mu}\left(1 - \frac{2}{3}v\right) - \frac{4}{3}e^{i\mu}(iw + 1) + \frac{1}{3} + \frac{2i}{3}w = 0, \ z = v + iw,$$

so that on ∂S

$$v = \frac{3}{2} + \frac{6\cos\mu - 9}{4\cos\mu - 2\cos 2\mu}, \quad w = \frac{\sin\mu(2 - \cos\mu)}{2\cos\mu - \cos 2\mu}.$$

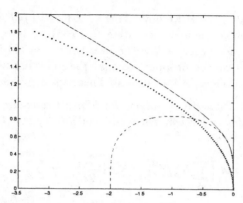

Fig. 5.7. Stability domain of extrapolated BDF (—) with oval (- - -) and parabola (···)

When $4\cos\mu - 2\cos 2\mu = 0$, i.e. for $\mu = \mu_0 = \arccos(\frac{1}{2} - \frac{1}{2}\sqrt{3}) \cong 1.95$, $z = v + iw$ goes to infinity. In Fig. 5.7 we plot ∂S for $0 \leq \mu \lesssim 1.73$. For $\mu > \mu_0, \partial S$ continues in the right half of the complex plane (not shown). The shape of ∂S is not unlike that for Adams-Bashforth-Crank-Nicolson, and to obtain stability conditions we again use the oval and the parabola. For $|\mu| \ll 1$ we have $v \cong -\frac{3}{4}w^4$ on ∂S, and the oval of Theorem 5.7.3 osculates with ∂S in $z = 0$ if $b = (2a/3)^{1/4}$. In order to stay inside ∂S we take b slightly smaller; with $a = 1$, $b = (1/2)^{1/4}$ we obtain the oval shown in Fig. 5.7. The parabola shown has parameter $b = 1$. For the same reason as for the Adams-Bashforth-Crank-Nicolson scheme we can allow $\gamma_1 \neq 0$. Application of Theorems 5.7.3 and 5.7.5 results in :

Theorem 5.8.7. *(Stability of extrapolated BDF scheme)*
A sufficient condition for von Neumann stability of the extrapolated BDF scheme is

$$\tilde{d} \leq 1 \text{ and } \left\{ \sum_\alpha (c_\alpha^4/d_\alpha)^{1/3} \leq q_1(\frac{1}{2})^{1/3} \quad or \quad \sum_\alpha c_\alpha \leq q_2(\frac{1}{2})^{1/2} \right\}$$

or

$$\sum_\alpha c_\alpha^2/d_\alpha \leq q_3 \,,$$

with q_1, q_2, q_3 *as in Theorems 5.7.3 and 5.7.5.*

Corresponding time step restrictions follow from (5.61)–(5.63) and (5.69), (5.70).

Runge-Kutta

Flow problems are often solved by a Runge-Kutta method. Applications to the Navier-Stokes equations are described in Meinke and Hänel (1990), Spalart, Moser, and Rogers (1991). As an example we take the method proposed by Wray as described in Spalart, Moser, and Rogers (1991), designed to save computer storage. This is a three-stage Runge-Kutta method, defined by

$$\begin{aligned}
\varphi^* &= \varphi^n - \tau L_h \gamma_1 \varphi^n , \\
\varphi^{**} &= \varphi^* - \tau L_h (\gamma_2 \varphi^* + \zeta_1 \varphi^n) , \\
\varphi^{n+1} &= \varphi^{**} - \tau L_h (\gamma_3 \varphi^{**} + \zeta_2 \varphi^*) ,
\end{aligned} \tag{5.80}$$

with parameters

$$\gamma_1 = \frac{8}{15}, \quad \gamma_2 = \frac{5}{12}, \quad \gamma_3 = \frac{3}{4}, \quad \zeta_1 = -\frac{17}{60}, \quad \zeta_2 = -\frac{5}{12} .$$

It suffices to store three instances of φ. We have

$$\varphi^{n+1} = P(-\tau L_h)\varphi^n, \; P(z) = 1 + z + \frac{1}{2}z^2 + \frac{1}{6}z^3 , \tag{5.81}$$

where P is called the *amplification polynomial*. The exact solution satisfies

$$\varphi(t + \tau) = e^{-\tau L_h} \varphi(t) .$$

Since $P(z) = e^z + \mathcal{O}(z^4)$, the temporal error per time step is $\mathcal{O}(\tau^4)$, so that the overall temporal accuracy of Wray's Runge-Kutta method is $\mathcal{O}(\tau^3)$.

Stability analysis of Runge-Kutta methods proceeds as follows. Taking the Fourier transform of (5.81) results in

$$\hat{\varphi}_\theta^{n+1} = P(-\tau \hat{L}_h(\theta))\hat{\varphi}_\theta^n .$$

The quantity $P(-\tau \hat{L}_h(\tau))$ is called the *amplification factor*. For von Neumann stability it is sufficient that

$$|P(-\tau \hat{L}_h(\theta))| \le 1, \quad \forall \theta . \tag{5.82}$$

The stability domain S of Runge-Kutta methods is defined as

$$S = \{z \in \mathbb{C} : |P(z)| \le 1\} .$$

We see that (5.82) is satisfied if (5.41) holds, as before.

The stability domain of Runge-Kutta methods is easily determined as follows. One solves numerically for r the equation $|P(-p+re^{i\mu})| = 1$, for given p and μ. Then the point $-p + re^{i\mu}$ belongs to ∂S. By varying μ between 0 and

2π, ∂S is traversed. The value p should be chosen such that the straight line $z = -p + re^{i\mu}$ intersects the upper half of ∂S in only one point and does not come close to being tangent to ∂S, in order to make the equation for r easy to solve numerically; $p = 2$ is found to be suitable in the present case. The stability domain is shown in Fig. 5.8, together with the ellipse of Theorem

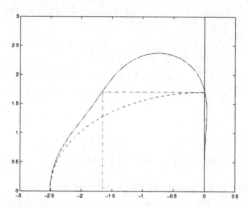

Fig. 5.8. Stability domain of Wray's Runge-Kutta method (—) with ellipse (- - -) and rectangle ($- \cdot -\cdot$)

5.7.4 (parameters: $a = 2.5127$, $b = \sqrt{3}$) and the rectangle of Theorem 5.7.1 (parameters $a = 1.65$, $b = 1.7$). The value $b = \sqrt{3}$ is chosen because ∂S intersects the imaginary axis in $z = i\sqrt{3}$, since $P(i\sqrt{3}) = -\frac{1}{2} + \frac{1}{2}i\sqrt{3}$. Thus we obtain

Theorem 5.8.8. *(Stability of Wray's Runge-Kutta method)*
A sufficient condition for von Neumann stability of Wray's Runge-Kutta method is

$$\tilde{d} \le \frac{a}{2} \quad and \quad \sum_\alpha c_\alpha \le b/(2 - \kappa)$$

with $a = 1.65$, $b = \sqrt{3}$, *or*

$$\tilde{d} \le \frac{a}{2} \quad and \quad \frac{2\tilde{d}}{b^2} \sum_\alpha c_\alpha^2/\tilde{d}_\alpha \le (2 - \kappa)^{-2}(1 + \sqrt{1 - 4\tilde{d}^2/a^2})$$

with $a = 2.5127$, $b = \sqrt{3}$.

Corresponding stability restrictions on the time step follow from (5.50), (5.52), (5.53) and (5.65)–(5.67). The advantage of using the rectangle is that the resulting stability conditions are simpler than for the ellipse.

Another Runge-Kutta method, especially designed for convection-diffusion problems, has been proposed in Sommeijer, van der Houwen, and Kok (1994). We will call it the SHK method. It is a four-stage method, and is given by

$$\varphi^* = \varphi^n - \alpha_1 \tau L_h \varphi^n ,$$
$$\varphi^{**} = \varphi^n - \alpha_2 \tau L_h \varphi^* ,$$
$$\varphi^{***} = \varphi^n - \frac{1}{2}\tau L_h \varphi^{**} ,$$
$$\varphi^{n+1} = \varphi^n - \tau L_h \varphi^{***} ,$$

where $\alpha_1 = 1/4$, $\alpha_2 = 1/3$. As for Wray's method, three instances of φ have to be stored. The amplification polynomial is given by

$$P(z) = 1 + z(1 + \frac{1}{2}z(1 + \alpha_2 z(1 + \alpha_1 z))) .$$

Since $P(z) = e^z + \mathcal{O}(z^5)$, the temporal accuracy of the SHK Runge-Kutta method is $\mathcal{O}(\tau^4)$.

The stability domain is determined in the same way as before, and is presented in Fig. 5.9. Because $|P(i\sqrt{8})| = 1$, ∂S intersects the imaginary

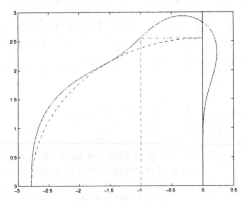

Fig. 5.9. Stability domain of SHK Runge-Kutta method (—) with ellipse (- - -) and rectangle (- · —·)

axis in $z = i\sqrt{8}$. Also shown are the ellipse of Theorem 5.7.4 (parameters: $a = 2.7853$, $b = 2.55$) and the rectangle of Theorem 5.7.1 (parameters: $a = 1$, $b = 2.55$). Again, the rectangle is included because the resulting stability conditions are simpler than for the ellipse. Stability conditions are obtained from Theorem 5.8.8 and equations (5.50), (5.52), (5.53) and (5.65)–(5.67) by inserting the appropriate values of a and b. The extra stage makes the SHK Runge-Kutta method about 33% more time-consuming than Wray's

Runge-Kutta method, but by comparing the values for b we see that if convection dominates, so that $\tau \hat{L}_h(\theta)$ is near the imaginary axis, the CFL numbers c_α and hence the time step τ can be about 50% larger.

One may also formulate implicit Runge-Kutta schemes to enhance stability. An application to the Navier-Stokes equations is presented in Marx (1994).

Mixed Runge-Kutta methods

The idea is to treat diffusion implicitly to enhance stability, and convection (which is often nonlinear) explicitly to decrease computational cost, and to have $\mathcal{O}(\tau^2)$ temporal accuracy for diffusion and $\mathcal{O}(\tau^3)$ for convection. Enhanced accuracy for convection is beneficial for convection dominated flows and for turbulence simulations, in which one wishes to have dissipation governed by physics and not by numerical errors in approximating convection. Such schemes have been applied to the Navier-Stokes equations in Baron and Quadrio (1994), Le and Moin (1991), Huser and Biringer (1993), Joslin, Streett, and Chang (1993), Rai and Moin (1991), Spalart (1988), Yang and Ferziger (1993). Let, for example, Wray's Runge-Kutta scheme be applied to convection, and the ω-scheme (to be discussed below) to diffusion. The resulting method, used by Le and Moin (1991), is given by

$$
\begin{aligned}
\varphi^* &= \varphi^n - D_h(\alpha_1\varphi^n + \beta_1\varphi^*) - C_h\sigma_1\varphi^n \,, \\
\varphi^{**} &= \varphi^* - D_h(\alpha_2\varphi^* + \beta_2\varphi^{**}) - C_h(\sigma_2\varphi^* + \zeta_2\varphi^n) \,, \\
\varphi^{n+1} &= \varphi^{**} - D_h(\alpha_3\varphi^{**} + \beta_3\varphi^{n+1}) - C_h(\sigma_3\varphi^{**} + \zeta_3\varphi^*) \,,
\end{aligned}
\tag{5.83}
$$

with

$$
\begin{aligned}
&\alpha_1 = \beta_1 = \tfrac{4}{15}, \ \alpha_2 = \beta_2 = \tfrac{1}{15}, \ \alpha_3 = \beta_3 = \tfrac{1}{6}, \ \sigma_1 = \tfrac{8}{15}, \ \sigma_2 = \tfrac{5}{12}, \\
&\sigma_3 = \tfrac{3}{4}, \ \zeta_2 = -\tfrac{17}{60}, \ \zeta_3 = -\tfrac{5}{12} \,.
\end{aligned}
\tag{5.84}
$$

We will call this the WLM Runge-Kutta method. Stability can be studied as follows. By taking the Fourier transformation of (5.83) one obtains

$$
\varphi_\theta^{n+1} = R(\gamma, \delta)\varphi_\theta^n \,,
$$

writing for brevity, as before, $\gamma = \hat{C}_h(\theta)$, $\delta = \hat{D}_h(\theta)$, with γ and δ given by (5.36) and (5.37). The amplification factor R is given by

$$
R(\gamma, \delta) = R_3\{R_2R_1 - \zeta_2\gamma/(1 + \beta_2\delta)\} - \zeta_3\gamma R_1/(1 + \beta_3\delta) \,,
$$

with

$$
R_k = (1 - \alpha_k\delta - \sigma_k\gamma)/(1 + \beta_k\delta), \quad k = 1, 2, 3 \,.
$$

This being a mixed scheme, we restrict ourselves to the case of purely imaginary $\gamma = i\gamma_2$, i.e. the central second or fourth order space discretization of the convection term, so that R is a function of the complex number $\delta + i\gamma_2$. To find the stability domain we write

$$
\begin{aligned}
R(z) &= R_3\{R_2R_1 + i\zeta_2 w/(1 - \beta_2 v)\} + i\zeta_3 w R_1/(1 - \beta_3 v)\,, \\
R_k &= (1 + \alpha_k v + i\sigma_k w)/(1 - \beta_k v), \quad k = 1, 2, 3\,.
\end{aligned}
\tag{5.85}
$$

The stability domain S is defined by

$$
S = \{z \in \mathbb{C} : |R(z)| \le 1\}\,.
$$

The locus of $|R(z)| = 1$ is identical to the boundary ∂S of S, and is easily determined numerically in the same way as for the preceding Runge-Kutta schemes. The result is shown in Fig. 5.10. Because with $\delta = 0$ the scheme is

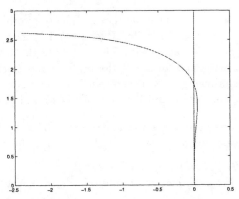

Fig. 5.10. Stability domain of WLM Runge-Kutta method.

identical to Wray's Runge-Kutta method, ∂S intersects the imaginary axis in $z = i\sqrt{3}$. Therefore we have stability if $|\gamma_2| \le \sqrt{3}$. Hence, we may apply Theorem 5.7.1 with $a = \infty$, $b = \sqrt{3}$. For the same reason as for Adams-Bashforth-Crank-Nicolson we may allow $\gamma_1 \ne 0$. We obtain:

Theorem 5.8.9. *(Stability of WLM Runge-Kutta method)*
A sufficient condition for von Neumann stability of the WLM Runge-Kutta method is

$$
\sum_\alpha c_\alpha \le \frac{\sqrt{3}}{2 - \kappa}\,.
$$

The corresponding time step restriction is

$$
\tau \le \frac{\sqrt{3}}{2 - \kappa}\bigg/ \sum_\alpha (|u_\alpha|/h_\alpha)\,.
$$

The fact that the temporal accuracy for convection is $\mathcal{O}(\tau^3)$ follows from the observation that if there is no diffusion we have Wray's Runge-Kutta scheme. On the other hand, if there is no convection, we have, taking $w = 0$ in (5.85),

$$R(z) = R_1 R_2 R_3 = e^z + \mathcal{O}(z^2) , \qquad (5.86)$$

so that the temporal accuracy of the approximation of diffusion is $\mathcal{O}(\tau^2)$.

The ω-scheme

This method is given by

$$\varphi^{n+1} - \varphi^n = -\tau L_h(\omega \varphi^{n+1} + (1 - \omega)\varphi^n) , \qquad (5.87)$$

where the parameter ω remains to be chosen. The temporal accuracy is $\mathcal{O}(\tau)$, but for $\omega = 1/2 + \mathcal{O}(\tau)$ it is $\mathcal{O}(\tau^2)$ (cf. Exercise 5.3.1). For $\omega = 0$ we have the explicit Euler scheme discussed before. For $\omega = 1/2$ the scheme is called the Crank-Nicolson scheme and for $\omega = 1$ the implicit Euler scheme is obtained. The Crank-Nicolson scheme has been applied to the Navier-Stokes equations in, for example, Choi, Moin, and Kim (1993), Choi and Moin (1994).

Stability can be analyzed for a particular value of ω by the method used before. But it is easy to treat the case $1/2 \leq \omega \leq 1$ as follows:

Theorem 5.8.10. *(Stability of ω-scheme)*
For $1/2 \leq \omega \leq 1$ the ω-scheme is unconditionally von Neumann stable.

Proof. By taking the Fourier transformation of (5.87) we obtain

$$\hat{\varphi}_\theta^{n+1} = R(\tau \hat{L}_h(\theta))\hat{\varphi}_\theta^n, \quad R(z) = \frac{1 - (1 - \omega)z}{1 + \omega z} .$$

We have $\tau \hat{L}_h(\theta) = v + iw$, $v = \delta + \gamma_1$, $w = \gamma_2$ with δ, γ_1 and γ_2 given by (5.35)–(5.37). We must show that $|R(v + iw)| \leq 1$. We have

$$|R(v + iw)|^2 = \frac{(1 - (1 - \omega)v)^2 + (1 - \omega)^2 w^2}{(1 + \omega v)^2 + \omega^2 w^2} .$$

Because $v \geq 0$ and $1/2 \leq \omega \leq 1$ the denominator is not smaller than the numerator. $\qquad \square$

The case $0 < \omega < 1/2$ is not of interest, because one has to pay the price of solving an implicit system, without the benefit of increased temporal accuracy, as for $\omega = 1/2 + \mathcal{O}(\tau)$. The case $\omega = 0$ has been treated in Theorem 5.8.1. The $\omega = 1$ scheme has certain advantages that will be discussed later, namely that it is strongly A-stable, and that the condition on τ for the scheme to be of positive type (Definition 5.4.1) is weaker than for $\omega < 1$. For example, with first order upwind discretization, (5.87) is of positive type if

$$\tau \le \frac{1}{1-\omega} / \sum_\alpha (|u_\alpha|/h_\alpha + 2D/h_\alpha^2) \,. \tag{5.88}$$

It is left to the reader to show this.

Further remarks

A large number of time discretization schemes has been presented, and the reader may wonder whether it would not have been better to confine our-selves to a smaller number of examples. However, almost all of these schemes are encountered in practice, usually without much analytic information on their stability. Such stability results as there are are scattered through the literature, and are frequently confined to one space dimension, one particular spatial discretization and/or absence of convection or diffusion terms, and are often more conservative than the conditions presented above. Therefore the preceding general and unifying stability analysis is useful. The reader may also wonder by which criteria a suitable method is to be selected from the many alternatives available. This will be our next subject. The present section is concluded with some further remarks on stability analysis.

Von Neumann stability analysis for multistep methods involves deriving conditions for the roots of the stability polynomial to be in the unit disk. Schur-Cohn theory, as described in Miller (1971), gives necessary and sufficient conditions on the coefficients. However, deriving stability restrictions on the time step from these conditions is usually an arduous task, that has to be undertaken anew for each scheme that one wishes to consider, and gets rapidly out of hand as the order of the multistep method increases. Hence, it is not surprising that few results have been published. Furthermore, Schur-Cohn theory is not applicable to Runge-Kutta methods, for which not the absolute value of the roots but of the amplification polynomial is not to exceed 1. In the foregoing we have presented an alternative approach. Based on Theorems 5.7.1–5.7.5, stability conditions are easy to find. Our approach does not get more complicated as the order of multistep methods increases, and applies equally to Runge-Kutta methods. Although not its principle, its ease of use in practice is affected if the stability polynomial depends on more than two parameters, so that the classical stability diagram of the time discretization method does not apply. This may happen if the terms in the partial differential equation under consideration are not all discretized in time by the same scheme, i.e. with hybrid schemes. Nevertheless, for such schemes the method may still work, for example, when central space discretization is applied to first order terms, as we have seen.

The principle of the method is applicable to general initial-boundary value problems in any number of dimensions, but Theorems 5.7.1–5.7.5 have been derived for the convection-diffusion equation, with the κ-scheme or the fourth order central scheme used for space discretization of the convection term. To illustrate the use of the method, von Neumann stability conditions are derived for a number of schemes. In most cases, sufficient conditions seem not to have been available before. The present method of von Neumann stability analysis has been proposed in Wesseling (1996c), Wesseling (1995).

Because Fourier transformation is used, von Neumann stability analysis is restricted to linear equations with constant coefficients, uniform grids and periodic boundary conditions. With different boundary conditions, von Neumann stability is necessary for explicit schemes, because it takes some time before the influence of (perhaps stabilizing) boundary conditions makes itself felt in the interior. Stability analysis in the presence of non-periodic boundary conditions is difficult. When the coefficients are variable we need to have local von Neumann stability with the local value of the coefficients, because usually instability is a local phenomenon, with the high wavenumber (short wavelength) Fourier modes being the most unstable. But it is shown in Widlund (1966) that this is not sufficient for stability in the variable coefficient case. For more background, see Thomeé (1990). Sometimes stability can be proved in the presence of variable coefficients, nonlinearity and non-periodic boundary conditions by means of the *energy method*, see Morton and Mayers (1994) Sect. 5.8. See Richtmyer and Morton (1967) for other methods of stability analysis, and Morton and Mayers (1994) Sect. 5.9 for a survey of the current situation, including the influence of boundary conditions.

Exercise 5.8.1. Consider the dimensionless one-dimensional convection-diffusion equation:

$$\frac{\partial \varphi}{\partial t} + \frac{\partial \varphi}{\partial x} - \mathrm{Pe}^{-1}\frac{\partial^2 \varphi}{\partial x^2} = 0 \ .$$

Show that Theorem 5.8.2 results in (replacing $\sqrt{1 - 4\tilde{d}^2/a^2}$ by 0) the following sufficient condition for von Neumann stability of the AB2 scheme in combination with the κ-scheme:

$$\tau \ \leq \ \min[h^2\mathrm{Pe}/(4 + (2 - 2\kappa)h\mathrm{Pe}), \ \max\{q_1(2h^2/\mathrm{Pe})^{1/3}, \ q_2h\sqrt{2}\},$$
$$h\sqrt{1/3}/(2 - \kappa)] \ .$$

Show that Theorem 5.8.1 gives for the explicit Euler scheme:

$$\tau \leq \min[\mathrm{Pe}h^2/(2 + (1 - \kappa)h\mathrm{Pe}), \ 2\mathrm{Pe}^{-1}/(2 - \kappa)^2] \ .$$

Show that for $h = 10^{-2}$, Pe $= 10^{-4}$ and $\kappa = 1/3$ this results in $\tau \leq 0.0035$ for AB2 and $\tau \leq 7.2 * 10^{-5}$ for Euler. We see that for this convection dominated case the stability condition is much more restrictive for explicit Euler than for AB2. Note that with $\tau = 0.0035$ the CFL number is 0.35.

Exercise 5.8.2. As Exercise 5.8.1, but for the FOC and first order upwind scheme.

Exercise 5.8.3. This exercise is about saving storage for the AB2 scheme. Asume a three-dimensional domain with a cubic $n_1 \times n_2 \times n_3$ grid. With lexicographic ordering, the matrix L_h is block-tridiagonal with $(n_1 n_2) \times (n_1 n_2)$ blocks B_j, D_j, C_j, $j = 1, ..., n_3$. Let $y^1 = \varphi^{n+1}$, $y^0 = \varphi^n$ and let y^p and y^q be two $n_1 n_2$-dimensional scratch vectors. Let $y^0 = (y_1^0, ..., y_{n_3}^0)$ be a partitioning in $n_1 n_2$-dimensional blocks, and similarly for y^1. Show that $L_h(3\varphi^{n+1} - \varphi^n)$ can be computed as follows, disregarding boundary modifications:

$$y^0 = 3y^1 - y^0$$
$$\text{for} \quad j = 1, ..., n_3 \quad \textbf{do}$$
$$y^p = y^q$$
$$y^q = y_j^0$$
$$y_j^0 = B_j y^p + D_j y^q + C_j y_{j+1}^0$$
$$\textbf{end}$$

Show how φ^{n+2} can be computed without using additional storage.

Exercise 5.8.4. Show that the parameters of the half ellipse of Theorem 5.7.4 that osculates with ∂S of the AB3 scheme in $z = -6/11$ are $a = 6/11$, $b = \frac{72}{11}\sqrt{2/235}$. (Hint: develop the stability polynomial with $\xi = e^{i\mu}$ in a power series around $\mu = \pi$; cf. our treatment of the AB2 scheme).

Exercise 5.8.5. As Exercise 5.8.1, but for AB3 scheme. Show that a sufficient stability condition is: $\tau \leq 0.0026$.

Exercise 5.8.6. Show that the leapfrog scheme applied to the convection-diffusion equation is unstable for every $\tau > 0$.

Exercise 5.8.7. Show that the temporal accuracy of the third order Adams-Bashforth scheme is $\mathcal{O}(\tau^3)$.

Exercise 5.8.8. Show that (5.78) implies (5.77).

Exercise 5.8.9. Prove (5.86).

Exercise 5.8.10. Prove (5.88).

5.9 Numerical experiments

Test problem

In order to motivate the introduction, besides their stability conditions, of additional criteria to judge the merits of discretization schemes, some numerical experiments will be presented for the following one-dimensional test problem:

$$\frac{\partial \varphi}{\partial t} + u \frac{\partial \varphi}{\partial x} - D \frac{\partial^2 \varphi}{\partial x^2} = q, \qquad 0 < x < 1, \quad 0 < t \le T, \tag{5.89}$$
$$q(t, x) = \beta^2 D \cos \beta(x - ut),$$

with D and $u > 0$ constant, and β a parameter. An exact solution is given by

$$\varphi(t, x) = \cos \beta(x - ut) + e^{-\alpha^2 Dt} \cos \alpha(x - ut), \tag{5.90}$$

with α arbitrary. Spatial discretization is done with the second order central ($\kappa = 1$) or with the first order upwind scheme on a uniform cell-centered grid with mesh size h. For temporal discretization the ω-scheme is used. The resulting scheme can be written as

$$\varphi_j^{n+1} - \varphi_j^n + \omega L_h \varphi_j^{n+1} + (1 - \omega) L_h \varphi_j^n = \tau q_j \tag{5.91}$$

with, in the interior,

$$L_h \varphi_j = \frac{1}{2} \sigma(\varphi_{j+1} - \varphi_{j-1}) + d(-\varphi_{j-1} + 2\varphi_j - \varphi_{j+1}), \quad j = 2, 3, ..., J - 1, \tag{5.92}$$

where $\sigma = u\tau/h$, $d = D\tau/h^2$. The grid is a uniform version of the grid depicted in Figure 4.1.

Choice of time step and mesh size and time scale

The length scale $\mathcal{L}$ and time scale $\mathcal{T}$ of the exact solution are given by

$$\mathcal{L} = \pi / \max(\alpha, \beta), \quad \mathcal{T} = \min\{\mathcal{L}/u, (D\alpha^2)^{-1}\},$$

where we take for the length scale of a harmonic function half its wavelength. We may expect accuracy to be sufficient if $\tau \ll \mathcal{T}$, $h \ll \mathcal{L}$. For efficiency, we would like to avoid more stringent restrictions on τ and h, such as might arise from stability or the positive type property. In the numerical experiments to be described we take $\alpha = 4\pi$, $\beta = 2\pi$. We take mostly $h = 1/30$, giving $h/\mathcal{L} \cong 0.13$.

We note that the discrete operator is of positive type (Definition 5.4.1) if, in the case of the second order central scheme,

$$\tau \leq \frac{h^2}{2(1-\omega)D} \quad \text{and} \quad p \equiv \frac{|u|h}{D} \leq 2 , \tag{5.93}$$

whereas in the case of the first order upwind scheme we must have

$$\tau \leq \frac{h^2}{1-\omega}/(2D + |u|h) . \tag{5.94}$$

The first conditions of (5.93) and (5.94) arise from the temporal discretization (and is different for other discretizations). If their violation causes spurious wiggles (the risk is greatest if the solution is nonsmooth), then τ should be decreased, or something should be done about the temporal discretization. This will lead us to the concept of strong stability in the next section. The second condition of (5.93) is the same as in the stationary case. How the situation may be handled is described in Chap. 4. For example, one may use the κ-scheme with limiting.

Dirichlet-Neumann boundary conditions

First we prescribe a Dirichlet condition at $x = 0$ and a Neumann condition at $x = 1$. Initial and boundary conditions are chosen conforming with the exact solution (5.90). With the second order central scheme, we have a discrete operator of positive type for $\omega = 1$ if the mesh Péclet number $p = uh/D$ satisfies $p \leq 2$. The left half of Fig. 5.11 shows a result. In this case we have $\mathcal{T} \cong 0.23$, $\tau \cong 0.15\mathcal{T}$, $p \cong 1.83$. The accuracy is disappointing. The cause is that with $\omega = 1$ the scheme is only first order in time. Making τ four times

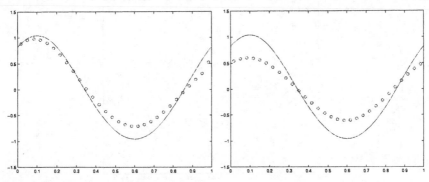

Fig. 5.11. Exact (—) and numerical solution (o) of (5.89). Second order central discretization; $\omega = 1$, $u = 1.1$, $D = 0.02$, $h = \tau = 1/30$, $t = 1$, $\alpha = 4\pi$, $\beta = 2\pi$. Left: Dirichlet-Neumann boundary conditions; right: periodic boundary condition.

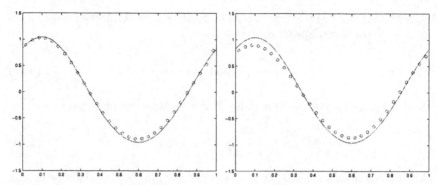

Fig. 5.12. As the preceding figure, but with $\tau = 1/120$.

as small gives the more accurate result of Fig. 5.12. Almost the same results (not shown) are obtained with the $\omega = 1/2$ scheme, but with τ much larger, namely $\tau = 1/30$. Hence, the $\omega = 1/2$ scheme is more efficient than the $\omega = 1$ scheme in this case. We have $\tau < h^2/\{2(1 - \omega)D\} = 1/18$ and $p = 1.83$, so that according to (5.93) the scheme is of positive type. The CFL number $c = |u|\tau/h$ and diffusion number $d = 2D\tau/h^2$ have values $c^2 = 1.21$ and $d = 1.2$, so that the $\omega = 0$ scheme is instable according to Theorem 5.8.1.

We continue with $\omega = 1/2$. In the preceding cases the discretization is of positive type, so that the maximum principle (Theorem 5.4.1)) holds, and spurious wiggles are excluded. In Fig. 5.13, where we have decreased D and where $p \cong 7.3$, the scheme is not of positive type, and we find spurious wiggles near $x = 1$. However, $\mathcal{L}$ and $\mathcal{T}$ are the same as before, so we would like to keep τ and h the same for efficiency. Just as in the stationary case discussed in Chap. 4, spurious wiggles may be avoided by local mesh refinement near

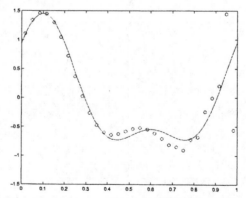

Fig. 5.13. As the left part of Fig. 5.11, but with $\omega = 1/2$ and $D = 0.005$.

$x = 1$, or by applying a homogeneous Neumann condition at $x = 1$, which gives the result presented in Fig. 5.14. Note that with the 'wrong' homoge-

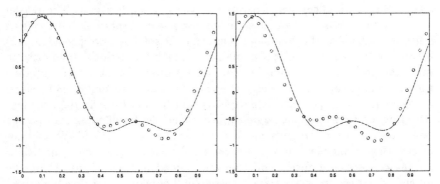

Fig. 5.14. As Fig. 5.11, but with a homogeneous Neumann condition at $x = 1$ for the left figure, and $\omega = 1/2$, $D = 0.005$.

neous Neumann condition the accuracy is much better than when $\partial\varphi(t, 1)/\partial x$ is derived from the exact solution. The reason is that an artificial numerical boundary layer at $x = 1$ is avoided, and that the upstream influence of the 'wrong' outflow boundary condition is restricted to a strip of width $\mathcal{O}(D)$, as explained in Sect. 2.5.

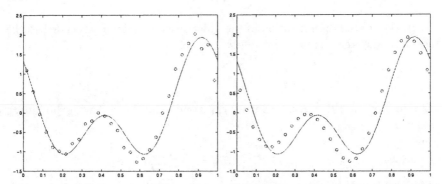

Fig. 5.15. Exact (—) and numerical (o) solution of equation (5.89). Second order central scheme; $\omega = 1/2$, $u = 1.1$, $D = 0.0005$, $h = \tau = 1/30$, $t = 0.8333$. Left: homogeneous Neumann outflow condition; right: periodic boundary conditions.

Transparent boundary condition

When D is decreased further, spurious wiggles gradually show up again, as seen in Fig. 5.15. They are generated at the boundary $x = 1$. The reason is that with D small we are close to solving a wave equation, and spurious reflection occurs at the Neumann boundary. In solving wave problems it is frequently the case that a large domain is cut off by an artificial boundary, to make the problem numerically tractable. Care has to be taken that outgoing waves travel unhampered through the artificial boundary, and do not reflect. To this end a *transparent boundary condition* (or weakly reflecting, or non-reflecting, or absorbing) has to be applied. Principles for deriving transparent boundary conditions are set forth in Engquist and Majda (1979). We will not discuss these principles, but use the results. Transparent boundary conditions for the convection-diffusion equation are studied in Hall (1986). There the following second order accurate transparent boundary condition is derived for (5.89):

$$\frac{\partial \varphi}{\partial t} + u \frac{\partial \varphi}{\partial x} = 0 \quad \text{at} \quad x = 1 . \tag{5.95}$$

This boundary condition is implemented numerically in the following way. We discretize (5.95) with the interior scheme (5.91), (5.92) in the boundary cell $j = J$, using a virtual point $J + 1$:

$$\frac{d\varphi_J}{dt} + \frac{\sigma}{2\tau}(\varphi_{J+1} - \varphi_{J-1}) = 0 . \tag{5.96}$$

Next, the spatial discretization of the differential equation is extended to the cell Ω_J using the virtual point $J + 1$:

$$\frac{d\varphi_J}{dt} + \frac{\sigma}{2\tau}(\varphi_{J+1} - \varphi_{J-1}) + \frac{d}{\tau}(-\varphi_{J-1} + 2\varphi_J - \varphi_{J+1}) = q_J . \tag{5.97}$$

Elimination of the virtual value φ_{J+1} results in the following boundary scheme:

$$\frac{d\varphi_J}{dt} + \frac{\sigma}{\tau}(\varphi_J - \varphi_{J-1}) = \frac{\sigma}{2d}q_J \tag{5.98}$$

This will be used as a transparent boundary condition.

Fig. 5.16 shows a result, which is much more satisfactory than that of Fig. 5.15. No boundary reflection is discernible. This shows the importance of using transparent outflow conditions for nonstationary convection dominated problems.

If diffusion dominates, the transparent outflow condition (5.98) gives wrong results. This is easily seen by taking $\sigma = 0$, in which case (5.97) determines φ_J^n completely, thus invoking a wrong Dirichlet boundary condition. But in

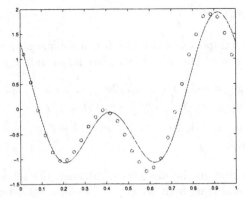

Fig. 5.16. Exact (—) and numerical solution (o) of 5.11. Second order central discretization; $\omega = 1/2, u = 1.1, D = 0.0005, h = \tau = 1/30, t = 0.8333$, transparent outflow condition.

a reasonable mathematical model, if diffusion dominates, a physical boundary condition should be available at an outflow boundary. So there should be no need for an artificial outflow condition. In any case, the transparent outflow condition (5.98) should be selected only if the mesh Péclet number $p = uh/D = \sigma/d$ is large enough, for example $p > 2$; otherwise, if no physical information is available, a homogeneous Neumann condition is to be preferred.

Periodic boundary conditions

Accuracy is enhanced in the preceding examples by the circumstance that at $x = 0$ the correct value is continuously fed in by the Dirichlet boundary condition. With a periodic boundary condition $\varphi(t, 1) = \varphi(t, 0)$ this is not the case, and the influence of discretization error is greater. The exact solution (5.90) is a combination of a damped and undamped traveling wave, and the numerical error may be broken down in *dissipation* (wrong amplitude) and *dispersion* (wrong propagation velocity). These concepts will be discussed more fully in Chap. 8.

Results with the periodic boundary condition are presented in the right parts of the preceding figures. In Fig. 5.11 we see that there is too much damping. This is due to the temporal discretization with $\omega = 1$; with τ smaller (Fig. 5.12) or with $\omega = 1/2$ (not shown) the numerical amplitude is more or less correct. Figs. 5.14 and 5.15 show that with the periodic boundary condition the scheme is not prone to spurious wiggles. The figures also show that the error is mainly due to the numerical propagation velocity being too small (dispersion).

Nonsmooth solution

In the following example the exact solution is not smooth. Again we consider (5.89), but with $q = 0$. The initial and boundary conditions are

$$
\begin{aligned}
\varphi(0,x) &= -1,\ 0 \le x < 1/2, \quad \varphi(0,x) = 1,\ 1/2 \le x \le 1\,; \\
\varphi(t,0) &= 1,\ 2k - 1/2 \le ut < 2k + 1/2\,; \\
\varphi(t,0) &= -1,\ 2k + 1/2 \le ut < 2k + 3/2\,; \\
\partial\varphi(t,1)/\partial x &= 0
\end{aligned}
\tag{5.99}
$$

for k integer, so that in the absence of diffusion the solution is a periodic step-function propagating with velocity u. We have not determined the exact solution for $D > 0$.

First the diffusion equation is considered ($u = 0$). Fig. 5.17 shows a result. When τ is increased beyond 0.01 the $\omega = 0$ scheme becomes unstable. With

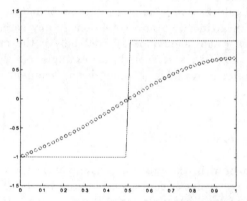

Fig. 5.17. Numerical solution (o) of equations (5.89), (5.99) with $q = u = 0$; —: initial condition; $\omega = 0$, $D = 1/50$, $h = 1/50$, $\tau = 1/110$, $t = 3$.

$\omega = 1$ and $\tau = 1/15$ almost the same result (not shown) is obtained as in Fig. 5.17. With $\omega = 1/2$ better temporal accuracy is obtained, but we have an operator of positive type only if $\tau \le h^2/D$ (cf. (5.93)), i.e. $d \le 2$. If this condition is not met numerical wiggles may occur. These indeed show up for small time, as seen in Fig. 5.18, where $d = 6.7$. Since here the convection term is absent, the temporal discretization is to blame. Improvement of the temporal discretization will be discussed in the next section.

Next, the convection-diffusion equation is considered. The transparent boundary condition is applied at $x = 1$. Although we do not know the exact solution, we can say that qualitatively the exact solution is a periodic step-function that is smoothed as time increases and propagates with velocity u. When

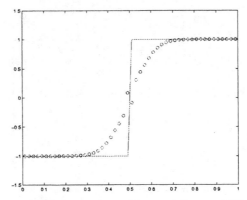

Fig. 5.18. As the preceding figure, but with $\omega = 1/2$, $\tau = 1/15$, $t = 0.2$

the discretization is of positive type there are no spurious wiggles, but with the present nonsmooth initial condition wiggles show up immediately when this not the case, as in Fig. 5.19, where the mesh Péclet number $p = 9 > 2$. Switching to first order upwind discretization restores monotonicity, but the

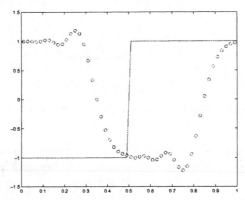

Fig. 5.19. Problem and symbols as in Fig. 5.17. Second order central discretization; $\omega = 1/2$, $D = 0.002$, $h = \tau = 1/50$, $t = 0.4$, $u = 0.9$.

result shows much false diffusion, cf. Fig. 5.20. A correct solution is shown in Fig. 5.21, obtained with $p = 1.8$. But this is an inefficient computation, because τ and h are very small compared to the physical scales $\mathcal{T}$ and $\mathcal{L}$.

The following conclusions may be drawn. If the second order central scheme is of positive type, good results are obtained. But conditions (5.93) are very restrictive. An excess of numerical damping can be caused by the time discretization ($\omega = 1$) or by first order upwind spatial discretization. Better temporal accuracy is obtained with $\omega = 1/2$, but if (5.93) is violated, numerical

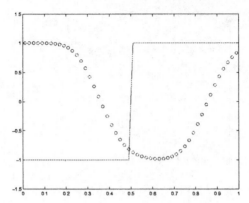

Fig. 5.20. As the preceding figure, but with first order upwind discretization

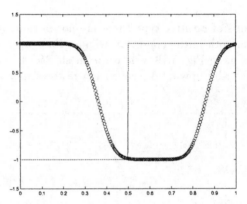

Fig. 5.21. As the preceding figure, but with $\tau = 0.008$ and $h = 0.004$

wiggles may occur if the exact solution is not smooth. In wave propagation problems the amplitude and the propagation velocity may be inaccurate due to spatial and/or temporal discretization errors.

To sum up, both temporal and spatial discretization need improvement relative to the methods discussed in this section. In the following, methods will be discussed that have gained acceptance in contemporary computational fluid dynamics.

Exercise 5.9.1. Carry out computational experiments for the problems for which the figures in this section give results. Experiment with defect correction.

5.10 Strong stability

As mentioned before, for efficiency the step sizes τ and h_α should be governed only by the physical scales $\mathcal{T}$ and $\mathcal{L}$ and preferably not by stability. This may be achieved by choosing methods whose stability domains satisfy certain criteria. These may depend on the application, for instance, on whether diffusion dominates or convection. A method is called *A-stable* if its stability domain S includes the second and third quadrant. Then the method is unconditionally stable for the convection-diffusion equation. An interesting theorem about A-stable schemes is *Dahlquist's second order barrier theorem* (see Hairer and Wanner (1991)), which says that the time accuracy of A-stable schemes is at best $\mathcal{O}(\tau^2)$. If the stability domain S includes the negative real axis, we will call the scheme *R-stable*. The advantage is that stability is not restricted by the diffusion term, because this term contributes only to the real part of the symbol of the scheme. As a consequence, instead of a stability condition of type $\tau = \mathcal{O}(h^2)$ one has $\tau = \mathcal{O}(h)$.

We will now formulate an additional requirement for A- or R-stable schemes. When this requirement is satisfied, spurious wiggles due to non-smoothness of the exact solution are damped. Such wiggles are seen in Fig. 5.18.

Strong stability

Let $q = 0$ in (5.89). By postulating a solution of the form $\varphi(t, x) = y(t)e^{i\theta x/h}$, the Fourier transform of (5.89) is obtained as

$$dy/dt = \lambda y, \quad \lambda = -D(\theta/h)^2 - iu\theta/h \ . \tag{5.100}$$

For our present purpose it is not necessary to discretize in space. We discretize (5.100) in time. Taking the $\omega = 1/2$ scheme as an example one obtains

$$y^{n+1} - y^n = \frac{1}{2}\lambda\tau(y^n + y^{n+1}) \ .$$

The stability polynomial is $(1 - \frac{1}{2}\lambda\tau)\xi - 1 - \frac{1}{2}\lambda\tau$, with root

$$\xi_1 = \frac{1 + \lambda\tau/2}{1 - \lambda\tau/2} \ . \tag{5.101}$$

The numerical solution is given by

$$y^n = \xi_1^n y^0 \ .$$

For temporal discretization schemes in general we have

$$y^n = \sum_{k=1}^{K} C_k \xi_k^n \ .$$

with C_k constants and $\xi_1, ..., \xi_K$ the roots of the stability polynomial. The exact solution is given by

$$y(n\tau) = e^{n\lambda\tau} y(0) .$$

We now make the following observation. We have with λ given by (5.100): $|e^{n\lambda\tau}| = e^{-\frac{1}{2}nd\theta^2}$ with $d = 2D\tau/h^2$. We see that Fourier components with θ large get damped quickly. However, for the $\omega = 1/2$ scheme this is not the case. On the contrary, if $|\theta| \to \infty$ we have $\lambda \to -\infty$ and according to (5.101) $\xi_1 \to -1$. This explains the wiggles in Fig. 5.18. Of course, there λ is different due to spatial discretization, and is given by, since $u = 0$,

$$\lambda\tau = -\tau \hat{L}_h(\theta) = -2d\sin^2\theta/2 ,$$

where (5.33) has been used. The Fourier mode with the shortest wavelength that can be represented on the grid has $\theta = \pi$. In Fig. 5.18 we have $d = 20/3$, so that for $\theta = \pi$ one finds $\xi_1 = -17/23 \cong -0.74$. This explains why spurious wiggles occur during the first time steps in the case of Fig. 5.18. These may be diminished or avoided altogether by requiring that the numerical scheme, like the differential equation, strongly damps short waves if diffusion is dominant. This leads to the following definition.

Definition 5.10.1. *Strong stability*
A discretization method is called *strongly A-stable* or *strongly R-stable* if it is A-stable or R-stable, respectively, and if when applied to $dy/dt = \lambda y$ the roots $\xi_1, ..., \xi_K$ of its stability polynomial satisfy

$$\lim_{\lambda \to -\infty} |\xi_k| < 1, \quad k = 1, ...K .$$

Of course, since the amplification factor $e^{\lambda\tau}$ of the exact solution satisfies $\lim_{\lambda \to -\infty} |e^{\lambda\tau}| = 0$, the smaller $\lim_{\lambda \to -\infty} |\xi_k|$, the better.

It is not difficult determine which of the A-stable or R-stable schemes discussed in Sect. 5.8 are strongly A-stable or R-stable. The stability polynomial of the BDF scheme was found in Sect. 5.8 to be $(1 - \frac{2}{3}z)\xi^2 - \frac{4}{3}\xi + \frac{1}{3}$ with in the notation of the present section $z = \tau\lambda$. As $z \to -\infty$ the roots satisfy $\xi_{1,2} \to 0$, so that BDF is strongly A-stable. Extrapolated BDF is strongly R-stable, to show this is left to the reader. From (5.101) we see that the $\omega = 1/2$ scheme is not strongly A-stable. The remaining R-stable schemes of Sect. 5.8 treat the diffusion term in a similar way as the $\omega = 1/2$ scheme, and turn out to be likewise not strongly R-stable. To show this for Adams-Bashforth-Crank-Nicolson is left to the reader. Although for Runge-Kutta schemes stability was studied by means of an amplification factor rather than the stability polynomial, the stability polynomial is readily defined, so that Definition 5.10.1 is easily extended to Runge-Kutta schemes. Application of Runge-Kutta to (5.100) gives

$$y^{n+1} = R(\tau\lambda)yn \ ,$$

with R the amplification factor introduced in Sect. 5.8. Hence the stability polynomial is $\xi - R$, with root $\xi_1 = R$. For the WLM Runge-Kutta method $R(z)$ is given by (5.85) with $z = v + iw = -\delta - i\gamma_2$. We have $\lim_{\delta\to\infty} R(z) = -1$, so that the method is not strongly R-stable.

The multistage ω-scheme

The ω-scheme is strongly A-stable if $\omega > 1/2$. But this reduces the temporal accuracy from $\mathcal{O}(\tau^2)$ to $\mathcal{O}(\tau)$. By using the ω-scheme in multistage fashion, $\mathcal{O}(\tau^2)$ accurate and strongly A-stable or R-stable schemes may be obtained, at similar computational cost as the $\omega = 1/2$ scheme. Such methods have been proposed and analyzed in Cash (1984), Gourlay and Morris (1980), Gourlay and Morris (1981), Lawson and Morris (1978), Glowinski and Périaux (1987), Bristeau, Glowinski, and Périaux (1987). The usefulness of these schemes for solving the Navier-Stokes equations is shown in Glowinski and Périaux (1987), Bristeau, Glowinski, and Périaux (1987), Rannacher (1989), Randall (1994), Turek (1994). A time step consists of three stages. Applied to $dy/dt = \lambda y + q$ the multistage ω-scheme is defined by, writing $z = \tau\lambda$,

$$(1 - \alpha\omega z)y^{n+\omega} = (1 + \alpha'\omega z)y^n + \omega\tau q^n \ ,$$
$$(1 - \alpha'\omega' z)y^{n+1-\omega} = (1 + \alpha\omega' z)y^{n+\omega} + \omega'\tau q^{n+1-\omega} \ , \qquad (5.102)$$
$$(1 - \alpha\omega z)y^{n+1} = (1 + \alpha'\omega z)y^{n+1-\omega} + \omega\tau q^{n+1} \ ,$$
$$\omega' = 1 - 2\omega, \qquad \alpha' = 1 - \alpha \ ,$$

where the parameters α and ω remain to be chosen. The three stages have time steps $\omega\tau$, $\omega'\tau$ and $\omega\tau$, respectively. Each stage is in fact an ω-scheme, so that the cost of a time step is three times the cost of the ω-scheme. However, this is compensated by the fact that τ can be chosen larger for the same accuracy. The amplification factor is

$$g(z) = \frac{(1 + \alpha'\omega z)^2(1 + \alpha\omega' z)}{(1 - \alpha\omega z)^2(1 - \alpha'\omega' z)} \ . \qquad (5.103)$$

For accuracy, $g(z)$ must approximate e^z for small z. We have

$$g(z) = 1 + z + \frac{1}{2}z^2\{1 + (1 - 2\alpha)(2\omega^2 - 4\omega + 1)\} + \mathcal{O}(z^3) \ .$$

Second order accuracy is obtained if $g(z) = e^z + \mathcal{O}(z^3)$. This is the case if either $\alpha = 1/2$, which brings us back to the Crank-Nicolson scheme, or

$$\omega = 1 - 1/\sqrt{2} \cong 0.292 \ , \qquad (5.104)$$

which is used in the multistage ω-scheme. If we choose α such that $\alpha\omega = \alpha'\omega'$, i.e.

$$\alpha = (1 - 2\omega)/(1 - \omega) = 2 - \sqrt{2} \cong 0.585 , \qquad (5.105)$$

then the coefficients in the left sides of (5.102) are the same, and hence also the matrices of the systems to be solved when applying the multistage ω-scheme to (5.89). We have the following theorem.

Theorem 5.10.1. *Strong A-stability*
If

$$\frac{1}{2} < \alpha \le 1 ,$$

then the multistage ω-scheme is strongly A-stable.

Proof. The amplification factor $g(z)$ given by (5.103) has poles only on the positive real axis, at $z = 1/\alpha\omega$ and $z = 1/(1 - \alpha)\omega'$. Furthermore,

$$\lim_{|z|\to\infty} |g(z)| = \alpha'/\alpha < 1 . \qquad (5.106)$$

Hence, $g(z)$ is analytic for $\mathrm{Re}(z) \le 0$. On the imaginary axis, writing $\alpha' = c\alpha$, $z = iy$, $i\alpha\omega y = iv$, we have

$$
\begin{aligned}
|g(iy)|^2 &= \frac{(1 + c^2v^2)^2(1 + 2v^2)}{(1 + v^2)^2(1 + 2c^2v^2)} \\
&= 1 + \frac{(c^4 - 1)v^4 + 2(c^4 - c^2)v^6}{(1 + v^2)^2(1 + 2c^2v^2)} .
\end{aligned}
$$

From (5.105) it follows that $c = \alpha/(1 - \alpha) < 1$, so that

$$|g(iy)|^2 \le 1 . \qquad (5.107)$$

Consider the closed contour consisting of the half-circle $z = re^{i\theta}$, $\pi/2 \le \theta \le 3\pi/2$ and the segment of the imaginary axis $z = iy$, $-r \le y \le r$. Inside this contour $g(z)$ is analytic, and according to the maximum modulus principle, inside the contour $|g(z)|$ is bounded by its maximum on the contour. Letting $r \to \infty$, we conclude from (5.106) and (5.107) that $g|(z)| \le 1$ for $\mathrm{Re}(z) \le 0$, which shows A-stability. Furthermore, according to (5.106),

$$\lim_{z\to-\infty} |g(z)| = \alpha'/\alpha < 1 ,$$

so that we have strong A-stability. $\qquad \square$

The choice $\alpha = 1$ (i.e. $\alpha' = 0$) seems attractive, because then

$$\lim_{z\to-\infty} |g(z)| = 0 ,$$

as for the exact amplification factor e^z of the differential equation. Then transient wiggles as those in Fig. 5.18 will surely be absent. But one would also like to approximate the exact amplification factor in another respect, namely undamped propagation of waves:

$$|g(iy)| \cong |g_e(iy)| = 1 .$$

With $\alpha = 1$ one finds

$$0.9 \leq |g(iy)| \leq 1 \quad \text{for} \quad 0 \leq |y| \lesssim 3 , \tag{5.108}$$

whereas with $\alpha = (1 - 2\omega)/(1 - \omega)$ one has

$$0.9 \leq |g(iy)| \leq 1 \quad \text{for} \quad 0 \leq |y| \lesssim 6 , \tag{5.109}$$

which is more attractive, For this value of α,

$$|g(-\infty)| = 1/\sqrt{2} \cong 0.702... ,$$

so that rough transients will be damped, but it will take a few time steps.

Application to the convection-diffusion equation

Application of the multistage ω-scheme to the spatial discretization of (5.89) gives

$$\varphi_j^{n+\omega} - \varphi_j^n + \alpha\omega\tau L_h\varphi_j^{n+\omega} + \alpha'\omega L_h\varphi_j^n = \omega\tau q_j^n ,$$
$$\varphi^{n+1-\omega} - \varphi_j^{n+\omega} + \alpha'\omega'\tau L_h\varphi_j^{n+1-\omega} + \alpha\omega' L_h\varphi_j^{n+\omega} = \tag{5.110}$$
$$\omega'\tau q^{n+1-\omega} ,$$
$$\varphi_j^{n+1} - \varphi^{n+1-\omega} + \alpha\omega\tau L_h\varphi^{n+1} + \alpha'\omega\tau L_h\varphi^{n+1-\omega} = \omega\tau q^{n+1} ,$$

where L_h is defined by (5.28) and (5.29). The amplification factor is given by

$$g(\theta) = g(z), \quad z = -\tau\hat{L}_h(\theta) ,$$

with $g(z)$ given by (5.103) and $\tau\hat{L}_h(\theta)$ by (5.34). Since $\text{Re}(-\tau\hat{L}_h(\theta)) \leq 0$ and the multistage ω-schema is A-stable, the scheme is unconditionally zero-stable and absolutely stable.

Let us now see whether the multistage ω-scheme is more attractive than the ω-scheme. Application to the test case of Fig. 5.18 gives the results of Fig. 5.22. In order to keep the cost of the multistage ω-scheme the same as for the $\omega = 1/2$ scheme, for Fig. 5.22 τ is three times as large as for Fig. 5.18. Although some wiggles occur in Fig. 5.22, they are smaller than in Fig. 5.18. If $d = 0$, $u \neq 0$ we have undamped wave propagation; therefore we would

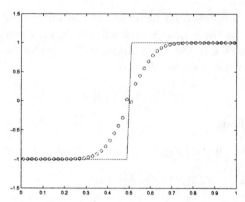

Fig. 5.22. Problem and symbols as in Fig. 5.18. Multistage ω-scheme; second order central discretization; $\omega = 1 - 1/\sqrt{2}$, $\alpha = 2 - \sqrt{2}$, $D = 0.02$; $u = 0$; $h = 1/50$, $\tau = 1/5$, $t = 0.2$.

like to have $|g(\theta)| = 1$ in this case. With $d = 0, \tau \hat{L}_h(\theta) = ic_1 \sin \theta$. From (5.109) it follows that $0.9 \leq |g(ic_1 \sin \theta)| \leq 1$ for $0 \leq c_1 \lesssim 6$ for all θ (for the $\omega = 1/2$ scheme, $|g(ic_1 \sin \theta)| = 1$ for all c_1 and θ, which is better, of course). The multistage ω-scheme gives results that are virtually indistinguishable from those of Figs. 5.14–5.18 with τ three times as large as for the $\omega = 1/2$ scheme. Figure 5.23 gives results for almost undamped wave propagation for long time. Here

$$q = 0 \,, \tag{5.111}$$

and the exact solution is given by

$$\varphi = \exp(-D\beta^2 t) \cos \beta(x - ut), \quad \beta = 3\pi \,, \tag{5.112}$$

with corresponding initial condition, Dirichlet condition at $x = 0$ and homogeneous Neumann condition at $x = 1$. Numerical dissipation is seen to be negligible, because the amplitude of the wave is faithfully represented. For the exact solution (5.112) the time and length scales are $\mathcal{T} = 1/3u \cong 0.3$ and $\mathcal{L} = 1/3$, so that τ and h are not unreasonably small, whereas the accuracy is reasonable. The wiggles are due to the central discretization of the convection term. Note that the mesh Péclet and CFL numbers satisfy $p = 73.3$ and $c_1 = 3.3$. The monotonicity of the solution will be improved later in this section.

We may conclude that the multistage ω-scheme is more attractive than the Crank-Nicolson ($\omega = 1/2$) scheme for temporal discretization.

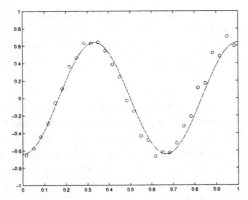

Fig. 5.23. Exact (—) and numerical (o) solution of (5.89), (5.111). Multistage ω-scheme, second order central discretization; $\omega = 1 - 1/\sqrt{2}$, $\alpha = 2 - \sqrt{2}$, $D = 0.0005$, $u = 1.1$, $h = 1/30$; $\tau = 0.1$; $t = 10$.

Choice of temporal discretization method

Because we have treated the ω-scheme and its multistage version rather extensively, the reader may be led to think that this is the method of choice. But the situation is not so clearcut. There are many possibilities for temporal discretization. For practical applications methods are mainly chosen on the basis of the following considerations.

(i) *Low artificial diffusion due to discretization of the convection term.* This is desirable if the temporal behavior of time-dependent flow is to be computed accurately. Physical oscillations should not be damped prematurely, and instable flows should not be stabilized artificially. This is an important consideration in large-eddy or direct simulation of turbulence. Since the convection term in the differential equation causes no dissipation and its symbol is purely imaginary (cf. (5.100)), it is desirable that the symbol of the spatial discretization of the convection term (given by (5.34) for the κ-scheme), is close to the imaginary axis. As far as the temporal discretization is concerned, we can say the following. For the differential equation the amplification factor for one time step is $\exp(\lambda\tau)$ with λ given by (5.100). In the absence of diffusion ($D = 0$) its modulus is 1. With $D = 0$ the symbol of the spatial discretization should be close to the imaginary axis, as argued above, and furthermore, the modulus of the amplification factor (for single step schemes) or of the dominant root of the stability polynomial (for multistep schemes) should be close to 1. That means that the boundary ∂S of the stability domain should be close to a sizeable segment of the imaginary axis. This rules out, for example, the use of the explicit and implicit ($\omega = 1$) Euler schemes for convection.

(ii)   *Accuracy.* Consideration (i) leads automatically to the use of $\mathcal{O}(\tau^2)$ accurate methods or better for convection. If diffusion is important (Péclet number not large) the diffusion term must also be approximated with $\mathcal{O}(\tau^2)$ accuracy. If diffusion is small (large Péclet number), $\mathcal{O}(\tau)$ accuracy may suffice for the diffusion term.

(iii)   *Stability and efficiency.* Explicit schemes generally lead to stability conditions of type $\tau = \mathcal{O}(h)$ for the convection term and $\tau = \mathcal{O}(h^2)$ for the diffusion term. If the latter condition is too severe, for example due to locally small h in realistic applications, then the diffusion term must be discretized implicitly, at additional cost per time step; this may be balanced by the fact that τ can be increased. As mentioned at the beginning of Sect. 5.9, for efficiency it is desirable that τ and h are constrained not by stability but by accuracy, i.e. related to $\mathcal{T}$ and $\mathcal{L}$, the time and length scales of the exact solution. If a stability condition of type $\tau = \mathcal{O}(h)$ is not too severe, then convection may be treated explicitly, using an IMEX scheme. Because in practice convection terms are usually nonlinear, this decreases the computing cost of implicit schemes. If spurious wiggles occur due to nonsmoothness of the solution, then a strongly stable scheme should be used, or in the explicit case, τ should be decreased in order to move away from ∂S (per definition, on ∂S we have no damping). If spurious wiggles are due to the spatial discretization (i.e. they persist when steady state is approached), then the measures discussed in Chap. 4 should be taken. In multidimensional problems with thin boundary layers, in which the spatial mesh size in a particular coordinate direction is much smaller than in the other directions, it may pay off to treat this direction implicitly and the other directions explicitly. In hyperbolic and convection domained problems, not only should numerical dissipation be small, but also dispersion, i.e. numerical wave propagation velocities should be close to the physical ones. We will come back to this in Chap. 8.

Defect correction

Defect correction is a way to improve the spatial discretization. For example, to improve the accuracy of the first order upwind scheme, while keeping spurious wiggles small, defect correction may be applied, just as in the stationary case (Sect. 4.8). As an example, consider one-step temporal discretization (such as the multistage ω-scheme). With first order upwind or second order central spatial discretization, this leads to systems of the following type:

$$B_u \varphi^{n+1} + A_u \varphi^n = q_u^n , \tag{5.113}$$
$$B_c \varphi^{n+1} + A_c \varphi^n = q_c^n , \tag{5.114}$$

where the subscript u refers to upwind and c to central discretization. Defect correction for time-dependent problems consists of the following steps, for each time step:

(i) Solve (5.113) and call the result $\varphi^{(0)}$;

(ii) For $k = 1, ..., K$:

$$B_u \delta\varphi^{(k)} = q_c^n - B_c\varphi^{(k-1)} - A_c\varphi^n, \quad \varphi^{(k)} = \varphi^{(k-1)} + \delta\varphi^{(k)} ; \quad (5.115)$$

(iii) $\varphi^{n+1} = \varphi^{(k)}$.

Step (ii) represents K defect corrections. If step (ii) is repeated until convergence so that $\delta\varphi^{(k)} \to 0$, (5.114) has been solved (in an expensive manner). But a middle course between (5.113) (low accuracy) and (5.114) (risk of spurious wiggles) is steered by taking $K = 1$. As shown in Sect. 4.6, this gives second order accuracy.

Applying defect correction with $K = 1$ to each successive stage of the multistage ω-scheme to the test case of Fig. 5.23 results in Fig. 5.24. A marked improvement is obtained compared to Fig. 5.23. One might think that the price for this is doubling of computing time, because now we have to solve twice per stage with matrix B_u, versus once with B_c without defect correction. But in realistic computational fluid dynamics problem solving with B_u (an M-matrix) is much faster than solving with B_c; more on this in Chap. 7.

Defect correction seems a good way to improve upon the accuracy of a first order scheme of positive type, without incurring too much of the non-monotonicity that may pollute results of non-positive second order schemes. In hyperbolic or convection dominated problems with nonsmooth solutions, the second order central scheme still introduces too much non-monotonicity. Then defect correction should be applied to the combination of the first order upwind and a flux-limited κ-scheme.

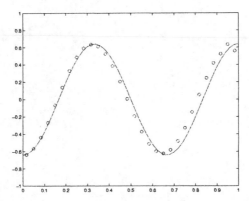

Fig. 5.24. Same case as Fig. 5.23, but with one defect correction based on first order upwind and second order central discretization

Exercise 5.10.1. Repeat the numerical experiments of this section. Does the accuracy improve when the number of defect corrections is increased?

Exercise 5.10.2. Choose $\alpha = 1$ in the multistage ω-scheme and recompute Figs. 5.22 and 5.23. Explain the differences that you observe.

Exercise 5.10.3. Verify Theorem 5.10.1 and equations (5.108), (5.109) by making a contour plot of $|g(z)|$ (using a contour plotting program from the website www.netlib.org or in MATLAB).

Exercise 5.10.4. Show that the extrapolated BDF scheme is strongly R-stable, and that the Adams-Bashforth-Crank-Nicolson scheme is not.

Exercise 5.10.5. Formulate defect correction for the second order Adams-Bashforth and BDF schemes.

6. The incompressible Navier-Stokes equations

6.1 Introduction

In this chapter, the incompressible Navier-Stokes equations in Cartesian coordinates discretized on uniform grids will be considered, discussing most of the basic numerical principles in a simple setting. In practical applications, of course, nonuniform grids and general coordinate systems are prevalent; these will be studied in later chapters, as will be the compressible case.

We can be relatively brief in discussing discretization of the Navier-Stokes equations, because we prepared the ground in our extensive discussion of the convection-diffusion equation in Chapters 3, 4 and 5. Therefore it wil not be necessary to discuss again the various possibilities for discretizing convection, or von Neumann stability conditions.

Only the *primitive variable formulation* will be discussed. This means that the velocity components and the pressure will be used as unknowns. This is because using other unknowns, such as the streamfunction and the vorticity, may be advantageous mainly in special circumstances, such as in two dimensions, or when vortices or vortex layers play a paramount role.

Other introductory texts on the subject are Ferziger and Perić (1996), Fletcher (1988), Peyret and Taylor (1985), Shyy (1994).

6.2 Equations of motion and boundary conditions

Equations of motion

The equations of motion have been discussed in Chap. 1. For ease of reference, the equations to be considered are repeated here. In this section the number of dimensions is three. We assume incompressible flow, i.e. $D\rho/Dt = 0$, so that

$$u^\alpha_{,\alpha} = 0 \,. \tag{6.1}$$

Conditions for incompressibility have been discussed in Sect. 1.12. We allow ρ to be variable, so that stratified flow or flow with large heat transfer can be taken into account.

The dimensionless incompressible Navier-Stokes equations are given by, in Cartesian tensor notation,

$$\frac{\partial \rho u^\alpha}{\partial t} + (\rho u^\alpha u^\beta)_{,\beta} = -p_{,\alpha} + \sigma^{\alpha\beta}_{,\beta} + f^\alpha, \quad \sigma^{\alpha\beta} = \mathrm{Re}^{-1}(u^\alpha_{,\beta} + u^\beta_{,\alpha}) , \quad (6.2)$$

where f is a body force. The following units are chosen: velocity: U; length: L; density: ρ_0; pressure: $\rho_0 U^2$. Then the Reynolds number is given by

$$\mathrm{Re} = \rho_0 U L / \mu ,$$

with μ the dynamic viscosity coefficient.

An example of a body force occurs in buoyant flow, where the dimensionless body force is given by (see Sect. 1.14)

$$f^\alpha = -\rho g^\alpha \mathrm{Ra} T .$$

Here g is the acceleration of gravity, T is temperature and the Rayleigh number is defined by

$$\mathrm{Ra} = \rho_0 |g| b \Delta T L / (\rho_0 U^2) ,$$

with ΔT, used as unit of temperature, a measure of the temperature variation, and b the thermal expansion coefficient of the fluid. Note that the Grashoff number of Sect. 1.14 is replaced by the Rayleigh number, because we use a different velocity unit.

The temperature is governed by equation (1.55). In dimensionless form:

$$\frac{\partial \rho T}{\partial t} + (\rho u^\alpha T)_{,\alpha} = (\mathrm{Pe}^{-1} T_{,\alpha})_{,\alpha} + \rho q , \quad (6.3)$$

where we have neglected the frictional heating term, and the Péclet number is given by

$$\mathrm{Pe} = U L / (c_p k) .$$

Usually μ and k are functions of T, but c_p is constant. Also $\rho = \rho(T)$, but the above equations also cover situations where ρ depends on the concentration of a solute, also designated by T. If ΔT is of moderate size, ρ may be taken constant, and for laminar flow the same holds for μ and k. But if a turbulence model is applied, or T varies strongly, μ and k vary.

Equation (6.3) is a convection-diffusion equation, and (6.2) is very similar. Hence, we will be able to profit a great deal from the insights gained in the preceding chapters concerning the numerical aspects of the convection-diffusion equation. The only, but significant, new complication is the lack of a time-evolution equation for the pressure p and the appearance of the kinematic constraint (6.1).

Boundary conditions

Both equations (6.2) and (6.3) are parabolic, and initial and boundary conditions are to be prescibed accordingly, cf. Chap. 2.

For the convection-diffusion equation (6.3) an initial condition

$$T(0, \boldsymbol{x}) = T_0(\boldsymbol{x}), \quad \boldsymbol{x} \in \Omega \tag{6.4}$$

is required, and on every point of the boundary $\partial \Omega$ of the domain Ω exactly one of the following conditions must be given:

$$T(t, \boldsymbol{x}) = f(t, \boldsymbol{x}) \quad \text{(Dirichlet)}, \tag{6.5}$$
$$\partial T(t, \boldsymbol{x})/\partial n = f(t, \boldsymbol{x}) \quad \text{(Neumann)}, \tag{6.6}$$
$$T(t, \boldsymbol{x}) + a\partial T(t, \boldsymbol{x})/\partial n = f(t, \boldsymbol{x}) \quad \text{(Robin)}, \tag{6.7}$$

with $a > 0$ and n the outward normal. If in a physical model nothing is known at an outflow boundary, it is advisable to prescribe a homogeneous Neumann condition, as argued in Sect. 4.2. If a singularity at $t = 0$ is to be avoided, the initial temperature field $T_0(\boldsymbol{x})$ in (6.4) should satisfy (6.5)–(6.7) at $t = 0$ at the corresponding parts of the boundary.

For the momentum equations (6.2) the following initial condition is required:

$$u^\alpha(0, \boldsymbol{x}) = w^\alpha(\boldsymbol{x}),$$

with the prescribed initial velocity field $\boldsymbol{w}$ satisfying the continuity equation (6.1):

$$w^\alpha_{,\alpha} = 0.$$

Note that there is no initial condition for the pressure, since $\partial p/\partial t$ does not occur.

No-slip condition

Viscous fluids cling to solid surfaces. This is called the *no-slip condition*. For the physical background of this condition, see Batchelor (1967). Hence, at a solid surface we have

$$u^\alpha(t, \boldsymbol{x}) = v^\alpha(t, \boldsymbol{x}), \tag{6.8}$$

with $v^\alpha(t, \boldsymbol{x})$ the local wall velocity. The Dirichlet condition (6.8) holds also at open parts of the boundary where the velocity is prescribed, which may be the case at an inflow boundary. But at an inflow boundary one may also prescribe condition (6.9) given below.

Free surface conditions

At a free surface the tangential stress components are zero. If the local (normal) displacement velocity of the surface $v(t, \boldsymbol{x})$ is known we have the following conditions:

$$u^\alpha n^\alpha = v(t, \boldsymbol{x}), \quad n^\alpha \sigma^{\alpha\beta} s^\beta = 0, \quad n^\alpha \sigma^{\alpha\beta} t^\beta = 0, \tag{6.9}$$

with $\boldsymbol{n}$ the unit outer normal, and $\boldsymbol{s}, \boldsymbol{t}$ linearly independent unit tangent vectors. One may also use (6.9) at an inflow boundary where the normal velocity is known. Conditions (6.9) may also arise at a plane of symmetry, in which case $v = 0$.

One can also have a free surface with unknown position and displacement velocity. Then the normal stress must also be prescribed, and (6.9) is replaced by

$$p - n^\alpha \sigma^{\alpha\beta} n^\beta = p_\infty, \quad n^\alpha \sigma^{\alpha\beta} s^\beta = 0, \quad n^\alpha \sigma^{\alpha\beta} t^\beta = 0. \tag{6.10}$$

In special cases one may wish to prescribe non-zero tangential stress in (6.9) or (6.10), for example, when one wishes to take the influence of wind shear on a water surface into account.

Outflow conditions

At an outflow boundary, often not enough physical information is available on which to base a sufficient number of boundary conditions. Usually only the pressure is known. This is not as serious as it may seem, because when $\mathrm{Re} \gg 1$ 'wrong' information generated by an artificial boundary condition propagates upstream only over a distance of $\mathcal{O}(\mathrm{Re}^{-1})$. This is plausible because of the resemblance of (6.2) to the convection-diffusion equation, and may in fact be shown directly by applying singular perturbation analysis to (6.2) in a similar manner as in Sect. 2.5. In order to avoid spurious numerical wiggles it is advisable to choose for artificial outflow conditions the Neumann conditions (6.10) in the stationary case. In the nonstationary case we expect this to be somewhat unsatisfactory, based on our experience with the one-dimensional convection-diffusion equation in Sect. 5.9, where we found the transparent boundary condition to be better. However, the question of what are good and easy to apply transparent boundary conditions for the incompressible Navier-Stokes equations seems to be largely unsettled, cf. the review presented in Sani and Gresho (1994). Outflow conditions can have a significant influence on results of instationary flow computations, as illustrated in Jin and Braza (1993). In Sect. 6.5 we will consider a generalization of the transparent outflow condition derived for the convection-diffusion equation in Sect. 5.9 to the incompressible Navier-Stokes equations, discretized on a staggered grid.

Compatibility condition

At every part of the boundary exactly one of the boundary conditions (6.8), (6.9) or (6.10) needs to be prescribed. If it is the case that along the whole of the boundary $\partial\Omega$ the normal velocity $v(t, \boldsymbol{x})$ is prescribed, then it follows from (6.1) and the divergence theorem that the following *compatibility condition* must be satisfied:

$$\int\limits_{\partial\Omega} v(t, \boldsymbol{x})dS = 0 \ . \tag{6.11}$$

It is shown in Temam (1985) that in order for (6.1), (6.2) to be well-posed for $t \geq 0$ the prescribed initial velocity field $\boldsymbol{w}(\boldsymbol{x})$ and the prescribed boundary velocity field must be related by

$$w_\alpha(\boldsymbol{x})n^\alpha = v_\alpha(0, \boldsymbol{x})n^\alpha$$

on parts of $\partial\Omega$ where the normal velocity is given. But the tangential components of the initial velocity need not be the same. Therefore, for example, a sliding wall may be set in motion instantaneously at $t = 0$ in a fluid originally at rest, but one should not let the speed of a body or of an inlet flow change discontinuously.

Theory

Much work has been done on the existence and uniqueness of (weak) solutions of the incompressible Navier-Stokes equations. For a survey, see Feistauer (1993) and Lions (1996); an introduction to the subject is given in Temam (1985). With certain restrictions on the boundary conditions, proofs have been given for the existence of solutions. If the Reynolds number is small enough the solution is unique, otherwise non-uniqueness may occur. This need not preclude numerical computation, because the possible solutions are often well separated. Some or all of them may be unstable. When the nonstationary equations are solved, one of the solutions is found, depending on the initial conditions. Interesting examples of computation of non-unique solutions are given in Roux (1990).

Exercise 6.2.1. Convince yourself that the Grashoff number measures the ratio of buoyancy and viscous forces, and that the Rayleigh number measures the ration of buoyancy and inertia.

6.3 Spatial discretization on colocated grid

Choice of grid

We restrict ourselves here to two dimensions. Generalization to three dimensions is straightforward. The domain Ω is a rectangle of size $L_1 \times L_2$. A cell-centered grid is used, as in Sect. 3.5. The domain Ω is subdivided in rectangular cells of size $h_1 \times h_2$. The computational grid G is the set of cell-centers:

$$G = \{\boldsymbol{x} \in \Omega : \boldsymbol{x} = \boldsymbol{x}_j, \ j = (j_1, j_2), \ j_\alpha = 1, 2, ..., m_\alpha, \ m_\alpha = L_\alpha / h_\alpha\} \,,$$

with $\boldsymbol{x}_j$ defined by

$$\boldsymbol{x}_j = (x_j^1, x_j^2), \ x_j^1 = (j_1 - 1/2)h_1, \ x_j^2 = (j_2 - 1/2)h_2 \,.$$

All discrete unknowns reside in the cell centers. This is called a *colocated*[1] *grid*. The cell with center at $\boldsymbol{x}_j$ is called Ω_j. We define

$$e_1 = (1/2, 0), \quad e_2 = (0, 1/2) \,.$$

The value of a quantity φ in $\boldsymbol{x}_j$ is designated by φ_j; φ_{j+e_1} is located at a cell face, namely at

$$\boldsymbol{x}_{j+e_1} = (j_1 h_1, (j_2 - 1/2)h_2) \,.$$

The cell at the 'east' side of Ω_j is designated by Ω_{j+2e_1}.

If one wishes, a vertex-centered grid with nodes on the boundary $\partial\Omega$ may be used instead. As shown in Chap. 3, this makes litte difference: despite differences in local error at boundaries, the global error will be of the same order.

Straightforward discretization of the continuity equation

Finite volume discretization of (6.1) gives

$$\int_{\Omega_j} u_{,\alpha}^\alpha d\Omega \cong h_2 u^1|_{j-e_1}^{j+e_1} + h_1 u^2|_{j-e_2}^{j+e_2} = 0 \,. \tag{6.12}$$

The cell face values in (6.12) need to be interpolated in terms of cell center values. The straightforward way to do this is to use linear interpolation, resulting in

$$h_2 u^1|_{j-2e_1}^{j+2e_1} + h_1 u^2|_{j-2e_2}^{j+2e_2} = 0 \,. \tag{6.13}$$

---

[1] Merriam-Webster dictionary: colocate: to locate together

Discretization of the momentum equation

Taking $\rho = 1$ and $\boldsymbol{f} = 0$ in (6.2), we have to discretize

$$\frac{\partial u^\alpha}{\partial t} + F^{\alpha\beta}_{,\beta} + p_{,\alpha} = 0, \quad F^{\alpha\beta} = u^\alpha u^\beta - \sigma^{\alpha\beta} . \tag{6.14}$$

Finite volume discretization gives

$$\int_{\Omega_j} \{\frac{\partial u^\alpha}{\partial t} + F^{\alpha\beta}_{,\beta} + p_{,\alpha}\} d\Omega \cong h_1 h_2 du^\alpha_j/dt + h_2 F^{\alpha 1}|^{j+e_1}_{j-e_1}$$

$$+ h_1 F^{\alpha 2}|^{j+e_2}_{j-e_\alpha} + h_\gamma p|^{j+e_\alpha}_{j-e_\alpha} = 0 , \tag{6.15}$$

with $\gamma \neq \alpha$. The cell face values have to be interpolated between cell center values. This is quite straightforward. For the pressure this is done as follows:

$$p_{j+e_\alpha} = \frac{1}{2}(p_j + p_{j+2e_\alpha}) . \tag{6.16}$$

The viscous stress $\sigma^{\alpha\beta}$ is approximated using

$$(Re^{-1}u^\alpha_{,1})_{j+e_1} \cong Re^{-1}_{j+e_1}(u^\alpha_{j+2e_1} - u^\alpha_j)/h_1 ,$$

$$(Re^{-1}u^\alpha_{,2})_{j+e_1} \cong \frac{1}{4}Re^{-1}_{j+e_1}(u^\alpha|^{j+2e_2}_{j-2e_2} + u^\alpha|^{j+2e_1+2e_2}_{j+2e_1-2e_2})/h_2 , \tag{6.17}$$

etcetera. For the inertia terms we may use

$$(u^\alpha u^\beta)_{j+e_\beta} \cong \frac{1}{2}(u^\alpha u^\beta)_j + \frac{1}{2}(u^\alpha u^\beta)_{j+2e_\beta} , \tag{6.18}$$

corresponding to second order central discretization.

The first order upwind scheme is obtained with (abandoning the summation convention for the remainder of this section)

$$(u^\alpha u^\beta)_{j+e_\beta} \cong (u^\alpha u^\beta)_j, \quad u^\beta_j + u^\beta_{j+2e_\beta} \geq 0 ;$$

$$(u^\alpha u^\beta)_{j+e_\beta} \cong (u^\alpha u^\beta)_{j+2e_\beta}, \quad u^\beta_j + u^\beta_{j+2e_\beta} < 0 . \tag{6.19}$$

The hybrid scheme easily follows from (6.18) and (6.19) using (4.36). For better accuracy one may use the κ-scheme (4.89) or the limited κ-scheme (4.94), which becomes in the present case, using (4.92),

$$(u^\alpha u^\beta)_{j+e_\beta} \cong f(\tilde{\varphi}_j), \quad \varphi = u^\alpha u^\beta,$$

$$\tilde{\varphi}_j = (\varphi_j - \varphi_{j-2e_\beta})/(\varphi_{j+2e_\beta} - \varphi_{j-2e_\beta}) , \tag{6.20}$$

where we assume $u^\beta_j + u^\beta_{j+2e_\beta} \geq 0$. In the contrary case we obtain using (4.95):

$$(u^\alpha u^\beta)_{j+e_\beta} \cong f(\tilde{\varphi}_{j+2e_\beta}), \quad \tilde{\varphi} = (\varphi - \varphi_{j+4e_\beta})/(\varphi_j - \varphi_{j+4e_\beta}) . \qquad (6.21)$$

For the κ-scheme without limiting, according to (4.93), f is given by

$$f(\tilde{\varphi}) = \frac{2-\kappa}{2}\tilde{\varphi} + \frac{1+\kappa}{4} .$$

Of course, if no limiting is used, for efficiency in programming one should not introduce the normalized variable $\tilde{\varphi}$, but work with u^α and u^β directly. For limiting, various possibilities for $f(\tilde{\varphi})$ have been discussed in Sect. 4.8.

Frequently, for the purpose of iterative solution, the nonlinear term $u^\alpha u^\beta$ is linearized, for example by $u^\alpha u^\beta \cong u^\alpha \bar{u}^\beta$ where $\bar{u}^\beta$ is known from a previous iteration or time level, or, with Newton's method, as $u^\alpha u^\beta \cong \bar{u}^\alpha u^\beta + u^\alpha \bar{u}^\beta - 2\bar{u}^\alpha \bar{u}^\beta$. It is a bad idea to use $u^\alpha u^\beta \cong \bar{u}^\alpha u^\beta$, because it is u^α that should be updated in the u^α-equation and not u^β; divergence results. With these linearizations, the following form of the κ-scheme is convenient:

$$\begin{aligned}
(u^\alpha u^\beta) &= u^\beta_{j+e_\beta} f(\tilde{u}^\alpha_j), \quad u^\beta_{j+e_\beta} = \tfrac{1}{2}(u^\beta_j + u^\beta_{j+2e_\beta}), \\
\tilde{u}^\alpha_j &= (u^\alpha_j - u^\alpha_{j-2e_\beta})/(u^\alpha_{j+2e_\beta} - u^\alpha_{j-2e_\beta}),
\end{aligned} \qquad (6.22)$$

assuming $u^\beta_{j+e_\beta} \geq 0$; the contrary case easily follows from (6.21).

Discretization of convection-diffusion equation

In a similar way, finite volume discretization of equation (6.3) gives

$$h_1 h_2 dT_j/dt + h_2(u^1 T)|^{j+e_1}_{j-e_1} + h_1(u^2 T)|^{j+e_2}_{j-e_2}$$
$$-h_2(Pe^{-1}T_{,1})|^{j+e_1}_{j-e_1} - h_1(Pe^{-1}T_{,2})|^{j+e_2}_{j-e_2} = h_1 h_2 q_j .$$

To approximate this further in terms of cell center values one may write

$$(Pe^{-1}T_{,\alpha})_{j+e_\alpha} \cong Pe^{-1}_{j+e_\alpha}(T_{j+2e_\alpha} - T_j)/h_\alpha .$$

Approximation of the convective flux $(u^\alpha T)_{j+e_\alpha}$ is quite analogous to the approximation of the inertia flux as given before. For example, the limited κ-scheme is obtained with (cf. (6.21))

$$(u^\alpha T)_{j+e_\alpha} \cong f(\tilde{\varphi}_j), \quad \varphi = u^\alpha T , \qquad (6.23)$$

assuming $u^\alpha_j + u^\alpha_{j+2e_\alpha} \geq 0$, or alternatively (cf. (6.22))

$$(u^\alpha T)_{j+e_\alpha} \cong u^\alpha_{j+e_\alpha} f(\tilde{T}_j), \quad u^\alpha_{j+e_\alpha} = \frac{1}{2}(u^\alpha_j + u^\alpha_{j+2e_\alpha}) .$$

Spurious checkerboard modes

The preceding straightforward spatial discretization method is completely analogous to what was done for the convection-diffusion equation in preceding chapters. But now a difficulty not encountered before crops up, that has to do with the lack of a time-evolution equation for the pressure p and the occurrence of the continuity equation (6.1). Assume $Re = $ constant, and neglect boundary conditions. We will show that

$$u_j^\alpha = (-1)^{j_1+j_2} g(t), \quad p = (-1)^{j_1+j_2} \tag{6.24}$$

is a solution of (6.15)–(6.18) for suitable $g(t)$. Equation (6.13) is trivially satisfied. The contributions of p and $u^\alpha u^\beta$ in (6.15) obviously cancel. We have

$$\mathrm{Re}\sigma_{j+e_1}^{11} \cong \frac{2}{h_1} u^1 |_j^{j+2e_1},$$

$$\mathrm{Re}\sigma_{j+e_1}^{12} \cong \frac{1}{h_1} u^2 |_j^{j+2e_1} + \frac{1}{4h_2} (u^1 |_{j-2e_2}^{j+2e_2} + u^1 |_{j+2e_1-2e_2}^{j+2e_1+2e_2}).$$

With u_j^α given by (6.24), we find

$$h_2 F^{\alpha 1} |_{j-e_1}^{j+e_1} = \frac{12h_1}{h_2 \mathrm{Re}} (-1)^{j_1+j_2} g(t) .$$

Similarly,

$$h_1 F^{\alpha 2} |_{j-e_2}^{j+e_2} = \frac{12h_1}{h_2 \mathrm{Re}} (-1)^{j_1+j_2} g(t) .$$

Substitution in (6.15) gives

$$\frac{dg}{dt} + (h_1^{-2} + h_2^{-2}) \frac{12}{\mathrm{Re}} g = 0 ,$$

so that (6.24) is a solution with

$$g(t) = \exp\{-\frac{12}{\mathrm{Re}} (h_1^{-2} + h_2^{-2}) t\} .$$

This shows that if $Re \gg 1$ the spurious checkerboard mode (6.24) is damped slowly for the velocity, but not at all for the pressure. In practice this spurious mode is found to occur to such an extent as to make the above spatial discretization useless.

One-sided discretization of div$u$ and grad$p$

The spurious checkerboard mode is made possible because the discretization of div$u = u_{,\alpha}^\alpha$ in (6.13) and of grad$p$ in (6.15), (6.16) straddles the central cell

Ω_j. This may be avoided by using forward differences for $u_{\alpha,\alpha}$ and backward differences for $p_{,\alpha}$, replacing (6.13) by

$$h_2 u^1 |_j^{j+2e_1} + h_1 u^2 |_j^{j+2e_2} = 0 \,,$$

and by approximating $p_{,\alpha}$ in $\boldsymbol{x}_j$ by

$$p_{,\alpha}(\boldsymbol{x}_j) \cong \frac{1}{h_\alpha} p|_{j-2e_\alpha}^j \,,$$

or the other way around. A proof of convergence is given in Ellison, Hall, and Porsching (1987). The method is applied in Fuchs and Zhao (1984), who find that the accuracy is $\mathcal{O}(h_1 + h_2)$.

Flux splitting

In Dick (1988), Dick (1989), Dick and Linden (1992) a flux splitting method is proposed for the incompressible Navier-Stokes equations. This method is similar to the flux splitting scheme of Roe for the compressible Euler equations, to be discussed in Sect. 10.3. This introduces a combination of forward and backward one-sided differences in a more sophisticated way than the preceding method, ruling out spurious modes as well, but at higher computing cost.

Pressure-weighted interpolation method

A colocated discretization method of accuracy $\mathcal{O}(h_1^2 + h_2^2)$ has been proposed by Rhie and Chow (1983). The idea is to prevent checkerboard oscillations by perturbing the continuity equation with pressure terms. Discretization of the momentum equations is as before. In Rhie and Chow (1983) the description of the discretization is intertwined with the description of an iterative solution method, and it is not easy to see which discretization is actually solved. Furthermore, depending on the choice for the iterative solution method, the final result may depend on certain relaxation factors, which would be undesirable. This is not the case for the version proposed in Miller and Schmidt (1988), where also the discretization that is solved is given. The continuity equation is discretized as in (6.12), with the cell face velocity components evaluated as follows:

$$u^\alpha_{j+e_\alpha} = \frac{1}{2}(u^\alpha_j + u^\alpha_{j+2e_\alpha}) + (\frac{h_\beta}{4a^\alpha}\Delta^\alpha p)|_j^{j+2e_\alpha} \quad \text{(no summation)} \,, \qquad (6.25)$$

where $\Delta^\alpha p_j = p_{j+2e_\alpha} - 2p_j + p_{j-2e_\alpha}$, $\beta \neq \alpha$ and a^α_j equals the negative sum of the coefficients of u^α_k, $k \neq j$ in equation (6.15). This way of expressing

$u^{\alpha}_{j+e_{\alpha}}$ in terms of cell center values is often called *pressure-weighted interpolation*, and the resulting colocated discretization is called the pressure-weighted interpolation (PWI) method.

The reasoning behing (6.25), see Rhie and Chow (1983), Miller and Schmidt (1988), is intuitive and somewhat involved. The basic mathematical principle is that a small regularizing term is added that excludes spurious modes. We have $\Delta^{\alpha}p = \mathcal{O}(h^{2}_{\alpha})$, so that, assuming a^{α}_{j} is smooth, the pressure term in (6.25) is $\mathcal{O}(h^{3}_{\alpha}h_{\beta})$. The error in linear interpolation between u^{α}_{j} and $u^{\alpha}_{j+2e_{\alpha}}$ is $\mathcal{O}(h^{2}_{\alpha})$, so that the regularizing pressure term in (6.25) does not affect the order of accuracy of the discretization error for the continuity equation. The main idea behind (6.25) is to use the momentum equations in an approximate way to interpolate cell face velocity components, in order to bring in the pressure. However, if there is a significant body force present, (6.25) does not work, and should be modified to bring in the body force, see Gu (1991).

Substitution of (6.25) in (6.12) results in the following discretization of div$\boldsymbol{u} = 0$ with the PWI method:

$$h_2 u^1|^{j+2e_1}_{j-2e_1} + h_1 u^2|^{j+2e_2}_{j-2e_2}$$
$$+ h^2_2\{(\frac{1}{2a^1}\Delta^1 p)_{j+2e_1} - (\frac{1}{a^1}\Delta^1 p)_j + (\frac{1}{2a^1}\Delta^1 p)_{j-2e_1}\} \tag{6.26}$$
$$+ h^2_1\{(\frac{1}{2a^2}\Delta^2 p)_{j+2e_2} - (\frac{1}{a^2}\Delta^2 p)_j + (\frac{1}{2a^2}\Delta^2 p)_{j-2e_2}\} = 0 \ .$$

Because in discretized form div$\boldsymbol{u} = 0$ only in the sense of (6.26), for the computation of mass flux through a cell face $u^{\alpha}_{j+e_{\alpha}}$ as given by (6.25) should be used in convection terms. This means that (6.23) and (6.18) must be replaced by

$$(u^{\alpha}T)_{j+e_{\alpha}} \cong \frac{1}{2}u^{\alpha}_{j+e_{\alpha}}(T_{j+2e_{\alpha}} + T_j)$$

and

$$(u^{\alpha}u^{\beta})_{j+e_{\alpha}} \cong \frac{1}{2}u^{\beta}_{j+e_{\alpha}}(u^{\alpha}_{j+2e_{\alpha}} + u^{\alpha}_j) \tag{6.27}$$

for second order central discretization. For other discretizations of convection, such as those discussed in Chapters 3 and 4 (first order upwind, hybrid or a (limited) κ-scheme), $T_{j+e_{\alpha}}$ and $u^{\alpha}_{j+e_{\alpha}}$ (in 6.15) are approximated by the appropriate formulae. This has been described in Sect. 4.8 and does not need to be discussed further.

Boundary conditions

Boundary conditions for $\boldsymbol{u}_j$ and T are implemented in the same way as for cell-centered discretization of the convection-diffusion equation, as described in Chapt. 3. The reader is reminded of the fact that, as seen in Chapt. 3, loss of order of accuracy in local error at boundaries does not cause the global error to deteriorate, so that, in spite of suggestions in the literature to the contrary, there is no need to introduce seemingly more accurate and more complicated discretizations at boundaries, or to prefer vertex-centered grids.

The PWI method requires some further specifications of conditions at boundaries, beyond what is given for the differential equations, which is a disadvantage of PWI. Let $(j = 1, j_2)$, so that the 'west' face of Ω_j is part of the boundary. In (6.15) (after using (6.16)), p_{j-2e_1} occurs, referring to a grid point outside the grid G. This value is approximated by extrapolation from the interior (Perić, Kessler, and Scheuerer (1988), Rhie and Chow (1983)):

$$p_{j-2e_1} = 2p_j - p_{j+2e_1} \,. \tag{6.28}$$

This is artificial, since the differential equations are not accompanied by a boundary condition for the pressure.

If the normal (outward) velocity $v(\boldsymbol{x}, t)$ at the 'west' face of Ω_j is prescribed, then one uses

$$u^1_{j-e_1} = -v(\boldsymbol{x}_{j-e_1}, t) \tag{6.29}$$

instead of (6.25) in (6.12). According to (6.25), in $u^1_{j+e_1}$ the value p_{j-2e_1} occurs. This is eliminated with (6.28). As a result, no pressure values outside G occur in (6.26). Of course, (6.29) is also used in (6.25) and (6.27) for normal cell face velocities in $\boldsymbol{x}_{j-e_1}$. If the normal velocity is prescribed over the whole boundary, then adding the discrete continuity equation (6.26) for each and every cell and substituting (6.29) at the boundaries results in

$$h_2\{\Sigma_E v_E - \Sigma_W v_W\} + h_1\{\Sigma_N v_N - \Sigma_S v_S\} = 0 \,, \tag{6.30}$$

where Σ_E indicates summation over all cell faces at the 'east' side of the boundary of Ω, etc. This is the discrete version of the compatibility relation (6.11). The boundary data must be specified such that (6.30) is satisfied within rounding error.

We found in the literature no discussion of the implementation of the free surface conditions (6.10) for the PWI scheme. Of course, in (6.15), $p_{j-e_1} - (\mathrm{Re}^{-1}\sigma^{11})_{j-e_1}$ is replaced by p_∞, and $(\mathrm{Re}^{-1}\sigma^{21})_{j-e_1}$ and $(\mathrm{Re}^{-1}\sigma^{12})_{j-e_1}$ are replaced by zero, so that no unknowns outside G occur in (6.15). In (6.26), in the expression (6.25) for $u^1_{j+e_1}$ we eliminate p_{j-2e_1} with (6.28). It remains to choose $u^1_{j-e_1}$. This has to be extrapolated from the interior. Two possibilities arise:

$$u^1_{j-e_1} = u^1_j, \quad u^1_{j-e_1} = \frac{3}{2}u^1_j - \frac{1}{2}u^1_{j+2e_1} \,.$$

We will not go into this further.

Often, the pressure distribution at boundaries needs to be computed. However, after the computation with the PWI scheme has been completed, the pressure is available only at cell centers. If required, the pressure at a boundary may be obtained by extrapolation from the interior (Perić, Kessler, and Scheuerer (1988), Rhie and Chow (1983)), for example by

$$p_{j-e_1} = \frac{3}{2}p_j - \frac{1}{2}p_{j+2e_1} .$$

Summary of equations

After spatial discretization, a system of ordinary differential equations in time is obtained. Putting all unknown velocity components is some order in an algebraic vector u and all pressure unknowns in an algebraic vector p, and dividing (6.15) and (6.26) by $h_1 h_2$, these equations can be written as

$$\frac{du}{dt} + N(u) + Gp = f(t) ,$$

$$Du + Bp = g(t) .$$

(6.31)

Here N is a nonlinear algebraic operator arising from the discretization of the inertia and viscous terms, G is a linear algebraic operator representing the discretization of the pressure gradient, D and B are linear algebraic operators corresponding to the velocity terms and the pressure terms, respectively, in the discretization (6.26) of the continuity equation, and f and g are known source terms arising from the boundary conditions and body forces. The system contains no time derivative for p, and contains both differential and algebraic equations. Therefore it is called a *differential-algebraic system*.

This completes our description of spatial discretization with the PWI method of Rhie and Chow (1983). Temporal discretization will be discussed later. A theory of convergence does not seem to be available. Numerical tests and comparisons with other discretization methods are given in Drikakis and Schäfer (1994), Kobayashi and Pereira (1991), Miller and Schmidt (1988), Perić, Kessler, and Scheuerer (1988). The accuracy seems to be the same as for the staggered discretization to be discussed next, and is $\mathcal{O}(h_1^2 + h_2^2)$.

Due to the relative ease with which colocated discretization can be extended to non-orthogonal grids, use of the PWI method has become widespread. Stationary flow computations are presented in Acharya and Moukalled (1989), Burns, Jones, Kightley, and Wilkes (1987), Deng (1989), Gu (1991), Kobayashi and Pereira (1991), Lapworth (1988), Lien and Leschziner (1994), Miller and Schmidt (1988), Moukalled and Acharya (1991), Perić, Kessler, and Scheuerer (1988), Rhie and Chow (1983), Rodi, Majumdar, and Schönung (1989); instationary flows are treated with the PWI method in Armfield (1994), Demirdžić and Perić (1990), Ho and Lakshminarayana (1993), Zang, Street, and Koseff

(1994). But if domain-decomposition is used (for greater geometric flexibility, or for parallel computing), then the coupling conditions at subdomain boundaries are more complicated than for staggered discretization, due to the artificial pressure terms.

Artificial compressibility method

Instead of using PWI to suppress the spurious modes, one may introduce an artificial compressibility term in the continuity equation. This has been proposed in Chorin (1967), and applied in Belov, Martinelli, and Jameson (1994), Chang and Kwak (1984), Drikakis and Schäfer (1994), Hartwich and Hsu (1988), Kwak and Chang (1984), Kwak, Chang, Shanks, and Chakravarthy (1986), Rogers and Kwak (1990), Rogers, Kwak, and Kiris (1991), Sheng, Taylor, and Whitfield (1995), Soh and Goodrich (1988), Tamamidis, Zhang, and Assanis (1996). For an introduction, see Peyret and Taylor (1985). The continuity equation (6.1) is replaced by the following artificial equation:

$$\beta \frac{\partial p}{\partial t} + u^\alpha_{,\alpha} = 0 \,, \tag{6.32}$$

with β a parameter to be chosen. When steady state is reached, (6.32) is equivalent with the original continuity equation (6.1), but the time dependence of the solution is falsified. This is a major reason why use of the artificial compressibility method is less widespread than the PWI method. However, time-dependence can be approximated by introducing a pseudo-time variable, see Merkle and Tsai (1986), Merkle and Athvale (1987), Soh and Goodrich (1988), Rogers and Kwak (1990), Rogers, Kwak, and Kiris (1991). Both for brevity and because it is less widely used than PWI, the artificial compressibility method will not be discussed further. In a comparative study of the PWI and artificial compressibility methods by Tamamidis, Zhang, and Assanis (1996), the PWI method is found to give better accuracy and mass conservation, and to require less memory.

6.4 Spatial discretization on staggered grid

This is the oldest and most straightforward approach to discretizing the Navier-Stokes equations. The method was first proposed in Harlow and Welch (1965), and is described in detail in Patankar (1980). On orthogonal grids it remains the method of choice. No artificial boundary conditions are necessary; the physical boundary conditions suffice. No special measures are required to prevent spurious modes. Theoretical convergence analysis is more developed than for the colocated scheme. For theoretical background, see Hou and Wetton (1993), Zhang and Kwok (1997) and references quoted there.

Staggered grid

Grid points for different unknowns are staggered with respect to each other. The pressure resides in the cell centers, whereas the cell face centers contain the normal velocity components, cf. Fig. 6.1. The same system of grid point

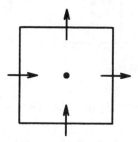

Fig. 6.1. Staggered placement of unknowns; →, ↑: velocity components; •: pressure.

indices is used as in the preceding section. Hence, x_j is a pressure point containing p_j and x_{j+e_α}, $\alpha = 1,2$ are velocity points containing $u^1_{j+e_1}$ or $u^2_{j+e_2}$ respectively.

Note that with a staggered grid we always have a mixture of vertex-centered and cell-centered discretization. Unavoidably, at a boundary, some unknowns will have nodes upon it, whereas other unknowns have no nodes on this boundary, but half a mesh size removed. Therefore it is fortunate, as seen in Chapt. 3, that vertex-centered and cell-centered discretization are on equal footing as far as global accuracy and ease of implementation of boundary conditions are concerned. It might be convenient to put a boundary where the normal stress is prescribed (first condition of (6.10)), through pressure points, but there is no real need to do so.

Discretization of continuity equation

As before, the continuity equation is integrated over Ω_j, resulting in (cf. (6.12))

$$h_2 u^1 |^{j+e_1}_{j-e_1} + h_1 u^2 |^{j+e_2}_{j-e_2} = 0 \ . \tag{6.33}$$

The advantage of the staggered placement of the unknowns is that no further approximation is necessary in this equation.

Discretization of momentum equations

Finite volume integration takes place over control volumes centered at u^α grid points. For example, for $\alpha = 1$ the momentum equation (6.14) is integrated over Ω_{j+e_1} (cf. Fig. 6.2). The treatment for $\alpha = 2$ is quite similar, and will

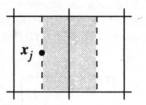

Fig. 6.2. Control volume for $u^1_{j+e_1}$.

not be given. Similarly to (6.15) one obtains

$$\int_{\Omega_{j+e_1}} \{\frac{\partial u^1}{\partial t} + F^{1\beta}_{,\beta} + p_{,1}\}d\Omega \cong h_1 h_2 du^1_{j+e_1}/dt + h_2 F^{11}|^{j+2e_1}_j$$

$$+ h_1 F^{12}|^{j+e_1+e_2}_{j+e_1-e_2} + h_2 p|^{j+2e_1}_j = 0 . \tag{6.34}$$

Now p occurs only in its own nodal points; interpolation is not necessary for p and spurious modes cannot occur. This is the advantage of staggered over colocated discretization. But a number of other quantities have to be interpolated between grid nodes. The viscous stress $\sigma^{1\beta}$ is approximated using

$$(\mathrm{Re}^{-1}u^1_{,1})_j \cong (h_1 \mathrm{Re}_j)^{-1}u^1|^{j+e_1}_{j-e_1} ,$$

$$(\mathrm{Re}^{-1}u^1_{,2})_{j+e_1+e_2} \cong (h_2 \mathrm{Re}_{j+e_1+e_2})^{-1}u^1|^{j+e_1+2e_2}_{j+e_1} ,$$

$$(\mathrm{Re}^{-1}u^2_{,1})_{j+e_1+e_2} \cong (h_1 \mathrm{Re}_{j+e_1+e_2})^{-1}u^2|^{j+2e_1+e_2}_{j+e_2} .$$

With the (limited) κ-scheme (4.94), the inertia terms are discretized as follows (cf. (6.20)):

$$(u^1u^1)_j \cong f(\tilde{\varphi}_{j-e_1}), \quad \varphi = u^1u^1 ,$$

$$\tilde{\varphi}_{j-e_1} = (\varphi_{j-e_1} - \varphi_{j-3e_1})/(\varphi_{j+e_1} - \varphi_{j-3e_1}) \tag{6.35}$$

if $u^1_{j-e_1} + u^1_{j+e_1} \geq 0$, whereas in the contrary case

$$(u^1u^1) \cong f(\tilde{\varphi}_{j+e_1}), \quad \tilde{\varphi}_{j+e_1} = (\varphi_{j+e_1} - \varphi_{j+3e_1})/(\varphi_{j-e_1} - \varphi_{j+3e_1}) , \tag{6.36}$$

or alternatively, in analogy with (6.22)

$$(u^1u^1)_j \cong u^1_j f(\tilde{u}_{j-e_1}), \quad u^1_j = \frac{1}{2}(u^1_{j-e_1} + u^1_{j+e_1}) , \tag{6.37}$$

assuming $u^1_j \geq 0$; the contrary case easily follows from (6.36). The second type of inertia term that has to be approximated in (6.34) is $(\rho u^1 u^2)_{j+e_1+e_2}$. Similarly to (6.35), we write

$$(u^1u^2)_{j+e_1+e_2} \cong f(\tilde{\varphi}_{j+e_1}), \quad \varphi = u^1u^2 ,$$
$$\tilde{\varphi}_{j+e_1} = (\varphi_{j+e_1} - \varphi_{j+e_1-2e_2})/(\varphi_{j+e_1+2e_2} - \varphi_{j+e_1-2e_2}) , \qquad (6.38)$$

assuming

$$u^2_{j+e_1+e_2} = \frac{1}{2}(u^2_{j+e_2} + u^2_{j+2e_1+e_2}) \ge 0 . \qquad (6.39)$$

The contrary case is left to the reader. In the order to evaluate φ_{j+e_1} etc. in (6.38), $u^2_{j+e_1}$ etc. has to be interpolated, which is done as in (6.39). This brings in quite a few grid points in the discretization, which is avoided by following (6.37) instead of using (6.38):

$$(u^1u^2)_{j+e_1+e_2} \cong u^2_{j+e_1+e_2} f(\tilde{u}^1_{j+e_1}) ,$$

assuming $u^2_{j+e_1+e_2}$ (given by (6.39)) to be positive; of course, $\tilde{u}^1_{j+e_1}$ is defined similar to $\tilde{\varphi}_{j+e_1}$ in (6.38).

Boundary conditions

On the staggered grid the implementation of the boundary conditions (6.4)–(6.10) is just as simple and done in the same way as for the convection-diffusion equation in Chapters 3 and 4. It is noteworthy that, unlike the colocated case, the boundary conditions that accompany the differential equations suffice to complete the discretization, and that no artificial extra boundary conditions for the pressure are required. This is easily seen as follows. Let $j = (1, j_2)$, so that the 'west' face of Ω_j is at the boundary. If the normal velocity is given, we know $u^1_{j-e_1}$, so that equation (6.34) is not used, and no reference to pressure or velocity nodes outside the domain is made. If the normal velocity is not prescribed, then, as in Sect. 3.4 for the convection-diffusion equation on a vertex-centered grid, the finite volume integration (6.34) is carried out over half a cell. For example, for $j = (1, j_2)$, the equation for $u^1_{j-e_1}$ (which is located at the boundary $x^1 = 0$) is obtained by integration over Ω_{j-e_1}, which is the rectangle with vertices at $\boldsymbol{x}_{j-e_1\pm e_2}$ and $\boldsymbol{x}_{j\pm e_2}$. This gives (cf. (6.34))

$$\int_{\Omega_{j-e_1}} \{\frac{\partial u^1}{\partial t} + F^{1\beta}_{,\beta} + p_{,1}\}d\Omega \cong \frac{1}{2}h_1h_2 du^1_{j-e_1}/dt$$

$$+ h_2 F^{11}|^j_{j-e_1} + \frac{1}{2}h_1 F^{12}|^{j-e_1+e_2}_{j-e_1-e_2} + h_2 p|^j_{j-e_1} = 0 . \qquad (6.40)$$

Since the normal velocity is not prescribed, the normal stress $p + \sigma^{11} = p_\infty$ is prescribed, according to the boundary conditions (6.10). This means that we can substitute

$$F^{11}_{j-e_1} + p_{j-e_1} = (u^1u^1)_{j-e_1} - \sigma^{11}_{j-e_1} + p_{j-e_1} = (u^1u^1)_{j-e_1} + p_\infty$$

in equation (6.40). For $F_{j-e_1+e_2}^{12}$ we need to approximate $(u^1 u^2 - \sigma^{12})_{j-e_1+e_2}$. We write

$$u_{j-e_1+e_2}^1 = \frac{1}{2}(u_{j-e_1}^1 + u_{j-e_1+2e_2}^1) . \tag{6.41}$$

If u^2 is given at the boundary under consideration, the prescribed value is substituted directly in $u^1 u^2$. Furthermore, in this case one has to approximate $\sigma_{j-e_1+e_2}^{12}$, for which we need to approximate $u_{,2}^1 + u_{,1}^2$. This can be done as follows:

$$(u_{,2}^1)_{j-e_1+e_2} \cong \frac{1}{h_2}(u_{j-e_1+2e_2}^1 - u_{j-e_1}^1) ,$$

$$(u_{,1}^2)_{j-e_1+e_2} \cong \frac{2}{h_1}(u_{j+e_2}^2 - u_{j-e_1+e_2}^2) ,$$

where $u_{j-e_1+e_2}^2$ is given by the boundary condition. If u^2 is not given, then the tangential stress is prescribed, according to the boundary conditions (6.9), (6.10). We substitute $\sigma^{12} = f$ in F^{12}, and it remains to determine $(u^1 u^2)_{j-e_1+e_2}$. For u^1 we use (6.41), and u^2 is approximated with a one-sided bias by $u_{j+e_2}^2$.

Summary of equations

Putting all unknown velocity components in some order in an algebraic vector u and all pressure unknowns in an algebraic vector p, and dividing by $h_1 h_2$, these equations go over in a differential-algebraic system of the following structure:

$$\frac{du}{dt} + N(u) + Gp = f(t) , \qquad Du = g(t) . \tag{6.42}$$

Here N is a nonlinear algebraic operator arising from the discretization of the inertia and viscous terms, G is a linear algebraic operator representing the discretization of the pressure gradient, D is a linear algebraic operator representing the discretization of the continuity equation, and f and g are known source terms, arising from the boundary conditions.

This completes our description of spatial discretization on the staggered grid.

6.5 On the choice of boundary conditions

Having completed our discussion of the principles of spatial discretization of the incompressible Navier-Stokes equations on Cartesian grids, we will now present further conditions on the choice of boundary conditions. This choice is not always clear-cut. At solid boundaries only one possibility presents itself, namely the no-slip condition, but at inflow and outflow boundaries choices may have to be made that influence physical realism and numerical accuracy. Some guidelines will be given by discussing two illustrative examples.

Flow over a backward facing step

Out first example is the flow over a backward facing step (Fig. 6.3). This is a frequently used test problem for testing numerical methods (Morgan, Périaux, and Thomasset (1984), Gartling (1990), Leone Jr. (1990)). The grid is not Cartesian, but this is irrelevant for what we want to discuss in this

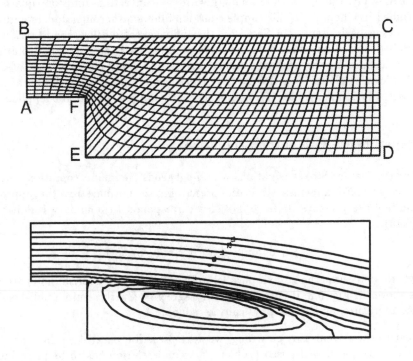

Fig. 6.3. Boundary-fitted grid and streamlines for flow over a backward facing step. Outflow conditions: $-p + \sigma'' = -p_\infty$, $\sigma^{12} = 0$.

section. In Chapt. 13 discretization on non-orthogonal structured grids will be considered. The results presented here were obtained using a staggered scheme.

The geometry of the domain is defined by (cf. Fig. 6.3): $|AB| = |AF| = |EF| = 1/2$, $|BC| = 3, |CD| = 1, |ED| = 2.5$. At AB, which is an inflow boundary, we prescribe $u^2 = 0$, and u^1 is given by a parabolic inflow profile. The Reynolds number based on the length $|AB|$ and the maximum inflow velocity is given by Re = 200. CD is an outflow boundary; the remainder of the boundary consists of solid walls.

Upstream influence of outflow conditions

The outflow conditions (at CD) are not known a priori. But nevertheless, we need to prescribe suitable conditions to make the problem determinate. Fortunately, there is a close analogy with the convection-diffusion equation, studied in Chap. 2–5. This simple equation allows a thorough understanding of many basic issues. In Sect. 2.5 the influence of outflow conditions was studied. Based on what we found there we may trust that the exact solution is influenced by the outflow conditions only in a small neigbourhood of CD of thickness $\mathcal{O}(\mathrm{Re}^{-1})$, if Re $\gg$ 1. Furthermore, rapid local variation of the exact solution is avoided if a homogeneous Neumann condition is applied.

Experiments with outflow conditions

For the Navier-Stokes equations a homogeneous Neumann condition corresponds to zero tangential stress and prescribing the normal stress (i.e. approximately the pressure if Re $\gg$ 1). This corresponds to boundary conditions (6.10):

$$-p + \sigma^{11} = -p_\infty, \quad \sigma^{21} = 0 \quad \text{on CD},$$

where we assume the x^1-axis parallel to BC. The resulting streamline pattern is shown in Fig. 6.3; it is in satisfactory agreement with results published in the literature, and seems physically reasonable.

Just to show what may go wrong, we next prescribe parallel outflow: $u^2 = 0$ at CD. To get a well-posed problem, a second relation has to be prescribed in relation to the equation for u^1. This could be the flux through CD, or u^1 itself. Starting with the first possibility, we prescribe

$$-p + \sigma^{11} = -p_\infty, \quad u^2 = 0 \quad \text{on CD}. \tag{6.43}$$

We now have a Dirichlet condition for u^2. We use the second order central discretization for the convection term. Keeping the analogy with the convection-diffusion equation in mind, we know that we now run the risk of spurious numerical wiggles, based on our experience with the convection-diffusion equation in Sect. 4.2. Spurious numerical wiggles indeed occur, as

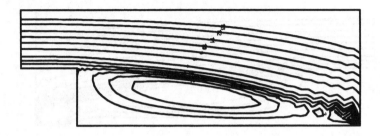

Fig. 6.4. Streamlines for flow over a backward facing step. Outflow conditions: $-p + \sigma^{11} = p_\infty$, $u^2 = 0$.

shown in Fig. 6.4. They get worse when we also prescribe a Dirichlet condition for u^1, namely a parabolic profile with the same mass flux as at the inlet. Again, $u^2 = 0$ at CD. The result is shown in Fig. 6.5. The results of Fig. 6.4 and 6.5 are unacceptable.

As for the convection-diffusion equation, we see that with the central scheme the type of outflow conditions has a large influence on the numerical results, and that the stress conditions (6.43) are to be preferred. With upwind discretization, there would have been no spurious numerical wiggles in Fig. 6.4 and 6.5, and the upstream influence of the outflow conditions would have been restricted to a few grid cells. But the length of the recirculation zone would have been too short, due to the numerical viscosity generated by the upwind scheme.

An analysis of outflow boundary conditions is given by Johansson (1993).

The chimney problem

It may also happen that inflow conditions are unknown a priori. What we shall call the chimney problem is an example. Consider a vertical pipe with a smooth entrance nozzle (Fig. 6.6). The wall of the pipe is hot, so that the fluid flows upward under the influence of body forces. The flow is governed by the Boussinesq equations (6.2), (6.3). We assume that the Rayleigh number is sufficiently large for the convection and inertia terms to be significant. This means that at the inflow boundary AB we must give Dirichlet bound-

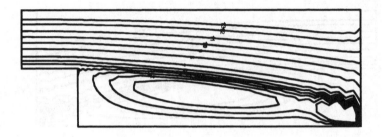

Fig. 6.5. Streamlines for flow over a backward facing step. Outflow conditions: u^1: parabolic profile; $u^2 = 0$.

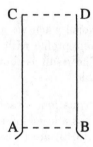

Fig. 6.6. The chimney problem.

ary conditions to obtain a well-posed problem. This has been demonstrated in Sect. 4.2 for the one-dimensional convection-diffusion problem. However, the inflow velocity is not known a priori. This difficulty can be resolved as follows. We assume a smooth inflow pattern without local separation, which is realistic because there is a smooth nozzle in front of the inflow boundary AB. As a consequence, Bernoulli's theorem holds approximately in front of AB. Furthermore, we assume parallel inflow with a constant velocity profile:

$$u^1 = 0 , \quad u^2 = \text{constant on AB}.$$

Bernoulli's theorem gives

$$\frac{1}{2}\rho u^2 + p = p_\infty \quad \text{on AB}, \tag{6.44}$$

where p_∞ is the pressure far away from the pipe. At CD the velocity is assumed to be vertical. Then the horizontal pressure gradient is zero, so that

$p = p_\infty$ on CD. We apply therefore the outflow conditions (6.10), assuming the viscous stress to be negligible. By means of an iterative procedure, u^2 on AB can be determined. We make a guess $u^2 = c^{(0)}$ on AB, solve and find p on AB, say $p = p^{(0)}$. Then we determine $c^{(1)}$ from

$$\frac{1}{2}\rho c^{(1)} c^{(1)} = p_\infty - p^{(0)} ,$$

put $u^{(2)} = c^{(1)}$, solve, and so on, until (6.44) is satisfied with sufficient accuracy.

6.6 Temporal discretization on staggered grid

We now discretize also in time, in order to obtain a fully discretized system. Spatial discretization on a staggered grid is assumed, so that (6.42) is our point of departure.

General formulation on staggered grid

As an example, the second order Adams-Bashforth scheme (5.74) is applied to (6.42), resulting in

$$u^n - u^{n-1} + \frac{1}{2}\tau\{3N(u^{n-1}) - N(u^{n-2}) + 3Gp^{n-1} - Gp^{n-2}\}$$
$$= \frac{1}{2}\tau\{3f(t^{n-1}) - f(t^{n-2})\} , \tag{6.45}$$
$$Du^n = g(t^n) .$$

In order to avoid an overdetermined system, u^n and p^{n-1} must be determined simultaneously. In order to obtain a general formulation that also applies to other time stepping schemes, we define $p^{n-1/2} = \frac{3}{2}p^{n-1} - \frac{1}{2}p^{n-2}$, and write

$$u^n - u^{n-1} + \frac{1}{2}\tau\{3N(u^{n-1}) - N(u^{n-2})\} + \tau Gp^{n-1/2}$$
$$= \frac{1}{2}\tau\{3f(t^{n-1}) - f(t^{n-2})\} , \tag{6.46}$$
$$Du^n = g(t^n) .$$

This formulation brings out more clearly than (6.45) the fact that the pressure acts as a Lagrange multiplier guaranteeing satisfaction of the continuity equation.

As a second example, the Crank-Nicolson scheme is applied to (6.42), which gives

$$u^n - u^{n-1} + \frac{1}{2}\tau\{N(u^n) + N(u^{n-1})\} + \tau G p^{n-1/2}$$
$$= \frac{1}{2}\tau\{f(t^n) + f(t^{n-1})\},$$
$$Du^n = g(t^n).$$

As seen from these examples, multistep methods applied to (6.42) can generally be written as

$$A(u^n) + \tau G p^{n-1/2} = r^n,$$
$$Du^n = g(t^n),$$
(6.47)

where r^n is known from previous time steps and the boundary conditions. For explicit methods, A is the identity I. The system (6.47) is an indefinite system for u^n and $p^{n-1/2}$, that can be solved iteratively with the methods to be described in Chap. 7 for the stationary case. However, less computing time is required if $p^{n-1/2}$ and u^n can be solved for separately. To this end the following method has been devised, which is the method of choice for nonstationary problems.

Pressure-correction method

Equation (6.47) is not solved as it stands, but first a prediction u^* is made that does not satisfy the continuity equation. Then a correction is computed involving the pressure such that the continuity equation is satisfied. The method is given by

$$A(u^*) + \tau G p^{n-3/2} = r^n,$$
(6.48)
$$u^n - u^* + \tau G(p^{n-1/2} - p^{n-3/2}) = 0,$$
(6.49)
$$Du^n = g(t^n).$$
(6.50)

Equation (6.48) more or less amounts to solving discretized convection-diffusion equations for the predicted velocity components. We use the best available guess for the pressure, namely $p^{n-3/2}$. Equation (6.49) is motivated by the fact, that if in the explicit case, where A is the identity, we eliminate u^* from (6.48) and (6.49), then the original system (6.47) is recovered. The pressure can be computed by applying the operator D to (6.49) and using (6.50), resulting in

$$DG\delta p = \frac{1}{\tau}\{Du^* - g(t^n)\}, \quad p^{n-1/2} = p^{n-3/2} + \delta p.$$
(6.51)

After $p^{n-1/2}$ has been computed, u^n follows from (6.49).

Equation (6.49) can be regarded as a correction of u^* for the change in pressure. Therefore (6.48)–(6.51) is called the *pressure-correction method*. It is an example of a *fractional step method*, in which a time step is split up in sub-steps, and different physical effects are accounted for separately in the sub-steps. Here pressure forces are accounted for in the second sub-step, and inertia and friction in the first sub-step. Confusingly, the term pressure-correction method is often also applied to various iterative methods to solve the stationary Navier-Stokes equations, in which velocity and pressure updates are carried out not simultaneously but successively. Such methods will be encountered in Chap. 7, where they will be called *distributive iteration methods*. These should not be confused with the pressure-correction method used in time accurate schemes as formulated above. This method may also be called a *projection method*, because in (6.49) the new velocity u^n is the projection of the intermediate velocity field u^* on the space of velocity fields with discretized divergence equal to zero.

Remembering that divgrad equals the Laplacian, we see that (6.51) looks very much like a discrete Poisson equation; it is frequently called the pressure Poisson equation. Note that no boundary condition needs to be invoked for δp (fortunately, for no such condition is given with the original equations), because the boundary conditions have already been taken into account in the construction of D, G and g; the operator DG works exclusively on pressure values in grid points in the interior of the domain. The issue of boundary conditions for the pressure Poisson equation (which does not arise with the approach followed here) is discussed extensively in Gresho and Sani (1987), Gresho (1991), Gresho (1992).

Even if the method is explicit ($A = I$), we still have to solve an implicit system for δp. This is an unavoidable consequence of the differential-algebraic nature of (6.42).

As we remarked before, by elimination of u^* it is easily seen that in the explicit case the pressure-correction method (6.48)–(6.51) is equivalent to (6.47), and that this remains true if $p^{n-3/2}$ is neglected in (6.48) and (6.49). But in the implicit case this does not hold, and inclusion of a sufficiently accurate first guess, such as $p^{n-3/2}$, for the pressure in (6.48) seems to be necessary to obtain full, i.e. $\mathcal{O}(\tau^2)$, temporal accuracy. This may make it necessary to compute the initial pressure field at the starting step ($n = 1$), to be used instead of $p^{-1/2}$. This may be done as follows. Application of D to (6.42) at $t = 0$ gives

$$dg(0)/dt + DN(u(0)) + DGp(0) = Df(0) . \tag{6.52}$$

After solving $p(0)$ from (6.52), we put $p^{-1/2} = p(0)$.

In Cartesian coordinates, D is the adjoint of G, but as this is not the case in general coordinates, this is not be assumed here.

Discrete compatibility condition

In case the pressure is not involved in any of the boundary conditions, it follows from the incompressible Navier-Stokes equations that the pressure is determined up to a constant. The system (6.51) for δp is singular in this case, and (6.51) has a solution only if the right-hand side satisfies a compatibility condition. The boundary conditions discussed in Sect. 6.2 are such that if the pressure is not involved in any of the boundary conditions, then the normal velocity component is prescribed all along the boundary, and the compatibility condition (6.11) is satisfied. Summing the discrete continuity equation (6.33) over all cells reduces, due to cancellation in the interior, to the following sum over boundary ponts:

$$\sum_{j_2} h_2 v|_{(1/2,j_2)}^{(J_1+1/2,j_2)} + \sum_{j_1} h_1 v|_{(j_1,1/2)}^{(j_1,J_2+1/2)} = 0 , \tag{6.53}$$

where v is the prescibed velocity component at the boundary. If (6.53) is not satisfied exactly one should adjust v, which should not be difficult, because of the compatibility condition (6.11). If (6.53) holds, then it turns out that the elements of the right-hand side vector of (6.51) sum to zero, which is precisely the compatibility condition required for existence of solutions of (6.51). If one desires to make the solution unique one can fix δp in some point, but iterative methods usually converge faster if one lets δp float.

Temporal accuracy

The pressure-correction method has been formulated and studied in Bell, Colella, and Glaz (1989), Chorin (1968), Chorin (1969), Chorin (1970), Dukowicz and Dvinsky (1992), Harlow and Welch (1965), Hou and Wetton (1993), Shen (1992a), Shen (1992b), Perot (1993), Temam (1977), van Kan (1986), including some rigorous convergence analysis. Indications are that the temporal accuracy of u^n is of the same order as rhe order of accuracy of the underlying time stepping method (for example, $\mathcal{O}(\tau^2)$ for Crank-Nicolson), but that the accuracy of $p^{n-1/2}$ is only $\mathcal{O}(\tau)$, irrespective of the time stepping method used. If one desires, a pressure field with improved accuracy can be obtained after u^n has been computed (with the pressure correction method) by proceeding in the same way as in the derivation of (6.52), leading to the following equation for $p^{n-1/2}$:

$$dg(t^n)/dt + DN(u^n) + DGp^{n-1/2} = Df(t^n) ,$$

which very likely results in a pressure field $p^{n-1/2}$ with the same order of temporal accuracy as the velocity field.

Stability

For stability of (6.48)–(6.51), it seems necessary that (6.48) is stable. It is conjectured that this is sufficient for the stability of (6.48)–(6.51); numerical evidence supports this conjecture, cf. Chorin (1968). We restrict ourselves to Fourier stability analysis of (6.48). To this end (6.48) is linearized, the coefficients are taken constant ('frozen'), the boundary conditions are assumed to be periodic, and the known source terms $\tau G p^{n-3/2}$ and r^n are neglected. Hence, carrying out Fourier stability analysis for (6.48) implies that the discretization of the following simplified and linearized version of (6.2) is considered:

$$\frac{\partial u^\alpha}{\partial t} + U^\beta u^\alpha_{,\beta} - \mathrm{Re}^{-1}(u^\alpha_{,\beta} + u^\beta_{,\alpha})_\beta = 0 \,,$$

with U^β and Re constant. For further simplification we use (6.1), from which it follows that $u^\beta_{,\alpha\beta} = 0$, and we obtain

$$\frac{\partial u^\alpha}{\partial t} + U^\beta u^\alpha_{,\beta} - \mathrm{Re}^{-1} u^\alpha_{,\beta\beta} = 0 \,. \tag{6.54}$$

Equation (6.54) consists of decoupled convection-diffusion equations, for which Fourier stability analysis is presented in Chap. 5. The computing cost of checking in every grid point whether a stability condition is satisfied is often not negligible, so in practice this is often done only every 5 or 10 time steps or so.

We now present a number of methods that are frequently used in practice.

Leapfrog-Euler

We write

$$N(u) = C(u) + Bu \,, \tag{6.55}$$

where the nonlinear operator $C(u)$ represents the spatial discretization of the convection terms, and the linear operator B is related to the viscous terms. Application of the leapfrog-Euler method (discussed in Sect. 5.8) to the velocity predictor version of (6.42) gives

$$\begin{aligned} u^* - u^{n-2} + 2\tau\{C(u^{n-1}) + Bu^{n-2}\} &= 2\tau f(t^{n-1}) \,, \\ u^n - u^* + Gp^{n-1} &= 0 \,, \quad DGp^{n-1} = Du^* - g(t^n) \,, \end{aligned} \tag{6.56}$$

which differs slightly from the general form (6.48)–(6.51). The use of the neutrally stable leapfrog method often causes numerical oscillations, which are commonly suppressed by the following smoothing operation, applied after u^n has been computed:

$$u^{n-1} := (1 - \gamma)u^{n-1} + \frac{1}{2}\gamma(u^n + u^{n-2}),   (6.57)$$

with the constant γ typically between 0.05 and 0.1. The smoothing (6.57) is called the Asselin filter (Asselin (1972)). After smoothing, the boundary conditions are reimposed on u^{n-1}. The global temporal accuracy seems to be $\mathcal{O}(\tau)$ for both velocity and pressure. Computation can be started with an explicit Euler step; in this step τ does not need to be subjected to a stability restriction.

This method has played an important role in the development of large-eddy simulation of turbulence (Arnal and Friedrich (1992), Boersma, Eggels, Pourquié, and Nieuwstadt (1994), Deardorff (1970), Eggels, Unger, Weiss, Westerweel, Adrian, Friedrich, and Nieuwstadt (1994), Friedrich and Unger (1991), Grötzbach (1982), Grötzbach and Wörner (1994), Manhart and Wengle (1994), Mason (1989), Schumann (1975), Schmitt and Friedrich (1988), Wagner and Friedrich (1994), Werner and Wengle (1989).

Usually the second order central scheme is used for spatial discretization, and the heuristic stability condition (5.78), put forward in Schumann (1975) and shown to be sufficient for von Neumann stability of the frozen coefficient version of the predictor step of (6.55) in Pourquié (1994), is commonly used. A weaker sufficient and necessary condition is given in Theorem 5.8.4, resulting in

$$\tau \le \text{Re}/\sum_\alpha h_\alpha^{-2}(2 + \sqrt{4 + qp_\alpha^2}), \quad p_\alpha = |u_\alpha|h_\alpha\text{Re},   (6.58)$$

with $q = 1$ for the second order central scheme and $q = 25/9$ for the fourth order central scheme (in which case (6.58) is merely sufficient).

Adams-Bashforth

Application of the second order Adams-Bashforth scheme (5.74) to (6.42) gives equation (6.46). The pressure-correction method results in

$$u^* - u^{n-1} + \frac{1}{2}\tau\{3N(u^{n-1}) - N(u^{n-2})\} = \frac{1}{2}\tau\{3f(t^{n-1}) - f(t^{n-2})\},$$
$$u^n - u^* + Gp^{n-1/2} = 0, \quad DGp^{n-1/2} = Du^* - g(t^n).$$
$$(6.59)$$

The global temporal accuracy of the velocity field is found to be $\mathcal{O}(\tau^2)$ in practice. Accuracy results on p seem not to have been published, but at least

$\mathcal{O}(\tau)$ is expected. Because it has better temporal accuracy and does not need the Asselin filter, this method now seems to be superseding the leapfrog-Euler scheme. Examples of application can be found in Andersson, Billdal, Eliasson, and Rizzi (1990), Balaras and Benocci (1994), Benocci and Pinelli (1990), Gao (1994), Gavrilakis, Tsai, Voke, and Leslie (1986), Hoffmann and Benocci (1994), Kajishima, Miyake, and Nishimoto (1990), Kobayashi and Kano (1986), Kobayashi, Morinishi, and Oh (1992), Krettenauer and Schumann (1992), Kristoffersen and Andersson (1993), Schmidt and Schumann (1989), Shimomura (1991), Thomas and Williams (1994), Voke and Gao (1995).

In practice one seems to proceed mainly by trial-and-error and by heuristics to determine a suitable size for the time step. But a dependable stability criterion may be deduced from Theorem 5.8.2, which gives for the κ-scheme:

$$\tau \leq \min\{\tau_1, \tau_2, \tau_3\} ,$$

with

$$\tau_1 = \frac{1}{2}\mathrm{Re}/\sum_\alpha h_\alpha^{-2}\{2 + (1 - \kappa)\mathrm{Re}|u_\alpha|h_\alpha\} ,$$

$$\tau_2 = \max\{q_1(2/\mathrm{Re})^{1/3}(\sum_\alpha u_\alpha^4/h_\alpha^2)^{-1/3}, \quad q_2\sqrt{2}/\sum_\alpha |u_\alpha|h_\alpha\} ,$$

$$\tau_3 = \frac{\sqrt{1/3}}{2 - \kappa}\left[\sum_\alpha h_\alpha^{-2}\{2 + (1 - \kappa)\mathrm{Re}|u_\alpha|h_\alpha\} \cdot \right.$$

$$\left. \cdot \sum_\alpha u_\alpha^2\{2 + (1 - \kappa)\mathrm{Re}|u_\alpha|h_\alpha\}^{-1}\right]^{-1/2} ,$$

where we have been a bit more strict than (5.65) for simplicity, and where we have used the fact that parts of (5.61) and (5.65) are identical in this case. Here q_1 and q_2 are given in Theorem 5.7.3. It is left to the reader to derive similar conditions for the fourth order central and first order upwind schemes, using Theorem 5.8.2 and (5.62), (5.66) or (5.63), 5.67), respectively.

Crank-Nicolson

Following ideas put forward in van Kan (1986), Bell, Colella, and Glaz (1989), application of the Crank-Nicolson scheme results in the following pressure-correction method:

$$u^* - u^{n-1} + \frac{1}{2}\tau\{N(u^*) + N(u^{n-1})\} + \tau G p^{n-3/2} = \frac{1}{2}\tau\{f(t^n) + f(t^{n-1})\},$$

$$u^n - u^* + G\delta p = 0,$$

$$DG\delta p = Du^* - g(t^n), \quad p^{n-1/2} = p^{n-3/2} + \delta p.$$

$$(6.60)$$

Unconditional stability for fixed τ and h_α is shown in van Kan (1986). Unconditional von Neumann stability of the velocity predictor step of (6.60) follows from Theorem 5.8.10. As argued in Sect. 5.10, the multistage ω-scheme is to be preferred over the Crank-Nicolson scheme in practice, because the latter scheme is not strongly stable. But it is useful as an ingredient in IMEX schemes, treating the nonlinear inertia terms with an explicit method. Two examples are given below.

However, the solution of nonlinear systems may also be avoided by linearizing the equation for u^*, for example in the way suggested in van Kan (1986). Inspection shows that $N(u^*) + N(u^{n-1})$ consists of terms of the type $\phi^*\psi^* + \phi^{n-1}\psi^{n-1}$, N being a discretization of $(u^\alpha u^\beta)_{,\beta}$. In van Kan (1986) it is shown that without loss of temporal accuracy the following Newton linearization may be applied: $\phi^*\psi^* + \phi^{n-1}\psi^{n-1} \cong \phi^*\psi^{n-1} + \phi^{n-1}\psi^*$. As a result $N(u^*) + N(u^{n-1})$ is replaced by $2E(u^{n-1})u^*$, with E a linear operator, so that the velocity predictor step becomes linear:

$$u^* - u^{n-1} + \tau E(u^{n-1})u^* + \tau G p^{n-3/2} = \frac{1}{2}\tau\{f(t^n) + f(t^{n-1})\}. \quad (6.61)$$

Numerical evidence and theoretical considerations (van Kan (1986), Bell, Colella, and Glaz (1989)) suggest that the temporal accuracy is $\mathcal{O}(\tau^2)$ for u and $\mathcal{O}(\tau)$ for p.

Adams-Bashforth-Crank-Nicolson

Less restrictive stability conditions than for Adams-Bashforth may be obtained if the viscous terms are treated implicitly with the Crank-Nicolson method, resulting in de Adams-Bashforth-Crank-Nicolson scheme. This results in the following pressure-correction method:

$$u^* - u^{n-1} + \frac{1}{2}\tau\{3C(u^{n-1}) - C(u^{n-2})\} + \frac{1}{2}\tau B(u^{n-1} + u^*) + \tau G p^{n-3/2}$$

$$= \tau f(t^{n-1/2}), \quad u^n - u^* + G\delta p = 0,$$

$$DG\delta p = Du^* - g(t^n), \quad p^{n-1/2} = p^{n-3/2} + \delta p.$$

This method has been applied in Esposito (1992), Gavrilakis (1993), Huser and Biringer (1992), Kim and Moin (1985). The accuracy for both velocity and pressure is found to be $\mathcal{O}(\tau^2)$ in a test example (Kim and Moin (1985)).

Suitable stability conditions follow from Theorem 5.8.6. Substitution of $b = 2/\sqrt{3}$ in (5.61), $a = 1/2$ and $b = 3^{1/4}/\sqrt{2}$ in (5.69) results in the following stability condition, if the κ-scheme is used for spatial discretization:

$$\tau \leq \max[\tau_1, \min\{\tau_2, \max(\tau_3, \tau_4)\}],$$

with

$$\tau_1 = \frac{8}{3}q_3\mathrm{Re}^{-1}/\sum_\alpha u_\alpha^2, \quad \tau_2 = \frac{1}{2}/\sum_\alpha h_\alpha^{-2}\{2\mathrm{Re}^{-1} + (1-\kappa)|u_\alpha|h\},$$

$$\tau_3 = q_1(3/\mathrm{Re})^{1/3}(\sum_\alpha u_\alpha^4/h_\alpha^2)^{-1/3}, \quad \tau_4 = q_2\sqrt{3}/\sum_\alpha |u_\alpha|/h_\alpha,$$

where q_1, q_2 are given in Theorem 5.7.3 and q_3 in Theorem 5.7.5. In a similar way, Theorem 5.8.6 can be used to obtain stability conditions for spatial discretization with the fourth order central and first order upwind schemes.

Runge-Kutta-Crank-Nicolson

In Le and Moin (1991) it is proposed to replace Adam-Bashforth by Wray's Runge-Kutta method (5.80) in order to reduce the stability restriction on the time step and the computing effort. For the convection-diffusion equation the resulting method is given by (5.83). Straightforward application of the pressure-correction method, applying the continuity equation to the velocity field at every stage, would require three pressure corrections per time step, making the method expensive. However, the pressure is treated with Crank-Nicolson, so that just as for (6.60) one expects that $\mathcal{O}(\tau^2)$ accuracy for the velocity is possible with just one pressure correction. In Le and Moin (1991) this is indeed found to be the case. The three stages of (5.83) are executed with the pressure from the previous time step and without invoking the continuity equation. Then a pressure correction step follows. This results in the following scheme:

$$u^* - u^{n-1} + \sigma_1\tau C(u^{n-1}) + \tau B(\alpha_1 u^n + \beta_1 u^*) + (\alpha_1 + \beta_1)\tau Gp^{n-3/2} = f^*,$$

$$u^{**} - u^* + \sigma_2\tau C(u^*) + \zeta_2\tau C(u^{n-1}) + \tau B(\alpha_2 u^* + \beta_2 u^{**})$$
$$+ (\alpha_2 + \beta_2)\tau Gp^{n-3/2} = f^{**},$$

$$u^{***} - u^{**} + \sigma_3\tau C(u^{**}) + \zeta_3\tau C(u^*) + \tau B(\alpha_3 u^{**} + \beta_3 u^{***})$$
$$+ (\alpha_3 + \beta_3)\tau Gp^{n-3/2} = f^{***},$$

$$u^n - u^{***} + G\delta p = 0,$$

$$DG\delta p = Du^{***} - g(t^n), \quad p^{n-1/2} = p^{n-3/2} + \delta p.$$

In the implementation of the pressure terms we have taken into account that the effective time step per Runge-Kutta stage is $(\alpha_k + \beta_k)\tau$. The coefficients $\alpha_k, \beta_k, \sigma_k$ and ζ_k are given by (5.84). The right-hand sides f^*, f^{**} and f^{***}

depend on the implementation of the boundary conditions and of body forces. For details on the treatment of time-dependent boundary conditions, see Kim and Moin (1985). If there is a body force s it can be treated with Crank-Nicolson or with Runge-Kutta. The latter option may be preferable if s depends on the solution. In that case the contribution to f^{**}, for example, becomes $\sigma_2 \tau s(t^*) + \zeta_2 \tau s(t^{n-1})$, with $t^* = t^{n-1} + (\alpha_1 + \beta_1)\tau$. Wray's Runge-Kutta method has been designed to economize on storage; only preceding velocity fields are needed to compute u^{**} and u^{***}. The accuracy of the velocity field is $\mathcal{O}(\tau^2)$.

Von Neumann stability of the three predictor stages leads with the use of Theorem 5.8.9 to the following stability condition:

$$\tau \leq \sqrt{3}/\{(2 - \kappa) \sum_\alpha |u_\alpha|/h_\alpha\}\ .$$

The multistage ω-scheme

The multistage ω-scheme (5.110) with α and ω given by (5.104) and (5.105) has better stability properties than the Crank-Nicolson scheme, because it is strongly stable. The computing cost is the same, because the cost of one stage is the same as the cost of one Crank-Nicolson step, whereas for the multistep ω-scheme τ can be three times as large as for Crank-Nicolson for the same accuracy, as illustrated in Sect. 5.10. Using the linearized version (6.61) in the multistage ω-scheme, only linear equations have to be solved, as for the IMEX versions of Crank-Nicolson discussed above. But the matrices of the systems to be solved change in time. However, this is also true for the IMEX schemes if the viscosity changes with time, as would be the case for turbulent flow computations. Despite its attractive properties, use of the multistage ω-scheme has been restricted to finite element methods (Glowinski and Périaux (1987), Bristeau, Glowinski, and Périaux (1987), Rannacher (1989), Rannacher (1993), Randall (1994), Turek (1994)), but probably it deserves to be used more widely.

Application of the multistage ω-scheme (5.110) as a pressure-correction method leads to the following scheme:

$$u^* - u^n + \alpha\omega\tau N(u^*) + \alpha^1\omega\tau N(u^n) + \omega\tau Gp^{n-\omega/2} = \omega\tau f^n \,,$$
$$u^{n+\omega} - u^* + G\delta p = 0, \quad DG\delta p = Du^* - g(t^n + \omega\tau) \,,$$
$$p^{n+\omega/2} = p^{n-\omega/2} + \delta p \,,$$
$$u^* - u^{n+\omega} + \alpha'\omega'\tau N(u^*) + \alpha\omega'\tau N(u^{n+\omega}) + \omega'\tau Gp^{n+\omega/2} = \omega'\tau f^{n+1-\omega} \,,$$
$$u^{n+1-\omega} - u^* + G\delta p = 0, \quad DG\delta p = Du^* - g(t^n + \tau - \omega\tau) \,,$$
$$p^{n+1/2} = p^{n+\omega/2} + \delta p \,,$$
$$u^* - u^{n+1-\omega} + \alpha\omega N(u^*) + \alpha'\omega N(u^{n+1-\omega}) + \omega\tau Gp^{n+\omega/2} = \omega\tau f^{n+1} \,,$$
$$u^{n+1} - u^* + G\delta p = 0, \quad DG\delta p = Du^* - g(t^{n+1}) \,,$$
$$p^{n+1-\omega/2} = p^{n-1/2} + \delta p \,.$$

The right-hand sides are to be computed from the body force s (if any) and the velocity boundary conditions at appropriate times. For example, for $f^{n+1-\omega}$ one uses $s(t^n + \tau - \omega\tau)$, the velocity boundary conditions on $u^{n+\omega}$ at $t^n + \omega\tau$ and on u^*, i.e. at $t^n + \tau - \omega\tau$. Without loss of accuracy, linear systems may be obtained by Newton linearization, replacing quantities like $\phi^*\psi^*$ by, in the second stage for example, $\phi^*\psi^{n+\omega} + \phi^{n+\omega}\psi^* - \phi^{n+\omega}\psi^{n+\omega}$.

The PISO method

The pressure-correction method (6.48)–(6.50) is not the only way, of course, to uncouple velocity and pressure by operator splitting. It is easy to think of variations, but no clear advantages over the basic method (6.48)–(6.50) seem to accrue. Just to show one example of how things can be done a little differently, and to give the reader a flavor of the considerations followed by practitioners, the PISO (Pressure Implicit with Splitting of Operators) method (Issa (1986)) will be discussed briefly. Equation (6.47) is approximated by the following steps:

$$u^* - u^{n-1} + \tau N(u^*) + \tau Gp^{n-3/2} = \tau f(t^n) \,, \tag{6.62}$$
$$u^n - u^{n-1} + \tau N(u^*) + \tau Gp^{n-1/2} = \tau f(t^n) \,. \tag{6.63}$$

By applying the discrete divergence operator D to (6.63) and imposing the continuity equation we obtain the following pressure equation:

$$DGp^{n-1/2} = \frac{1}{\tau}\{Du^{n-1} - g(t^n)\} - DN(u^*) + Df(t^n) \,. \tag{6.64}$$

After $p^{n-1/2}$ is solved from (6.64) it is substituted in (6.63) to determine u^n. In (6.64) we do not replace Du^{n-1} by $g(t^{n-1})$, in order to take account of the possibility that $Du^{n-1} \neq g(t^{n-1})$, due to inexact solution of the pressure equation for $p^{n-3/2}$; this is to avoid accumulation of errors.

For von Neumann stability analysis in the same way as before, the pressure terms and right-hand sides in (6.62), (6.63) are neglected. This makes (6.63) identical to (6.62), and we are left with the implicit Euler scheme, which is unconditionally stable. It also follows that the temporal accuracy of PISO is $\mathcal{O}(\tau)$, and this remains so if a second velocity-pressure correction is appended to (6.62), (6.63), as proposed in Issa (1986) (in order to improve the accuracy of the pressure, but found to yield no advantage in Issa, Gosman, and Watkins (1986)). To improve results for τ large (aiming for the steady state solution), equation (6.63) is made implicit in the velocity, while taking care that velocity and pressure may still be solved for separately. One writes

$$N(u) = N_0 u + N_1(u) ,$$

with N_0 a diagonal matrix containing the contribution of the central element of the discretization stencil, suitably linearized. Equation (6.63) is replaced by

$$(I + \tau N_0)u^n - u^{n-1} + \tau N_1(u^*) + \tau G p^{n-1/2} = \tau f(t^n) . \tag{6.65}$$

It follows that

$$u^n = (I + \tau N_0)^{-1}\{u^{n-1} - \tau N_1(u^*) - \tau G p^{n-1/2} + \tau f(t^n)\} , \tag{6.66}$$

where the inverse of $I + \tau N_0$ is readily available, because N_0 has been chosen diagonal. By invoking the continuity equation we obtain the following pressure equation:

$$D(I + \tau N_0)^{-1}G p^{n-1/2} =$$
$$D(I + \tau N_0)^{-1}\{\frac{1}{\tau}u^{n-1} - N_1(u^*) + f(t^n)\} - \frac{1}{\tau}g(t^n) . \tag{6.67}$$

Solution for $p^{n-1/2}$ and substitution in (6.66) gives u^n.

Finally, the method is made more efficient by computing increments instead of approximations. The predictor equation (6.62) remains. The corrector equation (6.65) is modified by subtracting (6.62). The pressure equation is obtained in the same way as before by invoking the continuity equation. This results in the PISO method as proposed in Issa (1986):

$$(I + \tau N_0)u^* - u^{n-1} + \tau N_1(u^*) + \tau G p^{n-3/2} = \tau f(t^n) ,$$
$$(I + \tau N_0)(u^n - u^*) + \tau G(p^{n-1/2} - p^{n-3/2}) = 0 , \tag{6.68}$$

$$D(I + \tau N_0)^{-1}G(p^{n-1/2} - p^{n-3/2}) = \frac{1}{\tau}\{Du^* - g(t^n)\} . \tag{6.69}$$

The advantage of solving for increments is seen to be, that the known terms in (6.68) and (6.69) are cheaper to evaluate than in (6.65) and (6.67).

The temporal accuracy of PISO is $\mathcal{O}(\tau)$, and there seems to be no advantage over the methods discussed before. But as a time stepping method to compute stationary solutions using large τ it is useful. It has much in common with iterative methods to solve the stationary equations that will be discussed in Chap. 7.

6.7 Temporal discretization on colocated grid

Equation (6.31) is our point of departure. Choosing a temporal discretization method, a system of equations of the following type results:

$$A(u^n) + \tau G p^{n-1/2} = r^n \,, \tag{6.70}$$

$$D u^n + B p^{n-1/2} = g(t^n) \,, \tag{6.71}$$

where r^n is known from previous time steps and the boundary conditions. The system has to be solved iteratively. The residual of (6.70) needs to be reduced to a size commensurate with the temporal discretization error, but (6.71) has to be solved accurately, to ensure mass conservation. A general framework and a theoretical basis seem to be lacking; in this respect the situation is more favorable for staggered grid methods.

Successive iterands approximating u^n and $p^{n-1/2}$ will be denoted by v^m and q^{m-1}, $m = 1, 2, \ldots$ respectively. We choose $q^0 = p^{n-3/2}$. One tries to avoid solving for v and q simultaneously. First, an iterative method that is reminiscent of pressure correction is formulated (cf. (6.48)–(6.50)):

$$A(v^*) + \tau G q^{m-1} = r^n \,, \tag{6.72}$$

$$v^m - v^* + \tau G(q^m - q^{m-1}) = 0 \,, \tag{6.73}$$

$$D v^m + B q^{m-1} = g(t^n) \,. \tag{6.74}$$

Application of D to (6.73) and substitution of (6.74) results in the following pressure equation:

$$DG\delta q = \frac{1}{\tau}(Dv^* + Bq^{m-1} - g(t^n)), \quad q^m = q^{m-1} + \delta q \,. \tag{6.75}$$

After solving for q^m, v^m follows from (6.73). Unlike the staggered grid method, we cannot stop after one iteration, writing $u^n = v^1$, $p^{n-1/2} = q^1$, because the residual in (6.71) is still too large, since q lags behind in (6.74). This is a distinct disadvantage of the colocated scheme for time-dependent flows, compensated somewhat by the fact that there is no reason anymore to solve (6.75) accurately.

In order to improve convergence, a PISO-like approach may be taken. We write

$$A(u) = A_0 u + A_1(u) \,,$$

with A_0 a diagonal matrix that may be chosen in various ways to enhance convergence. We write instead of (6.72)–(6.74)

$$A_0 v^* + A_1(v^*) + \tau G q^{m-1} = r^n \,, \tag{6.76}$$

$$A_0 v^m + A_1(v^*) + \tau G q^m = r^n \,, \tag{6.77}$$

$$D v^m + B q^{m-1} = g(t^n) \,. \tag{6.78}$$

Subtracting (6.76) from (6.77) gives

$$A_0(v^m - v^*) + \tau G\delta q = 0, \quad \delta q = q^m - q^{m-1} \ .$$

It follows that

$$v^m = v^* - \tau A_0^{-1} G\delta q \ .$$

Application of D and use of (6.78) results in the following pressure equation:

$$DA_0^{-1}G\delta q = \frac{1}{\tau}\{Dv^* + Bq^{m-1} - g(t^n)\} \ . \tag{6.79}$$

After convergence, we put $u^n = v^m$, $p^{n-1/2} = q^m$ and the next time step can be taken.

Equations (6.72) and (6.75), or (6.76) and (6.79), or, in the case of coupled solution of u and p, equations (6.70) and (6.71), can be solved with iterative methods developed for the stationary case, to be discussed in Chap. 7. Stability of the temporal discretization method chosen for (6.70), (6.72) or (6.76) can be studied in the same way as for staggered schemes. Examples of time-accurate computation with colocated spatial discretization using the PWI method (or something similar) may be found in Armfield (1994) (using Crank-Nicolson), Cantaloube and Lê (1992), Takemoto and Nakamura (1986), Takemoto and Nakamura (1989), Zang, Street, and Koseff (1994) (using Adams-Bashforth-Crank-Nicolson), Demirdžić and Perić (1990) (using implicit Euler), Ho and Lakshminarayana (1993), Kost, Bai, Mitra, and Fiebig (1992), Kost, Mitra, and Fiebig (1992) (using BDF), and Lê, Ryan, and Dang Tran (1992) (using second order Adams-Bashforth).

On the basis of what we have seen so far, for the computation of nonstationary flows, staggered discretization seems superior to colocated discretization, because of greater simplicity, efficiency and accuracy. But on general curvilinear grids, staggered spatial discretization is generally considered to be more complicated and less efficient than colocated discretization, which seems to be the more frequently used method on general grids.

7. Iterative methods

7.1 Introduction

An in-depth treatment of iterative methods in linear algebra will not be given, because this is a huge subject, that would require a separate volume. Instead, only a bird's eye view of the subject will be given, providing ample references to the literature, paying particular attention to difficulties that are peculiar to the incompressible Navier-Stokes equations. The uninitiated reader who wants to use these methods is advised to consult the references that we will quote.

The algebraic systems that arise by finite volume discretization are extremely sparse, and also very large, because many grid points are required for accuracy. Therefore iterative methods are more efficient and demand far less storage than direct methods, especially in three dimensions. In two dimensions, direct methods using sparse matrix techniques can still be useful. Hence, we will confine ourselves to iterative methods, with one exception: the equation for the pressure correction can often be solved very efficiently by so-called fast Poisson solvers, based on Fourier transformation and/or cyclic reduction.

Efficiency and robustness

When computing N unknowns, a method may be said to have optimal efficiency if the computing work $W = \mathcal{O}(N)$. This may be achieved by multigrid methods. The fast Poisson solvers mentioned above have $W = \mathcal{O}(N \ln N)$, which is also quite good. Generally, iterative methods have $W = \mathcal{O}(N^\alpha)$, $\alpha > 1$.

As the same time, the method needs to be *robust* (antonym: fragile), which means that the efficiency is insensitive to changes in the problem, such as variations in grid point distribution (especially cell aspect ratios), Reynolds number, flow direction or coordinate system (occurrence of mixed derivatives). In general, one should be wary of relaxation parameters that need to be adjusted to the problem at hand. Guidelines for improving robustness are given at the end of Sect. 7.2.

Obviously, in practice the turn-around time (i.e. the elapsed wall-clock time between problem input and completion of computation) is important. The turn-around time may be decreased by using a faster computer. The faster computers use parallel processing. An introduction to high performance computing in fluid dynamics is given in Wesseling (1996a).

The reader is assumed to be familiar with the basics of iterative methods in linear algebra. An elementary introduction is given in Golub and van Loan (1996), Chap. 10. Classics devoted to the subject are Faddeev and Faddeeva (1963), Hageman and Young (1981), Varga (1962) and Young (1971). More recent works are Axelsson (1994), Bruaset (1995), Greenbaum (1997), Hackbusch (1994), Saad (1996).

Software

Much standard software is available, easily accessible on the Internet. Try for instance: `math.nist.gov` or `www.netlib.org`.
This software can be used directly, or as examples or building blocks (templates: Barrett *et al.* (1994)) for the development of customized software. Of particular interest are the following libraries: BLAS (basic linear algebra subprograms), FISHPACK (fast Poisson solvers for the pressure equation), ITPACK (iterative methods) and LAPACK (numerical linear algebra).

7.2 Stationary iterative methods

Let the linear algebraic $n \times n$ system to be solved be denoted by

$$Ay = b .$$

A *stationary iterative method* is defined as follows:

$$y^{k+1} = By^k + c . \tag{7.1}$$

For consistency, we must have $y = By + c$, hence $c = (I - B)A^{-1}b$, so that (7.1) can be rewritten as

$$My^{k+1} = Ny^k + b , \tag{7.2}$$

with $M - N = A$ $(M = A(I - B)^{-1}, \ N = MB)$. So we see that every stationary iterative method corresponds to a splitting of A. Of course, M is chosen such that (7.2) can be solved with little work. We have

$$B = I - M^{-1}A = M^{-1}N .$$

The rate of convergence

The error $e^k = y - y^k$ satisfies

$$e^{k+1} = Be^k ,$$

and convergence is governed by B, called the *iteration matrix*. We have

$$\|e^k\| \leq \|B^k\| \, \|e^0\| . \tag{7.3}$$

Since the spectral radius satisfies

$$\rho(B) = \lim_{k \to \infty} \|B^k\|^{1/k} \tag{7.4}$$

(cf. Hackbusch (1994), Theorem 2.9.8) we see that we have convergence if and only if

$$\rho(B) < 1 .$$

The smaller $\rho(B)$, the faster convergence. This follows from (7.4), but becomes especially clear in the case that B is normal, and the spectral norm is used. Then

$$\|B^k\| = \rho(B)^k ,$$

and (7.3) becomes

$$\|e^k\| \leq \rho(B)^k \|e^0\| . \tag{7.5}$$

This inequality is sharp, since equality is obtained in the special case that e_0 is an eigenvector belonging to an eigenvalue with $|\lambda(B)| = \rho(B)$.

When to stop?

We will not present a thorough discussion of termination criteria for iterative processes. Good termination criteria are usually problem-dependent. Our main purpose here is to show that the difference between two successive iterates usually does not give a good indication of the precision achieved.

Let B have a single dominating real eigenvalue $\lambda = \rho(B)$, with corresponding eigenvector x. Then we have after many iterations

$$e^k \cong \alpha \lambda^k x , \tag{7.6}$$

with α some constant. It follows that

$$y^{k+1} - y^k \cong \alpha \lambda^k (\lambda - 1)x = (\lambda - 1)e^k ,$$

hence

$$e^k \cong \frac{y^{k+1} - y^k}{\lambda - 1} . \tag{7.7}$$

We see that if $\lambda = 1-\varepsilon$, $0 < \varepsilon \ll 1$, as frequently happens (slow convergence), then the error is much larger than the difference between successive iterations. If the assumption that B has a single dominating eigenvalue is satisfied, then an error estimate can be based on (7.7). To this end we need to estimate λ, which can be done by

$$\lambda \cong \|y^{k+1} - y^k\|/\|y^k - y^{k-1}\| . \tag{7.8}$$

Often, it is better to derive a termination criterion from the residual $r^k \equiv b - Ay^k$ instead of from the difference between two successive iterates. After k iterations we have solved exactly the following neighboring problem:

$$Ay^k = b - r^k ,$$

and frequently the perturbation of the right-hand side has a physical interpretation which enables one to decide what size of $\|r^k\|$ is tolerable.

Regular splittings and M- and K-matrices

We will now discuss an important class of matrices for which it is easy to obtain convergent stationary iterative methods of the type (7.2). All matrices to be considered have dimension $n \times n$, and by $A \geq 0$ we will mean $a_{ij} \geq 0$, $i, j = 1, ..., n$.

Definition 7.2.1. The splitting

$$A = M - N$$

is called *regular* if $M^{-1} \geq 0$ and $N \geq 0$. The splitting is called *convergent* if (7.2) converges.

Definition 7.2.2. (Varga (1962), Definition 3.3). The matrix A is called an *M-matrix* if $a_{ij} \leq 0$, $i \neq i$, $i, j = 1, ..., n$, A is non-singular and $A^{-1} \geq 0$.

Theorem 7.2.1. *A regular splitting of an M-matrix is convergent.*

Proof. See Varga (1962) Theorem 3.13.   □

Hence, it is worthwhile to try to discretize such that A is an M-matrix, because then convergent iterative methods are easy to obtain. As discussed in Chap. 4, it is also desirable that the discretization is of positive type (Definition 4.4.1), in order to exclude spurious wiggles. Fortunately, discretizations of positive type result in M-matrices. This we will now show.

Theorem 7.2.2. *If $a_{ii} > 0$, $a_{ij} \leq 0$, $i = 1, ..., n$, $j = 1, ..., n$, $j \neq i$, then A is an M-matrix if and only if $\rho(B) < 1$, $B = D^{-1}C$, $D = diag\,(A)$, $C = D - A$.*

Proof. See Young (1971), Theorem 2.7.2.   □

A matrix has *weak diagonal dominance* if

$$|a_{ii}| \geq \sum_{j \neq i} |a_{ij}|, \ i = 1, ..., n \,,$$

with strict inequality for at least one i. A matrix A called *irreducible* if the system $Ay = b$ does not consist of subsystems that are independent of each other. Of course, this is the usual case in practice. If not, one proceeds with the independent subsystems separately.

Theorem 7.2.3. *If A has weak diagonal dominance and is irreducible, then $det(A) \neq 0$ and $a_{ii} \neq 0$, $i = 1, ..., n$.*

Proof. See Young (1971), Theorem 2.5.3.   □

Theorem 7.2.4. *If A has weak diagonal dominance and is irreducible, then $\rho(B) < 1$, with B defined in Theorem 7.2.2.*

Proof. (see also Young (1971) p. 108). Assume $\rho(B) \geq 1$. Then B has an eigenvalue μ with $|\mu| \geq 1$. Furthermore, $det\,(Q) = 0$, $Q = I - \mu^{-1}B$. Since A is irreducible, so is Q. Since $|\mu^{-1}| \leq 1$, Q has weak diagonal dominance. From Theorem 7.2.3, $det\,(Q) \neq 0$, so that we have a contradiction.   □

The foregoing theorems allow us to formulate a sufficient condition for A to be an M-matrix, that can be verified simply by inspection of A.

Definition 7.2.3. A matrix A is called a K-*matrix* if

$$a_{ii} > 0, \quad i = 1, ..., n \,,$$
$$a_{ij} \leq 0, \quad i, j = 1, ..., n, \quad j \neq i \,,$$

and

$$\sum_j a_{ij} \geq 0, \quad i = 1, ..., n \,,$$

with strict inequality for at least one i.

Theorem 7.2.5. *An irreducible K-matrix is an M-matrix.*

Proof. According to Theorem 7.2.4, $\rho(B) < 1$. Then Theorem 7.2.2 gives the desired result.   □

It is easy to see, taking the boundary conditions into account, assuming that we have a Dirichlet or a Robin condition at least on part of the boundary, that a discretization of positive type (Definition 4.4.1) results in a K-matrix. This matrix is usually irreducible, otherwise the domain can be decomposed into subdomains that can be solved for independently. Hence we have an M-matrix.

If the discretization is of higher order, it cannot be of positive type (Theorem 4.4.2), and does not result in an M-matrix. Then one can use *defect correction* (Sect. 4.4) on a scheme of positive type. In each step of defect correction, one has to deal with an M-matrix, so that iterative solution methods have favorable convergence properties.

Regular (and hence convergent, Theorem 7.2.1) splittings are easy to obtain for M-matrices:

Theorem 7.2.6. *Let A be an M-matrix. If M is obtained by replacing some elements a_{ij}, $j \neq i$ by b_{ij} satisfying $a_{ij} \leq b_{ij} \leq 0$, then $A = M - N$ is a regular splitting.*

Proof. This theorem is an easy generalization of Theorem 3.14 in Varga (1962), suggested by Theorem 2.2 in Meijerink and van der Vorst (1977). □

Well-known examples of regular splittings are (block-) Jacobi and (block-) Gauss-Seidel methods, for which M is obtained by replacing certain elements a_{ij} by zero. Another class of methods is generated by *incomplete LU factorization* (ILU) in which one chooses

$$M = LU \ ,$$

with L and U sparse and easy to compute lower and upper triangular matrices. Examples and references to the literature are given in Axelsson (1994), Bruaset (1995), Hackbusch (1994), Wesseling (1992). For A an M-matrix, existence of L and U, numerical stability of the associated algorithms and regularity of the resulting splitting has been proven in Meijerink and van der Vorst (1977).

Acceleration of stationary iterative methods

In general, stationary iterative methods converge so slowly, that the ideal for the computing cost $W = \mathcal{O}(N)$ is far from reached. For example, for the standard 5-point finite difference discretization of the Laplacian on the unit square with uniform step-size h in both directions with Dirichlet boundary conditions, we have

$$\rho(B) = 1 - \mathcal{O}(h^2) \tag{7.9}$$

(cf. Hackbusch (1994), Sections 4.7 and 5.7) for methods of Jacobi and Gauss-Seidel type. Requiring an error reduction of $\mathcal{O}(h^2)$ (commensurate with the discretization error) the number of indications m must satisfy (cf. (7.5))

$$\rho^m(B) = \mathcal{O}(h^2) . \tag{7.10}$$

From (7.9) and (7.10) it follows that

$$m = \mathcal{O}(\ln h / \ln \rho(B)) = \mathcal{O}(-h^{-2} \ln h) .$$

Since the work per iteration is $\mathcal{O}(N)$ (N the number of unknowns) and $h = N^{-1/2}$ we find that the total computing work satisfies

$$W = \mathcal{O}(N^\alpha \ln N), \quad \alpha = 2 . \tag{7.11}$$

By introduction of relaxation parameters it turns out that α may be brought down to $\alpha = 3/2$ (provided the optimal relaxation parameter can be determined, which is difficult in general), but in general the computing work of basic iterative methods cannot be reduced further. However, these methods remain important, because they lend themselves to acceleration. This opens up the possibility to bring α down further and perhaps also to eliminate the factor $\ln N$ in (7.11). There are two classes of acceleration methods that when combined with a suitable stationary iterative method result in the most efficient and robust methods available at present. These are *Krylov subspace methods* and *multigrid methods*. The stationary iterative method is usually called preconditioner if a Krylov subspace method is used for acceleration, and is called smoother if multigrid is used.

Robustness

Robustness is only achieved with accelerated stationary iterative methods, if the preconditioner or smoother is suitably chosen. Guidelines can be derived from analysis of their performance for the convection-diffusion equation and the rotated anisotropic diffusion equation:

$$-\varepsilon(\varphi_{,11} + \varphi_{,22}) + cu_{,1} + su_{,2} = 0 , \tag{7.12}$$

$$-(\varepsilon c^2 + s^2)\varphi_{,11} - 2(\varepsilon - 1)csu_{,12} - (\varepsilon s^2 + c^2)u_{,22} = 0 , \tag{7.13}$$

where $s = \sin\beta$, $c = \cos\beta$. There are two parameters to be varied: $\varepsilon > 0$ and β. When discretized on a uniform grid, these equations model the effect of varying the Reynolds number, flow direction, cell aspect ratio and non-orthogonality of the coordinate system (occurrence of a mixed derivative). When efficiency is good for all ε and β, a method is called robust. Conclusions drawn from analysis of (7.12) and (7.13) carry over heuristically to more complicated systems, such as Navier-Stokes. Efficiency and robustness of many stationary iterative methods as smoothers for multigrid is investigated in Wesseling (1992) Chap. 7. It is general experience, that a good smoother for multigrid is also a good preconditioner for Krylov subspace methods. The following methods are found to be robust in two dimensions:

– Damped alternating line Jacobi;
– Alternating symmetric line Gauss-Seidel;
– Alternating damped zebra Gauss-Seidel;
– Certain ILU methods.

See Wesseling (1992) for remarks on robustness in three dimensions.

Exercise 7.2.1. Derive equations (7.6) and (7.8).

7.3 Krylov subspace methods

Basic idea

The system to be solved is denoted by $Ay = b$. When A is large and sparse it is attractive, because of efficiency and simplicity, to use A only to multiply with. This means that we can build polynomials in A. A general form of possible algorithms is

$$y^{m+1} = y^m + \alpha_m p^m, \quad p^m = \theta_m(A)r^0, \quad r^0 = b - Ay^0 ,$$

with θ_m an arbitrary polynomial of degree m, and α_m some real number. The special choice of r^0 as vector to be multiplied by $\theta_m(A)$ is made because we want to have $p^0 = 0$ in case y^0 is the exact solution. We have

$$\begin{aligned}
r^m &= b - Ay^m = b - Ay^{m-1} - \alpha_{m-1}Ap^{m-1} \\
&= r^{m-1} - \alpha_{m-1}A\theta_{m-1}(A)r^0 \\
&= r^0 - A\{\alpha_{m-1}\theta_{m-1}(A) + \alpha_{m-2}\theta_{m-2}(A) + ... + \alpha_0\theta_0(A)\}r^0 \\
&= \phi_m(A)r^0 ,
\end{aligned} \tag{7.14}$$

where ϕ_m is a polynomial degree m with the following property:

$$\phi_m(0) = 1 . \tag{7.15}$$

Since

$$y^m = A^{-1}(b - r^0) + \{\alpha_{m-1}\theta_{m-1}(A) + \alpha_{m-2}\theta_{m-2}(A) + ...\alpha_0\theta_0(A)\}r^0 ,$$

we have

$$y^m - y^0 = \{\alpha_{m-1}\theta_{m-1}(A) + \alpha_{m-2}\theta_{m-2}(A) + ... + \alpha_0\theta_0(A)\}r^0 .$$

We see that

$$y^m - y^0 \in K_m(A, r^0) ,$$

with the m-dimensional *Krylov subspace* K_m defined as

$$K_m(A, r^0) = \text{span } \{r^0, Ar^0, ..., A^{m-1}r^0\} .$$

Methods that look for optimal approximations to $y - y^0$ in $K_m(A, r^0)$ are called Krylov subspace methods. Introductions may be found in Bruaset (1995), Golub and van Loan (1996), Greenbaum (1997), Hackbusch (1994), Hageman and Young (1981), Saad (1996).

There are various ways to specify optimal approximations. Let $\|\cdot\|$ be some norm. Because of (7.14) and (7.15) we would like to construct ϕ_m such that $\|\phi_m(A)r^0\|$ is as small as possible, under the constraint $\phi_m(0) = 1$. More precisely, let us define the following set of polynomials:

$$\prod_m^1 = \{\psi(x) : \psi(0) = 1, \ \psi \text{ is polynomial of degree } \leq m\} .$$

We want to construct $\phi_m \in \prod_m^1$ such that $r^m = \phi_m(A)r^0$ satisfies

$$\|r^m\| = \min\{\|\psi(A)r^0\| : \psi \in \prod_m^1 \} . \tag{7.16}$$

Equivalent formulations of this optimal approximation property are obtained as follows. Equation (7.15) shows that

$$r^m = r^0 - v, \quad v \in V_m \equiv \text{span } \{Ar^0, ..., A^m r^0\} ,$$

and (7.16) is equivalent with

$$\|r^m\| = \min\{\|r^0 - v\| : v \in V_m\} = \min\{\|r^0 - Az\| : z \in K_m(A, r^0)\} . \tag{7.17}$$

Hence

$$\|A(y - y^m)\| = \min\{\|A(y - y_0 - z)\| : z \in K_m(A, r^0)\} ,$$

which gives the following formulation of the optimality property:

$$\|A(y - y^m)\| = \min\{\|A(y - w)\| : w - y_0 \in K_m(A, r^0)\} . \tag{7.18}$$

Rate of convergence

For a general matrix A the convergence behavior of Krylov subspace methods can be quite erratic and hard to predict, see Greenbaum, Ptak, and Strakos (1996). When A is symmetric we can say more, however. If A is symmetric there exists a unitary matrix Q (i.e. $Q^{-1} = Q^T$) such that

$$Q^T A Q = \Lambda , \quad \Lambda = \mathrm{diag}(\lambda_1, ..., \lambda_n) , \quad \lambda_k \text{ real.}$$

Until further notice we use the spectral norm. A unitary matrix Q satisfies

$$\|Q\| = 1 .$$

For Krylov subspace methods for symmetric A the convergence behavior depends on all eigenvalues of A. This follows from the following theorem.

Theorem 7.3.1. *Rate of convergence of Krylov subspace methods*
Let the $n \times n$ matrix A be symmetric with eigenvalues $\lambda_1, ..., \lambda_n$. Let r^m satisfy (7.16). Then

$$\|r^m\| \le \|r^0\| \min[\max\{\psi(\lambda) : \lambda \in \sigma(A)\}, \psi \in \prod_m^1] \qquad (7.19)$$

where $\sigma(A) = \{\lambda_1, ..., \lambda_n\}$.

Proof. We have $\psi(A) = Q\psi(\Lambda)Q^{-1}$. From (7.16) it follows that

$$\|r^m\| = \min\{\|Q\psi(\Lambda)Q^{-1}r^0\| : \psi \in \prod_m^1 \}$$

$$\le \|r^0\| \, \|Q\| \, \|Q^{-1}\| \min\{\|\psi(\Lambda)\| : \psi \in \prod_m^1 \}$$

$$= \|r^0\| \min[\max\{|\psi(\lambda)| : \lambda \in \sigma(A)\}, \psi \in \prod_m^1] .$$

$\square$

Note that it follows from (7.19) that convergence is monotone:

$$\|r^m\| \le \|r^{m-1}\| .$$

It remains to find efficient methods that generate residuals satisfying (7.16).

Preconditioning

From Theorem 7.3.1 it follows that it depends on the distribution of $\lambda(A)$ how large m must be for $\|r^m\|$ to be small. If the eigenvalues are clustered around a point $x \in \mathbb{R}$, then already $\|r^1\|$ is small, since we have

$$\min \left[\max\{|\psi(\lambda)| : \lambda \in \sigma(A)\} : \psi \in \prod_{1}^{1} \right] \leq \max\{|1 - \lambda/x| : \lambda \in \sigma(A)\} .$$

Preconditioning means changing the original problem $Ay = b$ such that the spectrum of the new problem is more clustered. The preconditioned problem is

$$\tilde{A}\tilde{y} = \tilde{b}, \quad \tilde{A} = C_L A C_R, \quad \tilde{y} = C_R^{-1}y, \quad \tilde{b} = C_L b . \tag{7.20}$$

C_L and C_R are called the left and right preconditioner, respectively; application of C_R is also called postconditioning. We may have $C_L = I$, or $C_R = I$, or both C_L and $C_R \neq I$. The aim is to have $\sigma(\tilde{A})$ much more clustered than $\sigma(A)$, giving much faster convergence of Krylov subspace methods.

Suitable preconditioning methods may be derived from stationary iterative methods. As discussed in Sect. 7.2, these can be written as

$$My^{k+1} = Ny^k + b ,$$

where y^{k+1} can be evaluated at little cost, i.e. evaluation of $M^{-1}x$ is cheap. We can take $C_L = M^{-1}$ or $C_R = M^{-1}$. With ILU methods (cf. Sect. 7.2) we have $M = LU$, and we can take $C_L = L^{-1}$, $C_R = U^{-1}$. Another possibility is to take $C_L = M^{-1}$, $C_R = M_2^{-1}$, with M_1 coresponding to a forward and M_2 corresponding to a backward Gauss-Seidel method. For remarks on the robustness of such preconditionings, see Sect. 7.2. It is preconditioning that makes Krylov subspace methods efficient. One may say that preconditioning accelerates Krylov subspace methods, or, equivalently, that Krylov subspace methods accelerate stationary iterative methods.

The GCR method

Of course, in computational fluid dynamics we have to deal with nonsymmetric matrices. The GCR (generalized conjugate residuals) method finds for general A the optimal Krylov subspace approximation in the sense of (7.16), (7.17) and (7.18).

We start from (7.17). Noting that V_m is a linear subspace, we can decompose r^0 uniquely as

$$r^0 = r_v^0 + r_p^0, \quad r_v^0 \in V_m, \quad r_p^0 \perp V_m .$$

Let us define orthogonality in the sense of the Euclidean inner product $(.,.)$:

$$r \perp q \iff (r, q) = 0 .$$

We write, with $v \in V_m$,

$$\|r^0 - v\|_2^2 = (r_v^0 - v + r_p^0, \ r_v^0 - v + r_p^0) = (r_v^0 - v, \ r_v^0 - v) + (r_p^0, \ r_p^0) ,$$

with $\| \cdot \|_2$ the Euclidean norm. Obviously, (7.17) is solved by $v = r_v^0$, which implies that

$$r^m \perp V_m .$$

This property enables us to determine $r^1, ..., r^m$ and $y^1, ..., y^m$. Let $\{v^1, ..., v^m\}$ be an orthonormal basis for V_m. Then

$$r^m = r^0 - r_v^0 = r^0 - \sum_1^m \alpha_j v^j, \quad \alpha_j = (r^0, v^j) .$$

This holds for $m = 1, 2, ...$ Note that α_j does not depend on m. We can write

$$\alpha_j = (r^0 - \sum_1^{j-1} (r^0, v^k) v^k, v^j) = (r^{j-1}, v^j) ,$$

so that

$$r^m = r^{m-1} - (r^{m-1}, v^m) v^m .$$

This shows what the residuals should be. Since $y^m = A^{-1}(b - r^m)$, the corresponding iterands satisfy

$$y^m = y^{m-1} + (r^{m-1}, v^m) s^m, \quad s^m = A^{-1} v^m .$$

It remains to determine $v^1, ..., v^m$ and $s^1, ..., s^m$ (avoiding the use of A^{-1}). This can be done with the Gram-Schmidt orthogonalization process. Given an orthonormal system $v^1, ..., v^{m-1}$, and given $s^j = A^{-1} v^j$, $j = 1, ..., m-1$, the following algorithm is executed to find v^m and s^m such that $\{v^1, ..., v^m\}$ is orthonormal and $s^m = A^{-1} v^m$:

Gram-Schmidt algorithm
　　Choose s^m;
　　$v^m = As^m$;
　　for $j = 1, ..., m-1$ **do**
　　　　$\alpha = (v^m, v^j); \ v^m = v^m - \alpha v^j; \ s^m = s^m - \alpha s^j;$
　　end for
　　$v^m = v^m / \|v^m\|_2; \ s^m = s^m / \|v^m\|_2;$

It remains to choose s^m. In a Krylov subspace method this should involve multiplication by A, so that we search for y^m in all of K_m, in order to profit from the optimality property (7.18). A suitable choice is $s^m = r^{m-1}$, leading to the GCR method as proposed originally in Eisenstat, Elman, and Schultz (1983):

GCR method
  Choose y^0; $r^0 = b - Ay^0$; $m = 0$;
  while stopcriterion not satisfied **do**
    $m = m + 1$;
    Choose s^m;   \\ For example: $s^m = r^{m-1}$
    $v^m = As^m$;
    for $j = 1, ...m - 1$ **do**   \\ Gram-Schmidt
    $\alpha = (v^m, v^j)$; $v^m = v^m - \alpha v^j$; $s^m = s^m - \alpha s^j$;
    end for
    $v^m = v^m / \|v^m\|_2$; $s^m = s^m / \|v^m\|_2$;
    $y^m = y^{m-1} + (r^{m-1}, v^m)s^m$; $r^m = r^{m-1} - (r^{m-1}, v^m)v^m$;
    inspect stopcriterion
  end while

Computing work

The vectors $v^1, ..., v^m$ and $s^1, ..., s^m$ have to be stored, and the work for Gram-Schmidt grows quadratically with m. Therefore, in practice, after a certain number of iterations, $\bar{m}$ say, the process is restarted or truncated. In a restart $v^1, ..., v^{\bar{m}}$ and $s^1, ..., s^{\bar{m}}$ are removed, and the algorithm is restarted with $y^0 = y^{\bar{m}}$. With truncation, we allow only $\bar{m}$ vectors s^j and v^j, and replace an old vector by a new one. Of course, the optimality property (7.18) gets lost when restart or truncation takes place.

Robustness and convergence

GCR satisfies the optimality property (7.16) (and, equivalently, (7.17) and (7.18)) with $\|\cdot\| = \|\cdot\|_2$. It can be shown that the only way in which GCR can lose the optimality property (apart from restart or truncation) is that $r^{m-1} = r^{m-2}$. In that case one way choose $s^m = As^{m-1}$, in order to search in the full space K_m. The rate of convergence can be estimated by Theorem 7.3.1, if something is known about $\sigma(A)$, and if A is symmetric. Inspection of the GCR algorithm shows that it can break down only if $\|v^m\|_2 = 0$. It can be shown that this can only happen if $y^{m-1} = y$, i.e. the exact solution is reached. From (7.17) it follows that convergence is monotone. We see that GCR is very robust. For efficiency, GCR usually needs to be preconditioned.

Preconditioning

The best choice for s^m in the GCR method is to make it equal to the error: $s^m = A^{-1}r^{m-1}$, so that y^m is exact. This is not practical, of course, but we could determine s^m by

$$\text{Solve approximately:}\quad As^m = r^{m-1} .\qquad(7.21)$$

As approximate solution method one can take, for example, an iteration with a stationary iterative method based on a splitting $A = M - N$, determining s^m by

$$s^m = M^{-1}r^{m-1} .$$

It is easy to see that this results in

$$y^m - y^0 \in K_m(M^{-1}A, M^{-1}r^0), \quad r^m \in K_m(AM^{-1}, r^0) ,$$

so that we can say that the stationary iterative method is used as preconditioner for GCR, or that GCR is used to accelerate the stationary iterative method. The optimality property (7.16) and Theorem 7.3.1 hold with A replaced by AM^{-1}. Of course, the method used for (7.21) can be anything; it may be chosen differently for every m, and it may be nonstationary. Then (7.16) and Theorem 7.3.1 do not apply, of course. In van der Vorst and Vuik (1994) it is proposed to use GMRES (to be discussed hereafter) for (7.21); the resulting method is called GMRESR.

The GMRES method

The GMRES (generalized minimal residual) method, proposed in Saad and Schultz (1986), generates the same iterates as GCR, but is cheaper. We will describe the method using both pre- and postconditioning, so that we solve

$$\tilde{A}\tilde{y} = \tilde{b}, \quad \tilde{A} = C_L A C_R, \quad \tilde{y} = C_R^{-1}y, \quad \tilde{b} = C_L b .$$

The GMRES method computes

$$\tilde{y} = \tilde{y}^0 + z^m, \quad z^m \in K_m(\tilde{A}, \tilde{r}^0) .$$

We have $\tilde{r}^m = \tilde{b} - \tilde{A}\tilde{y} = \tilde{r}^0 - \tilde{A}z^m$. GMRES finds z^m such that (7.17) holds in the $\|\cdot\|_2$ norm:

$$\|\tilde{r}^m\|_2 = \min\{\|\tilde{r}^0 - \tilde{A}z\|_2, \quad z \in K_m(\tilde{A}, \tilde{r}^0)\} .\qquad(7.22)$$

The principles of the method are as follows (cf. Bruaset (1995), Saad and Schultz (1986)). Let $\{v^1, ..., v^m\}$ be an orthonormal basis for $K_m(\tilde{A}, \tilde{r}^0)$, and let V_m be the $n \times m$ matrix with columns $v^1, ..., v^m$; $n \times n$ are the dimensions of $\tilde{A}$. Then $z^m = V_m u$ for some $u \in \mathbb{R}^m$, and we have to minimize $\|\tilde{r}^0 - \tilde{A}V_m u\|_2$. Define

$$h_{ij} = (\tilde{A}v^j, v^i), \quad i = 1, ..., j, \quad j = 1, ..., m,$$

$$h_{j+1,j} = \|\tilde{A}v^j - \sum_1^j h_{ij}v^i\|_2, \quad j = 1, ..., m.$$

The $(m+1) \times m$ matrix $H_{m+1,m}$ with entries h_{ij} is upper Hessenberg. It turns out that

$$\tilde{A}V_m = V_{m+1}H_{m+1,m}.$$

We can write $\tilde{r}^0 = \beta v^1$, $\beta \equiv \|\tilde{r}^0\|_2$, so that $V_{m+1}^T \tilde{r}^0 = \beta e^1$, with $(e^1)^T = (1, 0, ..., 0) \in \mathbb{R}^{m+1}$. We have

$$
\begin{aligned}
\|\tilde{r}^0 - \tilde{A}z\|_2 &= \|\tilde{r}^0 - \tilde{A}V_m u\|_2 = \|\beta v_1 - V_{m+1}H_{m+1,m}u\|_2 \\
&= \|V_{m+1}(\beta e_1 - H_{m+1,m}u)\|_2 \\
&= (V_{m+1}(\beta e_1 - H_{m+1,m}u), V_{m+1}(\beta e_1 - H_{m+1,m}u))^{1/2} \\
&= (V_{m+1}^T V_{m+1}(\beta e_1 - H_{m+1,m}u), \beta e_1 - H_{m+1,m}u)^{1/2} \\
&= \|\beta e_1 - H_{m+1,m}u\|_2,
\end{aligned}
$$

since $V_{m+1}^T V_{m+1}$ is the identity. Hence, the n-dimensional least squares problem (7.22) has been replaced by the following much smaller (m-dimensional) least squares problem: find $u \in \mathbb{R}^m$ such that

$$\|\beta e^1 - H_{m+1,m}u\|_2 = \min\{\|\beta e^1 - H_{m+1,m}u\|_2 : u \in \mathbb{R}^m\}. \quad (7.23)$$

The preconditioned GMRES method can be outlined as follows.

Preconditioned GMRES method
    Choose y^0; $r^0 = b - Ay^0$;
    $s^0 = C_L r^0$; $\beta = \|r^0\|_2$; $v^1 = s^0/\beta$; $j = 0$;
    while stopcriterion not satisfied **do**
        $j = j + 1$;
        $w = C_L A C_R v^j$;
        for $k = 1, ..., j$ **do**
            $h_{kj} = (v^k, w)$;
        end for
        $h_{j+1,j} = \|w\|_2$; $v^{j+1} = w/h_{j+1,j}$;
    end while
    Solve (7.23),
    $y^m = y^0 + C_R V_m u$;

The least squares problem (7.23) can be solved in the classical way, as suggested in Saad and Schultz (1986), by computing the QR factorization of

$H_{m+1,m}$ using plane rotations. The work involved is negligible if $m \ll n$.

Any stationary iterative method can be used as preconditioner. For example, as noted in Sect. 7.2, if A is an M-matrix, incomplete LU (ILU) factorizations exist:

$$A = M - N, \quad M = LU ,$$

and we may simply take $C_L = L^{-1}$, $C_R = U^{-1}$. ILU has proved to be very effective as preconditioning for Krylov subspace methods. For more information on ILU, see Chap. 3 of Barrett *et al.* (1994) and references quoted there.

Robustness and convergence

Since the minimization property (7.22) of GMRES is the same as the minimization property of GCR, the convergence behavior of the two methods is identical, apart from rounding errors. But GCR allows the use of a variable preconditioner in (7.21), which, as mentioned before, is taken advantage of in the GMRESR method (van der Vorst and Vuik (1994)). Obviously, GMRES requires less work and storage per step than GCR. A step of GMRES requires only one multiplication by A, and one iteration with the preconditioner. But in addition, the orthogonal basis and the matrix $H_{m+1,m}$ have to be constructed, and (7.23) has to be solved. The latter task is negligible if $m \ll n$. For a particular implementation, in Saad and Schultz (1986) it is estimated that construction of the basis and computation of y^m from u takes $m(m+2)n + m\bar{n}$ multiplications, with $\bar{n}$ the number of non-zero entries in A. Furthermore, storage of $\{v^1, ..., v^{m+1}\}$ takes $(m+1)n$ reals. As a consequence, in practice one is restricted to $m \ll n$. If the desired accuracy is not reached, one puts $y^0 = y^m$ and GMRES is restarted. Of course, this destroys the optimality property (7.22). Unlike GCR, there is no simple way to use truncation with GMRES.

The efficiency of GMRES, GMRESR and Bi-CGSTAB (to be discussed later) is compared in van der Vorst and Vuik (1994), Vuik (1993).

The conjugate gradient method

In the special case that A is symmetric positive definite (SPD), the minimization problem (7.16) can be solved by the *conjugate gradient method* (CG method) in a much cheaper way than with GCR or GMRES. About the only SPD matrix to be encountered in computational fluid dynamics is the pressure matrix in orthogonal coordinates, in which case fast Fourier/cyclic reduction Poisson solvers may be applicable, and if so, are to be preferred.

But if not, CG methods are a good choice, so they will be discussed.

Because A is SPD, $A^{1/2}$ exists. If the norm is chosen as follows:

$$\|r\|^2 = (A^{-1}r, r) = \|A^{-1/2}r\|_2^2 ,\qquad (7.24)$$

with $(\cdot, \cdot)$ the Euclidean inner product, then the following CG method solves (7.16):

Conjugate gradient method
  Choose y^0; $p^{-1} = 0$; $r^0 = b - Ay^0$; $\rho_{-1} = 1$; $m = -1$;
  while stopcriterion not satisfied **do**
    $m = m + 1$;
    $\rho_m = (r^m, r^m)$; $\beta_m = \rho_m/\rho_{m-1}$; $p^m = r^m + \beta_m p^{m-1}$;
    $\sigma_m = (Ap^m, p^m)$; $\alpha_m = \rho_m/\sigma_m$; $y^{m+1} = y^m + \alpha_m p^m$;
    $r^{m+1} = r^m - \alpha_m Ap^m$;
    Verify stopcriterion;
  end while
We will now show that this algorithm solves (7.16).

Theorem 7.3.2. *Optimality property of conjugate gradient method*
The conjugate gradient method gives a residual $r^m = b - Ay^m$ that satisfies

$$\|r^m\| = \min\{\|\psi(A)r^0\| : \psi \in \prod_m^1 \} ,$$

where $\|r\| = \|A^{-1/2}r\|_2$.

Proof. That $r^m = \phi_m(A)r^0$ with $\phi_m \in \prod_m^1$ is easy to see. We have

$$\|\psi(A)r^0\|^2 = (A^{-1}\psi(A)r_0, \psi(A)r^0) .$$

Let $\psi \in \prod_m^1$ be arbitrary. Then we can write

$$\psi(\xi) = \phi_m(\xi) + \xi\chi_{m-1}(\xi)$$

with χ_{m-1} a polynomial of degree $\leq m - 1$. It follows that

$$\|\psi(A)r^0\|^2 = \|\phi_m(A)r^0\|^2 + \|A\chi_{m-1}(A)r^0\|^2 + 2(\chi_{m-1}(A)r^0, \phi_m(A)r^0) .$$

We have proved the theorem if we can show that

$$(\chi_{m-1}(A)r^0, \phi_m(A)r^0) = 0, \quad \forall \chi_{m-1} .\qquad (7.25)$$

We have $r^k = \phi_k(A)r^0$. We assume $\alpha_1, \alpha_2, ..., \alpha_m \neq 0$ (if not, we have stumbled upon the solution: $\rho_k = 0$ for some $k \leq m$, $r_k = 0$ and (7.16) is obviously satisfied). Assuming A has full rank, $r^0, ..., r^m$ are linearly independent, and

$$\text{span } (r^0, ..., r^m) = K_m(A, r^0) \,.$$

Since $\chi_m(A)r^0 \in \text{span } (r^0, ..., r^m)$, (7.25) is satisfied if

$$(r^k, r^m) = 0, \quad k < m \,. \tag{7.26}$$

Hence, it suffices to establish (7.26). Using induction, assume

$$(p^k, Ap^{m-1}) = 0, \quad k < m-1 \tag{7.27}$$

and

$$(r^k, r^m) = 0, \quad k < m \,. \tag{7.28}$$

For $k < m - 1$ we have

$$(p^k, Ap^m) = (p^k, Ar^m + \beta_m Ap^{m-1}) = (Ap^k, r^m) = (r^k - r^{k+1}, r^m)/\alpha_k = 0 \,.$$

Furthermore,

$$(p^{m-1}, Ap^m) = (p^{m-1}, Ar^m + \beta_m Ap^{m-1}) = (Ap^{m-1}, r^m) + \beta_m \sigma_{m-1}$$
$$= (r^{m-1} - r^m, r^m)/\alpha_{m-1} + \beta_m \sigma_{m-1} = 0 \,,$$

so that (7.27) holds with m increased by one. Continuing with (7.28), for $k < m$ we have

$$(r^k, r^{m+1}) = (r^k, r^m - \alpha_m Ap^m) = -\alpha_m(r^k, Ap^m)$$
$$= -\alpha_m(p^k - \beta_k p^{k-1}, Ap^m) = 0 \,.$$

Furthermore,

$$(r^m, r^{m+1}) = \rho_m - \alpha_m(r^m, Ap^m) = \rho_m - \alpha_m(p^m - \beta_m p^{m-1}, Ap^m)$$
$$= \rho_m - \alpha_m \sigma_m = 0 \,.$$

so that (7.28) holds with m increased by one. Establishing (7.27) and (7.28) for $m = 1$ and $m = 2$ is left to the reader.

$\square$

Definitions of the norm other than (7.24) result in other variants of the CG method. The fact that the search directions are conjugate with respect to A (equation (7.27)) gives the method its name.

Convergence and work

The rate of convergence can be estimated by Theorem 7.3.1, which applies because of Theorem 7.3.2 with the corresponding norm.

In general, $\sigma(A)$ is unknown. An upper bound on the number of iterations required that uses only the condition number is given by the following theorem.

Theorem 7.3.3. *Rate of convergence of the CG method*
The residuals generated by the CG method satisfy

$$\|r^m\|/\|r^0\| \le 2\exp(-2m/\sqrt{\kappa}) \,,$$

with κ the spectral condition number of A, and $\|r\| = \|A^{-1/2}r\|_2$.

Proof. The eigenvalues of A are $\lambda_1 \le \lambda_2 \le \ldots \le \lambda_n$, $\lambda_1 > 0$. Define

$$\lambda(z) = \frac{1}{2}(\lambda_1 + \lambda_n) + \frac{1}{2}(\lambda_n - \lambda_1)z \,.$$

Then

$$\{\lambda_1, \ldots, \lambda_n\} \subset \{\lambda(z) : -1 \le z \le 1\} \,.$$

From Theorem 7.3.2 it follows that

$$\|r^m\| \le \|r^0\| \min \left[\max\{|\psi_m(\lambda(z))| : z \in [-1,1]\} : \psi_m \in \prod_m^1 \right] \,.$$

It is known that the sharpest bound is obtained by

$$\psi(\lambda(z)) = \bar{\psi}(z) = T_m(z)/T_m(z_0), \quad z_0 = (\lambda_1 + \lambda_n)/(\lambda_n - \lambda_1) \,,$$

with T_m the Chebyshev polynomial of degree m. This gives

$$\|r^m\|/\|r^0\| \le \max\{|\bar{\psi}(z)| : z \in [-1,1]\} = 1/T_m(z_0) \,.$$

T_m has the following representation:

$$T_m(z) = \frac{1}{2}(z + \sqrt{z^2-1})^m + \frac{1}{2}(z + \sqrt{z^2-1})^{-m} \,.$$

The spectral condition number is $\kappa = \lambda_n/\lambda_1$. We have

$$z_0 = \frac{\kappa+1}{\kappa-1}, \quad z_0 + \sqrt{z_0^2 - 1} = \frac{\sqrt{\kappa}+1}{\sqrt{\kappa}-1} \,,$$

$$\frac{1}{T_m(z_0)} = 2\left(\frac{\sqrt{\kappa}-1}{\sqrt{\kappa}+1}\right)^m \{1 + \left(\frac{\sqrt{\kappa}-1}{\sqrt{\kappa}+1}\right)^{2m}\}^{-1}$$
$$\leq 2\left(\frac{\sqrt{\kappa}-1}{\sqrt{\kappa}+1}\right)^m \leq 2e^{-2m/\sqrt{\kappa}} .$$

□

Obviously, a step with CG is much cheaper than a step with GCR or GMRES. There is no Gram-Schmidt orthogonalization. Each step requires only 3 vector updates, two inner products and one matrix-vector multiplication, requiring $\mathcal{O}(sn)$ work, with s the number of nonzero diagonals in A. Furthermore, only A and 3 vectors need to be stored: p^m (overwriting p^{m-1}), r^m (overwritten by r^{m+1}) and y^{m+1} (overwriting y^m). Hence, restarting is not necessary.

If we require a reduction of the residual by a factor ch^2 with h a measure of the mesh size, because the discretization error is $\mathcal{O}(h^2)$, then Theorem 7.3.3 tells us that a sufficient number of iterations is

$$m \geq -\frac{1}{2}\sqrt{\kappa}\ln ch^2/2 . \tag{7.29}$$

For elliptic problems we generally have $\kappa = \mathcal{O}(h^{-2})$. In d dimensions, the number of unknowns $n = \mathcal{O}(h^{-d})$. The work of one iteration with the conjugate gradient method is $\mathcal{O}(n)$, so the total work satisfies

$$W = \mathcal{O}(n^\alpha \ln n), \quad \alpha = 1 + 1/d . \tag{7.30}$$

This is not better than what can be achieved by stationary iterative methods using an optimal overrelaxation parameter. But the CG method derives its usefulness from the fact, that α can be brought down by preconditoning, or, equivalently, that the CG method can be used to accelerate stationary iterative methods, as first pointed out in Meijerink and van der Vorst (1977). No guessing of an optimal parameter will be required.

Implementation of the preconditioned conjugate gradient method

For application of CG, the preconditioned system must be SPD. Therefore (7.20) is replaced by

$$\tilde{A}\tilde{y} = \tilde{b}, \quad \tilde{A} = CAC^T, \quad \tilde{y} = C^{-T}, \quad \tilde{y} = C^{-T}y, \quad \tilde{b} = Cb, \tag{7.31}$$

so that $\tilde{A}$ is SPD if A is, assuming C to be nonsingular, of course. When CG is applied to (7.31), the matrix C can be eliminated by making the following substitutions: $r = C\tilde{r}$, $p = C^{-T}\tilde{p}$. Deleting the tildes, this results in the following algorithm (where $M = C^{-1}C^{-T}$):

Preconditioned CG method
  Choose y^0; $p^{-1} = 0$; $r^0 = b - Ay^0$; $\rho_{-1} = 1$; $m = 1$;
  while stopcriterion not satisfied **do**
    $m = m + 1$;
    Solve $Mz = r^m$; $\rho_m = (z^m, r^m)$; $\beta_m = \rho_m/\rho_{m-1}$;
    $p^m = z + \beta_m p^{m-1}$; $\sigma_m = (Ap^m, p^m)$; $\alpha_m = \rho_m/\sigma_m$;
    $y^{m+1} = y^m + \alpha_m p^m$, $r^{m+1} = r^m - \alpha_m Ap^m$;
    Evaluate stopcriterion;
  end while

It remainds to find a suitable $M = C^{-1}C^{-T}$ with M^{-1} approximating A^{-1}. The stationary iterative method (7.2) can be rewritten as

$$y^{k+1} = y^k + M^{-1}(b + Ny^k - My^k) = y^k + M^{-1}(b - Ay^k) \,,$$

from which it is clear that M^{-1} can be regarded as an approximate inverse of A. Solving $Mz = r^m$ in the preconditioned CG method is equivalent to executing one iteration with the stationary iterative method, with starting iterate zero and right hand side r^m. Because we must have $M = C^{-1}C^{-T}$ for some matrix C, M must be SPD. We give two examples.

Symmetric Gauss-Seidel

The forward Gauss-Seidel method is defined by $M = R+D$, $D = \text{diag}(A)$, R the lower triangular part of A excluding the main diagonal. Clearly, M is not SPD and cannot be used to precondition CG. For backward Gauss Seidel we have $M = R^T + D$, and an SPD matrix M is obtained by symmetric Gauss-Seidel (forward followed by backward), given by

$$y^{k+1/2} = y^k + M_1^{-1}(b - Ay^k), \quad M_1 = R + D,$$
$$y^{k+1} = y^{k+1/2} + M_1^{-T}(b - Ay^{k+1/2}) \,,$$

from which it follows that

$$y^{k+1} = y^k + M^{-1}(b - Ay^k), \quad M^{-1} = M_1^{-1} + M_1^{-T} - M_1^{-T}AM_1^{-1}$$
$$= M_1^{-1} + M_1^{-T} - M_1^{-T}(M_1 + M_1^T - D)M_1^{-1} = M_1^{-T}DM_1^{-1} \,.$$

We have, with $z = M_1^{-1}y$, if $y \neq 0$,

$$y^T M^{-1}y = z^T Dz > 0 \,,$$

because $d_{ii} > 0$, since A is SPD. Hence M is SPD, which means that there exists an SPD matrix C such that $C^{-2} = C^{-1}C^{-T} = M$. Note that in the preconditioned CG method the matrix C is not needed.

Incomplete Cholesky decomposition

This corresponds to the symmetric version of ILU:

$$A = M - N, \quad M = LL^T .$$

We now simply choose $C = L^{-1}$ in (7.31). This approach was proposed in Meijerink and van der Vorst (1977), leading to the well-known ICCG (incomplete Cholesky conjugate gradients) method. ICCG was found to be very effective in numerical experiments described in Meijerink and van der Vorst (1977), Kershaw (1978); it was also found numerically that (7.30) holds indeed. Subsequently, incomplete factorizations have found widespread use as preconditioners for Krylov subspace methods. The rate of convergence depends on the spectrum of CAC^T. Since $M^{-1}A = C^T(CAC^T)C^{-T}$, $M = C^{-1}C^{-T}$, $M^{-1}A$ has the same spectrum as CAC^T. Upper bounds for the required number of iterations can be obtained by means of Theorem 7.3.3, which asks for the condition number. Theoretical estimates of $\kappa(M^{-1}A)$ can be obtained only in special cases. For the standard 5-point discretization of the Laplacian on a uniform square grid and a special choice for L it was shown in Gustafsson (1978) that $\kappa(M^{-1}A) = \mathcal{O}(n^{1/2})$, so that for the two-dimensional case (7.29) gives $m = \mathcal{O}(n^{1/4} \ln n)$, leading to a total work of

$$W = \mathcal{O}(n^\alpha \ln n), \quad \alpha = 5/4 . \tag{7.32}$$

This value of α seems to be representative of what can be achieved in general with preconditioned Krylov subspace methods. If the eigenvalues of A are clustered, Theorem 7.3.3 is pessimistic. In Bruaset (1995) (Sect. 4.2) a survey of incomplete factorizations is given, and pictures of the spectra of A and $M^{-1}A$ are shown.

The Bi-CGSTAB method

A method that applies for general A, but with computing work and storage demands per step comparable to those of CG (hence, much less than for GCR and GMRES) is the bi-conjugate gradient stabilized (Bi-CGSTAB) method, proposed in van der Vorst (1992). However, for a general matrix A there is no minimization principle like (7.16), and convergence is not guaranteed. Although less robust, this method is frequently significantly more efficient than GMRES, because it is cheaper per step, and the restarts that GMRES may require (because of storage limitations) slow down its rate of convergence. A cheap version (Bruaset (1995)) is obtained if postconditioning rather than preconditioning is used: $AC\tilde{y} = b$, $y = C^{-1}\tilde{y}$. The algorithm is given by:

Postconditioned Bi-CGSTAB method

Choose y^0;

$r^0 = b - ay^0$; $p^0 = q^0 = 0$; $\alpha = \omega_0 = 1$;

while stopcriterion not satisfied **do**

$\quad \beta = \alpha(r^0, r^{m-1})/\{\omega_{m-1}(r^0, r^{m-2})\}$;

$\quad p^m = C(r^{m-1} - \beta\omega_{m-1}q^{m-1}) + \beta p^{m-1}$; $q^m = Ap^m$;

$\quad \alpha = (r^0, r^{m-1})/(r^0, q^m)$;

$\quad \tilde{s}^m = r^{m-1} - \alpha q^m$; $s^m = C\tilde{s}^m$; $t^m = As^m$;

$\quad \omega_m = (t^m, \tilde{s}^m)/(t^m, t^m)$;

$\quad y^m = y^{m-1} + \alpha p^m + \omega_m s^m$; $r^m = \tilde{s}^m - \omega_m t^m$;

$\quad$ Evaluate stopcriterion;

end while

Final remarks

A survey of Krylov subspace methods is given in Barrett *et al.* (1994). The choice of a good preconditioner seems to be more crucial for efficiency than the choice of a particular Krylov subspace method. Although this is not a theorem, equation (7.32) is representative of the computing work that is needed in the SPD case; in the general case, this will be more. An interesting comparison of various iterative methods is given in Botta *et al.* (1997). We will now discuss a class of methods that have the more optimal complexity $W = \mathcal{O}(n)$. The price to be paid is more complicated programming.

7.4 Multigrid methods

Here only a brief outline of multigrid methods will be sketched. Full introductions are given in Hackbusch (1985), Wesseling (1992). Furthermore, the elementary introduction Briggs (1987) and the pioneering paper Brandt (1977) can be recommended to those who want to get acquainted with the subject. The discovery of multigrid methods and (more general) multilevel methods (cf. Brandt (1977)) is the most significant development in numerical analysis in the last 25 years, and is having a strong impact on computational fluid dynamics. A review of the application of multigrid in computational fluid dynamics is given in Wesseling and Oosterlee (2000). The multigrid website, maintained by C.C. Douglas, is at www.mgnet.org.

The multigrid principle

Assume we have a discretization of a partial differential equation on the d-dimensional unit cube Ω. Let the computational grid Ω_l be uniform with stepsize h_l in all directions; the role of the index l (denoting the so-called level) will become clear shortly. The problem to be solved on the grid Ω_l is given by

$$A^l y^l = b^l . \tag{7.33}$$

To begin with, a small number of iterations is performed with some stationary iterative method. Let this be done with, in pseudo-programming language, by a call $S(A^l, b^l, y^l)$ to a subroutine S, called the *smoother*. Here y^l is both an input parameter (the initial iterate) and an output parameter (the approximation to the solution of (7.33) after smoothing); the dual use of the symbol y^l (solution of (7.33) and current iterate) should cause no confusion. As noted in Sect. 7.2, convergence is expected to be slow, with spectral radius $1 - \mathcal{O}(h_l^2)$. However, it frequently happens that the error is made smooth rapidly. This is called the smoothing property; for a precise mathematical definition see Hackbusch (1985) Sect. 6.1.3, Wesseling (1992) Sect. 6.5, and this property is assumed to hold here.

The basic idea is now to approximate the error on a coarser grid Ω_{l-1} with $h_{l-1} = 2h_l$; the number of cells in Ω_l is assumed to be even in all directions. Fig. 7.1 gives a one-dimensional example.

Prolongation and restriction

Let $U^k : \Omega_k \to \mathbb{R}$ be the set of grid functions on Ω_k, and let there be given

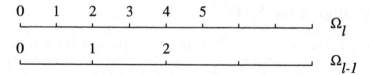

Fig. 7.1. Fine and coarse grid

a prolongation and restriction operator:

$$P^l : U^{l-1} \to U^l, \quad R^{l-1} : U^l \to U^{l-1} . \tag{7.34}$$

For example, in one dimension we could have, using linear interpolation, with the numbering system of Fig. 7.1,

$$(P^l y^{l-1})_{2j} = y_j^{l-1}, \quad (P^l y^{l-1})_{2j+1} = \frac{1}{2}(y_j^{l-1} + y_{j+1}^{l-1}) , \tag{7.35}$$

and a natural restriction would be

$$(R^{l-1} y^l)_j = y_{2j}^l . \tag{7.36}$$

It turns out that (7.36) is not good. For information on the requirements that P^l and R^{l-1} should satisfy, see Hackbusch (1985) (p. 149) or Wesseling (1992) (p.71). A good restriction is

$$(R^{l-1}y^l) = \frac{1}{4}(y^l_{2j-1} + 2y^l_{2j} + y^l_{2j+1}) \ .$$

Coarse grid correction

After smoothing, the second step in the algorithm is approximation of the error on the coarse grid. The error in the current iterand y^l, called e^l, satisfies

$$A^l e^l = b^l - A^l y^l \ . \tag{7.37}$$

Let A^{l-1} be an approximation of A^l on the coarse grid Ω^{l-1}, obtained for example by discretization of the underlying partial differential equation on Ω^{l-1}. Then we solve

$$A^{l-1}y^{l-1} = b^{l-1} \equiv R^{l-1}(b^l - A^l y^l) \ .$$

Next, an approximation to e^l (also called e^l) is obtained by interpolation to Ω^l:

$$e^l = P^l y^{l-1} \ . \tag{7.38}$$

If, as assumed here, the original error (satisfying (7.37)) is smooth, then (7.38) may be expected to give a good approximation to the error. This can be formulated mathematically in a precise way, and is called the *approximation property* (Hackbusch (1985) Sect. 6.1.3, Wesseling (1992) Sect. 6.5). The reason why (7.36) is not good is that the approximation property does not hold, as it turns out. The coarse grid correction is added to y^l:

$$y^l = y^l + e^l \ . \tag{7.39}$$

Finally, smoothing may be applied to y^l, and one two-grid iteration has been completed.

Two-grid algorithm

The foregoing leads us to the following iteration method. The coarse grid approximation to the error e^{l-1} is called y^{l-1}.

Two-grid method

Initialize y^l;

do until convergence

$S(A^l, b^l, y^l)$;

$b^{l-1} = R^{l-1}(b^l - A^l y^l)$;

Solve $A^{l-1} y^{l-1} = b^{l-1}$;

$y^l = y^l + P^l y^{l-1}$;

$S(A^l, b^l, y^l)$;

Check convergence criterion;

end

How to solve on Ω_{l-1} is discussed below.

Efficiency of multigrid method

The nice thing about the two-grid method is that its rate of convergence is independent of h_l, provided the smoothing and approximation properties hold. These requirements are not difficult to satisfy, as it turns out; for more information see Brandt (1977), Hackbusch (1985), Wesseling (1992).

Let N_k be the number of unknowns on Ω_k. Then the work required for one two-grid iteration excluding the solve step on the coarse grid is cN_l, with c some constant. Denoting the work of the solve step by W_{l-1}, the work W_l for one two-grid iteration is

$$W_l = cN_l + W_{l-1} . \tag{7.40}$$

We obtain a true multigrid method by replacing the solve step by γ two-grid iterations employing a still coarser grid Ω_{l-2}, and so on recursively, till a very coarse grid Ω_0 is reached, on which the cost of solving is negligible, and with N_0 independent of N_l. This means that if we refine Ω_l, the number of levels l is increased. Then we have, with d the dimension of Ω, for the number of grid points on Ω^k

$$N_k = 2^{kd} N_0 , \tag{7.41}$$

and the computing work satisfies the following recursion:

$$W_k = c2^{kd} N_0 + \gamma W_{k-1}, \quad k = 1, 2, ..., l . \tag{7.42}$$

To obtain a mathematically elegant formulation we assume $W_0 = cN_0$. This may be incorrect, but this is of no consequence because W_0 is very small. Then it follows that

$$\begin{aligned} W_l &= c2^{ld} N_0 (1 + \gamma(2^{-d} + \gamma(2^{-2d} + ... + \gamma 2^{-ld})...)) \\ &= cN_l (1 + \tilde{\gamma} + \tilde{\gamma}^2 + ... + \tilde{\gamma}^l), \quad \tilde{\gamma} = \gamma/2^d . \end{aligned} \tag{7.43}$$

Assume

$$\tilde{\gamma} < 1 . \tag{7.44}$$

Then

$$W_l < cN_l/(1 - \tilde{\gamma}) . \tag{7.45}$$

The following conclusions may be drawn from (7.45). The quantity cN_l is the work performed on the finest grid. In practice the bulk of this work is in the smoothing, which consists of a very small number of iterations with a simple stationary iterative method. The work on the coarse grids adds to this $100\tilde{\gamma}/(1 - \tilde{\gamma})$ percent. It turns out that usually $\gamma = 1$ or $\gamma = 2$ suffices to obtain a rate of convergence independent of h_l. Taking $\gamma = 2$, we get

$$W_l < 2cN_l \;\; (d = 2), \quad W_l < \frac{4}{3}cN_l \;\; (d = 3) .$$

So the cost of a complete multigrid cycle is equivalent to the cost of just a few steps with the smoothing method.

Because the rate of convergence is independent of h_l, $\mathcal{O}(|\ln h_k|) = \mathcal{O}(\ln N_l)$ iterations are required to obtain an error reduction of $\mathcal{O}(h_l^2)$. Since the work for one iteration is $\mathcal{O}(N_l)$ the total computing work satisfies

$$W = \mathcal{O}(N_l \ln N_l) . \tag{7.46}$$

The logarithmic factor in (7.46) may be removed by changing the multigrid algorithm, namely by starting computations on the coarsest grid Ω_0 (nested iteration or full multigrid), cf. Brandt (1980), Hackbusch (1985) Chap. 5, Wesseling (1992) Sect. 8.4, resulting in optimal computational complexity:

$$W = \mathcal{O}(N_l) .$$

This unique property has been proved rigorously under fairly general circumstances, see Hackbusch (1985). There is no other method known which achieves this. Consequently, multigrid methods have a strong impact on computational fluid dynamics.

Storage requirements

The additional storage requirements for the computations on the coarse grids are quite modest. A reasonable estimate for the storage required om grid Ω_k is cN_k, with c some constant which is the same on all grids. Total storage is therefore

$$cN_l(1 + 2^{-d} + 2^{-2d} + ...) < cN_l 2^d/(2^d - 1) ,$$

so that compared to single grid solution only about $1/3$ extra storage is required in two dimensions and $1/7$ in three dimensions.

Nonlinear multigrid method

A powerful property of multigrid is that it is directly applicable to nonlinear problems. Let the nonlinear discrete problem to be solved on Ω_l be denoted by

$$A^l(y^l) = b^l . \tag{7.47}$$

The nonlinear two-grid algorithm is given by

Nonlinear two-grid method
  Initialize y^l;
  do until convergence
      (1) $S(A^l, b^l, y^l)$;
      (2) $r^l = b^l - A^l(y^l)$;
      (3) Choose $\tilde{y}^{l-1}, s_{l-1}$;
      (4) $b^{l-1} = A^{l-1}(\tilde{y}^{l-1}) + s_{l-1} R^{l-1} r^l$;
      (5) Solve $A^{l-1}(y^{l-1}) = b^{l-1}$;
      (6) $y^l = y^l + (1/s_{l-1}) p^l (y^{l-1} - \tilde{y}^{l-1})$;
      (7) $S(A^l, b^l, y^l)$;
      Check convergence criterion;
  end

We proceed with a discussion of this algorithm. S represents some nonlinear smoothing iteration, for example nonlinear Gauss-Seidel. In (2) the residual is computed, which is going to steer the coarse grid correction. In this nonlinear case we cannot use r^l as right-hand side for the equation for the error; $A^l(e^l) = r^l$ might not even have a solution. For the same reason, $R^{l-1} r^l$ cannot be the right hand side for the coarse grid problem on Ω_{l-1}. Instead, it is added in (4) to $A^{l-1}(\tilde{y}^{l-1})$, with $\tilde{y}^{l-1}$ sufficiently close to the solution in some sense; choosing $\tilde{y}^{l-1} = R^{l-1} y^l$ would be a good idea. Obviously, $A^{l-1}(y^{l-1}) = A^{l-1}(\tilde{y}^{l-1})$ has a solution, and by choosing s_{l-1} small enough in (4), (5) also has a solution. In (6) the effect of s_{l-1} is undone again, and $y^{l-1} - \tilde{y}^{l-1}$ is a coarse grid approximation to the error. It is easy to see that in the linear case the nonlinear two-grid algorithm is equivalent to the linear two-grid algorithm. Extension from two-grid to multigrid is obvious.

Unlike the Krylov subspace methods discussed before, the nonlinear multigrid method does not need global linearization for application to nonlinear problems; only local linearization in the smoother is required. This is an advantage, because computation of a global Jacobian may be expensive, and it may be ill-conditioned, for example for transonic flows. On the other hand, programming multigrid methods is more complicated than programming Krylov subspace methods. For comparison of Krylov subspace and multigrid methods, see Wesseling (1992) Sect. 8.9 and references quoted there, and Zeng, Vuik, and Wesseling (1995).

Smoothing analysis

Usually, the smoother is the most computing intensive and critical part of multigrid. In general, the smoother is a simple stationary iterative method. Therefore, like Krylov subspace methods, multigrid may be viewed as a way to accelerate the convergence of stationary iterative methods. If multigrid does not converge rapidly, this is mostly because the smoother does not reduce all non-smooth components of the error effectively. Apart from changing the smoother, this can often be remedied by accelerating multigrid by a Krylov subspace method, by using a multigrid iteration as preconditioner. This has been done for example in Kettler (1982), Braess (1986), Zeng, Vuik, and Wesseling (1995), Washio and Oosterlee (1997).

Not every convergent iteration method has the smoothing property. But by introducing damping every convergent stationary iteration method can be made into a smoother, as noted in Wittum (1989a), cf. Wesseling (1992), p.98. Therefore it is desirable that A^l is an M-matrix, because then convergent methods and hence also smoothing methods are easy to find.

The smoothing property of stationary iterative methods can be investigated by *Fourier smoothing analysis*. To make Fourier analysis applicable, the coefficients in the differential equation are made constant (by 'freezing' local values just as for Fourier stability analysis in Sect. 5.5), and the grid is taken uniform and the boundary conditions periodic. Then discretizations of partial differential equations can be written as

$$L_h \psi_j = \sum_{k \in K} \alpha_k \psi_{j+k} = f_j, \quad j = (j_1, ..., j_d), \quad k = (k_1, ..., k_d) \,,$$

with d the number of dimensions. A stationary iterative method based on splitting (Sect. 7.2) corresponds to a splitting $K = K_1 \cup K_2$, and leads to the followig iteration scheme:

$$\sum_{k \in K_1} \alpha_k \psi_{j+k}^{m+1} = - \sum_{k \in K_2} \alpha_k \psi_{j+k}^m + f_j \,. \tag{7.48}$$

The error ε satisfies (7.48) with $f_j \equiv 0$. According to Theorem 5.5.1 we can write

$$\varepsilon_j^m = \sum_{\theta \in \Theta} c_\theta^m e^{ij\theta} \,.$$

Substitution in the homogeneous version of (7.48) gives, because the boundary conditions are periodic (cf. (5.20)):

$$c_\theta^{m+1} = g(\theta) c_\theta^m, \quad g(\theta) = - \sum_{K_2} \alpha_k e^{ik\theta} / \sum_{K_1} \alpha_k e^{ik\theta} \,. \tag{7.49}$$

We have the smoothing property if $|g(\theta)| < 1$ for all non-smooth wave-numbers θ. It is clear that all $|\theta| \ll 1$ are smooth wave-numbers. Somewhat arbitrarily, the sets Θ_s and Θ_r of smooth and rough wave-numbers are defined by

$$\Theta_s = \{\theta \in \Theta : |\theta_j| < \pi/2, \; j = 1, ..., d\}, \quad \Theta_r = \Theta \setminus \Theta_s \, . \tag{7.50}$$

The *smoothing factor* $\bar{\rho}$ is defined by

$$\bar{\rho} = \max\{|g(\theta)| : \; \theta \in \Theta_r\} \, . \tag{7.51}$$

We have the smoothing property if $\bar{\rho} < 1$. The smoothing factor $\bar{\rho}$ is easily determined numerically or analytically; see Wesseling (1992) (Chap. 7) for many examples and discussion of the influence of boundary conditions.

It turns out that in many cases $\bar{\rho}$ is much smaller than 1. For example, for Laplace's equation in two dimensions we have $\bar{\rho} = 1/2$ for the Gauss-Seidel method and $\bar{\rho} \cong 0.13$ for the 7-point ILU method. For a well-designed multigrid method the error reduction factor is often of about the same size as $\bar{\rho}$, so that the rate of convergence is not only independent of the mesh size, but also fast. Chap. 9 of Wesseling (1992) and Wesseling and Oosterlee (2000) may be consulted for application of multigrid to computational fluid dynamics.

7.5 Fast Poisson solvers

Certain methods developed especially for the Poisson equation have come to be known as *fast Poisson solvers*. Because they are so efficient, this class of direct methods gets an honorary position in this chapter on iterative methods.

Solving Poisson's equation for the pressure correction takes most of the time in computing nonstationary incompressible viscous flows on staggered grids, unless a fast Poisson solver is used. Without it large-eddy simulation of turbulence would not be feasible. Although fast Poisson solvers are restricted to orthogonal coordinates on rectangular domains, their importance for the applications just mentioned warrants a brief discussion here.

By fast Poisson solvers one means solution methods based on the *fast Fourier transform* and/or *cyclic reduction*. A survey of these methods is given in Swarztrauber (1984). They will not be described here, but we will indicate when they can be applied. Let us have a rectangular computational grid and let (j, k) be grid point indices in the usual way; $j, k \in \mathbb{N}$. Cyclic reduction is applicable if the stencil of the discretization is of the following form:

$$\begin{bmatrix} & e_k & \\ a_j & b_{jk} & c_j \\ & d_k & \end{bmatrix} \qquad (7.52)$$

That is, the coefficients referring to horizontal neighbours may depend only on the horizontal grid point index j, and similar in other directions; cyclic reduction is applicable in more than two dimensions. The fast Fourier transform can be brought into play when in one or more directions the coefficients are independent of the index in that direction; for example, if $a_j = a$, $b_{jk} = b_k$, $c_j = c$.

A stencil like (7.52) is possible only if the underlying partial differential equation is *separable*. That is, the coefficient of a derivative $\partial^m / \partial x_\alpha^m$ is allowed to depend only on x_α. Furthermore, no mixed derivatives may occur, which restricts us to orthogonal coordinate systems, so that general domains treated with boundary-fitted coordinates are out of bounds. The grid need not be uniform, but the step size in the horizontal direction may depend on j, and similarly for the other directions. We see that fast Poisson solvers are applicable to more general equations than the Poisson equation, but they were first developed for Poisson, and the name has stuck.

Fast Poisson solvers are of interest because they are fast. The computing work W satisfies $W = \mathcal{O}(N \ln N)$, or sometimes even $\mathcal{O}(N \ln \ln N)$, with reasonable constants of proportionality, and little storage is required. These methods are mature, and good computer codes are readily available on the Internet: `math.nist.gov/cgi-bin/gams-serve/class/I2b1a1a.html`.

Because they are so efficient and convenient, efforts have been made to somehow use fast Poisson solvers for more general problems (cf. Swarztrauber (1984)), such as problems on non-rectangular domains, or nonseparable equations. But efficiency suffers, and we feel that if more generality and flexibility are required, preconditioned Krylov subspace and multigrid methods are preferable. These have the potential to achieve similar efficiency in the nonseparable case.

7.6 Iterative methods for the incompressible Navier-Stokes equations

The algebraic system to be solved

We start with the discretization on a staggered grid in the stationary case. If we delete the time derivative in equation (6.42), the algebraic system to be solved may be denoted as

$$\begin{pmatrix} N & G \\ D & 0 \end{pmatrix} \begin{pmatrix} u \\ p \end{pmatrix} = \begin{pmatrix} f \\ g \end{pmatrix} . \qquad (7.53)$$

This can be looked upon as a linear system arising in an iterative procedure to solve (6.42) with nonlinear N, but we can for the time being also retain the possibility that N is a nonlinear operator on u. The nonlinearity is due to the inertia terms.

A crucial fact is that we do not have a K-matrix, nor an M-matrix (in the linear case), even if upwind discretization is used for convection, because we have a zero block on the main diagonal in (7.53). This precludes straightforward application of the iterative methods discussed before, and special methods have to be used. The literature is vast, and we can give only a few illustrative examples. First we present a general strategy to deal with non-M-matrices.

Distributive iteration

Suppose we have a linear system

$$Ay = b .\tag{7.54}$$

The system is postconditioned:

$$AB\hat{y} = b, \quad y = B\hat{y} .\tag{7.55}$$

The postconditioning matrix B is chosen such that (7.55) is easier to solve iteratively than (7.54). For example, AB is an M-matrix, while A is not. To (7.55) the iterative methods discussed before can be applied in straightforward fashion. For the iterative solution of (7.55) the splitting

$$AB = M - T\tag{7.56}$$

is introduced, to which corresponds the following splitting of the original matrix A:

$$A = MB^{-1} - TB^{-1} .$$

This leads to the following stationary iterative method for (7.54):

$$MB^{-1}y^{k+1} = TB^{-1}y^k + b ,$$

or

$$y^{k+1} = y^k + BM^{-1}(b - Ay^k) .\tag{7.57}$$

Note that the iterative method defined by (7.57) is obviously consistent for any nonsingular B and M: if it converges, it solves $Ay = b$. This means that when designing a distributive method, B and M in (7.57) need not be precisely the same as in (7.56). For example, one could use a complicated B to design a good M for (7.56), possibly even with $T = 0$, and then change M a little in (7.57) to make $My = c$ easily solvable, and also B may be simplified to make (7.57) easy to apply.

The method (7.57) is called *distributive iteration*. Equation (7.57) shows that the correction $M^{-1}(b - Ay^k)$ corresponding to non-distributive $(B = I)$ iteration is distributed, so to speak, over the elements of y^{k+1}, whence the name of the method, coined in Brandt and Dinar (1979).

Distributive iteration for Navier-Stokes

A general treatment of distributive iteration is given in Wittum (1989b), Wittum (1990a), Wittum (1990c), Wittum (1990b), where it is shown that a number of well-known iteration methods for the incompressible Navier-Stokes equations can be interpreted as distributive methods. Examples will follow. The distributive iteration framework permits us to give a unified description.

The system to be solved is (7.53). Define

$$A = \begin{pmatrix} N & G \\ D & 0 \end{pmatrix} . \tag{7.58}$$

A distribution matrix B such that AB is of block-triangular form:

$$AB = \begin{pmatrix} Q & 0 \\ R & S \end{pmatrix} \tag{7.59}$$

would be attractive. In the first place, splittings $AB = M - N$ are easily obtained by splitting Q and S, leading to simple separate updates for velocity and pressure. Secondly, in case one wants to use multigrid, smoothing analysis is simplified; in Wittum (1990a) it is shown that if the splitting is regular, then we have the smoothing property if we have it for the splittings of Q and S, so that only uncoupled equations for u^1, u^2 and p separately have to be analyzed. A possible choice for B is, with the same block structure as in (7.58):

$$B = \begin{pmatrix} I & B_{12} \\ 0 & B_{22} \end{pmatrix} ,$$

resulting in

$$AB = \begin{pmatrix} N & NB_{12} + GB_{22} \\ D & DB_{12} \end{pmatrix} .$$

Choosing B such that $NB_{12} + GB_{22} = 0$ results in the block triangular form (7.59). Therefore we choose

$$B_{12} = -N^{-1}GB_{22} ,$$

resulting in

$$AB = \begin{pmatrix} N & 0 \\ D & C \end{pmatrix} , \quad C = -DN^{-1}GB_{22} , \qquad (7.60)$$

with B_{22} still to be chosen. The main difficulty in the original formulation (7.53), inamely the zero block on the main diagonal, has disappeared. If upwind discretization is applied to convection, then N is an M-matrix. If B_{22} is chosen such that C is also an M-matrix, chances are that AB is an M-matrix, making the system suited for iterative solution. Various methods result from the choice of B_{22}.

SIMPLE method

A method widely known in the engineering literature as the SIMPLE method ('Semi-Implicit Method for Pressure-Linked Equations') is proposed in Patankar and Spalding (1972) and discussed in detail in Patankar (1980). This is perhaps the oldest and most widely used iterative method for the stationary Navier-Stokes equations. Many variations and improvements have been proposed, such as SIMPLER (Patankar (1980)), SIMPLEC (Van Doormaal and Raithby (1984)) and PISO (Issa (1986)). Without in any way disparaging the creativity that has gone into the design of these methods, nor denying the progress they have brought, it must be said that their formulation in the engineering literature is often ad-hoc and obscure; even their consistency with the system to be solved is not easy to verify. The distributive iteration framework, however, results in a unified and transparent formulation. To write some variants in a distributive formulation, not only postconditioning but also preconditioning must be applied.

The SIMPLE method is obtained by choosing

$$B_{22} = I ,$$

so that (7.60) becomes

$$AB = \begin{pmatrix} N & 0 \\ D & -DN^{-1}G \end{pmatrix} .$$

A splitting $AB = M - T$ is defined by

$$M = \begin{pmatrix} Q & 0 \\ D & R \end{pmatrix} ,$$

where Q and R are approximations to N and $-DN^{-1}G$ such that $My = b$ is easily solvable. For the distribution step in (7.57) B is approximated by

$$B = \begin{pmatrix} I & -\tilde{N}^{-1}G \\ 0 & I \end{pmatrix} ,$$

where $\tilde{N}^{-1}$ is an easy to evaluate approximate inverse of N (in the nonlinear case one may think of $\tilde{N}^{-1}(f)$ as giving an approximate solution of $N(u) = f$ by some iterative process). Depending on the choice of $\tilde{N}^{-1}, Q$ and R, various variants of the SIMPLE method are obtained. N just represents a set of convection-diffusion equations and it is easy to think of simple stationary iterative methods (linearizing globally or locally) to use, thus determining Q. In the original SIMPLE method, one chooses $\tilde{N} = \text{diag}(N)$. This makes $D\tilde{N}^{-1}G$ easy to determine. It has a simple 5-point stencil in two or 7-point stencil in three dimensions, and resembles the standard discrete Laplacian; it is a K-operator, and some simple stationary iterative method may be applied, thus determining R.

This results in the following algorithm. Its basic form is given by (7.57). In the present case we have

$$b - Ay^k = \begin{pmatrix} f \\ g \end{pmatrix} - \begin{pmatrix} N & G \\ D & 0 \end{pmatrix} \begin{pmatrix} u^k \\ p^k \end{pmatrix} = \begin{pmatrix} r_1^k \\ r_2^k \end{pmatrix} . \tag{7.61}$$

After computing the residuals r_1^k, r_l^k, a preliminary velocity correction δu is computed by solving

$$Q\delta u = r_1^k .$$

Next, a preliminary pressure correction is computed by solving

$$R\delta p = r_2^k - D\delta u .$$

In the distribution step new corrections are obtained by

$$\begin{pmatrix} \delta u \\ \delta p \end{pmatrix} := B \begin{pmatrix} \delta u \\ \delta p \end{pmatrix} = \begin{pmatrix} \delta u - \tilde{N}G\delta p \\ \delta p \end{pmatrix} .$$

Finally, in the SIMPLE method underrelaxation is applied with parameters ω_u and ω_p:

$$u^{k+1} = u^k + \omega_u \delta u , \quad p^{k+1} = p^k + \omega_p \delta p . \tag{7.62}$$

Robustness and efficiency properties of this type of methods are similar to those of the stationary iterative method implied by Q for the convection-diffusion model problem (7.12), combined with those of the stationary iterative method implied by R for the rotated anisotropic diffusion model problem (7.13). This means that convergence will slow down upon grid refinement. If there are grid cells with high aspect ratio, or if there are significant mixed derivatives due to the use of strongly non-orthogonal coordinates, or if convection dominates, one of the robust methods listed in Sect. 7.2 should be used. For high Reynolds numbers first order upwind needs to be used for convection in order to make N an M-matrix (roughly speaking), so that defect correction will be needed for better accuracy. The introduction of the relaxation parameters in (7.62), or using variants of SIMPLE such as SIMPLEC or PISO, do not give essential improvement of efficiency and robustness; see Van Doormaal and Raithby (1984) for a comparison. Like stationary iterative methods, this kind of method asks for acceleration by Krylov subspace or multigrid methods. Multigrid acceleration has been investigated, with favorable outcome. The smoothing factor of SIMPLE has been investigated in Shaw and Sivaloganathan (1988). Theory concerning the approximation and smoothing properties of distributive iterative methods for the Navier-Stokes equations is presented in Braess and Sarazin (1996) and Wittum (1990a).

Distributive Gauss-Seidel method

This method is proposed in Brandt and Dinar (1979) and put in distributive iteration formulation in Hackbusch (1985). It was designed right from the start as a smoothing method to be used with multigrid. This method does not bring AB exactly in the block triangular form (7.59). Choosing

$$B = \begin{pmatrix} I & -G \\ 0 & DG \end{pmatrix}$$

we obtain

$$AB = \begin{pmatrix} N & GDG - NG \\ D & -DG \end{pmatrix} . \tag{7.63}$$

The splitting $AB = M - N$ corresponds to the Gaus–Seidel method. The following algorithm results from (7.57). Residuals r_1^k, r_2^k are computed according to (7.61). Next, one Gauss-Seidel iteration is applied to

$$N\delta u = r_1^k$$

with initial guess $\delta u = 0$. This may be a nonlinear Gauss-Seidel iteration; previous linearization of N is not necessary. Then one Gauss-Seidel iteration with zero initial guess is applied to

$$DG\delta p = D\delta u .$$

Then the distribution step is carried out:

$$\begin{pmatrix} \delta u \\ \delta p \end{pmatrix} := \begin{pmatrix} I & -G \\ 0 & DG \end{pmatrix} \begin{pmatrix} \delta u \\ \delta p \end{pmatrix} ,$$

and finally

$$u^{k+1} = u^k + \delta u, \quad p^{k+1} = p^k + \delta p.$$

The following difficulties may arise. The Gauss-Seidel splitting of AB is dependable only if AB is an M-matrix. This requires using upwind discretization for convection for high Reynolds numbers, making N an M-matrix (roughly speaking). Furthermore, $GDG - NG$ must be small, so that M corresponding to Gauss-Seidel is a sufficiently accurate approximation of AB. For the Stokes equations (Re $= 1$, no convection) it may be shown that $GDG - NG = 0$ in the interior, but not at the boundaries (Hackbusch (1985), Wittum (1989b)). For high Reynolds numbers, however, $GDG - NG$ need not be small. For the Stokes equation a good smoothing factor (the same as for Gauss-Seidel applied to the Poisson equation) is found in Brandt and Dinar (1979), and in Wittum (1989b) numerical experiments show satisfactory multigrid convergence. But for high Reynolds numbers the method as it stands behaves less satisfactorily. Roughly speaking, for Navier-Stokes the method behaves very much like the Gauss-Seidel method for the convection-diffusion equation. The measures that can be taken to improve robustness in the latter case (cf. Wesseling (1992), Chapter 7) can be extended to the former case. The following method provides an example.

Distributive ILU method

This method is introduced in Wittum (1989b). The distribution matrix B is the same as in the preceding method, but M is derived from incomplete LU factorization:

$$AB = LU - N.$$

For an introduction to ILU and references to the literature, see Wesseling (1992) Chap. 4. Just as for the convection-diffusion equation, ILU gives better efficiency and robustness than the Gauss-Seidel variant disussed before. Furthermore, the block $GDG - NG$ in AB in (7.63) is taken better care of, and satisfactory multigrid convergence is obtained for high Reynolds number flows (Wittum (1989b), Wittum (1990c), Wittum (1990b)).

A distributive ILU method based on a SIMPLE type distribution matrix is proposed in Zeng and Wesseling (1995). We choose

$$B = \begin{pmatrix} I & -N^{-1}G \\ 0 & \zeta I \end{pmatrix}, \tag{7.64}$$

where the parameter ζ is introduced to enhance multigrid convergence. This gives

$$AB = \begin{pmatrix} N & (\zeta - 1)G \\ D & -DN^{-1}G \end{pmatrix}.$$

In order to construct a suitable splitting of AB, N^{-1} is replaced by $\tilde{N}^{-1}$, $\tilde{N} = \text{diag}(N)$. This changes AB in $A\tilde{B}$. With ILU one chooses

$$M = (L + D)D^{-1}(D + U),$$

with L and U strictly lower and upper triangular, respectively, and D diagonal. L, D and U are determined from

$$M_{kl} = (\tilde{A}B)_{kl}, \quad (k, l) \in G, \quad M_{kl} = 0, \quad (k, l) \notin G,$$

where the index set G is choosen as follows:

$$G = \{(k, k), (k, k \pm 1), (k, k \pm K \pm 1), (k, k \pm K \mp 1)\},$$

where K is the number of cells in the horizontal direction. ILU depends on the ordering of the unknowns. In Zeng and Wesseling (1995) $u^1_{j-e_1}, u^2_{j-e_2}$ and p_j are put behind each other, and j increases lexicographically. The iteration method becomes, according to (7.57):

$$y^{k+1} = y^k + \omega \tilde{B} M^{-1}(b - Ay^k),$$

where $\tilde{B}$ is a simplification of B in (7.64):

$$\tilde{B} = \begin{pmatrix} I & -\tilde{N}^{-1}G \\ 0 & \zeta I \end{pmatrix},$$

and ω is another relaxation factor. Numerical experiments with multigrid described in Zeng and Wesseling (1995) show that fixed values of $\omega = 0.7$ and $\zeta = 2$ result in rapid multigrid convergence for several flow problems in non-rectangular two- and three-dimensional domains, for medium and large Reynolds numbers. Galerkin coarse grid approximation is used as described in Zeng and Wesseling (1994).

Symmetric coupled Gauss-Seidel method

The SCGS method is proposed in Vanka (1986a) as a smoother to be used in multigrid. It is not of the distributive type (7.57). Each cell is visited in turn in some prescribed order, in Gauss-Seidel fashion. The five unknowns associated with a cell Ω_j (in the notation of Sect. 6.4), namely $u^1_{j\pm e_1}$, $u^2_{j\pm e_2}$ and p_j, are updated simultaneously, while keeping the other unknowns fixed. Note that the velocity variables belong to two cells; for example, $u^1_{j-e_1}$ belongs to Ω_j and Ω_{j-2e_1}. Hence, they are updated twice in a sweep over the domain. The convection terms are linearized. The system for the corrections to the five variables associated with Ω_j looks like

$$
\begin{pmatrix}
a_{11} & a_{12} & a_{13} & a_{14} & h_1^{-1} \\
a_{21} & a_{22} & a_{23} & a_{24} & -h_1^{-1} \\
a_{31} & a_{32} & a_{33} & a_{34} & h_2^{-1} \\
a_{41} & a_{42} & a_{43} & a_{44} & -h_2^{-1} \\
h_1^{-1} & -h_1^{-1} & h_2^{-1} & -h_2^{-1} & 0
\end{pmatrix}
\begin{pmatrix}
\delta u_{j-e_1}^1 \\
\delta u_{j+e_1}^1 \\
\delta u_{j-e_2}^2 \\
\delta u_{j+e_2}^2 \\
\delta p_j
\end{pmatrix}
= r \ .
$$

This system is approximated by dropping the off-diagonal a_{kl}, and damping is introduced by dividing the diagonal elements by damping factors, that hopefully can be fixed at problem-independent values. The result is

$$
\begin{pmatrix}
a_{11}/\sigma_1 & 0 & 0 & 0 & h_1^{-1} \\
0 & a_{22}/\sigma_1 & 0 & 0 & -h_1^{-1} \\
0 & 0 & a_{33}/\sigma_2 & 0 & h_2^{-1} \\
0 & 0 & 0 & a_{44}/\sigma_2 & -h_2^{-1} \\
h_1^{-1} & -h_1^{-1} & h_2^{-1} & -h_2^{-1} & 0
\end{pmatrix}
\begin{pmatrix}
\delta u_{j-e_1}^1 \\
\delta u_{j+e_1}^1 \\
\delta u_{j-e_2}^2 \\
\delta u_{j+e_2}^2 \\
\delta p_j
\end{pmatrix}
= r \ . \qquad (7.65)
$$

This system can be written in the following partioned form:

$$
\begin{pmatrix} A_1 & A_2 \\ A_2^T & 0 \end{pmatrix}
\begin{pmatrix} U_1 \\ U_2 \end{pmatrix}
= \begin{pmatrix} b_1 \\ b_2 \end{pmatrix} ,
$$

where U_1 contains the δu unknowns and $U_2 = \delta p_j$, and its solution is given by

$$
U_2 = (A_2^T A_1^{-1} b_1 - b_2)/A_2^T A_1^{-1} A_2, \quad U_1 = A_1^{-1}(b_1 - A_2 U_2) \ .
$$

Since A_1 is a diagonal matrix, this is easily evaluated. The procedure extends effortlessly to the three-dimensional case; Vanka (1986b) gives a three-dimensional application.

The method is called symmetric because the four velocity unknowns associated with a cell are treated the same way; it is called coupled because the velocity and pressure unknowns are updated simultaneously, and it is called a Gauss-Seidel method because the cells are visited sequentially in Gauss-Seidel fashion.

The method has been applied only in conjuction with multigrid. Suitable values for the relaxation factors σ_1, and σ_2 must be determined empirically. Usually, one can take $\sigma_1 = \sigma_2$, and optimum values are found to range between 0.3 and 0.8 (Vanka (1986a), Vanka (1986b), Oosterlee and Wesseling (1992b), Oosterlee and Wesseling (1992a), decreasing with increasing Reynolds number. In Wittum (1990c) it is found that with $\sigma_1 = \sigma_2 = 1$ SCGS is still an acceptable smoother. Fourier smoothing analysis of SCGS and numerical experiments are presented in Sivaloganathan, Shaw, Shah, and Mayers (1988), Sivaloganathan (1991), who find SCGS to be a more efficient smoother than SIMPLE, whereas in Wittum (1990c) distributive ILU is found to be a bit more efficient than SCGS. The method works for high Reynolds numbers. For robustness with respect to large cell aspect ratios it is necessary to update rows and columns of cells simultaneously, as is to be expected from the analysis of the model problems (7.12) and (7.13) in Wesseling (1992) Chap. 7. This can be kept relatively simple and cheap, although the systems to be solved are more complicated than (7.65). The method is described and applied in Thompson and Ferziger (1989), Oosterlee and Wesseling (1992c), Oosterlee and Wesseling (1993), Oosterlee and Wesseling (1994), and its robustness and efficiency are found to be satisfactory.

The colocated case

We will now discuss the case of spatial discretization on a colocated grid with the pressure-weighted interpolation method of Rhie and Chow (1983). The system to be solved may be written as, deleting the time derivative in (6.31):

$$\begin{pmatrix} N & G \\ D & C \end{pmatrix} \begin{pmatrix} u \\ p \end{pmatrix} = \begin{pmatrix} f \\ g \end{pmatrix} ,$$

where B in (6.31) is renamed C, reserving the symbol B for the distribution matrix. Distributive methods are called for here as well, although now there is no zero block on the main diagonal. But C is not a K-matrix, as is easily seen from (6.26). So, not surprisingly, iterative methods proposed in the literature are usually inspired by SIMPLE, and of distributive type, although the distributive formulation is not used to describe the methods.

The iterative method proposed in Rhie and Chow (1983) can be formulated as follows. The distribution matrix is of SIMPLE type, giving

$$AB = \begin{pmatrix} N & G \\ D & C \end{pmatrix} \begin{pmatrix} I & -N^{-1}G \\ 0 & I \end{pmatrix} = \begin{pmatrix} N & 0 \\ D & C-DN^{-1}G \end{pmatrix} .$$

A splitting $AB = M - N$ is defined by choosing

$$M = \begin{pmatrix} Q & 0 \\ D & R \end{pmatrix} ,$$

where Q is an approximation to N corresponding to some iterative method for the momentum equations without the pressure, and R is a similar approximation to $-D\tilde{N}^{-1}G$ with $\tilde{N} = \text{diag}(N)$. In the distribution step B is chosen as

$$B = \begin{pmatrix} I & -\tilde{N}^{-1}G \\ 0 & I \end{pmatrix} .$$

With (7.57) the following iterative method results. Compute:

$$\begin{pmatrix} r_1^k \\ r_2^k \end{pmatrix} = \begin{pmatrix} f \\ g \end{pmatrix} - \begin{pmatrix} N & G \\ D & C \end{pmatrix} \begin{pmatrix} u^k \\ p^k \end{pmatrix} .$$

Solve:

$$Q\delta u = r_1^k, \quad R\delta p = r_2^k - D\delta u \,.$$

Distribution step:

$$\delta u := \delta u - \tilde{N}^{-1}G\delta p \,.$$

Finally,

$$u^{k+1} = u^k + \omega_u \delta u, \quad p^{k+1} = p^k + \omega_p \delta p \,.$$

Of course, concerning robustness and efficiency the same remarks can be made as for the SIMPLE method on staggered grids; these properties depend on the choice of Q and R. In Rhie and Chow (1983) turbulent flow calculations on boundary-fitted grids around airfoils are reported to converge in from 500 to 1000 iterations. See Acharya and Moukalled (1989), Kobayashi and Pereira (1991), Miller and Schmidt (1988) for variations on the SIMPLE theme in the colocated case. Multigrid acceleration is described in Barcus, Perić, and Scheuerer (1988), Lien and Leschziner (1994), Moukalled and Acharya (1991), Zang, Street, and Koseff (1994).

The pressure-correction method

For nonstationary problems the pressure-correction method is efficient. The resulting algebraic systems are easier to solve than before, because the pressure-correction method uncouples velocity and pressure. With explicit schemes, the velocity update requires little effort, and computing the pressure correction is the computational bottleneck. Where applicable, a fast Poisson solver as described in Sect. 7.5 should be used. When this is not the case, as for instance in non-orthogonal coordinate systems, a Krylov subspace or multigrid method should be used. Efficiency and robustness are as for the anisotropic diffusion model problem (7.12). Application of preconditioned GMRES is described in Vuik (1993). For implicit schemes, systems have to be solved for the velocity components as well. How this can be done with preconditioned GMRES is described in Vuik (1993), Vuik (1996). Efficiency and robustness are as for the convection-diffusion model problem (7.12), discussed in Sect. 7.2. A comparison of Krylov subspace and multigrid methods and combination of the two (Krylov subspace preconditioned with multigrid) is presented in Zeng, Vuik, and Wesseling (1995).

8. The shallow-water equations

8.1 Introduction

We now enter the realm of hyperbolic equations. Instead of starting with an introduction to numerical methods for hyperbolic equations in general, we prefer to begin by plunging into the particular case of the shallow-water equations. Other hyperbolic equations will be considered in Chapters 9 and 10.

The derivation of the shallow-water equations from the Navier-Stokes equations is given in Sect. 1.16. The resulting equations (1.121) and (1.124) govern to a good approximation flows in channels, lakes, rivers and seas, under the shallowness assumptions mentioned in Sect. 1.16.

The core of mathematical models for flows on a planetary scale in the atmosphere and in the oceans consists of the shallow-water equations, but the computing methods used are specialized, and a sufficient introduction would exceed the confines of this book. We will therefore restrict ourselves mainly to applications in hydraulics, and not go into geophysical fluid dynamics.

More extensive introductions to computing methods for the shallow-water equations are given in Abbott (1979), Kowalik and Murty (1993), Vreugdenhil (1994).

8.2 The one-dimensional case

Governing equations

In one dimension the shallow-water equations (1.121) and (1.124) reduce to

$$\frac{\partial d}{\partial t} + \frac{\partial dU}{\partial x} = 0 \,, \tag{8.1}$$

$$\frac{\partial U}{\partial t} + U\frac{\partial U}{\partial x} + g\frac{\partial (d + H)}{\partial x} + \frac{gU|U|}{C^2 d} = 0 \,, \tag{8.2}$$

where U is the depth-averaged flow velocity, d the water depth, g the gravitational acceleration, H the height of the bottom, and C the Chézy friction coefficient.

Classification

Some basic mathematical properties of first order systems like (8.1) and (8.2) have been discussed in Chap. 2. The general form (2.15) becomes in the present case

$$\frac{\partial W}{\partial t} + F\frac{\partial W}{\partial x} = Q, \tag{8.3}$$

with

$$W = \begin{pmatrix} d \\ U \end{pmatrix}, \quad F = \begin{pmatrix} U & d \\ g & U \end{pmatrix}, \quad Q = \begin{pmatrix} 0 \\ -g\frac{\partial H}{\partial x} - g\frac{U|U|}{C^2 d} \end{pmatrix}.$$

According to Sect. 2.2, the type of (8.3) depends on the properties of the following generalized eigenproblem:

$$(n_0 I + n_1 F)\hat{W} = 0.$$

We have

$$\det(n_0 I + n_1 F) = 0 \tag{8.4}$$

if

$$\begin{pmatrix} n_0 \\ n_1 \end{pmatrix} = \begin{pmatrix} \pm c - U \\ 1 \end{pmatrix}, \quad c = \sqrt{gd}. \tag{8.5}$$

Since we have found two real linearly independent vectors (n_0, n_1), the system (8.3) is hyperbolic, according to Definition 2.2.2.

Eigenvectors corresponding to the plus and minus signs in (8.5) are, respectively,

$$\hat{W} = \begin{pmatrix} \mp c \\ g \end{pmatrix}. \tag{8.6}$$

Characteristics

According to Sect. 2.2 there exist in the frozen coefficient case wave-like solutions $W = \hat{W}e^{iw(t,x)}$ with

$$\frac{\partial w}{\partial t} = n_0 = \pm c - U, \quad \frac{\partial w}{\partial x} = n_1 = 1.$$

Since

$$\frac{\partial w}{\partial t} + (U \mp c)\frac{\partial w}{\partial x} = 0,$$

we see that $w = \text{constant}$ along trajectories $t = t(s)$, $x = x(s)$ with

$$\frac{dt}{ds} = 1, \quad \frac{dx}{ds} = U \mp c . \tag{8.7}$$

Manifolds on which $w(t, x) = \text{constant}$ were called characteristic surfaces in Sect. 2.2. In the present case these surfaces reduce to curves and are called *characteristics*. They are defined by (8.7), also in the nonlinear case.

Diagonalization

Let us introduce new unknowns Z such that $W = W(Z)$ and let B denote the Jacobian matrix

$$B = \{\partial W/\partial Z\} .$$

Then equation (8.3) becomes

$$\frac{\partial Z}{\partial t} + B^{-1}FB\frac{\partial Z}{\partial x} = B^{-1}Q . \tag{8.8}$$

It would be nice to choose Z such that $B^{-1}FB$ is a diagonal matrix, because then the left-hand side of (8.8) decouples into independent single equations. This can be achieved as follows. The eigenvalues and eigenvectors of F have already been determined (see (8.4)–(8.6)). Choosing the columns of B equal to the eigenvectors of F we get

$$B = \begin{pmatrix} c & -c \\ g & g \end{pmatrix}, \quad B^{-1} = \frac{1}{2}\begin{pmatrix} 1/c & 1/g \\ -1/c & 1/g \end{pmatrix}, \quad B^{-1}FB = \begin{pmatrix} U+c & 0 \\ 0 & U-c \end{pmatrix} .$$

The new unknowns Z satisfy

$$\frac{\partial z_i}{\partial w_j} = (B^{-1})_{ij} ,$$

which can be integrated to give

$$z_1 = \frac{1}{g}(c + U/2), \quad z_2 = \frac{1}{g}(-c + U/2) . \tag{8.9}$$

Equation (8.8) becomes

$$\frac{\partial z_1}{\partial t} + (U+c)\frac{\partial z_1}{\partial x} = (B^{-1}Q)_1, \quad \frac{\partial z_2}{\partial t} + (U-c)\frac{\partial z_2}{\partial x} = (B^{-1}Q)_2 . \tag{8.10}$$

Linearized equations

For future reference, we formulate a set of linear equations that approximate (8.1) and (8.2) in the case of small perturbations. Let $H = $ constant, $d = \bar{d} + d'$, $\bar{d} = $ constant, $U = \bar{U} + u$, $\bar{U} = $ constant. To keep things simple, the friction term is replaced by ru. Deleting primes, equations (8.1) and (8.2) become

$$\frac{\partial d}{\partial t} + \bar{d}\frac{\partial u}{\partial x} + \bar{U}\frac{\partial d}{\partial x} = 0 \,,$$

$$\frac{\partial u}{\partial t} + \bar{U}\frac{\partial u}{\partial x} + g\frac{\partial d}{\partial x} + ru = 0 \,.$$

This can be rewritten as

$$\frac{\partial W}{\partial t} + F\frac{\partial W}{\partial x} + GW = 0 \,, \tag{8.11}$$

where

$$W = \begin{pmatrix} d \\ u \end{pmatrix}, \quad F = \begin{pmatrix} \bar{U} & \bar{d} \\ g & \bar{U} \end{pmatrix}, \quad G = \begin{pmatrix} 0 & 0 \\ 0 & r \end{pmatrix} \,.$$

Instead of linearizing around a constant velocity $\bar{U}$ we can also assume that U is small, or more precisely

$$\mathrm{Fr} \equiv V/\sqrt{g\bar{d}} \ll 1 \,,$$

where V is a measure of the magnitude of $|U|$, and Fr is the *Froude number*. Neglecting terms involving U^2 and Ud', equations (8.1) and (8.2) again go over in (8.11), but with

$$W = \begin{pmatrix} d \\ U \end{pmatrix}, \quad F = \begin{pmatrix} 0 & \bar{d} \\ g & 0 \end{pmatrix}, \quad G = \begin{pmatrix} 0 & 0 \\ 0 & r \end{pmatrix} \,. \tag{8.12}$$

If we neglect friction, this gives us the simplest approximation of shallow water flow.

Waves

Various kinds of waves may be distinguished; for a survey see Vreugdenhil (1994).

Assume constant bottom height H and neglect the friction term, so that (8.10) becomes

$$\frac{\partial z_1}{\partial t} + (U + c)\frac{\partial z_1}{\partial x} = 0, \quad \frac{\partial z_2}{\partial t} + (U - c)\frac{\partial z_2}{\partial x} = 0. \tag{8.13}$$

Hence $z_1 =$ constant along the characteristic $dt/ds = 1$, $dx/ds = U + c$, and $z_2 =$ constant along the characteristic $dt/ds = 1$, $dx/ds = U - c$. The quantities z_1 and z_2 are called the *Riemann invariants*. We can look upon z_1 and z_2 as signals that travel with a speed $U + c$ and $U - c$, respectively. Superimposed on the mean flow is a signal velocity $\pm c$ that is caused by gravity only. We have wave-like solutions of a type called *gravity wave*. The velocity $c = \sqrt{gd}$ is called the *celerity*. The Froude number is $\mathrm{Fr} = U/c$.

Next, consider the linearized equation (8.11). Assume a harmonic wave of the form

$$W = \hat{W}e^{i(kx-\omega t)}, \quad \hat{W} = \text{constant} . \qquad (8.14)$$

This is a solution of (8.11) if

$$(-i\omega I + ikF + G)\hat{W} = 0 .$$

There are nontrivial solutions if the determinant is zero:

$$(ik\bar{U} - i\omega)(ik\bar{U} - i\omega + r) + k^2c^2 = 0 . \qquad (8.15)$$

This gives a functional dependence $\omega = \omega(k)$. This is called the dispersion relation. As will be clear from equation (8.14), a non-zero imaginary part $\mathrm{Im}(\omega)$ corresponds to damping or amplification, whereas the real part $\mathrm{Re}(\omega)$ corresponds to propagation. The *phase* and *group velocities* are defined by

$$c_p = \mathrm{Re}(\omega)/k , \quad c_g = \mathrm{Re}(d\omega/dk) . \qquad (8.16)$$

If the phase velocity of a harmonic wave depends on the wavenumber k, the wave is called *dispersive*. A dispersive wave, if it is composed of different harmonics, deforms while traveling. Energy travels with the group velocity. For a nondispersive wave, $c_g = c_p$.

Neglecting bottom friction ($r = 0$) in (8.15), we obtain the following dispersion relation:

$$\omega = k(\bar{U} \pm c) ,$$

so that

$$c_p = c_g = \bar{U} \pm c .$$

These waves are nondispersive, and are the gravity waves discussed before. The celerity c is seen to be the phase velocity in still water ($\bar{U} = 0$). Because ω is real, there is no damping.

Let us now include bottom friction. Solving (8.15) for ω gives the following dispersion relation:

$$\omega = k\bar{U} - ir/2 \pm \sqrt{k^2c^2 - r^2/4} \,. \tag{8.17}$$

If $r > 2kc$ then $\mathrm{Re}(\omega) = k\bar{U}$, and we have a nondispersive wave with

$$c_p = c_g = \bar{U} \,.$$

There is damping at a rate

$$\exp\{-\frac{1}{2}r \pm \sqrt{r^2/4 - k^2c^2}\}t \,. \tag{8.18}$$

Waves generated in rivers by heavy rainfall (called flood waves) have very long wavelength (i.e. $k \ll 1$). In this case

$$\sqrt{r^2/4 - k^2c^2} \cong \frac{r}{2}(1 - 2k^2c^2/r^2) \,.$$

Taking the plus sign in (8.18), corresponding to the most weakly damped wave, we see that flood waves are damped at a rate

$$\exp(-tk^2c^2/r) \,.$$

If $r \leq 2kc$ then (8.17) shows that

$$\mathrm{Re}(\omega) = k\bar{U} \pm \sqrt{k^2c^2 - r^2/4}$$

so that

$$c_p = \bar{U} \pm \sqrt{c^2 - r^2/4k^2}, \quad c_g = \bar{U} \pm c/\sqrt{1 - r^2/4k^2c^2} \,.$$

These waves are dispersive. As $r \downarrow 0$ the pure gravity waves are recovered.

Boundary and initial conditions

Boundary and initial conditions that lead to a well-posed problem for equations (8.1) and (8.2) are (obviously) precisely those which either explicitly or implicitly uniquely determine the Riemann invariants $z_{1,2}$ in equation (8.13). Such conditions may be determined by considering the geometry of the characteristics. Let the domain be $x \in (0, L)$, $t \in (0, T)$. Figure 8.1 gives a sketch of the two families of characteristics. The Riemann invariant z_1 is constant along characteristics inclining to the right, for example AC, whereas z_2 is constant along the other family of characteristics, for example BC. Suppose that d and U are given at the segment AB. Then, according to (8.9), $z_{1,2}$ are known at AB. Along their respective characteristics $z_{1,2}$ are constant. Consequently, in all of the shaded region ABC both z_1 and z_2, and hence also the solution (d, U), are known and uniquely determined by the data on AB. It follows that changes in d and/or U along AB (or anywhere else in

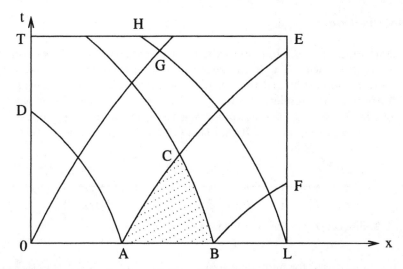

Fig. 8.1. Geometric structure of characteristics of equations (8.1, 8.2).

ABC) can influence the solution in C, but no influence is exerted in C by changes anywhere else. Therefore the shaded region ABC is called the *domain of dependence* of the point C. Furthermore, the data on AB propagate their influence along the characteristics emanating from AB. The domain swept out by these characteristics is called the *domain of influence* of AB. In the present case, this is the region ADTEFB.

Next, suppose (d, U) is given at $t = 0$. Then the solution is determined in OGL. In order to determine the solution outside OGL, boundary conditions are required at $x = 0$ and $x = L$ for $t > 0$. At $x = 0$, $t > 0$, z_2 is already determined by the data at $t = 0$. In order to determine the solution at $x = 0$ one additional quantity must be given. The general rule is, obviously:

> *The number of boundary conditions in a point on the boundary must be equal to the number of incoming characteristics at that point.*

Therefore, at $x = 0$ a condition independent of z_2 must be given, such that z_1 is determined. For example, suppose that $d(t, 0)$ is given. Hence, we know $c(t, 0) = \sqrt{gd(t, 0)}$. Then $z_1(t, 0) = \frac{2}{g}c(t, 0) + z_2(t, 0)$ is known, and the solution is now also uniquely determined in the region OGHT. It is left to the reader to show that giving one suitable quantity at (t, L) determines the solution in the remainder of the domain. No condition is to be given along TE, because both characteristics are outgoing there.

Discretization

We start with the simplest approximation to shallow water flow, namely equations (8.12) with $G = 0$. In order to obtain second order spatial accuracy with a compact stencil that excludes spurious solutions ('wiggles') and facilitates accurate implementation of boundary conditions, the staggered grid of Fig. 8.2 is used. The use of a staggered grid was already proposed in 1922 in the pioneering work of Richardson (1922). Staggered grids of this kind

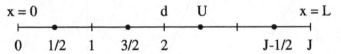

Fig. 8.2. Staggered grid for the shallow-water equations.

are commonly used for numerical solution of the shallow-water equations.

Let the mesh size be h = constant. We approximate $d(t, x)$ in $x_j = jh$ and $U(t, x)$ in $x_{j+1/2} = (j + 1/2)h$. We discretize (8.12) by finite differences in space in the following simple and natural way (postponing time discretization):

$$\frac{dd_j}{dt} + \bar{d}\frac{U_{j+1/2} - U_{j-1/2}}{h} = 0, \quad j = 1, 2, ..., J - 1, \tag{8.19}$$

$$\frac{dU_{j+1/2}}{dt} + g\frac{d_{j+1} - d_j}{h} = 0, \quad j = 0, 1, ..., J - 1. \tag{8.20}$$

The system has to be completed with appropriate boundary conditions. According to the rule given before, one boundary condition should be available at $x = 0$, and one at $x = L$. Consider $x = 0$. Implementation of a condition d_0 = given is trivial. If instead $U(t, 0)$ is given it is natural to redefine the grid such that $x_{1/2} = 0$, and again implementation is trivial. This is an advantage of the staggered grid. If instead one wants to stick to the grid of Fig. 8.2, then one may extend (8.19) to $j = 0$, and eliminate the virtual value $U_{-1/2}$ by means of

$$U_{-1/2} = 2U(t, 0) - U_{1/2} .$$

The boundary $x = L$ is handled similarly.

In order to develop some proficiency in stability analysis for the shallow-water equations, we will present a number of simple cases.

Explicit Euler scheme

By replacing the time derivatives in (8.19), (8.20) by a forward difference we obtain the following explicit scheme:

$$\frac{d_j^{n+1} - d_j^n}{\tau} + \bar{d}\frac{U_{j+1/2}^n - U_{j-1/2}^n}{h} = 0 , \tag{8.21}$$

$$\frac{U_{j+1/2}^{n+1} - U_{j+1/2}^n}{h} + g\frac{d_{j+1}^n - d_j^n}{h} = 0 , \tag{8.22}$$

where the superscript denotes the time level. In order to study von Neumann stability, we postulate

$$\begin{pmatrix} d_j^n \\ U_j^n \end{pmatrix} = \begin{pmatrix} \hat{d}^n \\ \hat{U}^n \end{pmatrix} e^{ij\theta} , \tag{8.23}$$

and find upon substitution in (8.21) and (8.22) (assuming the grid to be infinite)

$$\begin{pmatrix} \hat{d}^{n+1} \\ \hat{U}^{n+1} \end{pmatrix} = G \begin{pmatrix} \hat{d}^n \\ \hat{U}^n \end{pmatrix} , \tag{8.24}$$

with the amplification matrix G given by

$$G = \begin{pmatrix} 1 & -\frac{\bar{d}\tau}{h}2i\sin\theta/2 \\ -\frac{g\tau}{h}2i\sin\theta/2 & 1 \end{pmatrix} . \tag{8.25}$$

The eigenvalues of G are

$$\lambda_{1,2} = 1 \pm 2i\sigma|\sin\theta/2| ,$$

where

$$\sigma = c\tau/h$$

is called the CFL number, see Sect 5.6. Since there exist values of θ (in fact all $\theta \neq 0, 2\pi$) for which $|\lambda_{1,2}| > 1$, this scheme is instable.

Implicit Euler scheme

This scheme is given by

$$\frac{d_j^{n+1} - d_j^n}{\tau} + \bar{d}\frac{U_{j+1/2}^{n+1} - U_{j-1/2}^{n+1}}{h} = 0 ,$$

$$\frac{U_{j+1/2}^{n+1} - U_{j+1/2}^n}{\tau} + g\frac{d_{j+1}^{n+1} - d_j^{n+1}}{h} = 0 .$$

It is left to the reader to verify, that the amplification matrix is given by

$$G = \begin{pmatrix} 1 & \dfrac{\bar{d}\tau}{h} 2i \sin\theta/2 \\ \dfrac{g\tau}{h} 2i \sin\theta/2 & 1 \end{pmatrix}^{-1} \qquad (8.26)$$

with eigenvalues

$$\lambda_{1,2} = (1 \pm 2i\sigma \sin\theta/2)^{-1} \ .$$

Hence

$$|\lambda_{1,2}| = (1 + 4\sigma^2 \sin^2\theta/2)^{-1/2} \leq 1 \ ,$$

so that the scheme is unconditionally stable.

Leapfrog scheme

The leapfrog scheme (first used in Richardson (1922)) is given by

$$\frac{d_j^{n+1} - d_j^{n-1}}{2\tau} + \bar{d}\frac{U_{j+1/2}^n - U_{j-1/2}^n}{h} = 0 \ , \\ \frac{U_{j+1/2}^{n+1} - U_{j+1/2}^{n-1}}{2\tau} + g\frac{d_{j+1}^n - d_j^n}{h} = 0 \ . \qquad (8.27)$$

The temporal accuracy of this scheme is $\mathcal{O}(\tau^2)$, which is better than the Euler schemes. Because the scheme is explicit, the computing cost is low.

A straightforward way to investigate the stability of (8.27) is to substitute (8.23) and (8.24), which gives the following equation for the amplification matrix G:

$$G^2 - I + AG = 0 \ , \qquad (8.28)$$

with

$$A = \begin{pmatrix} 0 & \alpha \\ \beta & 0 \end{pmatrix}, \quad \alpha = 4i\frac{\tau\bar{d}}{h}\sin\theta/2, \quad \beta = 4i\frac{\tau g}{h}\sin\theta/2 \ .$$

It is a tedious affair to determine G from (8.28). However, since we do not need G, but only its eigenvalues, we can proceed as follows. If V is an eigenvector of G with corresponding eigenvalue λ, then

$$(\lambda^2 - 1)V + \lambda AV = 0 \ ,$$

so that V is also an eigenvector of A; let the corresponding eigenvalue be μ. Then the eigenvalues of G are related to those of A by

$$\lambda^2 - 1 + \mu\lambda = 0 . \tag{8.29}$$

We have

$$\mu_{1,2} = \pm 4i\sigma|\sin\theta/2| .$$

This gives us four eigenvalues λ (taking all possible permutations of the signs):

$$\lambda_{1,2,3,4} = \pm 2i\sigma|\sin\theta/2| \pm \sqrt{1 - 4\sigma^2 \sin^2\theta/2} .$$

The reason why we encounter four eigenvalues is that (8.28) has two solutions for G, each with two eigenvalues. If

$$2\sigma|\sin\theta/2| \le 1 \tag{8.30}$$

then

$$|\lambda_{1,2,3,4}| = 1 , \tag{8.31}$$

else there are λ's with $|\lambda| > 1$. Hence for stability we must have (8.30) for all θ, hence

$$\sigma \le 1/2 , \tag{8.32}$$

so that the leapfrog scheme is conditionally stable, with stability condition

$$\tau \le \frac{h}{2c} . \tag{8.33}$$

As it turns out, the pioneering calculations in Richardson (1922) violate this stability condition, so that wrong results were produced. L.F. Richardson was ahead of his time. Stability theory for numerical methods for initial value problems starts with the celebrated paper of Courant, Friedrichs, and Lewy (1928).

A somewhat different analysis of the stability of the leapfrog scheme is left to the reader in Exercise 8.2.4.

Equation (8.31) shows that the scheme does not contain any damping. As a consequence, perturbations in initial and boundary data persist, which may degrade the results. Inclusion of the bottom friction term has a damping effect, cf. Exercise 8.2.5.

Because it involves three time levels, the leapfrog scheme needs an extra initial condition compared to the differential equation. There are two linearly independent solutions, one of them *essential*, the other *spurious*. It depends on the implementation of the initial conditions how much the spurious solution makes itself felt. Because of (8.31) the spurious solution is not damped. Therefore in practice the next scheme is usually preferred.

Hansen scheme

This scheme has been popular for solving the shallow-water equations since the paper of Hansen (1956). It is given by

$$\frac{d_j^{n+1} - d_j^n}{\tau} + \bar{d}\frac{U_{j+1/2}^{n+1/2} - U_{j-1/2}^{n+1/2}}{h} = 0 , \qquad (8.34)$$

$$\frac{U_{j+1/2}^{n+3/2} - U_{j+1/2}^{n+1/2}}{\tau} + g\frac{d_{j+1}^{n+1} - d_j^{n+1}}{h} = 0 . \qquad (8.35)$$

Note that we now have also staggering in time. Because d^{n+1} can be evaluated before $U^{n+3/2}$, this scheme is effectively explicit. It shares with the leap-frog scheme the advantages of $\mathcal{O}(\tau^2)$ accuracy and explicitness, but it does not have spurious solutions, because no initial conditions additional to those given for the differential equations are required. Also it requires less storage, because the solution needs to be stored at only two time levels.

Implementation of initial conditions is easy. Let d^0 and U^0 be prescribed. We introduce the virtual values $U^{-1/2}$, use (8.35) for $n = -1$ and eliminate $U^{-1/2}$ with

$$U^{-1/2} = 2U^0 - U^{1/2} ,$$

so that $U^{1/2}$ is found from

$$U_{j+1/2}^{1/2} = U_{j+1/2}^0 + \frac{g\tau}{2h}(d_{j+1}^0 - d_j^0) .$$

A straightforward way to investigate the stability of (8.34), (8.35) is to substitute

$$\begin{pmatrix} d_j^n \\ U_{j+1/2}^{n+1/2} \end{pmatrix} = \begin{pmatrix} \hat{d}^n \\ \hat{U}^{n+1/2} \end{pmatrix} e^{ij\theta} ,$$

which gives

$$\begin{pmatrix} \hat{d}^{n+1} \\ \hat{U}^{n+3/2} \end{pmatrix} = G \begin{pmatrix} \hat{d}^n \\ \hat{U}^{n+1/2} \end{pmatrix} ,$$

with the amplification matrix G given by

$$G = G_2^{-1} G_1 ,$$

with

$$G_2 = \begin{pmatrix} 1 & 0 \\ \frac{g\tau}{h}(e^{i\theta} - 1) & 1 \end{pmatrix} , \quad G_1 = \begin{pmatrix} 1 & -\frac{\bar{d}\tau}{h}(1 - e^{-i\theta}) \\ 0 & 1 \end{pmatrix} .$$

The eigenvalue problem for G is

$$(G_2^{-1}G_1 - \lambda I)V = 0 \, ,$$

or

$$(G_1 - \lambda G_2)V = 0 \, .$$

From

$$\det(G_1 - \lambda G_2) = 0$$

we find

$$\lambda_{1,2} = 1 + \mu \pm \sqrt{\mu^2 + 2\mu}, \quad \mu = \sigma^2(\cos\theta - 1) \, .$$

For

$$-2 \leq \mu \leq 0 \tag{8.36}$$

we find

$$|\lambda_{1,2}| = 1 \, .$$

For $\mu \leq -2$ we have $|\lambda_2| > 1$ for certain values of θ, so that for stability it is necessary and sufficient that (8.36) holds for all θ. Hence we must have

$$\sigma \leq 1 \, , \tag{8.37}$$

so that the Hansen scheme is conditionally stable, with stability condition

$$\tau \leq h/c \, . \tag{8.38}$$

Sielecki scheme

This is another well-known scheme. It is similar to the Hansen scheme, but does not employ staggering in time. It has been proposed by Sielecki (1968), and is given by

$$\frac{d_j^{n+1} - d_j^n}{\tau} + \bar{d}\frac{(U_{j+1/2}^n - U_{j-1/2}^n)}{h} = 0$$

$$\frac{U_{j+1/2}^{n+1} - U_{j+1/2}^n}{\tau} + g\frac{d_{j+1}^{n+1} - d_j^{n+1}}{h} = 0$$

It is left to the reader to show that the amplification matrix G is the same as for the Hansen scheme. Hence, the Sielecki scheme is stable if (8.38) is satisfied. Because d^{n+1} can be evaluated before U^{n+1}, the scheme is effectively explicit. Because the scheme is not staggered in time, accuracy is only $\mathcal{O}(\tau)$, and is therefore less attractive than the Hansen scheme for the simple equations under consideration. But in the presence of Coriolis acceleration (in two dimensions) and atmospheric pressure effects, the Sielecki scheme has certain advantages, cf. Sielecki (1968).

An implicit scheme

Because explicit schemes usually have to obey a stability conditions, which may be too restrictive in practice, implicit schemes are often used. The implicit Euler scheme has $\mathcal{O}(\tau)$ temporal accuracy, which is generally insufficient. A straightforward example of a more accurate implicit scheme is:

$$
\begin{aligned}
\frac{d_j^{n+1} - d_j^n}{\tau} + \frac{\bar{d}}{2h}(U^{n+1} + U^n)|_{j-1/2}^{j+1/2} &= 0 , \\
\frac{U_{j+1/2}^{n+1} - U_{j+1/2}^n}{\tau} + \frac{g}{2h}(d^{n+1} + d^n)|_j^{j+1} &= 0 .
\end{aligned}
\tag{8.39}
$$

The temporal accuracy is $\mathcal{O}(\tau^2)$.

We now have a system of equations to solve for each time step. If we order the unknowns as follows:

$$
..., \; d_j^{n+1}, \; U_{j+1/2}^{n+1}, \; d_{j+1}^{n+1}, \; U_{j+3/2}^{n+1}, \; ...
$$

then the structure of the matrix A of the system to be solved is tridiagonal.

For the stability analysis we postulate (8.23), which results in

$$
G_1 \begin{pmatrix} \hat{d}^{n+1} \\ \hat{U}^{n+1} \end{pmatrix} = G_2 \begin{pmatrix} \hat{d}^n \\ \hat{U}^n \end{pmatrix} ,
$$

and the amplification matrix is given by

$$
G = G_1^{-1} G_2 .
$$

We find that G_1 and G_2 are given by

$$
G_1 = \begin{pmatrix} 1 & \alpha \\ \beta & 1 \end{pmatrix}, \quad G_2 = \begin{pmatrix} 1 & -\alpha \\ -\beta & 1 \end{pmatrix}
$$

with

$$
\alpha = i\frac{\bar{d}\tau}{h}\sin\theta/2, \quad \beta = i\frac{g\tau}{h}\sin\theta/2 .
$$

As before, $\lambda(G)$ follows from $\det(G_2 - \lambda G_1) = 0$, resulting in

$$
\lambda_{1,2} = \frac{(1 \pm i\sigma|\sin\theta/2|)^2}{1 + \sigma^2 \sin^2\theta/2} ,
$$

so that

$$
|\lambda_{1,2}| = 1 .
$$

Hence, the scheme is unconditionally stable and contains no damping.

Leendertse scheme

This scheme, proposed by Leendertse (1967) for the two-dimensional nonlinear shallow-water equations, is a two-step scheme. In the present simple case it reduces to:

$$\frac{d_j^{n+1/2} - d_j^n}{\tau/2} + \bar{d}\frac{U_{j+1/2}^{n+1/2} - U_{j-1/2}^{n+1/2}}{h} = 0 ,$$

$$\frac{U_{j+1/2}^{n+1/2} - U_{j+1/2}^n}{\tau/2} + g\frac{d_{j+1}^{n+1/2} - d_j^{n+1/2}}{h} = 0 ,$$

$$\frac{d_j^{n+1} - d_j^{n+1/2}}{\tau/2} + \bar{d}\frac{U_{j+1/2}^{n+1/2} - U_{j-1/2}^{n+1/2}}{h} = 0 ,$$

$$\frac{U_{j+1/2}^{n+1} - U_{j+1/2}^{n+1/2}}{\tau/2} + g\frac{d^{n+1/2} - d_j^{n+1/2}}{h} = 0 .$$

An implicit Euler step is followed by an explicit Euler step. The amplification matrix is the product of the two amplification matrices (8.25) and (8.26), with τ replaced by $\tau/2$:

$$G = G_2 G_1^{-1}, \quad G_2 = \begin{pmatrix} 1 & -i\bar{d}\frac{\tau}{h}\sin\theta/2 \\ -ig\frac{\tau}{h}\sin\theta/2 & 1 \end{pmatrix} ,$$

$$G_1 = \begin{pmatrix} 1 & i\bar{d}\frac{\tau}{h}\sin\theta/2 \\ ig\frac{\tau}{h}\sin\theta/2 & 1 \end{pmatrix} . \quad (8.40)$$

The eigenvalues $\lambda(G)$ follow from $\det(G_1 - \lambda G_2) = 0$, giving

$$(1 - \lambda)^2 + (1 + \lambda)^2 \sigma^2 \sin^2\theta/2 = 0 ,$$

so that

$$\lambda_{1,2} = \frac{1 \pm i\sigma\sin\theta/2}{1 \mp i\sigma\sin\theta/2} , \quad (8.41)$$

from which it follows that $|\lambda_{1,2}| = 1$, so that we have unconditional stability, and there is no damping.

Dissipation and dispersion

By postulating a harmonic wave $W = \hat{W}\exp\{i(kx - \omega t)\}$ for the linearized shallow water equations (8.12) without bottom friction we obtain the following dispersion relation:

$$\omega = \pm kc . \quad (8.42)$$

By assuming a harmonic wave for a numerical scheme we also get a dispersion relation. For example, for the Hansen scheme we may postulate

$$d_j^n = \hat{d}\exp\{i(kjh - \omega n\tau)\}, \quad U_{j+1/2}^{n+1/2} = \hat{U}\exp\{i(kjh - \omega n\tau)\} .$$

Substitution in (8.34) and (8.35) results in

$$\begin{pmatrix} e^{-i\omega\tau} - 1 & \frac{\tau\hat{d}}{h}(1 - e^{-ikh}) \\ \frac{\tau g}{h}(e^{ikh} - 1) & 1 - e^{i\omega\tau} \end{pmatrix} \begin{pmatrix} \hat{d} \\ \hat{U} \end{pmatrix} = 0 .$$

By requiring the determinant to be zero the following dispersion relation is obtained:

$$\omega = \pm\frac{2}{\tau}\arcsin(\sigma\sin kh/2) . \tag{8.43}$$

According to (8.37), for stability we must have $\sigma \leq 1$. This makes ω real, so that there is no numerical dissipation. In order to approximate a wave with wavenumber k we must have $h \ll 1/k$. Under this assumption (8.43) can be approximated by

$$\omega \cong \pm kc\{1 - \frac{1}{24}(1 - \sigma^2)k^2h^2\} . \tag{8.44}$$

Hence, using a superscript h to distinguish numerical from exact quantities, the numerical phase and group velocities (see (8.16)) are given by

$$c_p^h \cong \pm c\{1 - \frac{1}{24}(1 - \sigma^2)k^2h^2\}, \quad c_g^h = \pm c\{1 - \frac{1}{8}(1 - \sigma^2)k^2h^2\} , \tag{8.45}$$

showing $\mathcal{O}(h^2)$ accuracy, as expected.

Unlike the exact gravity waves, the numerical approximation shows dispersion, unless $\sigma = 1$, in which case the Hansen scheme has no error. The error for the group velocity is larger than for the phase velocity, which is a pity, because energy travels with the group velocity.

We briefly present dispersion analysis for the other (stable) schemes.

The dispersion relation for the Sielecki scheme is the same as for the Hansen scheme.

It is left to the reader to show the implicit Euler scheme has the following dispersion relation:

$$e^{i\omega\tau} = 1 \pm 2i\sigma\sin kh/2 . \tag{8.46}$$

Writing $\omega_r = \mathrm{Re}(\omega)$, $\omega_i = \mathrm{Im}(\omega)$, we see there is damping at a rate

$$e^{\omega_i\tau} = 1/\sqrt{1 + 4\sigma^2\sin^2 kh/2} .$$

Because the differential equation contains no damping (according to (8.42)), this numerical damping makes implicit Euler unattractive. According to (8.46),

$$\omega_r = \pm \frac{1}{\tau} \arctan(\sigma \sin kh/2) .$$

Approximation for $kh \ll 1$ gives

$$\omega_r \cong \pm kc\{1 - \frac{1}{24}(1 + 8\sigma^2)k^2h^2\} .$$

Hence

$$c_p^h \cong \pm c\{1 - \frac{1}{24}(1 + 8\sigma^2)k^2h^2\}, \quad c_g^h \cong \pm c\{1 - \frac{1}{8}(1 + 8\sigma^2)k^2h^2\} .$$

Comparison with (8.44) shows that the numerical dispersion of implicit Euler is significantly larger than for the Hansen scheme.

For the leapfrog scheme the dispersion relation is found to be

$$\omega = \pm \frac{1}{\tau} \arcsin(2\sigma \sin kh/2) . \tag{8.47}$$

We have stability only if $\sigma \leq 1$. We have propagation (ω real) only if $|2\sigma \sin kh/2| \leq 1$, in which case is no numerical dissipation. Approximation for $|kh| \ll 1$ gives

$$\omega \cong \pm kc\{1 - \frac{1}{24}(1 - 4\sigma^2)k^2h^2\} ,$$

so that

$$c_p^h \cong \pm c\{1 - \frac{1}{24}(1 - 4\sigma^2)k^2h^2\} , \quad c_g^h = \pm c\{1 - \frac{1}{8}(1 - 4\sigma^2)k^2h^2\} ,$$

which is a little less accurate than the Hansen scheme (cf. (8.45)).

For the implicit scheme (8.39) the following dispersion relation is obtained:

$$\omega = \pm \frac{2}{\tau} \arctan(\sigma \sin kh/2) . \tag{8.48}$$

Under all circumstances, ω is real, so there is no numerical dissipation. Expansion for $kh \ll 1$ gives

$$\omega \cong \pm kc\{1 - \frac{1}{24}(1 + 2\sigma^2)k^2h^2\} ,$$

from which it follows that

$$c_p^h \cong \pm c\{1 - \frac{1}{24}(1 + 2\sigma^2)k^2h^2\}, \quad c_g^h = \pm c\{1 - \frac{1}{8}(1 + 2\sigma^2)k^2h^2\} ,$$

which is less accurate than the Hansen scheme.

The dispersion relation of the Leendertse scheme may be determined in the following way. Postulate

$$d_j^n = \hat{d}^n e^{ijkh}, \quad U_j^n = \hat{U}^n e^{ijkh} .$$

Then we find

$$\begin{pmatrix} \hat{d}^{n+1} \\ \hat{U}^{n+1} \end{pmatrix} = G_2 G_1^{-1} \begin{pmatrix} \hat{d}^n \\ \hat{U}^n \end{pmatrix} ,$$

with G_1, G_2 given by (8.40) with θ replaced by kh. Assuming $\hat{d} = \hat{d} e^{-i\omega n\tau}$, $\hat{U}^n = \hat{U} e^{-i\omega n\tau}$, we find a solution if

$$(G_2 G_1^{-1} - e^{-i\omega\tau} I) \begin{pmatrix} \hat{d} \\ \hat{U} \end{pmatrix} = 0 .$$

The eigenvalues of $G_2 G_1^{-1}$ are given by (8.41), so that

$$e^{i\omega\tau} = \frac{1 \pm i\sigma \sin kh/2}{1 \mp i\sigma \sin kh/2} .$$

The modulus of the fraction equals 1, so that ω is real, and we have no numerical dissipation. The dispersion relation is

$$\omega = \pm \frac{2}{\tau} \arctan(\sigma \sin kh/2) , \tag{8.49}$$

which is the same as for the Hansen scheme (equation (8.43)). We may conclude that the accuracy of the Leendertse scheme is $\mathcal{O}(\tau^2 + h^2)$.

Exercise 8.2.1. At what speed do the waves travel caused by a stone thrown in a pond of depth 1 m?

Exercise 8.2.2. Derive the Riemann invariants z_1, z_2 for equations (8.11), (8.12). Give a discretization of the boundary condition $z_1(t, 0) =$ given for the explicit Euler scheme. Similarly for z_2. Why is it mathematically wrong to prescribe $z_2(t, 0)$? Perform numerical experiments and see what goes wrong with prescribing $z_2(t, 0)$.

Exercise 8.2.3. Solve equations (8.11), (8.12) numerically with the schemes discussed. Prescribe initial and boundary conditions corresponding to the following exact solution for the case without bottom friction:

$$d = \bar{d} \sin(t - x/c), \quad U = c \sin(t - x/c) ,$$

and study the numerical error. What is the numerical propagation velocity of gravity waves?

Exercise 8.2.4. *Alternative stability analysis of the leapfrog scheme*
Analyze the stability of the leapfrog scheme by postulating instead of (8.23)
and (8.24):

$$
\begin{pmatrix} \hat{d}^{n+1} \\ \hat{U}^{n+1} \\ \hat{d}^{n} \\ \hat{u}^{n} \end{pmatrix} = G \begin{pmatrix} \hat{d}^{n} \\ \hat{U}^{n} \\ \hat{d}^{n-1} \\ \hat{U}^{n-1} \end{pmatrix} .
$$

Determine the matrix G and its eigenvalues, and derive (8.32).
Hint: Consider a partitioned matrix

$$
G = \begin{pmatrix} A & I \\ I & 0 \end{pmatrix} ,
$$

where the blocks of G are square and of equal dimension. Let the corresponding partition of an eigenvector v be

$$
v = \begin{pmatrix} v_1 \\ v_2 \end{pmatrix} ,
$$

and let λ be the corresponding eigenvalue. Since $A v_1 + v_2 = \lambda v_1$, $v_1 = \lambda v_2$,
v_1 and v_2 are eigenvectors of A, so that λ is related to the eigenvalues $\mu(A)$
by (8.29).

Exercise 8.2.5. Consider the linearized shallow-water equations (8.11), (8.12)
including bottom friction. Carry out stability analysis for the leapfrog scheme,
and show that bottom friction has a damping effect.

Exercise 8.2.6. Make a plot of the dispersion relation (8.43), of the approximate dispersion relation (8.44) and compare with the exact dispersion
relation (8.42)

Exercise 8.2.7. Derive the dispersion relations (8.46) and (8.48).

8.3 The two-dimensional case

Governing equations

The two-dimensional shallow-water equations (1.121) and (1.124) can be written as

$$
\frac{\partial d}{\partial t} + \frac{\partial d U_1}{\partial x_1} + \frac{\partial d U_2}{\partial x_2} = 0 ,
$$

$$
\frac{\partial U_1}{\partial t} + U_1 \frac{\partial U_1}{\partial x_1} + U_2 \frac{\partial U_1}{\partial x_2} + g \frac{\partial (d + H)}{\partial x_1} - f U_2 + \frac{g U_1 |U|}{C^2 d} = 0 , \quad (8.50)
$$

$$
\frac{\partial U_2}{\partial t} + U_1 \frac{\partial U_2}{\partial x_1} + U_2 \frac{\partial U_2}{\partial x_2} + g \frac{\partial (d + H)}{\partial x_2} + f U_1 + \frac{g U_2 |U|}{C^2 d} = 0 ,
$$

where U_1 and U_2 are the depth-averaged velocity components, d the water depth, g the gravitational acceleration, f the Coriolis parameter and C the Chézy friction coefficient. The importance of the Coriolis acceleration is measured by the *Rossby number*, defined by

$$\mathrm{Ro} = \frac{V}{fL} \,,$$

where V is a measure of the magnitude of the velocity, and L a typical length scale.

Classification

As in the one-dimensional case, we determine the mathematical type of (8.50) along the lines set out in Chap. 2. The system (8.50) may be written as

$$\frac{\partial W}{\partial t} + F_1 \frac{\partial W}{\partial x_1} + F_2 \frac{\partial W}{\partial x_2} = Q \,, \tag{8.51}$$

with

$$W = \begin{pmatrix} d \\ U_1 \\ U_2 \end{pmatrix}, \quad F_1 = \begin{pmatrix} U_1 & d & 0 \\ g & U_1 & 0 \\ 0 & 0 & U_1 \end{pmatrix}, \quad F_2 = \begin{pmatrix} U_2 & 0 & d \\ 0 & U_2 & 0 \\ g & 0 & U_2 \end{pmatrix},$$

$$Q = \begin{pmatrix} 0 \\ fU_2 - gU_1|U|/C^2 d - g\partial H/\partial x_1 \\ -fU_1 - gU_2|U|/C^2 d - g\partial H/\partial x_2 \end{pmatrix}.$$

Following Sect. 2.2, the type of (8.51) depends on the properties of the eigenproblem given by

$$\det(n_0 I + n_1 F_1 + n_2 F_2) = 0 \,, \tag{8.52}$$

which gives

$$\tilde{n}_0\{\tilde{n}_0^2 - c^2(n_1^2 + n_2^2)\} = 0 \,, \tag{8.53}$$

where

$$\tilde{n}_0 = n_0 + n_1 U_1 + n_2 U_2, \quad c^2 = gd \,. \tag{8.54}$$

Three linearly independent solutions of (8.53) are given by

$$\tilde{n}_0 = 0, \quad \tilde{n}_0 = \pm c\sqrt{n_1^2 + n_2^2} \,. \tag{8.55}$$

Equation (8.52) arose in Chap. 2 from postulating a solution of the form $W = \hat{W}e^{iw(t,x)}$. The equation for $\hat{W}$ is

$$(n_0 I + n_1 F_1 + n_2 F_2)\hat{W} = 0 . \tag{8.56}$$

Corresponding to the three cases of (8.55) we find the following three solutions, choosing $n_1 = \cos\beta$, $n_2 = \sin\beta$ with β arbitrary:

$$\hat{W}_1 = \begin{pmatrix} 0 \\ \sin\beta \\ -g\cos\beta \end{pmatrix}, \quad \hat{W}_2 = \begin{pmatrix} c \\ -g\cos\beta \\ -g\sin\beta \end{pmatrix}, \quad \hat{W}_3 = \begin{pmatrix} -c \\ -g\cos\beta \\ -g\sin\beta \end{pmatrix}. \tag{8.57}$$

Since we have found a full set of eigenvectors of (8.56), the system (8.51) is hyperbolic according to the classification scheme discussed in Chap. 2.

Types of waves

In Sect. 8.2 we have, for the one-dimensional case, formulated a rule to determine appropriate boundary conditions based on the number of characteristics that enter the domain. In the two-dimensional case, appropriate boundary conditions can be selected by working with wave fronts (characteristic surfaces). We take a closer look at the solution $W = \hat{W} i e^{w(t,\boldsymbol{x})}$. With $\hat{W}$ given by (8.57) this is a solution of (8.51) with $Q = 0$ if

$$\frac{\partial w}{\partial t} = n_0, \quad \frac{\partial w}{\partial x_1} = n_1, \quad \frac{\partial w}{\partial x_2} = n_2, \tag{8.58}$$

with n_0, n_1, n_2 chosen according to (8.55), corresponding to the choice made in (8.57). Surfaces where $W = $ constant, i.e. $w(t, \boldsymbol{x}) = $ constant, are called wave fronts in Chap. 2.

According to (8.55), we have three kinds of waves. For $\tilde{n}_0 = 0$ it follows from (8.54) and (8.58) that

$$\frac{\partial w}{\partial t} + U_1 \frac{\partial w}{\partial x_1} + U_2 \frac{\partial w}{\partial x_2} = 0,$$

so that this wave moves with velocity (U_1, U_2). From (8.57) we see that for this type of wave, $d = 0$. The vorticity satisfies

$$\omega = \partial U_2/\partial x_1 - \partial U_1/\partial x_2 = i e^{iw}\left(-\cos\beta \frac{\partial w}{\partial x_1} - \sin\beta \frac{\partial w}{\partial x_2}\right) = -i e^{iw} .$$

Hence, the vorticity is constant along wave fronts. Therefore this type of wave is called a *vorticity wave*. It reprensents transport of vorticity by the flow.

The second kind of wave is obtained for $\tilde{n}_0 = c\sqrt{n_1^2 + n_2^2}$, or

$$n_0 = c - U_1 \cos\beta - U_2 \sin\beta, \quad n_1 = \cos\beta, \quad n_2 = \sin\beta .$$

From (8.58) it follows that

$$\frac{\partial w}{\partial t} + (U_1 - c\cos\beta)\frac{\partial w}{\partial x_1} + (U_2 - c\sin\beta)\frac{\partial w}{\partial x_2} = 0 \, ,$$

so that this wave moves with a velocity which is the sum of the flow velocity and a gravitational phase velocity vector in the direction $(-\cos\beta, -\sin\beta)$. The third kind of wave is obtained for $\tilde{n}_0 = -c\sqrt{n_1^2 + n_2^2}$. It is left to the reader to show that this wave moves with velocity $(U_1 + c\cos\beta, U_2 + c\sin\beta)$. Since β is arbitrary, there is no need to distinguish between the second and third kind of waves. These are called *gravity waves*. Summarizing, we have found two kinds of waves: vorticity waves, moving with velocity (U_1, U_2), and gravity waves, moving with velocity $(U_1 + c\cos\beta, U_2 + c\sin\beta)$, β arbitrary.

Waves in the linearized case

We now keep part of the right-hand side Q in (8.51), but linearize. Under the same assumptions as in Sect. 8.2, the two-dimensional linearized shallow-water equations are given by

$$\frac{\partial W}{\partial t} + F_1\frac{\partial W}{\partial x_1} + F_2\frac{\partial W}{\partial x_2} + GW = 0 \, , \tag{8.59}$$

where

$$\hat{W} = \begin{pmatrix} d \\ u_1 \\ u_2 \end{pmatrix}, \quad F_1 = \begin{pmatrix} \bar{U}_1 & \bar{d} & 0 \\ g & \bar{U}_1 & 0 \\ 0 & 0 & \bar{U}_1 \end{pmatrix}, \quad F_2 = \begin{pmatrix} \bar{U}_2 & 0 & \bar{d} \\ 0 & \bar{U}_2 & 0 \\ g & 0 & \bar{U}_2 \end{pmatrix},$$

$$G = \begin{pmatrix} 0 & 0 & 0 \\ 0 & r & -f \\ 0 & f & r \end{pmatrix} . \tag{8.60}$$

Assume a harmonic wave of the form

$$W = \hat{W}e^{i(\boldsymbol{k}\cdot\boldsymbol{x} - \omega t)}, \quad \hat{W} = \text{constant} \, ,$$

where $\boldsymbol{k} = (k_1, k_2)$, $\boldsymbol{x} = (x_1, x_2)$. This is a solution of (8.59) if

$$(-i\omega I + ik_1 F_1 + ik_2 F_2 + G)\hat{W} = 0 \, .$$

There are nontrivial solutions if the determinant is zero. Neglecting bottom friction $(r = 0)$, this results in:

$$i\tilde{\omega}(\tilde{\omega}^2 - k^2 c^2 - f^2) = 0, \quad k^2 = k_1^2 + k_2^2, \quad \tilde{\omega} = \omega - k_1\bar{U}_1 - k_2\bar{U}_2 \, .$$

The root $\tilde{\omega} = 0$ corresponds to the vorticity waves. The other roots are given by

$$\tilde{\omega} = \pm\sqrt{c^2 k^2 + f^2} \, .$$

The corresponding waves are called *Poincaré waves*. They are dispersive, with phase and group velocity components

$$c_p^\alpha = \omega/k_\alpha = -\bar{U}_\alpha \pm c\sqrt{1 + f^2/k^2c^2},$$
$$c_g^\alpha = \partial\omega/\partial k_\alpha = -\bar{U}_\alpha \pm c\sqrt{1 + f^2/k^2c^2}\ .$$

The pure gravity waves are recovered if $f/kc \ll 1$, or

$$\text{Bu} \equiv c^2 k^2/f^2 \gg 1\ ,$$

where Bu is called the *Burger number*.

At moderate latitudes, according to (1.122), the Coriolis parameter satisfies $f \cong 2\pi$ per 24 hrs. We have $f/kc = 1$ if the wavelength $L = 2\pi/k$ satisfies $L = 2\pi c/f$, which is, with the assumed value of f, the distance traveled by a gravity wave in 24 hours. In shallow water, for example $\bar{d} = 20$ m, we have $c \cong 14$ m/s, so that $L \cong 1200$ km; in deep water, for example $\bar{d} = 2$ km, we have $c \cong 140$ m/s and $L = 12,000$ km. We see that for large scale geophysical fluid dynamics the effect of Coriolis acceleration is not negligible. As mentioned before, its importance is measured by the Rossby number Ro. If Ro $\ll 1$, Coriolis acceleration dominates. The effect of bottom friction on wave motion will not be considered, because it has significant influence mainly in river flows, which can be modeled by the one-dimensional shallow-water equations; this case has already been examined in Sect. 8.2.

Boundary conditions

As initial condition, U must be given. The generalization to more dimensions of the rule for the boundary conditions found in Sect. 8.2 is:

> *The number of boundary conditions in a point on the boundary of the domain must be equal to the number of different types of waves that can enter the boundary at that point.*

As an illustration, consider a left boundary $x_1 = 0$. Two (of the many) possible gravity waves correspond to $\beta = 0, \pi$. These propagate with normal velocity component $U_1 \pm c$. The vorticity enters or leaves with normal velocity component U_1. This gives us the cases listed in Table 8.1, where n is the number of possible incoming waves and also the number of boundary conditions to be prescribed. The next question is, what to prescribe? For the gravity wave with $\beta = 0$, only d and U_1 are nonzero, according to (8.57). This reduces the governing equations to the one-dimensional case of Sect. 8.2, which tells us the following. In the subcritical cases, a quantity independent of $U_1 - 2\sqrt{gd}$ must be prescribed, such as U_1, d or $U_1 + 2\sqrt{gd}$. In the supercritical inflow case, both U_1 and d must be given. For the vorticity wave, only

Case	Comment	n
$U_1 \leq -c$	Supercritical outflow	0
$-c < U_1 \leq 0$	Subcritical outflow	1
$0 < U_1 \leq c$	Subcritical inflow	2
$c < U_1$	Supercritical inflow	3

Table 8.1. Number of boundary conditions (n) to be prescribed.

U_1 and U_2 are nonzero, according to (8.57). Since U_1 is already determined by the incoming and outgoing gravity waves, U_2 must be prescribed if the vorticity wave is incoming ($U_1 > 0$).

Often, a part of the domain boundary is artificial, and is introduced merely to restrict the domain of computation, for example, to separate the North Sea from the Atlantic. Then it is important that this boundary, as far as possible, does not generate artificial numerical wave reflections. It then becomes important what combinations of d, U_1 and U_2 are prescribed. Various *weakly reflecting boundary conditions* have been developed. We will not discuss these. Information may be found in Engquist and Majda (1979), Verboom and Slob (1984), Vreugdenhil (1994).

Choice of grids

Figure 8.3 shows the staggered grid proposed by Platzman (1959) and used by Hansen (1956) and Leendertse (1967). The velocity component U_1 is located

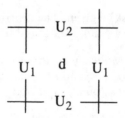

Fig. 8.3. Staggered grid

at grid points $(j+1/2, k)$, U_2 at $(j, k+1/2)$ and d at (j, k). Of course, there are different ways to assign unknowns to grid points, including colocated schemes. Five different configurations are studied in Arakawa and Lamb (1977), and are generally known as Arakawa grids A, B, C, D, E. The grid of Fig. 8.3 is the Arakawa C-grid. In a comparative study, Randall (1994) finds that the C-grid gives the best approximation of the dispersion relation. Here only the C-grid will be used. Several schemes will be presented with their stability analysis for the linearized shallow-water equations (8.59), (8.60) with $\bar{U}_1 =$

$\bar{U}_2 = 0$, $G = 0$. Numerical dispersion will not be analyzed, because the one-dimensional results of Sect. 8.2 give sufficient indication of the relative merits of the various schemes in this respect. Furthermore, a scheme that is frequently used in practice for the nonlinear equations (8.50) will be discussed. The linearized equations will be denoted for convenience by

$$\frac{\partial W}{\partial t} + F_1 \frac{\partial W}{\partial x_1} + F_2 \frac{\partial W}{\partial x_2} = 0,$$

$$W = \begin{pmatrix} d \\ U \\ V \end{pmatrix}, \quad F_1 = \begin{pmatrix} 0 & \bar{d} & 0 \\ g & 0 & 0 \\ 0 & 0 & 0 \end{pmatrix}, \quad F_2 = \begin{pmatrix} 0 & 0 & \bar{d} \\ 0 & 0 & 0 \\ g & 0 & 0 \end{pmatrix}.$$

The (uniform) mesh size in the x_α-direction is denoted by h_α.

Hansen scheme

The one-dimensional version of this scheme has been proposed in Hansen (1956). The scheme is staggered not only in space but also in time, and is given by

$$d_{jk}^{n+1} - d_{jk}^n + \frac{\tau \bar{d}}{h_1}(U_{j+1/2,k}^{n+1/2} - U_{j-1/2,k}^{n+1/2}) + \frac{\tau \bar{d}}{h_2}(V_{j,k+1/2}^{n+1/2} - V_{j,k-1/2}^{n+1/2}) = 0,$$

$$U_{j+1/2,k}^{n+3/2} - U_{j+1/2,k}^{n+1/2} + \frac{\tau g}{h_1}(d_{j+1,k}^{n+1} - d_{jk}^{n+1}) = 0,$$

$$V_{j,k+1/2}^{n+3/2} - V_{j,k+1/2}^{n+1/2} + \frac{\tau g}{h_2}(d_{j,k+1}^{n+1} - d_{jk}^{n+1}) = 0.$$

Stability analysis is standard, and similar to the analysis of the one-dimensional case in the preceding section. We postulate

$$\begin{pmatrix} d_{jk}^n \\ U_{j+1/2,k}^{n+1/2} \\ V_{j,k+1/2}^{n+1/2} \end{pmatrix} = \begin{pmatrix} \hat{d}^n \\ \hat{U}^{n+1/2} \\ \hat{V}^{n+1/2} \end{pmatrix} e^{i(j\theta_1 + k\theta_2)},$$

and find

$$\begin{pmatrix} \hat{d}^{n+1} \\ \hat{U}^{n+3/2} \\ \hat{V}^{n+3/2} \end{pmatrix} = G \begin{pmatrix} \hat{d}^n \\ \hat{U}^{n+1/2} \\ \hat{V}^{n+1/2} \end{pmatrix}, \quad G = G_2^{-1} G_1,$$

with

$$G_1 = \begin{pmatrix} 1 & -\beta_1 & -\beta_2 \\ 0 & 1 & 0 \\ 0 & 0 & 1 \end{pmatrix}, \quad G_2 = \begin{pmatrix} 1 & 0 & 0 \\ \alpha_1 & 1 & 0 \\ \alpha_2 & 0 & 1 \end{pmatrix},$$

where

$$\alpha_m = \frac{g\tau}{h_m}(e^{i\theta_m} - 1), \quad \beta_m = \frac{\bar{d}\tau}{h_m}(1 - e^{-i\theta_m}), \quad m = 1, 2.$$

From $\det(G_1 - \lambda G_2) = 0$ it follows that

$$(1 - \lambda)\{(1 - \lambda)^2 - 2\mu\lambda\} = 0, \quad \mu = \sigma_1^2(\cos\theta_1 - 1) + \sigma_2^2(\cos\theta_2 - 1), \quad (8.61)$$

where $\sigma_m = c\tau/h_m$, $m = 1, 2$. For

$$-2 \leq \mu \leq 0 \tag{8.62}$$

we find that the roots of (8.61) satisfy $|\lambda_{1,2,3}| = 1$, otherwise there are roots with modulus > 1. Hence (8.62) is sufficient and necessary for stability. The condition (8.62) is to hold for all θ_1, θ_2, so that the stability condition is

$$\sigma_1 + \sigma_2 \leq 1 \quad \text{or} \quad \tau \leq \frac{h_1 h_2}{c\sqrt{h_1^2 + h_2^2}}. \tag{8.63}$$

An implicit scheme

The two-dimensional version of the implicit scheme discussed in Sect. 8.2 is

$$d_{jk}^{n+1} - d_{jk}^n + \frac{\tau\bar{d}}{2h_1}(U^{n+1} + U^n)|_{j-1/2,k}^{j+1/2,k} + \frac{\tau\bar{d}}{2h_2}(V^{n+1} + V^n)|_{j,k-1/2}^{j,k+1/2} = 0,$$

$$U_{j+1/2,k}^{n+1} - U_{j+1/2,k}^n + \frac{\tau g}{2h_1}(d^{n+1} + d^n)|_{jk}^{j+1,k} = 0, \tag{8.64}$$

$$V_{j,k+1/2}^{n+1} - V_{j,k+1/2}^n + \frac{\tau g}{2h_2}(d^{n+1} + d^n)|_{jk}^{j,k+1} = 0.$$

Stability analysis proceeds as follows. We postulate

$$\begin{pmatrix} d_{jk}^n \\ U_{j+1/2,k}^n \\ V_{j,k+1/2}^n \end{pmatrix} = \begin{pmatrix} \hat{d}^n \\ \hat{U}^n \\ \hat{V}^n \end{pmatrix} e^{i(j\theta_1 + k\theta_2)}, \tag{8.65}$$

and find

$$\begin{pmatrix} \hat{d}^{n+1} \\ \hat{U}^{n+1} \\ \hat{V}^{n+1} \end{pmatrix} = G \begin{pmatrix} \hat{d}^n \\ \hat{U}^n \\ \hat{v}^n \end{pmatrix}, \tag{8.66}$$

where $G = G_2^{-1}G_1$, with

$$G_1 = \begin{pmatrix} 1 & -\alpha_1 & -\alpha_2 \\ -\beta_1 & 1 & 0 \\ -\beta_2 & 0 & 1 \end{pmatrix}, \quad G_2 = \begin{pmatrix} 1 & \alpha_1 & \alpha_2 \\ \beta_1 & 1 & 0 \\ \beta_2 & 0 & 1 \end{pmatrix}, \tag{8.67}$$

where $\alpha_m = (\tau\bar{d}/2h_m)(1 - \exp(-i\theta_m))$, $\beta_m = (\tau g/2h_m)(\exp(i\theta_m) - 1)$, $m = 1, 2$. Although the roots of $\det(G_1 - \lambda G_2)$ are easily determined, as before, we wish to demonstrate a different technique, that wil be useful later.

The matrices G_1 and G_2 can be symmetrized by a similarity transformation with the diagonal matrix D defined by

$$D = \begin{pmatrix} 1 & 0 & 0 \\ 0 & \sqrt{g/d}e^{i\theta_1/2} & 0 \\ 0 & 0 & \sqrt{g/d}e^{i\theta_2/2} \end{pmatrix}. \tag{8.68}$$

We find

$$D^{-1}G_1D = \tilde{G}_1 = \begin{pmatrix} 1 & -\gamma_1 & -\gamma_2 \\ -\gamma_1 & 1 & 0 \\ -\gamma_2 & 0 & 1 \end{pmatrix},$$

$$\tag{8.69}$$

$$D^{-1}G_2D = \tilde{G}_2 = \begin{pmatrix} 1 & \gamma_1 & \gamma_2 \\ \gamma_1 & 1 & 0 \\ \gamma_2 & 0 & 1 \end{pmatrix},$$

where $\gamma_m = i\sigma_m \sin\theta_m/2$, $\sigma_m = \tau c/h_m$, $m = 1, 2$. Because γ_m is imaginary, $\tilde{G}_1$ and $\tilde{G}_2$ are adjoint:

$$\tilde{G}_1 = \tilde{G}^H, \tag{8.70}$$

where $A = B^H$ means $a_{ij} = \bar{b}_{ji}$, $\mathrm{Re}(\bar{b}_{ji}) = \mathrm{Re}(b_{ji})$, $\mathrm{Im}(\bar{b}_{ji}) = -\mathrm{Im}(b_{ji})$. Furthermore, $\tilde{G}_2$ is normal:

$$\tilde{G}_2\tilde{G}_2^H = \tilde{G}_2^H\tilde{G}_2 . \tag{8.71}$$

With (8.70) and (8.71) we find that $\tilde{G} = \tilde{G}_2^{-1}\tilde{G}_1$ is unitary:

$$\tilde{G}\tilde{G}^H = I ,$$

so that $|\lambda(\tilde{G})| = 1$. Since $\lambda(G) = \lambda(\tilde{G})$, we have proved unconditional stability. The scheme has no damping.

Leendertse scheme

In order to reduce computing work, the implicit scheme may be approximated by stages that are implicit in one direction only. This results in the following ADI (alternating direction implicit) scheme:

$$d_{jk}^{n+1/2} - d_{jk}^n + \frac{\tau\bar{d}}{2h_1}U^{n+1/2}|_{j-1/2,k}^{j+1/2,k} + \frac{\tau\bar{d}}{2h_2}V^n|_{j,k-1/2}^{j,k+1/2} = 0,$$

$$U_{j+1/2,k}^{n+1/2} - U_{j+1/2,k}^n + \frac{\tau g}{2h_1}d^{n+1/2}|_{jk}^{j+1,k} = 0,$$

$$V_{j,k+1/2}^{n+1/2} - V_{j,k+1/2}^n + \frac{\tau g}{2h_2}d^n|_{jk}^{j,k+1} = 0,$$

$$d_{jk}^{n+1} - d_{jk}^{n+1/2} + \frac{\tau\bar{d}}{2h_1}U^{n+1/2}|_{j-1/2,k}^{j+1/2,k} + \frac{\tau\bar{d}}{2h_2}V^{n+1}|_{j,k-1/2}^{j,k+1/2} = 0, \qquad (8.72)$$

$$U_{j+1/2,k}^{n+1} - U_{j+1/2,k}^{n+1/2} + \frac{\tau g}{2h_1}d^{n+1/2}|_{jk}^{j+1,k} = 0,$$

$$V_{j,k+1/2}^{n+1} - V_{j,k+1/2}^{n+1/2} + \frac{\tau g}{2h_2}d^{n+1}|_{jk}^{j,k+1} = 0.$$

The time step is divided in two parts. In each part, only tridiagonal systems have to be solved, along horizontal or vertical grid lines. This is considerably cheaper than solving (8.64). The method has been designed such that $\mathcal{O}(\tau^2)$ accuracy is maintained. The scheme has been proposed in Leendertse (1967).

For the stability analysis of (8.72) we follow Stelling (1983). The principles are the same as before. By postulating (8.65) and (8.66) we find

$$G = G_4^{-1}G_3G_2^{-1}G_1,$$

with $G_2^{-1}G_1$ the amplification matrix of the first half step and $G_4^{-1}G_3$ that of the second. We find

$$G_1 = \begin{pmatrix} 1 & 0 & -\alpha_2 \\ 0 & 1 & 0 \\ -\beta_2 & 0 & 1 \end{pmatrix}, \quad G_2 = \begin{pmatrix} 1 & \alpha_1 & 0 \\ \beta_1 & 1 & 0 \\ 0 & 0 & 1 \end{pmatrix},$$

$$G_3 = \begin{pmatrix} 1 & -\alpha_1 & 0 \\ -\beta_1 & 1 & 0 \\ 0 & 0 & 1 \end{pmatrix}, \quad G_4 = \begin{pmatrix} 1 & 0 & \alpha_2 \\ 0 & 1 & 0 \\ \beta_2 & 0 & 1 \end{pmatrix},$$

with α_m and β_m defined after equation (8.67). We symmetrize with the matrix D given by (8.68) and find

$$\tilde{G} = D^{-1}GD = \tilde{G}_4^{-1}\tilde{G}_3\tilde{G}_2^{-1}\tilde{G}_1,$$

with

$$\tilde{G}_1 = \begin{pmatrix} 1 & 0 & -\gamma_2 \\ 0 & 1 & 0 \\ -\gamma_2 & 0 & 1 \end{pmatrix}, \quad \tilde{G}_2 = \begin{pmatrix} 1 & \gamma_1 & 0 \\ \gamma_1 & 1 & 0 \\ 0 & 0 & 1 \end{pmatrix}, \quad \tilde{G}_3 = \tilde{G}_2^H, \quad \tilde{G}_4 = \tilde{G}_1^H,$$

$$(8.73)$$

where γ_m is defined after equation (8.69). We have $\tilde{G} = \tilde{G}_1^{-H}\tilde{G}_2^H\tilde{G}_2^{-1}\tilde{G}_1$, hence

$$\tilde{G}_1^H \tilde{G} \tilde{G}_1^{-H} = \tilde{G}_2^H \tilde{G}_2^{-1} \tilde{G}_1 \tilde{G}_1^{-H}. \tag{8.74}$$

$\tilde{G}_1$ is normal, therefore

$$\begin{aligned}
(\tilde{G}_1 \tilde{G}_1^{-H})(\tilde{G}_1 \tilde{G}_1^{-H})^H &= \tilde{G}_1 \tilde{G}_1^{-H} \tilde{G}_1^{-1} \tilde{G}_1^H = \tilde{G}_1 (\tilde{G}_1 \tilde{G}_1^H)^{-1} \tilde{G}_1^H \\
&= \tilde{G}_1 (\tilde{G}_1^H \tilde{G}_1)^{-1} \tilde{G}_1^H = I
\end{aligned}$$

showing that $\tilde{G}_1 \tilde{G}_1^{-H}$ is unitary. Similarly, $\tilde{G}_2^H \tilde{G}_2^{-1}$ is unitary. Then it follows from (8.74) that $\tilde{G}_1^H \tilde{G} \tilde{G}_1^{-H}$ is unitary, so that $|\lambda(\tilde{G})| = 1$, proving unconditional stability. The scheme has no damping.

Disadvantage of ADI schemes

We see that without compromising the local truncation error or stability, we have replaced the implicit scheme (8.64) by the much cheaper ADI scheme (8.72). But ADI schemes have a disadvantage that the implicit scheme does not have. It turns out that, depending on the geometry of the problem, to achieve sufficient accuracy the time step τ must be much smaller for the ADI scheme than for the implicit scheme. This is a pity, because the main motivation for unconditionally stable schemes is to allow large τ. The efficiency advantage of ADI is lost if τ is severely constrained.

The loss of accuracy of ADI in domains of certain geometric shape may be explained as follows. Physical disturbances travel with the celerity c. For accuracy it is necessary that changes in the numerical solution propagate with a velocity that is approximately equal to c. That is, they should travel over a distance of approximately $c\tau$ in one time step, that is a distance $\sigma_m h_m$ in the x_m-direction. Now assume that the domain is the zig-zag channel of Fig. 8.4. A change in the numerical solution in P influences the whole hor-

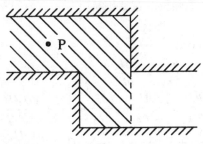

Fig. 8.4. Zig-zag channel

izontal row of cells in which P is located in the first half time step. In the second half step, all vertical rows of cells that intersect the previous horizontal row are affected. The numerical solution in other cells is not influenced.

Hence, a change in the solution in P affects only the shaded area in Fig. 8.4, in one time step. As a consequence, if σ_1 and σ_2 are so large that a physical disturbance in P travels much beyond the shaded are over a time-span τ, the ADI scheme is inaccurate. This is illustrated by applications worked out in Stelling, Wiersma, and Willemse (1986), in which the numerical propagation speed of tidal waves depends entirely on the geometry of the spatial and tidal flats, in combination with the spatial and temporal step sizes. This is the reason why in Wilders, van Stijn, Stelling, and Fokkema (1988) an implicit (non-ADI) scheme with an efficient iterative solver is developed. However, if the shape of the domain does not inhibit signal propagation, or if the time step has to be limited for other (accuracy) reasons anyway, then ADI schemes are attractive.

Stelling scheme

To illustrate how the non-simplified shallow-water equations may be solved numerically, we present details of a discretization scheme that has found widespread use in practice. It is a further development of the Leendertse scheme (Leendertse (1967)), and is described in Stelling (1983). The scheme is of ADI type. Coriolis acceleration, bottom friction, nonlinear advection terms and an interior viscous term are included. The spatial grid is staggered and of Arakawa C type, as before. The location of the unknowns is given in Fig. 8.5. The governing equations are (1.121) and (1.124) with inclusion of

$$
\begin{array}{ccc}
H & U_2 & \dot{H} \\[4pt]
U_1 & \xi & U_1 \\[4pt]
H & U2 & H
\end{array}
$$

Fig. 8.5. Location of unknowns in staggered grid

an interior viscous term:

$$
\frac{\partial \xi}{\partial t} + \frac{\partial dU}{\partial x_1} + \frac{\partial dV}{\partial x_2} = \frac{\partial H}{\partial t},
$$

$$
\frac{\partial U_1}{\partial t} + U\frac{\partial U}{\partial x_1} + V\frac{\partial U}{\partial x_2} + g\frac{\partial \xi}{\partial x_1} - fV + \frac{gU|U|}{C^2 d} - \nu\nabla^2 U = 0, \quad (8.75)
$$

$$
\frac{\partial V}{\partial t} + U\frac{\partial V}{\partial x_1} + V\frac{\partial V}{\partial x_2} + g\frac{\partial \xi}{\partial x_2} + fU + \frac{gV|U|}{C^2 d} - \nu\nabla^2 V = 0,
$$

where $\nabla^2 = \partial^2/\partial x_1^2 + \partial^2/\partial x_2^2$, $d = \xi - H$ and $|U| = \sqrt{U^2 + V^2}$. For simplicity we assume $\partial H/\partial t = 0$.

A time step is divided in two stages, in ADI fashion:

Stage 1:

$$\frac{2}{\tau}(\xi^{n+1/2} - \xi^n)_{jk} + \frac{1}{h_1}(d^{n+1/2}U^{n+1/2})|_{j-1/2,k}^{j+1/2,k} + \frac{1}{h_2}(d^n V^n)|_{j,k-1/2}^{j,k+1/2} = 0,$$

$$\frac{2}{\tau}(U^{n+1/2} - U^n)_{j+1/2,k} + \frac{1}{2h_1}U^{n+1/2}_{j+1/2,k}U^n|_{j-1/2,k}^{j+3/2,k} +$$

$$\frac{1}{12h_2}V^{n+1/2}_{j+1/2,k}(U^n|_{j+1/2,k-2}^{j+1/2,k+2} + 4U^n|_{j+1/2,k-1}^{j+1/2,k+1}) +$$

$$\frac{1}{h_1}g\xi^{n+1/2}|_{jk}^{j+1,k} - fV^{n+1/2}_{j+1/2,k} +$$

$$g[U^{n+1/2}\{(U^n)^2 + (V^{n+1/2})^2\}^{1/2}/(C^2 d^n)]_{j+1/2,k} - \nu\nabla_h^2 U^n_{j+1/2,k} = 0,$$

$$\frac{2}{\tau}(V^{n+1/2} - V^n)_{j,k+1/2} + A_{j,k+1/2}(U^n, V^{n+1/2}) +$$

$$\frac{1}{2h_2}V^n_{j,k+1/2}V^{n+1/2}|_{j,k-1/2}^{j,k+3/2} + \frac{1}{h_2}g\xi^n|_{jk}^{j,k+1} + fU^n_{j,k+1/2}$$

$$+g[V^{n+1/2}\{(U^n)^2 + (V^n)^2\}^{1/2}/(C^2 d^n)]_{j,k+1/2} - \nu\nabla_h^2 V^{n+1/2}_{j,k+1/2} = 0.$$

Stage 2:

$$\frac{2}{\tau}(\xi^{n+1} - \xi^{n+1/2})_{jk} + \frac{1}{h_1}(d^{n+1/2}U^{n+1/2})|_{j-1/2,k}^{j+1/2,k} +$$

$$\frac{1}{h_2}(d^{n+1}V^{n+1})|_{j,k-1/2}^{j,k+1/2} = 0,$$

$$\frac{2}{\tau}(U^{n+1} - U^{n+1/2})_{j+1/2,k} + \frac{1}{2h_1}U^{n+1/2}_{j+1/2,k}U^{n+1}|_{j-1/2,k}^{j+3/2,k}$$

$$+B_{j+1/2,k}(V^{n+1/2}, U^{n+1}) + \frac{1}{h_1}g\xi^{n+1/2}|_{jk}^{j+1,k} - fV^{n+1/2}_{j+1/2,k} +$$

$$g[U^{n+1}\{(U^{n+1/2})^2 + (V^{n+1/2})^2\}^{1/2}/(C^2 d^{n+1/2})]_{j+1/2,k} -$$

$$\nu\nabla_h^2 U^{n+1}_{j+1/2,k} = 0,$$

$$\frac{2}{\tau}(V^{n+1} - V^{n+1/2})_{j,k+1/2} +$$

$$\frac{1}{12h_1}U^{n+1}_{j,k+1/2}(V^{n+1/2}|_{j-2,k+1/2}^{j+2,k+1/2} + 4V^{n+1/2}|_{j-1,k+1/2}^{j+1,k+1/2}) +$$

$$\frac{1}{2h_2}V^{n+1}_{j,k+1/2}V^{n+1/2}|_{j,k-1/2}^{j,k+3/2}) + \frac{1}{h_2}g\xi^{n+1}|_{jk}^{j,k+1} + fU^{n+1}_{j,k+1/2}$$

$$+g[V^{n+1}\{(U^{n+1})^2 + (V^{n+1/2})^2\}^{1/2}/(C^2 d^{n+1/2})]_{j,k+1/2} -$$

$$\nu\nabla_h^2 V^{n+1/2}_{j,k+1/2} = 0.$$

Here ∇_h^2 stands for the standard discrete Laplacian. The operator A is defined by

$$A_{jk}(Y,Z) = Y_{jk}(3Z_{jk} - 4Z_{j-1,k} + Z_{j-2,k})/2h_1 \quad \text{if} \quad Y_{jk} > 0,$$

$$\text{(8.76)}$$

$$A_{jk}(Y,Z) = Y_{jk}(-3Z_{jk} + 4Z_{j+1,k} - Z_{j+2,k})/2h_1 \quad \text{if} \quad Y_{jk} \leq 0.$$

The operator B is defined similarly, but with differences in the x_2-direction. Quantities that have to be evaluated at points where they are not defined are obtained by straightforward averaging, for example,

$$d_{j+1/2,k} = \frac{1}{2}(\xi_{jk} + \xi_{j+1,k} - H_{j+1/2,k+1/2} - H_{j+1/2,k-1/2}). \qquad \text{(8.77)}$$

The implicitness is carefully distributed over the two stages, such that mainly tridiagonal systems have to be solved. In addition, a few iterations are carried out for the nonlinear terms, the operators A and B and the viscous terms. The convection terms have been discretized such that second order accuracy is obtained over a complete time step, with a certain amount of upwind bias to avoid spurious oscillations, while keeping numerical dissipation and dispersion small.

Implementation of boundary conditions is straightforward, except for the convection terms, which need special care. For details, see Stelling (1983), where also procedures to follow when d becomes zero (tidal flats) are described.

A non-ADI version, to be used when ADI is inaccurate due to the geometric effects discussed before, is presented in Wilders, van Stijn, Stelling, and Fokkema (1988), including discussion of efficient iterative solution methods.

Stability analysis of Stelling scheme

To give an example of Fourier stability analysis for a more difficult problem than before, we discuss stability of the Stelling scheme. The 'frozen coefficient' version of (8.75) is, assuming $H = 0$, and simplifying the bottom friction term as in Sect. 8.2,

$$\frac{\partial d}{\partial t} + \bar{d}\left(\frac{\partial U}{\partial x_1} + \frac{\partial V}{\partial x_2}\right) + \bar{U}\frac{\partial d}{\partial x_1} + \bar{V}\frac{\partial d}{\partial x_2} = 0,$$

$$\frac{\partial U}{\partial t} + \bar{U}\frac{\partial U}{\partial x_1} + \bar{V}\frac{\partial U}{\partial x_2} + g\frac{\partial d}{\partial x_1} - fV + rU - \nu\nabla^2 U = 0,$$

$$\frac{\partial V}{\partial t} + \bar{U}\frac{\partial V}{\partial x_1} + \bar{V}\frac{\partial V}{\partial x_2} + g\frac{\partial d}{\partial x_2} + fU + rV - \nu\nabla^2 V = 0.$$

The frozen coefficients are denoted by an overbar. Let $\bar{U} > 0$, $\bar{V} > 0$. By discretizing with the Stelling scheme and postulating (8.65) and (8.66) we find:

$$G_2\hat{W}^{n+1/2} = G_1\hat{W}^n, \quad G_4\hat{W}^{n+1} = G_3\hat{W}^{n+1/2}, \quad \hat{W} = (\hat{d}, \hat{U}, \hat{V})^T,$$

with

$$G_1 = \begin{pmatrix} 1 - \gamma_2 & 0 & -\alpha_2 \\ 0 & 1 - \gamma_1 - \delta_2 + \tilde{\nabla}^2 & 0 \\ -\beta_2 & -\tilde{f} & 1 \end{pmatrix},$$

$$G_2 = \begin{pmatrix} 1 + \gamma_1 & \alpha_1 & 0 \\ \beta_1 & 1 + \tau r/2 & -\tilde{f} \\ 0 & 0 & 1 + \tilde{A} + \gamma_2 + \tau r/2 - \tilde{\nabla}^2 \end{pmatrix},$$

(8.78)

$$G_3 = \begin{pmatrix} 1 - \gamma_1 & -\alpha_1 & 0 \\ -\beta_1 & 1 & \tilde{f} \\ 0 & 0 & 1 - \gamma_2 - \delta_1 + \tilde{\nabla}^2 \end{pmatrix},$$

$$G_4 = \begin{pmatrix} 1 + \gamma_2 & 0 & \alpha_2 \\ 0 & 1 + \tilde{B} + \gamma_1 + \tau r/2 - \tilde{\nabla}^2 & 0 \\ \beta_2 & \tilde{f} & 1 + \tau r/2 \end{pmatrix},$$

where

$$\alpha_m = \frac{\tau \bar{d}}{2h_m}(1 - e^{-i\theta_m}), \quad \beta_m = \frac{\tau \bar{g}}{2h_m}(e^{i\theta_m} - 1), \quad m = 1, 2;$$

$$\gamma_1 = \frac{i\tau \bar{U}}{2h_1}\sin\theta_1, \quad \gamma_2 = \frac{i\tau \bar{V}}{2h_2}\sin\theta_2, \quad \delta_1 = \frac{i\tau \bar{U}}{12h_1}(\sin 2\theta_1 + 4\sin\theta_1),$$

$$\delta_2 = \frac{i\tau \bar{V}}{12h_2}(\sin 2\theta_2 + 4\sin\theta_2), \quad \tilde{f} = \frac{\tau f}{8}(1 + e^{i\theta_1} + e^{-i\theta_2} + e^{i(\theta_1 - \theta_2)}),$$

$$\tilde{A} = \frac{\tau \bar{U}}{4h_1}(3 - 4e^{-i\theta_1} + e^{-2i\theta_1}), \quad \tilde{B} = \frac{\tau \bar{V}}{4h_2}(3 - 4e^{-i\theta_2} + e^{-2i\theta_2}),$$

$$\tilde{\nabla}^2 = \nu\tau\{(\cos\theta_1 - 1)/h_1^2 + (\cos\theta_2 - 1)/h_2^2\}.$$

Similarity transformation with the diagonal matrix D given by (8.68) results in

$$\tilde{G}_1 = \begin{pmatrix} 1 - \gamma_2 & 0 & -\mu_2 \\ 0 & g_1 & 0 \\ -\mu_2 & -\tilde{f} & 1 \end{pmatrix}, \quad \tilde{G}_2 = \begin{pmatrix} 1 + \gamma_1 & \mu_1 & 0 \\ \mu_1 & 1 + \tau r/2 & -\tilde{f} \\ 0 & 0 & g_2 \end{pmatrix},$$

$$\tilde{G}_3 = \begin{pmatrix} 1 - \gamma_1 & -\mu_1 & 0 \\ -\mu_1 & 1 & \tilde{f} \\ 0 & 0 & g_3 \end{pmatrix}, \quad \tilde{G}_4 = \begin{pmatrix} 1 + \gamma_2 & 0 & \mu_2 \\ 0 & g_4 & 0 \\ \mu_2 & \tilde{f} & 1 + \tau r/2 \end{pmatrix},$$

where $\tilde{G}_m = D^{-1}G_m D$, $m = 1, ..., 4$; $g_1 = 1 - \gamma_1 - \delta_2 + \tilde{\nabla}^2$, $g_2 = 1 + \tilde{A} + \gamma_2 + \tau r/2 - \tilde{\nabla}^2$, $g_3 = 1 - \gamma_2 - \delta_1 + \tilde{\nabla}^2$, $g_4 = 1 + \tilde{B} + \gamma_1 + \tau r/2 - \tilde{\nabla}^2$, $\mu_m = i\sigma_m \sin\theta_m/2$, $m = 1, 2$. The amplification matrix G satisfies

$$G = G_4^{-1}G_3 G_2^{-1}G_1 = D\tilde{G}_4^{-1}\tilde{G}_3\tilde{G}_2^{-1}\tilde{G}_1 D^{-1}$$
$$= (\tilde{G}_4 D^{-1})^{-1}(\tilde{G}_3\tilde{G}_2^{-1}\tilde{G}_1\tilde{G}_4^{-1})(\tilde{G}_4 D^{-1}).$$

Hence

$$\|G^n\| \le \|(\tilde{G}_4 D^{-1})^{-1}\| \, \|\tilde{G}_4 D^{-1}\| \, \|\tilde{G}_3 \tilde{G}_2^{-1}\|^n \|\tilde{G}_1 \tilde{G}_4^{-1}\|^n. \tag{8.79}$$

For simplicity we neglect bottom friction, i.e. $\bar{r} = 0$; in practice, bottom friction is found to have a stabilizing effect. The eigenvalues of $\tilde{G}_3 \tilde{G}_2^{-1}$ satisfy

$$\det(\tilde{G}_3 - \lambda \tilde{G}_2) = 0.$$

Expansion gives

$$(g_3 - \lambda g_2)\det(C - \lambda C^H) = 0,$$

where

$$C = \begin{pmatrix} 1 - \gamma_1 & -\mu_1 \\ -\mu_1 & 1 \end{pmatrix}.$$

The matrix C is normal, so that CC^{-H} is unitary. Hence, the roots of

$$\det(C - \lambda C^H) = 0$$

have modulus 1. It remains to inspect $|\lambda| = |g_3/g_2|$. Noting that $\tilde{\nabla}^2 \le 0$, it is easy to see that $|\mathrm{Re}(g_2)| \ge |\mathrm{Re}(g_3)|$. It is also easy to see that $|\mathrm{Im}(g_2)| \ge |\mathrm{Im}(g_3)|$, so that $|\lambda(\tilde{G}_3 \tilde{G}_2^{-1})| \le 1$, and $\|\tilde{G}_3 \tilde{G}_2^{-1}\|^n$ is bounded. In a similar way it can be shown that $\|\tilde{G}_1 \tilde{G}_4^{-1}\|^n$ is bounded. Furthermore,

$$\tilde{G}_4 D^{-1} = \begin{pmatrix} 1 + \gamma_2 & 0 & \mu_2 \sqrt{d/g} e^{-i\theta_2/2} \\ 0 & g_4 \sqrt{d/g} e^{-i\theta_1/2} & 0 \\ \mu_2 & \tilde{f}\sqrt{d/g} e^{-i\theta_1/2} & \sqrt{d/g} e^{-i\theta_2/2} \end{pmatrix}.$$

Since the elements are bounded, $\|\tilde{G}_4 D^{-1}\|$ is bounded. Finally,

$$\tilde{G}_4^{-1} = (1 + \gamma_2 - \mu_2^2)^{-1} \begin{pmatrix} 1 & \mu_2 \tilde{f}/g_4 & -\mu_2 \\ 0 & (1 + \gamma_2 - \mu_2^2)/g_4 & 0 \\ -\mu_2 & -(1 + \gamma_2)\tilde{f}/g_4 & 1 + \gamma_2 \end{pmatrix}.$$

Since g_4 is bounded away from zero, and

$$1 + \gamma_2 - \mu_2^2 = 1 + \sigma_2^2 \sin^2 \theta_2/2 + \frac{i\tau\bar{V}}{2h_2}\sin \theta_2$$

is also bounded away from zero, the elements of $(\tilde{G}_4 D^{-1})^{-1}$ are bounded, so that $\|(\tilde{G}_4 D^{-1})^{-1}\|$ is bounded. Hence, we have shown unconditional stability. The scheme remains stable when the interior viscous term is deleted.

Exercise 8.3.1. Derive equation (8.78).

9. Scalar conservation laws

9.1 Introduction

In Chap. 8 we have introduced systems of hyperbolic differential equations by discussing the particular case of the shallow-water equations. Especially in the linearized case, these form a very simple example of an hyperbolic system, permitting an easy introduction of basic concepts as characteristics, wave fronts and proper boundary conditions. In the nonlinear case, new phenomena occur, notably shocks and other types of discontinuity. These occur when the local flow velocity is larger than the phase velocity of waves. The discretizations discussed in Chap. 8 break down in this case. In order to prepare for the treatment of compressible flows, we give an introduction to scalar conservation laws in the present chapter.

We start by introducing the concept of monotonicity preservation and Godunov's order barrier theorem. Since all later work circles around these matters, we discuss this material in detail. This enables us later to clearly see the relation between various properties, notably the concepts of total variation diminishing (TVD) and positive schemes. It will be seen that the TVD property is harder to obtain than monotonicity preservation, and that in fact TVD is not necessary, but monotonicity preservation is.

More extensive introductions to the vast subject of scalar conservation laws and hyperbolic systems of conservation laws may be found in Godlewski and Raviart (1996), Kröner (1997), Laney (1998), LeVeque (1992), Majda (1984), Smoller (1983) and Toro (1997).

9.2 Godunov's order barrier theorem

The simplest hyperbolic equation is the simplest partial differential equation that one can think of:

$$\frac{\partial \varphi}{\partial t} + c\frac{\partial \varphi}{\partial x} = 0 , \quad t > 0 , \quad x \in \mathbb{R} , \tag{9.1}$$

with $c =$ constant and φ a scalar function. Without loss of generality we assume $c > 0$. The general solution is

$$\varphi(t, x) = \varphi(x - ct) , \qquad (9.2)$$

showing that φ is convected with the flow with velocity c. The simple problem (9.1) deserves special attention, because in a slightly more general form it is an essential ingredient in the equations governing fluid dynamics. Oddly enough, despite its simplicity, numerical solution of (9.1) is notoriously difficult, and a wealth of literature has appeared on the subject. The big difficulty is how to circumvent Godunov's order barrier theorem, which we shall present shortly.

Monotonicity preservation

A function $f(s)$ is called *monotone* if $f(s)$ is either non-decreasing or non-increasing as s increases. It follows from the exact solution (9.2) that if $\varphi(0, x)$ is monotone, than so is $\varphi(t, x)$, $t > 0$. Therefore (9.1) is called *monotonicity preserving*.

As we saw in our study of the convection-diffusion equation in Chapters 4 and 5, numerical solutions may exhibit spurious wiggles, which are to be avoided in most applications; this is also the case for the convection equation (9.1). Occurrence of such wiggles is unlikely if the numerical scheme shares the property of monotonicity preservation with the differential equation. Denote the numerical approximation to $\varphi(t^n, x_j)$ by φ_j^n, as usual, with $\{x_j\}$ the nodes of some computational grid with $x_j > x_{j-1}$, and $\{t^n\}$, $n = 0, 1, ...$ an increasing sequence of instants. For the present discussion it suffices to consider an infinite domain, so $j = 0, \pm 1, ...$, and no boundary conditions are involved. The sequence $\{\varphi_j^n\} = \{..., \varphi_{j-1}^n, \varphi_j^n, \varphi_{j+1}^n, ...\}$ is called monotone if φ_j^n is either non-decreasing or non-increasing as j increases. The desired property can be formulated as follows.

Definition 9.2.1. *Monotonicity preserving scheme*
A numerical scheme is called monotonicity preserving if for every non-decreasing (non-increasing) initial condition $\{\varphi_j^0\}$ the numerical solution at all later instants $\{\varphi_j^n\}$, $n = 1, 2, ...$, is non-decreasing (non-increasing).

This is the nonstationary version of the notion 'scheme of positive type', introduced in Sect. 4.4 for time-independent schemes.

Godunov's order barrier theorem

We follow the original work of Godunov (1959). Let the computational grid be uniform: $x_j = jh$, and let the time step be constant: $t^n = n\tau$. Then all (explicit and implicit) one-step numerical schemes can be written as

$$\sum_m \beta_m \varphi_{j+m}^{n+1} = \sum_m \alpha_m \varphi_{j+m}^n . \qquad (9.3)$$

It is assumed that $\{\beta_m\}$ is such that $\{\varphi_j^{n+1}\}$ is uniquely determined. Since (9.3) is a linear mapping between $\{\varphi_j^n\}$ and $\{\varphi_j^{n+1}\}$, we can write

$$\varphi_j^{n+1} = \sum_m \gamma_m \varphi_{j+m}^n \cdot \tag{9.4}$$

There is no need to express γ_m in terms of α_m and β_m.

Theorem 9.2.1. *Monotonicity preservation*
The scheme (9.4) is monotonicity preserving if and only if

$$\gamma_m \geq 0 \,, \quad \forall m \,. \tag{9.5}$$

Proof. (Godunov (1959))
If suffices to consider the case $n = 0$, and the case of non-decreasing $\{\varphi_j^0\}$.

Sufficiency.   Assume $\{\varphi_j^0\}$ to be non-decreasing. Then

$$\varphi_j^1 - \varphi_{j-1}^1 = \sum_m \gamma_m (\varphi_{j+m}^0 - \varphi_{j+m-1}^0) \geq 0 \,, \quad \forall j \,,$$

where we have used (9.5).

Necessity.   Assume $\gamma_p < 0$ for some p. Choose

$$\varphi_j^0 = 0 \,, \quad j \leq 0 \,; \quad \varphi_j^0 = 1 \,, \quad j > 0 \,.$$

Then

$$\varphi_{1-p}^1 - \varphi_{-p}^1 = \gamma_p < 0 \,,$$

so that $\{\varphi_j^1\}$ is not non-decreasing, and we arrive at a contradiction.   □

Godunov (1959) defines second order accurate schemes to be those that reproduce the exact solution if $\varphi(0, x)$ is a polynomial of second degree, and goes on to prove:

Theorem 9.2.2. *Godunov's order barrier theorem*
Linear one-step second-order accurate numerical schemes for the convection equation (9.1) cannot be monotonicity preserving, unless $|c|\tau/h \in \mathbb{N}$.

Proof. (Godunov (1959))
The general form of one-step schemes is (9.3), which can be rewritten in the form (9.4). Choose

$$\varphi(0, x) = (\frac{x}{h} - \frac{1}{2})^2 - \frac{1}{4} \,, \quad \varphi_j^0 = (j - \frac{1}{2})^2 - \frac{1}{4} \,.$$

The exact solution is

$$\varphi(t, x) = (\frac{x - ct}{h} - \frac{1}{2})^2 - \frac{1}{4} . \tag{9.6}$$

Assuming the scheme to be second order accurate, it reproduces this solution exactly:

$$\varphi_j^1 = (j - \sigma - \frac{1}{2})^2 - \frac{1}{4} , \quad \varphi_j^0 = (j - \frac{1}{2})^2 - \frac{1}{4} ,$$

where $\sigma = c\tau/h$ is the signed CFL number. Substitution in (9.4) gives:

$$(j - \sigma - \frac{1}{2})^2 - \frac{1}{4} = \sum_m \gamma_m \{(j + m - \frac{1}{2})^2 - \frac{1}{4}\} . \tag{9.7}$$

Now suppose that the scheme is monotonicity preserving. Then $\gamma_m \geq 0$, according to Theorem 9.2.1. Noting that $(j + m - \frac{1}{2})^2 - \frac{1}{4} \geq 0$, it follows from (9.7) that

$$(j - \sigma - \frac{1}{2})^2 - \frac{1}{4} \geq 0 , \quad \forall j . \tag{9.8}$$

Assume $\sigma > 0$, $\sigma \notin \mathbb{N}$ and choose j such that $j > \sigma > j - 1$. Then

$$(j - \sigma - \frac{1}{2})^2 - \frac{1}{4} = (j - \sigma)(j - \sigma - 1) < 0 ,$$

which contradicts (9.8). □

An alternative proof of Godunov's order barrier theorem is indicated in Exercise 9.2.2.

Of course, we are not interested in the exceptional case where the CFL number $\sigma = |c|\tau/h \in \mathbb{N}$, since this cannot be realized with variable coefficients.

The local discretization error of scheme (9.4) is defined in the usual way as

$$\tau_j^{n+1} = \varepsilon_j^{n+1} - \sum_m \gamma_m \varepsilon_{j+m}^n , \quad \varepsilon_j^n = \varphi(n\tau, jh) - \varphi_j^n , \tag{9.9}$$

where ε_j^n is the global discretization error. It is left to the reader to show that

$$\begin{aligned}
\tau_j^{n+1} &= \varphi(1 - \sum_m \gamma_m) + (\tau\varphi_t - h\varphi_x \sum_m m\gamma_m) \\
&\quad + \frac{1}{2}(\tau^2 \varphi_{tt} - h^2 \varphi_{xx} \sum_m m^2 \gamma_m) + \mathcal{O}(\tau^3 + h^3) \\
&= \varphi(1 - \sum_m \gamma_m) - h\varphi_x(\sigma + \sum_m m\gamma_m) + \\
&\quad \frac{1}{2}h^2 \varphi_{xx}(\sigma^2 - \sum_m m^2 \gamma_m) + \mathcal{O}(\tau^3 + h^3) ,
\end{aligned} \tag{9.10}$$

where φ and its partial derivatives φ_t, φ_x etc. are evaluated at $n\tau, jh$. It is left to the reader to show that if the scheme reproduces (9.6) exactly, then

$$\tau_j^n = \mathcal{O}(\tau^3 + h^3) \,. \tag{9.11}$$

The use of multistep schemes does not get us past Godunov's order barrier. This we will now show. Linear multistep schemes for (9.1), whether implicit or explicit, can be written in the following form (cf. (9.4)):

$$\varphi_j^n = \sum_{s=1}^{S} \sum_m \gamma_{ms} \varphi_{j+m}^{n-s} \,. \tag{9.12}$$

This is an S-step scheme. It needs specification of initial conditions at S time levels: $\varphi^0, ..., \varphi^{S-1}$. The scheme is called monotonicity preserving if for every set of non-decreasing (non-increasing) set of initial conditions $\{\varphi_j^0\}, ... \{\varphi_j^{S-1}\}$ the numerical solution at all later instants $\{\varphi_j^n\}$, $n = S, S+1, ...$, is non-decreasing (non-increasing).

Theorem 9.2.3. *Monotonicity preservation for multistep schemes*
The multistep scheme (9.12) is monotonicity preserving if and only if

$$\gamma_{ms} \geq 0 \,, \quad s = 1, ..., S \,, \quad \forall m \,.$$

Proof. Sufficiency follows as in the proof of Theorem 9.2.1. For necessity, take the non-decreasing case, and assume $\gamma_{pr} < 0$ for some p and r. Choose $\varphi_j^s = 0$, $s = 1, ..., S$, $\forall j$, except for $s = S - r$:

$$\varphi_j^{S-r} = 0 \,, \quad j \leq 0; \quad \varphi_j^{S-r} = 1 \,, \quad j > 0 \,.$$

Then

$$\varphi_{1-p}^S - \varphi_{-p}^S = \gamma_p^r < 0 \,,$$

which is a contradiction. $\qquad\qquad\square$

Theorem 9.2.4. *Order barrier theorem for multistep schemes*
Linear second-order accurate numerical schemes for the convection equation (9.1) cannot be monotonicity preserving, unless $|c|\tau/h \in \mathbb{N}$.

Proof. Second order accuracy requires that (9.11) holds, with the local truncation error defined as (cf. (9.9)):

$$\tau_j^n = \varepsilon_j^n - \sum_{s=1}^{S} \sum_m \gamma_{ms} \varepsilon_{j+m}^{n-s} \,.$$

It is left to the reader to show that this results in the following conditions:

$$\sum_{s=1}^{S}\sum_{m}\gamma_{ms} = 1\,,\quad \sum_{s=1}^{S}\sum_{m}(m+\sigma s)\gamma_{ms} = 0\,,\quad \sum_{s=1}^{S}\sum_{m}(m+\sigma s)^2\gamma_{ms} = 0\,.$$

$$(9.13)$$

It follows that

$$\sum_{s=1}^{S}\sum_{m}\{m+\sigma(s-1)\}\gamma_{ms} = -\sigma\,,\quad \sum_{s=1}^{S}\sum_{m}\{m+\sigma(s-1)\}^2\gamma_{ms} = \sigma^2\,.$$

$$(9.14)$$

Assuming the scheme to be monotonicity preserving, we have $\gamma_{ms} \geq 0$ according to Theorem 9.2.3, and we can write $\gamma_{ms} = \mu_{ms}^2$. From (9.13) and (9.14) it follows that

$$\{\sum_{s=1}^{S}\sum_{m}\mu_{ms}\eta_{ms}\}^2 = \{\sum_{s=1}^{S}\sum_{m}\mu_{ms}^2\}\{\sum_{s=1}^{S}\sum_{m}\eta_{ms}^2\}\,,\qquad (9.15)$$

with $\eta_{ms} = \{m+\sigma(s-1)\}\mu_{ms}$. The Cauchy-Schwarz inequality says that

$$(\boldsymbol{a}\cdot\boldsymbol{b})^2 \leq \|\boldsymbol{a}\|^2\|\boldsymbol{b}\|^2\,,\qquad (9.16)$$

with equality only if $\boldsymbol{a} = 0$ or $\boldsymbol{b} = 0$ or $\boldsymbol{a}$ and $\boldsymbol{b}$ parallel. Comparing this with (9.15) we see that the case of $\boldsymbol{a} = 0$ or $\boldsymbol{b} = 0$ cannot occur, because some $\mu_{ms} > 0$ and $m+\sigma(s-1) \neq 0$ in general if $|\sigma| \notin \mathbb{N}$. The case of $\boldsymbol{a}$ and $\boldsymbol{b}$ parallel applies only if $\gamma_{ms} > 0$ or if $m+\sigma(s-1) = $ constant. Starting with $s = 1$, we see that the constant must be an integer. Because $|\sigma| \notin \mathbb{N}$, no practical scheme is obtained. This means that inequality applies in (9.16), and we have a contradiction with (9.15).   □

The way to circumvent Godunov's order barrier is to use nonlinear schemes, as in Sect. 4.8. But first we will continue our discussion of linear schemes.

Properties equivalent to monotonicity preservation

The *total variation* $\mathrm{TV}(\varphi(t,\cdot))$ of the solution of (9.1) is defined as

$$\mathrm{TV}(\varphi(t,\cdot)) = \lim_{\varepsilon\to 0}\sup \frac{1}{\varepsilon}\int_{-\infty}^{\infty}|\varphi(t,x) - \varphi(t,x-\varepsilon)|dx\,.\qquad (9.17)$$

If the total variation of φ does not increase in time, φ is called *total variation diminishing* or TVD. It follows from the exact solution (9.2) that, writing $x' = x - ct$

$$\mathrm{TV}(\varphi(t,\cdot)) = \lim_{\varepsilon \to 0} \sup \frac{1}{\varepsilon} \int_{-\infty}^{\infty} |\varphi(-x') - \varphi(-x' - \varepsilon)| dx' \qquad (9.18)$$

is independent of t, so φ is TVD. It seems desirable that the scheme (9.3) mimicks this property. The total variation of the numerical solution is defined as

$$\mathrm{TV}(\varphi^n) = \sum_{j=-\infty}^{\infty} |\varphi_j^n - \varphi_{j-1}^n|, \qquad (9.19)$$

where we assume that the sum exists. A scheme is called TVD if $\mathrm{TV}(\varphi^n)$ does not increase with n, for all $\{\varphi_j^0\}$.

Theorem 9.2.5. *TVD property*
The scheme (9.4) is TVD if and only if

$$\gamma_m \geq 0, \quad \forall m, \quad \text{and} \quad \sum_m \gamma_m \leq 1. \qquad (9.20)$$

Proof. It suffices to consider the case $n = 0$.

Sufficiency. We have

$$\mathrm{TV}(\varphi^1) = \sum_j |\sum_m \gamma_m (\varphi_{j+m}^0 - \varphi_{j+m-1}^0)| \leq \sum_j \sum_m |\gamma_m| \, |\varphi_{j+m}^0 - \varphi_{j+m-1}^0|$$

$$= \sum_m \gamma_m \sum_{k=-\infty}^{\infty} |\varphi_k^0 - \varphi_{k-1}^0| \leq \mathrm{TV}(\varphi^0).$$

Necessity.
Let $\{\varphi_j^0\}$ be monotone. Assume $\gamma_m = 0$ for $|m| > M$. Choose

$$\begin{aligned}
\varphi_j^0 &= \varphi_- = \text{constant}, \quad j < -3M, \\
\varphi_j^0 &= \text{monotone}, \quad -3M \leq j \leq 3M, \qquad (9.21) \\
\varphi_j^0 &= \varphi_+ = \text{constant}, \quad j > 3M.
\end{aligned}$$

Assume that the resulting numerical solution $\{\varphi_j^1\}$ is not monotone. Then there is at least one local maximum φ_k^1 and minimum φ_m^1, and

$$\mathrm{TV}(\varphi^1) \geq |\varphi_+ - \varphi_-| + |\varphi_k^1 - \varphi_m^1| > |\varphi_+ - \varphi_-| = \mathrm{TV}(\varphi^0),$$

which is a contradiction. Hence, $\{\varphi_j^1\}$ is monotone. Because the support of the non-constant part of φ^0 is larger than that of $\{\gamma_m\}$, the initial condition (9.21) is sufficiently general to invoke Theorem 9.2.1, which proves the necessity of the first part of (9.20). Continuing with the second part, assume $\gamma_m \geq 0$

and $\sum_m \gamma_m > 1$. Then the scheme is monotonicity preserving. Choose φ^0 non-decreasing. Then φ^1 is non-decreasing, and we may write

$$
\begin{aligned}
\mathrm{TV}(\varphi^1) &= \sum_j (\varphi_j^1 - \varphi_{j-1}^1) = \sum_j \sum_m \gamma_m (\varphi_{j+m}^0 - \varphi_{j+m-1}^0) \\
&= \sum_m \gamma_m \sum_{k=-\infty}^{\infty} (\varphi_k^0 - \varphi_{k-1}^0) = \sum_m \gamma_m \mathrm{TV}(\varphi^0) > \mathrm{TV}(\varphi^0) \,,
\end{aligned}
$$

so that we have arrived at a contradiction, showing necessity of the second part of (9.20). □

We see that a TVD scheme is monotonicity preserving, but not vice-versa. The difference lies in the second condition of (9.20): $\sum_m \gamma_m \le 1$. Since consistency requires this condition anyway (cf. (9.10)), for practical purposes TVD and monotonicity preservation are equivalent in one dimension. But in more dimensions the gap between the two properties is wider, as we shall see. The concept of TVD solutions and TVD numerical schemes plays an important role in the theory of hyperbolic conservation laws.

Exercise 9.2.1. Prove (9.10).

Exercise 9.2.2. *Alternative proof of Godunov's order barrier theorem* (Roe (1986)).
Show that if the scheme (9.4) is exact when the exact solution is a second order polynomial, then τ_j^{n+1} as defined by (9.9) satisfies $\tau_j^{n+1} = \mathcal{O}(\tau^3 + h^3)$. Show that in that case (using (9.10)) and if the scheme is also monotonicity preserving, then

$$
\sum_m (\mu_m)^2 \sum_m (m\mu_m)^2 = \{ \sum_m m\mu_m^2 \}^2 \,,
$$

where $\mu_m = \sqrt{\gamma_m}$. Show that this violates the Cauchy-Schwarz inequality. Hint: if $a, b \in \mathbb{R}^p$ then $|a \cdot b| \le \|a\| \cdot \|b\|$, with equality only if a and b are parallel, or $a = 0$ or $b = 0$.

9.3 Linear schemes

Over the years, a great many schemes have been proposed, some of which are better than others, of course. It turns out that many schemes that have survived until today can be interpreted as members of a one-parameter family of schemes, namely the κ-scheme, introduced by van Leer (1977). We already encountered the κ-scheme in Sect. 4.8

A classical way to discretize in time is due to Lax and Wendroff (1960), as follows. Using (9.1), we have

$$
\begin{aligned}
\varphi(t + \tau) &= \varphi(t, x) + \tau\varphi_t(t, x) + \frac{1}{2}\tau^2\varphi_{tt}(t, x) + \mathcal{O}(\tau^3) \\
&= \varphi(t, x) - c\tau\varphi_x(t, x) + \frac{1}{2}c^2\tau^2\varphi_{xx}(t, x) + \mathcal{O}(\tau^3) ,
\end{aligned}
\tag{9.22}
$$

where $\varphi_t = \partial\varphi/\partial t$, $\varphi_x = \partial\varphi/\partial x$ etc. As in the preceding section, c is assumed constant. Approximation of φ_x with the κ-scheme of Sect. 4.8 and φ_{xx} with a weighted average of 3-point differences centered at x_j and x_{j-1} for $c \geq 0$ and at x_j and x_{j+1} for $c < 0$ results in, using a uniform spatial grid with step-size h,

$$
\begin{aligned}
\varphi_j^{n+1} &= \varphi_j^n - \frac{1}{4}\sigma\{(1 - \kappa)\varphi_{j-2}^n + (3\kappa - 5)\varphi_{j-1}^n + (3 - 3\kappa)\varphi_j^n + (1 + \kappa)\varphi_{j+1}^n\} \\
&\quad + \frac{1}{2}\sigma^2\{\mu(\varphi_{j-1}^n - 2\varphi_j^n + \varphi_{j+1}^n) + (1 - \mu)(\varphi_{j-2}^n - 2\varphi_{j-1}^n + \varphi_j^n)\} , \quad \sigma \geq 0 ,
\end{aligned}
\tag{9.23}
$$

$$
\begin{aligned}
\varphi_j^{n+1} &= \varphi_j^n + \frac{1}{4}\sigma\{(1 + \kappa)\varphi_{j-1}^n + (3 - 3\kappa)\varphi_j^n + (3\kappa - 5)\varphi_{j+1}^n + (1 - \kappa)\varphi_{j+2}^n\} \\
&\quad + \frac{1}{2}\sigma^2\{\mu(\varphi_{j-1}^n - 2\varphi_j^n + \varphi_{j+1}^n) + (1 - \mu)(\varphi_j^n - 2\varphi_{j+1}^n + \varphi_{j+2}^n)\} , \quad \sigma < 0 ,
\end{aligned}
$$

where $\sigma = c\tau/h$ is the signed CFL number. Scheme (9.23) can be rewritten as

$$
\begin{aligned}
\varphi_j^{n+1} &= \varphi_j^n + \frac{\sigma}{4}(\kappa - 2\mu\sigma - 1 + 2\sigma)\varphi_{j-2}^n \\
&\quad + \frac{\sigma}{4}(-3\kappa + 6\mu\sigma + 5 - 4\sigma)\varphi_{j-1}^n + \frac{\sigma}{4}(3\kappa - 6\mu\sigma - 3 + 2\sigma)\varphi_j^n \\
&\quad + \frac{\sigma}{4}(-\kappa + 2\mu\sigma - 1)\varphi_{j+1}^n , \quad \sigma \geq 0 , \\
\varphi_j^{n+1} &= \varphi_j^n + \frac{\sigma}{4}(\kappa + 2\mu\sigma + 1)\varphi_{j-1}^n + \frac{\sigma}{4}(-3\kappa - 6\mu\sigma + 3 + 2\sigma)\varphi_j^n \\
&\quad + \frac{\sigma}{4}(3\kappa + 6\mu\sigma - 5 - 4\sigma)\varphi_{j+1}^n + \frac{\sigma}{4}(-\kappa - 2\mu\sigma + 1 + 2\sigma)\varphi_{j+2}^n , \quad \sigma < 0 .
\end{aligned}
\tag{9.24}
$$

We see that effectively there is only one free parameter, namely $\kappa - 2\mu|\sigma|$. Defining

$$
\mu = \frac{1}{2} + \frac{3}{2}\kappa - \kappa|\sigma| ,
\tag{9.25}
$$

the κ-scheme as formulated by van Leer (1977) is recovered:

$$\varphi_j^{n+1} = \varphi_j^n + \frac{\sigma}{4}(-1 + \kappa + (1 - 3\kappa)\sigma + 2\kappa\sigma^2)\varphi_{j-2}^n$$
$$+ \frac{\sigma}{4}(5 - 3\kappa + (9\kappa - 1)\sigma - 6\kappa\sigma^2)\varphi_{j-1}^n$$
$$+ \frac{\sigma}{4}(-3 + 3\kappa - (1 + 9\kappa)\sigma + 6\kappa\sigma^2)\varphi_j^n$$
$$+ \frac{\sigma}{4}(-1 - \kappa + (1 + 3\kappa)\sigma - 2\kappa\sigma^2)\varphi_{j+1}^n, \quad \sigma \geq 0,$$
$$\varphi_j^{n+1} = \varphi_j^n + \frac{\sigma}{4}(1 + \kappa + (1 + 3\kappa)\sigma + 2\kappa\sigma^2)\varphi_{j-1}^n$$
$$+ \frac{\sigma}{4}(3 - 3\kappa - (1 + 9\kappa)\sigma - 6\kappa\sigma^2)\varphi_j^n$$
$$+ \frac{\sigma}{4}(-5 + 3\kappa - (1 - 9\kappa)\sigma + 6\kappa\sigma^2)\varphi_{j+1}^n$$
$$+ \frac{\sigma}{4}(1 - \kappa + (1 - 3\kappa)\sigma - 2\kappa\sigma^2)\varphi_{j+2}^n, \quad \sigma < 0. \tag{9.26}$$

By Taylor expansion and by using the differential equation to replace time derivatives by space derivatives we find for the local truncation error (dividing (9.26) by τ to make the scheme approximate the differential equation):

$$\tau_j^n = \frac{1}{12}ch^2(3\kappa - 1)(1 - 3\sigma + 2\sigma^2)\frac{\partial^3\varphi(n\tau, jh)}{\partial x^3}$$
$$+ \frac{1}{48}ch^3\{7\sigma(1 - \kappa) + (21\kappa - 8)\sigma^2 - 14\kappa\sigma^3 + \sigma^4\}\frac{\partial^4\varphi(n\tau, jh)}{\partial x^4} + \mathcal{O}(h^4), \tag{9.27}$$

which shows that the κ-scheme has third order accuracy for $\kappa = 1/3$ and second order accuracy otherwise.

The classical scheme of Lax and Wendroff (1960) is obtained by substituting $\mu = \kappa = 1$ in (9.24) (hence (9.25) is not valid). We obtain, both for positive and negative σ:

$$\varphi_j^{n+1} = \frac{1}{2}(\sigma^2 + \sigma)\varphi_{j-1}^n + (1 - \sigma^2)\varphi_j^n + \frac{1}{2}(\sigma^2 - \sigma)\varphi_{j+1}^n. \tag{9.28}$$

By Taylor expansion we find for the local truncation error:

$$\tau_j^n = \frac{1}{6}ch^2(1 - \sigma^2)\frac{\partial^3\varphi(n\tau, jh)}{\partial x^3} + \mathcal{O}(h^3). \tag{9.29}$$

For $\kappa = 1/3$, equation (9.26) gives the QUICKEST scheme proposed by Leonard (1979). As noted before, this value of κ gives the smallest truncation error. Note that the QUICK scheme, proposed in Leonard (1979) for the stationary case, corresponds to $\kappa = 1/2$.

For $\kappa = 0$, (9.26) gives the scheme proposed by Fromm (1968). It is designed to have a small phase error. In Wesseling (1973) it is shown that among all linear schemes that share the stencil of scheme (9.24), Fromm's scheme is optimal when the exact solution is a propagating step-function.

The fully one-sided upwind scheme proposed by Warming and Beam (1976) is obtained from (9.26) by choosing $\kappa - 2\mu\sigma = -1$.

With the Lax-Wendroff approach, temporal discretization takes place by replacing the time derivatives in (9.22) by space derivatives, using the differential equation (9.1). When a source term is present, or even more so when there are other differential terms, such as a diffusion term, this procedure becomes less straightforward. Another disadvantage of this way to do temporal discretization occurs when the sole aim of the computation is to compute a steady state. When steady state is reached, $\varphi_j^{n+1} = \varphi_j^n$, and equation (9.24) shows that the final solution depends on the time step τ through σ. This is unsatisfactory. These disadvantages are avoided by discretizing first only in space, as in Chap. 5. Using the κ-scheme, this results in the following system of ordinary differential equations:

$$c \geq 0:$$
$$d\varphi_j/dt = -\tfrac{c}{4h}\{(1-\kappa)\varphi_{j-2} + (3\kappa - 5)\varphi_{j-1} + (3 - 3\kappa)\varphi_j + (1+\kappa)\varphi_{j+1}\},$$

$$c < 0:$$
$$d\varphi_j/dt = \tfrac{c}{4h}\{(1+\kappa)\varphi_{j-1} + (3 - 3\kappa)\varphi_j + (3\kappa - 5)\varphi_{j+1} + (1-\kappa)\varphi_{j+2}\}.$$
$$(9.30)$$

Next, a discretization method for ordinary differential equations is applied. Examples have been given in Chap. 5.

Further schemes that should be mentioned are:

the first order upwind scheme:

$$d\varphi_j/dt = -\frac{1}{2h}\{-(c + |c|)\varphi_{j-1} + 2|c|\varphi_j - (c - |c|)\varphi_{j+1}\}, \qquad (9.31)$$

the fourth order central scheme (cf. Sect. 4.8):

$$d\varphi_j/dt = -\frac{1}{12h}\{\varphi_{j-2} - 8\varphi_{j-1} + 8\varphi_{j+1} - \varphi_{j+2}\},$$

and the Lax-Friedrichs scheme (Lax (1954)):

$$\varphi_j^{n+1} = \frac{1}{2}(\varphi_{j-1}^n + \varphi_{j+1}^n) - \frac{\sigma}{2}(\varphi_{j+1}^n - \varphi_{j-1}^n). \qquad (9.32)$$

With forward Euler temporal discretization, (9.31) goes over in the Courant-Isaacson-Rees scheme (Courant, Isaacson, and Rees (1952)):

$$\varphi_j^{n+1} = \varphi_j^n - \frac{1}{2}\{-(\sigma + |\sigma|)\varphi_{j-1}^n + 2|\sigma|\varphi_j^n - (\sigma - |\sigma|)\varphi_{j+1}^n\}. \qquad (9.33)$$

These schemes do not belong to the class of κ-schemes. The first order upwind and Lax-Friedrichs schemes are first order accurate. It is easy to see that the Lax-Friedrichs and Courant-Isaacson-Rees schemes are monotonicity preserving for $|\sigma| \leq 1$. This is also the case for the first order upwind scheme, for suitable temporal discretizations.

Boundary conditions

Consider the same equation as before, but now on a bounded domain:

$$\frac{\partial\varphi}{\partial t} + c\frac{\partial\varphi}{\partial x} = 0 , \quad T > 0 , \quad x \in (0,1) , \quad c = \text{constant} . \tag{9.34}$$

Recalling the discussion of characteristics and boundary conditions in Sect. 8.2, we see that equation (9.34) has characteristics parametrized by

$$\frac{dt}{dx} = 1 , \quad \frac{dx}{ds} = c . \tag{9.35}$$

Assuming that an initial condition is given for $\varphi(0, x)$, a boundary condition is required for $\varphi(t, 0)$ if $c > 0$ and for $\varphi(t, 1)$ if $c < 0$. The same is true for the first order upwind scheme. However, all other schemes discussed before need additional artificial boundary conditions, because the stencil of the spatial discretization contains three or more points. Assume $c > 0$, and let the grid points be $x_j = jh$, $j = 0, 1, ..., J$, $h = 1/J$. In (9.30), $\varphi_j(0)$ is given by the initial condition, and $\varphi_0(t)$ is specified by the boundary condition. In order to complete the equation for $d\varphi_1/dt$, we need an artificial boundary condition to specify $\varphi_{-1}(t)$. Similarly, to complete the equation for $d\varphi_J/dt$, an artificial boundary condition is needed for $\varphi_{J+1}(t)$. In the theoretical literature, the most common way to generate artificial boundary conditions is to extrapolate. For example, linear extrapolation gives

$$\varphi_{-1}(t) = 2\varphi_0(t) - \varphi_1(t) , \quad \varphi_{J+1}(t) = 2\varphi_J(t) - \varphi_{J-1}(t) . \tag{9.36}$$

Another option is to avoid the use of artificial boundary conditions by using the first order upwind scheme for $j = 1$, and a fully one-sided upwind scheme (first order, or $\kappa = -1$ or the Beam-Warming scheme) at $j = J$. In the present simple case, we found no significant differences between these approaches in numerical experiments.

Gustafsson (1975) has shown that the order of accuracy of the artificial boundary conditions can be one order lower than for the interior scheme, without impairing the overall order of accuracy. With (9.36), scheme (9.30) at the boundary becomes $(c > 0)$:

$$d\varphi_1/dt = -\frac{c}{4h}\{(\kappa - 3)\varphi_0 + (2 - 2\kappa)\varphi_1 + (1 + \kappa)\varphi_2\} .$$

It is left to the reader to show that

$$d\varphi_1/dt = -c\partial\varphi(t,0)/\partial x + \mathcal{O}(h^2)$$

so that the artificial boundary condition (9.36) is sufficiently accurate, even for $\kappa = 1/3$, in which case the discretization error in (9.30) is $\mathcal{O}(h^3)$ in the interior. It is of further interest to note that Kreiss and Lundqvist (1968) have shown that if the numerical scheme is dissipative (a concept that will be discussed shortly), then the influence of artificial boundary conditions is limited to an interval of length $\mathcal{O}(h|\ln h|)$.

As we saw in Sect. 5.9, artificial boundary conditions at an outflow boundary may give rise to spurious reflection. However, in the present simple case spurious reflection did not show up in the numerical experiments to be described later. Spurious reflection becomes an issue when other terms are added (e.g. diffusion), or when systems of equations are considered.

Stability

Stability analysis for the mixed initial-boundary value value problem (9.34) is presented by Richtmyer and Morton (1967) and Gustafsson, Kreiss, and Sundström (1972). This turns out to be a rather complicated affair, that we will not discuss. It is necessary that the scheme is stable for the pure initial value problem:

$$\frac{\partial\varphi}{\partial t} + c\frac{\partial\varphi}{\partial x} = 0 , \quad t > 0 , \quad -\infty < x < \infty .$$

Stability for the pure initial value problem can be analysed (in the case of constant coefficients) with the Fourier analysis method introduced in Sect. 5.5, and we will do this here. In Sect. 5.5, the domain is finite with periodic boundary conditions, and grid functions can be represented by the finite Fourier series (5.20). In the present case of a pure initial value problem, we have an infinite uniform grid, and grid functions can be represented by a Fourier integral:

Theorem 9.3.1. *Fourier integral*
If $\{\varphi_j : \varphi_j \in \mathbb{R}, j = 0, \pm 1, \pm 2, ...\}$ *satisfies*

$$\sum_{j=-\infty}^{\infty} \varphi_j^2 < \infty ,$$

then there exists a function $\hat{\varphi}(\theta) : (-\pi, \pi) \to \mathbb{R}$ *such that*

$$\varphi_j = \int_{-\pi}^{\pi} \hat{\varphi}(\theta)e^{ij\theta}\,d\theta , \quad \hat{\varphi}(\theta) = \frac{1}{2\pi}\sum_{j=-\infty}^{\infty}\varphi_j e^{-ij\theta} . \tag{9.37}$$

Proof. See Sect. 9.2 of Dahlquist and Björck (1974). □

As in Sect. 5.6, we study the behavior of a perturbation $\delta_j(t)$ caused by a perturbation $\delta_j(0)$ of the initial condition. Substitution in (9.30) gives

$$d\delta_j/dt + L_h\delta_j = 0 \,,$$

where L_h follows from comparison with (9.30). Substitution of (9.37) gives

$$\int_{-\pi}^{\pi} \{d\hat{\delta}(t,\theta)/dt + \hat{L}_h(\theta)\hat{\delta}(t,\theta)\}e^{ij\theta}d\theta = 0 \,, \tag{9.38}$$

where, as in Sect. 5.5, $\hat{L}_h$ is the symbol of L_h, defined by

$$\hat{L}_h(\theta) = e^{-ij\theta}L_h e^{ij\theta} \,.$$

From (9.38) it follows that

$$d\hat{\delta}(t,\theta)/dt + \hat{L}_h(\theta)\hat{\delta}(t,\theta) = 0 \,,$$

and, similar to Sect. 5.6, we have stability, i.e. $|\hat{\delta}(t,\theta)| \le |\hat{\delta}(0,\theta)|$, provided

$$-\hat{L}_h(\theta) \in S, \quad -\pi < \theta \le \pi \,,$$

with S the stability domain of the temporal discretization method used to approximate (9.30). The symbol $\hat{L}_h(\theta)$ follows from (5.34) by taking the one-dimensional case, replacing c_1 by σ and putting the viscosity $D = 0$, or $d_1 = 0$. The same substitutions transform the stability conditions derived for a number of temporal discretization methods in Sect. 5.7 and 5.8 into stability conditions for the initial value problem (9.30).

In a similar way, stability conditions can be derived for the Lax-Wendroff temporal discretizations for the κ-scheme (9.24) or (9.26).

We substitute

$$\delta_j^n = \int_{-\pi}^{\pi} \hat{\delta}^n(\theta)e^{ij\theta}d\theta \,,$$

and find

$$\hat{\delta}^{n+1}(\theta) = g(\theta)\hat{\delta}^n(\theta) \,.$$

We have stability if

$$|g(\theta)| \le 1, \quad \forall \theta \,. \tag{9.39}$$

Theorem 9.3.2. *Stability of Lax-Wendroff scheme*
A necessary and sufficient stability condition for the Lax-Wendroff scheme is

$$|\sigma| \leq 1 . \tag{9.40}$$

Proof. The Lax-Wendroff scheme is given by (9.28). We find for all σ:

$$g(\theta) = 1 - \sigma^2(1 - \cos\theta) - i\sigma\sin\theta . \tag{9.41}$$

By taking $\theta = \pi$ we see that (9.40) is necessary for (9.39). Furthermore, writing $s = \sin^2 \frac{1}{2}\theta$, we have

$$|g|^2 = 1 + 4s^2\sigma^2(\sigma^2 - 1) \leq 1 \tag{9.42}$$

where we have used (9.40). Hence, (9.40) is sufficient for (9.39). $\square$

Theorem 9.3.3. *Stability of κ-scheme*
A necessary and sufficient condition for the κ-scheme (9.26) to be stable for $-1 \leq \kappa \leq 1$ is

$$|\sigma| \leq 1 .$$

Proof. For $\sigma \geq 0$ the amplification factor is given by

$$\begin{aligned} g(\theta) &= 1 - 2\sigma^2 s + 2as^2 + i(as - \sigma)\sin\theta , \quad s = \sin^2 \tfrac{1}{2}\theta , \\ a &= \sigma(\sigma - 1)(1 - \kappa + 2\kappa\sigma) . \end{aligned} \tag{9.43}$$

We find

$$|g|^2 = 1 + 4s^2(a - 2a\sigma - \sigma^2 + \sigma^4) + 4as^3(a + 2\sigma - 2\sigma^2) . \tag{9.44}$$

Since $0 \leq s \leq 1$, $|g|^2 \leq 1$ for all θ is equivalent with

$$a - 2a\sigma - \sigma^2 + \sigma^4 + as(a + 2\sigma - \sigma^2) \leq 0 , \quad 0 \leq s \leq 1 ,$$

which is equivalent with

$$a - 2a\sigma - \sigma^2 + \sigma^4 \leq 0 \quad \text{and} \quad a(a + 2\sigma - \sigma^2) \leq 0 . \tag{9.45}$$

We start with sufficiency. We have

$$a - 2a\sigma - \sigma^2 + \sigma^4 = \sigma(\sigma - 1)\{1 - \sigma + \sigma^2 - \kappa(1 - 2\sigma)^2\} . \tag{9.46}$$

From $-1 \leq \kappa \leq 1$ and $0 \leq \sigma \leq 1$ it follows that

$$1 - \sigma + \sigma^2 - \kappa(1 - 2\sigma)^2 \geq 1 - \sigma + \sigma^2 - (1 - 2\sigma)^2 = 3(\sigma - \sigma^2) \geq 0 , \tag{9.47}$$

establishing the first part of (9.45). Continuing with the second part, we have $a \leq 0$, since for $-1 \leq \kappa \leq 1$ the extrema of $1 - \kappa + 2\kappa\sigma$ for fixed σ are attained for $\kappa = \pm 1$, and both are non-negative. It remains to show that in (9.45) we have $a + 2\sigma - \sigma^2 \geq 0$. We find, using $0 \leq \sigma \leq 1$,

$$a + 2\sigma - \sigma^2 = \sigma(\sigma - 1)(2\kappa\sigma - 1 - \kappa) \geq 0 ,$$

since the extrema of the last factor for σ fixed and $-1 \leq \kappa \leq 1$ are attained for $\kappa = \pm 1$, and both are non-positive.

Necessity is shown as follows. From (9.45) it follows that is necessary that

$$a - 2a\sigma - \sigma^2 + \sigma^4 \leq 0 .$$

We have

$$a - 2a\sigma - \sigma^2 + \sigma^4 = \sigma(\sigma - 1)\{1 - \kappa + (4\kappa - 1)\sigma + \sigma^2(1 - \kappa)\} .$$

The term between braces is positive for $-1 \leq \kappa \leq 1/4$, $\forall \sigma$ and changes sign for $\sigma > 0$ at

$$\sigma_1 = \frac{1}{2}(1 + \sqrt{3/(4\kappa - 1)}) , \quad \kappa > 1/4 .$$

Since $\sigma_1 > 1$ for $\kappa < 1$ the scheme is unstable for $1 < \sigma < \sigma_1$. It might be that the scheme is stable for $\sigma > \sigma_1$, but this does not need to be investigated, because a scheme that becomes instable if the time step τ (and hence σ) is decreased is impractical. For the case $\kappa = 1$ we look at the second part of (9.45). We find

$$a(a + 2\sigma - \sigma^2) = 2\sigma^3(\sigma - 1)(2\sigma^2 - 3\sigma + 2) ,$$

which is of wrong sign for $\sigma > 1$. This completes the proof for $\sigma \geq 0$. The case $\sigma < 0$ is left to the reader.   □

Theorem 9.3.4. *Stability of Lax-Friedrichs scheme*
A necessary and sufficient condition for the Lax-Friedrichs scheme (9.32) to be stable is

$$|\sigma| \leq 1 .$$

The proof is left to the reader.

Theorem 9.3.5. *Stability of Courant-Isaacson-Rees scheme*
A necessary and sufficient condition for the Courant-Isaacson-Rees scheme (9.33) to be stable is

$$|\sigma| \leq 1 .$$

The proof is left to the reader.

We can now state what a dissipative scheme is, a concept that was referred to earlier. Kreiss (1964) gives the following definition:

Definition 9.3.1. *Dissipative scheme*
A discretization scheme is dissipative of order $2r$, r a non-negative integer, if there exists a constant $\delta > 0$ such that

$$|g(\theta)| \leq 1 - \delta\theta^{2r} , \quad -\pi \leq \theta \leq \pi . \tag{9.48}$$

We show that the Lax-Wendroff and κ-schemes are dissipative. For the Lax-Wendroff scheme we have, from (9.42):

$$|g|^2 \leq 1 - \frac{1}{4}\theta^4\sigma^2(1 - \sigma^2) . \tag{9.49}$$

Since, obviously, we want τ to be greater than zero, it is reasonable to assume $\sigma^2 \geq \varepsilon > 0$. If also $\sigma^2 \leq 1 - \varepsilon$ then (9.49) gives

$$|g|^2 \leq 1 - \delta\theta^4 , \quad \delta = \frac{1}{4}\varepsilon(1 - \varepsilon), \quad \text{if} \quad \varepsilon \leq \sigma^2 \leq 1 - \varepsilon ,$$

so that the Lax-Wendroff scheme is dissipative of order 4. For the κ-scheme, (9.44) gives, assuming $0 \leq \sigma \leq 1$,

$$|g|^2 \leq 1 + 4s^2(a - 2a\sigma - \sigma^2 + \sigma^4) ,$$

since $a(a + 2\sigma - 2\sigma^2) \leq 0$, as shown in the proof of Theorem 9.3.3. From (9.46) and (9.47) it follows that

$$|g|^2 \leq 1 - \frac{3}{4}\theta^4\sigma^2(1 - \sigma^2) ,$$

hence

$$|g|^2 \leq 1 - \delta\theta^4 , \quad \delta = \frac{3}{4}\varepsilon(1 - \varepsilon), \quad \text{if} \quad \varepsilon \leq \sigma^2 \leq 1 - \varepsilon ,$$

so that the κ-scheme is also dissipative of order 4.

An advantage of dissipative schemes has been mentioned before, namely that the influence of artificial inflow boundary conditions is limited to a layer of thickness $\mathcal{O}(h|\ln h|)$. Another advantage is discussed by Richtmyer and Morton (1967) Sect. 5.4, namely that we have stability in the variable coefficients case.

It is left to the reader to show that with $\sigma = 1$ the Lax-Wendroff and κ-schemes reproduce the exact solution. But for $\sigma = 1$ the schemes are not dissipative. Nevertheless, artificial inflow boundary conditions have no influence. In fact, it turns out that a weaker condition than dissipativity is sufficient to restrain the influence of artificial boundary conditions; Kreiss and Lundqvist (1968) show that it is sufficient if the scheme is *contractive*:

$$|g(\theta)| \leq 1 , \quad |g(\theta + i\alpha)| \leq e^{-\beta} , \quad 0 \leq |\theta| \leq \pi ,$$

for some $\alpha > 0$ and a constant $\beta > 0$. For general σ, to prove contractivity is more difficult than stability, but the case $|\sigma| = 1$ is easy and is left to the reader.

Numerical dissipation and dispersion

The accuracy with which a numerical scheme approximates the propagation properties of wave-like solutions of hyperbolic equations depends on the dissipation and dispersion of the scheme, concepts that we encountered earlier, namely in Sect. 8.2.

Postulating a harmonic wave $\varphi = \hat{\varphi} \exp\{i(kx - \omega t)\}$ for the solution of (9.1) we find the following dispersion relation:

$$\omega = ck .$$

We recall from Sect. 8.2 that the phase velocity and group velocities are defined by

$$c_p = \mathrm{Re}(\omega)/k , \quad c_g = d\mathrm{Re}(\omega)/dk .$$

Hence, in the present case we have

$$c_p = c_g = c .$$

so that there is no dispersion: c_p does not depend on k. Assuming $c > 0$, substitution of a harmonic wave $\varphi_j^n = \hat{\varphi} \exp\{i(jkh - \omega_h t)\}$ in a difference scheme on a uniform grid with step-size h gives a relation of the form

$$\exp(-i\omega_h \tau) = g(\theta) , \quad \theta = kh . \tag{9.50}$$

For example, for the Lax-Wendroff scheme (9.28), g is given by (9.41), whereas for the κ-scheme, g is given by (9.43). It follows from (9.50) that

$$\omega_h = \frac{i}{\tau} \ln g ,$$

so that the numerical phase and group velocities are given by

$$c_p^h = -\mathrm{Im}\{\frac{1}{\tau k} \ln g\} , \quad c_g^h = -\mathrm{Im}\{\frac{1}{\tau g}\frac{dg}{dk}\} . \tag{9.51}$$

In general, numerical schemes will have dispersion. The fidelity of the scheme in reproducing the propagation properties of the exact solution can be assessed by comparing c_p^h and c_p, and c_g^h and c_g. This can also be done by comparing the numerical amplification factor $g(\theta)$ with the exact amplification factor $g_e(\theta)$ over one time step, given by

$$g_e(\theta) = e^{-i\omega\tau} = e^{-ick\tau} = e^{-i\sigma\theta} , \quad \sigma = c\tau/h .$$

The numerical dispersion of a wave with wavenumber k is defined as, writing $g = |g| \exp(i \arg(g))$,

$$\alpha(\theta) = \arg(g_e) - \arg(g) .$$

The numerical dispersion has the following physical interpretation. Since $\arg(g_e) = -ck\tau$ and $\ln g = \ln|g| + i\arg(g)$, we see from (9.51) that $\arg(g) = -c_p^h k\tau$, so that

$$\alpha(\theta) = (c_p^h - c)k\tau = (c_p^h - c)\tau 2\pi/L ,$$

where L is the wavelength, so that $\alpha(\theta)$ measures 2π times the distance that the numerical wave travels too far in one time step, divided by the wavelength.

The numerical dissipation is defined by

$$\varepsilon(\theta) = |g_e| - |g| .$$

This measures the amount by which the numerical scheme reduces the amplitude too much over one time step. In the present case, $|g_e| = 1$.

For the Lax-Wendroff scheme (9.28) we find from (9.41):

$$\arg(g) = -\arctan[\sigma \sin\theta/\{1 - \sigma^2(1 - \cos\theta)\}] .$$

Furthermore,

$$\arg(g_e) = -\sigma\theta .$$

Hence

$$\alpha(\theta) = \arctan[\sigma \sin(\theta)/\{1 - \sigma^2(1 - \cos\theta)\}] - \sigma\theta .$$

From (9.42) we find

$$\varepsilon(\theta) = 1 - \{1 + 4s^2\sigma^2(\sigma^2 - 1)\}^{1/2} , \quad s = \sin^2 \frac{1}{2}\theta .$$

For the κ-scheme (9.26) we find from (9.43) and (9.44)

$$\alpha(\theta) = \arctan\{(\sigma - as) \sin\theta/(1 - 2\sigma^2 s + 2as^2)\} - \sigma\theta , \qquad (9.52)$$

$$\varepsilon(\theta) = 1 - \{1 + 4s^2(a - 2a\sigma - \sigma^2 + \sigma^4) + 4as^3(a + 2\sigma - \sigma^2)\}^{1/2}, \quad (9.53)$$

where $a = \sigma(\sigma - 1)(1 - \kappa + 2\kappa\sigma)$.

We see that both the Lax-Wendroff and the κ-scheme contain dispersion as well as dissipation. The dispersion $\varepsilon(\theta)$ and dissipation $\alpha(\theta)$ for these schemes are plotted in Figs. 9.1–9.4. Comparison shows that the κ-scheme has less

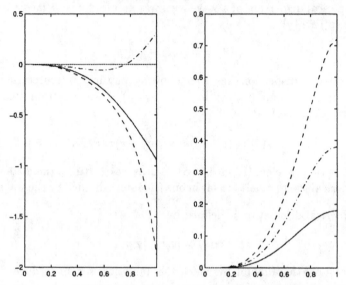

Fig. 9.1. Dispersion (left) and dissipation (right) versus θ/π for the Lax-Wendroff scheme. —— : $\sigma = 0.3$; - - - : $\sigma = 0.6$; — · — · — : $\sigma = 0.9$.

dispersion than the Lax-Wendroff scheme. The $\kappa = 0$ scheme was designed by Fromm (1968) for low dispersion, and indeed it comes out best in the dispersion plots. The $\kappa = 1/3$ scheme has third order spatial accuracy (cf. (9.27)), and has less damping than the other κ-schemes. Numerical damping is not all bad; in fact, it is desirable that modes that propagate at wrong phase velocity are damped. Therefore it is favorable that the schemes considered here have significant damping for values of θ for which dispersion is significant.

Numerical experiments

To see what happens in practice, the schemes discussed are applied to (9.34) with $c = 1$. The exact solution is $\varphi(t, x) = f(x - t)$. For f we choose

$$f(x) = 0, \quad x \notin (-0.3, -0.1), \quad f(x) = 1, \quad x \in [-0.3, -0.1], \quad (9.54)$$

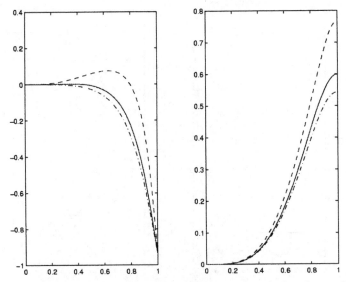

Fig. 9.2. Dispersion (left) and dissipation (right) versus θ/π for the κ-scheme, for $\sigma = 0.3$. - - - : $\kappa = -1$; —— : $\kappa = 0$; — · — · — : $\kappa = 1/3$.

or for $x \in [-0.3, -0.1]$,

$$f(x) = 1 - \cos^2 5\pi(x + 0.3) . \tag{9.55}$$

This exact solution is a pulse that travels with velocity c in the positive x-direction. The pulse enters the domain at $t = 0.1$, and has left at $t = 1.3$. In the case of (9.55) the pulse is twice differentiable, whereas in the case of (9.54) it consists of two step functions. We take the mesh size $h = 1/30$, so that the support of the pulse equals $6h$. For the CFL number we choose $\sigma = ch/\tau = 0.7$. Results for the Lax-Wendroff scheme (9.28) are given in Fig. 9.5 and for the κ-scheme (9.26) in Figs. 9.6 and 9.7. The pulse is shown after entering, after traveling through most of the domain and when leaving. The exact solution is represented by drawn curves, the numerical solution by circles. In both cases, we found almost no difference between the two ways to generate artificial boundary conditions that were discussed, and results are shown only for linear extrapolation (9.36). Spurious reflection from the outflow boundary appears to be absent. The Lax-Wendrofff scheme lets the pulse (especially the smooth one) travel too slowly, as predicted by the dispersion plot (Fig. 9.1). As is to be expected because of Godunov's order barrier theorem, both schemes show spurious wiggles, but these are much more severe for the Lax-Wendroff scheme than for the κ-scheme. Although both schemes are second order accurate, the $\kappa = 0$ scheme is significantly more accurate for the smooth pulse than the Lax-Wendroff scheme, for which not only the propagation velocity is too small but which also has more dissipation, as predicted by the dissipation plots in Figs. 9.1–9.4. Because of its third

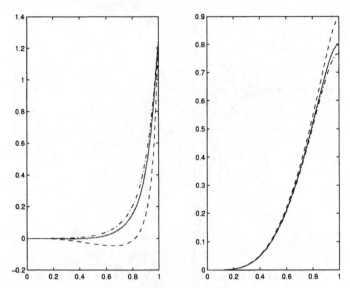

Fig. 9.3. Dispersion (left) and dissipation (right) versus θ/π for the κ-scheme, for $\sigma = 0.6$. - - - : $\kappa = -1$; —— : $\kappa = 0$; — · — · — : $\kappa = 1/3$.

order accuracy, the $\kappa = 1/3$ scheme is slightly more accurate than the $\kappa = 0$ scheme. For the step function pulse the exact solution is not smooth enough for the formal accuracy of the κ-schemes to pay off, althought they maintain their advantage of smaller spurious wiggles.

Often, spurious wiggles cannot be tolerated at all. Measures to obtain second order accurate wiggle-free schemes, such as flux limiting, were discussed in Sect. 4.8 for the stationary convection-diffusion equation. Before discussing this we abandon the linear case and consider the nonlinear case.

Exercise 9.3.1. Perhaps the simplest scheme that one can think of for the differential equation $\partial\varphi/\partial t + c\partial\varphi/\partial x = 0$ is

$$\varphi_j^{n+1} - \varphi_j^n + \frac{1}{2}\sigma(\varphi_{j+1}^n - \varphi_{j-1}^n) = 0 , \quad \sigma = c\tau/h .$$

Show that this scheme is unstable by means of Fourier stability analysis.

Exercise 9.3.2. Show that for $|\sigma| = 1$ the Lax-Wendroff and κ-schemes are contractive.

Exercise 9.3.3. Prove theorems 9.3.4 and 9.3.5.

Exercise 9.3.4. Prove that the Lax-Friedrichs and Courant-Isaacson-Rees schemes are dissipative.

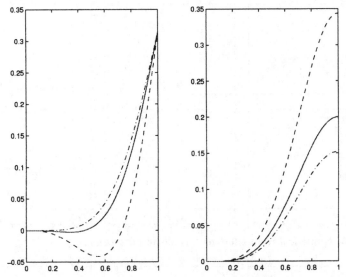

Fig. 9.4. Dispersion (left) and dissipation (right) versus θ/π for the κ-scheme, for $\sigma = 0.9$. - - - : $\kappa = -1$; —— : $\kappa = 0$; — · — · — : $\kappa = 1/3$.

9.4 Scalar conservation laws

The nonlinear generalization of (9.1) is

$$\frac{\partial \varphi}{\partial t} + \frac{\partial f(\varphi)}{\partial x} = 0 , \quad t > 0 , \quad x \in \mathbb{R} , \tag{9.56}$$

with φ a scalar (which in this context merely means single) function; this is why this is called a *scalar* conservation law. For the time being, boundary conditions will not be considered, which is why the domain is taken to be unbounded in x. An initial condition is required:

$$\varphi(0, x) = \varphi^0(x) . \tag{9.57}$$

Integration of (9.56) from $x = a$ to $x = b$ gives

$$\frac{d}{dt} \int\limits_a^b \varphi dx = f(\varphi)|_a^b .$$

This shows that the total amount of φ in (a, b) can change only by transport or flux through $x = a$ and $x = b$; f is called the flux function. If there is no transport, $\int\limits_a^b \varphi dx$ is constant or conserved, which is why (9.56) is called a conservation law.

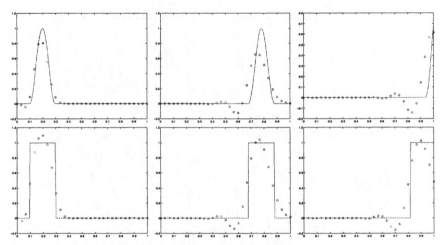

Fig. 9.5. Numerical results for the Lax-Wendroff scheme.

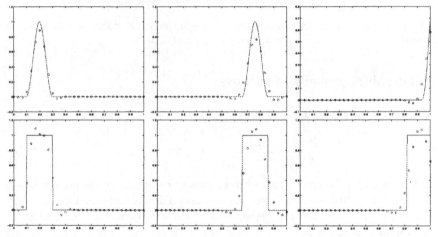

Fig. 9.6. Numerical results for $\kappa = 1/3$ scheme.

Because almost all (if not all) successful numerical methods for computing inviscid compressible flows are based on methods for scalar conservation laws in one dimension, and because there is a well-developed theory mainly for this case only, it is worthwhile to study properties of and numerical methods for scalar conservation laws, which we will do in this section. Despite the apparent simplicity of (9.56), this is a large subject, and the reader may wish to consult more extensive introductions to the subject. A good starting point is LeVeque (1992). Classics in the field are Courant and Friedrichs (1949) and Lax (1973). Recent expositions of the mathematical theory are Godlewski and Raviart (1996) and Smoller (1983). More emphasis on numerics is found in Sod (1985).

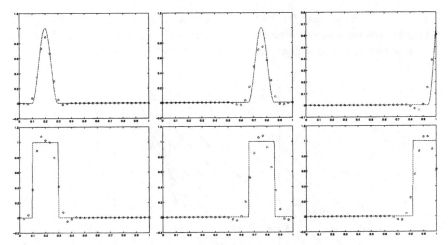

Fig. 9.7. Numerical results for $\kappa = 0$ scheme.

Characteristics

If f is differentiable, equation (9.56) can be rewritten as

$$\frac{\partial \varphi}{\partial t} + c(\varphi)\frac{\partial f}{\partial x} = 0 , \quad c(\varphi) = \frac{df}{d\varphi} .$$

This shows that φ is constant along characteristic curves defined by

$$t = t(s) , \quad x = x(s) , \quad \frac{dt}{ds} = 1 , \quad \frac{dx}{ds} = c(\varphi) .$$

Since φ and hence $c(\varphi)$ is constant along characteristics, the characteristics are straight. Depending on the initial condition $\varphi^0(t)$, it may easily happen that characteristics intersect. This would make the solution multi-valued, which is physically impossible. What happens is that a discontinuity, called a shock, develops. Even for smooth initial data $\varphi^0(t)$ the solution may become discontinuous in finite time, as illustrated in the following example.

Example 9.4.1. Inviscid Burgers equation

The Burgers equation (Burgers (1948)) is given by

$$\frac{\partial \varphi}{\partial t} + \frac{\partial f(\varphi)}{\partial x} = \nu \frac{\partial^2 \varphi}{\partial x^2} , \quad f(\varphi) = \frac{1}{2}\varphi^2 . \tag{9.58}$$

For $\nu = 0$ we obtain the inviscid Burgers equation, which is the simplest example of a nonlinear scalar conservation law. The initial condition is chosen as follows:

$$\varphi^0(x) = 1 , \quad x \leq 0 ; \quad \varphi^0(x) = 0 , \quad x \geq 1 ; \quad \varphi^0 \in C^\infty(\mathbb{R}) .$$

We leave φ^0 unspecified for $0 < x < 1$, except that we stipulate that φ^0 is infinitely differentiable and monotone. Characteristics are plotted in Fig. 9.8. Independent of the specification of φ^0 for $0 < x < 1$, the characteristics cer-

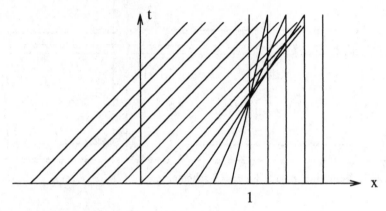

Fig. 9.8. Characteristics for inviscid Burgers equation.

tainly intersect, for $t > 1$ at the latest. Hence, although the initial condition $\varphi^0 \in C^\infty$, a discontinuity develops in finite time.   □

Weak solutions and shocks

As said before, a physically relevant solution is single-valued, and where characteristics intersect the solution becomes discontinuous. Since classical solutions of (9.56) are differentiable, the class of solutions has to be extended.

Definition 9.4.1. *Weak solution*
A bounded integrable function $\varphi(x,t)$ *is called a weak solution of the initial value problem (9.56), (9.57) if*

$$\int_0^\infty \int_{-\infty}^\infty (\varphi\psi_t + f(\varphi)\psi_x)dxdt + \int_{-\infty}^\infty \phi^0\psi(0, x)dx = 0 \qquad (9.59)$$

for all $\psi \in C_0^1(\mathbb{R}^+ \times \mathbb{R})$.

Here $C_0^1(\Omega)$ is the space of functions that are continuously differentiable, and that are identically zero outside some bounded set inside Ω. Note that (9.59) makes sense even if the solution φ is not continuous. It is left to the reader to show that if a weak solution is in C_0^1, then it is a classical solution.

We will now study the jump condition that holds at a discontinuity. Let $x = y(t)$ be a curve Γ along which there is a discontinuity, and let

$$\varphi_l(t) = \lim_{x \uparrow y(t)} \varphi(t, x) \, , \quad \varphi_r(t) = \lim_{x \downarrow y(t)} \varphi(t, x) \, .$$

Let φ be differentiable away from $x = y(t)$. Let $P = (t_0, y(t_0))$ be a point on Γ, and let D be a neighbourhood of P. Choose $\psi \in C_0^1(D)$. Let D_l and D_r be the parts of D that are to the left and to the right of Γ, respectively. Since φ is differentiable in D_l and D_r, we can apply the divergence theorem, and (9.59) gives

$$0 = \iint\limits_{D} (\varphi \psi_t + f \psi_x) dx dt = \{ \int\limits_{\partial D_l} + \int\limits_{\partial D_r} \} \psi(-\varphi dx + f dt) \, .$$

Since $\psi = 0$ on ∂D, the only contributions come from Γ:

$$0 = \int\limits_{P_1}^{P_2} \psi \{ [\varphi] dx - [f(\varphi)] dt \} \, ,$$

where P_1 and P_2 are the intersection points of Γ and D, and

$$[\varphi] = \varphi_r(t) - \varphi_l(t) \, , \quad [f(\varphi)] = f(\varphi_r(t)) - f(\varphi_l(t)) \, .$$

Since ψ is arbitrary, we must have

$$s[\varphi] = [f(\varphi)] \, , \tag{9.60}$$

where $s = dy/dt$ is the propagation speed of the discontinuity. Such a discontinuity is called a *shock*. Equation (9.60) is called *the jump condition*. It determines the location and the size of the discontinuity, and a single-valued weak solution is obtained, as illustrated in the following example, due to Lax (1973).

Example 9.4.2. Solution of inviscid Burgers equation
Consider the inviscid Burgers equation (9.58) with the following initial condition:

$$\varphi^0(x) = 1 \, , \quad x \le 0 \, ; \quad \varphi^0(x) = 1 - x \, , \quad 0 < x < 1 \, ; \quad \varphi^0(x) = 0 \, , \quad x \ge 1 \, .$$

The characteristics intersect for $t > 1$. It is left to the reader to show that for $t \le 1$ the solution is given by

$$\varphi(t, x) = 1 \, , \quad x < t \, ; \quad \varphi(t, x) = (1 - x)/(1 - t) \, , \quad t \le x \le 1 \, ; \\ \varphi(t, x) = 0 \, , \quad x > 1 \, . \tag{9.61}$$

For $t \ge 1$ we have a shock. Clearly, $\varphi_l = 1$ and $\varphi_r = 0$, so that the jump condition (9.60) gives $s = 1/2$. The shock starts at $(t, x) = (1, 1)$, so that for $t > 1$ a solution is given by

$$\varphi(t, x) = 1 \, , \quad x < 1 + \frac{1}{2}(t - 1) \, ; \quad \varphi(t, x) = 0 \, , \quad x > 1 + \frac{1}{2}(t - 1) \, .$$

The characteristics and shock are depicted in Fig. 9.9.   □

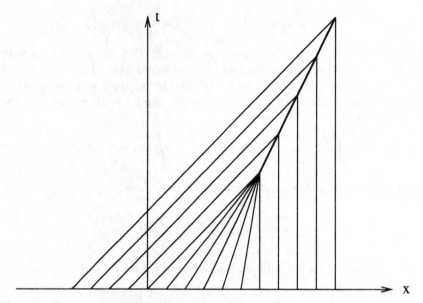

Fig. 9.9. Characteristics and shock for Example 9.4.2

The entropy condition and uniqueness

It turns out that the weak formulation (9.59) together with the jump condition (9.60) are not sufficient to determine a unique solution, as illustrated by the following example, again due to Lax (1973).

Example 9.4.3. Multiple solutions of inviscid Burgers equation
Consider the inviscid Burgers equation (9.58) with initial condition

$$\varphi^0(x) = 0 \,, \quad x < 0 \,; \quad \varphi^0(x) = 1 \,, \quad x > 0 \,.$$

By using the characteristics we see that

$$\varphi(t, x) = 0 \,, \quad x < 0 \,; \quad \varphi(t, x) = 1 \,, \quad x > t \,.$$

But what is the solution for $0 < x < t$? A solution satisfying the jump condition is

$$\varphi(t, x) = 0 \,, \quad x < t/2 \,; \quad \varphi(t, x) = 1 \,, \quad x > t/2 \,. \tag{9.62}$$

But another possible solution, not containing a shock, is given by

$$\varphi(t, x) = 0 \,, \quad x < 0 \,; \quad \varphi(t, x) = x/t \,, \quad 0 \le x \le t \,;$$
$$\varphi(t, x) = 1 \,, \quad x > t \,, \tag{9.63}$$

and there may be more solutions. Anyway, two is too many!   □

The question arises what additional condition should be imposed to make weak solutions unique. In fluid mechanics, an inviscid mathematical model is the vanishing viscosity limit of a viscous model. Analogously, in the case of the inviscid Burgers equation we require that the weak solution should be the $\nu \downarrow 0$ limit of a classical solution of the (viscous) Burgers equation. This, and other equivalent conditions, all go by the name of *entropy condition*. This name derives from the fact, that in the presence of viscosity (however small) a real physical flow causes the entropy of a fluid particle to be non-decreasing.

What is now needed is a mathematical formulation of the entropy condition that makes it (relatively) easy to check whether it is satisfied or not by a weak solution. One such formulation, due to Oleinik (1957), is the following:

Definition 9.4.2. *Entropy condition*
A weak solution is said to satisfy the entropy condition if in every point of every shock we have

- *(i) if $\varphi_r < \varphi_l$: The graph of $y = f(\varphi)$ over $[\varphi_r, \varphi_l]$ lies below the chord connecting $(\varphi_r, f(\varphi_r))$ to $(u_l, f(\varphi_l))$;*
- *(ii) if $\varphi_r > \varphi_l$: The graph of $y = f(\varphi)$ lies above the chord.*

A weak solution that satisfies the entropy condition is called an *entropy solution*. Note that the entropy condition says something about shocks only.

What does Definition 9.4.2 have to do with entropy? Suppose that f is convex: $f'' > 0$, as is the case, for example, for the inviscid Burgers equation. According to the jump condition (9.60) the shock speed equals the slope of the chord. Condition (i) is in the case $f'' > 0$ equivalent with

$$c(\varphi_r) < s < c(\varphi_l) , \quad c(\varphi) = f'(\varphi) , \qquad (9.64)$$

and this is obviously satisfied if $f'' > 0$. If $\varphi_r > \varphi_l$ we have for $f'' > 0$

$$c(\varphi_r) > s > c(\varphi_l) ,$$

and this is forbidden by condition (ii). Hence, in the case of convex f the entropy condition says that characteristics never emanate from a shock but always enter it. This means that from every point on a shock we can travel along characteristics backward in time only and not forward. In other words, information reaches the shock from the past only and not from the future. One may therefore regard Definition 9.4.2 as a causality principle. By distinguishing between past and future this entropy condition says something about the arrow of time: it implies time-irreversibility. Irreversibility is intimately related to the second law of thermodynamics, saying that entropy does not decrease in a closed system. Hence the connection between Definition 9.4.2 and entropy.

There are several equivalent formulations of the entropy condition, see Lax (1973) or Smoller (1983). The entropy condition guarantees that a weak solution is a vanishing viscosity solution and is unique, but we will not prove this.

It is easily seen that the entropy condition tells us how to choose between the two solutions (9.62) and (9.63) of the inviscid Burgers equation. In this case $f'' > 0$, and (9.64) should be satisfied by the discontinous solution (9.62), but it is not; hence, (9.62) is not an entropy solution. Since according to (9.63) φ is continuous, it is automatically an entropy solution; as asserted above, it is unique.

Non-equivalence of conservation laws

Two conservation laws that are equivalent for smooth solutions may not be equivalent for weak solutions. For example, the two conservation laws

$$\varphi_t + \frac{1}{2}(\varphi^2)_x = 0 \quad \text{and} \quad (\varphi^2)_t + \frac{2}{3}(\varphi^3)_x = 0$$

are obviously equivalent for differentiable φ. But the respective jump conditions are

$$s[\varphi] = \frac{1}{2}[\varphi^2] \quad \text{and} \quad s[\varphi^2] = \frac{2}{3}[\varphi^3] \,,$$

leading to the following shock propagation velocities, respectively:

$$s = \frac{1}{2}(\varphi_r + \varphi_l) \quad \text{and} \quad s = \frac{2}{3}(\varphi_r^2 + \varphi_r\varphi_l + \varphi_l^2)/(\varphi_r + \varphi_l) \,,$$

which are obviously different. In physical applications it is clear which form of a conservation law should be taken: the form that follows directly from the basic integral form of the physical conservation law.

Total variation and monotonicity

The solution of the conservation law (9.56) is *total variation diminishing* (TVD) and *monotonicity preserving*. It is desirable that numerical schemes mimick these (almost equivalent) properties, which we therefore discuss.

The *total variation* of a function $\varphi(t, x)$ is defined in equation (9.17).

Theorem 9.4.1. *Uniqueness, TVD and inverse monotonicity*
Assume that $\varphi^0 \in L^\infty(\mathbb{R})$ and let $f(\varphi)$ be differentiable. Then the problem (9.56), (9.57) has a unique entropy solution $\varphi \in L^\infty(\mathbb{R}^+ \times \mathbb{R})$. This solution satisfies for almost all $t > 0$

$$\|\varphi(t,\cdot)\|_{L^\infty} \leq \|\varphi^0\|_{L^\infty} .$$

If φ and ψ are entropy solutions associated with the initial conditions φ^0 and ψ^0, respectively, then we have

$$\varphi^0 \geq \psi^0 \implies \varphi(t,\cdot) \geq \psi(t,\cdot) \tag{9.65}$$

almost everywhere. If φ^0 has bounded total variation, then

$$TV(\varphi(t,\cdot)) \leq TV(\varphi^0) . \tag{9.66}$$

For a proof we refer to Godlewski and Raviart (1991), Chap.II (Sect. 5). Here L^∞ is the space of integrable functions with norm

$$\|\varphi(t,\cdot)\|_{L^\infty} = \inf\{\sup\{\varphi(t,x) : x \in \mathbb{R}\backslash A, \ A \text{ is set with zero measure}\} .$$

Equation (9.65) says that the equation is inverse monotone with respect to the initial condition. Equation (9.66) says that the solution is total variation diminishing or TVD. More precisely, the solution is total variation non-increasing, but we prefer the more crisp and commonly used terminology TVD.

Theorem 9.4.2. *Monotonicity preservation*
If φ^0 is monotone and is constant outside a bounded domain, then $\varphi(t,\cdot)$ is monotone.

Proof. If suffices to consider the case $x_2 > x_1 \implies \varphi^0(x_2) \geq \varphi^0(x_1)$. It follows that

$$\mathrm{TV}(\varphi^0) = \varphi^0(\infty) - \varphi^0(-\infty) .$$

Because φ^0 is constant outside a bounded domain, we have

$$\varphi(t,\pm\infty) = \varphi^0(\pm\infty) .$$

If $\varphi(t,\cdot)$ is monotone, then

$$\mathrm{TV}(\varphi(t,\cdot)) = \varphi(t,\infty) - \varphi(t,-\infty) = \mathrm{TV}(\varphi^0) . \tag{9.67}$$

If $\varphi(t,\cdot)$ is not monotone, its total variation is larger than given by (9.67), violating (9.66). □

This theorem says that TVD implies monotonicity preservation.

It is desirable that numerical solutions mimic these properties. This means that spurious numerical oscillations ('wiggles') are to be avoided. Especially near discontinuities in the solution these can be a nuisance.

The Riemann problem

The *Riemann problem* for (9.56), (9.57) is defined by the following special initial condition:

$$\varphi^0(x) = \varphi_l , \quad x < 0 ; \quad \varphi^0(x) = \varphi_r , \quad x > 0 , \tag{9.68}$$

with φ_l and φ_r constant. The solution of the Riemann problem plays an important role in the construction of many numerical methods. We will derive its solution.

It is easy to see that if $\varphi(t, x)$ is a solution of the Riemann problem, then so is $\varphi(at, ax)$ for every constant $a \neq 0$. So we can look for the solution in similarity form:

$$\varphi(t, x) = \varphi(x/t) .$$

Substitution in (9.56) gives

$$\varphi^{'}(\xi)\{f^{'}(\varphi(\xi)) - \xi\} = 0 . \tag{9.69}$$

Therefore we can construct a solution $\varphi(\xi)$, $\xi \in \mathbb{R}$, piece by piece by having the first or second factor in (9.69) zero. A constant state is obtained with $\varphi^{'}(\xi) = 0$. Putting the second factor zero gives

$$f^{'}(\varphi(\xi)) = \xi , \tag{9.70}$$

which means that φ is the inverse of $f^{'}$, and hence exists if $f^{'}$ is either strictly increasing or strictly decreasing, which is the case if f is either convex $(f^{''} > 0)$ or concave $(f^{''} < 0)$. If this is not the case, it is hard to solve the Riemann problem in general form; the case that $f^{''}$ changes sign will be illustrated shortly by an example. Because of an analogy with gasdynamics that will become apparent later, solutions of (9.70) are called *rarefaction fans* or *expansion fans*. Equation (9.69) assumes differentiability of φ. A similarity solution with a shock satisfying the jump condition (9.60) is

$$\varphi(x/t) = \varphi_l , \quad x < st ; \quad \varphi(x/t) = \varphi_r , \quad x > st ; \quad s = \frac{f(\varphi_r) - (f\varphi_l)}{\varphi_r - \varphi_l} .$$

Thus, we have obtained three building blocks for obtaining Riemann solutions: constant states, fans and shocks. These must be pieced together such that the initial condition (9.68) and the entropy condition (Def. 9.4.2) are satisfied.

Consider first the case $f^{''} > 0$. There are two possible cases to consider for the initial condition (9.68): $\varphi_r > \varphi_l$ and $\varphi_r < \varphi_l$. As entropy condition we can use (9.64). Hence, if $\varphi_r > \varphi_l$ there cannot be a shock, and the solution must be pieced together from solutions of (9.69), i.e. constant states and solutions of (9.70). Define

$$s_l = f'(\varphi_l) , \quad s_r = f'(\varphi_r) .$$

Then the solution is obviously given by

$$\varphi(x/t) = \varphi_l , \quad x < s_l t ; \quad \varphi(x/t) = \varphi_r , \quad x > s_r t ;$$
$$f'(\varphi(x/t)) = x/t , \quad s_l t < x < s_r t , \tag{9.71}$$

which can be solved for φ because $f'' > 0$. In the case of the inviscid Burgers equation, $f'(\varphi) = \varphi$, and the solution (9.63) is recovered. In the case $\varphi_r < \varphi_l$ the characteristics intersect (because $f'' > 0$) and we use a shock to solve. Using the jump condition (9.60) we find

$$\varphi(x/t) = \varphi_l , \quad x < st ; \quad \varphi(x/t) = \varphi_r , \quad x > st ;$$
$$s = \{f(\varphi_r) - f(\varphi_l)\}/(\varphi_r - \varphi_l) . \tag{9.72}$$

This completes the treatment of the case $f'' > 0$; the case $f'' < 0$ is similar and need not be discussed.

We see that in the cases $f'' > 0$ and $f'' < 0$ the discontinuity in the Riemann solution is separated either by a shock or by a fan, but not by both. However, this need not be so for f arbitrary.

The case that f is neither convex nor concave will be illustrated by the example of the scalar one-dimensional version of the Buckley-Leverett equation as given in Sect. 4.2 of LeVeque (1992). The flux function is given by

$$f(\varphi) = \frac{\varphi^2}{\varphi^2 + a(1 - \varphi)^2} .$$

The physical background is the flow of a mixture of oil and water through a porous medium. Here φ represents the saturation of water and lies between 0 and 1. The graph of f is shown in Fig. 9.10 for $a = 1/2$. Note that f'' changes sign at $\varphi = 1/2$. The solution of the Riemann problem can be established by geometric means in the following way. Because of the jump condition (9.60), a shock between two states φ_a and φ_b corresponds to a chord in the $f - \varphi$ diagram. This shock is admissible if its corresponding chord is in accordance with the entropy condition. If it is not we can go (part of) the way between φ_a and φ_b along the graph of f, corresponding to a fan. The procedure will become clear by taking an example. Let $\varphi_r = 0$, $\varphi_l = 1$. We can connect these states by an admissible shock between $\varphi = 0$ and $\varphi = \varphi_1$, where the chord coincides with the tangent, cf. Fig. 9.11. Between $\varphi = \varphi_1$ and φ_l we continue along the graph. This solution satisfies the entropy condition, and is the unique solution of the specified Riemann problem. In this particular case the solution can be obtained in a more or less explicit analytic form. From

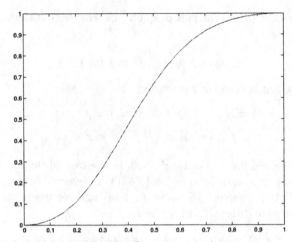

Fig. 9.10. Flux function for Buckley-Leverett equation

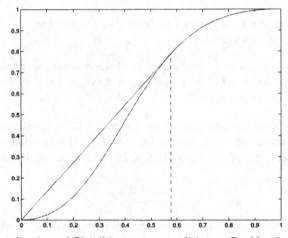

Fig. 9.11. Application of Oleinik's entropy condition to Buckley-Levertt problem

$$f(\varphi_1)/\varphi_1 = f'(\varphi_1)$$

we find $\varphi_1 = \frac{1}{3}\sqrt{3}$. So to begin with we have shock between $\varphi_r = 0$ and $\varphi_1 = \frac{1}{3}\sqrt{3}$ with

$$s = \frac{\varphi_r - \varphi_1}{f(\varphi_r - f(\varphi_1)} = \sqrt{3} - 1 .$$

Between φ_1 and $\varphi_l = 1$ we have a fan satisfying (9.70), i.e.

$$\frac{\varphi - \varphi^2}{\{\varphi^2 + \frac{1}{2}(1 - \varphi)^2\}^2} = \xi , \quad \xi = x/t . \tag{9.73}$$

This gives implicitly

$$\varphi = g(x/t) \, .$$

By plotting ξ as function of φ using (9.73) we obtain automatically the plot of φ versus $\xi = x/t$. Thus, we have obtained the following Riemann solution for the Buckley-Leverett equation:

$$\varphi(x/t) = 1 \, , \quad x < 0 \, ; \quad \varphi(x/t) = g(x/t) \, , \quad 0 < x < (\sqrt{3} - 1)t \, ; \qquad (9.74)$$
$$\varphi(x/t) = 0 \, , \quad x > (\sqrt{3} - 1)t \, .$$

Figure 9.12 shows the solution for $t = 1$. Note that we have a shock followed by a fan.

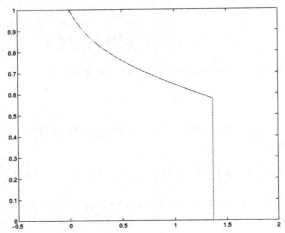

Fig. 9.12. Solution of Riemann problem for the Buckley-Leverett equation

Conservative and nonconservative numerical schemes

After this discussion of some important analytic aspects of scalar conservation laws, we now turn to numerical discretization methods. The occurrence of weak (discontinuous) solutions makes a significant difference compared to the case of strong (differentiable) solutions. In the latter case, consistency and stability imply convergence. But for convergence to weak solutions, these conditions do not suffice. Convergence to weak solutions satisfying the jump condition (genuine weak solutions) is guaranteed only if the numerical scheme is in conservation form (the definition of a conservative scheme is repeated below). Furthermore, just as for the differential condition, nonphysical weak solutions are excluded only if the numerical scheme satisfies an entropy condition.

The following example shows that a nonconservative scheme may fail to converge to a genuine weak solution.

Example 9.4.4. Nonconservative scheme for inviscid Burgers equation
This example is due to LeVeque (1992), Example 12.1. The nonconservative form of the inviscid Burgers equation (9.58) is

$$\varphi_t + \varphi \varphi_x = 0 .$$

The first order upwind scheme is given by, assuming $\varphi > 0$:

$$\varphi_j^{n+1} - \varphi_j^n + \lambda \varphi_j^n (\varphi_j^n - \varphi_{j-1}^n) = 0 , \quad \lambda = \tau/h . \tag{9.75}$$

Suppose

$$\varphi_j^0 = 1 , \quad j < 0 ; \quad \varphi_j^0 = 0 , \quad j \geq 0 .$$

It is easy to see that the numerical solution is given by

$$\varphi_j^n = \varphi_j^0 , \quad \forall n , \tag{9.76}$$

which is obviously wrong. It is left to the reader to show that the exact solution is given by

$$\varphi(t, x) = 1 , \quad x < t/2 ; \quad \varphi(t, x) = 0 , \quad x > t/2 . \tag{9.77}$$

We see that with the nonconservative scheme (9.75) the shock moves with the wrong speed, namely zero; as show by (9.77), the correct speed is $1/2$. □

Fortunately, convergence to a weak solution that satisfies the jump condition is guaranteed for conservative schemes. A conservative scheme for (9.56) is a scheme of the following form:

$$d\varphi_j/dt + \frac{1}{h}(F_{j+1/2} - F_{j-1/2}) = 0 , \tag{9.78}$$
$$F_{j+1/2} = F(\varphi_{j-p}, \varphi_{j-p+1}, ..., \varphi_{j+q}) ,$$

where the temporal discretization can still be chosen as one likes. Summing between $j = k$ and $j = m$ gives

$$\frac{d}{dt} \sum_{j=k}^{m} \varphi_j + \frac{1}{h}(F_{m+1/2} - F_{k-1/2}) = 0 ,$$

so that $\sum \varphi_j$ is conserved if the boundary fluxes $F_{m+1/2} = F_{k-1/2} = 0$, which is why (9.78) is called a conservative scheme.

We will now present a few examples of conservative schemes. Forward Euler temporal discretization of (9.78) gives

$$\varphi_j^{n+1} - \varphi_j^n + \lambda(F_{j+1/2}^n - F_{j-1/2}^n) = 0 \,. \tag{9.79}$$

A conservative nonlinear version of the Lax-Friedrichs scheme (9.32) is obtained with

$$F_{j+1/2}^n = \frac{1}{2\lambda}(\varphi_j^n - \varphi_{j+1}^n) + \frac{1}{2}\{f(\varphi_j^n + f(\varphi_{j+1}^n)\} \,. \tag{9.80}$$

In the linear case, $f(\varphi) = c\varphi$, and (9.32) is recovered. An obvious conservative nonlinear version of the Courant-Isaacson-Rees (or first order upwind) scheme (9.33) is obtained by choosing

$$\begin{aligned} F_{j+1/2}^n &= f(\varphi_j^n) \quad \text{if } f'(\varphi_{j+1/2}^n) \geq 0 \,; \\ F_{j+1/2}^n &= f(\varphi_{j+1}^n) \quad \text{if } f'(\varphi_{j+1/2}^n) < 0 \,, \end{aligned} \tag{9.81}$$

where $\varphi_{j+1/2}^n \equiv (\varphi_j^n + \varphi_j^n)/2$. We will see in Example 9.4.5 that this scheme has a serious flaw. A better nonlinear version of the Courant-Isaacson-Rees scheme is the Engquist-Osher scheme (Engquist and Osher (1981)). This scheme is obtained with

$$F_{j+1/2}^n = f^+(\varphi_j^n) + f^-(\varphi_{j+1}^n) \,,$$

$$f^+(\varphi) \equiv f(0) + \int_0^\varphi \max\{f'(\psi), 0\}d\psi \,, \quad f^-(\varphi) \equiv \int_0^\varphi \min\{f'(\psi), 0\}d\psi \,.$$

$$\tag{9.82}$$

That this scheme makes sense can be seen by noting that $f(\varphi) = f^+(\varphi) + f^-(\varphi)$. Furthermore, in the linear case $f(\varphi) = c\varphi$ the first order upwind scheme is recovered. Why (9.82) is a better nonlinear generalization of the first order upwind scheme than (9.81) will be seen shortly.

There are several conservative versions of the Lax-Wendroff scheme (9.26), all of two-stage form. MacCormack's scheme (MacCormack (1969)) is well-known; it is given by

$$\begin{aligned} \varphi_j^* - \varphi_j^n + \lambda\{f(\varphi_{j+1}^n) - f(\varphi_j^n)\} &= 0 \,, \\ \varphi_j^{n+1} - \frac{1}{2}(\varphi_j^n + \varphi_j^*) + \frac{\lambda}{2}\{f(\varphi_j^*) - f(\varphi_{j-1}^*)\} &= 0 \,. \end{aligned} \tag{9.83}$$

By elimination of φ^* it is easily seen that this scheme is conservative.

Schemes of the form (9.78) are called consistent if

$$F(\bar{\varphi}, ..., \bar{\varphi}) = f(\bar{\varphi}) \tag{9.84}$$

and if there exists a constant K such that

$$|F(\varphi_{j-p}, \varphi_{j-p+1}, ..., \varphi_{j+q}) - f(\bar{\varphi})| \leq K \max\{|\varphi_{j+m} - \bar{\varphi}| : m = -p, ..., q\} \tag{9.85}$$

for $\varphi_{j-p}, \varphi_{j-p+1}, ..., \varphi_{j+q}$ sufficiently close to $\bar{\varphi}$. It is left to the reader to show that this implies consistency of first order in the classical sense if the solution $\varphi(t, x)$ is sufficiently differentiable (cf. Kröner (1997) Lemma 2.2.4).

The celebrated Lax-Wendroff theorem (Lax and Wendroff (1960)) asserts that if the solution of a conservative and consistent scheme converges as $\tau \downarrow 0, h \downarrow 0$, then it converges to a genuine weak solution of the conservation law. In other words, we cannot get a shock that moves at a wrong speed. This is the merit of conservative schemes. For convergence we need stability, of course. An in-depth study of the reason why nonconservative schemes may converge to wrong solutions is given by Hou and Le Floch (1994).

The entropy condition

In order to ensure that a numerical solution converges to the physically relevant weak solution (the entropy solution), the numerical scheme must satisfy some form of the entropy condition, in addition to being consistent, conservative and stable. LeVeque (1992) (Example 12.3) gives an example of a conservative scheme that generates a numerical solution violating the entropy condition. We present a slight variation.

Example 9.4.5. Entropy condition violating scheme for inviscid Burgers equation
Consider the inviscid Burgers equation with initial condition

$$\varphi^0(x) = -1, \quad x < 0; \quad \varphi^0(x) = 1, \quad x > 0,$$

with the following corresponding initial data for the numerical solution:

$$\varphi_j^0 = -1, \quad j \leq 0; \quad \varphi_j^0 = 1, \quad j > 0.$$

Consider the nonlinear version of the first order upwind scheme defined by (9.79) and (9.81). Since $F_{j+1/2}^0 = F_{j-1/2}^0, \forall j$, we obtain $\varphi_j^1 = \varphi_j^0$, and hence

$$\varphi_j^n = \varphi_j^0, \quad \forall n.$$

However, this solution does not satisfy the entropy condition. The corresponding characteristics are given by

$$x/t = -1, \quad x < 0, \quad x/t = 1, \quad x > 1,$$

and emanate from the shock at $x = 0$ instead of converging into it. The entropy solution is given by

$$\varphi(t,x) = -1, \quad x \le -t; \quad \varphi(t,x) = x/t, \quad -t < x < t;$$
$$\varphi(t,x) = 1, \quad x \ge t.$$

We will show later that the Engquist-Osher scheme (9.79), (9.82) converges to the entropy solution.   □

We will not formulate entropy conditions for numerical schemes. For such a condition see LeVeque (1992) Sect. 13.4. It turns out that conservative monotone schemes converge to entropy solutions. Monotonicity is not difficult to check.

Monotone conservative schemes

As shown by Theorem 9.4.1, entropy solutions of scalar conservation laws have the following monotonicity property: if two initial conditions $\varphi^0(x)$ and $\psi^0(x)$ satisfy $\varphi^0 \ge \psi^0$, then $\varphi(t,\cdot) \ge \psi(t,\cdot)$. It is nice if numerical schemes mimick this property. We therefore define

Definition 9.4.3. *Monotone conservative schemes*
A conservative scheme

$$\varphi_j^{n+1} - \varphi_j^n + \lambda(F_{j+1/2}^n - F_{j-1/2}^n) = 0 \tag{9.86}$$

is called monotone if for any two numerical solutions φ_j^n and ψ_j^n we have the following property:

$$\varphi_j^n \ge \psi_j^n, \ \forall j \implies \varphi_j^{n+1} \ge \psi_j^{n+1}, \ \forall j.$$

By rewriting (9.86) in the following form

$$\varphi_j^{n+1} = H_j(\varphi^n) \tag{9.87}$$

we see that the property that H is a nondecreasing function of its arguments is equivalent with the scheme being monotone. We have

Theorem 9.4.3. *Convergence of monotone schemes to entropy solution*
The solution of a monotone conservative scheme converges to the entropy solution as $\tau \downarrow 0$ with τ/h fixed.

A proof has been given by Harten, Hyman, and Lax (1976) and and Crandall and Majda (1980).

Example 9.4.6. Nonmonotone upwind scheme
Because of the preceding theorem we expect that the upwind scheme of Example 9.4.5 is nonmonotone. That this is indeed the case may be shown as follows. Let

$$\varphi_j = -a \le 0, \quad j \le 0; \quad \varphi_j = b > 0, \quad j > 0.$$

Writing scheme (9.79), (9.81) in the form (9.87) we obtain for $b < a$:

$$H(\varphi) = \varphi_0 - \frac{1}{2}\lambda\varphi_1^2 + \frac{1}{2}\lambda\varphi_0^2 = -a + \frac{1}{2}\lambda(a^2 - b^2) > -a.$$

Increasing b such that $b > a$ gives

$$H_0(\varphi) = \varphi_0 - \frac{1}{2}\lambda\varphi_0^2 + \frac{1}{2}\lambda\varphi_0^2 = -a,$$

so that $H_0(\varphi)$ has decreased, whereas one of its arguments has increased. □

Since, as is easy to see, the first order upwind scheme is monotone in the linear case, it comes as a surprise that the nonlinear version (9.81) is nonmonotone. The following example shows that a monotone nonlinear generalization of the first order upwind scheme is provided by the Engquist-Osher scheme.

Example 9.4.7. Monotonicity of Engquist-Osher scheme
We will show that the Engquist-Osher scheme (9.79), (9.82) is monotone. The scheme can be written in the form (9.87) with

$$H_j(\varphi) = \lambda f^+(\varphi_{j-1}) + \varphi_j - \lambda f^+(\varphi_j) + \lambda f^-(\varphi_j) - \lambda f^-(\varphi_{j+1}).$$

Hence

$$\begin{aligned}
\partial H_j/\partial\varphi_{j-1} &= \lambda\max\{f'(\varphi_{j-1}), 0\} \ge 0, \\
\partial H_j/\partial\varphi_{j+1} &= -\lambda\min\{f'(\varphi_{j+1}), 0\} \ge 0, \\
\partial H_j/\partial\varphi_j &= 1 - \lambda|f'(\varphi_j)|.
\end{aligned}$$

We see that the Engquist-Osher scheme is monotone, provide the (nonlinear version of the) CFL number $\lambda|f'|$ satisfies

$$\lambda|f'(\varphi_j)| \le 1, \quad \forall j, \tag{9.88}$$

which can be taken care of by taking the time step τ small enough. □

It is left to the reader to show that the Lax-Friedrichs scheme (9.79), (9.80) is monotone, provided (9.88) is satisfied.

Just as in the linear case, monotone schemes are subject to Godunov's order barrier theorem: their order of consistency in regions where the solution is smooth is at most one. This is shown in Harten, Hyman, and Lax (1976).

The Godunov scheme

The Godunov scheme is of the following form:

$$\varphi_j^{n+1} - \varphi_j^n + \lambda(F_{j+1/2}^n - F_{j-1/2}^n) = 0 \, , \, F_{j+1/2}^n = F(\varphi_j^n, \varphi_{j+1}^n) \, .$$

Godunov (1959) proposes the following choice for F:

$$F(\varphi_j^n, \varphi_{j+1}^n) = \frac{1}{\tau} \int\limits_{t_n}^{t_{n+1}} f(\varphi^*(t, x_{j+1/2}))dt \, . \tag{9.89}$$

Here φ^* is the solution of the Riemann problem with left constant state φ_j^n and right constant state φ_{j+1}^n. This gives the exact result for τ small enough such that waves coming from adjacent cell faces do not interact, and if we define $\varphi(t_n, x)$ to be piecewise constant, with $\varphi(t_n, x) = \varphi_j^n$ for $x_j - h/2 < x < x_j + h/2$. Since φ^* is a similarity solution: $\varphi^*(t, x) = \varphi^*((x - x_{j+1/2})/t)$, we have $\varphi^*(t, x_{j+1/2}) = \varphi^*(0) = $ constant, so that (9.89) becomes

$$F(\varphi_j^n, \varphi_{j+1}^n) = f(\varphi^*(0)) \, .$$

The wave speed is bounded by $|f'(\varphi)|$, so that there will be no interaction between adjacent Riemann problem if

$$|\lambda f'(\varphi_j^n)| \leq 1 \, , \quad j = 0, \pm1, \pm2, \ldots$$

This is in fact a restriction on the CFL number, similar to the stability restrictions that we found for the linear case in Sect. 9.3.

Assume $f'' > 0$. Recalling the solution of the Riemann problem obtained earlier in this section, if $\varphi_{j+1}^n < \varphi_j^n$ then we have a shock propagating with velocity

$$s = \{f(\varphi_{j+1}^n) - f(\varphi_j^n)\}/(\varphi_{j+1}^n - \varphi_j^n) \, , \tag{9.90}$$

so that in (9.89)

$$\varphi^*(0) = \varphi_{j+1}^n \, , \quad s < 0 \, ; \quad \varphi^*(0) = \varphi_j^n \, , \quad s > 0 \, . \tag{9.91}$$

On the other hand, if $\varphi_{j+1}^n \geq \varphi_j^n$ then we have a fan satisfying (9.70). In our case, $\xi = (x - x_{j+1/2})/t$, $x = x_{j+1/2}$, so that $\xi = 0$, and $\varphi^*(0)$ follows from from the fan solution (9.71) with $x = 0$:

$$\varphi^*(0) = \varphi_j^n \, , \quad f'(\varphi_j^n) > 0 \, ; \tag{9.92}$$

$$\varphi^*(0) = \varphi_{j+1}^n \, , \quad f'(\varphi_{j+1}^n) < 0 \, ; \tag{9.93}$$

$$f'(\varphi^*(0)) = 0 \, , \quad f'(\varphi_j^n) < 0 < f'(\varphi_{j+1}^n) \, . \tag{9.94}$$

This completes the specification of the Godunov scheme.

Because $f'' > 0$, $f'(\varphi_j^n) > 0$ implies $s > 0$, and $f'(\varphi_{j+1}^n) < 0$ implies $s < 0$, with s given by (9.90). Hence, if we were to use instead of an exact Riemann solver (as Godunov) an approximate Riemann solver which violates the entropy condition and which uses an expansion shock rather than an expansion fan, so that we always use (9.90) and (9.91), then only in grid points where (9.94) occurs an error is made. Because of an analogy with gasdynamics, the case of (9.94) is called a transonic rarefaction.

In LeVeque (1992) Sect. 13.2 the following equivalent expression is given for the Godunov flux:

$$
\begin{aligned}
F(\varphi_j^n, \varphi_{j+1}^n) &= \min\{f(\varphi) : \varphi_j^n \leq \varphi \leq \varphi_{j+1}^n\} \, , \varphi_j^n \leq \varphi_{j+1}^n \, , \\
F(\varphi_j^n, \varphi_{j+1}^n) &= \max\{f(\varphi) : \varphi_{j+1}^n \leq \varphi \leq \varphi_j^n\} \, , \varphi_j^n > \varphi_{j+1}^n \, .
\end{aligned}
\tag{9.95}
$$

It turns out that this gives the correct flux for al scalar conservation laws, even nonconvex ones, for the Riemann solution that satisfies the entropy condition.

It has been shown by Crandall and Majda (1980) that the Godunov scheme is monotone. Hence, it satisfies the entropy condition, according to Theorem 9.4.3. A different proof that the Godunov scheme satisfies the entropy condition has been given by Harten and Lax (1981).

Higher order schemes

As is easily seen in the linear case, the Godunov scheme is only first order accurate. Lax-Wendroff type schemes, such as MacCormack's scheme (9.83), are second order accurate, but do not preserve monotonicity, and show spurious wiggles near discontinuities. For efficient computation of practical flow problems, second or higher order accuracy is desirable in smooth regions of the flow, with crisp resolution of discontinuities, while avoiding spurious wiggles near discontinuities. A way to achieve this is by extending the technique of flux limiting discussed in Sect. 4.8 to the nonlinear case.

First of all, schemes that are second order accurate in space in smooth parts of the flow have to be developed, generalizing the κ-scheme of Sect. 4.8 to the nonlinear case. A way to do this has been proposed by van Leer (1979), and is generally known as the MUSCL (monotone upwind schemes for conservation laws) approach. The idea is not to use the piecewise constant states φ_j and φ_{j+1} as input for the (approximate) Riemann solver $F(\varphi_j, \varphi_{j+1})$, but to extrapolate with a directional bias from the left and right to obtain right and left states $\varphi_{j+1/2}^+$ and $\varphi_{j+1/2}^-$, which are used as input for the (approximate) Rieman solver $F(\varphi_{j+1/2}^-, \varphi_{j+1/2}^+)$. This is to be done in such a way that in the linear case $(f(\varphi) = u\varphi)$ the numerical flux of the κ-scheme (9.30) is obtained. We therefore write

$$
\begin{aligned}
\varphi_{j+1/2}^- &= \frac{1}{2}(\varphi_j + \varphi_{j+1}) + \frac{1-\kappa}{4}(-\varphi_{j-1} + 2\varphi_j - \varphi_{j+1}) , \\
\varphi_{j+1/2}^+ &= \frac{1}{2}(\varphi_j + \varphi_{1+1}) + \frac{1-\kappa}{4}(-\varphi_j + 2\varphi_{j+1} - \varphi_{j+2}) .
\end{aligned}
\tag{9.96}
$$

In the numerical flux function F at $x_{j+1/2}$ the left and right states φ_j and φ_{j+1} are simply replaced by $\varphi_{j+1/2}^+$ and $\varphi_{j+1/2}^-$, respectively. This gives:

$$
\frac{d\varphi_j}{dt} + \frac{1}{h_j}\{F(\varphi_{j+1/2}^+, \varphi_{j+1/2}^-) - F(\varphi_{j-1/2}^+, \varphi_{j-1/2}^-)\} = 0 ,
\tag{9.97}
$$

with F a consistent numerical flux function, i.e. $F(\varphi, \varphi) = f(\varphi)$. This spatial discretization can be shown to be second order accurate (cf. Kröner (1997) Sect. 2.5). It is left to the reader to show that in the linear case $(f(\varphi) = c\varphi)$ the scheme (9.97) reduces to the κ-scheme (9.30).

Van Leer obtained (9.96) by replacing the piecewise constant per cell approximation of φ by a piecewise linear distribution. In order to avoid spurious wiggles, (9.96) is not used, but the slopes of the piecewise linear distribution are limited. Therefore this class of schemes, to be described next, may be called *slope limited schemes*.

Another way to obtain wiggle-free schemes is to put a limiter on the numerical flux function, resulting in *flux limited schemes*. This approach has been pioneered by Harten (1983), and put in a general framework by Sweby (1984). It turns out that the principles to ensure the absence of spurious wiggles are about the same as for slope limited schemes, and similar limiter functions are used. Numerical experiments by Anderson, Thomas, and van Leer (1986) show somewhat better accuracy for slope limited schemes as compared with flux limited schemes. We will not discuss flux limited schemes.

Still another way to realize second order schemes without spurious wiggles is to add explicitly a judicious amount of artificial viscosity to a second order central scheme. This results in *artificial viscosity schemes*, to be discussed later.

Slope limited schemes

In order to rule out spurious wiggles, the κ-scheme has to be modified. One way to do this follows easily from the linear case discussed in Sect. 4.8. As in (4.102) we replace (9.96) by

$$\varphi_{j+1/2}^{+} = \varphi_j + \frac{1}{2}\psi(r_j)(\varphi_{j+1} - \varphi_j) \,,$$

$$\varphi_{j+1/2}^{-} = \varphi_{j+1} + \frac{1}{2}\psi(\frac{1}{r_{j+1}})(\varphi_j - \varphi_{j+1}) \,, \qquad r_j = \frac{\varphi_j - \varphi_{j-1}}{\varphi_{j+1} - \varphi_j} \,. \tag{9.98}$$

We will now derive conditions on the limiter $\psi(r)$ for the resulting scheme to be monotonicity preserving. This is an easy generalization of the conditions obtained for the stationary linear case, provided the numerical flux function satisfies the following conditions:

$$F(\varphi_{j+1/2}^{+}, \varphi_{j+1/2}^{-}) = F^{-}(\varphi_{j+1/2}^{-}) + F^{+}(\varphi_{j+1/2}^{+}) \,,$$

$$\frac{dF^{-}(\varphi)}{d\varphi} \le 0 \,, \quad \frac{dF^{+}}{d\varphi} \ge 0 \,, \quad \forall \varphi \in \mathbb{R} \,. \tag{9.99}$$

The resulting semi-discetized scheme is given by

$$\frac{d\varphi_j}{dt} + \frac{1}{h}\{F^{-}(\varphi^{-}) + F^{+}(\varphi^{+})\}|_{j-1/2}^{j+1/2} = 0 \,. \tag{9.100}$$

In the case of explicit Euler discretization we get

$$\varphi_j^{n+1} - \varphi_j^{n} + \lambda\{F^{-}(\varphi^{-}) + F^{+}(\varphi^{+})\}|_{j-1/2}^{j+1/2} = 0 \,. \tag{9.101}$$

The basic idea here is to extrapolate a left state $\varphi_{j+1/2}^{+}$ and a right state $\varphi_{j+1/2}^{-}$ from neighboring grid points, to put limitations on $\varphi_{j+1/2}^{\pm}$ in order to suppress spurious wiggles, and to base the numerical flux $F_{j+1/2}$ on (approximate) solution of the Riemann problemRiemann!approximate solver corresponding to the left and right states $\varphi_{j+1/2}^{\pm}$.

Total variation decreasing schemes

In view of Theorem 9.4.3 one would ideally like to have monotone schemes. However, this seems to be asking too much for second order schemes. Instead, we will be content with a weaker requirement that excludes spurious wiggles. Harten (1983) shows (in the one-dimensional scalar case) that a TVD scheme is monotonicity preserving. The discrete TVD property has been defined after equation (9.19). Hence, TVD schemes share the properties put forward in Theorems 9.4.1 and 9.4.2 with the exact solution. For a TVD scheme the total variation is obviously bounded (TVB), and TVB schemes are convergent (see LeVeque (1992) Sect. 15.3, for example). Hence, TVD schemes are stable.

We will give conditions for explicit schemes to be TVD. Let the scheme be denoted by

$$\varphi_j^{n+1} = \varphi_j^{n} + a_j(\varphi_{j+1}^{n} - \varphi_j^{n}) + b_j(\varphi_{j-1}^{n} - \varphi_j^{n}) \,, \tag{9.102}$$

where a_j and b_j may depend on $\varphi_j^{n}, \varphi_{j\pm1}^{n}, \varphi_{j\pm2}^{n}, \dots$ We have the following theorem.

Theorem 9.4.4. *(Harten (1983)).*
If the coefficients in scheme (9.102) satisfy for all j

$$a_j \geq 0, \quad b_j \geq 0, \quad a_j + b_{j+1} \leq 1, \qquad (9.103)$$

then the scheme is TVD.

Proof. It follows from (9.102) that

$$\varphi_{j+1}^{n+1} - \varphi_j^{n+1} = b_j(\varphi_j^n - \varphi_{j-1}^n) + (1 - b_{j+1} - a_j)(\varphi_{j+1}^n - \varphi_j^n) \\ + a_{j+1}(\varphi_{j+2}^n - \varphi_{j+1}^n). \qquad (9.104)$$

Because of (9.103) the coefficients in (9.104) are nonnegative, so that

$$|\varphi_{j+1}^{n+1} - \varphi_j^{n+1}| \leq b_j|\varphi_j^n - \varphi_{j-1}^n| + (1 - b_{j+1} - a_j)|\varphi_{j+1}^n - \varphi_j^n| \\ + a_{j+1}|\varphi_{j+2}^n - \varphi_{j+1}^n|.$$

Summation over j gives

$$\mathrm{TV}(\varphi^{n+1}) = \sum_j |\varphi_{j+1}^{n+1} - \varphi_j^{n+1}| \leq \\ \sum_j (b_{j+1} + 1 - b_{j+1} - a_j + a_j)|\varphi_{j+1}^n - \varphi_j^n| = \mathrm{TV}(\varphi^n).$$

□

We now give conditions for the slope limited scheme (9.101) to be TVD. The scheme can be written in the form (9.102) with

$$a_j = -\lambda \frac{F^-(\varphi_{j+1/2}^-) - F^-(\varphi_{j-1/2}^-)}{\varphi_{j+1/2}^- - \varphi_{j-1/2}^-} \cdot \frac{\varphi_{j+1/2}^- - \varphi_{j-1/2}^-}{\varphi_{j+1}^n - \varphi_j^n} \\ b_j = \lambda \frac{F^+(\varphi_{j+1/2}^+) - F^+(\varphi_{j-1/2}^+)}{\varphi_{j+1/2}^+ - \varphi_{j-1/2}^+} \cdot \frac{\varphi_{j+1/2}^+ - \varphi_{j-1/2}^+}{\varphi_j^n - \varphi_{j-1}^n}. \qquad (9.105)$$

Since according to (9.99) we have

$$\frac{dF^-(\varphi)}{d\varphi} \leq 0, \quad \frac{dF^+}{d\varphi} \geq 0, \quad \forall \varphi \in \mathbb{R},$$

the first fraction in a_j is nonpositive and the first fraction in b_j is nonnegative. In order to have $a_j \geq 0$, $b \geq 0$ according to (9.103), we must therefore have

$$\frac{\varphi_{j+1/2}^- - \varphi_{j-1/2}^-}{\varphi_{j+1}^n - \varphi_j^n} \geq 0, \quad \frac{\varphi_{j+1/2}^+ - \varphi_{j-1/2}^+}{\varphi_j^n - \varphi_{j-1}^n} \geq 0. \qquad (9.106)$$

This is equivalent to (4.109), and conditions on the limiter $\psi(r)$ to satisfy (9.106) have already been studied in Sect. 4.8. The graph of $\psi(r)$ must lie in the admissible region of Fig. 4.11 It remains to satisfy the last part of (9.103).

Assume

$$\frac{\varphi^-_{j+1/2} - \varphi^-_{j-1/2}}{\varphi^n_{j+1} - \varphi^n_j} \leq M , \quad \frac{\varphi^+_{j+1/2} - \varphi^+_{j-1/2}}{\varphi^n_j - \varphi^n_{j-1}} \leq M$$

(this is the case if the graph of $\psi(r)$ satisfies the condition just stated). Define

$$c_{j+1/2} \equiv \frac{F^+(\varphi^+_{j+3/2}) - F^+(\varphi^+_{j+1/2})}{\varphi^+_{j+3/2} - \varphi^+_{j+1/2}} - \frac{F^-(\varphi^-_{j+1/2}) - F^-(\varphi^-_{j-1/2})}{\varphi^-_{j+1/2} - \varphi^-_{j-1/2}} \leq c , \quad \forall j .$$

Then the last part of (9.103) is satisfied if the time step is chosen small enough, i.e.

$$\lambda \leq \frac{1}{cM} . \tag{9.107}$$

This is similar to a stability condition on the CFL number, since it turns out that for practical schemes $c_{j+1/2}$ is related to the wave propagation velocity $f'(\varphi)$. For example, for the Engquist-Osher scheme we have

$$c_{j+1/2} = \frac{1}{\varphi^+_{j+3/2} - \varphi^+_{j+1/2}} \int_{\varphi^+_{j+1/2}}^{\varphi^+_{j+3/2}} \max\{f'(\psi), 0\} d\psi$$

$$- \frac{1}{\varphi^-_{j+1/2} - \varphi^-_{j-1/2}} \int_{\varphi^-_{j-1/2}}^{\varphi^-_{j+1/2}} \min\{f'(\psi), 0\} d\psi .$$

Hence

$$0 \leq c_{j+1/2} \leq c = \max\{|f'(\varphi)|\} .$$

It now becomes clear why it is reasonable to choose $M = 3$, leading to the admissible region for the graph of $\psi(r)$ given in Fig. 4.11. In Sect. 4.8 it has been shown that for second order accuracy we must have $\psi(1) = 1$, leading to $M \geq 2$. In order to give limiters some leeway we choose $M > 2$; in order to make (9.107) not overly strict we do not want to choose M large, leaving us with $M = 3$ as a reasonable compromise.

For first order flux splitting schemes that do not use extrapolation and limiting the analysis is easier. For the Enquist-Osher scheme we have already obtained condition (9.88) for the scheme to be monotone, hence TVD.

As we remarked before, TVD schemes are stable. Furthermore, scheme (9.101) is in conservation form, so that genuine weak solutions satisfying the jump condition will be obtained. In order to demonstrate satisfaction of the entropy condition it would suffice that the scheme is monotone, but this is not the case. But that correct solutions will be obtained can be made plausible as follows. The main defect of schemes that do not satisfy the entropy condition is that they allow an expansion shock where the exact solution has an expansion fan, and is smooth and monotone. In the case $f' > 0$, $f'' > 0$ (for example) the exact solution would be smoothly increasing. Now suppose the numerical solution shows locally an expansion shock, with $0 < \varphi_{j-1} < \varphi_j \ll \varphi_{j+1}$. This brings r_j close to zero, hence $\psi(r_j)$ will be close to zero (cf. Fig. 4.12), and the slope limiter switches the scheme to the underlying first order scheme. But this scheme does not allow an expansion shock, if it satisfies the entropy condition. Hence, we may be confident that slope limited second order schemes will not develop expansion shocks, of the underlying first order scheme satisfies the entropy condition. It suffices that the first order scheme is monotone, which is easy to check.

Other temporal discretizations

It is frequently desirable to replace the $\mathcal{O}(\tau)$ temporal discretization (9.101) by an $\mathcal{O}(\tau^2)$ scheme for better accuracy, or by an implicit scheme, in order to relax the time step restriction (9.107). We start with the following implicit scheme:

$$
\begin{aligned}
&\varphi_j^{n+1} + c_j(\varphi_{j+1}^{n+1} - \varphi_j^{n+1}) + d_j(\varphi_{j-1}^{n+1} - \varphi_j^{n+1}) = \\
&\varphi_j^n + a_j(\varphi_{j+1}^n - \varphi_j^n) + b_j(\varphi_{j-1}^n - \varphi_j^n) \,,
\end{aligned}
\tag{9.108}
$$

where a_j and φ_j may depend on $\varphi_j^n, \varphi_{j\pm1}^n, \varphi_{j\pm2}^n, \dots$ and c_j and d_j may depend on $\varphi_j^{n+1}, \varphi_{j\pm1}^{n+1}, \varphi_{j\pm2}^{n+1} \dots$. The following theorem gives conditions for scheme (9.108) to be TVD.

Theorem 9.4.5. *(Harten (1984)).*
If the coefficients in scheme (9.108) satisfy for all j

$$
a_j \geq 0\,, \quad b_j \geq 0\,, \quad a_j + b_j \leq 1\,, \quad -\infty < C \leq c_j\,, \ d_j \leq 0
$$

then the scheme is TVD.

Proof. Let us denote the scheme (9.108) as

$$
L\varphi^{n+1} = R\varphi^n \,.
$$

First we show that $\mathrm{TV}(\varphi) \leq \mathrm{TV}(L\varphi)$. Writing $\psi = L\varphi$, we have

$$
\begin{aligned}
&\psi_{j+1} - \psi_j = d_j(\varphi_j - \varphi_{j-1}) + (1 - c_j - d_j)(\varphi_{j+1} - \varphi_j) \\
&\quad + c_{j+1}(\varphi_{j+2} - \varphi_{j+1}) \,, \\
&\psi_{j+1} - \psi_j - d_j(\varphi_j - \varphi_{j-1}) - c_{j+1}(\varphi_{j+2} - \varphi_{j+1}) = \\
&(1 - c_j - d_{j+1})(\varphi_{j+1} - \varphi_j) \,.
\end{aligned}
$$

Taking absolute values, taking the signs of the coefficients into account and using the triangle inequality we obtain

$$|\psi_{j+1} - \psi_j| - d_j|\varphi_j - \varphi_{j-1}| - c_{j+1}|\varphi_{j+2} - \varphi_{j+1}| \geq$$
$$(1 - c_j - d_{j+1})|\varphi_{j+1} - \varphi_j| .$$

Summation over j gives

$$\mathrm{TV}(\psi) - \sum(d_{j+1} + c_j)|\varphi_{j+1} - \varphi_j| \geq \mathrm{TV}(\varphi) - \sum(d_{j+1} + c_j)|\varphi_{j+1} - \varphi_j| .$$

Hence $\mathrm{TV}(\psi) \geq \mathrm{TV}(\varphi)$, or

$$\mathrm{TV}(\varphi) \leq \mathrm{TV}(L\varphi) .$$

From Theorem 9.4.4 it follows that

$$\mathrm{TV}(R\varphi) \leq \mathrm{TV}(\varphi) .$$

Combination of the last two equations gives the desired result:

$$\mathrm{TV}(\varphi^{n+1}) \leq \mathrm{TV}(L\varphi^{n+1}) = \mathrm{TV}(R\varphi^n) \leq \mathrm{TV}(\varphi^n) .$$

$\square$

Let us apply this theorem to the temporal discretization of (9.100) with the ω-method:

$$\varphi_j^{n+1} + \lambda\{\omega F^-(\varphi^{-,n+1}) + \omega F^+(\varphi^{+,n+1}) +$$
$$(1 - \omega)F^-(\varphi^{-,n}) + (1 - \omega)F^+(\varphi^{+,n})\}|_{j-1/2}^{j+1/2} = 0 .$$

with $F^\pm$ satisfying (9.99). This results in

$$\varphi_j^{n+1} - \omega a_j^{n+1}(\varphi_{j+1}^{n+1} - \varphi_j^{n+1}) - \omega b_j^{n+1}(\varphi_{j-1}^{n+1} - \varphi_j^{n+1}) =$$
$$\varphi_j^n + (1 - \omega)a_j^n(\varphi_{j+1}^n - \varphi_j^n) + (1 - \omega)b_j^n(\varphi n_{j-1} - \varphi_j^n) ,$$

where a_j^n, b_j^n, a_j^{n+1} and b_j^{n+1} are given by (9.105) by putting in the appropriate time level superscripts. According to the preceding theorem, this scheme is TVD if, assuming $0 \leq \omega \leq 1$,

$$a_j^{n+1} \geq 0 , \quad b_j^{n+1} \geq 0 , \quad a_j^n \geq 0 , \quad b_j^n \geq 0 , \tag{9.109}$$

and

$$(1 - \omega)(a_j^n + b_j^n) \leq 1 . \tag{9.110}$$

The explicit case $\omega = 0$ has already been studied. Conditions (9.109) also have to be satisfied in the explicit case, leading to (9.106), which results in the usual restrictions on the limiter $\psi(r)$. But for $\omega > 0$ (9.110) is weaker than the last condition of (9.103), and it is easy to see that (9.107) is replaced by the weaker time step restriction

$$\lambda \le \frac{1}{(1-\omega)cM} \cdot$$

The price to pay is of course the need to solve a nonlinear system per time step. If we linearize a^{n+1} and b^{n+1} in some fashion, the system to be solved becomes linear, but is no longer conservative. This makes it less suitable for time-accurate computation of time-varying flows, but for time-stepping to steady state with large time-steps (taking ω close to 1) a linearized scheme is fine, since the steady-state version of the scheme is conservative. Furthermore, the matrix of the system to be solved is obviously diagonally dominant, provided the inequalities (9.109) are maintained after linearization.

The preceding implicit scheme is $\mathcal{O}(\tau^2)$ accurate for $\omega = 1/2$. Higher order accuracy may also be obtained with explicit schemes. For brevity we rewrite the semi-discretized scheme (9.100) as

$$\frac{d\varphi}{dt} = \frac{1}{h}R(\varphi) \,,$$
$$R(\varphi)_j \equiv -\{F^-(\varphi^-) + F^+(\varphi^+)\}|_{j-1/2}^{j+1/2} \,,$$

where φ is the algebraic vector with elements φ_j. Let us consider Runge-Kutta methods for temporal discretization. A two-stage Runge-Kutta method (known as the method of Heun) is given by:

$$\varphi^* = \varphi^n + \lambda R(\varphi^n) \,,$$
$$\varphi^{n+1} = \varphi^n + \frac{1}{2}\lambda\{R(\varphi^n) + R(\varphi^*)\} \,. \tag{9.111}$$

We will show that this scheme is TVD (hence stable) under certain conditions. The scheme can be rewritten as a succession of two explicit Euler steps:

$$\varphi^* = \varphi^n + \lambda R(\varphi^n) \,,$$
$$\varphi^{**} = \varphi^* + \lambda R(\varphi^*) \,,$$
$$\varphi^{n+1} = \frac{1}{2}(\varphi^n + \varphi^{**}) \,.$$

Conditions for the explicit Euler scheme to be TVD have already been given. If these conditions are satisfied, then

$$TV(\varphi^{**}) \le TV(\varphi^*) \,, \quad TV(\varphi^*) \le TV(\varphi^n) \,.$$

Since

$$TV(\varphi^{n+1}) \le \frac{1}{2}TV(\varphi^n) + \frac{1}{2}TV(\varphi^{**})$$

the Runge-Kutta scheme (9.111) is TVD under the same conditions as the explicit Euler scheme (9.101). If the spatial accuracy is $\mathcal{O}(h^2)$, the temporal accuracy need not be better than $\mathcal{O}(\tau^2)$, so (9.111) is an attractive time-accurate scheme.

If the spatial accuracy is third order, for example when a limited $\kappa = 1/3$ scheme is used, then it may be attractive to have third order accuracy in time as well. The following Runge-Kutta method

$$\varphi^* = \varphi^n + \lambda R(\varphi^n) ,$$

$$\varphi^{**} = \frac{3}{4}\varphi^n + \frac{1}{4}\varphi^* + \frac{\lambda}{4}R(\varphi^*) , \qquad (9.112)$$

$$\varphi^{n+1} = \frac{1}{3}\varphi^n + \frac{2}{3}\varphi^{**} + \frac{2\lambda}{3}R(\varphi^{**}) ,$$

is shown in Shu and Osher (1988) to be third order accurate in time and TVD under the same condition under which the explicit Euler scheme is TVD, i.e. (9.107). A general study of TVD Runge-Kutta schemes is presented in Shu (1988), and the subject is also considered in Hundsdorfer, Koren, van Loon, and Verwer (1995).

This completes our general discussion of higher order TVD schemes for scalar conservation laws. Specific schemes may now be formulated by combining an approximate Riemann solver (i.e. a numerical flux function) with a limiter function, corresponding with some value of κ; a number of limiter functions are discussed in Sect. 4.8. But we will discuss particular schemes and numerical results not for the scalar case, but for the Euler equations, in Chap. 10.

Artificial viscosity schemes

The TVD schemes discussed before belong to the class of approximate Riemann solvers: the numerical flux $F^n_{j+1/2}$ is related to the approximate solution of the Riemann problem based on two constant states φ^n_j and φ^n_{j+1}, or $\varphi^+_{j+1/2}$ and $\varphi^-_{j+1/2}$ in the case of second order slope limited schemes. In essence, these are generalizations of upwind biased schemes to the nonlinear case. To combine second order accuracy and the TVD property, one may also try to modify central schemes, such as the Lax-Wendroff scheme, by adding a solution-dependent amount of artificial viscosity. A widely used scheme of this type has been developed in Jameson, Schmidt, and Turkel (1981), Jameson (1985b), Jameson (1985a), Jameson (1988), and is generally known as the Jameson-Schmidt-Turkel scheme or JST scheme for short. The artificial viscosity has to be judiciously chosen, and has evolved over the years. Jameson (1985a) proposes the following scheme. The conservation law (9.56) is semi-discretized as follows:

$$\frac{d\varphi_j}{dt} + \frac{1}{h}\{F(\varphi) + d\}|^{j+1/2}_{j-1/2} = 0 \,. \qquad (9.113)$$

Here $F(\varphi)$ is a central approximation of the flux:

$$F(\varphi_{j+1/2}) = \frac{1}{2}\{f(\varphi_j) + f(\varphi_{j+1})\} \,.$$

The term d represents artificial dissipation, needed to suppress spurious wiggles. The term $d(\varphi)$ must be carefully chosen to ensure crisp resolution of discontinuities. Jameson, Schmidt, and Turkel (1981) define d as follows:

$$d_{j+1/2} = -\varepsilon^{(2)}_{j+1/2}R_{j+1/2}(\varphi_{j+1} - \varphi_j) +$$
$$\varepsilon^{(4)}_{j+1/2}R_{j+1/2}(\varphi_{j+2} - 3\varphi_{j+1} + 3\varphi_j - \varphi_{j-1}) \,.$$

Here $\varepsilon^{(2)}$ and $\varepsilon^{(4)}$ are solution-dependent coefficients, and $R_{j+1/2}$ is a factor to give $d_{j+1/2}$ the proper dimension and scale:

$$R_{j+1/2} = \frac{1}{2}\{f'(\varphi_j) + f'(\varphi_{j+1})\} \,, \quad \varepsilon^{(2)}_{j+1/2} = \min\{\frac{1}{2}, k^{(2)}\bar{\nu}_{j+1/2}\} \,,$$

$$\varepsilon^{(4)}_{j+1/2} = \max\{0, k^{(4)} - \alpha\bar{\nu}_{j+1/2}\} \,, \quad \bar{\nu}_{j+1/2} = \max\{\nu_{j+2}, \nu_{j+1}, \nu_j, \nu_{j-1}\} \,,$$

$$\nu_j = \left|\frac{\varphi_{j+1} - 2\varphi_j + \varphi_{j-1}}{\varphi_{j+1} + 2\varphi_j + \varphi_{j-1}}\right| \,, \quad k^{(2)} = 1 \,, \quad k^{(4)} = \frac{1}{32} \,, \quad \alpha = 2 \,.$$

The quantity ν_j is meant to be a sensor of shocks. When it is small, $\varepsilon^{(2)}$ is small, and $\varepsilon^{(4)} \approx k^{(4)}$, giving only an $\mathcal{O}(h^3)$ dissipative flux d. Near a shock $\nu_j \approx 1$ and $\varepsilon^{(2)}_{j+1/2} \approx 1$, and the artificial viscosity makes the scheme first order accurate.

Although the preceding spatial discretization appears to work well in practice, it does not seem to be TVD. Jameson (1988) presents an artificial viscosity scheme that is TVD, based on the following train of thought.

Let the artificial dissipation be given by

$$d_{j+1/2} = e_{j+3/2} - 2e_{j+1/2} + e_{j-1/2} \,,$$

where

$$e_{j+1/2} = \alpha_{j+1/2}(\varphi_{j+1} - \varphi_j) \,, \quad \alpha_{j+1/2} > 0 \,.$$

Noting that

$$F(\varphi)|^{j+1/2}_{j-1/2} = \frac{1}{2}f(\varphi)|^{j+1}_j + \frac{1}{2}f(\varphi)|^j_{j-1} = \frac{1}{2}c_{j+1/2}\varphi|^{j+1}_j - \frac{1}{2}c_{j-1/2}\varphi|^j_{j-1} \,,$$

with

$$c_{j+1/2} = \frac{f(\varphi_{j+1}) - f(\varphi_j)}{\varphi_{j+1} - \varphi_j} , \quad \varphi_{j+1} \neq \varphi_j ; \quad c_{j+1/2} = f'(\varphi_j) , \quad \varphi_{j+1} = \varphi_j ,$$

we can rewrite the scheme (9.113) as follows:

$$h \frac{d\varphi_j}{dt} = -\frac{1}{2} c_{j+1/2} \varphi|_j^{j+1} + \frac{1}{2} c_{j-1/2} \varphi|_{j-1}^{j} - \alpha_{j+3/2} \varphi|_j^{j+2} +$$
$$3\alpha_{j+1/2} \varphi|_j^{j+1} - 3\alpha_{j-1/2} \varphi|_{j-1}^{j} + \alpha_{j-3/2} \varphi|_{j-2}^{j-1} .$$

It can be shown that for $\{\varphi_j(t)\}$ to be TVD it is necessary that $\alpha_{j\pm3/2} \leq 0$, which is not the case. Therefore a limiter $\psi(r)$ is introduced in the artificial dissipation, redefining $d_{j+1/2}$ as

$$d_{j+1/2} = \psi(r_{j+1}) e_{j+3/2} - 2 e_{j+1/2} + \psi(\frac{1}{r_j}) e_{j-1/2} , \quad r_j = \frac{e_{j-1/2}}{e_{j+1/2}} .$$

Noting that $e_{j+3/2} = e_{j+1/2}/r_{j+1}$ and $e_{j-3/2} = r_{j-1} e_{j-1/2}$, the scheme becomes

$$h \frac{d\varphi_j}{dt} = -\frac{1}{2} c_{j+1/2} \varphi|_j^{j+1} + \frac{1}{2} c_{j-1/2} \varphi|_{j-1}^{j} + \alpha_{j+1/2} \{2 + \psi(r_j) -$$
$$\frac{1}{r_{j+1}} \psi(r_{j+1}) \} \varphi|_j^{j+1} - \alpha_{j-1/2} \{2 + \psi(\frac{1}{r_j}) - r_{j-1} \psi(\frac{1}{r_{j-1}}) \} \varphi|_{j-1}^{j} .$$

$$(9.114)$$

It is easily shown that $\{\varphi_j(t)\}$ is TVD if the coefficient of $\varphi|_j^{j+1}$ is nonnegative and the coefficient of $\varphi|_{j-1}^{j}$ is nonpositive. This gives

$$\alpha_{j+1/2} \{2 + \psi(r_j) - \frac{1}{r_{j+1}} \psi(r_{j+1}) \} - \frac{1}{2} c_{j+1/2} \geq 0 ,$$

$$\alpha_{j-1/2} \{2 + \psi(\frac{1}{r_j}) - r_{j-1} \psi(\frac{1}{r_{j-1}}) \} - \frac{1}{2} c_{j-1/2} \geq 0 .$$

Choosing

$$\alpha_{j+1/2} \geq \frac{1}{2} |c_{j+1/2}| ,$$

we must have

$$2 + \psi(r) - \frac{1}{s} \psi(s) \geq 0 , \quad \forall r, s \in \mathbb{R} .$$

This satisfied if

$$\psi(r) \geq M - 1 \geq -2 , \quad \psi(r) \leq (1 + M) r . \quad (9.115)$$

Hence, it is sufficient if the graph of $\psi(r)$ lies in the admissible region of Fig. 4.12. Jameson (1988) proposes:

$$\psi(r) = 0 \, , \quad r < 0 \, ; \quad \psi(r) = r \, , \quad 0 \geq r \geq 1 \, ; \quad \psi(r) = 1 \, , \quad r > 1 \, .$$

Note that this makes the semi-discretized system (9.113) TVD, but the TVD property can still be destroyed by the temporal discretization. Further developments of TVD versions of the JST scheme are described in Jameson (1995a), Jameson (1995b), Kim and Jameson (1995).

Jameson, Schmidt, and Turkel (1981), Jameson (1985b), Jameson (1985a), Jameson (1988) consider only time-stepping schemes designed to reach steady state quickly, without regard for temporal accuracy. The semi-discretized scheme (9.113) can be denoted as

$$\frac{d\varphi}{dt} + R(\varphi) = 0 \, ,$$

where $\varphi(t)$ is an algebraic vector containing the unknowns φ_j. Runge-Kutta schemes are selected for temporal discretization. The following class of m-stage Runge-Kutta schemes is considered, to step from $t = n\tau$ to $t = (n+1)\tau$:

$$
\begin{aligned}
\varphi^{(0)} &= \varphi^n \, , \\
\varphi^{(1)} &= \varphi^{(0)} - \beta_1 \tau R^{(0)} \, , \\
&\cdots\cdots\cdots \\
\varphi^{(m-1)} &= \varphi^{(0)} - \beta_{m-1} \tau R^{(m-2)} \, , \\
\varphi^{(m)} &= \varphi^{(0)} - \tau R^{(m-1)} \, , \\
\varphi^{n+1} &= \varphi^{(m)} \, ,
\end{aligned}
\tag{9.116}
$$

where

$$R^{(k)} = \sum_{r=0}^{k} \gamma_{kr} R(\varphi^{(r)}) \, , \quad \sum_{r=0}^{k} \gamma_{kr} = 1 \, .$$

The coefficients β_k are chosen not to achieve temporal accuracy, but to tailor the stability domain such that a large time step τ can be taken. Because the symbol of the linearized differential equation is purely imaginary (it is left to the reader to check this), β_r is chosen to maximize the stability interval along the imaginary axis (Kinmark (1984)). Because of the artificial dissipation, the symbol of the (linearized) scheme is not purely imaginary, and stability cannot be completely guaranteed, the more so because the scheme is nonlinear. When we have reached steady state, we have solved

$$R(\varphi) = 0 \, ,$$

which can be written as (cf. (9.114))

$$a_j(\varphi_{j+1} - \varphi_j) - b_j(\varphi_{j-1} - \varphi_j) = 0 \, , \quad j = 1, ..., J-1 \, . \tag{9.117}$$

Boundary conditions are left unspecified. The following theorem shows why the limiter introduced in (9.114) is useful.

Theorem 9.4.6. *If*

$$a_j \geq 0 , \quad b_j \geq 0 \tag{9.118}$$

then the solution of (9.117) is monotone.

Proof. We can write

$$\varphi_j = \frac{a_j}{a_j + b_j} \varphi_{j+1} + \frac{b_j}{a_j + b_j} \varphi_{j-1} , \quad j = 1, ..., J - 1 .$$

Since $0 \leq \frac{a_j}{a_j + b_j}, \frac{b_j}{a_j + b_j} \leq 1$ we have

$$\min\{\varphi_{j-1}, \varphi_{j+1}\} \leq \varphi_j \leq \max\{\varphi_{j-1}, \varphi_{j+1}\} ,$$

therefore φ_j cannot be a local extremum. This applies for all j, so that $\{\varphi_j\}$ is monotone. □

Hence, if (9.118) is satisfied then spurious wiggles are excluded. Conditions (9.118) are satisfied if the limiter $\psi(r)$ in (9.114) satisfies conditions (9.115).

Further specifications of the time-stepping scheme are given in Jameson (1985b). For efficiency, the dissipative part of the flux is treated different from the convective part, defining $R^{(k)}$ as follows:

$$R(\varphi)_j = Q(\varphi)_j + D(\varphi)_j ,$$

$$Q(\varphi_j) = -\frac{1}{h} F(\varphi)\big|_{j-1/2}^{j+1/2} , \quad D = -\frac{1}{h} d\big|_{j-1/2}^{j+1/2} ,$$

$$R^{(k)} = \sum_{r=0}^{k} \{\gamma_{kr} Q(\varphi^{(r)}) + \delta_{kr} D(\varphi^r)\} ,$$

$$\sum_{r=0}^{k} \gamma_{kr} = 1 , \quad \sum_{r=0}^{k} \delta_{kr} = 1 .$$

In Jameson (1985b) it is made plausible that we have stability in the case of four stages ($m = 4$) with

$$\gamma_{kk} = \delta_{kk} = 1 , \quad \beta_1 = \frac{1}{3} , \quad \beta_2 = \frac{4}{15} , \quad \beta_3 = \frac{5}{9} .$$

These coefficients β maximize the stability interval along the imaginary axis, as shown by Sonneveld and van Leer (1985). For efficiency, the dissipative term is computed only once per time-step, leading to

$$R^{(k)} = Q(\varphi^{(k)}) + D(\varphi^{(0)}) .$$

A plausible stability condition is found to be that the CFL number satisfies

$$\sigma = \max\{\frac{\tau}{h}|f'(\varphi_j^n)|\} \leq 2.6 \ .$$

In order to increase τ further, so that fewer time-steps are needed to reach steady state, Jameson and Baker (1984) introduce *residual averaging*. That is, at each stage the residual $R^{(k)}$ is replaced by $\bar{R}^{(k)}$, defined by

$$-\varepsilon\bar{R}^{(k)}(\varphi)_{j-1} + (1+2\varepsilon)\bar{R}^{(k)}(\varphi)_j - \varepsilon\bar{R}^{(k)}(\varphi)_{j+1} = R^{(k)}(\varphi)_j \ . \qquad (9.119)$$

The scheme can be made stable by choosing

$$\varepsilon \geq \frac{1}{4}\{(\frac{\sigma}{\sigma^*})^2 - 1\} \ ,$$

where σ^* is the CFL number limit for the scheme without residual averaging. Solving (9.119) is not expensive. After convergence, $\bar{R} = 0$, hence $R = 0$, so a solution of the original scheme is obtained.

For time-accurate solutions, the time-stepping scheme needs to be TVD. This has the additional advantage that stability is guaranteed. Examples of TVD Runge-Kutta methods have already been given in (9.111) and (9.112). TVD multistage schemes have been studied by Shu (1988) in general, and for the JST scheme by Jameson (1995a), Jameson (1995b), Kim and Jameson (1995). We will not go into this further.

More dimensions

The multi-dimensional case offers no essentially new aspects compared to the one-dimensional case. The scalar conservation law is now given by

$$\frac{\partial\varphi}{\partial t} + \sum_\alpha \frac{\partial f^\alpha(\varphi)}{\partial x^\alpha} = 0 \ , \quad t > 0 \ , \quad x^\alpha \in \mathbb{R} \ , \quad \alpha = 1, ..., d \ ,$$

where d is the number of space dimensions. The summation convention does not apply in the remainder of this section. The total variation $\text{TV}(\varphi)$ is defined as follows, in the two-dimensional case (for brevity):

$$\text{TV}(\varphi(t,\cdot)) = \limsup_{\varepsilon\downarrow 0} \frac{1}{\varepsilon} \int\limits_{-\infty}^{\infty}\int |\varphi(x^1+\varepsilon, x^2) - \varphi(x^1, x^2)|dx^1dx^2 \ .$$

$$+ \limsup_{\varepsilon\downarrow 0} \frac{1}{\varepsilon} \int\limits_{-\infty}^{\infty}\int |\varphi(x^1, x^2+\varepsilon) - \varphi(x^1, x^2)|dx^1dx^2 \ .$$

The extension to three dimensions is obvious. It can be shown (Crandall and Majda (1980)) that φ is TVD. The total variation of a grid function φ_j, $j = (j_1, ..., j_d)$ on a d-dimensional grid is defined analogously:

$$\text{TV}(\varphi) = \sum_{j_1} \cdots \sum_{j_d} \sum_{\alpha} |\varphi_{j+2e_\alpha} - \varphi_j| \,,$$

where $e_1 = (\frac{1}{2}, 0, ..., 0)$, $e_2 = (0, \frac{1}{2}, 0, ..., 0)$ etc. Just as in the one-dimensional case, experience shows that we get satisfactory results if the numerical scheme is TVD, and again the TVD property implies stability. We will therefore analyze conditions under which numerical schemes are TVD.

We will study slope limited schemes, and generalize (9.100) to more dimensions as follows:

$$\frac{d\varphi_j}{dt} + \sum_{\alpha} \frac{1}{h_\alpha} \{F_\alpha^-(\varphi^-) + F_\alpha^+(\varphi^+)\}|_{j-e_\alpha}^{j+e_\alpha} = 0 \,,$$

where $\varphi_{j+e_\alpha}^{\pm}$ are extrapolated values obtained from φ in neigboring grid points in the x^α-direction in exactly the same way as in the one-dimensional case. In the case of explicit Euler discretization we obtain

$$\varphi_j^{n+1} - \varphi_j^n + \sum_{\alpha} \lambda_\alpha \{F_\alpha^-(\varphi^-) + F_\alpha^+(\varphi^+)\}|_{j-e_\alpha}^{j+e_\alpha} = 0 \,, \quad \lambda_\alpha = \tau/h_\alpha \,. \quad (9.120)$$

We will now give conditions for this scheme to be TVD, in a similar way as in the one-dimensional case. To this end we rewrite the scheme as follows (cf. (9.102)):

$$\varphi_j^{n+1} = \varphi_j^n + \sum_{\alpha} \{a_j^\alpha(\varphi_{j+2e_\alpha}^n - \varphi_j^n) + b_j^\alpha(\varphi_{j-2e_\alpha}^n - \varphi_j^n)\} \,. \quad (9.121)$$

In the same way as Theorem 9.4.4 it can be proved that this scheme is TVD if

$$a_j^\alpha \geq 0 \,, \quad b_j^\alpha \geq 0 \,, \quad \sum_{\alpha}(a_j^\alpha + b_{j+2e_\alpha}^\alpha) \leq 1 \,. \quad (9.122)$$

Comparison of (9.120) and (9.121) shows that we can write

$$a_j^\alpha = -\lambda_\alpha \frac{F_\alpha^-(\varphi_{j+e_\alpha}^-) - F_\alpha^-(\varphi_{j-e_\alpha}^-)}{\varphi_{j+e_\alpha}^- - \varphi_{j-e_\alpha}^-} \cdot \frac{\varphi_{j+e_\alpha}^- - \varphi_{j-e_\alpha}^-}{\varphi_{j+2e_\alpha}^n - \varphi_j^n} \,,$$

$$b_j^\alpha = \lambda_\alpha \frac{F_\alpha^+(\varphi_{j+e_\alpha}^+) - F_\alpha^+(\varphi_{j-e_\alpha}^+)}{\varphi_{j+e_\alpha}^+ - \varphi_{j-e_\alpha}^+} \cdot \frac{\varphi_{j+e_\alpha}^+ - \varphi_{j-e_\alpha}^+}{\varphi_j^n - \varphi_{j-2e_\alpha}^n} \,.$$

We assume that the numerical fluxes satisfy (cf. (9.99)):

$$\frac{dF_\alpha^-(\varphi)}{d\varphi} \leq 0 \,, \quad \frac{dF_\alpha^+(\varphi)}{d\varphi} \geq 0 \,, \quad \forall \varphi \in \mathbb{R} \,.$$

In order to have $a_j^\alpha \geq 0$, $b_j^\alpha \geq 0$ we must therefore have

$$\frac{\varphi^-_{j+e_\alpha} - \varphi^-_{j-e_\alpha}}{\varphi^n_{j+2e_\alpha} - \varphi^n_j} \geq 0 \, , \quad \frac{\varphi^+_{j+e_\alpha} - \varphi^+_{j-e_\alpha}}{\varphi^n_j - \varphi^n_{j-2e_\alpha}} \geq 0$$

This will be satisfied if we use flux limiting with a limiter $\psi(r)$ whose graph is inside the admissible region of Fig. 4.11. It remains to satisfy the last part of (9.122). Define

$$\mu^\alpha_{j+e_\alpha} \equiv \frac{F^+_\alpha(\varphi^+_{j+3e_\alpha}) - F^+(\varphi^+_{j+e_\alpha})}{\varphi^+_{j+3e_\alpha} - \varphi^+_{j+e_\alpha}} - \frac{F^-_\alpha(\varphi^-_{j+e_\alpha}) - F^-(\varphi^-_{j-e_\alpha})}{\varphi^-_{j+e_\alpha} - \varphi^-_{j-e_\alpha}} \, .$$

Assume

$$\frac{\varphi^-_{j+e_\alpha} - \varphi^-_{j-e_\alpha}}{\varphi^n_{j+2e_\alpha} - \varphi^n_j} \leq M \, , \quad \frac{\varphi^+_{j+e_\alpha} - \varphi^+_{j-e_\alpha}}{\varphi^n_j - \varphi^n_{j-2e_\alpha}} \leq M \, ,$$

which is the case if the limiter satisfies the usual conditions. Then the last part of (9.122) is satifsied if

$$\sum_\alpha \lambda^\alpha \mu_{j+e_\alpha} \leq 1/M \, . \tag{9.123}$$

As in the one-dimensional case we usually have

$$\mu^\alpha_{j+e_\alpha} \cong |\frac{df^\alpha(\varphi)}{d\varphi}|$$

for some φ, so that (9.123) can be interpreted as a condition on the CFL numbers in the various coordinate directions, which can be satisfied by taking the time step τ small enough.

It has been shown that, unlike the one-dimensional case, in two dimensions TVD schemes cannot be formally second order accurate (Goodman and Leveque (1985)), and presumably this holds also in the three-dimensional case. Nevertheless, schemes of the type considered above using a one-dimensional slope limited flux splitting method in each coordinate direction usually give second order accurate results in smooth parts of the flow and give non-oscillatory and sharp solutions near discontinuities.

Exercise 9.4.1. Show that if a weak solution is in C^1, then it is a classical solution.

Exercise 9.4.2. Prove (9.61).

Exercise 9.4.3. Show that (9.62) satisfies the jump condition.

Exercise 9.4.4. Solve the Riemann problem for the Buckley-Leverett equation with $\varphi_r = 1$, $\varphi_l = 0$.

Exercise 9.4.5. Prove (9.77).

Exercise 9.4.6. Show that the local truncation error for (9.78) tends to zero with the mesh size, if $F_{j+1/2}$ satisfies the consistency conditions (9.84) and (9.85), assuming the solution $\varphi(t, x)$ to be differentiable (cf. Kröner (1997) Lemma 2.2.4).

Exercise 9.4.7. Show that under suitable restrictions on λ (which?) the Lax-Friedrichs scheme (9.79), (9.80) is monotone.

Exercise 9.4.8. Show that in the linear case $(f(\varphi) = c\varphi)$ the scheme (9.97) reduces to the first order upwind scheme, if for F the Engquist-Osher flux (9.82) or the Godunov flux (9.95) is chosen.

Exercise 9.4.9. For what reason can the proof that the Engquist-Osher scheme is monotone (Example 9.4.7) not be extended to the scheme (9.81)?

10. The Euler equations in one space dimension

10.1 Introduction

Since almost all computing methods for the Euler equations in two and three space dimensions rely heavily on techniques developed for the one-dimensional case, we devote a full chapter to the one-dimensional Euler equations. In the one-dimensional case many interesting analytic aspects can be brought to light, and this we do first. The shock tube problem is a very useful test problem for discretization schemes; we will present its analytic solution. Then we turn to discretization methods.

More extentive introductions to numerical methods for the Euler equations are given by Godlewski and Raviart (1996), Kröner (1997), Laney (1998), Majda (1984), Toro (1997), Smoller (1983), Hirsch (1990).

10.2 Analytic aspects

Useful books about analytic and physical aspects of the Euler equations are Liepmann and Roshko (1957) and Landau and Lifshitz (1959).

In one dimension, the Euler equations (1.76)–(1.78) become, without heat addition and body force,

$$
U_t + f(U)_x = 0 \,, \quad U = \begin{pmatrix} \rho \\ m \\ \rho E \end{pmatrix} \,, \quad f = \begin{pmatrix} m \\ m^2/\rho + p(\rho, e) \\ mH \end{pmatrix} \,, \qquad (10.1)
$$

where $m = \rho u$ with u the flow velocity, $E = e + \frac{1}{2}u^2$, $H = h + \frac{1}{2}u^2$, $h = e + p/\rho$. We assume a perfect gas, and recall that in this case we have $p = \rho RT$, $e = e(T)$, $c_v = de/dT$. Hence $h = h(T)$, $c_p = dh/dT$, $R = c_p - c_v$, $p = (\gamma - 1)\rho e$, $\gamma = c_p/c_v$. These relations will be used frequently.

The difference with the preceding chapter is, that now we have a *system* of conservation laws.

Homogeneity of the flux function

Suppose the equation of state is given by

$$p = \rho g(e) .$$

Note that the perfect gas law is a special case, since $e = e(T)$. For the flux function f we can write

$$f(U) = \begin{pmatrix} U_2 \\ U_2^2/U_1 + U_1 g(e) \\ U_2 U_3/U_1 + U_2 g(e) \end{pmatrix} . \tag{10.2}$$

Since $e = e(U) = (U_3 - \frac{1}{2}U_2^2/U_1)/U_1$, we have have $e(\mu U) = e(U)$, with μ an arbitrary real number. Then (10.2) shows that

$$f(\mu U) = \mu f(U) .$$

A function with this property is called *homogeneous of degree one*. Differentiation with respect to μ and taking $\mu = 1$ results in the following interesting consequence:

$$f(U) = f'(U)U ,$$

where f' is the Jacobian of f, i.e. the matrix with elements $\partial f_k/\partial U_j$.

The Jacobian of the flux function

If U_x exists, equation (10.1) can be rewritten as

$$U_t + f'(U)U_x = 0 . \tag{10.3}$$

From Sect. 2.2 we recall that the system is hyperbolic if f' has a full set of eigenvectors. We will therefore solve the eigenproblem for f'. Assume a perfect gas, so that

$$p = (\gamma - 1)(U_3 - \frac{1}{2}U_2^2/U_1) ,$$

and f' is easily found to be

$$f' = \begin{pmatrix} 0 & 1 & 0 \\ \frac{\gamma-3}{2}u^2 & (3-\gamma)u & \gamma - 1 \\ (\gamma-1)u^3 - \gamma uE & \gamma E - \frac{3}{2}(\gamma-1)u^2 & \gamma u \end{pmatrix} .$$

For later reference, we note that f' can also be written as

$$f' = \begin{pmatrix} 0 & 1 & 0 \\ \frac{\gamma-3}{2}u^2 & (3-\gamma)u & \gamma-1 \\ \frac{\gamma-1}{2}u^3 - uH & H-(\gamma-1)u^2 & \gamma u \end{pmatrix} . \tag{10.4}$$

This follows from the identity

$$\gamma E = H + \frac{\gamma-1}{2}u^2 .$$

Transformation to nonconservative variables

In order to facilitate solving the eigenproblem for f' we transform to new dependent variables, namely

$$W = \begin{pmatrix} \rho \\ u \\ p \end{pmatrix} . \tag{10.5}$$

We have

$$U = \begin{pmatrix} W_1 \\ W_1 W_2 \\ \frac{1}{\gamma-1}W_3 + \frac{1}{2}W_1 W_2^2 \end{pmatrix} .$$

The Jacobian $Q = \partial U/\partial W$ and its inverse are found to be

$$Q = \begin{pmatrix} 1 & 0 & 0 \\ u & \rho & 0 \\ \frac{1}{2}u^2 & \rho u & \frac{1}{\gamma-1} \end{pmatrix} , \quad Q^{-1} = \begin{pmatrix} 1 & 0 & 0 \\ -u/p & 1/\rho & 0 \\ \frac{\gamma-1}{2}u^2 & (1-\gamma)u & \gamma-1 \end{pmatrix} . \tag{10.6}$$

Equation (10.3) can be rewritten as

$$QW_t + f'QW_x = 0 . \tag{10.7}$$

Premultiplication with Q^{-1} gives the following nonconservative form of the Euler equations:

$$W_t + \tilde{f}'W_x = 0 , \tag{10.8}$$

$$\tilde{f}' = Q^{-1}f'Q = \begin{pmatrix} u & \rho & 0 \\ 0 & u & 1/\rho \\ 0 & \rho c^2 & u \end{pmatrix} , \quad c^2 = \gamma p/\rho ,$$

where c is known to be the speed of sound waves (see Sect. 1.12). The eigenvalues of $\tilde{f}'$ (and hence of f') are easily found to be

$$\lambda_1 = u - c , \quad \lambda_2 = u , \quad \lambda_3 = u + c . \tag{10.9}$$

The corresponding eigenvectors of $\tilde{f}'$ are

$$\tilde{R}_1 = \frac{1}{2}\begin{pmatrix} -\rho/c \\ 1 \\ -\rho c \end{pmatrix}, \quad \tilde{R}_2 = \begin{pmatrix} 1 \\ 0 \\ 0 \end{pmatrix}, \quad \tilde{R}_3 = \frac{1}{2}\begin{pmatrix} \rho/c \\ 1 \\ \rho c \end{pmatrix}. \tag{10.10}$$

These are linearly independent. Hence, the system is hyperbolic. If the eigenvalues are distinct (as in our case), the system is also called strictly hyperbolic.

Characteristics and Riemann invariants

Define the matrix $\tilde{R}$ by choosing its columns to be $\tilde{R}_{1,2,3}$:

$$\tilde{R} = (\tilde{R}_1 \ \tilde{R}_2 \ \tilde{R}_3).$$

We find

$$\tilde{R}^{-1} = \begin{pmatrix} 0 & 1 & -1/\rho c \\ 1 & 0 & -1/c^2 \\ 0 & 1 & 1/\rho c \end{pmatrix}. \tag{10.11}$$

Premultiplication of (10.8) by $\tilde{R}^{-1}$ gives:

$$u_t + (u-c)u_x - \frac{1}{\rho c}\{p_t + (u-c)p_x\} = 0, \tag{10.12}$$

$$\rho_t + u\rho_x - \frac{1}{c^2}(p_t + up_x) = 0, \tag{10.13}$$

$$u_t + (u+c)u_x + \frac{1}{\rho c}\{p_t + (u+c)p_x\} = 0. \tag{10.14}$$

The entropy is defined by relating small changes in the entropy (here to be denoted by S) to small changes in e and ρ by (1.50):

$$T\delta S = \delta e + p\delta(1/\rho).$$

For a perfect gas this can be rewritten as

$$\delta S = c_v \delta \ln \frac{p}{\rho} - R\delta \ln \rho = c_v \delta \ln p - c_p \delta \ln \rho = c_v \delta \ln p\rho^{-\gamma}. \tag{10.15}$$

This defines S up to a constant, which is chosen such that, assuming $c_v = $ constant,

$$S = c_v \ln p\rho^{-\gamma}. \tag{10.16}$$

Furthermore, continuing with (10.15),

$$\delta S = \frac{\gamma c_v}{\rho}(\frac{\rho}{\gamma p}\delta p - \delta \rho) = \frac{\gamma c_v}{\rho}(\frac{1}{c^2}\delta p - \delta \rho) \,. \tag{10.17}$$

Hence, using (10.13),

$$S_t + u S_x = \frac{\gamma c_v}{\rho}\{\frac{1}{c^2}(p_t + u p_x) - \rho_t - u \rho_x\} = 0 \,.$$

Therefore, as long as the unknowns are differentiable, the entropy S is constant along particle paths. When a shock is crossed a fluid particle undergoes an increase in entropy. If we assume that we are locally in a region where the particle paths carry the same entropy, so that S is constant locally, then equations (10.12) and (10.14) can be rewritten as follows. From (10.16) it follows that

$$p = k\rho^\gamma \,, \quad k = \exp(S/c_v) \,. \tag{10.18}$$

Hence $c^2 = k\gamma\rho^{\gamma-1}$ and if S, hence k, is constant we have

$$\int \frac{1}{\rho c}dp = \frac{2c}{\gamma - 1} \,.$$

This means that (10.12) and (10.14) can be rewritten as

$$(u - \frac{2c}{\gamma - 1})_t + (u - c)(u - \frac{2c}{\gamma - 1})_x = 0 \,,$$

$$(u + \frac{2c}{\gamma - 1})_t + (u + c)(u + \frac{2c}{\gamma - 1})_x = 0 \,.$$

Hence, we have found that $u \pm 2c/(\gamma - 1)$ is constant along the characteristics defined by

$$dx/dt = u \pm c \,, \tag{10.19}$$

and S is constant along characteristics defined by

$$dx/dt = u \,. \tag{10.20}$$

The quantities $u \pm 2c/(\gamma - 1)$ and S are called *Riemann-invariants*. As we shall see, these are useful for constructing analytic solutions.

Eigenvectors of Jacobian

For later reference, we write down the eigenvectors R_p, $p = 1, 2, 3$ of the untransformed Jacobian $f'(U)$. They follow from $R_p = Q\tilde{R}_p$:

$$R_1 = \tfrac{1}{2}\rho \begin{pmatrix} -1/c \\ 1 - u/c \\ u - \tfrac{1}{2}u^2/c - \tfrac{1}{\gamma-1}c \end{pmatrix}, \quad R_2 = \begin{pmatrix} 1 \\ u \\ \tfrac{1}{2}u^2 \end{pmatrix},$$

$$R_3 = \tfrac{1}{2}\rho \begin{pmatrix} 1/c \\ 1 + u/c \\ u + \tfrac{1}{2}u^2/c + \tfrac{1}{\gamma-1}c \end{pmatrix}. \tag{10.21}$$

Let R be the matrix with columns R_1, R_2, R_3. For later reference, we determine R^{-1}. We have $R = Q\tilde{R}$, hence $R^{-1} = \tilde{R}^{-1}Q^{-1}$. From (10.11) and (10.6) we find:

$$R^{-1} = \begin{pmatrix} -\tfrac{u}{\rho}(1 + \tfrac{\gamma-1}{2}\tfrac{u}{c}) & \tfrac{1}{\rho}(1 + (\gamma-1)\tfrac{u}{c}) & -(\gamma-1)\tfrac{1}{\rho c} \\ 1 - \tfrac{\gamma-1}{2}\tfrac{u^2}{c^2} & (\gamma-1)\tfrac{u}{c^2} & -(\gamma-1)\tfrac{1}{c^2} \\ -\tfrac{u}{\rho}(1 - \tfrac{\gamma-1}{2}\tfrac{u}{c}) & \tfrac{1}{\rho}(1 - (\gamma-1)\tfrac{u}{c}) & (\gamma-1)\tfrac{1}{\rho c} \end{pmatrix}. \tag{10.22}$$

Boundary conditions

As in Chap. 8, boundary and initial conditions that lead to a well-posed problem for the Euler equations are precisely those that determine implicitly or explicitly the Riemann invariants. As in Chap. 8, the general rule is, that the number of boundary conditions in a point on the boundary must be equal to the number of incoming characteristics at that point. Let $U(0, x)$ be given, and let the domain be $x \in (0, 1)$. Consider the boundary $x = 0$. If $u > 0$ and $u \pm c > 0$ (supersonic inflow) three conditions must be given that determine $U(t, 0)$. If $u > 0$ and $u - c < 0$ (subsonic inflow), two conditions must be given that determine the Riemann invariants corresponding to $dx/dt = u$ and $dx/dt = u + c$, i.e. S and $u + 2c/(\gamma - 1)$. If $u < 0$ and $u + c > 0$ (subsonic outflow), then $u + 2c/(\gamma - 1)$ must be given. If $u < 0$ and $u + c < 0$ (supersonic outflow) then no boundary condition is to be given. The numerical implementation of these conditions is not straightforward, since in general u, S and c are not used as unknowns, but ρ, u and e. A good approach in practice is to prescribe u and e at an inflow boundary, and the pressure p at a supersonic inflow boundary and at a subsonic outflow boundary.

This leaves us with the following difficulty. In many numerical schemes the density is required at an inflow boundary, since together with the inflow velocity this determines the mass flux. However, as just said, at a subsonic inflow boundary we prescribe only the temperature; the pressure and the density have to follow from the solution. This difficulty may be resolved by taking the boundary density from a previous time step or iteration, in a time stepping or iterative solution procedure.

Shocks and Rankine-Hugoniot conditions

As we saw in the preceding chapter on the scalar case, when characteristics intersect, a shock is formed. Obviously, the characteristics of type (10.20) cannot intersect, since they are particle lines. But the characteristics of type (10.19) can intersect, so that shocks may be expected. Weak solutions may be defined in the same way as for the scalar case (Sect. 9.4), and exactly in the same way as in Sect. 9.4 it can be shown that the following jump condition must hold across a shock:

$$s[U] = [f(U)] , \tag{10.23}$$

where s is the speed of the shock and $[\cdot]$ is the size of the jump. This gives, denoting quantities immediately to the left and right of the shock by subscripts 1 and 2, respectively,

$$s(\rho_2 - \rho_1) = m_2 - m_1 , \tag{10.24}$$
$$s(m_2 - m_1) = m_2^2/\rho_2 + p_2 - m_1^2/\rho_1 - p_1 , \tag{10.25}$$
$$s(\rho_2 E_2 - \rho_1 E_1) = m_2 H_2 - m_1 H_1 . $$

These are known as the Rankine-Hugoniot conditions (Hugoniot (1889), Rankine (1870)).

Several useful properties and relations can be derived from these conditions; see Liepmann and Roshko (1957) (Sect. 2.13) for more details. We start with the case of a stationary shock: $s = 0$. We have $m_2 = m_1$. Dividing (10.25) by $m_1 = m_2$ we get

$$u_1 - u_2 = c_2^2/\gamma u_2 - c_1^2/\gamma u_1 . \tag{10.26}$$

Since $h = c_p T = c^2/(\gamma - 1)$, $H_1 = H_2$ can be rewritten as

$$\frac{1}{2}u_1^2 + \frac{1}{\gamma - 1}c_1^2 = \frac{1}{2}u_2^2 + \frac{1}{\gamma - 1}c_2^2 . \tag{10.27}$$

The sonic condition is defined as the condition in which $u = c$, and is denoted by an asterisk. Hence (10.27) can be written as

$$\frac{1}{2}u_1^2 + \frac{1}{\gamma - 1}c_1^2 = \frac{1}{2}u_2^2 + \frac{1}{\gamma - 1}c_2^2 = \frac{\gamma + 1}{\gamma - 1}c_*^2 . \tag{10.28}$$

Solution of c_1 and c_2 from (10.28) and substitution in (10.26) gives after some manipulation

$$u_1 u_2 = c_*^2 . \tag{10.29}$$

This interesting relation (called the Prandtl-Meyer relation) shows that either $|u_1| > c_*$ and $|u_2| < c_*$, or vice-versa, so that the flow at one side of a shock is always supersonic ($|u| > c$) and at the other side subsonic ($|u| < c$). (In more space dimensions, this remains true for the velocity component normal to the shock).

We now show that the ratios of all quantities depend on only one parameter, for which we first take the Mach number $M_1 = u_1/c_1$. Defining $M^* = u/c^*$, we get from (10.28):

$$M_1^{*2} + \frac{2}{\gamma - 1}\left(\frac{M_1^*}{M_1}\right)^2 = \frac{\gamma + 1}{\gamma - 1},$$

hence

$$M_1^{*2} = (\gamma + 1)M_1^2/\{(\gamma - 1)M_1^2 + 2\}. \tag{10.30}$$

We have

$$\frac{u_1}{u_2} = \frac{u_1^2}{u_1 u_2} = M_1^{*2},$$

so that (10.30) gives

$$\frac{u_1}{u_2} = (\gamma + 1)M_1^2/\{(\gamma - 1)M_1^2 + 2\}. \tag{10.31}$$

From $m_1 = m_2$ it then follows that

$$\frac{\rho_1}{\rho_2} = \{(\gamma - 1)M_1^2 + 2\}/(\gamma + 1)M_1^2. \tag{10.32}$$

From (10.25) with $s = 0$ we obtain

$$p_2 - p_1 = \rho_1 u_1 (u_1 - u_2),$$

or

$$\frac{p_2}{p_1} - 1 = \frac{\rho_1 u_1^2}{p_1}\left(1 - \frac{u_2}{u_1}\right) = \gamma M_1^2\left(1 - \frac{u_2}{u_1}\right).$$

Substitution of (10.31) gives

$$\frac{p_2}{p_1} - 1 = \frac{2\gamma}{\gamma + 1}(M_1^2 - 1),$$

or

$$\frac{p_1}{p_2} = \frac{\gamma + 1}{1 - \gamma + 2\gamma M_1^2}. \tag{10.33}$$

We see the sizes of the jumps depend only on one parameter M_1. Solving M_1^2 from (10.33), we can also express the jumps in ρ and u in terms of the parameter p_1/p_2. The result is:

$$\frac{u_1}{u_2} = \frac{\rho_2}{\rho_1} = \frac{P + \alpha}{\alpha P + 1}, \quad P = \frac{p_1}{p_2}, \quad \alpha = \frac{\gamma + 1}{\gamma - 1}. \tag{10.34}$$

In the case of a moving shock ($s \neq 0$) there are two free parameters. This case can be obtained from the above case of a stationary shock by a Galilean transformation $x \longrightarrow x - st$. That is, we switch to an observer who travels with speed $-s$, and who sees the shock move with speed s. The thermodynamic quantities do not change under this transformation, but u_1 and u_2 have to be taken relative to the moving shock, so that u_1 and u_2 have to be replaced by $u_1 - s$ and $u_2 - s$ in the jump conditions above:

$$\frac{u_1 - s}{u_2 - s} = \frac{P + \alpha}{\alpha P + 1}, \quad P = \frac{p_1}{p_2}. \tag{10.35}$$

We now have two free parameters, for which we can take P and s. Alternatively, s can be obtained from (10.35) as

$$\frac{s}{u_2} = 1 + \frac{\gamma - 1}{2} \frac{1 + \alpha P}{1 - P} (1 - \frac{u_1}{u_2}).$$

Another expression for the shock speed can be obtained as follows. For the case of a stationary shock we have (10.33), which can be rewritten as

$$\mathrm{M}_1^2 = 1 + \frac{\gamma + 1}{2\gamma} (\frac{p_2}{p_1} - 1), \tag{10.36}$$

with $\mathrm{M}_1^2 = u_1^2/c_1^2$. For a moving shock this becomes $\mathrm{M}_1^2 = (u_1 - s)^2/c_1^2$. Hence

$$(s - u_1)^2 = c_1^2 \{1 + \frac{\gamma + 1}{2\gamma} (\frac{p_2}{p_1} - 1)\}. \tag{10.37}$$

Note that subscript 1 refers to the low pressure side of the shock.

The entropy condition

As in the scalar case, discontinuities do not only have to satisfy the jump condition (10.23), but also the entropy condition, in order to be physically realizable. In the present case we can check the behavior of the entropy directly, and require that the entropy of a fluid particle does not decrease as it traverses a shock. From (10.16) we have

$$S = c_v \ln p - c_p \ln \rho = R\{\frac{1}{\gamma - 1} \ln p - \frac{\gamma}{\gamma - 1} \ln \rho\},$$

hence

$$\frac{\gamma - 1}{R} (S_2 - S_1) = \ln \frac{p_2}{p_1} - \gamma \ln \frac{\rho_2}{\rho_1}.$$

First, consider the case of a stationary shock. Substitution of (10.32) and (10.33) and writing $M_1^2 = 1 + \varepsilon$ gives

$$\frac{\gamma - 1}{R}(S_2 - S_1) = \ln(1 + z_1) - \gamma \ln(1 + \varepsilon) + \gamma \ln(1 + z_2) \,,$$

where

$$z_1 = \frac{2\gamma}{\gamma + 1}\varepsilon \,, \quad z_2 = \frac{\gamma - 1}{\gamma + 1}\varepsilon \,.$$

Expansion for $|\varepsilon| \ll 1$ gives

$$\frac{\gamma - 1}{R}(S_2 - S_1) = \frac{2\gamma(\gamma - 1)}{3(\gamma + 1)^2}\varepsilon^3 + \mathcal{O}(\varepsilon^4) \,.$$

Now assume $u_1 > 0$, and let subscript 1 denote the state immediately to the left of the shock. Hence, before entering the shock a fluid particle has entropy S_1, and immediately upon leaving it has entropy S_2. The entropy condition requires that $S_2 \geq S_1$. Hence we must have $\varepsilon \geq 0$, or

$$M_1^2 \geq 1 \,. \tag{10.38}$$

We see that the entropy condition is equivalent to the statement, that a fluid particle can only enter a shock from the supersonic side. (We saw earlier that a shock always separates a supersonic and a subsonic state). In the case of a moving shock this remains true for the velocity of a fluid particle relative to the shock. This means that an observer that moves with a fluid particle with the flow velocity u sees a shock coming on always at supersonic speed. If this is not the case, transition from one state to another does not take place through a shock but through a *contact discontinuity* or an *expansion wave*.

Equations (10.38) and (10.36) imply $p_2 \geq p_1$. Hence a fluid particle that crosses a shock never undergoes a drop in pressure.

Contact discontinuities

Since the characteristics of type (10.20) (particle paths) cannot intersect themselves, they do not give rise to shocks. But since the corresponding Riemann invariant S can be different for different particle paths, they can support a jump in S. The discontinuity travels with speed

$$s = u_1 = u_2 \,, \tag{10.39}$$

with subscripts 1 and 2 denoting the states adjacent to the discontinuity. It is clear that a fluid particle cannot cross this type of discontinuity, and two bodies of fluid that are initially in contact at the discontinuity remain so at later times; hence the name contact discontinuity. Substitution of (10.39) in the Rankine-Hugoniot conditions (10.24), (10.25) gives

$$\rho_2 - \rho_1 = \rho_2 - \rho_1 \,,$$
$$s^2(\rho_2 - \rho_1) = s^2(\rho_2 - \rho_1) + p_2 - p_1 \,.$$

It follows that $p_1 = p_2$, but ρ can jump. The corresponding jump in entropy follows from (10.16):

$$S_2 - S_1 = c_p \ln \rho_1/\rho_2 \,.$$

Simple waves and expansion fans

A simple wave is a part of the domain where the solution is smooth and where S and one of the other two Riemann invariants are constant. For example, let

$$S = S_1 \,, \quad u + \frac{2c}{\gamma - 1} = u_1 + \frac{2c_1}{\gamma - 1} \,. \tag{10.40}$$

Because the solution is smooth we have

$$u - \frac{2c}{\gamma - 1} = \text{constant} \quad \text{along} \quad \frac{dx}{dt} = u - c \,.$$

Therefore along a characteristic $u \pm 2c/(\gamma - 1) = $ constant, and hence both u and c are constant. As a consequence, the characteristics are straight in a simple wave.

A simple wave which is a similarity solution, i.e. a solution of the form

$$U = U(\xi) \,, \quad \xi = x/t$$

is called an expansion fan. We will give an explicit solution. Suppose to the left of the expansion fan we have a constant state, denoted by subscript 1:

$$U(\xi) = U_1 \,, \quad \xi \le \xi_1 \,.$$

Because u and c are constant along $\frac{dx}{dt} = u - c$, we have along a characteristic

$$u - c = x/t \,. \tag{10.41}$$

By solving u and c from (10.41) and (10.40) we get

$$u = \frac{2}{\gamma + 1}(\frac{x}{t} + c_1) + \frac{\gamma - 1}{\gamma + 1}u_1 \,,$$
$$c = \frac{\gamma - 1}{\gamma + 1}(u_1 - \frac{x}{t}) + \frac{2}{\gamma + 1}c_1 \,.$$

By using $\xi_1 = u_1 - c_1$ this can be rewritten as

$$u - u_1 = \frac{2}{\gamma + 1}(\xi - \xi_1) \ , \quad c - c_1 = \frac{\gamma - 1}{\gamma + 1}(\xi_1 - \xi) \ . \qquad (10.42)$$

Pressure and density can be determined as follows. Because $S = $ constant, it follows from (10.16) that

$$p\rho^{-\gamma} = p_1 \rho_1^{-\gamma} \ .$$

From this and from $c^2 = \gamma p / \rho$ it follows that

$$(c/c_1)^2 = (p/p_1)^{(\gamma-1)/\gamma} = (\rho/\rho_1)^{\gamma - 1} \ . \qquad (10.43)$$

Substitution in (10.42) gives

$$\frac{p}{p_1} = \{1 + \frac{\gamma - 1}{\gamma + 1}(\xi_1 - \xi)/c_1\}^{2\gamma/(\gamma-1)}, \quad \frac{\rho}{\rho_1} = \{1 + \frac{\gamma - 1}{\gamma + 1}(\xi_1 - \xi)/c_1\}^{2/(\gamma-1)} \ . \qquad (10.44)$$

Suppose the expansion fan separates two constant states (denoted by subscripts 1 and 2). Then the relation between these states can be expressed with p_2/p_1 as parameter as follows. Equation (10.43) gives

$$c_2/c_1 = (p_2/p_1)^{(\gamma-1)/2\gamma} \ , \quad \rho_2/\rho_1 = (p_2/p_1)^{1/\gamma} \ . \qquad (10.45)$$

Equation (10.40) gives

$$u_2 - u_1 = \frac{2}{\gamma - 1}(c_1 - c_2) \ ,$$

which by using (10.45) can be rewritten as

$$u_2 - u_1 = \frac{2}{\gamma - 1} c_1 \{1 - (p_2/p_1)^{(\gamma-1)/2\gamma}\} \qquad (10.46)$$

The Riemann problem

The Riemann problem consists of solving the one-dimensional Euler equations for a special class of initial conditions, namely a constant (left) state for $-\infty < x < 0$ and a different constant (right) state for $0 < x < \infty$. The left state will be indicated by subscript 1 and the right state will be indicated by subscript 4. So the initial conditions are:

$$u(0, x) = u_1 \ , \quad p(0, x) = p_1 \ , \quad \rho(0, x) = \rho_1 \ , \quad x < 0 \ ; $$
$$u(0, x) = u_4 \ , \quad p(0, x) = p_4 \ , \quad \rho(0, x) = \rho_4 \ , \quad x > 0 \ . \qquad (10.47)$$

For the time being we take $u_4 = 0$. If also $u_1 = 0$, then the problem is called the shock tube problem. In Sect. 18B of Smoller (1983) it is shown that for the Euler equations, the Riemann problem always has a solution (if $\rho_{1,4} \geq 0$).

It is easy to see that if $U(t, x)$ is a solution of (10.1), with initial condition (10.47), then so is $U(at, ax)$, with a constant. Therefore the solution is of similarity form:

$$U = U(\xi), \quad \xi = x/t .$$

We restrict ourselves to the case $p_1 \geq p_4$. We will use the following building blocks to bridge the gap between states 1 and 4 for $t > 0$:

 (i) a shock separating states 4 and 3;
(ii) a contact discontinuity separating states 3 and 2;
(iii) an expansion fan separating states 2 and 1.

The situation is sketched in Fig. 10.1. If we succeed in solving the Riemann problem with these building blocks, then the structure of the solution assumed here is correct.

We start with the shock. Equation (10.37) becomes, with appropriate changes of indices,

$$(s - u_4)^2 = c_4^2 \{1 + \frac{\gamma + 1}{2\gamma}(P - 1)\} , \quad P = p_3/p_4 .$$

We assume $p_3 > p_4$, otherwise the entropy condition (10.38) would be violated, according to (10.36). As noted before, a fluid particle that traverses a shock cannot undergo a drop in pressure, hence $s > u_4$, so that

$$s = u_4 + c_4(\frac{\gamma - 1}{2\gamma} + \frac{\gamma + 1}{2\gamma}P)^{1/2} . \tag{10.48}$$

Equation (10.35) gives, with $u_4 = 0$:

$$u_3 = s(1 - \frac{P + \alpha}{\alpha P + 1}) = c_4(P - 1)\sqrt{\frac{2}{\gamma}}\{\gamma - 1 + (\gamma + 1)P\}^{-1/2} .$$

For the density we have equation (10.34):

$$\frac{\rho_3}{\rho_4} = \frac{1 + \alpha P}{\alpha + P} ,$$

Next, we consider the contact discontinuity separating states 3 and 2. We have

$$u_2 = u_3 , \quad p_2 = p_3 , \quad \rho_2 \neq \rho_3 . \tag{10.49}$$

The contact discontinuity moves with speed $u_2 = u_3$.

Finally, we consider the expansion fan separating states 1 and 2. Equation (10.46) gives

$$u_2 = u_1 + \frac{2}{\gamma - 1} c_1 \{1 - (p_2/p_1)^{(\gamma-1)/2\gamma}\} \,. \tag{10.50}$$

Equating u_2 and u_3 and p_2 and p_3 gives a nonlinear equation for $P = p_3/p_4$ in terms of the known pressure ratio:

$$\frac{1}{P} \left\{ 1 + \frac{\gamma-1}{2} \frac{u_1}{c_1} - \frac{(\gamma-1)c_4(P-1)}{c_1\sqrt{2\gamma(\gamma-1+(\gamma+1)P)}} \right\}^{\frac{2\gamma}{\gamma-1}} = \frac{p_4}{p_1} \,. \tag{10.51}$$

With $u_1 = 0$, this is the basic shock tube equation. After P has been determined numerically, the solution can be constructed. It is assumed that (10.51) has a real solution for P. For this it is necessary that the term between braces is positive. If this is not the case, then the solution has a different structure than assumed. This is also the case when $p_4 > p_1$. We will not go into this further, and assume that states 1 and 4 are specified such that (10.51) has a solution.

When P has been obtained, equations (10.50) and (10.49) give

$$u_2 = u_3 = u_1 + \frac{2}{\gamma - 1} c_1 \{1 - (Pp_4/p_1)^{\frac{\gamma-1}{2\gamma}}\} \,.$$

Equations (10.45) and (10.34) give

$$p_2 = p_1(Pp_4/p_1)^{1/\gamma} \,, \quad p_3 = p_4 \frac{1 + \alpha P}{\alpha + P} \,.$$

The speed of sound follows from $c^2 = \gamma p/\rho$. The shock moves with velocity s given by (10.48). The contact discontinuity moves with speed $u_2 = u_3$. The left and right sides of the expansion fan move with velocity

$$\xi_1 = -c_1 \,, \quad \xi_2 = u_2 - c_2 \,.$$

The variation of u, p and ρ in the expansion fan is given by equations (10.42) and (10.44). Fig. 10.1 gives a qualitative picture of the flow.

We have constructed a weak solution that satisfies the entropy condition. It can be shown that such solutions are unique; see for example Taylor (1996) Chap. 17.

The case $u_4 \neq 0$ is easily handled by a Galilei transformation. Define new velocities:

$$\tilde{u}_1 = u_1 - u_4 \,, \quad \tilde{u}_4 = u_4 - u_4 = 0 \,.$$

The Riemann problem for $(\tilde{u}, p, \rho)$ can be solved in the way outlined above, provided the data are such that a solution of the assumed structure exists.

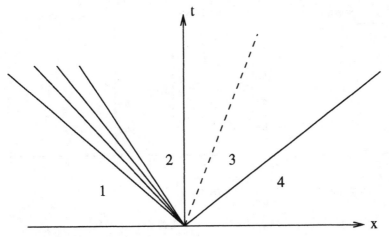

Fig. 10.1. Shock tube problem: $x - t$ diagram of shock (—), contact discontinuity (- - -) and expansion fan.

After the solution has been obtained one puts $u = \tilde{u} + u_4$.

We will not present the general solution of the Riemann problem. The general solution can be obtained in the same spirit as above, but more possibilities have to be considered. For instance, there could be two shocks, or none; or there could be two expansions fans, or none. See Sect. 93 of Landau and Lifshitz (1959) for the various possibilities. Approximate solutions of the Riemann problem play a dominant role in an important class of numerical methods for the Euler equations, as we will see.

The *shock tube problem* is a special kind of Riemann problem. Its solution is in the class of solutions that we have constructed above. We have an infinitely long tube along the x-axis, with a diaphragm at $x = 0$, which separates two bodies of gas at rest (i.e. $u_1 = u_4 = 0$). We assume here that gas is the same at both sides of the diaphragm, but that ρ and p are different. The shock tube problem consists of determining the flow that develops when at $t = 0$ the diagram is suddenly removed.

A well-known test problem is the shock tube problem proposed by Sod (1978). The initial conditions are given by

$$u_1 = 0 , \quad p_1 = 1 , \quad \rho_1 = 1 , \quad u_4 = 0 , \quad p_4 = 0.1 , \quad \rho_4 = 0.125 .$$

The solution can be determined in the way described above, and is presented in Fig. 10.2. Going from right to left we observe a shock, a contact discontinuity and an expansion fan. In this and the following figures, the quantity called entropy is $\ln p\rho^{-\gamma}$, which is in fact proportional to the physical entropy (cf. (10.16)). There can be a jump of either sign across a contact discontinuity,

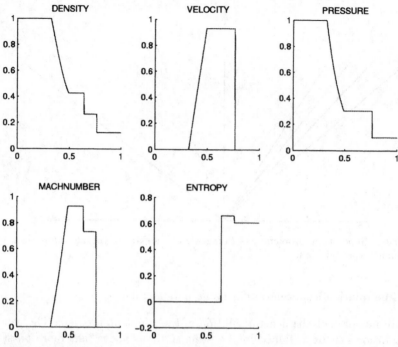

Fig. 10.2. Solution of Sod's shock tube problem at $t = 0.15$

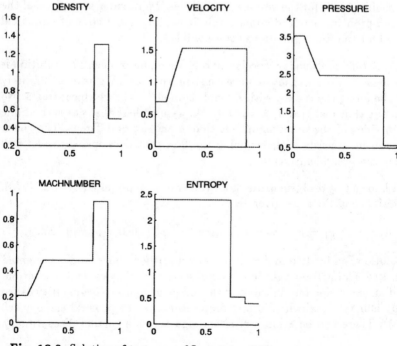

Fig. 10.3. Solution of test case of Lax at $t = 0.15$

depending on the choice of the left and right states of the Riemann problem. Fluid particles do not cross a contact discontinuity. Figure 10.2 shows that the entropy of fluid particles is constant in time, except when they cross a shock, in which case the entropy increases.

Some publications showing numerical results for this test case are Sod (1978), Harten, Engquist, Osher, and Chakravarty (1987), Arora and Roe (1997). Another frequently used test problem is the Lax test case, see Lax (1954), Harten (1983) and the papers just quoted. The initial state is specified by

$$u_1 = 0.698 , \quad p_1 = 3.528 , \quad \rho_1 = 0.445 , \quad u_4 = 0 , \quad p_4 = 0.571 , \quad \rho_4 = 0.5 .$$

The solution is shown in Fig. 10.3. The contact discontinuinity and the shock are stronger than in Sod's shock tube problem.

In these two test cases the flow remains subsonic. Supersonic flow may bring additional numerical difficulties. A test case in which supersonic flow occurs

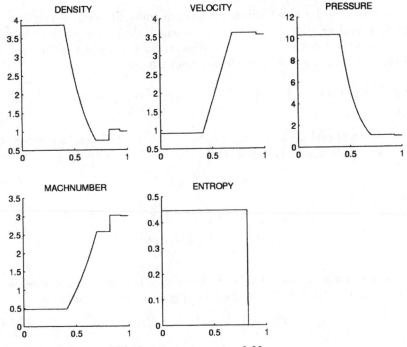

Fig. 10.4. Solution of Mach 3 test case at $t = 0.09$.

is considered by Arora and Roe (1997), and will be called the Mach 3 test case. The data are:

$$u_1 = 0.92 , \quad p_1 = 10.333 , \quad \rho_1 = 3.857 , \quad u_4 = 3.55 , \quad p_4 = 1 , \quad \rho_4 = 1 .$$

The solution is shown in Fig. 10.4. The flow is dominated by a strong expansion fan.

10.3 The approximate Riemann solver of Roe

The usual discretization method for hyperbolic systems of conservation laws is the finite volume method. A number of well-known finite volume schemes will be presented. Treatment of boundary conditions is postponed till our discussion of the three-dimensional case in Chap. 12.

For more information on finite volume methods for hyperbolic systems of conservation laws and the Euler equations in particular, see Godlewski and Raviart (1991), Hirsch (1990), Kröner (1997), Laney (1998) and Toro (1997).

Conservative schemes

Just as for the scalar case discussed in the preceding chapter, convergence to a genuine weak solution can be guaranteed only if the scheme is conservative. Therefore we will discuss only conservative schemes, i.e. schemes that can be written as (assuming a uniform spatial mesh size h):

$$dU_j/dt + \frac{1}{h}(F_{j+1/2} - F_{j-1/2}) = 0 , \quad F_{j+1/2} = F(U_{j-p}, U_{j-p+1}, ..., U_{j+q}) ,$$
(10.52)

where the temporal discretization can be chosen as one likes. Integration of the Euler equations (10.1) over a cell results in (10.52) with $F_{j+1/2}$ given by

$$F_{j+1/2} = f(U_{j+1/2})$$
(10.53)

and

$$U_j(t) = \frac{1}{h} \int\limits_{x_{j-1/2}}^{x_{j+1/2}} U(t, x) dx .$$

Requiring the spatial discretization to be exact if $U = $ constant, comparison of (10.52) and (10.53) results in the following consistency condition:

$$F_{j+1/2}(U, U, ..., U) = f(U) .$$
(10.54)

The Godunov scheme

The Godunov scheme has been described in Chap. 9 for a single conservation law. Its principle extends to systems of conservation laws; in fact, it was originally developed by Godunov (1959) for the Euler equations. The Godunov scheme is described in Sect. 12.15 of Richtmyer and Morton (1967).

Suppose that at time $t = t^n$ we have available the values U_j^n in the cell centers. These values are extended to the whole cell, resulting in a piecewise constant function $U(t^n, x)$. This is used as initial condition to solve

$$\frac{\partial U}{\partial t} + \frac{\partial f(U)}{\partial x} = 0$$

for $t > t^n$. For $t - t^n < \frac{h}{2}\max|\lambda|$, with λ ranging over the set of eigenvalues of the Jacobian of f, the exact solution consists of solutions of local Riemann problems at the cell interfaces. For t larger the waves emanating from the cell interfaces start to interact, and the solution is no longer composed of Riemann solutions. The Riemann problem at the cell interface $x = x_{j+1/2}$ has left state U_j^n and right state U_{j+1}^n. Denote the solution of the Riemann problem by $\hat{U}\left(\frac{x - x_{j+1/2}}{t}; U_j^n, U_{j+1}^n\right)$. For $t - t_n$ small enought we have

$$f(U(t, x_{j+1/2})) = f(\hat{U}(0; U_j^n, U_{j+1}^n)) .$$

In the Godunov scheme this is used for the numerical flux function $F_{j+1/2}$ in the semi-discretized system (10.52):

$$F_{j+1/2} = F(U_j, U_{j+1}) = f(\hat{U}(0; U_j^n, U_{j+1}^n)) . \tag{10.55}$$

Thus, evaluation of $F_{j+1/2}$ requires the exact solution of a Riemann problem. The Riemann problem for the Euler equations has been solved for the shock tube problem in the preceding section. Its general solution is somewhat complicated and will not be discussed.

Because at every time step the solution is approximated by a piecewise constant distribution, the Godunov scheme is not exact. Therefore one may as well replace the exact solution of the Riemann problem in (10.55) by a simpler approximate solution. This has given rise to a class of discretization methods called *approximate Riemann solvers*. These are the subject of the present section. As a preliminary, we first discuss the Riemann problem for a linear system.

The Riemann problem for a linear system

Consider the following Riemann problem:

$$\frac{\partial U}{\partial t} + A\frac{\partial U}{\partial x} = 0 , \quad A \in \mathbb{R}^{m \times m} \text{ constant} , $$
$$U(0, x) = U_L , \quad x < 0 ; \quad U(0, x) = U_R , \quad x > 0 . \tag{10.56}$$

It is assumed that A has m linearly independent eigenvectors $R_1, ..., R_m$ with corresponding eigenvalues $\lambda_1 < \lambda_2 < ... < \lambda_m$. Define $\alpha_1, ..., \alpha_m$ implicitly by

$$U_R - U_L = \sum_{p=1}^{m} \alpha_p R_p . \tag{10.57}$$

We diagonalize (10.56) in the usual way. Let R be the matrix with columns $R_1, ..., R_m$. Then (10.56) can be rewritten as

$$\frac{\partial W}{\partial t} + \Lambda \frac{\partial W}{\partial x} = 0 , \quad W = R^{-1}U , \quad \Lambda = \text{diag}(\lambda_1, ..., \lambda_m) ,$$
$$W(0, x) = W_L , \quad x < 0 ; \quad W(0, x) = W_R , \quad x > 0 .$$

Writting $W = (w_1, ..., w_m)$, the solution is, with $\xi = x/t$:

$$w_k(\xi) = w_{Lk} , \quad \xi < \lambda_k ; \quad w_k(\xi) = w_{Rk} , \quad \xi > \lambda_k , \quad k = 1, ..., m ,$$

where W_L and W_R are determined as follows. Let

$$U_L = \sum_{p=1}^{m} \beta_p R_p , \quad U_R = \sum_{p=1}^{m} \gamma_p R_p .$$

From $W = R^{-1}U$ it follows that

$$W_L = (\beta_1, ..., \beta_m)^T , \quad W_R = (\gamma_1, ..., \gamma_m)^T .$$

The solution follows from $U(\xi) = RW(\xi)$. We find: $U(\xi) = U_L, \xi < \lambda_1$; $U(\xi) = U_L + \alpha_1 R_1, \lambda_1 < \xi < \lambda_2$; etc., resulting in

$$U(\xi) = U_L , \quad \xi < \lambda_1 ;$$
$$U(\xi) = U_L + \sum_{p=1}^{k} \alpha_p R_p , \quad \lambda_k < \xi < \lambda_{k+1} , \quad k = 1, ..., m-1 ; \tag{10.58}$$
$$U(\xi) = U_L + \sum_{p=1}^{m} \alpha_p R_p = U_R , \quad \lambda_m < \xi .$$

The solution is seen to be piecewise constant. It is left as an exercise to show that the jump at $x/t = \lambda_k$ satisfies the jump condition (10.23), verifying that (10.58) is the correct weak solution.

The Roe scheme

The exact Riemann problem is:

$$\frac{\partial U}{\partial t} + \frac{\partial f(U)}{\partial x} = 0\,,$$

$$U(0, x) = U_L\,, \quad x < 0\,; \quad U(0, x) = U_R\,, \quad x > 0\,. \tag{10.59}$$

Roe (1981) approximates this by the following linear Riemann problem:

$$\frac{\partial U}{\partial t} + A\frac{\partial U}{\partial x} = 0\,, \quad A \in \mathbb{R}^{m \times m} \text{ constant},$$

$$U(0, x) = U_L\,, \quad x < 0\,; \quad U(0, x) = U_R\,, \quad x > 0\,. \tag{10.60}$$

The matrix A depends on U_L and U_R : $A = A(U_L, U_R)$, and is assumed to satisfy the following conditions:

$$f(V) - f(W) = A(V, W)(V - W)\,, \tag{10.61}$$

$$A(V, W) \longrightarrow f'(V) \quad \text{as} \quad W \longrightarrow V\,, \tag{10.62}$$

$$A(V, W) \quad \text{has only real eigenvalues,} \tag{10.63}$$

$$A(V, W) \quad \text{has a complete system of eigenvectors.} \tag{10.64}$$

Condition (10.61) ensures that the exact solution of the original Riemann problem (10.59) is obtained in the particular case when the solution consists of a single shock or contact discontinuity. Condition (10.62) ensures consistency with the original equation. Conditions (10.63) and (10.64) ensure solvability of the linear Riemann problem (10.60).

Roe's scheme is given by

$$\frac{dU_j}{dt} + \frac{1}{h}(F_{j+1/2} - F_{j-1/2}) = 0\,,$$

$$F_{j+1/2} = F(U_j, U_{j+1}) = A_{j+1/2}(U_j, U_{j+1})\hat{U}(0; U_j, U_{j+1})\,, \tag{10.65}$$

where $\hat{U}$ is the exact solution of the linear Riemann problem (10.60) with $U_L = U_j$ and $U_R = U_{j+1}$, and A is now named $A_{j+1/2}$. We call $A_{j+1/2}$ the Roe matrix, and $F_{j+1/2}$ as defined in (10.65) is called the Roe flux.

A useful expression for the Roe flux can be derived as follows. From the exact solution of the linear Rieman problem (10.58) it follows that we have for $\hat{U}$ in (10.65):

$$\hat{U} = U_{j+1}\,, \quad \lambda_m < 0\,; \quad \hat{U} = U_j\,, \quad \lambda_1 > 0\,;$$

whereas in the case that for some k we have $\lambda_k < 0 < \lambda_{k+1}$

$$\hat{U} = U_j + \sum_{p=1}^{k} \alpha_p R_p = U_{j+1} - \sum_{p=k+1}^{m} \alpha_p R_p$$

$$= \tfrac{1}{2}(U_j + U_{j+1}) + \tfrac{1}{2}\{\sum_{p=1}^{k} - \sum_{p=k+1}^{m}\}\alpha_p R_p\,.$$

Hence

$$
\begin{aligned}
F_{j+1/2} &= A_{j+1/2}\hat{U} = \tfrac{1}{2}A_{j+1/2}(U_j + U_{j+1}) - \tfrac{1}{2}\sum_{p=1}^{m}|\lambda_p|\alpha_p R_p \\
&= \tfrac{1}{2}\{f(U_j) + f(U_{j+1})\} - \tfrac{1}{2}\sum_{p=1}^{m}|\lambda_p|\alpha_p R_p \,,
\end{aligned}
\tag{10.66}
$$

where we have used (10.61). In this expression we may obviously allow $\lambda_j = 0$; this has to be excluded in (10.58) and hence in the derivation leading up to (10.66), because it makes no sense to write $\hat{U}(0)$ if $\hat{U}(\xi)$ has a jump at $\xi = 0$, but the flux is perfectly well defined (namely zero) in this case. Noting that according to (10.58)

$$
\sum_{p=1}^{m}\alpha_p R_p = U_{j+1} - U_j \,,
\tag{10.67}
$$

we see that the Roe flux satisfies

$$
F_{j+1/2} = f(U_j)\,, \quad \lambda_1 \geq 0\,; \quad F_{j+1/2} = f(U_{j+1})\,, \quad \lambda_m \leq 0\,,
$$

so that the Roe scheme is identical to the first order upwind scheme in these cases.

A matrix $A_{j+1/2}$ satisfying Roe's conditions (10.61)–(10.64) can be constructed as follows. For the Euler equations (10.1) the state and flux vectors

$$
U = \begin{pmatrix} \rho \\ \rho u \\ \rho E \end{pmatrix}\,, \quad f = \begin{pmatrix} \rho u \\ \rho u^2 + p \\ \rho u H \end{pmatrix}\,,
$$

where $H = E + p/\rho$, are expressed as functions of the vector

$$
Z = \sqrt{\rho}\begin{pmatrix} 1 \\ u \\ H \end{pmatrix}
$$

as follows. Assuming a perfect gas, the equation of state gives

$$
p = (\gamma - 1)\rho e = (\gamma - 1)\left(\rho E - \frac{1}{2}\rho u^2\right)\,,
$$

so that

$$
p = \frac{\gamma - 1}{\gamma}\left(\rho H - \frac{1}{2}\rho u^2\right)\,, \quad \rho E = \frac{1}{\gamma}\rho H + \frac{\gamma - 1}{2\gamma}\rho u^2 \,.
$$

We find

$$
U = \begin{pmatrix} z_1 z_1 \\ z_1 z_2 \\ \frac{1}{\gamma}z_1 z_3 + \frac{\gamma-1}{2\gamma}z_2 z_2 \end{pmatrix}\,, \quad f = \begin{pmatrix} z_1 z_2 \\ \frac{\gamma+1}{2\gamma}z_2 z_2 + \frac{\gamma-1}{\gamma}z_1 z_3 \\ z_2 z_3 \end{pmatrix}\,.
$$

Observe that the elements of U and f are homogeneous quadratic functions of the elements of Z. We define the difference $\delta a = a_{j+1} - a_j$ and the average $\bar{a} = \frac{1}{2}(a_j + a_{j+1})$. Then we have $\delta(ab) = \bar{a}\delta b + \bar{b}\delta a$, $\bar{a}\delta a = \frac{1}{2}\delta(aa)$. Using these identities, the following equalities are easily verified:

$$\delta U = \bar{B}\delta Z , \quad \bar{B} = \begin{pmatrix} 2\bar{z}_1 & 0 & 0 \\ \bar{z}_2 & \bar{z}_1 & 0 \\ \frac{1}{\gamma}\bar{z}_3 & \frac{\gamma-1}{\gamma}\bar{z}_2 & \frac{1}{\gamma}\bar{z}_1 \end{pmatrix} ,$$

$$\delta f = \bar{C}\delta Z , \quad \bar{C} = \begin{pmatrix} \bar{z}_2 & \bar{z}_1 & 0 \\ \frac{\gamma-1}{\gamma}\bar{z}_3 & \frac{\gamma+1}{\gamma}\bar{z}_2 & \frac{\gamma-1}{\gamma}\bar{z}_1 \\ 0 & \bar{z}_3 & \bar{z}_2 \end{pmatrix} .$$

Hence

$$\bar{C}\bar{B}^{-1}\delta U = \bar{C}\delta Z = \delta f ,$$

so that

$$A_{j+1/2} = \bar{C}\bar{B}^{-1}$$

satisfies condition (10.61). We find:

$$A_{j+1/2} = \begin{pmatrix} 0 & 1 & 0 \\ \frac{\gamma-3}{2}(\frac{\bar{z}_2}{\bar{z}_1})^2 & (3-\gamma)\frac{\bar{z}_2}{\bar{z}_1} & \gamma-1 \\ \frac{\gamma-1}{2}(\frac{\bar{z}_2}{\bar{z}_1})^3 - \frac{\bar{z}_2\bar{z}_3}{\bar{z}_1\bar{z}_1} & \frac{\bar{z}_3}{\bar{z}_1} - (\gamma-1)(\frac{\bar{z}_2}{\bar{z}_1})^2 & \gamma\frac{\bar{z}_2}{\bar{z}_1} \end{pmatrix} .$$

Define the following averages:

$$\bar{u} = \frac{\sqrt{\rho_j}u_j + \sqrt{\rho_{j+1}}u_{j+1}}{\sqrt{\rho_j} + \sqrt{\rho_{j+1}}} , \quad \bar{H} = \frac{\sqrt{\rho_j}H_j + \sqrt{\rho_{j+1}}H_{j+1}}{\sqrt{\rho_j} + \sqrt{\rho_{j+1}}} . \tag{10.68}$$

These are called the Roe averages. It is easily seen that the Roe matrix can be rewritten in terms of the Roe averages as

$$A_{j+1/2} = \begin{pmatrix} 0 & 1 & 0 \\ \frac{\gamma-3}{2}\bar{u}^2 & (3-\gamma)\bar{u} & \gamma-1 \\ \frac{\gamma-1}{2}\bar{u}^3 - \bar{u}\bar{H} & \bar{H} - (\gamma-1)\bar{u}^2 & \gamma\bar{u} \end{pmatrix} .$$

Comparison with (10.4) shows that $A_{j+1/2}$ equals the Jacobian evaluated at the Roe-averaged state (10.68):

$$A_{j+1/2} = f'(\bar{U}) . \tag{10.69}$$

This immediately establishes that $A_{j+1/2}$ satisfies the properties (10.62)–(10.64). Roe and Pike (1984) show that for the Euler equations the matrix $A_{j+1/2}$ satisfying conditions (10.61)–(10.64) is unique.

For the Roe flux we have the expression (10.66). Because of (10.69), λ_p follows immediately from (10.9):

$$\lambda_1 = \bar{u} - \bar{c}, \quad \lambda_2 = \bar{u}, \quad \lambda_3 = \bar{u} + \bar{c},$$

where $\bar{c}$ is the sound speed associated with $\bar{u}$ and $\bar{H}$:

$$\bar{c}^2 = (\gamma - 1)(\bar{H} - \frac{1}{2}\bar{u}^2). \tag{10.70}$$

It is to be noted that $\bar{c}$ is not a Roe-average of c_j and c_{j+1}. The eigenvectors R_p follow from (10.21). For simplicity we rescale R_1 and R_3; R_1 is multiplied by $-2c/\rho$ and R_3 is multiplied by $2c/\rho$. This gives, using (10.70),

$$R_1 = \begin{pmatrix} 1 \\ \bar{u} - \bar{c} \\ \bar{H} - \bar{u}\bar{c} \end{pmatrix}, \quad R_2 = \begin{pmatrix} 1 \\ \bar{u} \\ \frac{1}{2}\bar{u}^2 \end{pmatrix}, \quad R_3 = \begin{pmatrix} 1 \\ \bar{u} + \bar{c} \\ \bar{H} + \bar{u}\bar{c} \end{pmatrix}.$$

The coefficients α_p in the Roe flux (10.66) follow from (10.67) as follows. Let R be the matrix with columns R_1, R_2, R_3. The rows of R^{-1} (let us call them R^1, R^2, R^3) constitute an orthogonal system with the columns of R:

$$R^p \cdot R_q = \delta^p_q,$$

with δ^p_q the Kronecker delta. Taking the inner product of (10.67) with R^q gives:

$$\alpha_q = R^q \cdot (U_{j+1} - U_j). \tag{10.71}$$

The vectors R^1, R^2, R^3 equal the rows of (10.22), scaled appropriately: the first row is to be multiplied by $-\rho/2c$ and the third by $\rho/2c$. This gives:

$$R^1 = \left(\frac{\bar{u}}{4\bar{c}}(2 + (\gamma - 1)\frac{\bar{u}}{\bar{c}}), \quad -\frac{1}{2\bar{c}}(1 + (\gamma - 1)\frac{\bar{u}}{\bar{c}}), \quad \frac{\gamma - 1}{2}\frac{1}{\bar{c}^2} \right),$$

$$R^2 = \left(1 - \frac{\gamma - 1}{2}\frac{\bar{u}^2}{\bar{c}^2}, \quad (\gamma - 1)\frac{\bar{u}}{\bar{c}^2}, \quad -(\gamma - 1)\frac{1}{\bar{c}^2} \right),$$

$$R^3 = \left(-\frac{\bar{u}}{4\bar{c}}(2 - (\gamma - 1)\frac{\bar{u}}{\bar{c}}), \quad \frac{1}{2\bar{c}}(1 - (\gamma - 1)\frac{\bar{u}}{\bar{c}}), \quad \frac{\gamma - 1}{2}\frac{1}{\bar{c}^2} \right).$$

This completes our description of how to determine λ_p, α_p and R_p for computing the flux $F_{j+1/2}$ for the Roe scheme according to (10.66).

We end our description of the Roe scheme with deriving a useful reformulation. Let R be the matrix with columns R_p. Then the inverse R^{-1} is the matrix with rows R^p. It is not difficult to see that we have, with α_p given by (10.71),

$$\sum_{p=1}^m |\lambda_p|\alpha_p R_p = R^{-1}|\Lambda|R(U_{j+1} - U_j),$$

where $|\Lambda| = \text{diag}(|\lambda_1|, ..., |\lambda_m|)$. We have $R^{-1}\Lambda R = A$, and write therefore $|A| = R^{-1}|\Lambda|R$. This results in the following compact expression for the Roe flux:

$$F_{j+1/2} = \frac{1}{2}\{f(U_j) + f(U_{j+1})\} - \frac{1}{2}|A_{j+1/2}|(U_{j+1} - U_j) .$$

Numerical tests of Roe scheme

We will apply the Roe scheme to the three test cases presented in the preceding section. We discretize (10.65) in time by the explicit Euler method, and obtain

$$U_j^{n+1} = U_j^n - \lambda(F_{j+1/2}^n - F_{j-1/2}^n) , \quad \lambda = \tau/h , \quad j = 2, \ldots, J - 1 . \quad (10.72)$$

The exact solution is used to prescribe U_1^n and U_J^n. The time step τ and mesh size h are constant. A discussion of stability conditions on τ is deferred to the next section. In all cases we take $h = 1/48$. Results are shown in

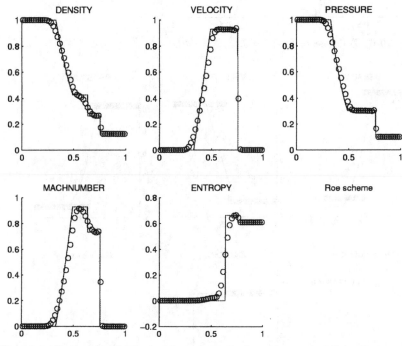

Fig. 10.5. Sod's shock tube problem; exact (—) and numerical (o) solution with Roe scheme; $t = 0.1458$, $\lambda = 0.5$.

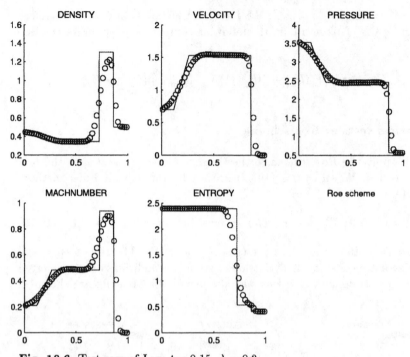

Fig. 10.6. Test case of Lax; $t = 0.15$, $\lambda = 0.2$.

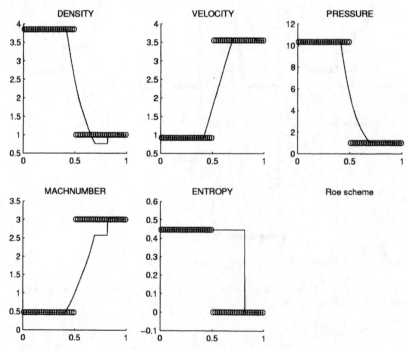

Fig. 10.7. Mach 3 test case; $t = 0.0875$, $\lambda = 0.2$.

Figs. 10.5–10.7. As explained in Sect. 10.2, the entropy jump acros the contact discontinuity is a function of the initial conditions. Across the shock the numerical entropy increases and in the expansion fan it is constant, as it should be, in Figs. 10.5 and 10.6. Because we have a first order scheme, the smearing of the contact discontinuities in Figs. 10.5 and 10.6 is to be expected. Near shocks, the characteristics converge; this produces a steepening effect, so that shocks are less smeared than contact discontinuities. In fact, shock resolution is very crisp with the Roe scheme. An important observation is that spurious wiggles are absent. In the test cases of Figs. 10.5 and 10.6 the accuracy is satisfactory for a first order scheme. For the Mach 3 test case (Fig. 10.7), however, the numerical solution is completely wrong. This is a beautiful example of violation of the entropy condition. The expansion fan is replaced by an expansion shock in the sonic point. According to (10.41), the sonic point in an expansion fan does not move. Hence fluid particles cross the sonic point from left to right. The figure shows they undergo a decrease of entropy, in violation of the entropy condition. This violation of the entropy condition in sonic points of expansion fans by the Roe scheme will be explained later. Note that the expansion fans occuring in Figs. 10.5 and 10.6 do not contain a sonic point, and are approximated satisfactorily.

When the expansion fan is less strong than in the Mach 3 test case, the Roe scheme does not necessarily violate the entropy condition. This is illustrated by the following shock tube problem, which contains a supersonic zone, and which we will refer to as the supersonic shock tube problem:

$$p_L = \rho_L = 8 \,, \quad p_R = \rho_R = 0.2 \,, \quad u_L = u_R = 0 \,. \tag{10.73}$$

The exact and numerical solutions are shown in Fig. 10.8. A small jump is visible near the sonic point in the expansion fan. This is called a sonic glitch. It looks like a small expansion shock, violating the entropy condition. However, the entropy is seen not to decrease as a fluid crosses the stationary sonic point, but to increase a little. Hence there is no violation of the entropy condition. We see that a sonic glitch may appear even if the entropy condition is satisfied. The lower part of Fig. 10.8 shows that the sonic glitch becomes smaller when the mesh size is decreased. But it is found that for the Mach 3 test case mesh refinement does not help. We will now consider a remedy for this shortcoming of the Roe scheme.

Sonic entropy fix for the Roe scheme

As seen in Sect. 10.2, expansion fans are associated with λ_1 or λ_3. In order to analyze the way in which an expansion fan is approximated by the Roe scheme, let us consider the special case in which $U_{j+1} - U_j$ is such that $\alpha_2 = \alpha_3 = 0$ in (10.57). If we also assume that

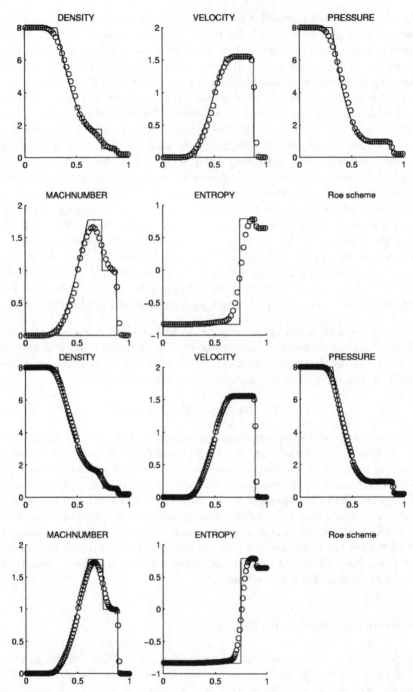

Fig. 10.8. Supersonic shock tube problem; $t = 0.1562$, $\lambda = 0.3$, $h = 1/48$ (above), $h = 1/96$ (below).

$$\lambda_1 = \bar{u} - \bar{c} = 0 \,, \tag{10.74}$$

then equation (10.66) gives for the Roe flux

$$F_{j+1/2} = \frac{1}{2}\{f(U_j) + f(U_{j+1})\} \,.$$

This means that central discretization is used, without artificial viscosity, so there is a danger that entropy decreases, and that discontinuities are insufficiently smeared out. This is what we see in Fig. 10.7. But if (10.74) does not hold, i.e. the flow is not sonic, then we do not have central discretization, and there may be enough dissipation to prevent violation of the entropy condition. This is confirmed by the correct approximation of the subsonic expansion fans in Fig. 10.5 and 10.6.

An often used artifice to add dissipation to the Roe scheme near sonic conditions has been proposed by Harten (1984). The eigenvalues λ_1 and λ_3 are slightly increased in the vicinity of zero, replacing them in (10.66) by

$$\begin{aligned}
|\tilde{\lambda}_p| &= |\lambda_p| \quad \text{if} \quad |\lambda_p| \geq \varepsilon\,, \quad p = 1,3; \\
|\tilde{\lambda}_p| &= \tfrac{1}{2}(\tfrac{\lambda_p^2}{\varepsilon} + \varepsilon) \quad \text{if} \quad |\lambda_p| < \varepsilon\,, \quad p = 1,3 \,.
\end{aligned} \tag{10.75}$$

for some small value of ε. Since (10.75) comes only into play when $|u| \approx c$, this is called a sonic entropy fix. Fig. 10.9 presents numerical results for the supersonic shock tube test case, with $\varepsilon = 0.5$. Compared with Fig. 10.8, the sonic glitch has almost disappeared. But the strong expansion shock of Fig. 10.7 is not completely replaced by an expansion fan by this sonic entropy fix, as shown in Fig 10.10, at least for $\varepsilon = 0.5$; we think this is rather large value of ε. Nevertheless, the entropy condition is now satisfied, as shown by the entropy plot.

Summarizing, the Roe scheme performs well for flows without strong expansion fans. The sonic entropy fix of Harten cures the violation of the entropy condition, but significant sonic glitches may remain.

Exercise 10.3.1. Show that the solution (10.58) satisfies the jump condition (10.23).

10.4 The Osher scheme

The Osher scheme (Osher (1981), Osher and Solomon (1982), Osher and Chakravarthy (1983)) can be regarded as a generalization of the Engquist-Osher scheme (9.82) from the scalar case to the systems case. In Sect. 9.4 we saw that the Engquist-Osher scheme is an upwind scheme that satisfies the

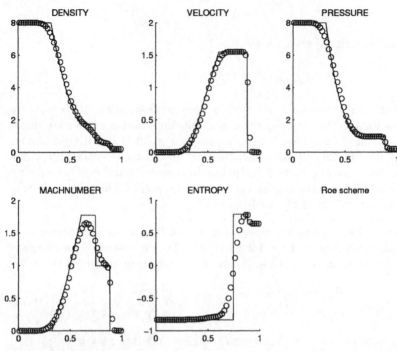

Fig. 10.9. Supersonic shock tube problem with sonic entropy fix; $\varepsilon = 0.5$.

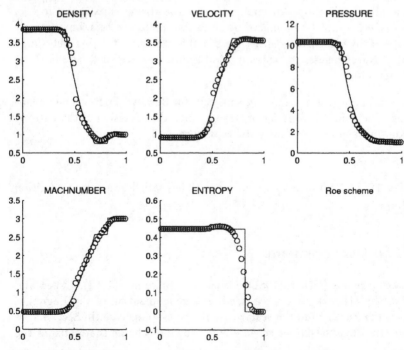

Fig. 10.10. Mach 3 test case with sonic entropy fix; $\varepsilon = 0.5$.

entropy condition. This was seen to be not the case for the Courant-Isaacson-Rees upwind scheme. It turns out that the Osher scheme also satisfies the entropy condition.

We first rewrite the Engquist-Osher flux as follows. Let $\mu(\psi)$ be the Heaviside function:

$$\mu(\psi) = 1 \, , \quad \psi \geq 0 \, ; \quad \mu(\psi) = 0 \, , \quad \psi < 0 \, .$$

Then the Engquist-Osher flux can be written as

$$F_{j+1/2} \; = \; f(0) + \int_0^{\varphi_j} \mu(\psi) f'(\psi) d\psi + \int_0^{\varphi_{j+1}} (1 - \mu(\psi)) f'(\psi) d\psi$$

$$= \; f(\varphi_j) + \int_{\varphi_j}^{\varphi_{j+1}} (1 - \mu(\psi)) f'(\psi) d\psi = f(\varphi_j) + \int_{\varphi_j}^{\varphi_{j+1}} \min\{f'(\psi), 0\} d\psi \, .$$

We can also write

$$F_{j+1/2} = f(\varphi_{j+1}) - \int_{\varphi_j}^{\varphi_{j+1}} \mu(\psi) f'(\psi) d\psi = f(\varphi_{j+1}) - \int_{\varphi_j}^{\varphi_{j+1}} \max\{f'(\psi), 0\} d\psi \, .$$

Hence, we can write for the Engquist-Osher flux:

$$F_{j+1/2} = \frac{1}{2}\{f(\varphi_j) + f(\varphi_{j+1})\} - \frac{1}{2} \int_{\varphi_j}^{\varphi_{j+1}} |f'(\psi)| d\psi \, .$$

This is generalized to systems by interpreting the absolute value of the Jacobian as any other matrix function, namely as

$$|f'| = R|\Lambda|R^{-1} \, ,$$

where $|\Lambda|$ is the diagonal matrix with elements the absolute values of the eigenvalues of f' (given by (10.9)) and R the matrix with the eigenvectors (given by (10.21)) as columns. This gives the following generalization of the Engquist-Osher flux for systems:

$$F_{1/2} = \frac{1}{2}\{f(U_0) + f(U_1)\} - \frac{1}{2} \int_{U_0}^{U_1} |f'(U)| dU \, , \tag{10.76}$$

where for brevity we have put $j = 0$; U_0 is the state to the left and U_1 is the state to the right of the cell interface. The path is parametrized by $U = U(\sigma)$, $0 \leq \sigma \leq 1$, $U(0) = U_0$, $U(1) = U_1$, so that (10.76) becomes

$$F_{1/2} = \frac{1}{2}\{f(U_0) + f(U_1)\} - \frac{1}{2}\int_0^1 |f'(U)|\frac{dU}{d\sigma}d\sigma \ . \tag{10.77}$$

The integral depends on the path. In Osher's scheme the path is chosen such that the integral is easy to evaluate and the computing cost is small. The path is divided in three parts $\Gamma_1 \, (0 \leq \sigma < 1/3), \quad \Gamma_2 \, (1/3 \leq \sigma < 2/3)$ and $\Gamma_3 \, (2/3 \leq \sigma \leq 1)$. The integration path is chosen such that

$$\frac{dU}{d\sigma} = R_3 \ \ \text{on } \Gamma_1 \, , \quad \frac{dU}{d\sigma} = R_2 \ \ \text{on } \Gamma_2 \, , \quad \frac{dU}{d\sigma} = R_1 \ \ \text{on } \Gamma_3 \, . \tag{10.78}$$

This choice has the following consequences. Let R^1, R^2, R^3 be the rows of R^{-1} (given by (10.22)), so that $R^p R_q = \delta_q^p$ (Kronecker delta). Then it follows from (10.78) that we have on each subpath

$$R^q \frac{dU}{d\sigma} = 0 \tag{10.79}$$

for two values of q. We show that this implies that two Riemann invariants are constant on Γ_p. To this end we transform to the nonconservative variables W defined by (10.5), so that (10.79) becomes

$$R^q Q \frac{dW}{d\sigma} = 0 \, ,$$

where $Q = \partial U/\partial W$ is given by (10.6). We obtain

$$R^1 Q dW = du - \frac{1}{\rho c}dp \, ,$$

$$R^2 Q dW = d\rho - \frac{1}{c^2}dp = \frac{\rho}{\gamma c_v}dS \, ,$$

$$R^3 Q dW = du + \frac{1}{\rho c}dp \, .$$

where we have used (10.17). On Γ_1 equation (10.78) holds for $q = 1, 2$, so that

$$S = \text{constant} \, , \quad du - \frac{1}{\rho c}dp = 0 \, . \tag{10.80}$$

This can be rewritten as follows. Since the entropy is constant, equation (10.17) gives with $c^2 = k\gamma\rho^{\gamma-1}$

$$\frac{1}{\rho c}dp = \frac{2}{\gamma - 1}dc \, ,$$

and (10.80) becomes, using (10.16):

$$pp^{-\gamma} = \text{constant} , \quad u - \frac{2c}{\gamma - 1} = \text{constant}.$$

Similarly, on Γ_3:

$$pp^{-\gamma} = \text{constant} , \quad u + \frac{2c}{\gamma - 1} = \text{constant} .$$

Finally, on Γ_2 we have $du \pm dp/\rho c = 0$, so that on Γ_2

$$u = \text{constant} , \quad p = \text{constant} .$$

By means of these relations we can determine the end points of Γ_1, Γ_2 and Γ_3 in the following way. Defining

$$z = \ln(pp^{-\gamma}) ,$$

we can write the perfect gas law as follows, remembering that $c^2 = \gamma RT$:

$$p = p(c, z) = (\gamma^\gamma \frac{e^z}{c^{2\gamma}})^{\frac{1}{1-\gamma}} . \tag{10.81}$$

The conditions along the subpaths give the following relations between the states at the end points:

$$z_{1/3} = z_0 , \quad u_{1/3} - \frac{2}{\gamma - 1}c_{1/3} = u_0 - \frac{2}{\gamma - 1}c_0 \equiv \Psi_0 ,$$
$$u_{1/3} = u_{2/3} , \quad p_{2/3} = p_{1/3} ,$$
$$z_{2/3} = z_1 , \quad u_{2/3} + \frac{2}{\gamma - 1}c_{2/3} = u_1 + \frac{2}{\gamma - 1}c_1 \equiv \Psi_1 .$$

From $p_{2/3} = p_{1/3}$ it follows that, using (10.81),

$$c_{2/3}/c_{1/3} = e^{(z_{2/3}-z_{1/3})/2\gamma} = e^{(z_1-z_0)/2\gamma} = c_1/c_0 \equiv \alpha .$$

We obtain the following linear system:

$$u_{1/3} - \frac{2}{\gamma - 1}c_{1/3} = \Psi_0 , \quad u_{1/3} + \frac{2\alpha}{\gamma - 1}c_{1/3} = \Psi_1 .$$

The solution is

$$c_{1/3} = \frac{\gamma - 1}{2 + 2\alpha}(\Psi_1 - \Psi_0) , \quad u_{1/3} = \frac{\alpha \Psi_0 + \Psi_1}{\alpha + 1} . \tag{10.82}$$

Since $c_{1/3}$ connot be negative, a meaningful solution does not exist if $\Psi_1 < \Psi_0$. This corresponds to $u_1 \ll u_0$. We will not worry here about this uncommon case.

Having determined the integration path, we now have to evaluate the integral in (10.77). It turns out that for this we need to determine points on Γ_1 where λ_1 changes sign and on Γ_3 where λ_3 changes sign (sonic points). Since $\lambda_2 = u = $ constant on Γ_2, λ_2 does not change sign on Γ_2. It is a fact — not shown here — that for the Euler equations there cannot be more than one sonic point on Γ_1; the same holds for Γ_3. Hence, a sonic point occurs on Γ_1 if and only if

$$(u_0 + c_0)(u_{1/3} + c_{1/3}) \leq 0 .$$

Let in the sonic point on Γ_1 (if any) $u = u_1^s$, $c = c_1^s = -u_1^s$. Then we have

$$z_0^s = z_0 , \quad u_1^s = \frac{\gamma - 1}{\gamma + 1}\Psi_0 , \quad c_1^s = -u_1^s .$$

Similarly, on Γ_3 a sonic point occurs if and only if

$$(u_1 - c_1)(u_{2/3} - c_{2/3}) \leq 0 ,$$

and satisfies

$$z_3^s = z_1 , \quad u_3^s = c_3^s = \frac{\gamma - 1}{\gamma + 1}\Psi_1 .$$

We are now ready to evaluate the integral in (10.77). We denote the value of σ on Γ_1 or Γ_3 where a sonic point occurs (if any) by σ_1 or σ_3, respectively, and define the sign function by

$$\text{sign}(\lambda) = 1 , \quad \lambda > 0 ; \quad \text{sign}(\lambda) = 0 , \quad \lambda = 0 ; \quad \text{sign}(\lambda) = -1 , \quad \lambda < 0 .$$

We have

$$\int_0^{\sigma_1} |f'(U)|dU = \int_0^{\sigma_1} |f'(U)|R_3 d\sigma = \int_0^{\sigma_1} |\lambda_3|R_3 d\sigma$$

$$= \text{sign}(\lambda_3(0)) \int_0^{\sigma_1} \lambda_3 R_3 d\sigma = \text{sign}(\lambda_3(0)) \int_0^{\sigma_1} f'(U)dU$$

$$= \text{sign}(\lambda_3(0))(f_{\sigma_1} - f_0) ,$$

where we abbreviate

$$f(U(\sigma_1)) = f_{\sigma_1} , \quad f(U(0)) = f_0 .$$

The remaining parts of the integral are evaluated similarly. Equation (10.77) results in the following flux for the Osher scheme:

$$F_{1/2} = \frac{1}{2}\{1 + \text{sign}(\lambda_3(0))\}f_0 + \frac{1}{2}\{\text{sign}(\lambda_3(\frac{1}{3})) - \text{sign}(\lambda_3(0))\}f_{\sigma_1}$$

$$+ \frac{1}{2}\{\text{sign}(u_{1/3}) - \text{sign}(\lambda_3(\frac{1}{3}))\}f_{1/3} + \frac{1}{2}\{\text{sign}(\lambda_1(\frac{2}{3})) - \text{sign}(u_{1/3})\}f_{2/3}$$

$$+ \frac{1}{2}\{\text{sign}(\lambda_1(1)) - \text{sign}(\lambda_1(\frac{2}{3}))\}f_{\sigma_3} + \frac{1}{2}\{1 - \text{sign}(\lambda_1(1))\}f_1 .$$

In order to avoid unnecessary evaluations of f, which are expensive, this can be programmed as follows:

$$s_1 = 1 + \text{sign}(\lambda_3(0)) ; \quad s_2 = \text{sign}(\lambda_3(\tfrac{1}{3})) - \text{sign}(\lambda_3(0)) ;$$
$$s_3 = \text{sign}(u_{1/3}) - \text{sign}(\lambda_3(\tfrac{1}{3})) ; \quad s_4 = \text{sign}(\lambda_1(\tfrac{2}{3})) - \text{sign}(u_{1/3}) ;$$
$$s_5 = \text{sign}(\lambda_1(1)) - \text{sign}(\lambda_1(\tfrac{2}{3})) ; \quad s_6 = 1 - \text{sign}(\lambda_1(1)) ;$$
$$F_{1/2} = 0 ;$$

if $s_1 \neq 0$ then
$$F_{1/2} = F_{1/2} + \tfrac{1}{2} s_1 f_0;$$
if $s_2 \neq 0$ then
$$F_{1/2} = F_{1/2} + \tfrac{1}{2} s_2 f_{\sigma_1};$$
if $s_3 \neq 0$ then
$$F_{1/2} = F_{1/2} + \tfrac{1}{2} s_3 f_{1/3};$$
if $s_4 \neq 0$ then
$$F_{1/2} = F_{1/2} + \tfrac{1}{2} s_4 f_{2/3};$$
if $s_5 \neq 0$ then
$$F_{1/2} = F_{1/2} + \tfrac{1}{2} s_5 f_{\sigma_3};$$
if $s_6 \neq 0$ then
$$F_{1/2} = F_{1/2} + \tfrac{1}{2} s_6 f_1;$$

This may still result in some superfluous evaluations of f, since it can happen that some contributions cancel. To avoid this the various possibilities can be put in a logical table; see for instance Sect. 12.3.3 of Toro (1997).

Hemker and Spekreijse (1986) have proposed a version of the Osher scheme in which the order of the subpaths is reversed. That is, Γ_p is parallel to R_p, $p = 1, 2, 3$. It turns out that this usually results in significantly fewer evaluations of the flux $f(U)$, so that this version requires less computing time.

It has been proven (Osher and Solomon (1982)) that the Osher scheme in the semi-discrete form (10.52) converges as $h \downarrow 0$ to a solution satisfying the entropy condition, under the assumption that the numerical solution converges. This proof does not hold for the version of Hemker and Spekreijse.

Another nice property of the Osher scheme (in both versions) is that the flux $F_{1/2}$ is a differentiable function of U_0 and U_1, as shown by Hemker and Spekreijse (1986), who also give expressions for the derivatives concerned. This is beneficial for the convergence behavior of iterative solution methods for implicit versions of the scheme. The Roe flux is not a differentiable function of the two neighbouring states.

Numerical experiments

We apply the Osher scheme to the test cases of Sect. 10.3. The temporal discretization is given by (10.72). We start with the test case that is most demanding with respect to the entropy condition, namely the Mach 3 test case. The original Osher scheme is called the O-variant; the version of Hemker and Spekreijse is called the H-variant. Fig. 10.11 gives results for the O-variant; results for the H-variant are found to be identical in this case. Although near the sonic point there is a slight discontinuity that looks like an expansion shock, the results are not in disagreement with entropy condition, because the entropy plot shows that the entropy of a fluid particle does not decrease. As shown by the lower half of the figure, the sonic glitch becomes smaller with mesh refinement. Fig. 10.12 gives results for the supersonic shock tube problem. Unless stated otherwise, $h = 1/48$. With the O-variant, there is almost no sonic glitch in the expansion zone. The H-variant shows a larger sonic glitch, but the entropy plot shows there is no violation of the entropy condition.

Results for Sod's shock tube problem are shown in Fig. 10.13. The O- and H-variants give good and almost indistinguishable results. For the test case of Lax (Fig. 10.14) results for the O- and H-variants are found to be indistinguishable.

Summarizing, these tests show that the O- and H-variants of the Osher scheme give almost the same results, except near sonic points; it may happen that the H-variant gives a larger sonic glitch than the O-variant. Both variants seem to satisfy the entropy condition, a fact that has been proven for the O-variant.

In Table 10.1 the number of flux evaluations required to compute these test cases is listed. The H-variant is seen to bring significant savings.

	O-variant	H-variant
Mach 3	1995	987
Supersonic s.t.	3622	1682
Sod	3360	1480
Lax	5356	1968

Table 10.1. Number of flux evaluations required for test case computations.

Comparing with the results obtained with the Roe scheme in Sect. 10.3, we see that for the test cases of Sod and Lax (both without sonic points) the Osher scheme and the Roe scheme give almost the same results. When a

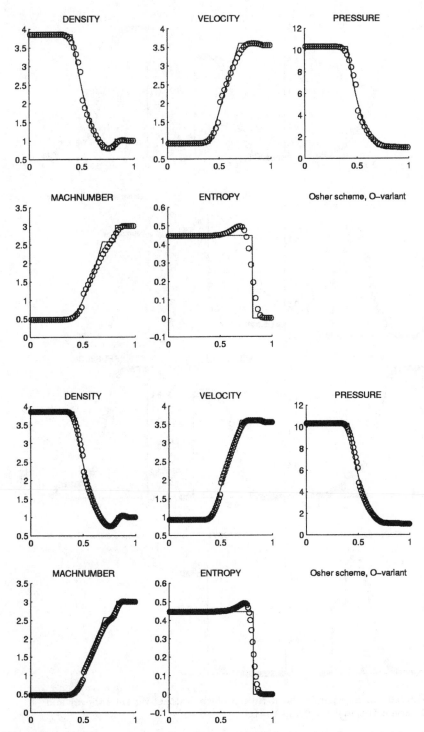

Fig. 10.11. Mach 3 test case, O-variant; $t = 0.0875$, $\lambda = 0.2$, $h = 1/48$ (above), $h = 1/96$ (below); —: exact solution, o: numerical solution.

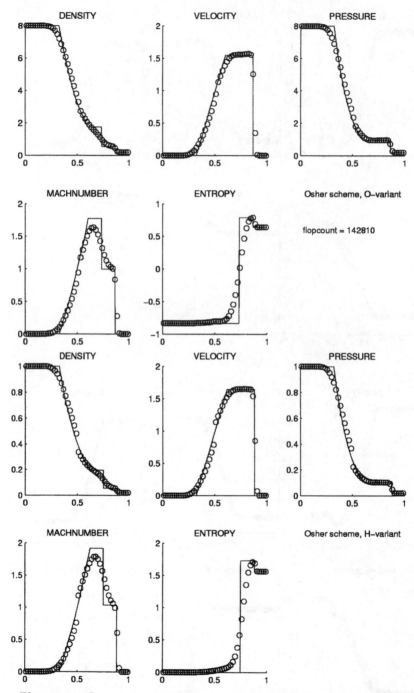

Fig. 10.12. Supersonic shock tube problem with O-variant (above) and H-variant (below); $t = 0.15$, $\lambda = 0.3$.

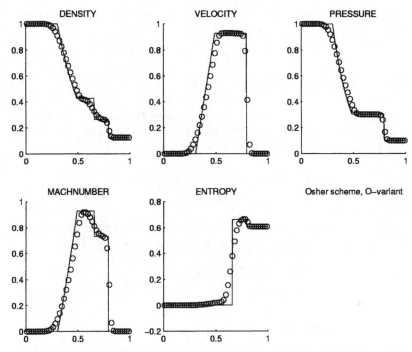

Fig. 10.13. Sod's shock tube problem with O-variant; $t = 0.1667$, $\lambda = 0.4$.

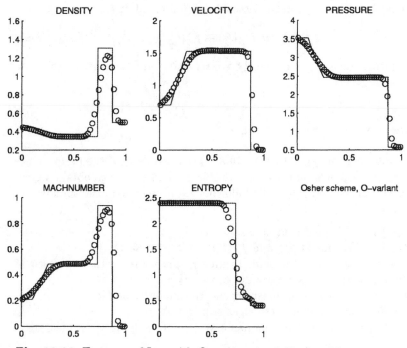

Fig. 10.14. Test case of Lax with O-variant; $t = 0.15$, $\lambda = 0.2$.

strong expansion fan is present, as in the Mach 3 test case, the Roe scheme suffers from violation of the entropy condition.

10.5 Flux splitting schemes

Flux splitting schemes can be regarded as a generalization of the Courant-Isaacson-Rees scheme (9.81) from the scalar to the systems case. We recall that in the scalar case the numerical flux is given by

$$F_{j+1/2} = f(\varphi_j) \quad \text{if} \quad f'(\varphi_{j+1/2}) \geq 0 \,,$$
$$F_{j+1/2} = f(\varphi_{j+1}) \quad \text{if} \quad f'(\varphi_{j+1/2}) < 0 \,,$$

where $\varphi_{j+1/2} \equiv (\varphi_j + \varphi_{j+1})/2$. Steger and Warming (1981) have generalized this to systems in the following way. The Euler flux is split as follows:

$$f(U) = f^+(U) + f^-(U) \tag{10.83}$$

such that

$$\lambda(\partial f^+/\partial U) \geq 0 \,, \quad \lambda(\partial f^-/\partial U) < 0 \,, \tag{10.84}$$

i.e. the Jacobian of f^+ has only nonnegative eigenvalues and the Jacobian of f^- has only negative eigenvalues. Then we can write down the following generalization of the Courant-Isaacson-Rees numerical flux:

$$F_{j+1/2} = f^+(U_j) + f^-(U_{j+1}) \,. \tag{10.85}$$

The resulting scheme is obviously conservative and consistent (because (10.54) is satisfied).

The van Leer scheme

Various splittings of type (10.83), (10.84) are possible, and the performance of the scheme depends on the choice that is made. In order to obtain a good scheme, van Leer (1982) puts the following requirements on the splitting, in addition to (10.84):

(i) $f^{\pm}(U)$ must depend continuously on U;
(ii) $f^+(U) = f(U)$ for $M \geq 1$ and $f^-(U) = f(U)$ for $M \leq -1$;
(iii) $f^+ + f^-$ must have the same symmetry properties with respect to M (keeping all other state variables constant) as f;
(iv) The Jacobians $\partial f^{\pm}/\partial U$ must depend continuously on U;
(v) $\partial f^{\pm}/\partial U$ must have one zero eigenvalue for $|M| < 1$;
(vi) Like f, $f^{\pm}$ must be a polynomial in M, and of the lowest possible degree.

We will now clarify these requirements. We note that we can write

$$
f(U) = \begin{pmatrix} \rho c M \\ \rho c^2 (M^2 + 1/\gamma) \\ \rho c^3 M(\frac{1}{2}M^2 + 1/(\gamma - 1)) \end{pmatrix} , \quad M = u/c .
$$

Hence, keeping ρ and c constant,

$$
f_{1,3}(M) = -f_{1,3}(-M) , \quad f_2(M) = f_2(-M) .
$$

Requirement (iii) therefore becomes:

$$
f_{1,3}^+(M) + f_{1,3}^-(M) = -f_{1,3}^+(-M) + f_{1,3}^-(M) ,
$$
$$
f_2^+(M) + f_2^-(M) = f_2^+(-M) + f_2^-(-M) .
$$

Requirements (i) and (iv) make the numerical solution smooth , especially near sonic points ($M = 1$) and stagnation points ($u = 0$), where eigenvalues change sign. Furthermore, these requirements enhance the convergence behavior of iterative methods for the solution of implicit versions of the scheme. Requirement (ii) ensures that the numerical scheme has the same domain of dependence as the differential equation. Requirement (iii) makes the numerical flux share an important property with the exact flux. It turns out that requirement (v) enables the scheme to capture stationary shocks in two cells. Finally, requirement (vi) makes the splitting unique.

From these requirements van Leer derives the following splitting:

$$
f^+ = \begin{pmatrix} \frac{1}{4}\rho c(1 + M)^2 \\ \frac{c}{\gamma} f_1^+ (2 + (\gamma - 1)M) \\ \frac{\gamma^2}{2(\gamma^2 - 1)}(f_2^+)^2 / f_1^+ \end{pmatrix} ,
$$

$$
\tag{10.86}
$$

$$
f^- = \begin{pmatrix} -\frac{1}{4}\rho c(1 - M)^2 \\ \frac{c}{\gamma} f_1^- ((\gamma - 1)M - 2) \\ \frac{\gamma^2}{2(\gamma^2 - 1)}(f_2^-)^2 / f_1^- \end{pmatrix}
$$

for $|M| < 1$; for $|M| \geq 1$ the splitting is given by requirement (ii). The only requirement that is not trivially satisfied is (10.84). It is left as an exercise to show that (10.84) is satisfied.

The modification of Hänel, Schwane and Seider

According to Bernoulli's law, for stationary inviscid flow the total enthalpy satisfies $H = h + \frac{1}{2}u^2 = $ constant along streamlines. If the streamlines emanate from a region of constant H, then $H = $ constant everywhere. We can also see immediately from the stationary mass and energy equations

$$\mathrm{div}\rho\boldsymbol{u} = 0 , \quad \mathrm{div}\rho\boldsymbol{u}H = 0$$

that $H = $ constant is a solution. In general, for numerical schemes $H = $ constant is not an exact solution, but gives a residual of discretization error size. Hänel, Schwane, and Seider (1987) report that this may cause significant errors in regions of strong Mach number changes. They propose to modify van Leer's flux splitting for the energy equation by:

$$\rho u H = f_1^+ H + f_1^- H , \tag{10.87}$$

with $f_1 = f_1^+ + f_1^-$ the van Leer flux splitting for the mass conservation equation. That the flux splitting (10.87) has the desired effect can be seen as follows. In the stationary case the discretized energy equation becomes with the splitting (10.87):

$$f_1^+ (U_{j-1})H_{j-1} + \{f_1^- (U_j) - f_1^+ (U_j)\}H_j - f_1^- (U_{j+1})H_{j+1} = 0 . \tag{10.88}$$

The stationary discrete mass conservation equation gives

$$f_1^+ (U_{j-1}) + f_1^- (U_j) - f_1^+ (U_j) - f_1^- (U_{j+1}) = 0 ,$$

so that $H_j = $ constant is indeed a solution of (10.88).

It turns out that of the six requirements that determine van Leer's flux splitting, the modification of Hänel *et al.* violates only requirement (v), the purpose of which is (as stated before) to ensure capturing of stationary shocks in two cells. Fig. 10.15 shows an example with a slowly moving weak shock. We see that the modified scheme indeed needs a few more cells to capture the shock. But it turns out that in most cases the shock resolution of the modified scheme is about as crisp as that of the original scheme. Results obtained with the modified scheme for the four test cases discussed before will not be shown, because they resemble closely the results obtained with the original scheme, except that the modified scheme gives a somewhat smaller sonic glitch for the supersonic shock tube problem.

The Osher scheme allows $H_j = $ constant in the stationary case. As for the scheme of Hänel *et al.*, this follows from the fact that the energy flux equals H times the mass flux. In the stationary case the scheme gives

$$F_{j+1/2} - F_{j-1/2} = 0 . \tag{10.89}$$

For the mass conservation equation this gives with the Osher scheme an expression of the type

$$\sum_s \alpha_s m_s = 0 , \quad m = \rho u , \tag{10.90}$$

where s indicates the various states occurring along the integration paths used to determine $F_{j\pm1/2}$, and α_s are coefficients that need not be specified. The Osher scheme gives for the stationary energy equation

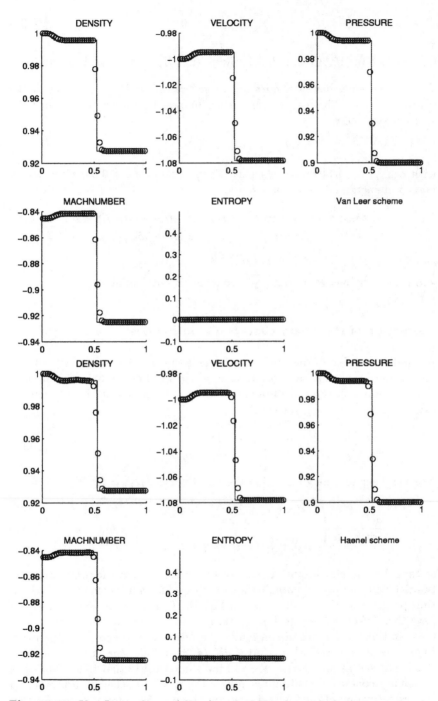

Fig. 10.15. Van Leer scheme (above) and of Hänel *c.s.* (below); $p_L = 1$, $p_R = 0.9$, $\rho_L = 1$, $\rho_R = 0.9275$, $u_L = -1$, $u_R = -1.0781$, $t = 0.175$, $\lambda = 0.4$.

$$\sum_s \alpha_s m_s H_s = 0 . \tag{10.91}$$

Comparison of (10.90) and (10.91) shows that $H_s = $ constant is a solution.

The Roe scheme, however, does not in general allow $H_j = $ constant in the stationary case. This can be shown as follows. It suffices to consider the special case where

$$U_{j+1} - U_j = (\alpha R_1)_{j+1/2} .$$

with $\alpha_{j+1/2}$ an arbitrary coefficient. Then (10.66) and (10.89) give for the mass and energy conservation equations

$$m_{j+1} - m_{j-1} - (\alpha|\bar{u} - \bar{c}|)_{j+1/2} + (\alpha|\bar{u} - \bar{c}|)_{j-1/2} = 0 ,$$
$$(mH)_{j+1} - (mH)_{j-1} - (\alpha|\bar{u} - \bar{c}|(\bar{H} - \bar{u}\bar{c}))_{j+1/2} +$$
$$(\alpha|\bar{u} - \bar{c}|(H - \bar{u}\bar{c}))_{j-1/2} = 0 ,$$

and it is clear that $H_j = \bar{H}_{j+1/2} = $ constant is not a solution.

Resolution of stationary contact discontinuities

A significant defect of the van Leer scheme is the fact that stationary contact discontinuities cannot be resolved, as noted in van Leer (1982). This we now show. Suppose the flow is stationary, and that there is a stationary contact discontinuity at $x_{j+1/2}$. Hence

$$u_k = 0 , \quad p_k = p_{k+1} , \; \forall k ;$$
$$\rho_k = \rho_j , \; k \leq j ; \quad \rho_k = \rho_{j+1} , \; k \geq j+1 . \tag{10.92}$$

The van Leer scheme gives, using the fact that $(\rho c^2)_{j+1} = (\rho c^2)_j$,

$$\frac{dU_j}{dt} = \frac{1}{h} \begin{pmatrix} \frac{1}{4}\{(\rho c)_{j+1} - 2(\rho c)_j + (\rho c)_{j-1}\} \\ 0 \\ \frac{(\rho c^2)_j}{2(\gamma^2-1)}(c_{j+1} - 2c_j + c_{j-1}) \end{pmatrix} \neq 0 , \tag{10.93}$$

so that the numerical solution will not be stationary, but the contact discontinutity will be approximated by a smooth profile that widens diffusively as time progresses. This is illustrated in Fig. 10.16. The Osher scheme does not have this defect, as seen in Fig. 10.16. It is left as an exercise to show that for a stationary contact discontinuity the Osher flux is zero, like the exact Euler flux. Hence the scheme gives $dU_j/dt = 0$. The Roe flux is also zero; this is also left as an exercise. We will show in Chap. 12 that this smearing of stationary contact discontinuities makes the van Leer scheme unsuitable for the discretization of the inviscid terms in viscous flow computations. Further numerical results will be presented only for the improved version below.

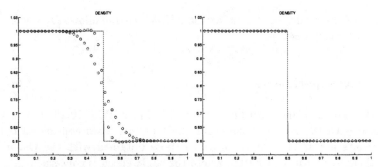

Fig. 10.16. Stationary contact discontinuity at $t = 0.1$ and $t = 1$. Left: van Leer scheme; right: Osher scheme. $\lambda = 0.4$, $p_L = p_R = 0.5$, $u_L = u_R = 0$, $\rho_L = 1$, $\rho_R = 0.6$.

The AUSM scheme

In Liou and Steffen (1993) a flux splitting scheme is proposed that is similar to the van Leer scheme, but that has a satisfactory resolution of stationary contact discontinuities. This scheme is often called the AUSM (advection upstream splitting method) scheme, and represents the state of the art of flux splitting schemes. It rivals the accuracy of the Osher and Roe schemes, but requires less computing.

The AUSM scheme is defined as follows:

$$F_{j+1/2} = \frac{1}{2}(M + |M|)_{j+1/2}\tilde{U}_j + \frac{1}{2}(M - |M|)_{j+1/2}\tilde{U}_{j+1} + \tilde{P}_j^+ + \tilde{P}_{j+1}^- \,,$$

$$M_{j+1/2} = M_j^+ + M_{j+1}^- \,,$$

$$M^\pm = \pm\frac{1}{4}(M \pm 1)^2 \,, \ |M| \le 1 \,; \ M^\pm = \frac{1}{2}(M \pm |M|) \,, \ |M| > 1 \,,$$

$$\tilde{U} = \rho c \begin{pmatrix} 1 \\ u \\ H \end{pmatrix} \,, \ \tilde{P}^\pm = \begin{pmatrix} 0 \\ p^\pm \\ 0 \end{pmatrix} \,,$$

$$p^\pm = \frac{p}{2}(1 \pm M) \,, \ |M| \le 1 \,; \ p^\pm = \frac{p}{2}(1 \pm |M|/M) \,, \ |M| > 1 \,.$$

Note that this scheme is not quite of the form (10.85).

Suppose we have a stationary contact discontinuity at $x_{j+1/2}$, with U_k given by (10.92). It is left to the reader to show that in this case

$$\frac{dU_k}{dt} = 0 \,, \quad \forall k \,,$$

so that a stationary contact discontinuity is resolved exactly by the AUSM scheme.

It is left to the reader to show that the AUSM scheme allows $H_j = \text{constant}$ in the stationary case.

Numerical experiments

We discretize in time according to (10.72). Figs. 10.17–10.20 show results obtained with the AUSM scheme for the four test cases considered before. The results are at least as accurate as those obtained with the Osher and Roe schemes, and in fact better where the sonic glitch is concerned.

Exercise 10.5.1. Show that van Leer's flux splitting (10.86) satisfies condition (10.84).
Hint: one eigenvalue is zero. Show that the other two satisfiy for $|M| < 1$:

$$\lambda^2 - \lambda\frac{3c}{2}(1+M)[1 + \frac{(\gamma-1)(1-M)}{12\gamma(\gamma+1)}\{\gamma(1-M)^2 - 2\gamma(1-M) - 2(\gamma+3)\}]$$
$$+ \frac{c^2}{4}(1+M)^3[1 - \frac{1-M}{8\gamma(\gamma+1)}\{4\gamma(\gamma-1)(1-M) + (\gamma+1)(\gamma-3)\}] = 0$$

the roots of which are positive for $1 \le \gamma \le 3$.

Exercise 10.5.2. Derive equation (10.93).

Exercise 10.5.3. Show that the AUSM scheme allows $H_j = \text{constant}$ in the stationary case.

Exercise 10.5.4. Show that for the AUSM scheme we have $dU_j/dt = 0$ in the case of a stationary contact discontinuity.
Hint: show that $M_{j+1/2} = 0$.

Exercise 10.5.5. Show that for the Osher scheme we have $dU_j/dt = 0$ in the case of a stationary contact discontinuity.
Hint: show that (10.82) gives $u_{1/3} = 0$. Since $u_0 = u_1 = 0$, all contributions to the Osher flux are zero.

Exercise 10.5.6. Show that for the Roe scheme we have $dU_j/dt = 0$ in the case of a stationary contact discontinuity.
Hint: show that $\alpha_1 = \alpha_3 = \lambda_2 = 0$.

10.6 Numerical stability

Rigorous stability analysis of numerical schemes for nonlinear hyperbolic systems is difficult, and has rarely been attempted. With von Neumann stability analysis, we linearize, freeze the coefficients and study the growth or decay

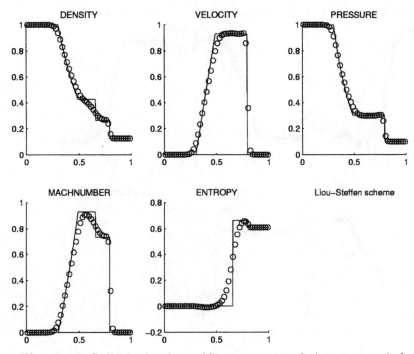

Fig. 10.17. Sod's shock tube problem; —: exact solution; o: numerical solution; $t = 0.168$, $\lambda = 0.35$.

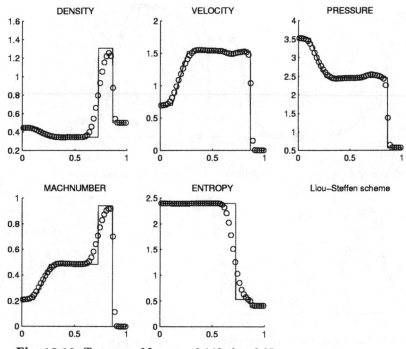

Fig. 10.18. Test case of Lax: $t = 0.146$, $\lambda = 0.25$.

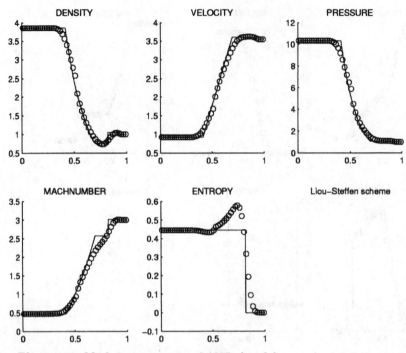

Fig. 10.19. Mach 3 test case; $t = 0.0875$, $\lambda = 0.2$.

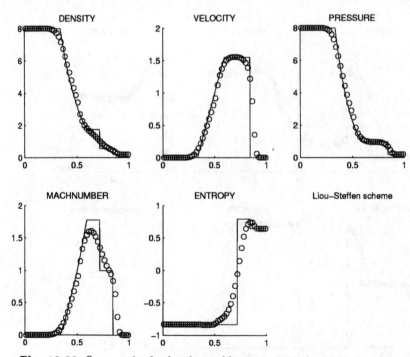

Fig. 10.20. Supersonic shock tube problem; $t = 0.14$, $\lambda = 0.11$.

of the solution by means of Fourier analysis. As an example, consider the explicit Euler method for a flux splitting scheme. The numerical flux is given by (10.85), and we get

$$U_j^{n+1} - U_j^n + \lambda\{f^+(U_j^n) - f^+(U_{j-1}^n) + f^-(U_{j+1}^n) - f^-(U_j^n)\} = 0 .$$

Linearization and freezing of the coefficients gives

$$U_j^{n+1} - U_j^n + \lambda F^+(U_j^n - U_{j-1}^n) + \lambda F^-(U_{j+1}^n - U_j^n) = 0 ,$$

where $F^\pm$ are the Jacobians of $f^\pm$, assumed constant. Assuming an harmonic wave $U_j^n = \hat{U}^n e^{ij\theta}$ gives $\hat{U}^{n+1} = G\hat{U}^n$, with the amplification matrix G given by

$$\begin{aligned} G &= I - \lambda F^+(1 - e^{-i\theta}) - \lambda F^-(e^{i\theta} - 1) \\ &= I - \lambda(F^+ - F^-)(1 - \cos\theta) - i\lambda f' \sin\theta , \end{aligned}$$

where we have used $F^+ + F^- = f'$, with f' the Jacobian of f (cf. (10.83)). We have stability if the eigenvalues μ of G satisfy

$$|\mu(G)| \le 1 , \quad \forall\theta .$$

However, because the eigenvectors of $F^+ - F^-$ are in general different from those of f', G is not easily diagonalizable, if at all, and $\mu(G)$ is hard to determine. Furthermore, freezing the coefficients is unrealistic where the solution is not smooth, for instance near shocks.

For simplicity we proceed in an even less rigorous manner, and linearize the system of differential equations instead of the numerical scheme, and freeze the coefficients. This leads us to the following system:

$$\frac{\partial U}{\partial t} + f'\frac{\partial U}{\partial x} = 0 ,$$

with the Jacobian f' assumed constant. Let R be the matrix with columns R_1, R_2, R_3, given by (10.21). Diagonalization gives

$$\frac{\partial V}{\partial t} + \Lambda\frac{\partial V}{\partial x} = 0 , \quad V = R^{-1}U ,$$

with Λ a diagonal matrix containing the eigenvalues $\mu = u, u \pm c$ of f'. This is an uncoupled system of scalar equations of the type

$$\frac{\partial v}{\partial t} + \mu\frac{\partial v}{\partial x} = 0 . \tag{10.94}$$

We now require that the scheme used for the Euler equations is stable when applied to (10.94). The schemes discussed before are generalizations of the first order upwind scheme to the systems case. Application of the first order upwind scheme to (10.94) gives, assuming $\mu > 0$,

$$\frac{dv_j}{dt} + \frac{\mu}{h}(v_j - v_{j-1}) = 0 . \tag{10.95}$$

Von Neumann stability analysis for temporal discretizations of (10.95) is easily carried out following the principles explained in Sect. 5.6. The symbol of the scheme is given by

$$\hat{L}_h(\theta) = \frac{\sigma}{\tau}(1 - e^{-i\theta}) , \quad \sigma = \mu\tau/h .$$

We have stability if $-\tau\hat{L}_h$ is in the stability domain S of the time stepping method to be used. Stability is required for all values μ that occur in the flow domain. This means that we have to take

$$\mu = \bar{\mu} \equiv \max\{|u| + c\} .$$

The corresponding value of σ, i.e.

$$\bar{\sigma} = \bar{\mu}\tau/h$$

is usually called the CFL (Courant-Friedrichs-Lewy) number.

For the forward Euler method we have stability if

$$\bar{\sigma} \leq 1 . \tag{10.96}$$

From the exact solutions, we obtain for the four test cases considered in the preceding section the approximate estimates for $\bar{\mu}$ listed in Table 10.2, which also gives the stability bounds λ_1 for $\lambda = \tau/h$ that follow from (10.96). The

	$\bar{\mu}$	λ_1
Sod	2.2	0.46
Lax	4.7	0.21
Mach 3	5.0	0.20
Supers. shock tube	3.1	0.32

Table 10.2. Estimates for $\bar{\mu}$ and stability bounds for λ.

numerical results in the preceding sections were obtained with values of λ close to λ_1.

Exercise 10.6.1. Derive equation (10.96).

10.7 The Jameson-Schmidt-Turkel scheme

Second order central discretization gives the following numerical flux:

$$F_{j+1/2} = \frac{1}{2}\{f(U_j) + f(U_{j+1})\} .$$

This scheme is unusable because it gives serious oscillations near discontinuities, and may also generate oscillations elsewhere. The schemes of Roe, Osher and van Leer have a numerical flux that can be written as

$$F_{j+1/2} = \frac{1}{2}\{f(U_j) + f(U_{j+1})\} + \text{extra term} .$$

The extra term is given explicitly for the Roe scheme in (10.66) and for the Osher scheme in (10.76). The extra term provides just enough dissipation to obliterate numerical wiggles, while still maintaining crisp resolution of discontinuities; moving contact discontinuities are smeared, however. In the schemes of Roe, Osher and in flux splitting schemes the extra term is an implicit consequnce of the way in which the Riemann problem at the cell face is approximated. Instead, one may try to design the extra term explicitly. This is the approach of the Jameson-Schmidt-Turkel or JST scheme, developed by Jameson, Schmidt, and Turkel (1981), Jameson (1985b), Jameson (1985a), Jameson (1988). Of course, the extra term has to be designed judiciously. The aim is to obtain a scheme with greater simplicity and economy of computation than the schemes described before, allowing if necessary a slight increase of numerical shock thickness.

Artificial viscosity

For the scalar case, the JST scheme has been presented in Sect. 9.4. In Jameson (1985a) the following numerical flux is proposed for the Euler equations:

$$F_{j+1/2} = \frac{1}{2}\{f(U_j) + f(U_{j+1})\} - d_{j+1/2} ,$$

where $d_{j+1/2}$ is an artificial viscosity term. Writing $d = (d^1, d^2, d^3)^T$, for the mass conservation equation d^1 is chosen as follows:

$$d^1_{j+1/2} = r_{j+1/2}\{\varepsilon^{(2)}_{j+1/2}(\rho_{j+1} - \rho_j) - \varepsilon^{(4)}_{j+1/2}(\rho_{j+2} - 3\rho_{j+1} + 3\rho_j - \rho_{j-1})\} . \tag{10.97}$$

Here $r_{j+1/2}$ is a coefficient chosen to give the artificial viscosity term the proper scale. On dimensional grounds, it must have the dimension of velocity, and is taken to be an estimate of the special radius of the Jacobian $\partial f/\partial U$ at the cell face:

$$r_{j+1/2} = \frac{1}{2}(|u_j| + c_j + |u_{j+1}| + c_{j+1}) .$$

The term with $\varepsilon^{(2)}$ represents artificial diffusion proportional to the second derivative of ρ. Its purpose is to damp oscillations near discontinuities. This term is turned off adaptively in regions where the solution is smooth; there the term with $\varepsilon^{(4)}$ takes over. This term is proportional to the fourth derivative of ρ. A sensor that signals the presence of a shock is needed. This shock sensor is chosen as follows:

$$\nu_j = \left| \frac{p_{j+1} - 2p_j + p_{j-1}}{p_{j+1} + 2p_j + p_{j-1}} \right| .$$

We set

$$\bar{\nu}_{j+1/2} = \max\{\nu_{j+2}, \nu_{j+1}, \nu_j, \nu_{j-1}\} .$$

Then we take

$$\varepsilon^{(2)}_{j+1/2} = \min\{\frac{1}{2}, k^{(2)}\bar{\nu}_{j+1/2}\} ,$$
$$\varepsilon^{(4)} = \max\{0, k^{(4)} - \alpha\bar{\nu}_{j+1/2}\} ,$$

where $k^{(2)}$, $k^{(4)}$ and α are constants, typically chosen as follows:

$$k^{(2)} = 1 , \quad k^{(4)} = 1/32 , \quad \alpha = 2 . \tag{10.98}$$

We see that near shocks, where $\bar{\nu}_{j+1/2}$ becomes of order one, $\varepsilon^{(4)}_{j+1/2}$ is turned off; this is necessary because the fourth order dissipation term is found to produce overshoots near shocks. This completes the spatial discretization of the mass conservation equation. In regions where the solution is smooth we have $\bar{\nu}_{j+1/2} \ll 1$, so that $\varepsilon^{(2)}_{j+1/2}$ is switched of. Hence $d^1_{j+1/2} - d^1_{j-1/2} = \mathcal{O}(h^4)$, giving the scheme the order of accuracy of a second order central scheme, namely $\mathcal{O}(h^2)$. Near shocks the solution is not differentiable, and it makes little sense to estimate the local truncation error with Taylor expansion.

The spatial discretization of the momentum equation is obtained by replacing ρ by $m = \rho u$ in (10.97):

$$d^2_{j+1/2} = r_{j+1/2}\{\varepsilon^{(2)}_{j+1/2}(m_{j+1} - m_j)$$
$$-\varepsilon^{(4)}_{j+1/2}(m_{j+2} - 3m_{j+1} + 3m_j - m_{j-1})\} ,$$

where the coefficients are defined as before. The spatial discretization of the energy equation is obtained by replacing ρ not by ρE, as one might expect, but by ρH. This is done to make $H = $ constant a solution of the stationary energy equation, making the scheme conform to Bernoulli's law. So we obtain

$$d^3_{j+1/2} = r_{j+1/2}\{\varepsilon^{(2)}_{j+1/2}((\rho H)_{j+1} - (\rho H)_j) -$$
$$\varepsilon^4_{j+1/2}((\rho H)_{j+2} - 3(\rho H)_{j+1} + 3(\rho H)_j - (\rho H)_{j-1})\} .$$

Temporal discretization

Since in regions where the solution is smooth the JST scheme is virtually identical to the second order central scheme, the explicit Euler scheme cannot be used for time stepping. In the linearized scalar case, the Fourier symbol of the spatial discretization is (almost) purely imaginary. Therefore a temporal discretization method is needed with a stability domain that includes part of the imaginary axis. Certain Runge-Kutta methods are suitable. The system of ordinary differential equations that results after spatial discretization can be written as

$$\frac{dV}{dt} + R(V) = 0\,,$$

where the algebraic vector $V(t)$ contains all unknowns. Consider the following explicit four-stage Runge-Kutta method:

$$
\begin{aligned}
V^{(0)} &= V^n\,, \\
V^{(1)} &= V^{(0)} - \alpha_1 \tau R(V^{(0)})\,, \\
V^{(2)} &= V^{(0)} - \alpha_2 \tau R(V^{(1)})\,, \\
V^{(3)} &= V^{(0)} - \alpha_3 \tau R(V^{(2)})\,, \\
V^{n+1} &= V^{(0)} - \tau R(V^{(3)})\,.
\end{aligned}
\tag{10.99}
$$

The coefficients are chosed to maximize the stability interval along the imaginary axis. Temporal accuracy is given low priority, because the JST scheme is designed primarily for stationary flows. The general problem of finding the coefficients that define Runge-Kutta methods such that for a given temporal accuracy the stability interval along the imaginary axis is maximized has been considered by van der Houwen (1977). For temporal accuracy of order one, Sonneveld and van Leer (1985) give a solution, namely

$$\alpha_1 = 1/3\,, \quad \alpha_2 = 4/15\,, \quad \alpha_3 = 5/9\,. \tag{10.100}$$

The classical four-stage Runge-Kutta method (maximizing order of accuracy) has

$$\alpha_1 = 1/4\,, \quad \alpha_2 = 1/3\,, \quad \alpha_3 = 1/2\,. \tag{10.101}$$

The stability domains of these two methods are presented in Fig. 10.22. We see that if the symbol of the (linearized) spatial discretization is purely imaginary, then the version of Sonneveld and van Leer allows a time step that is 5% larger than allowed by the classical version. Hence, this sacrifice of accuracy for stability hardly pays off.

A substantial saving in computing time can be realized by evaluating the artificial dissipation term only once. This can be done by writing

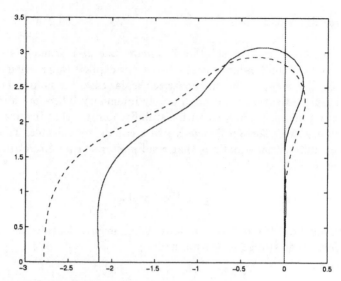

Fig. 10.21. Stability domains of four-stage Runge-Kutta method with coefficients given by (10.100) (—) and (10.101) (- - -).

$R(V) = Q(V) - D(V)$ where D arises from the artificial dissipation term, and by defining in (10.99):

$$R(V^{(q)}) = Q(V^{(q)}) - D(V^{(0)}) \,.$$

We have applied the JST scheme to the test cases considered before. The coefficients are chosen as proposed by Jameson (1985a), and are given by (10.98) and (10.100). We will also present results with a larger value of $k^{(2)}$, in order to improve results near shocks. The step size is $h = 1/48$ in all cases. The time step is chosen close to the stability limit on an empirical basis. The upper half of Fig. 10.22 shows shock-induced wiggles. As shown in the lower half of the figure, these disappear if the second order artificial viscosity is increased sufficiently ($k^{(2)} = 8$); the time step has to be decreased in order to maintain stability. We see that shock capturing is less crisp than with the schemes discussed earlier, but still the results of the lower half of the figure are good, although there is a slight violation of the entropy condition .

Fig. 10.23 also shows wiggles for $k^{(2)} = 1$. Here $k^{(2)} = 6$ was found to be sufficient to obtain smooth results, but in order to see whether good results can be obtained with the parameters fixed for all problems, we have used $k^{(2)} = 8$ for the lower half of Fig. 10.23, as in Fig. 10.22 (it turns out there is very little difference between results with $k^{(2)} = 6$ and $k^{(2)} = 8$ in this case). Again, the shock is somewhat more smeared than with the schemes discussed earlier, but nevertheless the accuracy of the results with $k^{(2)} = 8$ is good.

Fig. 10.24 shows that the JST scheme does not give a sonic glitch (unlike the

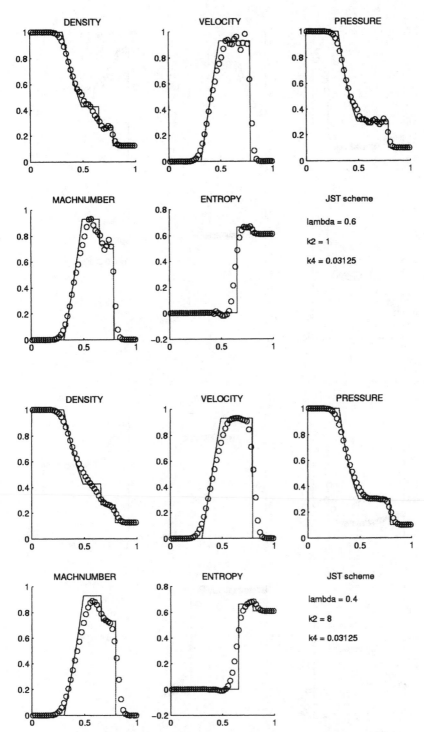

Fig. 10.22. Sod's shock tube problem at $t = 0.1667$. Above: $\lambda = 0.6$, $k^{(2)} = 1$ below: $\lambda = 0.4$, $k^{(2)} = 8$.

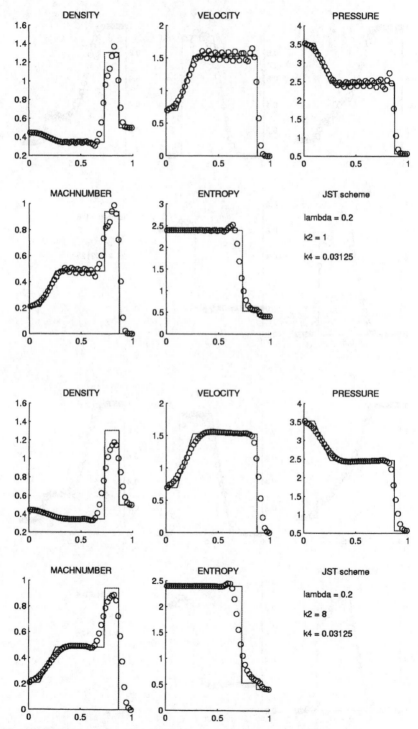

Fig. 10.23. Test case of Lax at $t = 0.15$. Above: $\lambda = 0.2$, $k^{(2)} = 1$; below: $\lambda = 0.2$, $k^{(2)} = 8$.

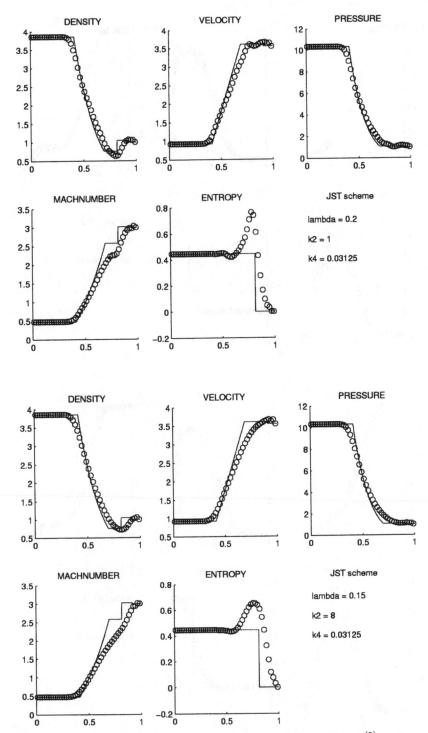

Fig. 10.24. Mach 3 test case at $t = 0.0875$. Above: $\lambda = 0.2$, $k^{(2)} = 1$; below: $\lambda = 0.15$, $k^{(2)} = 8$.

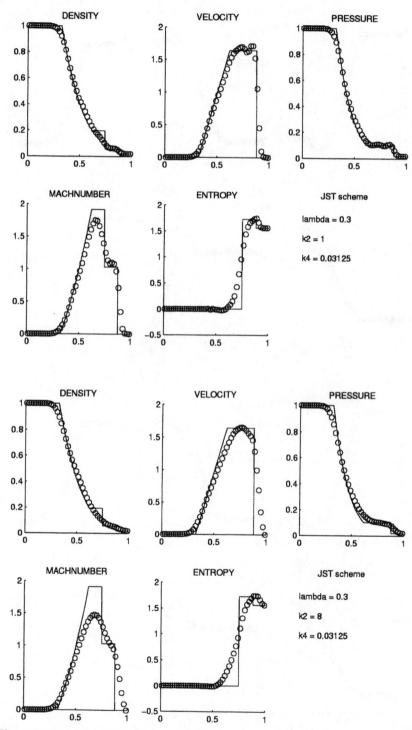

Fig. 10.25. Supersonic shock tube problem at $t = 0.15$. Above: $\lambda = 0.3$, $k^{(2)} = 1$; below: $\lambda = 0.3$, $k^{(1)} = 8$.

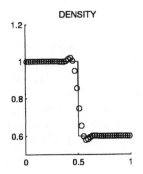
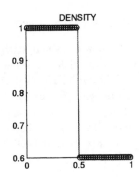

Fig. 10.26. Stationary contact discontinuity at $t = 1$ with $\lambda = 2$. Left: $k^{(4)} = 1/32$; right : $k^{(4)} = 0$.

schemes discussed before), despite the fact that there is a slight violation of the entropy condition near the sonic point ($x = 0.5$; note that fluid particles cross this point from left to right). The small velocity wiggles can be made to disappear by increasing $k^{(2)}$ to $k^{(2)} = 4$, but again we show results for $k^{(2)} = 8$, for which it is found to be necessary to decrease the time step. The difference between results for $k^{(2)} = 4$ and $k^{(2)} = 8$ was found to be negligible, but compared to $k^{(2)} = 1$, Fig. 10.24 shows a slight decrease in accuracy, apart from the removal of wiggles, of course. Nevertheless, the results compare well with those obtained with the Osher and van Leer schemes in this case.

As seen in Fig. 10.25, also for the supersonic shock tube no sonic glitch occurs, although again these is a slight violation of the entropy condition near the sonic point. The wiggles in the velocity in the upper half in the figure dissappear when $k^{(2)}$ is increased to $k^{(2)} = 4$ (not shown); almost the same results are obtained with $k^{(2)} = 8$. Results for a stationary contact discontinuity (same data as for Fig. 10.16) are shown in Fig. 10.26. There is some smearing, but the numerical width of the contact discountinuity does not increase with time, as for the van Leer scheme. Since the pressure is constant, the second order artificial viscosity is zero. The small overshoot that occurs is caused by the fourth order artificial viscosity; when it is turned off the result of the scheme is exact. But for a moving contact discontinuity, with a vanishing artificial viscosity unacceptable wiggles occur (results not shown). Hence, the fourth order artificial viscosity has to be turned on, leading to the nonmonotone approximation shown in the left half of Fig. 10.26. This lack of monotonicity corresponds to the fact that the JST scheme is not TVD, as shown for the scalar case in Sect. 9.4. In Jameson (1988), Kim and Jameson (1995) a fourth order artificial dissipation term is presented that makes the scheme TVD; this TVD version was described in Sect. 9.4 for the scalar case, and will not be further considered here.

Summarizing, the results obtained with the JST scheme are of the same quality as those obtained with the schemes of Roe and Osher and the flux splitting schemes. Although slight violations of the entropy condition occur, no sonic glitches are present. Shock resolution is less crisp. It is necessary to choose the parameters governing the artifical dissipation right. The parameters given by Jameson (1985a) and listed in (10.98) are not adequate for the one-dimensional problems treated here; $k^{(2)}$ should be changed to $k^{(2)} = 8$.

Exercise 10.7.1. Show that in the stationary case the JST scheme allows the solution $H_j = \text{constant}$.

10.8 Higher order schemes

The Roe, Osher and flux splitting schemes are generalizations of first order upwind schemes from the scalar case to the case of systems. As a consequence, these schemes are first order accurate in space. In order to improve accuracy and resolution of contact discontinuities, it is necessary to make these schemes second or higher order accurate in regions where the solution is smooth, while avoiding spurious wiggles. In Sect. 9.4 we discussed a good way to do this for the scalar case, namely the MUSCL (monotone upwind schemes for conservation laws) approach proposed in van Leer (1979), van Leer (1984). This approach generalizes easily to the case of systems.

The MUSCL approach

Suppose we have a first order spatial discretization, which leads to the following semi-discrete system:

$$\frac{dU_j}{dt} + \frac{1}{h}(F_{j+1/2} - F_{j-1/2}) = 0 , \quad F_{j+1/2} = F(U_j, U_{j+1}) .$$

Just as in the scalar case discussed in Sect. 9.4, the accuracy may be improved by inserting extrapolated left and right states in the numerical flux function:

$$F_{j+1/2} = F(U_{j+1/2}^L, U_{j+1/2}^R) .$$

To avoid spurious wiggles, the slope limited extrapolation (9.98) is applied to each of the state variables separately. For instance,

$$\rho_{j+1/2}^L = \rho_j + \frac{1}{2}\psi(r_j)(\rho_{j+1} - \rho_j) , \quad \rho_{j+1/2}^R = \rho_{j+1} + \frac{1}{2}\psi(\frac{1}{r_{j+1}})(\rho_j - \rho_{j+1}) ,$$

$$r_j = \frac{\rho_j - \rho_{j-1}}{\rho_{j+1} - \rho_j} .$$

The limiter $\psi(r)$ is chosen such that where the solution is smooth a higher order scheme, for instance the κ-scheme with a specified value of κ, is emulated.

Numerical stability

In order to derive stability conditions, we use the same heuristic reasoning as in Sect. 10.6, and require the scheme to be stable for the scalar conservation law

$$\frac{\partial \varphi}{\partial t} + \frac{\partial f(\varphi)}{\partial x} = 0 , \tag{10.102}$$

where we do not specify $f(\varphi)$, except that the wave speed $df/d\varphi$ is assumed to be equal to the largest wave speed for the hyperbolic system under consideration. For the Euler equations this leads to

$$\frac{df}{d\varphi} = |u| + c . \tag{10.103}$$

In Sect. 9.4 conditions were found for schemes for (10.102) to be TVD and hence stable. Although the TVD concept does not carry over to systems, it is found in practice that a scheme that is TVD for the scalar case is usually free from spurious oscillations in the systems case.

As an example, we choose the van Albada limiter (van Albada, van Leer, and Roberts (1982)):

$$\psi(r) = (r^2 + r)/(1 + r^2) , \ r \geq 0 ; \ \ \psi(r) = 0 , \ r < 0 .$$

Since $\psi'(1) = 1/2$, this gives us the $\kappa = 1/2$ scheme in smooth parts of the flow. Simple analysis shows that we have

$$0 \leq \frac{\psi(r)}{r} \leq (1 + \sqrt{2})/2 \cong 0.85 , \tag{10.104}$$
$$0 \leq \psi(r) \leq (1 + \sqrt{2})/2 .$$

The graph of $\psi(r)$ is shown in Fig. 10.27, together with the lines $\psi(r) = r(1 + \sqrt{2})/2$ and $\psi(r) = r$. It is clear that the graph of the van Albada limiter is in the admissibility region of Fig. 4.12.

It was shown in Sect. 9.4 that the explicit Euler scheme is TVD if the graph of $\psi(r)$ is in the admissibility region of Fig. 4.12, which is the case, and if the time step is small enough (cf. (9.107)):

$$\lambda = \frac{\tau}{h} \leq \frac{1}{\bar{\mu} M} , \ \ M > 1 + \max\{\psi(r)/r : \ r > 0\} , \tag{10.105}$$
$$\bar{\mu} = \sup(|u| + c) .$$

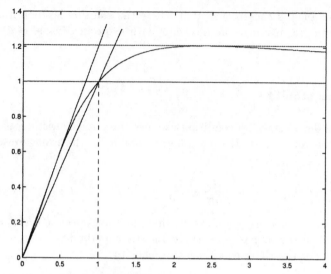

Fig. 10.27. Graph of van Albada limiter.

It follows from (10.104) that a suitable value of the constant M (not to be confused with the Mach number) is

$$M = (3 + \sqrt{2})/2 \approx 1.85 \, .$$

It was shown in Sect. 9.4 that the Runge-Kutta schemes of Heun (9.111) and of Shu and Osher (9.112) are TVD under the same conditions as the explicit Euler scheme. The bounds λ_1 on λ that follow from (10.105) for the four test problems discussed before are listed in Table 10.3. We see that these bounds are much more strict than those of Table 10.2.

	$\bar{\mu}$	λ_1
Sod	2.2	0.25
Lax	4.7	0.11
Mach 3	5.0	0.11
Supers. shock tube	3.1	0.17

Table 10.3. Estimates for $\bar{\mu}$ and TVD bounds λ_1 for λ for the explicit Euler scheme with the MUSCL method using the van Albada limiter.

If the time stepping scheme is not TVD, or if one wishes to increase the time step, the TVD requirement may be replaced by the requirement of stability, which is often weaker. A heuristic stability criterion is obtained for MUSCL schemes as follows. We look for stability of the scheme for the scalar advection equation (10.94). In smooth parts of the flow the scheme switches to the underlying κ-scheme, otherwise the scheme switches to the first order upwind scheme. We therefore require the time stepping scheme to be stable for both schemes. The symbol of the first order upwind scheme is given by (10.94). The symbol of the κ-scheme applied to (10.94) is given by (cf. Sect. 5.6):

$$\hat{L}_h(\theta) = \frac{1}{\tau}(\gamma_1(\theta) + i\gamma_2(\theta)) , \quad \gamma_1(\theta) = 2\sigma(1-\kappa)s^2 ,$$

$$\gamma_2(\theta) = \sigma\{(1-\kappa)s + 1\}\sin\theta , \quad s = \sin^2\frac{1}{2}\theta , \quad \sigma = \bar{\mu}\tau/h . \tag{10.106}$$

Since the locus of $-\tau\hat{L}_h(\theta)$ as given by (10.106) is close to the imaginary axis, it is desirable to use a time stepping scheme with a stability domain that contains part of the imaginary axis. To see what happens if we use a time stepping scheme that is not (known to be) TVD, we select the SHK Runge-Kutta method described in Sect. 5.8. The stability domain S is depicted in Fig. 5.9. The intersections with the real and imaginary axes are at -2.7853 and $i\sqrt{8}$. For the first order upwind scheme, the locus of $-\tau\hat{L}_h(\theta)$ is a circle with center at $-\sigma$ and radius $\sigma = \bar{\mu}\lambda$; this circle is inside S if $\sigma < 1.3926$, which leads to the following stability condition:

$$\lambda < 1.3926/\bar{\mu} . \tag{10.107}$$

Using Theorem 5.7.4, we find that for the κ-scheme $-\tau\hat{L}_h(\theta)$ is inside the ellipse of Fig. 5.9 if

$$\sigma < \frac{a}{2(1-\kappa)} \quad \text{and} \quad \sigma < \frac{1}{\sqrt{2}}\frac{b}{2-\kappa} ,$$

with $a = 2.7853$ and $b = 2.55$. This gives, with $\kappa = 1/2$,

$$\lambda < \lambda_1 = 1.2/\bar{\mu} . \tag{10.108}$$

It is left to the reader to show that the use of the rectangle of Fig. 5.9 leads to a more restrictive stability condition. By comparision of (10.107) and (10.108) we see that both the first order upwind and the $\kappa = 1/2$ scheme are stable if (10.108) is satisfied. The resulting values of λ_1 for the four test cases are listed in Table 10.4. These bounds were indeed found to be sufficient in the numerical tests to be described next. Note that the stability bounds of Table 10.4 are much weaker than the TVD bounds of Table 10.3, but the TVD property is not guaranteed. Nevertheless, spurious oscillations were found to be absent in the numerical experiments presented below.

Numerical experiments

In all cases, the spatial step size satisfies $h = 1/48$. Figs. 10.28–10.31 show results for the MUSCL version of the Roe scheme, without a sonic entropy fix.

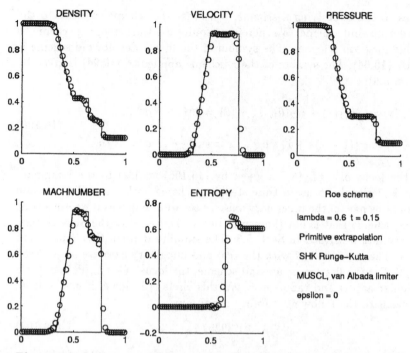

Fig. 10.28. Sod's test problem with Roe scheme; $t = 0.15$, $\lambda = 0.6$.

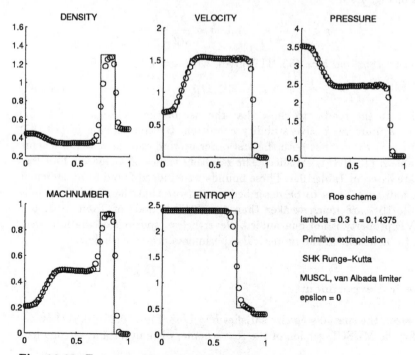

Fig. 10.29. Test case of Lax with Roe scheme; $t = 0.144$, $\lambda = 0.3$.

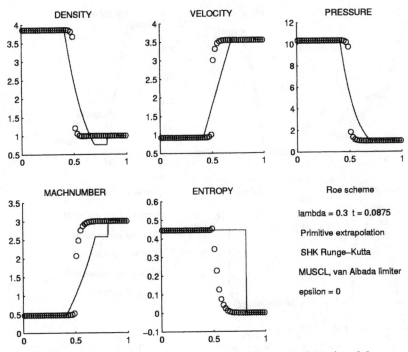

Fig. 10.30. Mach 3 test case with Roe scheme; $t = 0.0875$, $\lambda = 0.3$.

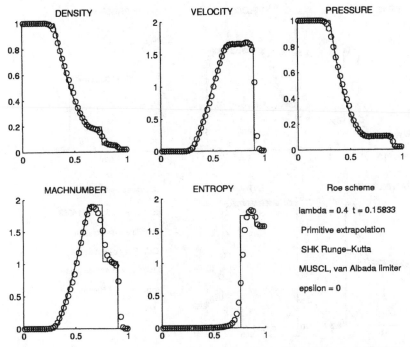

Fig. 10.31. Supersonic shock tube problem with Roe scheme; $t = 0.158$, $\lambda = 0.4$.

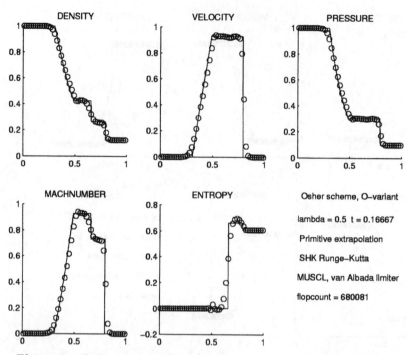

Fig. 10.32. Sod's test problem with Osher scheme; $t = 0.167$, $\lambda = 0.5$

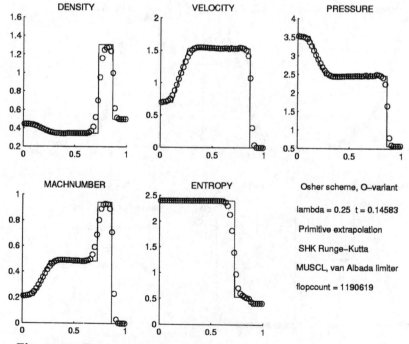

Fig. 10.33. Test case of Lax with Osher scheme; $t = 0.146$, $\lambda = 0.25$.

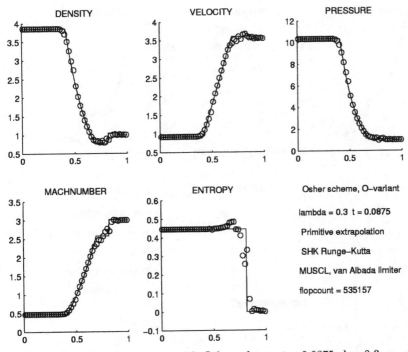

Fig. 10.34. Mach 3 test case with Osher scheme; $t = 0.0875$, $\lambda = 0.3$.

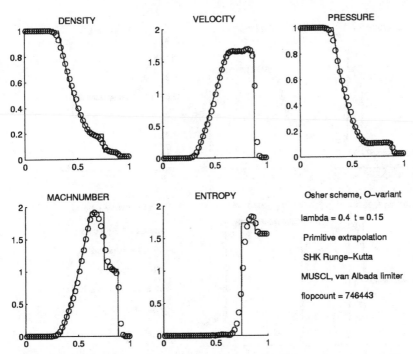

Fig. 10.35. Supersonic shock tube problem with Osher scheme; $t = 0.15$; $\lambda = 0.4$.

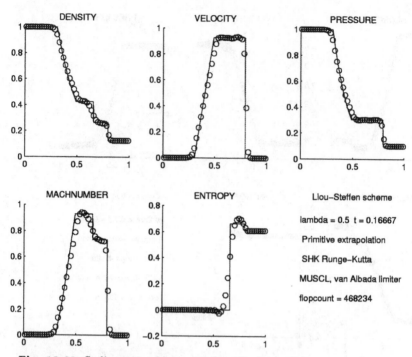

Fig. 10.36. Sod's test problem with Liou-Steffen scheme; $t = 0.167$, $\lambda = 0.5$

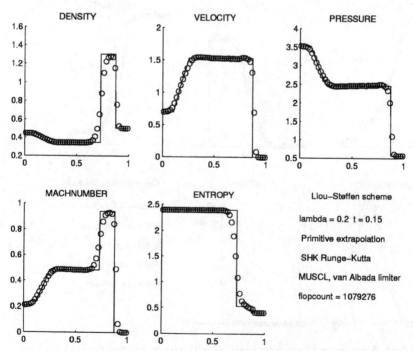

Fig. 10.37. Test case of Lax with Liou-Steffen scheme; $t = 0.15$, $\lambda = 0.2$.

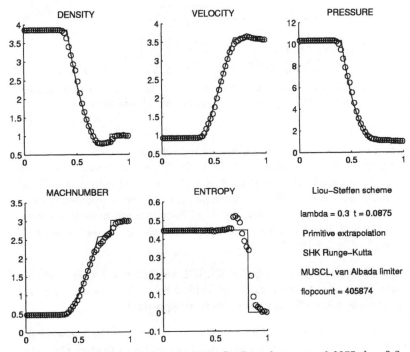

Fig. 10.38. Mach 3 test case with Liou-Steffen scheme; $t = 0.0875$, $\lambda = 0.3$.

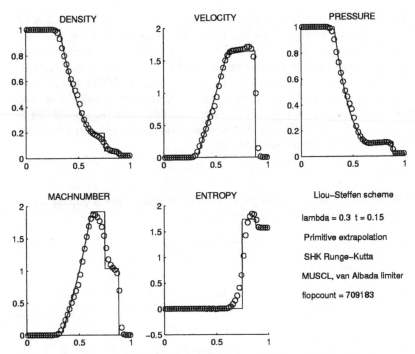

Fig. 10.39. Supersonic shock tube problem with Liou-Steffen scheme; $t = 0.15$, $\lambda = 0.3$.

	$\bar{\mu}$	λ_1
Sod	2.2	0.55
Lax	4.7	0.26
Mach 3	5.0	0.24
Supers.shock tube	3.1	0.39

Table 10.4. Estimates for $\bar{\mu}$ and stability bounds λ_1 for λ for the SHK Runge-Kutta method with the MUSCL method using the van Albada limiter.

Comparing with the corresponding figures in Sect. 10.3, we see a better resolution of contact discontinuities, and the sonic glitch has almost disapeared for the supersonic shock tube problem. The violation of the entropy condition for the Mach 3 test case is still unacceptable, but has become somewhat less severe. With MUSCL, the sonic entropy fix of Harten needs to be less drastic than when MUSCL is not applied; with $\varepsilon = 0.2$ we obtain better results (not shown) than in Fig. 10.10.

Figs. 10.32–10.35 show results for the MUSCL version of the Osher scheme (O-variant). Compared to the results obtained in Sect. 10.4, the accuracy has improved and the sonic glitches have almost disappeared.

Figs. 10.36–10.39 show results for the MUSCL version of the Liou-Steffen (AUSM) scheme. The accuracy improvement is similar to that of the other schemes. Sonic glitches disappear.

Summarizing, the accuracy of the three schemes improves by application of MUSCL. The Liou-Steffen scheme has the best accuracy and is the cheapest of the three. The Roe scheme is as accurate as the Osher scheme, except when a strong expansion fan is present, as in the Mach 3 test case; this can be cured by a sonic entropy fix. The price we pay here for MUSCL is a more expensive time stepping scheme; because the CFL number $(u+c)\tau/h$ cannot be much larger than for explicit Euler time stepping with the non-MUSCL schemes, and because each Runge-Kutta step consists of four Euler steps, the amount of computing almost quadruples.

Exercise 10.8.1. Use the rectangle of Fig. 5.9 to obtain the following sufficient stability condition for the κ-scheme :

$$\sigma < 1(2 - 2\kappa) \,,$$

where σ is the CFL number.

11. Discretization in general domains

11.1 Introduction

In the foregoing chapters we have restricted ourselves to Cartesian computational grids, in order to avoid unnecessary technicalities in discussing the basic principles of computational fluid dynamics. In the present and in the following two chapters, we will go into the complications brought about by geometric complexity of the domain in which the flow takes place. These complications are mainly of a technical nature, and no new basic principles will emerge. Nevertheless, for successful computation of flows in general domains, the technicalities involved in computation on general grids must be considered carefully.

We will use boundary-fitted coordinates for creating grids in general domains. Simple and exact expressions will be derived for cell volumes and cell face areas. Tensor analysis is a valuable tool for expressing physical laws in general coordinates. Therefore we include an introduction to tensor analysis for the uninitiated reader. In the following two chapters, discretization of the governing equations of fluid dynamics on general (structured) grids will be presented, both for the staggered and the colocated placement of the dependent variables. We will see how the basic principles established in earlier chapters using simple grids can be used for developing discretizations on general grids. The issue of how to deal with rough (i.e. strongly non-uniform) grids and the effect of grid roughness on accuracy, already discussed in sections 3.7 and 4.7, will come up again, this time in a more general context.

11.2 Three types of grid

One of the distinguishing features of practical computational fluid dynamics is the geometric complexity of the domains in which flows take place. Examples are flows around aircraft, in turbines, in estuaries, and physiological flows, such as in hearts and arteries. There are many ways to discretize in general domains. We will give a brief survey of the various possibilities, and go into more detail about structured grids.

Discretization always entails subdivision of the domain in small cells, or, which amounts to the same thing, distribution of a finite set of points in the domain. The resulting set of cells or points is called the (computational) grid. There are basically three types of grid:

– Cartesian
– Structured boundary-fitted
– Unstructured

These three types are illustrated in Fig. 11.1. In a Cartesian grid the cells are rectangular. A grid is called structured if all interior cell vertices belong to the same number of cells. Hence, grids (a) and (b) of Fig.11.1 are both structured.

If the boundary consists of cell faces the grid is called *boundary-fitted*. Unstructured grids are invariably boundary fitted. In Cartesian grids the domain boundary usually does not coincide with cell faces, which makes accurate implementation of boundary conditions complicated. However, in applications where the precise location of the boundary is unknown or of little importance, the boundary may be approximated by Cartesian cell faces. Fig. 11.2 gives an example. This example concerns the tidal flow in an estuary with

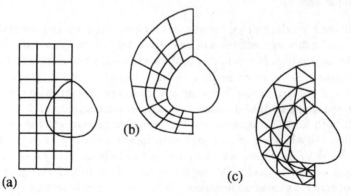

(a)   (b)   (c)

Fig. 11.1. Three types of grid: (a) Cartesian; (b) structured boundary-fitted; (c) unstructured.

tidal flats. In this case the boundary between land and water is vague, and no harm is done by using the 'staircase' approximation for the land-water interface shown in the figure. But Cartesian grids may also be employed when the domain boundary is given precisely, and cuts in an arbitrary way through the grid. An application of the Cartesian approach in the hyperbolic case is given by Coirier and Powell (1995), Deister, Rocher, Hirschel, and Monnoyer (1998) and Ye, Mittal, Udaykumar, and Shyy (1999), and reviews may be found in Henshaw (1996) and Berger and Aftosmis (1998).

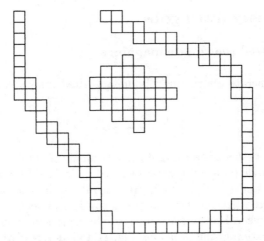

Fig. 11.2. Cartesian grid for tidal flow in estuary with tidal flats. Only dry cells are shown.

Advantages and disadvantages

The advantage of structured grids over unstructured grids is, that datastructured are simpler and no indirect addressing is required, leading to smaller computing times. Furthermore, fast multigrid iterative solution methods are easier to use. On the other hand, in complicated three-dimensional domains the construction of unstructured grids is far easier and requires far less time from the user than the construction of boundary-fitted structured grids, which may require weeks of work. This seems to be leading at present to a growing popularity of unstructured grids. In such grids the cells need not be triangles, as in Fig 11.1(c), but can be quadrilaterals as well. In three dimensions we may have tetrahedrons, hexahedrons, prisms etc.

Of the three kinds of grids listed above, Cartesian grids are least popular. But they are easiest to generate and away from the boundaries discretization is also the easiest. Accurate discretization near the boundaries is more difficult than for the other two options. Nevertheless, interest in Cartesian grids seems to be growing; see for example the references just quoted.

Structured boundary-fitted grids seem to dominate at present, and we will concentrate exclusively on this approach. Much of the analysis that follows is also useful for the unstructured case.

11.3 Boundary-fitted grids

Boundary-fitted coordinate mappings

In order to generate a structured boundary-fitted grid, a coordinate transformation

$$T : \boldsymbol{x} = \boldsymbol{x}(\boldsymbol{\xi}) \tag{11.1}$$

is constructed, that maps a computational domain G onto a physical domain Ω. G is a rectangle (or a rectangular hexahedron in 3D). Here $\boldsymbol{x} = (x^1, x^2, x^3)$ are Cartesian, and $\boldsymbol{\xi} = (\xi^1, \xi^2, \xi^3)$ are general curvilinear coordinates. The mapping is such that the boundary $\partial\Omega$ consists of segments where $\xi^\alpha =$ constant for some α. How to generate T will be remarked upon later. Fig. 11.3 gives an illustration of the principle. Fig. 11.4 gives a realistic example.

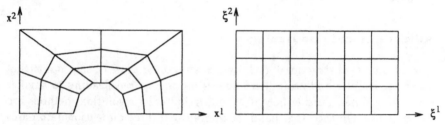

Fig. 11.3. Mapping of physical domain on computational domain.

Orthogonal coordinates

By a ξ^α-coordinate curve we mean a curve $\boldsymbol{x} = \boldsymbol{x}(\boldsymbol{\xi})$ in Ω along which ξ^β = constant, $\beta \neq \alpha$. In other words, along a ξ^α-coordinate line ξ^α is the only $\boldsymbol{\xi}$-component that varies. If the angle of intersection of every ξ^α- and ξ^β-coordinate curve with $\beta \neq \alpha$ is orthogonal, the $\boldsymbol{\xi}$-coordinate system is called orthogonal. Note that here the $\boldsymbol{\xi}$-coordinate curves are considered in Ω; in G they are orthogonal per definition.

Conformal mapping

We consider the two-dimensional case. Let $z = x^1 + ix^2, w = \xi^1 + i\xi^2$, with $i = \sqrt{-1}$. A mapping

$$z = f(w) \tag{11.2}$$

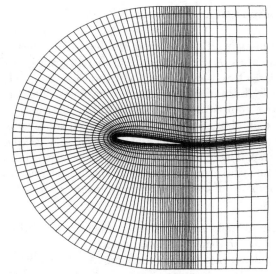

Fig. 11.4. A boundary-fitted grid around an airfoil.

i.e.

$$x^1 = \mathrm{Re}\{f(\xi^1 + i\xi^2)\}, \ \ x^2 = \mathrm{Im}\{f(\xi^1 + i\xi)\}$$

is called *conformal* if f is analytic. The name conformal derives from the fact that in the small figures retain their shape under (11.2); they are merely rotated and stretched a bit. In particular, angles between intersecting curves are preserved. As a result, the $\boldsymbol{\xi}$-coordinates are orthogonal, like $\boldsymbol{x}$. Conformal mapping is a classic topic in complex function theory, and was an important tool in fluid dynamics before the arrival of the computer. It is restricted to two dimensions.

In general, the equations of motion become more complicated when ξ^α are introduced as independent variables, but this is hardly the case if the $\boldsymbol{\xi}$-system is orthogonal. Hence one might think that conformal mapping is an attractive way to generate boundary-fitted grids. But this is not so, because it leaves little freedom to put grid points where we want them. So conformal mapping is little used.

Orthogonality is a weaker condition than conformality, as illustrated in Exercise 11.3.1. However, it turns out that mappings that are merely orthogonal are also not sufficiently flexible in practice. But it is quite feasible to generate grids that are orthogonal at the domain boundary only. This facilities accurate implementation of the boundary conditions. But we will consider only the general case in what follows.

Domain decomposition

A non-singular mapping T (11.1) exists if Ω and G are *topologically equivalent*. This means that one can obtain G from Ω by continuous deformation without cutting. But even when this is not the case, a suitable mapping can be constructed by introducing suitable cuts in Ω, cf. Exercise 11.3.1. If the mapping remains complicated and the resulting distribution of grid point undesirable, the number of cuts may be increased, subdividing Ω in subdomains, each of which is mapped separately on its own rectangle G. The domains may overlap. The conditions by which the solutions in the subdomains are connected to each other are subject of much research. Apart from grid generation in domains of complicated shape, other reasons to apply domain decomposition are:

- Zonal method: different mathematical models are applied in different subdomains, in order to reduce computing cost.
- Parallel computing: each subdomain is handled by a different processor.

Domain decomposition will not be further discussed.

Generation of boundary-fitted grids

Grid generation is a large subject by itself. We will say little about it, but refer the reader to the following monographs: George (1991), Knupp and Steinberg (1993) and Thompson, Warsi, and Mastin (1985). Recent reviews are given by Henshaw (1996) and Spekreijse (1995). Excellent sources of information on the Internet are R. Schneiders' *Mesh generation & grid generation on the Web* site:

> www-users.informatik.rwth-aachen.de/~roberts
> /meshgeneration.html

and P. Heckbert's collection of mesh generation links:

> www.cs.cmu.edu/~ph/mesh.html

and the *Grid technology* site of Mississippi State University:

> www.erc.msstate.edu/research/thrusts/grid/

Here pointers to the literature and software packages may be found.

Piecewise multilinear coordinate mapping

Let the physical domain Ω be topologically equivalent to the unit cube G. In G we have Cartesian coordinates $\boldsymbol{\xi} = (\xi^1, \xi^2, \xi^3)$ and a uniform grid G_h consisting of grid points located at $\boldsymbol{\xi}_j$, $j = (j_1, j_2, j_3)$:

$$G_h = \{ \boldsymbol{\xi}_j : \xi_j^\alpha = j_\alpha \Delta \xi^\alpha, \ j_\alpha = 0, 1, ..., 1/\Delta \xi^\alpha, \ \alpha = 1, 2, 3 \} . \tag{11.3}$$

where $1/\Delta\xi^\alpha \in \mathbb{N}$. Greek indices are used exclusively to refer to coordinate directions, and vice-versa. Unless stated otherwise, summation is implied exclusively over pairs of equal Greek sub- and superscripts in terms and products. This summation convention does not apply to (11.3). The two-dimensional case is subsumed under the three-dimensional case by restricting the range of α to $\{1, 2\}$; in this case G is of course the unit square.

Boundary-fitted grid generation is a numerical process that generates the coordinate mapping

$$T : x_j = x(\xi_j), \quad x_j \in \Omega, \quad \xi_j \in G_h \tag{11.4}$$

only in the discrete set of points $G_h \subset G$. In order to obtain accurate discretizations, we have to be precise about how the mapping (11.4) is extended to all of Ω and G. This is necessary in order to discretize correctly near grid nonuniformities, and to compute cell volumes and cell face areas exactly and in a consistent manner. The extension should preserve the one-to-one property of the mapping, and allow exact and efficient computation of cell volumes and cell face areas.

We recall from Sect. 3.7 that if the mesh size jumps, the local discretization error for second derivatives is of first order for vertex-centered and of zeroth order (which makes the scheme inconsistent in the maximum norm) for cell-centered schemes for the convection-diffusion equation (Manteuffel and White, Jr. (1986)). This has made some believe that grids need to be smooth for accurate results. But this is not so. In Forsyth, Jr. and Sammon (1988), Manteuffel and White, Jr. (1986), Weiser and Wheeler (1988), Wesseling (1996b) it is shown that the global discretization error is second order on strongly nonuniform grids. These results for the convection-diffusion equation may be expected to carry over to the Navier-Stokes equations. This is fortunate, because it allows us to switch abruptly from a fine mesh in thin boundary or shear layers to a coarse mesh outside. For such grids, in Farrell, Miller, O'Riordan, and Shishkin (1996), Roos, Stynes, and Tobiska (1996), Wesseling (1996b) it is shown that the accuracy is uniform in the Reynolds number, for the convection-diffusion equation, as we saw in Sect. 4.7. In order to allow abrupt changes in mesh size, we will assume the coordinate mapping to be merely piecewise differentiable. We will therefore extend the mapping by multilinear interpolation.

Exercise 11.3.1. Let Ω be the concentric annulus between $r = 1$ and $r = 2$, with $r = (x^1 x^1 + x^2 x^2)^{1/2}$. Let $\xi^1 = r$, $\xi^2 = \theta$ with (r, θ) cylindrical coordinates. Determine the form that relation (11.1) takes in this case, and show that the image G of Ω is a rectangle. Determine where Ω has been cut. Show that the mapping is not conformal by comparing the angle of intersection φ of two curves $\theta = $ constant, $x^1 = 1$ in the x-plane (answer: $\varphi = \pi/2 - \theta$) and in the ξ-plane (answer: $\varphi = \arctan(\cos\theta/\tan\theta)$).

Exercise 11.3.2. Show that for piecewise multilinear coordinate mappings, cell edges are straight in physical space.
Hint: a cell edge has the property that only one component of ξ varies along it.

11.4 Basic geometric properties of grid cells

A detailed account will be given of accurate and efficient computation of cell volumes, cell face areas and cell face normals. This is of obvious importance for finite volume discretization. This elementary topic has of course been discussed before; see the excellent review of discretization in physical space by Vinokur (1989), and references quoted there. But the topic still warrants attention, because (in three dimensions) many authors define a cell face to consist of two plane triangles, see for example He and Salcudean (1994). This is less satisfactory than the doubly-ruled surface that we will use. The resulting formulae are not more economical, and a face can be subdivided into triangles in two ways, and two adjacent cells have to fit, which complicates grid generation; furthermore, the normal is ill-defined. Therefore we feel that the approach taken here, in which the mapping is extended by multilinear interpolation, so that cell faces turn out to be doubly-ruled surfaces, is to be preferred; the more so because the resulting expressions for areas and volumes are cheap, as will be seen.

Two-dimensional case

Let $\Omega_j \in \Omega$ be a geometric figure to be further defined shortly with vertices

$$\boldsymbol{x}_{j-e_1-e_2}, \quad \boldsymbol{x}_{j+e_1-e_2}, \quad \boldsymbol{x}_{j-e_1+e_2}, \quad \boldsymbol{x}_{j+e_1+e_2}, \quad e_1 = (1/2, 0), \quad e_2 = (0, 1/2)$$

(since the vertices have integer indices, this implies that j_1 and j_2 are fractional). The image of Ω_j in G is a rectangle called G_j; G_j and Ω_j are called cells. The cell G_j is defined by (cf. (11.3)) (no summation):

$$G_j = \{\boldsymbol{\xi} = (\xi^1, \xi^2) : \xi^\alpha \in [(j_\alpha - \frac{1}{2})\Delta\xi^\alpha, (j_\alpha + \frac{1}{2})\Delta\xi^\alpha)]\} \ .$$

Let $\boldsymbol{\xi}_0$ be some point in G_j. Bilinear interpolation in Ω_j gives the following relation between $\boldsymbol{x}$ and $\boldsymbol{\xi}$:

$$\boldsymbol{x} = \boldsymbol{x}_0 + c_\alpha(\xi^\alpha - \xi_0^\alpha) + c_{12}(\xi^1 - \xi_0^1)(\xi^2 - \xi_0^2) \ . \tag{11.5}$$

Since this mapping is different in every cell, it is piecewise bilinear. It is also continuous. The edges of G_j are straight and on every edge, ξ^1 or ξ^2 is constant. This makes the mapping (11.5) linear at the edges, so that the edges of Ω_j are also straight, as indicated in Fig. 11.5. The coefficients x_0, c_α

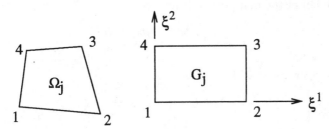

Fig. 11.5. Grid cell in physical space (Ω_j) and in computational space (G_j).

and c_{12} follow from the requirement $x(\xi_m) = x_m$, $m = 1, 2, 3, 4$, where the subscript m refers to the vertices enumerated in Fig. 11.5, and are easily determined as follows. Choose ξ_0 in the center of G_j, and define new variables s^α by

$$\xi^\alpha = \xi_0^\alpha + \frac{1}{2}\Delta\xi^\alpha s^\alpha \quad \text{(no summation)},$$

so that (11.5) becomes

$$x = x_0 + b_\alpha s^\alpha + b_{12}s^1 s^2 , \tag{11.6}$$

with

$$b_\alpha = \frac{1}{2}\Delta\xi^\alpha c_\alpha, \quad b_{12} = \frac{1}{4}\Delta\xi^1 \Delta\xi^2 c_{12} \quad \text{(no summation)} .$$

By requiring $x(\xi_m) = x_m$, $m = 1, 2, 3, 4$, we obtain the following system of equations:

$$Ay = r , \tag{11.7}$$

with $y = (b_1, b_2, b_{12}, x_0)^T$, $r = (x_1, x_2, x_3, x_4)^T$ and

$$A = \begin{pmatrix} -1 & -1 & 1 & 1 \\ 1 & -1 & -1 & 1 \\ 1 & 1 & 1 & 1 \\ -1 & 1 & -1 & 1 \end{pmatrix} .$$

The columns of A are orthogonal. Premultiplication of (11.7) by A^T gives $y = \frac{1}{4}A^T r$, so that we have, defining $x_{mn} \equiv x_m + x_n$:

$$b_1 = \frac{1}{4}(x_{23} - x_{14}), \quad b_2 = \frac{1}{4}(x_{34} - x_{12}) ,$$
$$b_{12} = \frac{1}{4}(x_{13} - x_{24}), \quad x_0 = \frac{1}{4}(x_{12} + x_{34}) .$$

Geometric quantities

For later use, we need to determine various geometric quantities related to the coordinate mapping.

The Jacobian of the mapping is defined by

$$J \equiv \det\{\frac{\partial \boldsymbol{x}}{\partial \boldsymbol{\xi}}\} = \frac{\partial \boldsymbol{x}}{\partial \xi^1} \otimes \frac{\partial \boldsymbol{x}}{\partial \xi^2} \,, \tag{11.8}$$

where the operator $\otimes$ is defined as

$$\boldsymbol{a} \otimes \boldsymbol{b} = a^1 b^2 - a^2 b^1 \,.$$

It is assumed that

$$J > 0 \,, \tag{11.9}$$

i.e. the $\boldsymbol{x}$- and $\boldsymbol{\xi}$-coordinate systems are right-handed.

By the definition of the outer product, in three dimensions the vector $\boldsymbol{a} \times \boldsymbol{b}$ has magnitude equal to twice the area of the triangle spanned by $\boldsymbol{a}$ and $\boldsymbol{b}$; hence, in two dimensions $\boldsymbol{a} \otimes \boldsymbol{b}$ with $\boldsymbol{a}$ and $\boldsymbol{b}$ ordered counterclockwise equals twice the area of the triangle spanned by $\boldsymbol{a}$ and $\boldsymbol{b}$. It follows that the cell area is, with the vertex enumeration of Fig. 11.5,

$$|\Omega_j| = \frac{1}{2}(\boldsymbol{x}_3 - \boldsymbol{x}_1) \otimes (\boldsymbol{x}_4 - \boldsymbol{x}_2) \,, \tag{11.10}$$

where we have taken advantage of the identities $\boldsymbol{a} \otimes \boldsymbol{a} = 0$ and $\boldsymbol{a} \otimes \boldsymbol{b} = -\boldsymbol{b} \otimes \boldsymbol{a}$. The vector $\boldsymbol{s}_{mn}$ normal to a cell edge $\boldsymbol{x}_m - \boldsymbol{x}_n$ with length $|\boldsymbol{x}_m - \boldsymbol{x}_n|$ will be called the cell edge vector. For complete specification, we add the requirement, assuming $\xi^\alpha = $ constant on $\boldsymbol{x}_m - \boldsymbol{x}_n$, that $\boldsymbol{s}_{mn}$ points into the cell where $\xi^\beta (\beta \neq \alpha)$ is largest. It follows that (cf. Fig. 11.5)

$$\begin{aligned} \boldsymbol{s}_{12} &= \begin{pmatrix} x_1^2 - x_2^2 \\ x_2^1 - x_1^1 \end{pmatrix}, & \boldsymbol{s}_{23} &= \begin{pmatrix} x_3^2 - x_2^2 \\ x_2^1 - x_3^1 \end{pmatrix}, \\ \boldsymbol{s}_{34} &= \begin{pmatrix} x_4^2 - x_3^2 \\ x_3^1 - x_4^1 \end{pmatrix}, & \boldsymbol{s}_{14} &= \begin{pmatrix} x_4^2 - x_1^2 \\ x_1^1 - x_4^1 \end{pmatrix}. \end{aligned} \tag{11.11}$$

Later, we will need the so-called contravariant and covariant base vectors. The covariant base vectors $\boldsymbol{a}_{(\alpha)}$ are defined by

$$\boldsymbol{a}_{(\alpha)} \equiv \frac{\partial \boldsymbol{x}}{\partial \xi^\alpha} \quad \text{or} \quad a^\beta_{(\alpha)} \equiv \frac{\partial x^\beta}{\partial \xi^\alpha} \,,$$

with $a^\beta_{(\alpha)}$ the Cartesian x^β-component of the vector $\boldsymbol{a}_{(\alpha)}$. We put parentheses around α to emphasize that no component is intended. Because $\boldsymbol{x} = \boldsymbol{x}(\boldsymbol{\xi})$

is piecewise bilinear, $a_{(\alpha)}$ is piecewise continuous. In the cell interior and on edges $\xi^\beta =$ constant, $\beta \neq \alpha$, $a_{(\alpha)}$ is continuous, but $a_{(\alpha)}$ is discontinuous at cell edges $\xi^\alpha =$ constant. For instance, $a_{(1)}$ is discontinuous at the cell edge $x_3 - x_2$ (cf. Fig. 11.5). We will need $a_{(\alpha)}$ only in the cell center $x_C = \frac{1}{4}(x_1 + x_2 + x_3 + x_4)$ and at those cell edge centers where $a_{(\alpha)}$ is continuous. For mnemonic convenience, cell edge centers are indicated by subscripts N, W, S, E, where N stands for north etc., so that for instance $x_N = \frac{1}{2}(x_3 + x_4)$. In x_C we have in (11.6) $s^1 = s^2 = 0$, in x_N we have $s^1 = 0$, $s^2 = 1$ etc. It is easily seen that we have exactly

$$
\begin{aligned}
a_{(1)C} &= (x_E - x_W)/\Delta\xi^1, & a_{(2)C} &= (x_N - x_S)/\Delta\xi^2, \\
a_{(1)N} &= (x_3 - x_4)/\Delta\xi^1, & a_{(1)S} &= (x_2 - x_1)/\Delta\xi^1, \\
a_{(2)E} &= (x_3 - x_2)/\Delta\xi^2, & a_{(2)W} &= (x_4 - x_1)/\Delta\xi^2.
\end{aligned}
\tag{11.12}
$$

The contravariant base vectors are defined by

$$
a^{(\alpha)} \equiv \nabla\xi^\alpha \quad \text{or} \quad a^{(\alpha)}_\beta \equiv \frac{\partial\xi^\alpha}{\partial x^\beta} ,
$$

where $a^{(\alpha)}_\beta$ is the Cartesian x^β-component of the vector $a^{(\alpha)}$. We have

$$
a^{(\alpha)} \cdot a_{(\beta)} = \delta^\alpha_\beta ,
$$

with δ^α_β the Kronecker delta. Solving this equation gives

$$
a^{(1)} = \frac{1}{\sqrt{g}} \begin{pmatrix} a^2_{(2)} \\ -a^1_{(2)} \end{pmatrix} , \quad a^{(2)} = \frac{1}{\sqrt{g}} \begin{pmatrix} -a^2_{(1)} \\ a^1_{(1)} \end{pmatrix} ,
\tag{11.13}
$$

where $\sqrt{g}$ is the common designation in tensor analysis for the Jacobian J defined in (11.8), i.e.

$$
\sqrt{g} \equiv a_{(1)} \otimes a_{(2)} .
\tag{11.14}
$$

From the smoothness properties of $a_{(\alpha)}$ it follows that $\sqrt{g}$, $a^{(\alpha)}$ and $\sqrt{g}a^{(\alpha)}$ are discontinuous at cell edges, except that $\sqrt{g}a^{(\alpha)}$ is continuous at cell edges of the type $\xi^\alpha =$ constant. From (11.13), (11.12) and (11.11) it follows that $\sqrt{g}a^{(\alpha)}$ is closely related to cell edge vectors:

$$
\begin{aligned}
(\sqrt{g}a^{(1)})_E &= s_E/\Delta\xi^2, & (\sqrt{g}a^{(1)})_W &= s_W/\Delta\xi^2, \\
(\sqrt{g}a^{(2)})_N &= s_N/\Delta\xi^1, & (\sqrt{g}a^{(2)})_S &= s_S/\Delta\xi^1,
\end{aligned}
$$

where $s_N = s_{34}$ etc.

Not surprisingly, $\sqrt{g}$ is closely related to the cell area. We have, using (11.14), (11.13) and (11.10),

$$
\begin{aligned}
\sqrt{g_C}\Delta\xi^1\Delta\xi^2 &= (x_E - x_W) \otimes (x_N - x_S) = \\
&= \frac{1}{4}(x_2 + x_3 - x_1 - x_4) \otimes (x_3 + x_4 - x_1 - x_2) = \Omega_j .
\end{aligned}
$$

Geometric identity

According to the divergence theorem, we have for a function $f(\boldsymbol{x})$:

$$\int_{\Omega_j} \operatorname{grad} f d\Omega = \int_{\Gamma_j} f \boldsymbol{n} d\Gamma ,$$

where $\boldsymbol{n}$ is the outward unit normal. Choosing $f(\boldsymbol{x})$ to be constant, this gives

$$\int_{\Gamma_j} \boldsymbol{n} d\Gamma = 0 . \tag{11.15}$$

This is called the *geometric identity*. It expresses the fact that the contour Γ_j is closed. It is desirable that (11.15) is satisfied exactly when Γ_j is approximated numerically. We find indeed from (11.11):

$$\int_{\Gamma_{ij}} \boldsymbol{n} d\Gamma = s_{43} + s_{23} - s_{21} - s_{14} = 0 .$$

This is a necessary condition for uniform flow to be an exact solution of the numerical scheme.

Three-dimensional case

In three dimensions, the mapping (11.1) is extended by piecewise trilinear interpolation. Let Ω_j be a geometric figure to be further defined shortly with vertices

$$\boldsymbol{x}_{j\pm(e_1+e_2+e_3)}, \; \boldsymbol{x}_{j\pm(e_1-e_2+e_3)}, \; \boldsymbol{x}_{j\pm(e_1+e_2-e_3)} ,$$
$$\boldsymbol{x}_{j\pm(-e_1+e_2+e_3)}, \; e_1 = (\frac{1}{2},0,0), \; e_2 = (0,\frac{1}{2},0), \; e_3 = (0,0,\frac{1}{2}) .$$

Since the vertices have integer indices, this implies that j_α is fractional. In the following, $\boldsymbol{x}_j$ will be a cell center, $\boldsymbol{x}_{j+e_\alpha}$ will be a cell face center, $\boldsymbol{x}_{j+e_\alpha+e_\beta}$ $(\beta \neq \alpha)$ will be a cell edge center, and $\boldsymbol{x}_{j+e_1+e_2+e_3}$ will be a cell vertex. The image of Ω_j in G is a rectangular hexahedron called G_j. The cell G_j is defined by (no summation):

$$G_j = \{\boldsymbol{\xi} = (\xi^1,\xi^2,\xi^3) : \xi^\alpha \in [(j_\alpha - \frac{1}{2})\Delta\xi^\alpha, (j_\alpha + \frac{1}{2})\Delta\xi^\alpha]\} .$$

Let $\boldsymbol{\xi}_0$ be some point in G_j. Then trilinear interpolation inside Ω_j gives the following relation between $\boldsymbol{x}$ and $\boldsymbol{\xi}$:

$$\begin{aligned}
\boldsymbol{x} = \; & \boldsymbol{x}_0 + c_\alpha(\xi^\alpha - \xi_0^\alpha) + c_{12}(\xi^1 - \xi_0^1)(\xi^2 - \xi_0^2) + c_{23}(\xi^2 - \xi_0^2)(\xi^3 - \xi_0^3) \\
& + c_{13}(\xi^3 - \xi_0^3)(\xi^1 - \xi_0^1) + c_{123}(\xi^1 - \xi_0^1)(\xi^2 - \xi_0^2)(\xi^3 - \xi_0^3) .
\end{aligned} \tag{11.16}$$

It follows that the edges of the cell Ω_j are straight. Consider a face of Ω_j with $\xi^3 = $ constant. In this face the mapping (11.16) becomes, choosing $\boldsymbol{\xi}_0$ in the corresponding face of G_j so that we have $\xi^3 = \xi_0^3$:

$$\boldsymbol{x} = \boldsymbol{x}_0 + \boldsymbol{c}_1(\xi^1 - \xi_0^1) + \boldsymbol{c}_2(\xi^2 - \xi_0^2) + \boldsymbol{c}_{12}(\xi^1 - \xi_0^1)(\xi^2 - \xi_0^2) , \qquad (11.17)$$

with coefficients $\boldsymbol{x}_0$ and $\boldsymbol{c}$ in general different from those in (11.16), since $\boldsymbol{\xi}_0$ is changed. With $\xi^1 = $ constant or $\xi^2 = $ constant equation (11.17) describes straight lines, so that the cell face contains two families of straight lines, and is therefore a doubly-ruled surface.

The following change of variables is convenient:

$$\xi^\alpha = \xi_0^\alpha + \frac{1}{2}\Delta\xi^\alpha s^\alpha \quad \text{(no summation)} , \qquad (11.18)$$

with $\boldsymbol{\xi}_0$ in the center of G_j. By rewriting the mapping (11.16) in the local s-coordinates we obtain:

$$\boldsymbol{x} = \boldsymbol{x}_0 + \boldsymbol{b}_\alpha s^\alpha + \boldsymbol{b}_{12}s^1 s^2 + \boldsymbol{b}_{13}s^1 s^3 + \boldsymbol{b}_{23}s^2 s^3 + \boldsymbol{b}_{123}s^1 s^2 s^3 ,$$

with (no summation)

$$\boldsymbol{b}_\alpha = \frac{1}{2}\Delta\xi^\alpha \boldsymbol{c}_\alpha, \quad \boldsymbol{b}_{\alpha\beta} = \frac{1}{4}\Delta\xi^\alpha \Delta\xi^\beta \boldsymbol{c}_{\alpha\beta},$$

$$\boldsymbol{b}_{123} = \frac{1}{8}\Delta\xi^1 \Delta\xi^2 \Delta\xi^3 \boldsymbol{c}_{123} . \qquad (11.19)$$

Requiring $\boldsymbol{x} = \boldsymbol{x}_m$, $m = 1, 2, ..., 8$ (see Fig. 11.6) gives the following system of equations:

$$Ay = r , \qquad (11.20)$$

with $y = (\boldsymbol{b}_1, \boldsymbol{b}_2, \boldsymbol{b}_3, \boldsymbol{b}_{12}, \boldsymbol{b}_{13}, \boldsymbol{b}_{23}, \boldsymbol{b}_{123}, \boldsymbol{x}_0)^T$, $r = (\boldsymbol{x}_1, \boldsymbol{x}_2, ..., \boldsymbol{x}_8)^T$ and

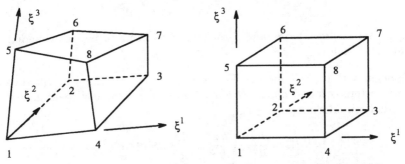

Fig. 11.6. Three-dimensional cell. Its image in G is a rectangular hexahedron.

$$A = \begin{pmatrix} -1 & -1 & -1 & 1 & 1 & 1 & -1 & 1 \\ -1 & 1 & -1 & -1 & 1 & -1 & 1 & 1 \\ 1 & 1 & -1 & 1 & -1 & -1 & -1 & 1 \\ 1 & -1 & -1 & -1 & -1 & 1 & 1 & 1 \\ -1 & -1 & 1 & 1 & -1 & -1 & 1 & 1 \\ -1 & 1 & 1 & -1 & -1 & 1 & -1 & 1 \\ 1 & 1 & 1 & 1 & 1 & 1 & 1 & 1 \\ 1 & -1 & 1 & -1 & 1 & -1 & -1 & 1 \end{pmatrix} .$$

The columns of A are orthogonal. Premultiplication of (11.20) by A^T gives

$$y = \tfrac{1}{8} A^T r ,$$

so that

$$
\begin{aligned}
b_1 &= \tfrac{1}{8}(x_{3478} - x_{1256}), & b_2 &= \tfrac{1}{8}(x_{2367} - x_{1458}), \\
b_3 &= \tfrac{1}{8}(x_{5678} - x_{1234}), & b_{12} &= \tfrac{1}{8}(x_{1357} - x_{2468}), \\
b_{13} &= \tfrac{1}{8}(x_{1278} - x_{3456}), & b_{23} &= \tfrac{1}{8}(x_{1467} - x_{2358}), \\
b_{123} &= \tfrac{1}{8}(x_{2457} - x_{1368}),
\end{aligned}
\tag{11.21}
$$

where

$$x_{jkmn} \equiv x_j + x_k + x_m + x_n .$$

This completes the specification of the coordinate mapping in three dimensions.

Geometric quantities

For the Jacobian we can write

$$J \equiv \det\{\frac{\partial x}{\partial \xi}\} = \frac{\partial x}{\partial \xi^1} \cdot (\frac{\partial x}{\partial \xi^2} \times \frac{\partial x}{\partial \xi^3}) . \tag{11.22}$$

Again, we assume the x- and ξ-coordinate systems to be right-handed, so that (11.9) holds.

We begin with the cell volume. The volume of the parallelepiped spanned by three vectors a, b and c is $a \cdot (b \times c)$. Hence, the volume element generated by infinitesimal changes in ξ is

$$d\Omega = \frac{\partial x}{\partial \xi^1} \cdot (\frac{\partial x}{\partial \xi^2} \times \frac{\partial x}{\partial \xi^3}) d\xi^1 d\xi^2 d\xi^3 .$$

This is positive, because right-handed coordinate systems are assumed, cf. (11.22) and (11.9). It follows that the cell volume is given by

$$|\Omega_j| = \int\limits_{\Omega_j} d\Omega = \int\limits_{\Omega_j} J d\xi^1 d\xi^2 d\xi^3 \ . \tag{11.23}$$

Let us choose $\boldsymbol{\xi}_0$ in (11.18) in the center of the cell G_j. Then the ranges of integration in this triple integral are $(\xi_0^\alpha - \frac{1}{2}\Delta\xi^\alpha, \ \xi_0^\alpha + \frac{1}{2}\Delta\xi^\alpha)$, $\alpha = 1, 2, 3$, with $\Delta\xi^\alpha$ the length of the edges of G_j. The following change of variables is convenient:

$$\xi^\alpha = \xi_0^\alpha + \tfrac{1}{2}\Delta\xi^\alpha s^\alpha \quad \text{(no summation)} \ , \tag{11.24}$$

so that (11.23) becomes

$$|\Omega_j| = \tfrac{1}{8}\Delta\xi^1 \Delta\xi^2 \Delta\xi^3 \int\limits_{-1}^{1}\int\limits_{-1}^{1}\int\limits_{-1}^{1} J ds^1 ds^2 ds^3 \ .$$

According to (11.16) we have

$$\begin{aligned}
\frac{\partial \boldsymbol{x}}{\partial \xi^1} &= \boldsymbol{c}_1 + \boldsymbol{d}_{12}s^2 + \boldsymbol{d}_{13}s^3 + \boldsymbol{d}_{123}s^2 s^3 \ , \\
\frac{\partial \boldsymbol{x}}{\partial \xi^2} &= \boldsymbol{c}_2 + \boldsymbol{d}_{21}s^1 + \boldsymbol{d}_{23}s^3 + \boldsymbol{d}_{231}s^3 s^1 \ , \\
\frac{\partial \boldsymbol{x}}{\partial \xi^3} &= \boldsymbol{c}_3 + \boldsymbol{d}_{31}s^1 + \boldsymbol{d}_{32}s^2 + \boldsymbol{d}_{312}s^1 s^2 \ ,
\end{aligned} \tag{11.25}$$

where

$$\begin{aligned}
\boldsymbol{d}_{12} &= \tfrac{1}{2}\Delta\xi^2 \boldsymbol{c}_{12}, \ \ \boldsymbol{d}_{13} = \tfrac{1}{2}\Delta\xi^3 \boldsymbol{c}_{13}, \ \ \boldsymbol{d}_{123} = \tfrac{1}{4}\Delta\xi^2 \Delta\xi^3 \boldsymbol{c}_{123} \ , \\
\boldsymbol{d}_{21} &= \tfrac{1}{2}\Delta\xi^1 \boldsymbol{c}_{12}, \ \ \boldsymbol{d}_{23} = \tfrac{1}{2}\Delta\xi^3 \boldsymbol{c}_{23}, \ \ \boldsymbol{d}_{231} = \tfrac{1}{4}\Delta\xi^3 \Delta\xi^1 \boldsymbol{c}_{123} , \\
\boldsymbol{d}_{31} &= \tfrac{1}{2}\Delta\xi^1 \boldsymbol{c}_{13}, \ \ \boldsymbol{d}_{32} = \tfrac{1}{2}\Delta\xi^2 \boldsymbol{c}_{23}, \ \ \boldsymbol{d}_{312} = \tfrac{1}{4}\Delta\xi^1 \Delta\xi^2 \boldsymbol{c}_{123} \ .
\end{aligned} \tag{11.26}$$

Because of symmetry (which is why $\boldsymbol{\xi}_0$ was chosen in the center), terms that contain odd powers of s^α do not contribute. Neglecting these terms, equation (11.22) gives

$$\begin{aligned}
J \ \sim \ & \boldsymbol{c}_1 \cdot (\boldsymbol{c}_2 \times \boldsymbol{c}_3) + s^1 s^1 \boldsymbol{c}_1 \cdot (\boldsymbol{d}_{21} \times \boldsymbol{d}_{31}) \\
& + s^2 s^2 \boldsymbol{d}_{12} \cdot (\boldsymbol{c}_2 \times \boldsymbol{d}_{32}) + s^3 s^3 \boldsymbol{d}_{13} \cdot (\boldsymbol{d}_{23} \times \boldsymbol{c}_3) \\
& + s^1 s^1 s^2 s^2 s^3 s^3 \boldsymbol{d}_{123} \cdot (\boldsymbol{d}_{231} \times \boldsymbol{d}_{312}) \ ,
\end{aligned}$$

where the symbol $\sim$ indicates that terms have been neglected. Because the $\boldsymbol{d}_{\alpha\beta\gamma}$ vectors are parallel, the last term is zero. Integration gives, using (11.26),

$$\begin{aligned}
|\Omega_j| = \Delta\xi^1 \Delta\xi^2 \Delta\xi^3 \{ & \boldsymbol{c}_1 \cdot (\boldsymbol{c}_2 \times \boldsymbol{c}_3) + \tfrac{2}{3}\Delta\xi^1 \Delta\xi^1 \boldsymbol{c}_1 \cdot (\boldsymbol{c}_{12} \times \boldsymbol{c}_{13}) \\
& + \tfrac{2}{3}\Delta\xi^2 \Delta\xi^2 \boldsymbol{c}_{12} \cdot (\boldsymbol{c}_2 \times \boldsymbol{c}_{23}) + \tfrac{2}{3}\Delta\xi^3 \Delta\xi^3 \boldsymbol{c}_{13} \cdot (\boldsymbol{c}_{23} \times \boldsymbol{c}_3) \} \ .
\end{aligned} \tag{11.27}$$

We want to write (11.27) in terms of the location of the cell vertices. Substitution of (11.19) in (11.27) gives

$$|\Omega_j| = 8b_1 \cdot (b_2 \times b_3) + \tfrac{8}{3}\{b_1 \cdot (b_{12} \times b_{13})$$
$$+ b_{12} \cdot (b_2 \times b_{23}) + b_{13} \cdot (b_{23} \times b_3)\} \,. \tag{11.28}$$

This formula is exact. Before giving a more efficient expression we first give a formula for the area of a cell face multiplied by its outward unit normal, which will call a cell face vector.

Consider the cell face 1234 (cf. Fig. 11.6). In this cell face the coordinate mapping can be written in the form given by (11.17). The vector normal to the face in a point x in the positive ξ^3-direction with length equal to the area of the parallelogram spanned by two infinitesimal displacement vectors $dx_{(1)}$ and $dx_{(2)}$ is given by $dx_{(1)} \times dx_{(2)}$, so that the average normal vector with length equal to the area of the face 1432 (which we call the cell face vector) is given by

$$s_{1432} = \int_{1432} n d\Gamma = \int_{1432} dx_{(1)} \times dx_{(2)} \,. \tag{11.29}$$

The indices are put in parentheses to emphasize that no components are intended. Introducing s^α defined by (11.24) gives

$$dx_{(\alpha)} = \frac{\partial x}{\partial \xi^\alpha} d\xi^\alpha \quad \text{(no summation)} \,,$$

with

$$\frac{\partial x}{\partial \xi^1} = c_1 + d_{12} s^2, \quad \frac{\partial x}{\partial \xi^2} = c_2 + d_{21} s^1 \,,$$

where d_{12} and d_{21} are defined by (11.26). Substitution in (11.29) gives

$$s_{1432} = \tfrac{1}{4} \Delta\xi^1 \Delta\xi^2 \int_{-1}^{1}\int_{-1}^{1} \frac{\partial x}{\partial \xi^1} \times \frac{\partial x}{\partial \xi^2} ds^1 ds^2 = \Delta\xi^1 \Delta\xi^2 c_1 \times c_2 \,.$$

The vectors c_1 and c_2 are easily determined as follows. The transformation (11.17) can be rewritten as

$$x = x_0 + \tfrac{1}{2}\Delta\xi^1 c_1 s^1 + \tfrac{1}{2}\Delta\xi^2 c_2 s^2 + \tfrac{1}{4}\Delta\xi^1 \Delta\xi^2 c_{12} s^1 s^2 \,.$$

Requiring $x = x_m$, $m = 1, 2, 3, 4$, gives a system of equations that is easily solved in the same way as (11.20). The result is

$$c_1 = \frac{1}{2\Delta\xi^1}(x_{34} - x_{12}), \quad c_2 = \frac{1}{2\Delta\xi^2}(x_{23} - x_{14}) \,,$$

where $\boldsymbol{x}_{jm} \equiv \boldsymbol{x}_j + \boldsymbol{x}_m$. Using $\boldsymbol{a} \times \boldsymbol{a} = 0$ this gives the following efficient form:

$$s_{1432} = \tfrac{1}{2}(\boldsymbol{x}_4 - \boldsymbol{x}_2) \times (\boldsymbol{x}_3 - \boldsymbol{x}_1) \, , \qquad (11.30)$$

which is identical to the formula for a plane quadrilateral. This equivalence holds only for doubly-ruled surfaces with straight edges. The general formula for a cell face vector is

$$s_{mnpq} = \tfrac{1}{2}(\boldsymbol{x}_n - \boldsymbol{x}_q) \times (\boldsymbol{x}_p - \boldsymbol{x}_m) \, , \qquad (11.31)$$

assuming the cell vertices are numbered such that the diagonals are mp and nq and that the vector has a positive component in a direction of increasing ξ^α (cf. Fig. 11.6).

It may be shown that

$$s_{mnpq} // \boldsymbol{n}_0 \, , \qquad (11.32)$$

with $\boldsymbol{n}_0$ the normal in the cell face center $\boldsymbol{x}_0$. It is easily seen from (11.17) that

$$\boldsymbol{x}_0 = \tfrac{1}{4}(\boldsymbol{x}_m + \boldsymbol{x}_n + \boldsymbol{x}_p + \boldsymbol{x}_q) \, .$$

Geometric identity

Let Γ_j be the surface of the cell Ω_j. Since Γ_j is closed we must have the geometric identity (11.15). As mentioned before, for uniform flow to be an exact solution of discretization schemes it is necessary that the geometric identity is satisfied exactly. Since our cell faces are smooth, $\boldsymbol{n}$ is defined unambiguously, and it is easily shown that

$$\int_{\Gamma_j} \boldsymbol{n} d\Gamma = \sum_{\Gamma_j} s_{mnpq} = 0 \, ,$$

where summation takes place over the cell faces constituting Γ_j. We see that the geometric identity is satisfied.

Economic formula for cell volume

We can now write down a more economical formula for the cell volume. The cell face vectors are needed in the discretization. It would be nice if they could be reused to compute cell volumes. It would not be unexpected if the volume is related to face area times face distance. The vector $2\boldsymbol{b}_\alpha$ (defined in (11.21)) connects face centers. We rewrite (11.28) as

$$|\Omega_j| = \tfrac{8}{3}\{b_1 \cdot (b_2 \times b_3 + b_{12} \times b_{13}) + b_2 \cdot (b_3 \times b_1 + b_{23} \times b_{12})$$
$$+ b_3 \cdot (b_1 \times b_2 + b_{13} \times b_{23})\} \,. \tag{11.33}$$

After some manipulation we find

$$b_2 \times b_3 = \tfrac{1}{32}(x_{23} \times x_{67} + x_{67} \times x_{58} + x_{14} \times x_{23} + x_{58} \times x_{14})$$

and

$$b_{12} \times b_{13} = \tfrac{1}{32}(x_{17} \times x_{28} + x_{35} \times x_{17} + x_{28} \times x_{46} + x_{46} \times x_{35}) \,,$$

so that

$$b_1 \cdot (b_2 \times b_3 + b_{12} \times b_{13}) = \tfrac{1}{8} b_1 \cdot (s_{1265} + s_{4378}) \,.$$

Not unexpectedly, this is the average area of two opposing faces times their distance divided by 8. The other two terms in (11.33) can be handled in the same way, resulting in the following efficient formula:

$$|\Omega_j| = \tfrac{1}{3}\{b_1 \cdot (s_{1265} + s_{4378}) + b_2 \cdot (s_{1584} + s_{2673}) + b_3 \cdot (s_{1432} + s_{8765})\} \,. \tag{11.34}$$

Since s_{mnpq} is needed anyway, as noted before, this is the most efficient formula for the cell volume. Equation (11.34) consists of parts that can be reused for adjacent volumes. For example, $b_1 = \tfrac{1}{8}(x_{3478} - x_{1256})$, and x_{3478} can be used both in Ω_j and Ω_{j+2e_1}. The same holds for s_{4378}, and the other vectors in (11.34) can also be used more than once. A more complicated exact formula for the cell volume has been given by Davies and Salmond (1985).

Exercise 11.4.1. Show that the volume of the parallelepiped spanned by three vectors a, b, c is $a \cdot (b \times c)$.
Hint: $|b \times c|$ is the area of the parallelogram spanned by b, c.

Exercise 11.4.2. Prove equation (11.32).
Hint: on a cell face , $n \,//\, \tfrac{\partial x}{\partial \xi^1} \times \tfrac{\partial x}{\partial \xi^2}$.

11.5 Introduction to tensor analysis

The reader who is only interested in discretization schemes on colocated grids can skip this section.

When working in a general coordinate system it is desirable to formulate the equations in an *invariant form*, i.e. a form that is independent of the coordinate system. For this purpose vector and tensor notation have been developed. Vector notation is a special case of tensor notation, and is less handy for expressing higher order tensors. We therefore prefer tensor notation.

For an introduction to tensor analysis, see Aris (1962), Sedov (1971) or Sokolnikoff (1964). Here we will give an introduction that covers the essentials required for formulation and discretization of physical laws in general coordinates. The reader is assumed to be familiar with vector analysis.

11.5.1 Invariance

At the basis of tensor analysis is the fact that a physical entity remains the same, no matter what coordinate system is used for its description. In other words, it is *invariant* under coordinate transformations. This means there must be a relation between descriptions in two different coordinate systems. This relation is studied in tensor analysis. Furthermore, physical laws do not depend on coordinate systems. They can therefore be expressed in a form that does not depend on the coordinate system used, and is invariant under coordinate transformations. Such invariant formulations can be derived with tensor analysis. A law that cannot be put in invariant (tensor) form cannot be physical.

So invariance refers to independence of coordinate systems of *physical objects*, and, at the same time, of the *formulation* of their description and of the physical laws which govern them. To give an example, we define a vector $\boldsymbol{u}$ in Cartesian coordinates by

$$\boldsymbol{u} = u_1 \boldsymbol{e}^{(1)} + u_2 \boldsymbol{e}^{(2)} + u_3 \boldsymbol{e}^{(3)} \,,$$

with $\boldsymbol{e}^{(\alpha)}$ the unit vector in the x^α-direction. This expression obviously holds only in Cartesian coordinates. With tensor analysis we will find the following description:

$$\boldsymbol{u} = U_1 \boldsymbol{a}^{(1)} + U_2 \boldsymbol{a}^{(2)} + U_3 \boldsymbol{a}^{(3)} \,.$$

This formula holds in any coordinate system, with suitable definitions of U_α and $\boldsymbol{a}^{(\alpha)}$.

Coordinate transformations

We will restrict ourselves to Euclidean space. This means that the space can be described by a Cartesian coordinate system $\boldsymbol{x} = (x^1, x^2, x^3)$. We will always use $\boldsymbol{x}$ to denote a Cartesian coordinate system and Greek letters, such as $\boldsymbol{\xi}$, to denote general coordinates. All coordinate systems will be right-handed and coordinate transformations will be one-to-one, so that the Jacobian of a coordinate transformation $\boldsymbol{\eta} = \boldsymbol{\eta}(\boldsymbol{\xi})$ satisfies, as already discussed in Sect. 11.4,

$$J > 0,$$

(11.35)

with

$$J \equiv \left| \frac{\partial(\eta^1, \eta^2, \eta^3)}{\partial(\xi^1, \xi^2, \xi^3)} \right| \equiv \left| \frac{\partial \boldsymbol{\eta}}{\partial \boldsymbol{\xi}} \right| \equiv \frac{\partial \boldsymbol{\eta}}{\partial \xi^1} \cdot \left(\frac{\partial \boldsymbol{\eta}}{\partial \xi^2} \times \frac{\partial \boldsymbol{\eta}}{\partial \xi^3} \right)$$

$$= \begin{vmatrix} \frac{\partial \eta^1}{\partial \xi^1} & \frac{\partial \eta^1}{\partial \xi^2} & \frac{\partial \eta^1}{\partial \xi^3} \\ \frac{\partial \eta^2}{\partial \xi^1} & \frac{\partial \eta^2}{\partial \xi^2} & \frac{\partial \eta^2}{\partial \xi^3} \\ \frac{\partial \eta^3}{\partial \xi^1} & \frac{\partial \eta^3}{\partial \xi^2} & \frac{\partial \eta^3}{\partial \xi^3} \end{vmatrix}.$$

(11.36)

Coordinate transformations satisfying (11.35) are called *admissible*.

Summation convention and notation

In order to simplify notation, tensor analysis always uses the *summation convention*. This was used before but is repeated here:

> *Summation takes place over Greek indices that occur once as a subscript and once as a superscript in a term or product.*

Since nothing changes when an index over which summation takes place is replaced by another letter, it is called a *dummy index*. Other indices are called *free indices*.

A vector will be denoted by lower case Latin letters, e.g. $\boldsymbol{u}$, its components in a Cartesian coordinate system are denoted by u_α or u^α ($u_\alpha = u^\alpha$, of course), and its components in a general coordinate system are denoted by U^α or U_α. We will see how U^α and U_α may be deduced from $\boldsymbol{u}$ shortly, and find $U_\alpha \neq U^\alpha$.

When a vector (or, more generally, a tensor) is defined everywhere in the domain as a function of location we have a *vector field*. For brevity we will just call this a vector, and similarly for tensors and scalars.

When we write an expression or equation with free indices, for example $\delta^{\alpha\beta} U_\beta = V^\alpha$, we mean $\delta^{\alpha\beta} U_\beta = V^\alpha$, $\alpha = 1, 2, 3$. Free indices in terms in an equation must correspond. For example, $U^\alpha = V_\alpha$ is not allowed, neither is $U^\alpha = V^\beta$. Any free indices must be present in every term in an equation. Furthermore, the only way in which an index may be repeated in a term or product is as a single superscript-subscript pair. Hence, $U^\alpha U^\alpha$ is not allowed. These syntactic rules help to guard against mistakes. If they are violated in an equation, it is usually not a tensor equation.

Contravariant and covariant vectors

Let $\boldsymbol{\eta} = \boldsymbol{\eta}(\boldsymbol{\xi})$ (i.e $\eta^\alpha = \eta^\alpha(\xi^1, \xi^2, \xi^3)$) be an admissible transformation from a general coordinate system $\boldsymbol{\xi}$ to a general coordinate system $\boldsymbol{\eta}$. Let $U^\alpha(\boldsymbol{\xi})$ be a set of three functions of ξ^1, ξ^2, ξ^3, which transforms to another set of three functions $\tilde{U}^\beta(\boldsymbol{\eta})$ in $\boldsymbol{\eta}$-coordinates. If we have the following relations

$$\tilde{U}^\beta = \frac{\partial \eta^\beta}{\partial \xi^\alpha} U^\alpha , \tag{11.37}$$

then U^α is said to be a *contravariant vector* (or contravariant tensor of order 1). The contravariant nature is expressed by the index being a superscript. The order is the number of indices. The quantities $\tilde{U}^\beta, \partial \eta_\beta / \partial \xi^\alpha$ and U^α are functions of position in the underlying space. It is understood that they are evaluated in the same position in equation (11.37) and all other equations.

If $U_\alpha(\boldsymbol{\xi})$ is again a set of three functions, but it transforms according to

$$\tilde{U}_\beta = \frac{\partial \xi^\alpha}{\partial \eta^\beta} U_\alpha , \tag{11.38}$$

then U_α is said to be a *covariant vector*. The covariant nature is expressed by the index being a subscript.

The special thing about these transformation rules is that they are *linear*, *homogeneous* and *transitive*. We show this for (11.37). Linearity follows from

$$\frac{\partial \eta^\beta}{\partial \xi^\alpha}(U^\alpha + V^\alpha) = \frac{\partial \eta^\beta}{\partial \xi^\alpha}U^\alpha + \frac{\partial \eta^\beta}{\partial \xi^\alpha}V^\alpha .$$

Homogeneity follows from, with c a constant,

$$\frac{\partial \eta^\beta}{\partial \xi^\alpha}(cU^\alpha) = c\frac{\partial \eta^\beta}{\partial \xi^\alpha}U^\alpha .$$

These two properties show that a linear combination of tensors is a tensor, making it possible to formulate tensor equations. For transitivity, consider a third coordinate system $\boldsymbol{\zeta}$. Suppose rule (11.37) is satisfied also when we go from $\boldsymbol{\eta}$- to $\boldsymbol{\zeta}$-coordinates. Transitivity means that the transformation rule (11.37) is also satisfied when we go directly from $\boldsymbol{\xi}$ to $\boldsymbol{\zeta}$. This follows easily, since we have, denoting by $\overset{\approx}{U}{}^\alpha$ the components in $\boldsymbol{\zeta}$-coordinates:

$$\overset{\approx}{U}{}^\alpha = \frac{\partial \zeta^\alpha}{\partial \eta^\beta}\tilde{U}^\beta = \frac{\partial \zeta^\alpha}{\partial \eta^\beta}\frac{\partial \eta^\beta}{\partial \xi^\gamma}U^\gamma = \frac{\partial \zeta^\alpha}{\partial \xi^\gamma}U^\gamma .$$

Invariance of tensor laws

We assume that all transformations are admissible, so that the corresponding Jacobians are positive. Hence the determinant

$$\det \{\frac{\partial \eta^\beta}{\partial \xi^\alpha}\} = J \neq 0 ,$$

so that in (11.37) we have

$$U^\alpha = 0 \iff \tilde{U}^\beta = 0 .$$

Let two contravariant tensors satisfy

$$cU^\alpha + dV^\alpha = 0 , \tag{11.39}$$

with c and d constants. It follows from linearity and homogeneity that the left-hand-side is a tensor. From $J \neq 0$ it follows that in another coordinate system

$$c\tilde{U}^\alpha + d\tilde{V}^\alpha = 0 ,$$

and from transitivity that (11.39) holds in every admissible coordinate system. Thus we have obtained the invariance concept that we want:

Lemma 11.5.1. *(Ricci)*
If a tensor law is valid in one coordinate system, then it is valid in every coordinate system.

This means that it is not necessary to say what the coordinate system is when we write down an equation like (11.39). Ricci's lemma is a powerful tool, because it allows us to verify the correctness of a tensor law by inspecting it in a special coordinate system, such as a Cartesian system.

General tensors

The transformation laws (11.37) and (11.38) are not the only laws that are linear, homogeneous and transitive. More generally we have: $T_{\beta_1\beta_2...\beta_q}^{\alpha_1\alpha_2...\alpha_p}$ is a mixed tensor of rank $p + q$ and weight w if its components $\tilde{T}_{\beta_1...\beta_q}^{\alpha_1...\alpha_p}$ in η-coordinates and $T_{\beta_1...\beta_q}^{\alpha_1...\alpha_p}$ in ξ-coordinates are related by:

$$\tilde{T}_{\beta_1...\beta_q}^{\alpha_1...\alpha_p} = \left|\frac{\partial \xi}{\partial \eta}\right|^w \frac{\partial \eta^{\alpha_1}}{\partial \xi^{\gamma_1}} \cdots \frac{\partial \eta^{\alpha_p}}{\partial \xi^{\gamma_p}} \frac{\partial \xi^{\delta_1}}{\partial \eta^{\beta_1}} \cdots \frac{\partial \xi^{\delta_q}}{\partial \eta^{\beta_q}} T_{\delta_1...\delta_q}^{\gamma_1...\gamma_p} , \tag{11.40}$$

where w is a real number, and $\left|\frac{\partial \xi}{\partial \eta}\right|$ is the Jacobian defined in (11.36). This is called a mixed tensor, because it has both covariant (subscripts) and contravariant (superscripts) indices. It has order $p + q$, and contravariant order p and covariant order q. If $w \neq 0$, the tensor is called a *relative tensor* with weight w; if $w = 0$ the tensor is called *absolute*.

The product of two tensors is a tensor, for example $U^\alpha U^\beta$ is a contravariant tensor of order 2. A vector is a tensor of order 1.

Scalars

A tensor of order zero (no indices at all) is called a *scalar*. From (11.40) follows the transformation law for a relative scalar of weight w:

$$\tilde{\varphi} = \left| \frac{\partial \boldsymbol{\xi}}{\partial \boldsymbol{\eta}} \right|^{w} \varphi . \qquad (11.41)$$

The simplest example of a scalar is a function of position in space. Neither the space nor the position of a point P change as we change coordinates from $\boldsymbol{\xi}$ to $\boldsymbol{\eta}$, so that

$$\tilde{\varphi}(\boldsymbol{\eta}_p) = \varphi(\boldsymbol{\xi}_p) ,$$

or

$$\tilde{\varphi}(\eta) = \tilde{\varphi}(\eta(\xi)) = \varphi(\xi) ,$$

which is what is meant when we write

$$\tilde{\varphi} = \varphi ,$$

so that according to the transformation law (11.41) this is an (absolute) scalar. It follows immediately that the value of a relative ($w \neq 0$) scalar in a point in space changes as we change coordinates. One might think that such a quantity has no physical meaning, because its value depends on the coordinate system chosen. However, this is not so. We will see later that the mass flux in a flow field is relative vector. The reason is that the length scales associated with coordinate increments change under coordinate transformation.

The Kronecker tensor

From now on we write δ_{β}^{α} for the Kronecker delta. It is defined by

$$\delta_{\beta}^{\alpha} = 1 \text{ if } \alpha = \beta, \quad \delta_{\beta}^{\alpha} = 0 \text{ if } \alpha \neq \beta \qquad (11.42)$$

in *every coordinate system*. Let (11.42) hold in $\boldsymbol{\xi}$-coordinates. We try to see if the transformation law (11.40) holds. We have

$$\tilde{\delta}_{\beta}^{\alpha} = \frac{\partial \eta^{\alpha}}{\partial \xi^{\gamma}} \frac{\partial \xi^{\sigma}}{\partial \eta^{\beta}} \delta_{\sigma}^{\gamma} = \frac{\partial \eta^{\alpha}}{\partial \xi^{\gamma}} \frac{\partial \xi^{\gamma}}{\partial \eta^{\beta}} = \frac{\partial \eta^{\alpha}}{\partial \eta^{\beta}} = \delta_{\beta}^{\alpha} ,$$

since $\eta^{1}, \eta^{2}, \eta^{3}$ vary independently, so that $\partial \eta^{\alpha} / \partial \eta^{\beta} = 0$ if $\alpha \neq \beta$. Comparison with (11.42) shows that δ_{β}^{α} is indeed an absolute mixed tensor of order 2. According to Exercise 11.5.3, $\delta^{\alpha\beta}$ and $\delta_{\alpha\beta}$ are not tensors.

The infinitesimal displacement vector

Let $d\boldsymbol{x}$ be an infinitesimal displacement vector. Its Cartesian components dx^α and its components in a general $\boldsymbol{\xi}$-coordinate system $d\xi^\alpha$ are related by

$$dx^\alpha = \frac{\partial x^\alpha}{\partial \xi^\beta} d\xi^\beta \ . \tag{11.43}$$

We see that the components of an infinitesimal displacement vector $d\xi^\alpha$ transform like a contravariant vector (or tensor).

The gradient

Let $\varphi(\boldsymbol{x})$ be a differentiable function. The Cartesian components of its gradient are $\partial\varphi/\partial x^\alpha$. We have

$$\frac{\partial\varphi}{\partial x^\alpha} = \frac{\partial\xi^\beta}{\partial x^\alpha}\frac{\partial\varphi}{\partial\xi^\beta} \ ,$$

so that $\frac{\partial\varphi}{\partial\xi^\beta}$ is a covariant vector. This shows why it is natural to use subscripts for covariant components.

Exercise 11.5.1. Show that the Jacobian of the product of two transformations equals the product of the Jacobians of these transformations.

Exercise 11.5.2. Show that the following transformation law

$$\tilde{V}^\alpha = J^w \frac{\partial\eta^\alpha}{\partial\xi^\beta} V^\beta$$

is linear, homogeneous and transitive. (Hint: use Exercise 11.5.1).

Exercise 11.5.3. Show that $\delta_{\alpha\beta}$ and $\delta^{\alpha\beta}$ are not tensors.

Exercise 11.5.4. Express U^α in terms of u^α for cylindrical polar coordinates.

Exercise 11.5.5. Show that $\partial^2\varphi/\partial\xi^\alpha\partial\xi^\beta$ is not a tensor.

11.5.2 The geometric quantities

The geometric quantities are fundamental quantities related to the transformation from Cartesian to general coordinates.

The base vectors

The base vectors have already been introduced in the preceding section; we repeat their definition here. The *contravariant base vectors* are defined by

$$\boldsymbol{a}^{(\alpha)} \equiv \operatorname{grad} \xi^{\alpha} \ . \tag{11.44}$$

In other words, the Cartesian components of $\boldsymbol{a}^{(\alpha)}$ are given by

$$a_{\beta}^{(\alpha)} \equiv \frac{\partial \xi^{\alpha}}{\partial x^{\beta}} \ .$$

Note that $\boldsymbol{a}^{(\alpha)}$ is a vector and not a vector component. The parentheses serve to distinguish between $a_{\beta}^{(\alpha)}$ and $a_{(\beta)}^{\alpha}$; $\boldsymbol{a}_{(\alpha)}$ will be introduced presently. For the same reason $\boldsymbol{a}^{(\alpha)}$ is *not* a contravariant vector; $a_{\beta}^{(\alpha)}$ are its Cartesian components. It is called the contravariant base vector because it is written with a superscript, which is natural because of its definition (11.44). It follows from (11.44) that $\boldsymbol{a}^{(\alpha)}$ is perpendicular to coordinate surfaces $\xi^{\alpha} = $ constant.

The *covariant base vectors* are defined by

$$\boldsymbol{a}_{(\alpha)} \equiv \frac{\partial \boldsymbol{x}}{\partial \xi^{\alpha}} \ , \tag{11.45}$$

which means that its Cartesian components are given by

$$a_{(\alpha)}^{\beta} = \frac{\partial x^{\beta}}{\partial \xi^{\alpha}} \ .$$

It follows from (11.45) that $\boldsymbol{a}_{(\alpha)}$ is tangential to coordinate lines along which ξ^{α} varies and $\xi^{\beta}, \beta \neq \alpha$ is constant. Fig. 11.7 gives a geometric interpretation of the two kinds of base vectors in two dimensions. Note that, unlike the customary base vectors in a Cartesian coordinate system, the length of $\boldsymbol{a}^{(\alpha)}$ and $\boldsymbol{a}_{(\alpha)}$ need not be unity.

It follows that the contravariant transformation law (11.37) in the case of a transformation from Cartesian to general coordinates can be written as

$$U^{\beta} = a_{\alpha}^{(\beta)} u^{\alpha} = \boldsymbol{a}^{(\beta)} \cdot \boldsymbol{u} \ , \tag{11.46}$$

$$u^{\beta} = a_{(\alpha)}^{\beta} U^{\alpha} \quad \text{or} \quad \boldsymbol{u} = \boldsymbol{a}_{(\alpha)} U^{\alpha} \ , \tag{11.47}$$

which tells us how the Cartesian and contravariant components of a vector $\boldsymbol{u}$ are related. In fact, we could have taken (11.46) as the definition of the contravariant components of the vector field $\boldsymbol{u}$ in general coordinate system, from which the transformation law (11.37) would have followed. Similarly, for a transformation from Cartesian to general coordinates the covariant transformation law (11.38) can be written as

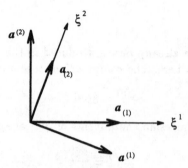

Fig. 11.7. Covariant and contravariant base vectors.

$$U_\beta = a^\alpha_{(\beta)} u_\alpha = a_{(\beta)} \cdot u \,, \tag{11.48}$$

$$u_\beta = a^{(\alpha)}_\beta U_\alpha \quad \text{or} \quad u = a^{(\alpha)} U_\alpha \,. \tag{11.49}$$

We see that U^β and U_β are two different representations of the same vector u; all that is involved is a change in base vectors in which u is expressed, according to (11.47) and (11.49).

As suggested by Fig. 11.7, the contravariant and covariant base vectors are mutually orthogonal. We have

$$a^{(\alpha)} \cdot a_{(\beta)} = \delta^\alpha_\beta \,, \tag{11.50}$$

since

$$\frac{\partial \xi^\alpha}{\partial x^\gamma} \frac{\partial x^\gamma}{\partial \xi^\beta} = \frac{\partial \xi^\alpha}{\partial \xi^\beta} \,.$$

Equation (11.50) allows us to determine $a^{(\alpha)}$ if $a_{(\alpha)}$ is given, and vice-versa. For example, let $a_{(\alpha)}$ be given. Then we have the following three equations for the three components of $a^{(1)}$:

$$a^{(1)} \cdot a_{(\beta)} = \delta^1_\beta \,, \quad \beta = 1, 2, 3 \,. \tag{11.51}$$

The matrix of this system is $(a_{(1)} \; a_{(2)} \; a_{(3)})^T$, where $(a_{(1)} \; a_{(2)} \; a_{(3)})$ is understood to be the matrix with $a_{(\alpha)}$ as columns, and its determinant is the Jacobian for a mapping $x = x(\xi)$ given by (11.22):

$$\sqrt{g} = \left| \frac{\partial x}{\partial \xi} \right| = a_{(1)} \cdot (a_{(2)} \times a_{(3)}) = |a_{(1)} \; a_{(2)} \; a_{(3)}| \,. \tag{11.52}$$

This quantity was formerly called J and was assumed to be positive; why we now prefer to call it $\sqrt{g}$ will becomes clear later. Using Cramer's rule, the solution of (11.51) is easily found to be

$$a^{(1)} = \frac{1}{\sqrt{g}} (a_{(2)} \times a_{(3)}) \,.$$

The general relation is

$$a^{(\alpha)} = \frac{1}{\sqrt{g}}(a_{(\beta)} \times a_{(\gamma)}) , \quad \alpha, \beta, \gamma \ \text{cyclic} .$$

(11.53)

Here cyclic means that α, β, γ should be an even permutation of 1, 2, 3, i.e. 1, 2, 3 or 2, 3, 1 or 3, 1, 2.

The permutation symbol

The *permutation symbol* $\varepsilon_{\alpha\beta\gamma}$ is defined as follows:

$$\varepsilon_{\alpha\beta\gamma} \equiv \begin{cases} 0, & \text{if any two of } \alpha, \beta, \gamma \text{ are equal} \\ 1, & \text{if } \alpha, \beta, \gamma \text{ is an even permutation of 1, 2, 3} \\ -1, & \text{if } \alpha, \beta, \gamma \text{ is an odd permutation of 1, 2, 3} \end{cases}$$

(11.54)

For example, $\varepsilon_{122} = 0$, $\varepsilon_{231} = 1$, $\varepsilon_{213} = -1$.

The vector product $c = a \times b$ of two vectors a and b can be expressed as, in Cartesian coordinates,

$$c_\alpha = \varepsilon_{\alpha\beta\gamma} a^\beta b^\gamma .$$

Hence

$$a_{(\alpha)} \cdot (a_{(\beta)} \times a_{(\gamma)}) = \varepsilon_{\delta\sigma\mu} a^\delta_{(\alpha)} a^\sigma_{(\beta)} a^\mu_{(\gamma)} .$$

Let us define $\varepsilon_{\alpha\beta\gamma}$ to be a tensor with components defined by (11.54) in Cartesian coordinates. If this quantity is indeed to be a tensor we must have according to (11.40) in $\boldsymbol{\xi}$-coordinates

$$\tilde{\varepsilon}_{\alpha\beta\gamma} = |\frac{\partial \boldsymbol{x}}{\partial \boldsymbol{\xi}}|^w \frac{\partial x^\delta}{\partial \xi^\alpha} \frac{\partial x^\sigma}{\partial \xi^\beta} \frac{\partial x^\mu}{\partial \xi^\gamma} \varepsilon_{\delta\sigma\mu} = (\sqrt{g})^w a_{(\alpha)} \cdot (a_{(\beta)} \times a_{(\gamma)}) .$$

(11.55)

According to (11.52), the right-hand-side equals $\pm\sqrt{g}^{1+w}$ if α, β, γ are an even or odd permutation of 1, 2, 3, respectively, whereas it equals zero if any two vectors in (11.55) are the same, so that if $w = -1$ we have indeed, as desired,

$$\tilde{\varepsilon}_{\alpha\beta\gamma} = \varepsilon_{\alpha\beta\gamma} ,$$

which shows that $\varepsilon_{\alpha\beta\gamma}$ is a relative tensor of weight -1. Similarly, $\varepsilon^{\alpha\beta\gamma}$ is a relative tensor of weight 1. Absolute variants are $\sqrt{g}\varepsilon_{\alpha\beta\gamma}$ and $\frac{1}{\sqrt{g}}\varepsilon^{\alpha\beta\gamma}$.

The metric tensor

The length ds of an incremental displacement vector $d\boldsymbol{x}$ satisfies, using (11.43):

$$ds^2 = d\boldsymbol{x} \cdot d\boldsymbol{x} = \sum_{\alpha=1}^{3} \frac{\partial x^\alpha}{\partial \xi^\beta} \frac{\partial x^\alpha}{\partial \xi^\gamma} d\xi^\beta d\xi^\gamma = g_{\beta\gamma} d\xi^\beta d\xi^\gamma \,, \tag{11.56}$$

with

$$g_{\alpha\beta} \equiv \sum_{\gamma=1}^{3} \frac{\partial x^\gamma}{\partial \xi^\alpha} \frac{\partial x^\gamma}{\partial \xi^\beta} = \boldsymbol{a}_{(\alpha)} \cdot \boldsymbol{a}_{(\beta)} \,. \tag{11.57}$$

We cannot use the summation convention in one member of (11.56) and (11.57) because we sum over two superscripts. It is easy to see that $g_{\alpha\beta}$ is an absolute covariant tensor of order 2:

$$\tilde{g}_{\alpha\beta} \equiv \sum_{\gamma=1}^{3} \frac{\partial x^\gamma}{\partial \eta^\alpha} \frac{\partial x^\gamma}{\partial \eta^\beta} = \sum_{\gamma=1}^{3} \frac{\partial x^\gamma}{\partial \xi^\mu} \frac{\partial \xi^\mu}{\partial \eta^\alpha} \frac{\partial \xi^\sigma}{\partial \xi^\sigma} \frac{\partial \xi^\sigma}{\partial \eta^\beta} = \frac{\partial \xi^\mu}{\partial \eta^\alpha} \frac{\partial \xi^\sigma}{\partial \eta^\beta} g_{\mu\sigma} \,,$$

so that the transformation law (11.40) holds when we go from $\boldsymbol{\xi}$- to $\boldsymbol{\eta}$-coordinates. The tensor $g_{\alpha\beta}$ is called the *metric tensor*, because it relates distance to infinitesimal coordinate increments, according to (11.56). Note that $g_{\alpha\beta}$ is symmetric:

$$g_{\alpha\beta} = g_{\beta\alpha} \,.$$

Contravariant and mixed versions are defined by

$$g^{\alpha\beta} \equiv \boldsymbol{a}^{(\alpha)} \cdot \boldsymbol{a}^{(\beta)}, \quad g_\beta^\alpha \equiv \boldsymbol{a}^{(\alpha)} \cdot \boldsymbol{a}_{(\beta)} \,.$$

Because of (11.50),

$$g_\beta^\alpha = \delta_\beta^\alpha \,.$$

Cartesian coordinates correspond to the mapping $\boldsymbol{x} = \boldsymbol{\xi}$, so that in Cartesian coordinates

$$\begin{aligned} g_{\alpha\alpha} &= g^{\alpha\alpha} = 1 \quad \text{(no summation)}, \\ g_{\alpha\beta} &= g^{\alpha\beta} = 0, \ \alpha \neq \beta \,. \end{aligned} \tag{11.58}$$

The determinant with elements $g_{\alpha\beta}$ is called g. We have, from (11.57),

$$g = |(\boldsymbol{a}_{(1)} \ \boldsymbol{a}_{(2)} \ \boldsymbol{a}_{(3)})^T (\boldsymbol{a}_{(1)} \ \boldsymbol{a}_{(2)} \ \boldsymbol{a}_{(3)})| = |\boldsymbol{a}_{(1)} \ \boldsymbol{a}_{(2)} \ \boldsymbol{a}_{(3)}|^2 \,, \tag{11.59}$$

where $|(\boldsymbol{a}_{(1)} \ \boldsymbol{a}_{(2)} \ \boldsymbol{a}_{(3)}|$ is understood to be the matrix with $\boldsymbol{a}_{(\alpha)}$ as columns. Equations (11.59) and (11.22) explain why the Jacobian of a mapping $\boldsymbol{x} = \boldsymbol{x}(\boldsymbol{\xi})$ (from general to Cartesian coordinates) is called $\sqrt{g}$.

Contraction

Contraction is summing over a pair of indices of a tensor or tensor product. It can be shown that this gives a new tensor with the same weight but the order reduced by two. For example,

$$T^{\alpha\beta}_{\beta} = V^{\alpha} , \quad g^{\alpha\beta} T_{\beta\gamma} = S^{\alpha}_{\gamma} ,$$

with V^{α} and S^{α}_{γ} tensors. Note that $g^{\alpha\beta} T_{\delta\gamma}$ is a tensor of order 4. It is left to one of the exercises to show that

$$g^{\alpha\gamma} g_{\gamma\beta} = \delta^{\alpha}_{\beta} . \tag{11.60}$$

Raising and lowering indices

From (11.46) and (11.49) it follows that

$$U^{\beta} = \boldsymbol{a}^{(\beta)} \cdot \boldsymbol{a}^{(\alpha)} U_{\alpha} = g^{\beta\alpha} U_{\alpha} .$$

Similarly,

$$U_{\beta} = g_{\alpha\beta} U^{\alpha} .$$

This shows how the metric tensor is used to raise or lower and index. More generally,

$$g^{\beta_1\gamma_1} T^{\alpha_1...\alpha_p}_{\beta_1...\beta_q} = T^{\gamma_1\alpha_1...\alpha_p}_{\beta_2...\beta_q} .$$

It follows that dummy indices can be lowered and raised at will, e.g.

$$T^{\alpha\beta} N_{\beta} = T^{\alpha}_{\beta} N^{\beta} . \tag{11.61}$$

The inner product

Using (11.47) and (11.49) we can write

$$\boldsymbol{u} \cdot \boldsymbol{v} = \boldsymbol{a}_{(\alpha)} \cdot \boldsymbol{a}^{(\beta)} U^{\alpha} V_{\beta} .$$

Using (11.50) this becomes

$$\boldsymbol{u} \cdot \boldsymbol{v} = U^{\alpha} V_{\alpha} = U_{\alpha} V^{\alpha} .$$

Physical components

If $\boldsymbol{n}$ is some unit vector (length one, direction arbitrary) then the projection of a vector $\boldsymbol{u}$ on $\boldsymbol{n}$ (i.e. $\boldsymbol{u} \cdot \boldsymbol{n}$) is called the physical component of $\boldsymbol{u}$ in the direction $\boldsymbol{n}$, and denoted by $u_p(\boldsymbol{n})$. We have

$$u_p(\boldsymbol{n}) = U^\alpha N_\alpha = U_\alpha N^\alpha \; .$$

In a Cartesian system u_α is the physical component in the x^α-direction, but in general U_α is not the physical component in the ξ^α-direction. The unit vector parallel to $\boldsymbol{a}_{(\alpha)}$ for some α is

$$\boldsymbol{n} = \boldsymbol{a}_{(\alpha)}/|\boldsymbol{a}_{(\alpha)}| = \boldsymbol{a}_{(\alpha)}/\sqrt{g_{\alpha\alpha}} \quad \text{(no summation)} \,,$$

so that

$$N^\beta = g_\alpha^\beta/\sqrt{g_{\alpha\alpha}} \quad \text{(no summation)} \,, \tag{11.62}$$

and

$$u_p(\boldsymbol{n}) = U_\alpha/\sqrt{g_{\alpha\alpha}} \quad \text{(no summation)}.$$

Similarly, the covariant representation of the unit vector parallel to $\boldsymbol{a}^{(\alpha)}$ for some α is

$$N_\beta = g_\beta^\alpha/\sqrt{g^{\alpha\alpha}} \quad \text{(no summation)} \,, \tag{11.63}$$

and

$$u_p(\boldsymbol{n}) = U^\alpha/\sqrt{g^{\alpha\alpha}} \quad \text{(no summation)} \,.$$

We will also need physical components of second order tensors, in order to evaluate the shear stress at a wall, for example. In Cartesian coordinates, in a fluid the stress component in the x^α-direction on a surface element with unit normal $\boldsymbol{n}$ is given by

$$f^\alpha = t^{\alpha\beta} n_\beta \,,$$

with $t^{\alpha\beta}$ the (symmetric) stress tensor. The physical stress component in the direction of a unit vector $\boldsymbol{m}$ is given by

$$\boldsymbol{f} \cdot \boldsymbol{m} = m_\alpha t^{\alpha\beta} n_\beta \,.$$

Because this is a tensor equation, equating two scalars, according to Ricci's lemma it holds in every coordinate system:

$$\boldsymbol{f} \cdot \boldsymbol{m} = M_\alpha T^{\alpha\beta} N_\beta \,.$$

Suppose we want to determine the physical shear stress at a wall along which $\xi^3 = $ constant. The unit normal to the surface is given by (11.63) with $\alpha = 3$. This gives

$$\boldsymbol{f} \cdot \boldsymbol{m} = M_\alpha T^{\alpha 3}/\sqrt{g^{33}} \,.$$

For the component perpendicular to the wall, M_α is given by (11.63) with $\alpha = 3$, resulting in

$$\boldsymbol{f} \cdot \boldsymbol{m} = T^{33}/\sqrt{g^{33}} \,.$$

Christoffel symbol

The remaining geometric quantity of importance is

$$\left\{{\alpha \atop \beta\gamma}\right\} \equiv \boldsymbol{a}^{(\alpha)} \cdot \frac{\partial \boldsymbol{a}_{(\beta)}}{\partial \xi^\gamma} \ ,$$

which is called the *Christoffel symbol* (in the ξ-coordinate system). We have

$$\left\{{\alpha \atop \beta\gamma}\right\} = \frac{\partial \xi^\alpha}{\partial x^\sigma} \frac{\partial^2 x^\sigma}{\partial \xi^\beta \partial \xi^\gamma} \ ,$$

so that

$$\left\{{\alpha \atop \beta\gamma}\right\} = \left\{{\alpha \atop \gamma\beta}\right\} \ .$$

There are 18 Christoffel symbols in three dimensions and 6 in two. The Christoffel symbol is *not* a tensor (we will not show this), which is why we use the notation with curly braces and not something like $\Gamma^\alpha_{\beta\gamma}$, which would suggest a third order tensor. The role played by the Christoffel symbol will become clear later.

Exercise 11.5.6. Prove (11.61).

Exercise 11.5.7. Prove (11.60).

Exercise 11.5.8. Prove (11.62) and (11.63).

Exercise 11.5.9. Prove that a coordinate system is orthogonal if and only if $g_{\alpha\beta} = 0$ for $\beta \neq \alpha$.

Exercise 11.5.10. Show that the angle θ between ξ^α and ξ^β coordinate lines satisfies $\cos \theta = g_{\alpha\beta}/\sqrt{g_{\alpha\alpha}g_{\beta\beta}}$ (no summation).

Exercise 11.5.11. Determine $\sqrt{g}$ for cylindrical coordinates.

Exercise 11.5.12. Find the relation between U^α and $\boldsymbol{u}$ in cylindrical coordinates, and determine the physical components.

Exercise 11.5.13. Cylindrical coordinates ξ^α are defined by $x^1 = \xi^1$, $x^2 = \xi^2 \cos \xi^3$, $x^3 = \xi^2 \sin \xi^3$. Show that the only non-zero Christoffel symbols are:

$$\left\{{2 \atop 33}\right\} = -\xi^2 \ , \quad \left\{{3 \atop 22}\right\} = 1/\xi^2 \ .$$

11.5.3 Tensor calculus

The covariant derivative

Operations such as taking the divergence or the curl of a vector field, or the formula for the stress tensor in a fluid, involve taking derivatives of vector fields. In Cartesian coordinates we define the following notation:

$$u_{\alpha,\beta} = \partial u_\alpha / \partial x_\beta . \tag{11.64}$$

In general coordinates we want to define $U_{\alpha,\beta}$ and $U^\alpha_{,\beta}$ such that these quantities are tensors and reduce to (11.64) in the case of the identity mapping $\boldsymbol{x} = \boldsymbol{\xi}$. It turns out that $\partial U_\alpha / \partial \xi^\beta$ is not a tensor (cf. Exercise 11.5.14). We have, assuming that $U_{\alpha,\beta}$ will be defined such that it is a tensor, using (11.40),

$$
\begin{aligned}
U_{\alpha,\beta} &= \frac{\partial x^\sigma}{\partial \xi^\alpha} \frac{\partial x^\mu}{\partial \xi^\beta} u_{\sigma,\mu} \\
&= \frac{\partial x^\sigma}{\partial \xi^\alpha} \frac{\partial}{\partial \xi^\beta} \left(\frac{\partial \xi^\gamma}{\partial x^\sigma} U_\gamma \right) \\
&= \frac{\partial x^\sigma}{\partial \xi^\alpha} \frac{\partial \xi^\gamma}{\partial x^\sigma} \frac{\partial U_\gamma}{\partial \xi^\beta} + \frac{\partial x^\sigma}{\partial \xi^\alpha} \frac{\partial}{\partial \xi^\beta} \left(\frac{\partial \xi^\gamma}{\partial x^\sigma} \right) U_\gamma = \\
&= \frac{\partial U_\alpha}{\partial \xi^\beta} + \frac{\partial x^\sigma}{\partial \xi^\beta} \frac{\partial}{\partial \xi^\beta} \left(\frac{\partial \xi^\gamma}{\partial x^\sigma} \right) U_\gamma ,
\end{aligned} \tag{11.65}
$$

where we have assumed that U_α is an absolute tensor. Equation (11.65) is not very elegant, because the second term involves a mixed derivative with respect to both $\boldsymbol{x}$ and $\boldsymbol{\xi}$. We can eliminate this mixed derivative as follows:

$$
\begin{aligned}
\frac{\partial x^\sigma}{\partial \xi^\alpha} \frac{\partial}{\partial \xi^\beta} \left(\frac{\partial \xi^\gamma}{\partial x^\sigma} \right) &= \frac{\partial}{\partial \xi^\beta} \left(\frac{\partial x^\sigma}{\partial \xi^\alpha} \frac{\partial \xi^\gamma}{\partial x^\sigma} \right) - \frac{\partial \xi^\gamma}{\partial x^\sigma} \frac{\partial^2 x^\sigma}{\partial \xi^\alpha \partial \xi^\beta} \\
&= -\boldsymbol{a}^{(\gamma)} \cdot \frac{\partial \boldsymbol{a}_{(\alpha)}}{\partial \xi^\beta} = -\{^\gamma_{\alpha\beta}\} ,
\end{aligned}
$$

so that we find

$$U_{\alpha,\beta} = \frac{\partial U_\alpha}{\partial \xi^\beta} - \{^\gamma_{\alpha\beta}\} U_\gamma . \tag{11.66}$$

This is called the *covariant derivative* of the absolute tensor U^α. In a similar way we find

$$U^\alpha_{,\beta} = \frac{\partial U^\alpha}{\partial \xi^\beta} + \{^\alpha_{\beta\gamma}\} U^\gamma , \tag{11.67}$$

and for absolute second order tensors:

$$T^{\alpha\beta}_{,\gamma} = \frac{\partial T^{\alpha\beta}}{\partial \xi^\gamma} + \{^{\alpha}_{\gamma\delta}\}T^{\delta\beta} + \{^{\beta}_{\gamma\delta}\}T^{\alpha\delta} \,, \tag{11.68}$$

$$T^{\alpha}_{\beta,\gamma} = \frac{\partial T^{\alpha}_{\beta}}{\partial \xi^\gamma} + \{^{\alpha}_{\gamma\delta}\}T^{\delta}_{\beta} - \{^{\delta}_{\beta\gamma}\}T^{\alpha}_{\delta} \,, \tag{11.69}$$

$$T_{\alpha\beta,\gamma} = \frac{\partial T_{\alpha\beta}}{\partial \xi^\gamma} - \{^{\delta}_{\alpha\gamma}\}T_{\delta\beta} - \{^{\delta}_{\beta\gamma}\}T_{\alpha\delta} \,. \tag{11.70}$$

The usual rules that hold for differentiation also hold for the covariant derivative, such as

$$(U^\alpha U^\beta)_{,\gamma} = U^\beta U^{\alpha}_{,\gamma} + U^\alpha U^{\beta}_{,\gamma} \,, \tag{11.71}$$

as is easily seen. In fact, this follows immediately from Ricci's lemma, since (11.71) obviously holds in Cartesian coordinates.

The covariant derivative of the metric tensor

In Cartesian coordinates $g^{\alpha\beta}$ is given by (11.58), so that in these coordinates

$$g^{\alpha\beta}_{,\gamma} = 0 \,. \tag{11.72}$$

Because this is a tensor equation, equation (11.72) holds in every coordinate system, according to Ricci's lemma. Proving (11.72) without Ricci's lemma would be a tedious affair.

Suppose $U^{\alpha}_{,\beta} = 0$. We have, using (11.72),

$$U_{\alpha,\beta} = (g_{\alpha\gamma}U^\gamma)_{,\beta} = g_{\alpha\gamma}U^{\gamma}_{,\beta} \,,$$

hence

$$U_{\alpha,\beta} = 0 \,.$$

The covariant derivative of a constant vector

Let c be a constant vector field. Then

$$c^{\alpha}_{,\beta} = 0$$

holds in Cartesian coordinates, and hence in general coordinates we have

$$C^{\alpha}_{,\beta} = 0 \,.$$

The derivative of a scalar

For an absolute scalar φ we have $\varphi(\boldsymbol{\xi}) = \varphi(\boldsymbol{\xi}(\boldsymbol{x}))$, so that

$$\varphi_{,\alpha} \equiv \partial\varphi/\partial\xi^\alpha$$

is a covariant vector.

A useful identity that we will not demonstrate here is the following:

$$\frac{\partial\sqrt{g}}{\partial\xi^\alpha} = \sqrt{g}\{^\beta_{\beta\alpha}\} . \tag{11.73}$$

This clearly does not transform like a tensor, so that the covariant derivative of relative scalars is not given by (11.73) but by something else. More generally, the covariant derivative of relative tensors is defined differently from (11.66)–(11.70), but will not be needed here.

Divergence, curl and second order differential operator

We have, using (11.73), for the divergence of a vectorfield:

$$\text{div}\boldsymbol{u} = U^\alpha_{,\alpha} = \partial U^\alpha/\partial\xi^\alpha + \{^\alpha_{\alpha\beta}\}U^\beta = \frac{1}{\sqrt{g}}\frac{\partial}{\partial\xi^\alpha}(\sqrt{g}U^\alpha) . \tag{11.74}$$

Similarly,

$$T^{\alpha\beta}_{,\beta} = \frac{1}{\sqrt{g}}\frac{\partial\sqrt{g}T^{\alpha\beta}}{\partial\xi^\beta} + \{^\alpha_{\beta\gamma}\}T^{\beta\gamma} .$$

In Cartesian coordinates, the curl of a vectorfield $\boldsymbol{u}$, called $\boldsymbol{\omega} = \text{curl}\boldsymbol{u}$, is given by $\omega^\alpha = \varepsilon^{\alpha\beta\gamma}u_{\beta,\gamma}$. In general coordinates this becomes

$$\Omega^\alpha \equiv \frac{1}{\sqrt{g}}\varepsilon^{\alpha\beta\gamma}U_{\beta,\gamma} , \tag{11.75}$$

where we have inserted the factor $1/\sqrt{g}$ ($=1$ in Cartesian coordinates) in order to make Ω^α an absolute tensor if U_α is absolute. Because this notation is common in fluid dynamics we make an exception here, denoting a vector by the Greek letter Ω. Because (11.75) is a tensor equation and is true in the Cartesian case, it holds in general coordinates. Equivalent forms of (11.75) are

$$\Omega^\alpha = \frac{1}{\sqrt{g}}\varepsilon^{\alpha\beta\gamma}g_{\beta\delta}U^\delta_{,\gamma} = \frac{1}{\sqrt{g}}\varepsilon^{\alpha\beta\gamma}(g_{\beta\delta}U^\delta)_{,\gamma} .$$

In Cartesian coordinates the general second order differential operator $L\varphi$ working on an absolute scalar φ is given by

$$L\varphi \equiv (a^{\alpha\beta}\varphi_{,\alpha})_{,\beta} .$$ (11.76)

Define the vector field b by

$$b^\beta = a^{\alpha\beta}\varphi_{,\alpha} ,$$

then

$$L\varphi = \text{div } b ,$$

which shows that $L\varphi$ is a scalar. Hence (11.76) is a tensor equation, and holds in general coordinates. Using (11.74) we can write

$$L\varphi = \frac{1}{\sqrt{g}}\frac{\partial}{\partial\xi^\beta}(\sqrt{g}A^{\alpha\beta}\frac{\partial\varphi}{\partial\xi^\alpha}) .$$

In order to obtain the Laplacian $\nabla^2\varphi$ in general coordinates we cannot replace $a^{\alpha\beta}$ by $\delta^{\alpha\beta}$, because $\delta^{\alpha\beta}$ is not a tensor, as shown in Sect. 11.5.1. But $a^{\alpha\beta} = g^{\alpha\beta}$ does the job in Cartesian and hence in general coordinates, so that in general coordinates the Laplacian is given by

$$\nabla^2\varphi = \frac{1}{\sqrt{g}}\frac{\partial}{\partial\xi^\beta}(\sqrt{g}g^{\alpha\beta}\frac{\partial\varphi}{\partial\xi^\alpha}) .$$ (11.77)

11.5.4 The equations of motion in general coordinates

Using the rules of tensor notation, tensor calculus and Ricci's lemma, the formulation of the governing equations of fluid dynamics in general coordinates is surprisingly easy. It suffices to write down syntactically correct tensor laws, that are valid in the special case of a Cartesian system. The equations in a Cartesian system have been presented in Chap. 1. They can almost be copied verbatim, except that here or there an index has to be raised or lowered.

General coordinate systems

The invariant form of the mass conservation equation (1.13) is found to be:

$$\frac{\partial\rho}{\partial t} + (\rho U^\alpha)_{,\alpha} = 0 .$$

Using (11.74) this can be rewritten as

$$\frac{\partial\rho}{\partial t} + \frac{1}{\sqrt{g}}\frac{\partial}{\partial\xi^\alpha}(\sqrt{g}\rho U^\alpha) = 0 .$$

The momentum equation (1.24) takes the following general form:

$$\frac{\partial \rho U^\alpha}{\partial t} + (\rho U^\alpha U^\beta)_{,\beta} = \tau_{,\beta}^{\alpha\beta} + \rho F_b^\alpha , \qquad (11.78)$$

with F_b^α the contravariant representation of the body force. For the invariant formulation of (1.25) some raising of indices is required, and the Kronecker symbol $\delta_{\alpha\beta}$ (not a tensor) must be replaced by $g^{\alpha\beta}$, cf. Sect. 11.5.1. The invariant expression for $\tau^{\alpha\beta}$ follows from (1.25):

$$\tau^{\alpha\beta} = -pg^{\alpha\beta} + 2\mu(e^{\alpha\beta} - \frac{1}{3}\Delta g^{\alpha\beta}) ,$$
$$e^{\alpha\beta} = \frac{1}{2}(g^{\beta\gamma}U_{,\gamma}^\alpha + g^{\alpha\gamma}U_{,\gamma}^\beta) , \quad \Delta = \frac{1}{\sqrt{g}}\frac{\partial}{\partial\xi^\alpha}(\sqrt{g}U^\alpha) .$$

The invariant form of the energy equation (1.40) is found to be:

$$\frac{\partial \rho E}{\partial t} + \frac{1}{\sqrt{g}}\frac{\partial}{\partial\xi^\alpha}(\sqrt{g}\rho U^\alpha E) = (U_\alpha \tau^{\alpha\beta})_{,\beta} + (kg^{\alpha\beta}T_{,\beta})_{,\alpha} + \rho U_\alpha F_b^\alpha + \rho q .$$

$$(11.79)$$

Exercise 11.5.14. Prove that $\partial U_\alpha / \partial \xi^\beta$ is not a tensor.

Exercise 11.5.15. Prove (11.67).

Exercise 11.5.16. Show that the 3×3 determinant with elements $b_{\alpha\beta}$ equals $\varepsilon_{\alpha\beta\gamma}b_{1\alpha}b_{2\beta}b_{3\gamma}$, with $\varepsilon_{\alpha\beta\gamma}$ the permutation symbol.

Exercise 11.5.17. Derive the Laplacian in cylindrical and spherical coordinates, starting from (11.77).

Exercise 11.5.18. Show that the heat conduction term in the energy equation (11.79) can be written as

$$\frac{\partial}{\partial\xi^\alpha}(k\sqrt{g}g^{\alpha\beta}\frac{\partial T}{\partial\xi^\beta}) .$$

Exercise 11.5.19. Take $\boldsymbol{u} = $ constant and use (11.74) to show that

$$\frac{\partial}{\partial\xi^\alpha}(\sqrt{g}a^{(\alpha)}) = 0 .$$

12. Numerical solution of the Euler equations in general domains

12.1 Introduction

The purpose of this chapter is to generalize the methods described in Chap. 10 for computing inviscid compressible flows in one dimension to more dimensions and general domains. Although no new principles will emerge, this generalization is not completely straightforward, except for the JST scheme (which is its great charm), which will therefore not be discussed here.

We will mostly discuss the three-dimensional case, leaving the two-dimensional case for the reader to work out. The domain $\Omega \subset \mathbb{R}^3$ is arbitrary, and is assumed to contain a single block boundary-fitted structured grid with hexahedrons as cells, as discussed in Chap. 11. In practice the shape of the domain is often so complicated, that domain decomposition (cf. Sect. 11.3) has to be used. Domain decomposition is also frequently used to create subtasks for parallel processing. Conditions to couple subdomain solutions and iterative methods to iterate to global convergence have to be devised. Introductions to the mathematical background of domain decomposition are given in Chan and Mathew (1994) and Smith, Bjørstad, and Gropp (1996). We will not discuss domain decomposition.

For reasons to be explained in Chap. 14, the methods to be disussed in the present chapter do not work well in the (almost) incompressible case. A unified method for the compressible and the incompressible case will be presented in Chap. 14.

12.2 Analytic aspects

This section runs parallel to Sect. 10.2; here we discuss only extensions brought about by multi-dimensionality.

Euler equations

The Euler equations in Cartesian coordinates (1.76)–(1.78) can be written as

$$U_t + \partial f^\alpha(U)/\partial x^\alpha = Q , \quad \boldsymbol{x} \in \Omega , \quad t \in (0, T] , \qquad (12.1)$$

where we sum over $\alpha \in \{1, 2, 3\}$, and where

$$U = \begin{pmatrix} \rho \\ m^1 \\ m^2 \\ m^3 \\ \rho E \end{pmatrix} , \quad f^\alpha = \begin{pmatrix} m^\alpha \\ u^\alpha m^1 + p\delta_1^\alpha \\ u^\alpha m^2 + p\delta_2^\alpha \\ u^\alpha m^3 + p\delta_3^\alpha \\ m^\alpha H \end{pmatrix} , \quad Q = \begin{pmatrix} 0 \\ \rho f_b^1 \\ \rho f_b^2 \\ \rho f_b^3 \\ \rho q + \rho u^\beta f_b^\beta \end{pmatrix} , \qquad (12.2)$$

where $m^\alpha = \rho u^\alpha$, and δ_β^α is the Kronecker delta. This is the conservative formulation. We will also need the nonconservative formulation with dependent variables $\rho, \boldsymbol{u}$ and p. For the momentum and energy equations we take the inviscid and nondiffusive versions of (1.27) and (1.43), respectively. We use the perfect gas law $e = \frac{1}{\gamma-1}\frac{p}{\rho}$ to eliminate e from the energy equation, which becomes:

$$\frac{Dp}{Dt} - \frac{p}{\rho}\frac{D\rho}{Dt} = (1 - \gamma)p\mathrm{div}\boldsymbol{u} + (\gamma - 1)\rho q .$$

Using the mass conservation equation, $D\rho/Dt$ is replaced by $-\rho\mathrm{div}\boldsymbol{u}$. The nonconservate formulation of the Euler equations becomes:

$$\frac{D\rho}{Dt} + \rho\mathrm{div}\boldsymbol{u} = 0 , \qquad (12.3)$$

$$\frac{D\boldsymbol{u}}{Dt} + \frac{1}{\rho}\mathrm{grad}p = \boldsymbol{f}_b , \qquad (12.4)$$

$$\frac{Dp}{Dt} + \gamma p\mathrm{div}\boldsymbol{u} = (\gamma - 1)\rho q . \qquad (12.5)$$

Homogeneity of the flux function

Suppose the equation of state is given by

$$p = \rho g(e) .$$

Note that the perfect gas law is a special case. Since $e = e(U) = \{U_5 - \frac{1}{2}(U_2^2 + U_3^2 + U_4^2)/U_1\}/U_1$, we have $e(\mu U) = e(U)$, with μ an arbitrary real number. In the same way as in Sect. 10.2 it is easily shown that the flux functions are homogeneous of degree one:

$$f^\alpha(\mu U) = \mu f^\alpha(U) ,$$

and that we have

$$f^\alpha(U) = F^\alpha(U)U , \quad F^\alpha \equiv \partial f^\alpha(U)/\partial U .$$

The Jacobians of the flux functions

The Euler equations (12.1) can be rewritten in nonconservative form as

$$U_t + F^\alpha \partial U/\partial x^\alpha = Q \ . \tag{12.6}$$

According to Sect. 2.2 this is a hyperbolic system if there are 5 linearly independent real eigenvectors R with real eigenvalues λ for the following generalized eigenproblem:

$$(n_\alpha F^\alpha - \lambda I)R = 0 \ . \tag{12.7}$$

To settle this we will study the generalized eigenproblem (12.7).

It turns out that the eigenvalue problem is more tractable if we transform to the following nonconservative variables:

$$W = \begin{pmatrix} \rho \\ u \\ p \end{pmatrix} \ , \quad u = \begin{pmatrix} u^1 \\ u^2 \\ u^3 \end{pmatrix} \ . \tag{12.8}$$

By defining $Q = \partial U/\partial W$ and writing

$$U = \begin{pmatrix} \rho \\ \rho u \\ \frac{p}{\gamma-1} + \frac{1}{2}\rho u \cdot u \end{pmatrix}$$

(where we have used the perfect gas law $\rho e = p/(\gamma - 1)$) we obtain:

$$Q = \begin{pmatrix} 1 & 0 & 0 \\ u & \rho I & 0 \\ \frac{1}{2}u \cdot u & \rho u^T & \frac{1}{\gamma-1} \end{pmatrix} \ , \quad I \equiv \begin{pmatrix} 1 & 0 & 0 \\ 0 & 1 & 0 \\ 0 & 0 & 1 \end{pmatrix} \ . \tag{12.9}$$

The inverse is found to be:

$$Q^{-1} = \begin{pmatrix} 1 & 0 & 0 \\ -u/\rho & I/\rho & 0 \\ \frac{\gamma-1}{2}u \cdot u & -(\gamma-1)u^T & \gamma-1 \end{pmatrix} \ .$$

We expect that the eigenvalue problem (12.7) is simplified by similarity transformation with Q, as in the one-dimensional case (cf. Sect. 10.2). The transformed eigenvalue problem becomes:

$$(n_\alpha \tilde{F}^\alpha - \lambda I)\tilde{R} = 0 \ , \quad \tilde{F}^\alpha = Q^{-1}F^\alpha Q \ , \quad \tilde{R} = Q^{-1}R \ . \tag{12.10}$$

However, we do not wish to undertake the labor of carrying out the matrix multiplications $Q^{-1}F^\alpha Q$. It is easier to obtain $\tilde{F}^\alpha$ by inspection of the nonconservative Euler equations (12.3)–(12.5). These can be written as

$$\frac{\partial V}{\partial t} + \tilde{F}^\alpha \frac{\partial V}{\partial x^\alpha} = \tilde{Q} \, ,$$

with

$$\tilde{F}^1 = \begin{pmatrix} u^1 & \rho & 0 & 0 & 0 \\ 0 & u^1 & 0 & 0 & 1/\rho \\ 0 & 0 & u^1 & 0 & 0 \\ 0 & 0 & 0 & u^1 & 0 \\ 0 & \gamma p & 0 & 0 & u^1 \end{pmatrix} \, ,$$

$$\tilde{F}^2 = \begin{pmatrix} u^2 & 0 & \rho & 0 & 0 \\ 0 & u^2 & 0 & 0 & 0 \\ 0 & 0 & u^2 & 0 & 1/\rho \\ 0 & 0 & 0 & u^2 & 0 \\ 0 & 0 & \gamma p & 0 & u^2 \end{pmatrix} \, ,$$

$$\tilde{F}^3 = \begin{pmatrix} u^3 & 0 & 0 & \rho & 0 \\ 0 & u^3 & 0 & 0 & 0 \\ 0 & 0 & u^3 & 0 & 0 \\ 0 & 0 & 0 & u^3 & 1/\rho \\ 0 & 0 & 0 & \gamma p & u^3 \end{pmatrix} \, .$$

The eigenvalue problem (12.10) can be conveniently formulated as

$$(A - \lambda I)\tilde{R} = 0 \, ,$$

with

$$A = \begin{pmatrix} \boldsymbol{n \cdot u} & \rho n_1 & \rho n_2 & \rho n_3 & 0 \\ 0 & \boldsymbol{n \cdot u} & 0 & 0 & n_1/\rho \\ 0 & 0 & \boldsymbol{n \cdot u} & 0 & n_2/\rho \\ 0 & 0 & 0 & \boldsymbol{n \cdot u} & n_3/\rho \\ 0 & \gamma p n_1 & \gamma p n_2 & \gamma p n_3 & \boldsymbol{n \cdot u} \end{pmatrix} \, .$$

The characteristic polynomial is easily found to be

$$(\boldsymbol{n \cdot u} - \lambda)^3 \{ (\boldsymbol{n \cdot u} - \lambda)^2 - c^2 \boldsymbol{n \cdot n} \} \, , \quad c^2 = \gamma p / \rho \, ,$$

with c the speed of sound. This gives us

$$\lambda_1 = \boldsymbol{n \cdot u} - c|\boldsymbol{n}| \, , \quad \lambda_2 = \lambda_3 = \lambda_4 = \boldsymbol{n \cdot u} \, , \quad \lambda_5 = \boldsymbol{n \cdot u} + c|\boldsymbol{n}| \, . \qquad (12.11)$$

The eigenvectors

For $\lambda = \lambda_2, \lambda_3, \lambda_4$ we find the following equations for the eigenvector $\tilde{R} = (\tilde{r}_1, ..., \tilde{r}_5)^T$:

$$\tilde{r}_5 = 0 \, , \quad n_1 \tilde{r}_2 + n_2 \tilde{r}_3 + n_3 \tilde{r}_4 = 0 \, .$$

Three linearly independent solutions are:

$$\tilde{R}_2 = \begin{pmatrix} n_1 \\ 0 \\ n_3 \\ -n_2 \\ 0 \end{pmatrix} , \quad \tilde{R}_3 = \begin{pmatrix} n_3 \\ -n_2 \\ n_1 \\ 0 \\ 0 \end{pmatrix} , \quad \tilde{R}_4 = \begin{pmatrix} n_2 \\ -n_3 \\ 0 \\ n_1 \\ 0 \end{pmatrix} . \quad (12.12)$$

For $\lambda = \lambda_1$ and $\lambda = \lambda_5$ solutions are easily found by choosing a nonzero value for the fifth component and solving for the other components by backsubstitution. In analogy with the one-dimensional case (cf. (10.10)) we choose the values $-\rho c/2$ and $\rho c/2$ for the fifth component of $\tilde{R}_1$ and $\tilde{R}_5$, respectively, and we obtain:

$$\tilde{R}_1 = \begin{pmatrix} -\rho/c \\ n/|n| \\ -\rho c \end{pmatrix} , \quad \tilde{R}_5 = \frac{1}{2} \begin{pmatrix} \rho/c \\ n/|n| \\ \rho c \end{pmatrix} . \quad (12.13)$$

Because we have obtained a full set of eigenvectors for the generalized eigenvalue problem (12.10), the system is hyperbolic.

Characteristic surfaces and bicharacteristics

As seen in Sect. 2.2, if we freeze F^α locally then there are plane wave solutions of the type

$$U = \hat{U} e^{i(n \cdot x - \lambda t)} , \quad \hat{U} = \text{constant} .$$

A plane defined by $w(x, t) \equiv n \cdot x - \lambda t = \text{constant}$ is called a *characteristic surface*. The normal to the characteristic surface in (t, x) space satisfies $\text{grad} w = n$, $w_t = -\lambda$. Now we take F^α variable, so that n and λ are not constant, and define characteristic surfaces by

$$w(t, x) = \text{constant} , \quad \text{grad} w = n , \quad w_t = -\lambda .$$

As seen from (12.11), $\lambda = \lambda(n)$, and n is arbitrary. In every point there is an infinite family of characteristic surfaces. Their (local) envelope is called the Monge or Mach conoid, and the curve of tangency between a characteristic surface and the Monge conoid is called a bicharacteristic. The bicharacteristics sweep out the Monge conoid. We will determine the bicharacteristics. Let (s_0, s) be a tangent vector to a bicharacteristic in (t, x) space. Since this vector is tangential to the characteristic surface, we have $(s_0, s) \cdot (-\lambda, n) = 0$, or

$$s \cdot n = \lambda(n) s_0 .$$

Since (s_0, s) is also tangential to the Monge conoid, it is tangential to the characteristic with n changed infinitesimally:

$$s \cdot (n + \delta n) = \lambda(n + \delta n)s_0 \ .$$

Obviously, in these equations we can choose $s_0 = 1$. For $\lambda = n \cdot u$ these equations for the bicharacteristic direction become:

$$s \cdot n = u \cdot n \ , \quad s \cdot \delta n = u \cdot \delta n$$

with solution $s = u$, so that

$$(s_0, s) = (1, u) \ .$$

In other words, the bicharacteristics corresponding to λ_1, λ_2 and λ_3 are the particle paths. For $\lambda = n \cdot u \pm c|n|$ we obtain

$$s \cdot n = u \cdot n \pm c|n| \ , \quad s \cdot (n + \delta n) = (n + \delta n) \cdot u \pm c|n + \delta n| \ ,$$

with solution $s = u \pm cn/|n|$. Hence, tangent vectors to a bicharacteristic corresponding to λ_1 and λ_5 are, respectively,

$$(s_0, s) = (1, u - cn/|n|) \ , \quad (s_0, s) = (1, u + cn/|n|) \ .$$

Boundary conditions

Information travels along bicharacteristics. The number of boundary conditions to apply in a point P at the boundary is therefore equal to the number of different types of bicharacteristics that enter the domain at P for every possible choice of $n(\lambda)$. Let the inward normal to the boundary at P be m. If $m \cdot u > 0$, the bicharacteristics corresponding to λ_2, λ_3 and λ_4 enter; these are counted three times, so the particle path bicharacteristics bring along three boundary conditions. If P is at a solid wall $(m \cdot u = 0)$ or if there is outflow at P $(m \cdot u < 0)$ at most two boundary conditions are to be applied. Bicharacteristics corresponding to λ_1 enter for all n if

$$m \cdot u - cm \cdot n/|n| > 0 \ , \quad \forall n \ ,$$

which is the case if the normal inflow velocity is supersonic:

$$m \cdot u > c \ .$$

Finally, bicharacteristics corresponding to λ_5 enter for all n if

$$m \cdot u + cm \cdot n/|n| > 0 \ , \quad \forall n \ ,$$

which is the case if there is inflow or if there is outflow with subsonic normal component:

$$m \cdot u > -c \ .$$

Having determined the number of boundary conditions to prescribe, the question arises what kind of boundary conditions to prescribe. This is restricted by the so-called *compatibility relations*. These hold in characteristics, and give rise to the Riemann invariants in the one-dimensional case. The multidimensional case is more complicated. Compatibility relations will not be considered here; see Hirsch (1990), Chap. 16 for an extensive discussion of compatibility relations and boundary conditions. Some guidance in selecting boundary conditions may be obtained from the one-dimensional case (Chap. 10), where the availability of Riemann invariants makes it easy to determine which combinations of boundary conditions are correct.

For a solid wall, it follows from the criteria given above that precisely one condition is to be given. This should be the normal velocity component. At a supersonic inflow boundary all unknowns must be prescribed. At a subsonic inflow boundary, a good possibility is to prescribe the velocity and one thermodynamic quantity, for example temperature. At a subsonic outflow boundary, for the single condition to be prescribed one may select another thermodynamic quantity, for example pressure.

Rotation of coordinate system

In the next section we will find it convenient to rotate coordinates, for easy evaluation of the expression $f^\alpha(U)n_\alpha$, with n a unit vector. We start by studying rotation of coordinates.

Let $x = (x^1, x^2, x^3)$ be a Cartesian coordinate system and let $y = (y^1, y^2, y^3)$ be another coordinate system, defined by

$$y = L^{-1}x, \quad x = Ly \qquad (12.14)$$

with L a nonsingular constant matrix. It is well-known that the transformation (12.14) represents a rotation if and only if L is an orthogonal matrix, i.e.

$$L^{-1} = L^T.$$

This we now show. Obviously, the coordinate systems are share the same origin, and because the transformation is linear, the y-system is rectilinear. Denote the base vector along the y^α-axis by $a_{(\alpha)}$. We have

$$a_{(\alpha)} = \frac{\partial x}{\partial y^\alpha}, \quad a_{(\alpha)}^\beta = \frac{\partial x^\beta}{\partial y^\alpha} = l_{\beta\alpha},$$

i.e. $l_{\beta\alpha}$ is the β-component of $a_{(\alpha)}$ in the x-system. We have a rotation if the y-system is again Cartesian, i.e. if the base vectors are orthogonal:

$$a_{(\alpha)} \cdot a_{(\beta)} = \delta_{\alpha\beta} \ .$$

This gives (the position of indices is irrelevant in this section, and the summation convention is invoked for pairs of equal Greek indices)

$$l_{\gamma\alpha} \cdot l_{\gamma\beta} = \delta_{\alpha\beta} \ ,$$

or

$$LL^T = I \ ,$$

which is what we wanted to show.

We now turn to the Euler equations. Denote the velocity components in the y-system by $v = (v^1, v^2, v^3)$. We have

$$v^\alpha = u \cdot a_{(\alpha)} = u^\beta a^\beta_{(\alpha)} = u^\beta l_{\beta\alpha} \ , \tag{12.15}$$

hence

$$v = L^T u \ , \quad u = Lv \ . \tag{12.16}$$

We wish to consider what happens to the expression $f^\alpha(U)n_\alpha$, given in the x-coordinate system, under a specific rotation of coordinates. Let the rotated y-coordinate system have the y^1-axis aligned with n:

$$a_{(1)} = n \ .$$

We do not care about the direction of the y^2- and y^3-axes, but if one wants to make a choice a possibility would be to take, denoting the base vectors of the x-coordinate system by $e_{(\alpha)}$,

$$a_{(2)} = e_{(1)} \times a_{(1)} \ , \quad a_{(3)} = a_{(1)} \times a_{(2)} \ ,$$

provided $e_{(1)} \neq a_{(1)}$; otherwise no rotation is applied. Using (12.15) and (12.16) we obtain

$$f^\alpha(U)n_\alpha = f^\alpha(U)a^\alpha_{(1)} = \begin{pmatrix} \rho v^1 \\ \rho v^1 u + p a_{(1)} \\ \rho v^1 H \end{pmatrix} ,$$

so that we can write

$$f^\alpha(U)n_\alpha = \tilde{L} f^1(V) \ , \tag{12.17}$$

with

$$\tilde{L} = \begin{pmatrix} 1 & 0 & 0 \\ 0 & L & 0 \\ 0 & 0 & 1 \end{pmatrix} , \quad V = \begin{pmatrix} \rho \\ \rho v \\ \rho E \end{pmatrix} .$$

Equation (12.17) will be used later.

Exercise 12.2.1. Show that the coordinate system defined by (12.14) is rectilinear.

Exercise 12.2.2. Show that if (12.14) is a rotation, then $l_{\alpha\beta}$ is the cosine of the angle between the base vectors $e_{(\alpha)}$ and $a_{(\beta)}$ of the original and the rotated coordinate system, respectively.

12.3 Cell-centered finite volume discretization on boundary-fitted grids

Integration over finite volume

Finite volume discretization is carried out in the usual way. The Euler equations (12.1) are integrated over the cell Ω_j depicted in Fig. 11.6. By using the divergence theorem we obtain, using the summation convection for Greek indices and the notation of Chap. 11,

$$|\Omega_j|\frac{dU_j}{dt} + \int_{\Gamma_j} f^\alpha n_\alpha dS = |\Omega_j|Q_j ,$$

with n the outer normal, and U_j and Q_j the volume averages of U and Q, respectively. The surface integral is evaluated by approximating f^α by its value at the center of the cell face. Taking as an example the integral over the 'east' cell face, i.e. the cell face that separates Ω_j and Ω_{j+2e_1} (let us denote this part of Γ_j by Γ_E), we get

$$\int_{\Gamma_E} f^\alpha n_\alpha dS \cong F_E^\alpha \int_{\Gamma_E} n_\alpha dS = F_E^\alpha s_{\alpha E} ,$$

with the cell face vector s_E defined by (11.30), and F_E^α an approximation of f_E^α. For the 'west' face we get, taking the sign convection for the cell face vectors s into account,

$$\int_{\Gamma_W} f^\alpha n_\alpha dS \cong -F_W^\alpha s_{\alpha W} .$$

First, we assume a Cartesian grid, and generalize the schemes discussed in Chap. 10 to three dimensions. Then general structured grids will be considered.

Cartesian grid

Consider the cell-face Γ_E, supposed to be perpendicular to the x^1-axis with outward normal in the positive direction. We obtain

$$\int_{\Gamma_E} f^\alpha n_\alpha dS = \int_{\Gamma_E} f^1 dS \cong F_E^1 |s_E| , \qquad (12.18)$$

with $|s_E|$ the area of Γ_E, and F_E^1 the numerical flux, which remains to be specified. This can be done in a way that is closely related to the one-dimensional case, because f^2 and f^3 do not occur in (12.18).

With flux splitting schemes, F_E^1 is obtained by splitting the flux f^1, analogous to (10.83):

$$F_E^1 = f^+(U_L) + f^-(U_R) ,$$

with U_L and U_R states to the left and right of Γ_E. These states may be the states in the centers of the two neighbor cells of Γ_E, or extrapolated states at Γ_E in MUSCL fashion.

With approximate Riemann solvers, F^1 is obtained by specifying a Riemann problem at Γ_E. Without loss of generality we assume $t = 0$. Then the Riemann problem to be used is given by

$$\frac{\partial U}{\partial t} + \frac{\partial f^1}{\partial x^1} = 0 ,$$
$$U(0, \boldsymbol{x}) = U_L , \quad x^1 < x_E^1 , \quad U(0, \boldsymbol{x}) = U_R , \quad x^1 > x_E^1 . \qquad (12.19)$$

This problem is similar to the one we had in the one-dimensional case, and extension to the multi-dimensional case is relatively straightforward. The above Riemann problem is solved exactly (Godunov) or approximately (Osher) or is approximated by a linear problem which is solved exactly (Roe). Denote the solution obtained by $\hat{U}(\frac{x^1 - x_E^1}{t})$. For the numerical flux we take

$$F_E^1 = f^1(\hat{U}(0)) .$$

The Roe scheme

Similarly to the one-dimensional case, with the Roe scheme the Riemann problem (12.19) is approximated by the following linear problem:

$$\frac{\partial U}{\partial t} + A \frac{\partial U}{\partial x} = 0 , \quad A = \frac{\partial f^1(\bar{U})}{\partial U} . \qquad (12.20)$$

The constant matrix A is the Jacobian of f^1 evaluated at the Roe-averaged state $\bar{U}$. It suffices, as it turns out, to specify the averages of $\boldsymbol{u}$ and H:

$$\bar{\boldsymbol{u}} = \frac{\sqrt{\rho_L}\boldsymbol{u}_L + \sqrt{\rho_R}\boldsymbol{u}_R}{\sqrt{\rho_L} + \sqrt{\rho_R}} , \quad \bar{H} = \frac{\sqrt{\rho_L}H_L + \sqrt{\rho_R}H_R}{\sqrt{\rho_L} + \sqrt{\rho_R}} . \qquad (12.21)$$

The Riemann problem for (12.20) is solved in the same way as in the one-dimensional case. Let $R_1, ..., R_5$ be the eigenvectors of A, and let

$$U_R - U_L = \sum_{p=1}^{5} \alpha_p R_p . \tag{12.22}$$

Then we get (cf. (10.66)):

$$F_E^1 = \frac{1}{2}\{f^1(U_L) + f^1(U_R)\} - \frac{1}{2}\sum_{p=1}^{5} |\lambda_p| \alpha_p R_p .$$

The quantities λ_p, α_p and R_p are determined as follows. The eigenproblem for the Jacobian of f^1 follows from the eigenproblem (12.7) by taking $n = (1, 0, 0)$. From (12.11) we find

$$\lambda_1 = \bar{u}^1 - \bar{c}, \quad \lambda_2 = \lambda_3 = \lambda_4 = \bar{u}^1, \quad \lambda_5 = \bar{u}^1 + \bar{c} .$$

The averaged velocity $\bar{u}^1$ follows from (12.21). The averaged sound speed $\bar{c}$ is given by (10.70):

$$\bar{c}^2 = (\gamma - 1)(\bar{H} - \frac{1}{2}\bar{u} \cdot \bar{u}) .$$

The eigenvectors follow from $R_p = Q\tilde{R}_p$ with Q given by (12.9) and $\tilde{R}_p$ by (12.12) and (12.13). In order to beautify R_p we apply a suitable scaling to $\tilde{R}_p$, and choose, with $\tilde{R}$ the matrix with columns $\tilde{R}_1, ..., \tilde{R}_5$:

$$\tilde{R} = \begin{pmatrix} 1 & 1 & 0 & 0 & 1 \\ -c/\rho & 0 & 0 & 0 & c/\rho \\ 0 & 0 & 1/\rho & 0 & 0 \\ 0 & 0 & 0 & 1/\rho & 0 \\ c^2 & 0 & 0 & 0 & c^2 \end{pmatrix} .$$

Let R be the matrix with columns $R_1, ..., R_5$. From $R = Q\tilde{R}$ we obtain, evaluating in the Roe-averaged state:

$$R = \begin{pmatrix} 1 & 1 & 0 & 0 & 1 \\ \bar{u}^1 - \bar{c} & \bar{u}^1 & 0 & 0 & \bar{u}^1 + \bar{c} \\ \bar{u}^2 & \bar{u}^2 & 1 & 0 & \bar{u}^2 \\ \bar{u}^3 & \bar{u}^3 & 0 & 1 & \bar{u}^3 \\ \bar{H} - \bar{c}\bar{u}^1 & \frac{1}{2}\bar{u}\cdot\bar{u} & \bar{u}^2 & \bar{u}^3 & \bar{H} + \bar{c}\bar{u}^1 \end{pmatrix} . \tag{12.23}$$

Let $\{R^p\}$ be the orthogonal set of row vectors satisfying $R^p R_q = \delta_q^p$. Then we get from (12.22) the following expression for α_p:

$$\alpha_p = R^p \cdot (U_R - U_L) .$$

To evaluate α_p we need $R^1, ..., R^5$, which are the rows of R^{-1}. We note that

$$\tilde{R}^{-1} = \begin{pmatrix} 0 & -\frac{\rho}{2c} & 0 & 0 & \frac{1}{2c^2} \\ 1 & 0 & 0 & 0 & -\frac{1}{c^2} \\ 0 & 0 & \rho & 0 & 0 \\ 0 & 0 & 0 & \rho & 0 \\ 0 & \frac{\rho}{2c} & 0 & 0 & \frac{1}{2c^2} \end{pmatrix} . \tag{12.24}$$

From $R^{-1} = \tilde{R}^{-1}Q^{-1}$ we obtain, evaluating in the Roe-averaged state,

$$R^{-1} = \begin{pmatrix} \frac{\bar{u}^1}{2\bar{c}} + \frac{\tilde{\gamma}}{4}\bar{M}^2 & -\frac{1}{2\bar{c}} - \frac{\tilde{\gamma}\bar{u}^1}{2\bar{c}^2} & -\frac{\tilde{\gamma}\bar{u}^2}{2\bar{c}^2} & -\frac{\tilde{\gamma}\bar{u}^3}{2\bar{c}^2} & \frac{\tilde{\gamma}}{2\bar{c}^2} \\ 1 - \frac{\tilde{\gamma}}{2}\bar{M}^2 & \frac{\tilde{\gamma}\bar{u}^1}{\bar{c}^2} & \frac{\tilde{\gamma}\bar{u}^2}{\bar{c}^2} & \frac{\tilde{\gamma}\bar{u}^3}{\bar{c}^2} & -\frac{\tilde{\gamma}}{\bar{c}^2} \\ -\bar{u}^2 & 0 & 1 & 0 & 0 \\ -\bar{u}^3 & 0 & 0 & 1 & 0 \\ -\frac{\bar{u}^1}{2\bar{c}} + \frac{\tilde{\gamma}}{4}\bar{M}^2 & \frac{1}{2\bar{c}} - \frac{\tilde{\gamma}\bar{u}^1}{2\bar{c}^2} & -\frac{\tilde{\gamma}\bar{u}^2}{2\bar{c}^2} & -\frac{\tilde{\gamma}\bar{u}^3}{2\bar{c}^2} & \frac{\tilde{\gamma}}{2\bar{c}^2} \end{pmatrix} , \tag{12.25}$$

where $\tilde{\gamma} = \gamma - 1$ and $\bar{M}^2 = \bar{u} \cdot \bar{u}/\bar{c}^2$. This completes the specification of the Roe scheme in the three-dimensional Cartesian case. The application of the sonic entropy fix of Harten (1984) follows easily from the one-dimensional case (cf. Sect. 10.3).

The Osher scheme

Application of the Osher scheme to (12.18) gives the following numerical flux (cf. (10.77)):

$$F_E^1 = \frac{1}{2}\{f^1(U_L) + f^1(U_R)\} - \frac{1}{2}\int_0^1 |A(U)|\frac{dU}{d\sigma}d\sigma , \tag{12.26}$$

with $A(U) = \partial f^1(U)/\partial U$. The procedure is similar to the one-dimensional case in Sect. 10.4. For an arbitrary hyperbolic system, the path would be divided in five parts $\Gamma_1, ..., \Gamma_5$, with

$$\frac{dU}{d\sigma} = R_5 \quad \text{on} \quad \Gamma_1 , \qquad \frac{dU}{d\sigma} = R_4 \quad \text{on} \quad \Gamma_2 \quad \text{etc.,}$$

with R_p the eigenvectors of A. In the particular case of the Euler equations, where we have $\lambda_2 = \lambda_3 = \lambda_4 = u^1$, we need to divide the path in only three parts, as we will see. We choose Γ_1 ($0 \leq \sigma < 1/3$), Γ_2 ($1/3 \leq \sigma < 2/3$), Γ_3 ($2/3 \leq \sigma \leq 1$) according to

$$\frac{dU}{d\sigma} = R_5 \quad \text{on} \quad \Gamma_1, \qquad \frac{dU}{d\sigma} \in \text{Span}\{R_2, R_3, R_4\} \quad \text{on} \quad \Gamma_2 , \qquad \frac{dU}{d\sigma} = R_1 \quad \text{on} \quad \Gamma_3 .$$

The eigenvectors R_p are the columns of R in (12.23), omitting the overbars. The rows of R^{-1} are called R^p; R^{-1} is given by (12.25), omitting the overbars. We have

$$R^q \frac{dU}{d\sigma} = 0, \quad q = 1, 2, 3, 4 \quad \text{on} \quad \Gamma_1,$$

$$R^q \frac{dU}{d\sigma} = 0, \quad q = 1, 5 \quad \text{on} \quad \Gamma_2, \tag{12.27}$$

$$R^q \frac{dU}{d\sigma} = 0, \quad q = 2, 3, 4, 5 \quad \text{on} \quad \Gamma_3.$$

To draw conclusions from this we transform to the nonconservative variables W defined by (12.8), so that (12.27) gives for suitable values of q

$$R^q Q \frac{dW}{d\sigma} = 0,$$

with Q given by (12.9)). Remembering that $R^1, ..., R^5$ are the rows of R^{-1}, and noting that $R^{-1}Q = \tilde{R}^{-1}$ with $\tilde{R}^{-1}$ given by (12.24), we easily obtain

$$R^{-1}Q dW = \tilde{R}^{-1} dW = \begin{pmatrix} -\frac{\rho}{2c}(du^1 - \frac{1}{\rho c}dp) \\ d\rho - \frac{1}{c^2}dp \\ \rho du^2 \\ \rho du^3 \\ \frac{\rho}{2c}(du^1 + \frac{1}{\rho}dp) \end{pmatrix}.$$

In the same way as in the one-dimensional case we see that

$$p\rho^{-\gamma} = \text{const.}, \quad u^1 - \frac{2c}{\gamma - 1} = \text{const.} \quad \text{on} \quad \Gamma_1.$$

Furthermore, obviously,

$$u^2 = \text{const.}, \quad u^3 = \text{const.} \quad \text{on} \quad \Gamma_1.$$

Similarly,

$$p\rho^{-\gamma} = \text{const.}, \quad u^1 + \frac{2c}{\gamma - 1} = \text{const.}, \quad u^2 = \text{const.}, \quad u^3 = \text{const.} \quad \text{on} \quad \Gamma_3.$$

On Γ_2 (12.27) gives

$$u^1 = \text{const.}, \quad p = \text{const.}$$

From these relations the states at the end points of the subpaths can be determined. The relations between u^1, p, ρ and c are the same as in the one-dimensional case. Using results from Sect. 10.4 we get

$$u^1_{2/3} = u^1_{1/3} = \frac{\alpha\psi_0 + \psi_1}{\alpha + 1},$$

$$c_{1/3} = \frac{\gamma - 1}{2 + 2\alpha}(\psi_1 - \psi_0), \quad c_{2/3} = \alpha c_{1/3},$$

$$z_{1/3} = z_0, \quad z_{2/3} = z_1, \quad p_{2/3} = p_{1/3}.$$

Here $\alpha = c_1/c_0$, $\psi_0 = u_0 - 2c_0/(\gamma-1)$, $\psi_1 = u_1 + 2c_1/(\gamma-1)$, $z = \ln(p\rho^{-\gamma})$. The pressure follows from (10.81). Furthermore, we have

$$u^q_{1/3} = u^q_0, \quad u^q_{2/3} = u^q_1, \quad q = 1, 2.$$

This determines the integration path. The integral in (12.26) is evaluated in exactly the same manner as in Sect. 10.4.

The van Leer scheme

For the van Leer scheme we write

$$F^1_E = f^+(U_L) + f^-(U_R).$$

It easily seen that in the three-dimensional case a splitting $f^1 = f^+ + f^-$ satisfying (10.84) and the six requirements listed in Sect. 10.5 is given by

$$f^+ = \begin{pmatrix} \frac{\rho c}{4}(1 + M)^2 \\ \frac{c}{\gamma}f^+_1((\gamma - 1)M + 2) \\ u^2 f^+_1 \\ u^3 f^+_1 \\ (\gamma f^+_2)^2/\{2(\gamma^2 - 1)f^+_1\} + \frac{1}{2}f^+_1(u^2 u^2 + u^3 u^3) \end{pmatrix},$$

$$f^- = \begin{pmatrix} -\frac{\rho c}{4}(1 - M)^2 \\ \frac{c}{\gamma}f^-_1((\gamma - 1)M - 2) \\ u^2 f^-_1 \\ u^3 f^-_1 \\ (\gamma f^-_2)^2/\{2(\gamma^2 - 1)f^-_1\} + \frac{1}{2}f^-_1(u^2 u^2 + u^3 u^3) \end{pmatrix},$$

where $M = u^1/c$.

The AUSM scheme

The three-dimensional version of the AUSM scheme proposed by Liou and Steffen (1993) is given by

$$F_E^1 = \frac{1}{2}(M + |M|)_E \tilde{U}_L + \frac{1}{2}(M - |M|)_E \tilde{U}_R + \tilde{P}_L^+ + \tilde{P}_R^- ,$$

$$M_E = M_L^+ + M_R^- ,$$

$$M^\pm = \pm\frac{1}{4}(M \pm 1)^2 , \quad |M| \le 1; \quad M^\pm = \frac{1}{2}(M + |M|) , \quad |M| > 1 ,$$

$$\tilde{U} = \rho c \begin{pmatrix} 1 \\ u^1 \\ u^2 \\ u^3 \\ H \end{pmatrix} , \quad P^\pm = \begin{pmatrix} 0 \\ p^\pm \\ 0 \\ 0 \\ 0 \end{pmatrix} ,$$

$$p^\pm = \frac{p}{2}(1 \pm M) , \quad |M| \le 1; , \quad p^\pm = \frac{p}{2}(1 \pm |M|/M) , \quad M > 1 .$$

Boundary-fitted grids

On boundary-fitted grids the flux at a cell face Γ_E is approximated as follows:

$$\int_{\Gamma_E} f^\alpha n_\alpha dS \cong n_{\alpha E} \int_{\Gamma_E} f^\alpha dS ,$$

where n_E is the outward unit normal at the center of Γ_E. This can be rewritten in a convenient form by introduction of a local Cartesian y-coordinate system, with the y^1-axis aligned with n_E. By using (12.17) we obtain

$$n_{\alpha E} \int_{\Gamma_E} f^\alpha dS = \tilde{L}_E \int_{\Gamma_E} f^1(V)dS ,$$

where we have appended a subscript E to $\tilde{L}$ to emphasize that the coordinate rotation is associated with Γ_E; for each cell face a different coordinate rotation is used. The advantage of coordinate rotation is, that f^2 and f^3 disappear, so that we have a close analogy with the Cartesian case, already discussed. The integral over Γ_E is aproximated in the same way as (12.18) for the Cartesian case:

$$\int_{\Gamma_E} f^1(V)dS \cong F_E^1 |s_E| ,$$

and $F_E^1(V_L, V_R)$ is evaluated in the same way as in the Cartesian case. The final result is:

$$\int_{\Gamma_E} f^\alpha n_\alpha ds \cong \tilde{L}_E F_E^1(V_L, V_R)|s_E| .$$

Of course, the flux over the same face for the neighbor cell (Γ_W in that case) does not need to be evaluated separately, but equals minus the flux just obtained, as in any conservative scheme.

The semi-discretized scheme that results after spatial discretization can be summarized as follows:

$$|\Omega_J|\frac{dU_j}{dt} = \sum_{\alpha}\{\tilde{L}|s|F^{\alpha}(V^+, V^-)\}|_{j-e_{\alpha}}^{j+e_{\alpha}},$$

$$V_{j+e_{\alpha}}^{\pm} = \tilde{L}_{j+e_{\alpha}}^T U_{j+e_{\alpha}}^{\pm},$$

(12.28)

where $j = (j_1, j_2, j_3)$, $e_1 = (\frac{1}{2}, 0, 0)$, $e_2 = (0, \frac{1}{2}, 0)$, $e_3 = (0, 0, \frac{1}{2})$. The index $j \pm e_{\alpha}$ refers to the face of Ω_j in the $\pm x^{\alpha}$-direction. If MUSCL is not used then $U_{j+e_{\alpha}}^+ = U_{j+2e_{\alpha}}$, $U_{j+e_{\alpha}}^- = U_j$, otherwise these are the slope limited extrapolations from the neighboring cells to the cell face at $j + e_{\alpha}$ in the obvious way.

Exercise 12.3.1. Show that the Riemann invariants associated with the hyperbolic system

$$\frac{\partial U}{\partial t} + \partial f^1(U)/\partial x^1 = 0$$

are the entropy S, the velocity components u^2 and u^3, and $u^1 \pm 2c/(\gamma - 1)$. Show that S, u^2 and u^3 correspond to a contact discontinuity.

12.4 Numerical boundary conditions

As we have seen, the derivation of numerical fluxes in the general case is effectively reduced to the Cartesian case by rotation of coordinates. Therefore it suffices to discuss numerical boundary conditions for the Cartesian case.

Number of analytic boundary conditions

Suppose the cell face Γ_E is at the boundary of the domain. The numerical flux at Γ_E is given by

$$F_E^1 = F^1(U_C, U_E),$$

where U_C is the state in the interior cell of which Γ_E is a face, and it remains to specify the state U_E at the outer side of Γ_E using the analytic boundary conditions and the scheme. Analytic boundary conditions have been discussed in Sect. 12.2. An extensive treatment of analytic and numerical boundary conditions is given in Chap. 19. of Hirsch (1990).

The Riemann invariants relevant to the choice of boundary conditions are the Riemann invariants of the hyperbolic system (12.3.1). It is left to the reader to show that these Riemann invariants and their corresponding eigenvalues λ are given by

$$
\begin{aligned}
u^1 - 2c/(\gamma - 1)\,, &\quad \lambda = u^1 - c\,, \\
S\,, &\quad \lambda = u^1\,, \\
u^2\,, &\quad \lambda = u^1\,, \\
u^3\,, &\quad \lambda = u^1\,, \\
u^1 + 2c/(\gamma - 1)\,, &\quad \lambda = u^1 + c\,.
\end{aligned}
$$

We assume that the outward normal on Γ_E points in the positive x^1-direction. Then if $\lambda < 0$ the corresponding Riemann invariant is incoming, and if $\lambda \geq 0$ it is not incoming, and will be called outgoing. As seen in Sect. 12.2, the number of analytic boundary conditions equals the number of incoming Riemann invariants.

Numerical boundary conditions

As said before, U_E is the state at the outer side of Γ_E, and there may be some vagueness as to whether the desired boundary state is effectively enforced at Γ_E or in the cell center of a virtual cell at the outer side of Γ_E. In the case of a solid boundary it is important to ensure that the boundary condition is enforced precisely at the boundary, but in the case of an inflow or an outflow boundary the precise location of the boundary is usually of little consequence, so the precise location where the boundary condition is effectively enforced is also of little consequence.

In the case of *supersonic inflow*, i.e. $u_E^1 < -c$, the above difficulty does not occur, since schemes based on approximate Rieman solvers or flux splitting all give

$$
F_E^1 = f^1(U_E)\,,
$$

and it suffices to prescribe the full state vector U_E, which in this case is given by the analytic boundary conditions.

In the case of *subsonic inflow*, i.e. $-c \leq u_E^1 < 0$, there are four incoming Riemann invariants and four analytic boundary conditions are given. First, assume that the four incoming Riemann invariants are given:

$$
S_E = S_0\,, \quad u_E^2 = u_0^2\,, \quad u_E^3 = u_0^3\,, \quad u_E^1 - 2c_E/(\gamma - 1) = r_1\,.
$$

As said before, the state in the cell adjacent to Γ_E is denoted by U_C. To obtain a fifth condition, we may extrapolate the outgoing Riemann invariant:

$$u_E^1 + 2c_E/(\gamma - 1) = u_C^1 + 2c_C/(\gamma - 1) \equiv r_2 \ . \tag{12.29}$$

We find

$$u_E^1 = \frac{1}{2}(r_1 + r_2) \ , \quad c_E = \frac{\gamma - 1}{4}(r_2 - r_1) \ . \tag{12.30}$$

Using (10.18) we have

$$c^2 = \gamma \rho^{\gamma - 1} \exp(S/c_v) \ , \tag{12.31}$$

which enables us to find ρ from c and S. The following boundary state is obtained:

$$\rho_E = \{\frac{1}{\gamma}c_E \exp(-S_0/c_v)\}^{\frac{1}{1-\gamma}} \ ,$$
$$m_E^1 = \rho_E u_E^1 \ , \quad m_E^2 = \rho_E u_0^2 \ , \quad m_E^3 = \rho_E u_0^3 \ ,$$
$$(\rho E)_E = \rho_E(\frac{1}{\gamma(\gamma - 1)}c_E^2 + \frac{1}{2}u_E \cdot u_E) \ . \tag{12.32}$$

In practice, usually the incoming Riemann invariants are not given, but four other quantities are prescribed, for example ρ_0 and u_0. If again we extrapolate the outgoing Riemann invariant, we obtain

$$\rho_E = \rho_0 \ , \quad m_E = \rho_0 u_0 \ , \quad c_E = \frac{\gamma - 1}{2}(r_2 - u_0^1) \ ,$$

and $(\rho E)_E$ follows from (12.32).

At a *solid boundary* the analytic boundary condition is $u^1 = 0$. To enforce this precisely at Γ_E, reflection is used:

$$u_E^1 = -u_C^1 \ , \quad u_E^2 = u_C^2 \ , \quad u_E^3 = u_C^3 \ , \quad S_E = S_C \ .$$

A fifth condition is obtained by extrapolation of the outgoing Riemann invariant, according to (12.29). This results in

$$c_E = \frac{\gamma - 1}{2}(r_2 + u_C^1) \ .$$

Equation (12.31) gives

$$\rho_E = \{\frac{1}{\gamma}c_E \exp(-S_C/c_v)\}^{\frac{1}{1-\gamma}} \ ,$$

and $(\rho E)_E$ follows from (12.32).

In the case of *subsonic outflow*, i.e. $0 < u_E^1 < c$, there is one incoming Riemann invariant, and one analytic boundary condition is given. Ideally, the incoming Riemann invariant is specified:

$$u_E^1 - 2c_E/(\gamma - 1) = r_1 \, .$$

Four additional conditions may be generated by extrapolating the outgoing Riemann invariants:

$$S_E = S_C \, , \quad u_E^1 + 2c_E/(\gamma - 1) = u_C^1 + 2c_C/(\gamma - 1) \equiv r_2 \, , \quad u_E^2 = u_C^2 \, ,$$
$$u_E^3 = u_C^3 \, .$$

Then u_E^1 and c_E follow from (12.30), and the boundary state is obtained as

$$\rho_E = \{\frac{1}{\gamma} c_E \exp(-S_C/c_v)\}^{\frac{1}{1-\gamma}} \, ,$$
$$m_E^1 = \rho_E u_E^1 \, , \quad m_E^2 = \rho_E u_C^2 \, , \quad m_E^3 = \rho_E u_C^3 \, ,$$

and $(\rho E)_E$ follows from (12.32). In practice, often some other quantity is given instead of the incoming Riemann invariant, for example the pressure: $p_E = p_0$. If we extrapolate the outgoing Riemann invariants we get in a similar way as above

$$\rho_E = \{p_0 \exp(-S_C/c_v)\}^{\frac{1}{1-\gamma}} \, , \quad c_E = \sqrt{\gamma p_0/\rho_E} \, ,$$
$$u_E^1 = r_2 - 2c_E/(\gamma - 1) \, , \quad u_E^2 = u_C^2 \, , \quad u_E^3 = u_C^2 \, ,$$

and $(\rho E)_E$ follows from (12.32).

If, as sometimes happens, u_E^1 is not known a priori, one does not know which of the above cases applies. One may then test on u_C^1 instead of u_E^1.

Transparent boundary conditions

If the domain is infinite, it has to be restricted artificially for numerical purposes by the introduction of an artificial boundary. Suppose that no analytic information whatsoever is available on the artificial boundary. Artificial numerical boundary conditions have to be given at the artificial boundary. In order to avoid artificial reflections of waves at the artificial boundary as much as possible, *transparent boundary conditions* may be specified. The simplest kind of transparent boundary conditions are:

$$\rho_E = \rho_C \, , \quad m_E = m_C \, , \quad p_E = p_C \, .$$

For more discussion on transparent boundary conditions, see Bayliss and Turkel (1980), Giles (1990), Karni (1992), Roe (1989).

Far-field boundary conditions for two-dimensional flow around airfoils

Let us consider the flow around a two-dimensional airfoil in an unbounded domain. The undisturbed flow velocity is denoted by U_∞ and the x^1-axis is taken in the undisturbed flow direction. There may be shocks at the airfoil, but sufficiently far from the airfoil there are no shocks and we have potential flow. Let (u^1, u^2) be the perturbation of the velocity with respect to the undisturbed flow. It is well-known (see for example Thomas and Salas (1986)) that if the airfoil carries lift, then far from the airfoil the perturbation velocity is asymptotically equal to

$$\delta u^1 = \frac{\kappa}{2\pi} \frac{\beta x^2}{x^1 x^1 + \beta^2 x^2 x^2} , \quad \delta u^2 = -\frac{\kappa}{2\pi} \frac{\beta x^1}{x^1 x^1 + \beta^2 x^2 x^2} , \quad \beta = \sqrt{1 - M_\infty^2} ,$$

$$(12.33)$$

where κ is the circulation. The relation between the circulation and the lift force L working in the x^2-direction is

$$L = \kappa \rho_\infty U_\infty ,$$

as shown by Blasius in 1910 (cf. Sect. 72b of Lamb (1945)).

It is clear from (12.33) that when $\beta \ll 1$ the influence of the airfoil extends over a large distance in the x^2-direction. When β is very close to zero the asymptotic analysis on which (12.33) is based needs to be refined (cf. Murman and Cole (1971)), but the preceding statement about the long-distance influence of airfoils with lift in transonic flow remains true. As a consequence, the distance from the airfoil at which the infinite domain is truncated needs to be sufficiently large. To give a rough idea, experience shows that if free-stream conditions are applied at the part of the artificial boundary where there is no outflow, then the boundary has to be at a distance of a least fifty chords from the airfoil, especially in the x^2-direction.

Since already at significantly smaller distances from the airfoil the solution shows little variation, choosing the computational domain so large implies a waste of computing resources, unless the size of the grid cells is increased drastically at increasing distance from the airfoil. But this usually has a detrimental effect on the rate of convergence of iterative solution methods.

An alternative is to choose the computational domain smaller, and not to apply free-stream conditions on the non-outflow part of the artificial boundary, but to prescribe

$$u^1 = u_\infty^1 + \delta u^1 , \quad u^2 = \delta u^2 ,$$

with δu^α given by (12.33). Of course, this implies iteration on the lift force L. For further information, see Thomas and Salas (1986) and Sect. 19.3 of Hirsch (1990).

Evaluation of the pressure at a solid boundary

With cell-centered finite volume discretization, the state vector is obtained in cell centers, which are not located at the boundary. If for post-processing purposes certain quantities are required at the boundary, extrapolation has to be applied. This is the case, for instance, when lift or drag on a solid body are to be determined. For this the pressure distribution on the solid boundary is needed.

The simplest method is piecewise constant extrapolation, in which the pressure at a cell edge on a solid boundary is taken equal to the pressure in the cell-center. However, this often implies wasting some of the accuracy that has been realized in the solution in the cell centers, especially when a higher order scheme (for example, of MUSCL type) has been used. The normal derivative of the pressure may be significant, making piecewise constant extrapolation inaccurate. As a simple example, consider the two-dimensional flow around a circular cylinder. Using cylindrical coordinates (r, θ), the non-conservative form of the stationary momentum equation in r-direction is given by

$$\rho v_r \frac{\partial v_r}{\partial r} + \frac{1}{r} \rho v_\theta \frac{\partial v_r}{\partial \theta} - \frac{1}{r} \rho v_\theta^2 = -\frac{\partial p}{\partial r} .$$

At the cylinder $v_r = 0$, so that

$$\frac{\partial p}{\partial r} = \frac{1}{r} \rho v_\theta^2 .$$

Hence, where a solid body is strongly curved (r small), such as near the leading edge of an airfoil, the normal derivative $\partial p / \partial n$ may be significant.

Extrapolation of the pressure to a solid boundary may be done by using an estimate of $\partial p / \partial n$ obtained from the equations of motion. This is described in Rizzi (1978) and Sect. 19.2.4 of Hirsch (1990). An alternative that is somewhat simpler, especially in three dimensions, is multilinear extrapolation. We describe this in two dimensions with the aid of Fig. 12.1. With a piecewise bilinear boundary-fitted coordinate mapping, the solid boundary is approximated by a piecewise linear curve. The pressure is assumed known in A, B, C, D. The points of intersection of AD and BC with the solid boundary are called E and F, respectively. The pressure is assumed to be bilinear in the patch $EFCD$. We take advantage of the fact that p varies linearly along straight lines. This gives

$$p_E = p_A \frac{|DE|}{|AD|} - p_D \frac{|AE|}{|AD|} , \quad p_F = p_B \frac{|CF|}{|BC|} - p_C \frac{|BF|}{|BC|} .$$

Let H be the center of AB, and let the point of intersection of GH with CD be I. Linear extrapolation gives

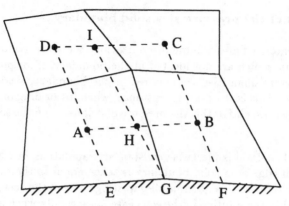

Fig. 12.1. Bilinear extrapolation of the pressure

$$p_G = p_H \frac{|GI|}{|HI|} - p_I \frac{|GH|}{|HI|} .$$

Between E and G, G and F the pressure varies linearly. The three-dimensional case can be handled similarly, but is a bit technical, and will not be discussed for brevity.

Free slip at solid boundary

In an inviscid flow model, such as the Euler equations, at a solid boundary the normal velocity component is zero, but the fluid should slip freely along the boundary in a tangential direction. A condition of no-slip should arise only due to viscous effects, and not because of numerical effects in the numerical approximations to inviscid terms. Numerical schemes for the Euler equations that are to be used for the approximation of the inviscid part of the Navier-Stokes equations should therefore allow free slip. We will show that the van Leer scheme does not satisfy this requirement, which is why this scheme was developed further by Liou and Steffen (1993) into the AUSM scheme.

We consider a two-dimensional flow in an unbounded domain. At time $t = 0$ we have flow in the x^2-direction with a slip line along the x^2-axis:

$$u^1(0, \boldsymbol{x}) = 0 , \quad \rho(0, \boldsymbol{x}) = \rho_0 = \text{constant} , \quad e(0, \boldsymbol{x}) = e_0 = \text{constant} ,$$
$$u^2(0, \boldsymbol{x}) = u_L^2 , \quad x^1 < 0 ; \quad u^2(0, \boldsymbol{x}) = u_R^2 , \quad x^1 > 0 ,$$
$$(12.34)$$

with u_L^2 and u_R^2 constant. From the equation of state it follows that $p(0, \boldsymbol{x}) = p_0 = \text{constant}$. It is easy to see that the exact solution is given by

$$U(t, \boldsymbol{x}) = U(0, \boldsymbol{x}) .$$

Let us choose a uniform Cartesian grid with step size h in the x^1-direction. Since there is no x^2-dependence there is no need to specify a step size or grid point labeling in the x^2-direction. The cell-centers x_j are chosen at $x_j = (j - 1/2)h$, and cell-centered finite volume schemes take the following form:

$$\frac{dU_j}{dt} + \frac{1}{h}(F^1_{j+1/2} - F^1_{j-1/2}) = 0 \,.$$

At time $t = 0$ we have $c_j = c_0 = $ constant, and $M_j = u^1_j/c_j = 0$. The van Leer flux is found to be at time $t = 0$:

$$F^1_{j+1/2} = \frac{1}{4}\rho_0 c_0 \begin{pmatrix} 0 \\ 4c_0/\gamma \\ u^2_j - u^2_{j+1} \\ \frac{1}{2}(u^2_j)^2 - \frac{1}{2}(u^2_{j+1})^2 \end{pmatrix} \,,$$

so that at $t = 0$ we have

$$\frac{dU_1}{dt} = -\frac{dU_0}{dt} = \frac{1}{4h}\rho_0 c_0 \begin{pmatrix} 0 \\ 0 \\ u^2_L - u^2_R \\ \frac{1}{2}(u^2_L)^2 - \frac{1}{2}(u^2_R)^2 \end{pmatrix} \,.$$

Hence, the exact solution is not recovered, and the contact discontinuity gets smeared as time increases, in a similar way as in Fig. 10.16, corresponding to a large amount $(\mathcal{O}(h))$ of artificial viscosity. A flux splitting scheme that gives no smearing of stationary contact discontinuities is the AUSM scheme. It is left to the reader to show this. The Roe and Osher schemes also give no smearing of stationary contact discontinuities.

With the JST scheme, the second order artificial viscosity term is switched off at a contact discontinuity, because it is triggered by pressure variations only; we now see why the scheme has been designed this way. The remaining fourth order artificial viscosity is sufficiently small for the JST scheme to be applicable to computation of viscous flows.

Exercise 12.4.1. Show that for the AUSM scheme we have $dU_j/dt = 0$ when the initial conditions are given by (12.34).

12.5 Temporal discretization

The semi-discretized system (12.28) is easily discretized in time by the same methods as considered for the one-dimensional case in Sect. 10.6 and 10.8.

Numerical stability

For the derivation of heuristic numerical stability criteria, we are led by the same reasoning as in Sect. 10.6 to consider the following linear scalar conservation law:

$$\frac{\partial v}{\partial t} + \sum_\alpha \mu_\alpha \frac{\partial v}{\partial x^\alpha} = 0 . \tag{12.35}$$

In the spirit of von Neumann stability analysis, we have to assume a uniform grid and constant coefficients. Considering flux splitting schemes to be generalizations of the first order upwind scheme to the systems case, we discretize (12.35) in space with the first order upwind scheme, assuming $\mu_\alpha \geq 0$:

$$\frac{dv_j}{dt} + \sum_\alpha \frac{\mu_\alpha}{h_\alpha}(v_j - v_{j-1}) = 0 . \tag{12.36}$$

We carry out von Neumann stability analysis according to the principles presented in Sect. 5.6. The symbol of scheme (12.36) is given by

$$\hat{L}_h(\theta) = \frac{1}{\tau} \sum_\alpha c_\alpha (1 - e^{-i\theta_\alpha}) , \quad c_\alpha = \mu_\alpha \tau / h_\alpha . \tag{12.37}$$

Stability requires that $-\tau \hat{L}_h(\theta) \in S$ for all $\theta = (\theta_1, \theta_2, ..., \theta_d)$, with S the stability domain of the time stepping scheme to be used, for all values of μ_α that occur in the flow domain. We must interpret μ_α as being the largest eigenvalue of the Jacobian $\partial f^\alpha(U)/\partial U$, i.e.

$$\mu_\alpha = |u_\alpha| + c ,$$

with c the speed of sound.

The stability results obtained in Chap. 5 are easily applied to the present case. Equation (12.37) can be rewritten as

$$\tau \hat{L}_h(\theta) = \delta(\theta) + i\gamma_2(\theta) ,$$

$$\delta(\theta) = 2 \sum_\alpha c_\alpha s_\alpha , \quad \gamma_2(\theta) = \sum_\alpha c_\alpha \sin \theta_\alpha , \quad s_\alpha = \sin^2 \frac{1}{2} \theta_\alpha . \tag{12.38}$$

Comparison shows that this is equivalent with (5.34) if we define $d_\alpha = c_\alpha$, $\kappa = 1$. This enables us to apply directly the stability results of Sect. 5.8.

For explicit Euler discretization in time, Theorem 5.8.1 gives the following sufficient stability condition:

$$\sum_\alpha c_\alpha \leq 1 . \tag{12.39}$$

It is left to the reader to show that this condition is also necessary (cf. Exercise 12.5.1).

In practice we do not have a uniform grid and a scalar equation, but a boundary-fitted grid and a system of equations. The question arises what to take for $|\mu_\alpha|$ and h_α in (12.37). The general form of the scheme is (12.28). Inspection of this equation shows that we have two candidates for h_α, namely

$$h_{j\pm e_\alpha} = |s_{j\pm e_\alpha}|/|\Omega_j| \ .$$

Furthermore, $(\tilde{L}F^\alpha)_{j+e_\alpha}$ corresponds to a discretization of a one-dimensional Euler equation with x^α-axis perpendicular to the cell face Γ_{j+e_j}. The corresponding maximum wave speed is $|u_{j+e_\alpha}|+c_{j+e_\alpha}$, where u_{j+e_α} is the velocity component perpendicular to Γ_{j+e_α}. We define

$$c_{\alpha j} = ((|u| + c)/h)_{j+e_\alpha} \ ,$$

and we replace (12.39) by

$$\tau \sum_\alpha c_{\alpha j} \leq 1 \ , \quad \forall j \ . \tag{12.40}$$

Higher order schemes

Just as in one dimension, for better spatial accuracy than given by first order approximate Riemann solvers or flux splitting schemes, we may approximate the numerical fluxes $F^\alpha(V^+, V^-)|_{j-e_\alpha}^{j+e_\alpha}$ in the scheme (12.28) by the MUSCL approach. That is, we may use a slope limited scheme. The formulation of such schemes follows immediately from the one-dimensional case, and will not be given.

In order to derive stability conditions, we follow the same heuristic reasoning as in the one-dimensional case (Sect. 10.8), and require the scheme to be stable for the following scalar conservation law:

$$\frac{\partial \varphi}{\partial t} + \sum_\alpha \frac{\partial f^\alpha(\varphi)}{\partial x^\alpha} = 0 \ .$$

If we discretize in time by the explicit Euler method, then in Sect. 9.4 the TVD condition (9.123) is found, which is sufficient for stability:

$$\sum_\alpha \lambda^\alpha \mu_{j+e_\alpha} \leq 1/M \ , \quad \lambda^\alpha = \tau/h^\alpha \ , \tag{12.41}$$

where M is a parameter that depends on the flux limiter used. For example, in Sect. 10.8 we found $M \approx 1.85$ for the van Albada limiter. Furthermore,

μ is the wave speed. Equation (12.41) has been derived for a uniform grid. Heuristic extension to systems is done in a way that is by now familiar by replacing μ by the maximum wave speed:

$$\sum_\alpha \lambda^\alpha (|u| + c)_{j+e_\alpha} \leq 1/M \ .$$

As before, this is heuristically extended to boundary-fitted grids as follows:

$$\tau \sum_\alpha c_{\alpha j} \leq 1/M \ , \quad \forall j \ , \tag{12.42}$$

with $c_{\alpha j}$ defined before (12.40).

As we saw in Sect. 9.4, some higher order Runge-Kutta time stepping schemes inherit the TVD property from the explicit Euler scheme, and (12.42) continues to hold. But if a time stepping scheme is used for which this is not the case, or if one wishes to work with a weaker stability condition, then one can forego TVD considerations and rely solely on von Neumann stability analysis, as in the one-dimensional case of Sect. 10.8. There a heuristic stability criterion for MUSCL schemes, that works well in practice, is obtained by requiring von Neumann stability for spatial discretization by the first order upwind scheme and by the κ-scheme that corresponds to the limiter selected, for the scalar equation (12.35). As an illustration we extend the example of Sect. 10.8 to the multi-dimensional case. The symbol of the first order upwind scheme is given by (12.38). The symbol of the κ-scheme applied to (12.35) is given by (cf. Sect. 5.6):

$$\tau \hat{L}_h(\theta) = \gamma_1(\theta) + i\gamma_2(\theta) \ , \quad \gamma_1(\theta) = 2(1 - \kappa) \sum_\alpha c_\alpha s_\alpha^2 \ ,$$

$$\gamma_2(\theta) = \sum_\alpha c_\alpha \{(1 - \kappa)s_\alpha + 1\} \sin \theta_\alpha \ , \quad s_\alpha = \sin^2 \frac{1}{2}\theta_\alpha \ , \quad c_\alpha = |\mu_\alpha|\tau/h_\alpha \ .$$

Comparision shows that the symbol of the first order upwind scheme is given by (5.34) if we take $d_\alpha = c_\alpha$ and $\kappa = 1$. The symbol of the κ-scheme is given by (5.34) if we take $d_\alpha = 0$. This enables us to apply the theorems of Sect. 5.7 and 5.8.

As in Sect. 10.8 we select as an example the SHK Runge-Kutta method given in Sect. 5.8. Theorem 5.7.2 shows that the symbol of the first order upwind scheme is within a circle with center at $-a$ and radius a if $\sum_\alpha c_\alpha \leq a$. As noted in Sect. 10.8 such a circle is inside the stability domain of the SHK method for $a \leq 1.3926$, so that we obtain the following sufficient stability condition for the first order upwind scheme:

$$\sum_\alpha c_\alpha \leq 1.3926 \ , \tag{12.43}$$

which is weaker than (12.39). We continue with the κ-scheme. Using Theorem 5.7.4 we see that for the κ-scheme $-\tau \hat{L}_h(\theta)$ is inside the ellipse of Fig. 5.9 if

$$\sum_\alpha c_\alpha \leq \frac{a}{2(1-\kappa)} \quad \text{and} \quad \sum_\alpha c_\alpha \leq \frac{b}{(2-\kappa)\sqrt{2}} ,$$

where $a = 2.7853$, $b = 2.55$. With the van Albada limiter we have $\kappa = 1/2$, so that we obtain the following stability condition:

$$\sum_\alpha c_\alpha \leq 1.2 . \tag{12.44}$$

Comparing (12.43) and (12.44), we see that both the first order upwind and $\kappa = 1/2$ scheme are stable if (12.44) is satisfied. As before, we extend this condition heuristically to the case of systems on boundary-fitted grids by replacing it by

$$\tau \sum_\alpha c_{\alpha j} \leq 1.2 , \quad \forall j ,$$

with $c_{\alpha j}$ defined before (12.40).

Final remarks

We have discussed explicit temporal discretization. Sometimes implicit schemes are more efficient than explicit schemes. For example, if the mesh size of the grid is locally very small, stability may require an extremely small time step for explicit schemes, making implicit schemes more efficient. Furthermore, if a stationary problem is to be solved, iterative solution of the stationary problem may be more efficient than time stepping to steady state. For a stationary problem or a time step with an implicit scheme, a complicated nonlinear algebraic system has to be solved. Doing this efficiently is not a straightforward matter. Much research is going on in this area. Due to the large number of unknows arising in typical applications and the sparsity of the associated matrices, iterative methods are to be preferred over direct methods. Advances in efficiency and turn-around time are being made by algorithmic improvements and the efficient use of high performance parallel computers. To limit the scope of the present work, we will not go into this subject. In Chap. 7 a brief survey of iterative methods has been given. The application of multigrid methods is computational fluid dynamics is reviewed in Chap. 9 of Wesseling (1992) and in Wesseling and Oosterlee (2000). An introduction to high performance parallel computing in fluid dynamics is given in Wesseling (1996a). Iterative methods related to domain decomposition are reviewed in Chan and Mathew (1994) and Smith, Bjørstad, and

Gropp (1996). In these publications the reader will find ample references to the literature.

Exercise 12.5.1. Show that (12.39) is necessary. Hint: take $\theta_\alpha = \pi$.

13. Numerical solution of the Navier-Stokes equations in general domains

13.1 Introduction

Most of the introductory remarks made in Sect. 12.1 apply to the present chapter as well, and will not be repeated. First, the compressible case will be discussed. The inviscid terms can be discretized as in Chap. 12, so that only the spatial discretization of the viscous terms needs to be given. Then the incompressible Navier-Stokes equations will be considered. The methods discussed in Chap. 6 will be generalized from Cartesian grids to structured boundary-fitted grids. A unified method for both the compressible and the incompressible case will be presented in the following chapter.

13.2 Analytic aspects

Governing equations

The compressible Navier-Stokes equations have been derived in Chap. 1. They consist of the equation of mass conservation (1.13), the momentum equations (1.24) and the energy equation (1.40). These equations can be gathered together in the following system, which is nothing but the Euler equations (1.76)–(1.78) with viscous and heat diffusion terms added:

$$U_t + \partial f^{(\alpha)}(U)/\partial x_\alpha + \partial g^{(\alpha)}(U)/\partial x_\alpha = Q , \quad \boldsymbol{x} \in \Omega , \quad t \in (0, T) , \quad (13.1)$$

where we sum over $\alpha \in \{1, 2, 3\}$, where U, f^α and Q are given by (12.2), and where the viscous flux $g^{(\alpha)}$ is given by

$$g^{(\alpha)} = - \begin{pmatrix} 0 \\ \sigma^{\alpha 1} \\ \sigma^{\alpha 2} \\ \sigma^{\alpha 3} \\ u^\beta \sigma^{\alpha \beta} + k T_{,\alpha} \end{pmatrix} .$$

Here $T_{,\alpha}$ stands for $\partial T / \partial x^\alpha$, and $\sigma^{\alpha\beta}$ is the stress tensor, defined as (cf.(1.25)):

$$\sigma^{\alpha\beta} = 2\mu(e^{\alpha\beta} - \frac{1}{3}\delta^{\alpha\beta}u^{\gamma}_{,\gamma}) , \quad e^{\alpha\beta} = \frac{1}{2}(u^{\alpha}_{,\beta} + u^{\beta}_{,\alpha}) .$$

We see that $\partial g^{(\alpha)}/\partial x^{\alpha}$ contains second derivatives of velocity and temperature, and it turns out that the spatial operator in (13.1) is elliptic, making (13.1) parabolic.

Boundary conditions

Usually the Reynolds number is large, and the viscous terms come into play only in small regions (e.g. boundary layers) with high gradients. As a consequence, the compressible Navier-Stokes equations behave very much like the (hyperbolic) Euler equations, and boundary conditions should be chosen such that we retain a well-posed problem when $\mu \downarrow 0$ and $k \downarrow 0$. We therefore apply as far as possible the same boundary conditions as for the Euler equations, plus some additional conditions to make the parabolic case well-posed. This implies that precisely four conditions are to be given at every point of the boundary.

At an inflow boundary we have to prescribe u and two thermodynamic quantities, for example ρ and p; T follows from the equation of state. Based on our experience with the convection-diffusion equation in Sect. 4.2 we apply homogeneous Neumann conditions on the stress and the temperature at an outflow boundary:

$$\tau^{\alpha\beta}n^{\beta} = -p_{\infty}n^{\alpha} , \quad n^{\alpha}T_{,\alpha} = 0 , \tag{13.2}$$

where n is the unit normal on the outflow boundary, p_{∞} is the outflow pressure, and

$$\tau^{\alpha\beta} = \sigma^{\alpha\beta} - p\delta^{\alpha\beta} .$$

At a solid boundary the no-slip condition is applied:

$$u = u_b ,$$

where u_b is the local speed with which the boundary moves; on a fixed object, $u_b = 0$. Furthermore, at a solid boundary a temperature condition is to be given. If the solid boundary is a perfect heat conductor, then the temperature of the fluid equals the temperature of the solid boundary:

$$T = T_b .$$

If the solid boundary is a perfect heat insulator (adiabatic wall), then we get a homogeneous Neumann condition at the boundary:

$$\partial T/\partial n = 0 .$$

If neither of the two above conditions apply, then the heat flow problem in the solid body has to be solved simultaneously with the flow problem, and the two problems are coupled by the condition of continuity of heat flux:

$$(\mu \partial T/\partial n)_f = (\mu \partial T/\partial n)_b \, ,$$

where the subscripts f and b refer to the fluid side and the body side of the solid boundary, respectively.

At first sight, the outflow condition (13.2) seems to present a problem in the case of supersonic outflow, since when $\mu \ll 1$ we effectively prescribe the pressure, which is forbidden in the Euler case. However, if the inviscid part of the equations is discretized with an approximate Riemann solver or a flux splitting scheme this presents no problem, because then the scheme is of fully upwind type near a supersonic outflow boundary, which implies that the scheme automatically disregards the outflow boundary condition.

For further remarks on boundary conditions, see Sect. 6.5.

13.3 Colocated scheme for the compressible Navier-Stokes equations

Integration over finite volume

Finite volume discretization is carried out in the same way as in Sect. 12.3. Integration over the cell Ω_j shown in Fig. 11.6 gives

$$|\Omega_j| \frac{dU_j}{dt} + \int_{\Gamma_j} (f^\alpha + g^\alpha) n_\alpha dS = |\Omega_j| Q_j \, .$$

The surface integral of f^α can be approximated in the same way as in Chap. 12, and does not need to be further considered here. We continue with the surface integral of g^α. Taking as an example the integral over the 'east' cell face, we write

$$\int_{\Gamma_E} g^\alpha n_\alpha dS \cong G_E^\alpha s_{\alpha E} \, ,$$

with s_E the cell face vector defined by (11.30), and G_E^α an approximation of g_E^α.

The path integral method

G_E^α has to be expressed in terms of surrounding cell center values of U. For this we need approximations of first order spatial derivatives of u^α and T

in terms of surrounding cell center values. A flexible and accurate way to do this is provided by the path integral method (van Beek, van Nooyen, and Wesseling (1995), Wesseling, Segal, Kassels, and Bijl (1998), Wesseling, Segal, and Kassels (1999)). Denoting the gradient of T by ∇T we have

$$b_1 \equiv T|_j^{j+2e_1} = \int_{\boldsymbol{x}_j}^{\boldsymbol{x}_{j+2e_1}} \nabla T \cdot d\boldsymbol{x} \cong \nabla T_{j+e_1} \cdot \boldsymbol{c}_{(1)}\,, \quad \boldsymbol{c}_{(1)} \equiv \boldsymbol{x}|_j^{j+2e_1}\,. \qquad (13.3)$$

Remembering that Γ_E is between Ω_j and Ω_{j+2e_1}, we see that ∇T_{j+e_1} is the quantity that we wish to approximate. We see that (13.3) gives a relation between ∇T and surrounding cell center values T_j and T_{j+2e_1}. In order to determine ∇T in three dimensions, we need two more such relations. These may be obtained by choosing two additional non-parallel integration paths. For accuracy, these are chosen symmetric with respect to $\boldsymbol{x}_{j+e_1}$. We choose:

$$b_2 \equiv T|_{j-2e_2}^{j+2e_2} + T|_{j+2e_1-2e_2}^{j+2e_1+2e_2} = \left\{ \int_{\boldsymbol{x}_{j-2e_2}}^{\boldsymbol{x}_{j+2e_2}} + \int_{\boldsymbol{x}_{j+2e_1-2e_2}}^{\boldsymbol{x}_{j+2e_1+2e_2}} \right\} \nabla T \cdot d\boldsymbol{x}$$

$$\cong \nabla T_{j+e_1} \cdot \boldsymbol{c}_{(2)}\,, \quad \boldsymbol{c}_{(2)} \equiv \boldsymbol{x}|_{j-2e_2}^{j+2e_2} + \boldsymbol{x}|_{j+2e_1-2e_2}^{j+2e_1+2e_2}\,.$$

Similarly,

$$b_3 \equiv T_{j-2e_3}^{j+2e_3} + T|_{j+2e_1-2e_3}^{j+2e_1+2e_3} = \left\{ \int_{\boldsymbol{x}_{j-2e_3}}^{\boldsymbol{x}_{j+2e_3}} + \int_{\boldsymbol{x}_{j+2e_1-2e_3}}^{\boldsymbol{x}_{j+2e_1-2e_3}} \right\} \nabla T \cdot d\boldsymbol{x}$$

$$\cong \nabla T_{j+e_1} \cdot \boldsymbol{c}_{(3)}\,, \quad \boldsymbol{c}_{(3)} \equiv \boldsymbol{x}|_{j-2e_3}^{j+2e_3} + \boldsymbol{x}|_{j+2e_1-2e_3}^{j+2e_1+2e_3}\,.$$

We now have three relations expressing ∇T in terms of surrounding cell center values of T. This system of three equations is solved for ∇T by defining

$$\boldsymbol{c}^{(\alpha)} \equiv \boldsymbol{c}_{(\beta)} \times \boldsymbol{c}_{(\gamma)}/C\,, \quad C \equiv \boldsymbol{c}_{(1)} \cdot (\boldsymbol{c}_{(2)} \times \boldsymbol{c}_{(3)})\,,$$

with α, β and γ cyclic. Then the solution for ∇T is given by

$$\frac{\partial T}{\partial x^\alpha} \cong c_\alpha^{(\beta)} b_\beta\,.$$

The same procedure is followed for the derivatives of u^β, and in fact the coefficients in the resulting 3×3 system are the same. If we redefine

$$b_1 \equiv u^\beta|_j^{j+2e_1}\,, \quad b_\gamma \equiv u^\beta|_{j-2e_\gamma}^{j+2e_\gamma} + u^\beta|_{j+2e_1-2e_\gamma}^{j+2e_1+2e_\gamma}\,, \quad \gamma = 2, 3,$$

then the solution is given by

$$\frac{\partial u^\beta}{\partial x^\alpha} \cong c_\alpha^{(\beta)} b_\beta\,.$$

For the viscosity and heat conduction coefficients one may take

$$\mu_{j+e_1} \cong \frac{1}{2}(\mu_j + u_{j+2e_1}) , \quad k_{j+e_1} \cong \frac{1}{2}(k_j + k_{j+2e_1}) .$$

This in essence completes the definition of the discretization of the compressible Navier-Stokes equations in the interior of the domain. Numerical implementation of boundary conditions presents no particular difficulties. Suppose for example that Γ_E is part of a boundary where $\xi^1 = $ constant. If $\partial T/\partial n$ is given at this boundary, then $\int_{\Gamma_e} \frac{\partial T}{\partial n} dS$ is given, and nothing needs to be done. If T is given at this boundary, then T_{j+e_1} is given, and the integration paths are modified by replacing $2e_1$ by e_1 in the above definitions of integration paths. Similar procedures can be followed for the momentum equations.

Time stepping

Stability of explicit schemes can only be analized in a approximate maner, in a similar way as in Sect. 12.5 for the inviscid case. A representative viscous term is added to the model equation (12.35), and stability of the resulting scheme is analysed in the same way as in Sect. 12.5. The model equation now is a convection-diffusion equation, and methods and results of Chap. 5 can be used.

The remarks made at the end of Chap. 12 about iterative solution methods for implicit schemes apply here also.

13.4 Colocated scheme for the incompressible Navier-Stokes equations

Governing equations

The governing equations can be written as follows in Cartesian tensor notation (cf. (6.1) and (6.14)):

$$u^{\alpha}_{,\alpha} = 0 , \tag{13.4}$$

$$\frac{\partial u^{\alpha}}{\partial t} + F^{\alpha\beta}_{,\beta} + p_{,\alpha} = 0 , \tag{13.5}$$

$$F^{\alpha\beta} = u^{\alpha}u^{\beta} - \sigma^{\alpha\beta} , \quad \sigma^{\alpha\beta} = \nu(u^{\alpha}_{,\beta} + u^{\beta}_{,\alpha}) .$$

Integration over finite volumes

All unknowns reside in the centers of the cells, as before. Integration of the continuity equation (13.4) over the cell Ω_j depicted in Fig. 11.6 gives

$$\int_{\Omega_j} u^\alpha_{,\alpha} d\Omega = \int_{\Gamma_j} u^\alpha n_\alpha dS = 0 \,,$$

with n the unit outer normal. The surface integral is evaluated by approximating u^α by its value at the center of the cell face. As an example, we take the integral over the 'east' cell face, i.e. the cell face that separates Ω_j and Ω_{j+2e_1} (in the notation of Chap. 11). This part of Γ_j is denoted by Γ_E. We get

$$\int_{\Gamma_E} u^\alpha n_\alpha dS \cong u^\alpha_E \int_{\Gamma_E} n_\alpha dS = u^\alpha_E s_{\alpha E} \,, \tag{13.6}$$

with the cell face vector s_E defined by (11.31). For the 'west' face we get, taking the sign convection for the cell face vectors s into account,

$$\int_{\Gamma_W} u^\alpha n_\alpha dS \cong -u^\alpha_W s_{\alpha W} \,.$$

The cell face values of u^α have to be approximated in terms of surrounding cel center values u^α_j. We will come back to this later.

For the momentum equations (13.5) we obtain

$$|\Omega_j| \frac{du^\alpha_j}{dt} + \int_{\Gamma_j} F^{\alpha\beta} n_\beta dS + \int_{\Gamma_j} p n_\alpha dS = 0 \,.$$

The pressure integral over Γ_E is approximated as follows:

$$\int_{\Gamma_E} p n_\alpha dS \cong p_E s^\alpha_E \,,$$

and p_E is evaluated as follows:

$$p_E \cong \frac{1}{2}(p_j + p_{j+2e_1}) \,.$$

For the viscous terms we have to approximate integrals, as for example

$$\int_{\Gamma_E} \nu u^\alpha_{,\beta} n_\beta d\Gamma \,.$$

This is done as follows:

$$\int_{\Gamma_E} \nu u^{\alpha}_{,\beta} n_\beta d\Gamma \cong \nu_E(u^{\alpha}_{,\beta})_E s^{\beta}_E \ .$$

Derivatives of u^α are approximated in terms of surrounding cell center values by the path integral method explained in Sect. 13.3.

The inertia terms are approximated as follows:

$$\int_{\Gamma_E} u^\alpha u^\beta n_\beta dS \cong u^{\alpha}_E u^{\beta}_E s^{\beta}_E \ . \tag{13.7}$$

Here u^{β}_E must be expressed in terms of surrounding cell center values by the same method that is to be used for u^{α}_E in the continuity equation (13.6). For u^{α}_E in (13.7) we can use the usual approximations for convection. For example, the second order central scheme is obtained with

$$u^{\alpha}_E \cong \frac{1}{2}(u^{\alpha}_j + u^{\alpha}_{j+2e_1}) \ .$$

The first order upwind scheme is obtained with

$$u^{\alpha}_E \cong u^{\alpha}_j \quad \text{if} \quad u^{\beta}_E s^{\beta}_E \geq 0 \ ,$$
$$u^{\alpha}_E \cong u^{\alpha}_{j+2e_1} \quad \text{if} \quad u^{\beta}_E s^{\beta}_E < 0 \ .$$

The κ-scheme is obtained with (cf. (9.96)):

$$u^{\alpha}_E \cong \frac{1}{2}(u^{\alpha}_j + u^{\alpha}_{j+2e_1}) + \frac{1-\kappa}{4}(-u^{\alpha}_{j-2e_1} + 2u^{\alpha}_j - u^{\alpha}_{j+2e_1})$$

if $u^{\beta}_E s^{\beta}_E \geq 0$, and in the opposite case

$$u^{\alpha}_E \cong \frac{1}{2}(u^{\alpha}_j + u^{\alpha}_{j+2e_1}) + \frac{1-\kappa}{4}(-u^{\alpha}_j + 2u^{\alpha}_{j+2e_1} - u^{\alpha}_{j+4e_1}) \ .$$

In the case of a Cartesian grid, this scheme gives rise to spurious checkerboard modes, as shown in Sect. 6.3, where various remedies are discussed. The most popular of these is the pressure-weighted interpolation method of Rhie and Chow (1983), which is discussed in great detail for the case of arbitrary grids in Sect. 8.8 of Ferziger and Perić (1996). This implies approximation of u^{β}_E in terms of surrounding cell center values including pressure values, analogous to (6.25). We will not further discuss this method here. Time stepping methods have been discussed in Sect. 6.7. At the end of Sect. 6.3, references have been given to a large number of publications in which applications of colocated schemes to computation of incompressible flows are discussed.

13.5 Staggered scheme for the incompressible Navier-Stokes equations

We think that nobody will dispute, that on Cartesian grids, computation of incompressible flows is best performed with the staggered scheme proposed by Harlow and Welch (1965) and discussed in Chap. 6. In combination with the pressure-correction method (Sect. 6.6) an efficient and accurate method to compute nonstationary flows is obtained. The physical boundary conditions suffice. The main advantage over colocated schemes is, that no artificial measures are needed to rule out spurious oscillations.

However, there is no such consensus when the grid is arbitrary. In fact, the use of colocated schemes is more prevalent. The reason is that generalization from the Cartesian to the general case is relatively straightforward, as seen in the preceding section. Staggered schemes are more complicated, and may suffer from accuracy problems on nonuniform grids. If the smoothness properties of the boundary-fitted coordinate mapping are not carefully taken into account, or if too much smoothness is assumed, the accuracy can become bad, even on mildly nonuniform grids. This experience has led to a widespread opinion that on curvilinear grids staggered schemes are inherently les accurate than colocated schemes. But this is not so.

On colocated grids it is customary to discretize in physical space, as we did in the preceding section, and no reference is made to the coordinate mapping, so that its smoothness properties do not come into play, and no serious degradation of accuracy is observed as the grid becomes less smooth. Staggered discretization may also be carried out in physical space; this is done in Rosenfeld, Kwak, and Vinokur (1991). But staggered discretization in physical space puts a heavy demand on geometric insight and pictorial representation, which is why we will develop an algebraic formulation (cf. Wesseling, Segal, and Kassels (1999)). Furthermore, we think it desirable to bring out explicitly the role of the smoothness properties of the coordinate mapping.

Geometric quantities and their smoothness properties

A described in Sect. 11.3, the grid is generated by a piecewise trilinear coordinate mapping

$$\boldsymbol{x}_j = \boldsymbol{x}(\boldsymbol{\xi}) , \quad \boldsymbol{x}_j \in \Omega , \quad \boldsymbol{\xi} \in G_h .$$

Fig. 13.1 shows a cell in the physical space Ω and the computational space G. This is the same as Fig. 11.6, repeated here for convenience. For mnemonic convenience, the cell centers and cell face centers ara indicated by subscripts C, N, S, E, W, F, R, with C for center, F for front, R for rear, N for north etc. Hence, with the indexing system of Sect. 11.3,

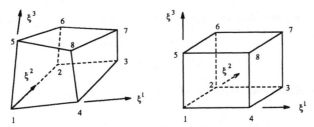

Fig. 13.1. Three-dimensional cel. Its image in G is rectangular hexahedron.

$$\boldsymbol{x}_C = \boldsymbol{x}_j , \quad \boldsymbol{x}_F = \boldsymbol{x}_{j-e_2} , \quad \boldsymbol{x}_E = \boldsymbol{x}_{j+e_1} ,$$

for example. With the vertex numbering of Fig. 13.1 we have for example

$$\boldsymbol{x}_C = \frac{1}{8} \sum_1^8 \boldsymbol{x}_m , \quad \boldsymbol{x}_F = \frac{1}{4}(\boldsymbol{x}_1 + \boldsymbol{x}_4 + \boldsymbol{x}_5 + \boldsymbol{x}_8) .$$

We know from Sect. 11.4 that the points $\boldsymbol{x}_N, ...$ belong to the doubly-ruled cell faces.

The contravariant and covariant base vectors $\boldsymbol{a}^{(\alpha)}$ and $\boldsymbol{a}_{(\alpha)}$ have been defined in Sect. 11.5. We repeat for convenience:

$$\boldsymbol{a}^{(\alpha)} \equiv \nabla \xi^\alpha , \quad \boldsymbol{a}_{(\alpha)} = \frac{\partial \boldsymbol{x}}{\partial \xi^\alpha} .$$

It follows that the Cartesian components of these vectors are given by

$$a_\beta^{(\alpha)} = \frac{\partial \xi^\alpha}{\partial x^\beta} , \quad a_{(\alpha)}^\beta = \frac{\partial x^\beta}{\partial \xi^\alpha} .$$

In this section the language of tensor analysis (Sect. 11.5) will be used. Therefore the location of an index (subscript or superscript) matters, and summation is implied over pairs of Greek indices, one of which must be a superscript and one a subscript.

Because $\boldsymbol{x} = \boldsymbol{x}(\boldsymbol{\xi})$ is piecewise trilinear, $\boldsymbol{a}_{(\alpha)}$ is piecewise continuous and does not exist everywhere. At a cell face $\xi^\alpha =$constant, $\boldsymbol{a}_{(\alpha)}$ is discontinuous and will not be used, but $\boldsymbol{a}_{(\beta)}$, $\beta \neq \alpha$ is continuous. For economy we need to express $\boldsymbol{a}_{(\alpha)}$ is terms of the cell vertex locations $x_1, ..., x_8$ in an efficient way. The derivatives $\partial \boldsymbol{x}/\partial \xi^\alpha$ are given by (11.25). In C we have $(s^1, s^2, s^3) = (0,0,0)$; in N we have $(s^1, s^2, s^3) = (0,0,1)$ etc. After some algebra we find from (11.25), (11.26), (11.19) and (11.21)

$$\boldsymbol{a}_{(1)C} = (\boldsymbol{x}_E - \boldsymbol{x}_W)/\Delta\xi^1 , \quad \boldsymbol{a}_{(2)C} = (\boldsymbol{x}_R - \boldsymbol{x}_F)/\Delta\xi^2 ,$$
$$\boldsymbol{a}_{(3)C} = (\boldsymbol{x}_N - \boldsymbol{x}_S)/\Delta\xi^3 ,$$
$$\boldsymbol{a}_{(2)W} = \frac{1}{2\Delta\xi^2}(\boldsymbol{x}_{26} - \boldsymbol{x}_{15}) , \quad \boldsymbol{a}_{(3)W} = \frac{1}{2\Delta\xi^3}(\boldsymbol{x}_{56} - \boldsymbol{x}_{12}) ,$$
$$\boldsymbol{a}_{(1)F} = \frac{1}{2\Delta\xi^1}(\boldsymbol{x}_{48} - \boldsymbol{x}_{15}) , \quad \boldsymbol{a}_{(3)F} = \frac{1}{2\Delta\xi^3}(\boldsymbol{x}_{58} - \boldsymbol{x}_{14}) ,$$
$$\boldsymbol{a}_{(1)S} = \frac{1}{2\Delta\xi^1}(\boldsymbol{x}_{34} - \boldsymbol{x}_{12}) , \quad \boldsymbol{a}_{(2)S} = \frac{1}{2\Delta\xi^2}(\boldsymbol{x}_{23} - \boldsymbol{x}_{14})$$

etc., where we have used the notation $\boldsymbol{x}_{mn} = \boldsymbol{x}_m + \boldsymbol{x}_n$. For the contravariant base vectors we have

$$\boldsymbol{a}^{(\alpha)} \cdot \boldsymbol{a}_{(\beta)} = \delta_\beta^\alpha,$$

with δ_β^α the Kronecker delta. Solving gives (cf. (11.53)):

$$\boldsymbol{a}^{(\alpha)} = \frac{1}{\sqrt{g}}(\boldsymbol{a}_{(\beta)} \times \boldsymbol{a}_{(\gamma)}), \quad \alpha, \beta, \gamma \ \text{cyclic},$$

where $\sqrt{g}$ is given by

$$\sqrt{g} = \boldsymbol{a}_{(\alpha)} \cdot (\boldsymbol{a}_{(\beta)} \times \boldsymbol{a}_{(\gamma)}), \quad \alpha, \beta, \gamma \ \text{cyclic}.$$

From the smoothness properties of $\boldsymbol{a}_{(\alpha)}$ it follows that $\sqrt{g}$ and $\boldsymbol{a}^{(\alpha)}$ are discontinuous at cell faces. But $\sqrt{g}\boldsymbol{a}^{(\alpha)}$ turns out to be continuous at cell faces of the type $\xi^\alpha = $ constant. After some algebra we find that in the cell face centers where $\sqrt{g}a^\alpha$ is continuous, $\sqrt{g}\boldsymbol{a}^{(\alpha)}$ equals the corresponding cell face vector times a scaling factor:

$$
\begin{aligned}
(\sqrt{g}\boldsymbol{a}^{(1)})_{W,E} &= \tfrac{1}{\Delta\xi^2\Delta\xi^3}\boldsymbol{s}_{W,E} , \quad (\sqrt{g}\boldsymbol{a}^{(2)})_{R,F} = \tfrac{1}{\Delta\xi^1\Delta\xi^3}\boldsymbol{s}_{R,F} , \\
(\sqrt{g}\boldsymbol{a}^{(3)})_{N,S} &= \tfrac{1}{\Delta\xi^1\Delta\xi^2}\boldsymbol{s}_{N,S} .
\end{aligned}
\tag{13.8}
$$

Staggered representation of the velocity field

We want to generalize the classical staggered Marker-and-Cell (MAC) scheme proposed in Harlow and Welch (1965) from Cartesian to general coordinates. This means that we wish to compute the pressure in cell centers, and normal velocity components in cell face centers. It follows from (13.8) that the contravariant base vector $\boldsymbol{a}^{(\alpha)}$ is perpendicular to faces of the type $\xi^\alpha = $ constant, so at first sight the contravariant velocity components U^α, defined as (cf. (11.46))

$$U^\alpha = \boldsymbol{u} \cdot \boldsymbol{a}^{(\alpha)}$$

would seem to be suitable for representing the velocity field $\boldsymbol{u}$. But we just saw that $\boldsymbol{a}^{(\alpha)}$ is discontinuous at cell faces. As a consequence, the use of U^α leads to bad accuracy on rough grids. But, as we saw, $\sqrt{g}\boldsymbol{a}^{(\alpha)}$ is continuous at cell faces where $\xi^\alpha = $ constant. Therefore the following coordinate-invariant staggered representation of the velocity field $\boldsymbol{u}$ will be employed (from now on denoting $\boldsymbol{x}_E$ by $\boldsymbol{x}_{j+e_1}$, etc.):

$$V_{j+e_\alpha}^\alpha = (\sqrt{g}\boldsymbol{a}^{(\alpha)} \cdot \boldsymbol{u})_{j+e_\alpha} \quad \text{(no summation)}.$$

From (13.8) it follows that we may also write

$$V_{j+e_\alpha}^\alpha = (s \cdot u)_{j+e_\alpha} / \Delta\xi^\beta \Delta\xi^\gamma, \quad \alpha, \beta, \gamma \text{ cyclic.}$$

This shows that $V^\alpha \Delta\xi^\beta \Delta\xi^\gamma$ approximates the volume flux through the cell face. Therefore V^α will be called the volume flux components. Equation (11.31) gives us an efficient formula for $\sqrt{g}a^{(\alpha)}$, for example (cf. Fig. 13.1):

$$(\sqrt{g}a^{(1)})_{j+e_1} = \tfrac{1}{2}(x_7 - x_4) \times (x_8 - x_3)/\Delta\xi^2 \Delta\xi^3 .$$

We will need to aproximate V^α and u not only in the cell face centers, but in other points as well. The general relation between V^α and u is (using (11.46), (11.47)):

$$V^\alpha = \sqrt{g}a^{(\alpha)} \cdot u, \quad u = a_{(\alpha)}V^\alpha/\sqrt{g}. \tag{13.9}$$

Finding u requires evaluation of V^α in points other than its proper grid nodes. Because of lack of smoothness of the geometric quantities, evaluation of V^α in other points than its proper grid nodes, and evaluation of u needs to be done carefully. A certain tedium is unavoidable. We impose the following accuracy requirement on formulas for defining V^α in other points than its proper grid nodes x_{j+e_α}: constant velocity fields u must be invariant under transformation to V^α-representation and back. More precisely, if u is given, and $V_{j+e_\alpha}^\alpha$ is computed with (13.9), and V^α is determined in another point by some interpolation recipe, and u is computed in this point with (13.9), then the original u is recovered exactly in the special case that $u = $ constant. Furthermore, in the case of the identity mapping $x = \xi$, we will have multilinear interpolation. We have found these requirements to be essential for maintaining accuracy on rough grids.

In the cell centers x_j we define (no summation):

$$V_j^\alpha \equiv \tfrac{1}{2}(V_{j-e_\alpha}^\alpha + V_{j+e_\alpha}^\alpha), \tag{13.10}$$

$$(\sqrt{g}a^{(\alpha)})_j \equiv \tfrac{1}{2}\{(\sqrt{g}a^{(\alpha)})_{j-e_\alpha} + (\sqrt{g}a^{(\alpha)})_{j+e_\alpha}\}, \tag{13.11}$$

$$(\frac{1}{\sqrt{g}}a_{(\alpha)})_j \equiv \left\{ \frac{\sqrt{g}a^{(\beta)} \times \sqrt{g}a^{(\gamma)}}{\sqrt{g}a^{(1)} \cdot (\sqrt{g}a^{(2)} \times \sqrt{g}a^{(3)})} \right\}_j, \quad \alpha, \beta, \gamma \text{ cyclic.} \tag{13.12}$$

We show that the above invariance requirement is met. Suppose $u = $ constant. Then (13.10), (13.11) and (13.12) give

$$V_j^\alpha = (\sqrt{g}a^{(\alpha)})_j \cdot u .$$

Recomputing u_j from V_j^α gives, using (13.9) and (13.12):

$$u_j = (\frac{1}{\sqrt{g}}a_{(\alpha)})_j(\sqrt{g}a^{(\alpha)})_j \cdot u = u \tag{13.13}$$

(cf. Exercise 13.5.1), which we wanted to show.

In the cell face centers x_{j+e_α} we define for $\beta \neq \alpha$:

$$V_{j+e_\alpha}^\beta \equiv \tfrac{1}{4}(V_{j+e_\beta}^\beta + V_{j-e_\beta}^\beta + V_{j+2e_\alpha+e_\beta}^\beta + V_{j+2e_\alpha-e_\beta}^\beta) , \qquad (13.14)$$

$$\begin{aligned}
(\sqrt{g}a^{(\beta)})_{j+e_\alpha} \equiv \tfrac{1}{4}\{ & (\sqrt{g}a^{(\beta)})_{j+e_\beta} + (\sqrt{g}a^{(\beta)})_{j-e_\beta} \\
& + (\sqrt{g}a^{(\beta)})_{j+2e_\alpha+e_\beta} + (\sqrt{g}a^{(\beta)})_{j+2e_\alpha-e_\beta}\}
\end{aligned} \qquad (13.15)$$

(no summation), and $(a_{(\alpha)}/\sqrt{g})_{j+e_\alpha}$ is related to $(\sqrt{g}a^{(\beta)})_{j+e_\alpha}$ by replacing j by $j + e_\alpha$ in (13.12). In a similar way as before it is easily seen that the invariance requirement is satisfied.

In the cell edge centers $x_{j+e_\alpha+e_\beta}$, $\beta \neq \alpha$ we define (no summation):

$$\begin{aligned}
V_{j+e_\alpha+e_\beta}^\alpha &\equiv \tfrac{1}{2}(V_{j+e_\alpha}^\alpha + V_{j+e_\alpha+2e_\beta}^\alpha) , \\
(\sqrt{g}a^{(\alpha)})_{j+e_\alpha+e_\beta} &\equiv \tfrac{1}{2}\{(\sqrt{g}a^{(\alpha)})_{j+e_\alpha} + (\sqrt{g}a^{(\alpha)})_{j+e_\alpha+2e_\beta}\} , \\
V_{j+e_\alpha+e_\beta}^\beta &\equiv \tfrac{1}{2}(V_{j+e_\beta}^\beta + V_{j+2e_\alpha+e_\beta}^\beta) , \\
(\sqrt{g}a^{(\beta)})_{j+e_\alpha+e_\beta} &\equiv \tfrac{1}{2}\{(\sqrt{g}a^{(\beta)})_{j+e_\beta} + (\sqrt{g}a^{(\beta)})_{j+2e_\alpha+e_\beta}\} ,
\end{aligned} \qquad (13.16)$$

and for $\gamma \neq \alpha, \beta$

$$\begin{aligned}
V_{j+e_\alpha+e_\beta}^\gamma \equiv \tfrac{1}{8}(& V_{j-e_\gamma}^\gamma + V_{j+e_\gamma}^\gamma + V_{j-e_\gamma+2e_\alpha}^\gamma + V_{j+e_\gamma+2e_\alpha}^\gamma + V_{j-e_\gamma+2e_\beta}^\gamma \\
& + V_{j+e_\gamma+2e_\beta}^\gamma + V_{j-e_\gamma+2e_\alpha+2e_\beta}^\gamma + V_{j+e_\gamma+2e_\alpha+2e_\beta}^\gamma) , \\
(\sqrt{g}a^{(\gamma)})_{j+e_\alpha+e_\beta} \equiv \tfrac{1}{8}\{ & (\sqrt{g}a^{(\gamma)})_{j-e_\gamma} + (\sqrt{g}a^{(\gamma)})_{j+e_\gamma} + (\sqrt{g}a^{(\gamma)})_{j-e_\gamma+2e_\alpha} \\
& + (\sqrt{g}a^{(\gamma)})_{j+e_\gamma+2e_\alpha} + (\sqrt{g}a^{(\gamma)})_{j-e_\gamma+2e_\beta} \\
& + (\sqrt{g}a^{(\gamma)})_{j+e_\gamma+2e_\beta} \\
& + (\sqrt{g}a^{(\gamma)})_{j-e_\gamma+2e_\alpha+2e_\beta} + (\sqrt{g}a^{(\gamma)})_{j+e_\gamma+2e_\alpha+2e_\beta}\} ,
\end{aligned}$$

and $(a_{(\gamma)}/\sqrt{g})_{j+e_\alpha+e_\beta}$ is related to $(\sqrt{g}a^{(\gamma)})_{j+e_\alpha+e_\beta}$ by replacing j by $j+e_\alpha + e_\beta$ in (13.12). Again, it is easily seen that the invariance requirement is satisfied.

Furthermore, u is required at the cell vertices $x_{j+e_1+e_2+e_3}$. The way to proceed is clear from the preceding cases. We define ($\alpha \neq \beta$, $\alpha \neq \gamma$, $\beta \neq \gamma$, no summation):

$$\begin{aligned}
V_{j+e_1+e_2+e_3}^\alpha &\equiv \tfrac{1}{4}(V_{j+e_\alpha}^\alpha + V_{j+e_\alpha+2e_\beta}^\alpha + V_{j+e_\alpha+2e_\gamma}^\alpha + V_{j+e_\alpha+2e_\beta+2e_\gamma}^\alpha) , \\
(\sqrt{g}a^{(\alpha)})_{j+e_2+e_3} &\equiv \tfrac{1}{4}\{(\sqrt{g}a^{(\alpha)})_{j+e_\alpha} + (\sqrt{g}a^{(\alpha)})_{j+e_\alpha+2e_\beta} \\
& \quad + (\sqrt{g}a^{(\alpha)})_{j+e_\alpha+2e_\gamma} \\
& \quad + (\sqrt{g}a^{(\alpha)})_{j+e_\alpha+2e_\beta+2e_\gamma}\} ,
\end{aligned}$$

and $(a_{(\alpha)}/\sqrt{g})$ is related to $(\sqrt{g}a^{(\alpha)})$ in $x_{j+e_1+e_2+e_3}$ by (13.12) with appropriate grid point indices.

Coordinate invariant discretization of the incompressible Navier-Stokes equations

Finite volume integration of the continuity equation div $\boldsymbol{u} = 0$ over Ω_j gives

$$
\begin{aligned}
0 &= \int\limits_{\Omega_j} \operatorname{div}\boldsymbol{u}d\Omega = \int\limits_{\Gamma_j} \boldsymbol{u}\cdot\boldsymbol{n}d\Gamma = \sum_{\alpha=1}^{3}(\boldsymbol{u}\cdot\boldsymbol{s})|_{j-e_\alpha}^{j+e_\alpha} \\
&= \sum_{\alpha=1}^{3} \Delta\xi^\beta \Delta\xi^\gamma V^\alpha|_{j-e_\alpha}^{j+e_\alpha}, \quad \alpha,\beta,\gamma \text{ cyclic}.
\end{aligned}
\tag{13.17}
$$

For robustness, the discretization scheme should be coordinate invariant. This is the case for (13.17), because it contains a contravariant representation of the velocity field.

A straightforward way to proceed would be to discretize the momentum equations written in coordinate invariant form, using tensor analysis. However, in this formulation the so-called Christoffel symbols occur, encountered in Sect. 11.5. These involve second derivatives of the mapping $\boldsymbol{x} = \boldsymbol{x}(\boldsymbol{\xi})$. Because the mapping is piecewise bilinear, the Christoffel symbols are 'infinite' at cell edges. Approximation of the Christoffel symbols by straightforward finite differences gives reasonable results on smooth grids only. This is perhaps what has led to a widespread belief that staggered discretization is inaccurate in general coordinates. In order to avoid this difficulty with the Christoffel symbols we first transform only the independent variables, obtaining a form that is not coordinate invariant, that is discretized and used as a stepping stone to arrive at a coordinate invariant discretization.

The derivative of some quantity φ transforms according to

$$
\frac{\partial\varphi}{\partial x^\alpha} = a_\alpha^{(\beta)} \frac{\partial\varphi}{\partial\xi^\beta} .
\tag{13.18}
$$

Using the identity (cf. Exercise 11.5.19)

$$
\frac{\partial}{\partial\xi^\alpha}(\sqrt{g}\boldsymbol{a}^{(\alpha)}) = 0 ,
$$

this can be rewritten as

$$
\frac{\partial\varphi}{\partial x^\alpha} = \frac{1}{\sqrt{g}} \frac{\partial}{\partial\xi^\beta}(\sqrt{g}a_\alpha^{(\beta)}\varphi) .
\tag{13.19}
$$

The momentum equations can be written as

$$
\frac{\partial\boldsymbol{u}}{\partial t} + \frac{\partial}{\partial x^\alpha}(u^\alpha\boldsymbol{u}) = -\nabla p + \frac{\partial\nu e^{(\alpha)}}{\partial x^\alpha} ,
\tag{13.20}
$$

where

$$e^{(\alpha)} \equiv \begin{pmatrix} \dfrac{\partial u^1}{\partial x^\alpha} + \dfrac{\partial u^\alpha}{\partial x^1} \\[2mm] \dfrac{\partial u^2}{\partial x^\alpha} + \dfrac{\partial u^\alpha}{\partial x^2} \\[2mm] \dfrac{\partial u^3}{\partial x^\alpha} + \dfrac{\partial u^\alpha}{\partial x^3} \end{pmatrix} .$$

By applying (13.19), equation (13.20) can be rewritten as

$$N(\boldsymbol{u},p) \equiv \frac{\partial \boldsymbol{u}}{\partial t} + \frac{1}{\sqrt{g}} \frac{\partial}{\partial \xi^\alpha}(\boldsymbol{u}V^\alpha) + \nabla p - \frac{1}{\sqrt{g}} \frac{\partial}{\partial \xi^\alpha}(\nu \sqrt{g} a_\beta^{(\alpha)} e^{(\beta)}) = 0 .$$

$$(13.21)$$

This equation is integrated over the shifted finite volume Ω_{j+e_1} depicted in Fig. 13.2. Treating each term successively, we obtain

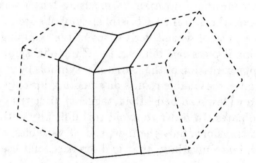

Fig. 13.2. Shifted finite volume Ω_{j+e_1}

$$\int_{\Omega_{j+e_1}} \frac{\partial \boldsymbol{u}}{\partial t} d\Omega = |\Omega_{j+e_1}| d\boldsymbol{u}_{j+e_1}/dt , \qquad (13.22)$$

$$\int_{\Omega_{j+e_1}} \frac{1}{\sqrt{g}} \frac{\partial}{\partial \xi^\alpha}(\boldsymbol{u}V^\alpha) d\Omega = \int_{G_{j+e_1}} \frac{\partial}{\partial \xi^\alpha}(\boldsymbol{u}V^\alpha) d\xi^1 d\xi^2 d\xi^3 .$$

We have

$$\boldsymbol{I}_{11} \equiv \int_{G_{j+e_1}} \frac{\partial}{\partial \xi^1}(\boldsymbol{u}V^1) d\xi^1 d\xi^2 d\xi^3 = \int_{\xi_{j_2-1/2}^2}^{\xi_{j_2+1/2}^2} \int_{\xi_{j_3-1/2}^3}^{\xi_{j_3+1/2}^3} (\boldsymbol{u}V^1)|_{j_1}^{j_1+1} d\xi^2 d\xi^3 ,$$

where the fact that $\boldsymbol{u}V^1$ is continuous in G_{j+e_1} has been used. This integral is approximated by the midpoint rule:

$$\boldsymbol{I}_{11} \cong \Delta\xi^2 \Delta\xi^3 (\boldsymbol{u}V^1)|_j^{j+2e_1} . \tag{13.23}$$

This is approximated further by using

$$V_j^1 \cong \tfrac{1}{2}(V_{j-e_1}^1 + V_{j+e_1}^1) .$$

which contains V^1 in its proper nodes. If a second order central scheme is to be used, then $\boldsymbol{u}$ is expressed in terms of V^α in the manner described before. If a first order upwind scheme is to be used, then we write

$$\boldsymbol{u}_j = \boldsymbol{u}_{j-e_1} \quad \text{if} \quad V_j^1 \geq 0 , \quad \boldsymbol{u}_j = \boldsymbol{u}_{j+e_1} \quad \text{if} \quad V_j^1 < 0 .$$

Similarly,

$$\boldsymbol{I}_{12} \equiv \int\limits_{G_{j+e_1}} \frac{\partial}{\partial\xi^2}(\boldsymbol{u}V^2)d\xi^1 d\xi^2 d\xi^3 \cong \Delta\xi^1 \Delta\xi^3 (\boldsymbol{u}V^2)|_{j+e_1-e_2}^{j+e_1+e_2} .$$

This is further approximated using

$$V_{j+e_1+e_2}^2 \cong \tfrac{1}{2}(V_{j+e_2}^2 + V_{j+2e_1+e_2}^2) ,$$

and $\boldsymbol{u}$ is handled in a similar way as before, using a shift in the ξ^2-direction in the upwind case. Analogously,

$$\boldsymbol{I}_{13} \equiv \int\limits_{G_{j+e_1}} \frac{\partial}{\partial\xi^3}(\boldsymbol{u}V^3)d\xi^1 d\xi^2 d\xi^3 \cong \Delta\xi^1 \Delta\xi^2 (\boldsymbol{u}V^3)|_{j+e_1-e_3}^{j+e_1+e_3} , \tag{13.24}$$

$$V_{j+e_1+e_3}^3 \cong \tfrac{1}{2}(V_{j+e_3}^3 + V_{j+2e_1+e_3}^3) ,$$

and $\boldsymbol{u}$ is handled as above. Integration of the pressure term is done as follows:

$$\boldsymbol{I}_{14} \equiv \int\limits_{\Omega_{j+e_1}} \nabla p \, d\Omega \cong \nabla p_{j+e_1} |\Omega_{j+e_1}| ,$$

where we have used smoothness of ∇p. The term ∇p_{j+e_1} is expressed in terms of surrounding nodal values using the path integral method (Sect. 13.3). We write, using smoothness of ∇p,

$$b_1 \equiv p|_j^{j+2e_1} = \int\limits_{\boldsymbol{x}_j}^{\boldsymbol{x}_{j+2e_1}} \nabla p \cdot d\boldsymbol{x} \cong \nabla p_{j+e_1} \cdot \boldsymbol{c}_{(1)}, \quad \boldsymbol{c}_{(1)} \equiv \boldsymbol{x}|_j^{j+2e_1} . \tag{13.25}$$

Similarly,

$$b_2 \equiv p|_{j-2e_2}^{j+2e_2} + p|_{j+2e_1-2e_2}^{j+2e_1+2e_2} = \left\{ \int\limits_{\boldsymbol{x}_{j-2e_2}}^{\boldsymbol{x}_{j+2e_2}} + \int\limits_{\boldsymbol{x}_{j+2e_1-2e_2}}^{\boldsymbol{x}_{j+2e_1+2e_2}} \right\} \nabla p \cdot d\boldsymbol{x} \cong \nabla p_{j+e_1} \cdot \boldsymbol{c}_{(2)} ,$$

$$c_{(2)} \equiv x|_{j-2e_2}^{j+2e_2} + x|_{j+2e_1-2e_2}^{j+2e_1+2e_2} .$$ (13.26)

Using the third direction we obtain similarly

$$b_3 \equiv p|_{j-2e_3}^{j+2e_3} + p|_{j+2e_1-2e_3}^{j+2e_1+2e_3} \cong \nabla p_{j+e_1} \cdot c_{(3)} ,$$

$$c_{(3)} \equiv x|_{j-2e_3}^{j+2e_3} + x|_{j+2e_1-2e_3}^{j+2e_1+2e_3} .$$ (13.27)

We now have three equations expressing ∇p in terms of surrounding nodal values, without any assumption about the smoothness of the coordinate mapping. The system is solved for ∇p by defining

$$c^{(\alpha)} \equiv c_{(\beta)} \times c_{(\gamma)}/C, \quad C \equiv c_{(1)} \cdot (c_{(2)} \times c_{(3)}) .$$ (13.28)

Then the solution of the above three equations for ∇p is

$$\frac{\partial p}{\partial x_\alpha} \cong c_\alpha^{(\beta)} b_\beta .$$

Integration of the viscous term over Ω_{j+e_1} gives three contributions. The first is

$$I_{15} \equiv \int\limits_{G_{j+e_1}} \frac{\partial}{\partial \xi_1}(\sqrt{g}a_\beta^{(1)}\nu e^{(\beta)})d\xi^1 d\xi^2 d\xi^3$$

$$= \int\limits_{\xi_{j_2-1/2}^2}^{\xi_{j_2+1/2}^2} \int\limits_{\xi_{j_3-1/2}^3}^{\xi_{j_3+1/2}^3} (\sqrt{g}a_\beta^{(1)}\nu e^{(\beta)})|_{j_1}^{j_1+1}d\xi^2 d\xi^3 ,$$

where $\sqrt{g}a_\beta^{(1)}$ is constant. We write

$$I_{15} \cong \Delta\xi^2 \Delta\xi^3(\sqrt{g}a_\beta^{(1)}\nu e^{(\beta)})|_j^{j+2e_1} .$$

The second contribution is

$$I_{16} \equiv \int\limits_{\xi_{j_1}^1}^{\xi_{j_1+1}^1} \int\limits_{\xi_{j_3-1/2}^3}^{\xi_{j_3+1/2}^3} (\sqrt{g}a_\beta^{(2)}\nu e^{(\beta)})|_{j_2-1/2}^{j_2+1/2}d\xi^1 d\xi^3 ,$$

where $\sqrt{g}a^{(2)}$ is piecewise constant. We make the following approximation:

$$I_{16} \cong \Delta\xi^1 \Delta\xi^3(\sqrt{g}a_\beta^{(2)}\nu e^{(\beta)})|_{j+e_1-e_2}^{j+e_1+e_2} ,$$

where we define

$$(\sqrt{g}a^{(2)})_{j+e_1\pm e_2} \equiv \tfrac{1}{2}\{(\sqrt{g}a^{(2)})_{j\pm e_2} + (\sqrt{g}a^{(2)})_{j+2e_1\pm e_2}\} .$$

The third contribution is I_{17}, which is handled just like I_{16}:

$$I_{17} \equiv \int_{\xi_{j_1}^1}^{\xi_{j_1+1}^1} \int_{\xi_{j_2-1/2}^2}^{\xi_{j_2+1/2}^2} (\sqrt{g}a_\beta^{(3)}\nu e^{(\beta)})|_{j_3-1/2}^{j_3+1/2} d\xi^1 d\xi^2$$
$$\cong \Delta\xi^1 \Delta\xi^2 (\sqrt{g}a_\beta^{(3)}\nu e^{(\beta)})|_{j+e_1-e_3}^{j+e_1+e_3} ,$$

where

$$(\sqrt{g}a^{(3)})_{j+e_1\pm e_3} \equiv \tfrac{1}{2}\{(\sqrt{g}a^{(3)})_{j\pm e_3} + (\sqrt{g}a^{(3)})_{j+2e_1\pm e_3}\} .$$

In I_{15}, I_{16} and I_{17}, $e^{(\beta)}$ has to be approximated, requiring discretizations of derivatives of $\boldsymbol{u}$. We start with I_{15} and write, using (13.18):

$$\left(\frac{\partial \boldsymbol{u}}{\partial x^\beta}\right)_j = \left(a_\beta^{(\alpha)} \frac{\partial \boldsymbol{u}}{\partial \xi^\alpha}\right)_j .$$

In a cell-center we are not hampered by non-smoothness of the mapping $\boldsymbol{x} = \boldsymbol{x}(\boldsymbol{\xi})$, and the following straightforward approximation can be used:

$$\left(\frac{\partial \boldsymbol{u}}{\partial x^\beta}\right)_j \cong \sum_{\alpha=1}^{3} \frac{1}{\Delta\xi^\alpha} \left(a_\beta^{(\alpha)}\right)_j \boldsymbol{u} \,|_{j-e_\alpha}^{j+e_\alpha} .$$

The same procedure cannot be followed for I_{16} and I_{17}, because we are at cell edges, where the geometric quantities are discontinuous. Instead, we proceed in a similar way as for ∇p. An approximation for $\partial u^\alpha/\partial x^\beta$ in $\boldsymbol{x}_{j+e_1+e_2}$, for example, is derived by writing (no summation):

$$b_\gamma \equiv u^\alpha|_{j+e_1+e_2-e_\gamma}^{j+e_1+e_2+e_\gamma} = \int_{\boldsymbol{x}_{j+e_1+e_2-e_\gamma}}^{\boldsymbol{x}_{j+e_1+e_2+e_\gamma}} \nabla u^\alpha \cdot d\boldsymbol{x} \cong \nabla u_{j+e_1+e_2}^\alpha \cdot \boldsymbol{c}_{(\gamma)}, \tag{13.29}$$

$$\boldsymbol{c}_{(\gamma)} \equiv \boldsymbol{x}|_{j+e_1+e_2-e_\gamma}^{j+e_1+e_2+e_\gamma}, \quad \gamma = 1, 2, 3 .$$

This system of three equations for ∇u^α is solved in similar manner as the system for ∇p. We define $\boldsymbol{c}^{(\alpha)}$ by (13.28), taking $\boldsymbol{c}_{(\alpha)}$ from (13.29), and find

$$(\frac{\partial u^\alpha}{\partial x^\beta})_{j+e_1+e_2} \cong c_\beta^{(\gamma)} b_\gamma .$$

In $c_\beta^{(\gamma)}$ values of $\boldsymbol{u}$ in cell face centers, cell edge centers and cell vertices occur; these are expanded in terms of V^α by means of (13.9), (13.12) and (13.14)–(13.16).

The momentum equation (13.21) is integrated also over shifted finite volumes Ω_{j+e_2} and Ω_{j+e_3}, and discretized in a completely similar manner. We may consider the resulting finite volume discretization of (13.21), briefly denoted as

$$\int_{\Omega_{j+e_\alpha}} N(\boldsymbol{u}, p)d\Omega = 0 \qquad (13.30)$$

with the integral approximated in the way indicated above, as semi-discretized evolution equations for $\boldsymbol{u}$ at the cell face centers (by semi-discretization we mean discretization in space but not in time). This is not what we want, because we wish to generalize the staggered scheme of Harlow and Welch (1965) from Cartesian to general grids, which implies that we want evolution equations for the volume flux components at the cell face centers. Using (13.9), this is achieved by replacing (13.30) by

$$(\sqrt{g}\boldsymbol{a}^{(\alpha)})_{j+e_\alpha} \cdot \int_{\Omega_{j+e_\alpha}} N(\boldsymbol{u}, p)d\Omega = 0 \quad \text{(no summation)},$$

with the integral approximated in the way described above. This gives for the various terms in $N(\boldsymbol{u}, p)$ the following results. From (13.22) and (13.9) it follows that

$$(\sqrt{g}\boldsymbol{a}^{(\alpha)})_{j+e_\alpha} \cdot \int_{\Omega_{j+e_\alpha}} \frac{\partial \boldsymbol{u}}{\partial t}d\Omega = |\Omega_{j+e_\alpha}|dV^\alpha_{j+e_\alpha}/dt \quad \text{(no summation)},$$

which is precisely what we want. Furthermore, using (13.23),

$$(\sqrt{g}\boldsymbol{a}^{(1)})_{j+e_1} \cdot \boldsymbol{I}_{11} \cong \frac{1}{2}\Delta\xi^2\Delta\xi^3(\sqrt{g}\boldsymbol{a}^{(1)})_{j+e_1} \cdot (\boldsymbol{u}_{j+2e_1}V^1|^{j+3e_1}_{j+e_1} - \boldsymbol{u}_j V^1|^{j+e_1}_{j-e_1})$$

$$(13.31)$$

(no summation). Here $\boldsymbol{u}$ is expressed in terms of V^α in the way described after equation (13.23). The remaining terms are handled in a completely similar manner. A similar method to avoid Christoffel symbols is followed in He and Salcudean (1994), Karki and Patankar (1988), Melaaen (1992), Wesseling, Segal, Kassels, and Bijl (1998), Wesseling, Segal, and Kassels (1999).

In this way a coordinate invariant discretization is obtained, that in the case of the identity mapping $\boldsymbol{x} = \boldsymbol{\xi}$ reduces to the classical Cartesian staggered discretization of Harlow and Welch (1965). Furthermore, the discretization error is zero for $\boldsymbol{u}$ and ∇p constant, regardless of roughness and non-orthogonality of the grid. It would be too tedious to show this here, but verification by numerical experiment is easy.

One may wonder how it is possible that a coordinate invariant discretization has been obtained without encountering Christoffel symbols. This may, to a certain extent, be elucidated as follows. The Christoffel symbols are defined by

$$\left\{ {\alpha \atop \beta\gamma} \right\} \equiv \boldsymbol{a}^{(\alpha)} \cdot \frac{\partial \boldsymbol{a}_{(\beta)}}{\partial \xi^\gamma} . \qquad (13.32)$$

In (13.31), for example, if u is expressed in terms of V^α, products of $a^{(\alpha)}$ in one point and $a_{(\beta)}$ in neighbouring points are hidden, similar to what one would get by discretizing (13.32) (assuming differentiability of $a_{(\beta)}$). Another way of looking at our circumvention of Christoffel symbols is to note that finite volume integration precedes transformation to invariant form, so that integrals of the Christoffel symbols are required, removing the derivative in (13.32).

The structure of the stencil

Fig. 13.3 shows at which points u, V^α and p are used in the discretization of the inertia and pressure terms in the equation for V^1; values of V^α needed to express u are included. For clarity, we show only the two-dimensional case. Fig. 13.4 shows the points where $e^{(\alpha)}$, u and V^α occur in the viscous terms. The u points are those needed to express $e^{(\beta)}$ in terms of u with the path

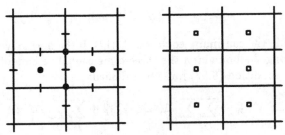

Fig. 13.3. Stencils of inertia and pressure terms; $\bullet$: u, $-$: V^1, $|$: V^2, $\square$: p.

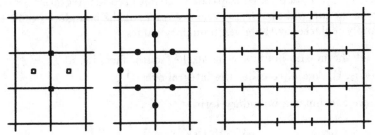

Fig. 13.4. Stencil of viscous term: $\square$: $e^{(\beta)}$, $\bullet$: u, $|$: V^1, $-$: V^2.

integral method, and the V^α points are those needed to express u. Note that the viscous term contains 9 V^1 points and 12 V^2 points; this does not seem exorbitant in view of the fact, that mixed derivatives have to be approximated in the case of non-orthogonal grids.

Boundary conditions

Numerical implementation of boundary conditions has been discussed for Cartesian grids in Sect. 6.4. Generalization to general structured grids is relatively straightforward. A guiding principle is that the numerical boundary conditions of Sect. 6.4 should be recovered in the special case of a Cartesian grid.

No-slip condition

Let at the 'west' face of the domain $\xi^1 = \xi^1_{1/2}$ the following boundary condition be given:

$$u(t)_{j-e_1} = v_b(t, x_{j-e_1}) , \quad j = (1, j_2, j_3) , \quad e_1 = (\tfrac{1}{2}, 0, 0) .$$

Then

$$V^1_{j-e_1} = -(\sqrt{g}a^{(1)} \cdot v_b)_{j-e_1}$$

is substituted in the continuity equation (13.17). If the normal velocity is prescribed all along the boundary, the three-dimensional version of the discrete compatibility condition (6.53) must be satisfied:

$$\sum_{j_2} \sum_{j_3} \Delta\xi^2 \Delta\xi^3 V^1|^E_W + \sum_{j_3} \sum_{j_1} \Delta\xi^3 \Delta\xi^2 V^2|^R_F + \sum_{j_1} \sum_{j_2} \Delta\xi^1 \Delta\xi^2 V^3|^N_S = 0 ,$$

where E stands for the 'east' face of the domain, etc., in the same way as in the beginning of this section for cells.

Because $V^1_{j-e_1}$, $j = (1, j_2, j_3)$ is given, we do not need to integrate over Ω_{j-e_1}. Where in the integral over Ω_{j+e_1} velocity values at the boundary occur, these are simply replaced by the given boundary values.

Next, let the no-slip boundary be at the 'south' face, i.e. at $\xi^3 = \xi^3_{1/2}$. Let $j = (j_1, j_2, 1)$. Again, consider the integral over Ω_{j+e_1}.

In I_{13} we encounter a boundary term

$$\Delta\xi^1 \Delta\xi^2 (uV^3)_{j+e_1-e_3}$$

which is known from the boundary condition.

In I_{14} we cannot evaluate ∇p_{j+e_1} with the path integral method in the same way as in the interior, because this would involve p_{j-2e_3}, referring to a grid point outside the domain. Therefore the integration path in the ξ^3-direction is chosen in a one-sided way. We write

$$b_3 \equiv p|_j^{j+2e_3} + p|_{j+2e_1}^{j+2e_1+2e_3} \cong \nabla p_{j+e_1} \cdot c_{(3)} \, ,$$

$$c_{(3)} \equiv x|_j^{j+2e_3} + x|_{j+2e_1}^{j+2e_1+2e_3} \, .$$

The two other integration paths for the approximation of ∇p_{j+e_1} are chosen in the same way as in the interior.

In I_{17} we encounter a boundary term

$$\Delta\xi^1\Delta\xi^2(\sqrt{g}a_\beta^{(3)}\nu e^{(\beta)})_{j+e_1-e_3} \, .$$

Evaluation of $e^{(\beta)}$ by the path integral method in the same way as in the interior would lead to the use of grid points outside the domain. We therefore again choose a one-sided integration path in the ξ^3-direction. In the case of a Cartesian grid also a one-sided approximation is necessary. We write for the approximation of $\nabla u_{j+e_1-e_3}^\alpha$:

$$b_3 \equiv u^\alpha|_{j+e_1-e_3}^{j+e_1} \cong \nabla u_{j+e_1-e_3}^\alpha \cdot c_{(3)} \, ,$$

$$c_{(3)} \equiv x|_{j+e_1-e_3}^{j+e_1} \, .$$

The other two integration paths are chosen as in the interior.

This covers the principles of the modifications required near a no-slip boundary.

Free surface conditions

We recall from Sect. 6.2 that the free surface conditions are given in Cartesian tensor notation by:

$$u^\alpha n^\alpha = v_b(t, x) \, , \quad n_\alpha\sigma^{\alpha\beta}s_\beta = 0 \, , \quad n_\alpha\sigma^{\alpha\beta}t_\beta = 0 \, , \tag{13.33}$$

with n the unit outer normal, and s and t two linearly independent tangent vectors in the surface. It is assumed that the grid near the free surface has the following local orthogonality property: grid lines that go from the free surface into the interior are orthogonal to the free surface. The situation is sketched in Fig. 13.5 for the two-dimensional case. If this local orthogonality property is not satisfied, numerical implementation of the free surface conditions becomes complicated.

Local grid adjustment to satisfy the above local orthogonality property is easily done as follows. The (in general non-orthogonal) cell faces on the free surface are left unchanged. Take a vertex P on the free surface. The local normal direction is taken to be the direction of the average of the four surrounding cell face vectors. In this direction a line is drawn into the interior. Its first intersection with an interior cell face is selected as a new vertex P',

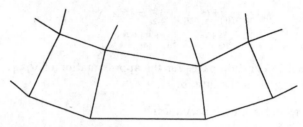

Fig. 13.5. Locally orthogonal grid

which replaces the interior vertex that was originally connected to P. This procedure is clarified in Fig. 13.6 for the two-dimensional case.

Let the free boundary be at the 'west' face, i.e. at $\xi^1 = \xi^1_{1/2}$. Then

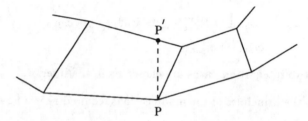

Fig. 13.6. Local grid adjustment.

$$V^1_{j-e_1} = -(|\sqrt{g}a^{(1)}|v_b)_{j-e_1}, \quad j = (1, j_1, j_2)$$

is given, so that we do not need to integrate over Ω_{j-e_1}.

The pressure integrals are treated in the same way as for the no-slip condition, and need not be discussed again.

From Fig. 13.4 we see that the integral over Ω_{j+e_1} requires u_{j-e_1}. However, if $j = (1, j_1, j_2)$, x_{j-e_1} is at the free surface, and u_{j-e_1} is unknown; only the normal component is given. In the special case of a Cartesian grid, the normal component of u_{j-e_1} is all that is needed. Due to the condition of local orthogonality of the grid that we have imposed above, the Cartesian situation is approximated to a certain extent, so that the tangential components of u_{j-e_1} have little influence. Therefore we venture to express u_{j-e_1} in terms of interior values of V^α by means of a one-sided evaluation. We write

$$V^1_{j-e_1} = -(|\sqrt{g}a^{(1)}|v_b)_{j-e_1},$$

$$V^2_{j-e_1} \cong \frac{1}{2}(V^2_{j+e_2} + V^2_{j-e_2}),$$

$$V^3_{j-e_1} \cong \frac{1}{2}(V^3_{j+e_3} + V^3_{j-e_3}).$$

We write, furthermore,

$$V^\alpha_{j-e_1} = (\sqrt{g}\boldsymbol{a}^{(\alpha)} \cdot \boldsymbol{u})_{j-e_1} , \tag{13.34}$$

where we define (no summation)

$$(\sqrt{g}\boldsymbol{a}^{(\alpha)})_{j-e_1} = \frac{1}{2}\{(\sqrt{g}\boldsymbol{a}^{(\alpha)})_{j+e_\alpha} + (\sqrt{g}\boldsymbol{a}^{(\alpha)})_{j-e_\alpha}\} , \quad \alpha = 2,3 .$$

Using the dual vectors $\tilde{\boldsymbol{a}}_{(\alpha)}$ defined by

$$\tilde{\boldsymbol{a}}_{(\alpha)} \cdot \sqrt{g}\boldsymbol{a}^{(\beta)} = \delta^\beta_\alpha ,$$

equation (13.34) is solved by

$$\boldsymbol{u}_{j-e_1} = (\tilde{\boldsymbol{a}}_{(\alpha)}V^\alpha)_{j-e_1} .$$

The dual vectors $\tilde{\boldsymbol{a}}_{(\alpha)}$ follow from

$$\tilde{\boldsymbol{a}}_{(\alpha)} = \frac{\sqrt{g}\boldsymbol{a}^{(\beta)} \times \sqrt{g}\boldsymbol{a}^{(\gamma)}}{\sqrt{g}\boldsymbol{a}^{(1)} \cdot \sqrt{g}\boldsymbol{a}^{(2)} \times \sqrt{g}\boldsymbol{a}^{(3)}} , \quad \alpha,\beta,\gamma \ \ \text{cyclic.}$$

In this way $\boldsymbol{u}_{j-e_1}$ is constructed using both information from the boundary conditions and from the interior.

Next, let the free surface be at the 'south' face, i.e. at $\xi^3 = \xi^3_{1/2}$. Now $V^3_{j-e_3}$ is given by the free surface condition:

$$V^3_{j-e_3} = -(|\sqrt{g}\boldsymbol{a}^{(3)}|v_b)_{j-e_3} , \quad j = (j_1, j_2, 1) .$$

Consider the integral over Ω_{j+e_1}. In the integral I_{13}, given by (13.24), we encounter a boundary term $(\boldsymbol{u}V^3)_{j+e_1-e_3}$. We make the following approximation:

$$V^3_{j+e_1-e_3} \simeq \frac{1}{2}(V^3_{j-e_3} + V^3_{j+2e_1-e_3}) ,$$

and $\boldsymbol{u}_{j+e_1-e_3}$ is evaluated in similar one-sided fashion as before, using

$$V^1_{j+e_1-e_3} \simeq V^1_{j+e_1} ,$$

$$V^2_{j+e_1-e_3} \simeq \frac{1}{4}(V^2_{j+e_2} + V^2_{j-e_2} + V^2_{j+2e_1+e_2} + V^2_{j+2e_1-e_2}) .$$

We write

$$V^\alpha_{j+e_1-e_3} = (\sqrt{g}\boldsymbol{a}^{(\alpha)} \cdot \boldsymbol{u})_{j+e_1-e_3}$$

$$(\sqrt{g}\boldsymbol{a}^{(1)})_{j+e_1-e_3} \equiv (\sqrt{g}\boldsymbol{a}^{(1)})_{j+e_1} ,$$

$$(\sqrt{g}\boldsymbol{a}^{(2)})_{j+e_1-e_3} \equiv \frac{1}{4}\{(\sqrt{g}\boldsymbol{a}^{(2)})_{j+e_2} + (\sqrt{g}\boldsymbol{a}^{(2)})_{j-e_2}$$

$$+ (\sqrt{g}\boldsymbol{a}^{(2)})_{j+2e_1+e_2} + (\sqrt{g}\boldsymbol{a}^{(2)})_{j+2e_1-e_2}\} .$$

We obtain

$$\boldsymbol{u}_{j+e_1-e_3} = (\tilde{\boldsymbol{a}}_{(\alpha)}V^\alpha)_{j+e_1-e_3} \, ,$$

with the dual vectors $\tilde{\boldsymbol{a}}_{(\alpha)}$ related to $\sqrt{g}\boldsymbol{a}^{(\alpha)}$ as before.

It remains to consider the integral over the viscous term. The evaluation of $\boldsymbol{e}^{(\beta)}$ in $\boldsymbol{I}_{15}$ and $\boldsymbol{I}_{16}$ requires $\boldsymbol{u}$ at the boundary, which is approximated in a similar way as above. In $\boldsymbol{I}_{17}$ the boundary term

$$\Delta\xi^1\Delta\xi^2(\sqrt{g}a_\beta^{(3)}\nu\boldsymbol{e}^{(\beta)})_{j+e_1-e_3}$$

occurs. This term is not zero in general, but nevertheless it can be neglected. The reason is that what is really needed is not $\boldsymbol{I}_{17}$ but $(\sqrt{g}\boldsymbol{a}^{(1)})_{j+e_1} \cdot \boldsymbol{I}_{17}$. Taking the inner product of the above boundary term with $\sqrt{g}\boldsymbol{a}^{(1)}$, we get

$$\begin{aligned} B_{17} &\equiv \Delta\xi^1\Delta\xi^2(\sqrt{g}\boldsymbol{a}^{(1)})_{j+e_1} \cdot (\sqrt{g}a_\beta^{(3)}\nu\boldsymbol{e}^{(\beta)})_{j+e_1-e_3} \\ &= \Delta\xi^1\Delta\xi^2(\sqrt{g}a_\gamma^{(1)})_{j+e_1}(\sqrt{g}a_\beta^{(3)}\sigma^{\beta\gamma})_{j+e_1-e_3} \, , \end{aligned}$$

with $\sigma^{\beta\gamma}$ the Cartesian shear stress components. Per definition, $\boldsymbol{a}^{(3)}$ is in the direction of $-\boldsymbol{n}$ (with $\boldsymbol{n}$ the outward normal) and because of the local orthogonality property of the grid that we have postulated, $\boldsymbol{a}^{(1)}$ is in a direction $\boldsymbol{s}$ tangential to the free surface. It follows that we can write

$$B_{17} = -\Delta\xi^1\Delta\xi^2|\sqrt{g}\boldsymbol{a}^{(1)}|_{j+e_1}|\sqrt{g}\boldsymbol{a}^{(3)}|_{j+e_1-e_3}(s_\gamma n_\beta\sigma^{\beta\gamma})_{j+e_1-e_3} = 0 \, ,$$

according to the free surface boundary condition. So the boundary term can be neglected. This concludes our discussion of the numerical implementation of the boundary conditions (13.33).

If we have a truly free surface, it moves with the flow, and the normal velocity cemponent is unknown. All three stress components are prescribed:

$$p - n_\alpha\sigma_{\alpha\beta}n_\beta = p_\infty \, , \quad n_\alpha\sigma_{\alpha\beta}s_\beta = 0 \, , \quad n_\alpha\sigma_{\alpha\beta}t_\beta = 0 \, , \tag{13.35}$$

where we have used Cartesian tensor notation.

Let the free surface be at the 'west' face of the domain, i.e. at $\xi^1 = \xi^1_{1/2}$. Since the normal velocity at the free surface is unknown, we have to compute $V^1_{j-e_1}$, $j = (1, j_2, j_3)$. We therefore integrate over the cell Ω_{j-e_1}, which has half the normal size, cf. Fig. 13.7. The various integrals that we encountered in the discretization in the interior can now be approximated as follows.

First of all we have

$$\boldsymbol{I}_{11} \cong \Delta\xi^2\Delta\xi^3(\boldsymbol{u}V^1)|^j_{j-e_1} \, .$$

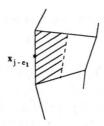

Fig. 13.7. The cell $\Omega_{j-e_1}, j = (1, j_2, j_3)$

When the inner product with $(\sqrt{g}a^{(1)})_{j-e_1}$ is taken, u_{j-e_1} is replaced by $V^1_{j-e_1}$, which is the unknown for which the integral over Ω_{j-e_1} provides an equation, so no further measures are required.

Next, we write

$$I_{12} \cong \Delta\xi^1\Delta\xi^3(u_{j-e_1+e_2}V^2_{j+e_2} - u_{j-e_1-e_2}V^2_{j-e_2}) \, ,$$

where u will be replaced by V^1 when the inner product with $(\sqrt{g}a^{(1)})_{j-e_1}$ is taken, and no further modifications are necessary. I_{13} is handled similarly.

In order to take advantage of the fact that the total stress is prescribed on the boundary, we lump the pressure and viscous terms together, and write, in Cartesian tensor notation,

$$I^1_\alpha \equiv \int_{\Omega_{j-e_1}} (p_{,\alpha} - \sigma_{\alpha\beta,\beta})d\Omega \, .$$

Using (13.19) this is rewritten as

$$I^1_\alpha = \int_{G_{j-e_1}} \frac{\partial}{\partial\xi^\gamma}(\sqrt{g}a^{(\gamma)}_\alpha p - \sqrt{g}a^{(\gamma)}_\beta \sigma_{\alpha\beta})d\xi^1 d\xi^2 d\xi^3$$

This gives several cell face integrals. To begin with,

$$I^{11}_\alpha \cong \Delta\xi^2\Delta\xi^3(\sqrt{g}a^{(1)}_\alpha p - \sqrt{g}a^{(1)}_\beta \sigma_{\alpha\beta})|^j_{j-e_1} \, .$$

The inner product with $(\sqrt{g}a^{(1)})_{j-e_1}$ is required, which is in the direction of $-n$, with n the unit outer normal, so that $\sqrt{g}a^{(1)} = -|\sqrt{g}a^{(1)}|n$. Therefore we can write, at $x = x_{j-e_1}$:

$$\sqrt{g}a^{(1)}_\alpha(\sqrt{g}a^{(1)}_\alpha p - \sqrt{g}a^{(1)}_\beta \sigma_{\alpha\beta}) =$$
$$= |\sqrt{g}a^{(1)}|^2(p - n_\alpha\sigma_{\alpha\beta}n_\beta) = |\sqrt{g}a^{(1)}|^2 p_\infty \, ,$$

which takes care of the contribution at $x = x_{j-e_1}$. For the contribution at x_j we can let p_j stand, and the derivatives of u in $\sigma_{\alpha\beta}$ are evaluated with the path integral method as usual, but expressing $\boldsymbol{u}_{j-e_1}$ in terms of V^α with a one-sided bias, as before, using $V^1_{j-e_1}$, $V^2_{j\pm e_2}$, $V^3_{j\pm e_3}$. Next, we encounter

$$I^{12}_\alpha \cong \frac{1}{2}\Delta\xi^1\Delta\xi^3(\sqrt{g}a^{(2)}_\alpha p - \sqrt{g}a^{(2)}_\beta\sigma_{\alpha\beta})|^{j-e_1+e_2}_{j-e_1-e_2} \, .$$

Because of the local orthogonality condition of the grid, $\boldsymbol{a}^{(2)}$ is perpendicular to $\boldsymbol{a}^{(1)}$ at the boundary. Since the inner product with $(\sqrt{g}a^{(1)})_{j-e_1} \cong (\sqrt{g}a^{(1)})_{j-e_1\pm e_2}$ is required, we may just put $p = 0$ in I^{12}_α. Furthermore, again because of the orthogonality condition, $(\sqrt{g}a^{(2)})_{j-e_1\pm e_2}$ is in a direction tangential to the free surface, so that we may write

$$(\sqrt{g}a^{(1)}_\alpha)_{j-e_1}(\sqrt{g}a^{(2)}_\beta\sigma_{\alpha\beta})_{j-e_1\pm e_2}$$
$$\cong -|\sqrt{g}a^{(1)}|_{j-e_1}|\sqrt{g}a^{(2)}|_{j-e_1\pm e_2}(n_\alpha\sigma_{\alpha\beta}s_\beta)_{j-e_1\pm e_2}) = 0 \, ,$$

where s is some vector tangential to the surface. Similarly, the contribution in the third direction $I^{13}_\alpha \cong 0$. This completes our discussion of the finite volume discretization for the cell Ω_{j-e_1} of Fig. 13.7 with boundary conditions (13.35).

Next, we have to consider the integral over Ω_{j+e_1} in the case where the boundary conditions (13.35) are applied at the 'south' face. This can be done in identical fashion as described for the boundary conditions (13.33), with one small difference: $V^3_{j-e_3}$ is not replaced by a known boundary value, but is left to stand, because now it belongs to the set of unknowns.

Outflow conditions

The purpose of outflow conditions is to let the fluid flow out of the domain while disturbing it as little as possible. For reasons given in Sect. 6.2, prescribing the free surface conditions (13.35) is a good option.

Solution methods

The structure of the system of ordinary differential equations that results from the spatial discretization described above is the same as in the case of a Cartesian grid as presented in Sect. 6.4, and is given by equation (6.42). Temporal discretization can be done in the same way as in Sect. 6.6, using the pressure-correction method. For the iterative solution of the pressure-correction equation and of algebraic systems resulting from implicit time stepping schemes the same methods can be used as for the Cartesian case, presented in Chap. 7.

Concluding remarks

We have given a description of a coordinate-invariant spatial discretization using a staggered scheme on a boundary-fitted grid. When the grid is Cartesian, the classical scheme of Harlow and Welch (1965) is recovered. As will be illustrated by an example in the next section, accuracy is maintained on nonsmooth grids. To achieve this we have carefully taken into account the smoothness proporties of the boundary-fitted coordinate mapping, avoided the use of Christoffel symbols in the way described above, and by implementing transformations from V^α to u and vice-versa in a special way.

What is the price to pay for the above properties? In a comparitive study, in Manhart *et al.* (1998) it is found that a three-dimensional time-implicit implementation of the method described here uses 243 words per cell computer memory. A time-explicit Cartesian staggered code is found to require only 12 words per cell, whereas a colocated code on a curvilinear grid using a time stepping scheme of IMEX type (Sect. 5.8) is found to require about 200 words per node. So a staggered scheme does not require much more memory than a colocated scheme.

Exercise 13.5.1. Prove (13.13). Hint: note that $(\frac{1}{\sqrt{g}} a_{(\alpha)})_j$ has been defined such that $(a_{(\alpha)})_j \cdot (a^{(\beta)})_j = \delta_\alpha^\beta$. Take the inner product of (13.13) with $(\sqrt{g} a^{(\beta)})_j$.

13.6 An application

In order to illustrate that the general coordinates staggered discretization described before is at least as accurate as the discretization methods that are mostly used at present in codes to compute Navier-Stokes solutions in complicated domains, namely colocated finite volume methods using pressure-weighted interpolation (Rhie and Chow (1983)) and finite element methods, we approximate a simple exact solution on a rough grid. To make our point, it suffices to consider a simple two-dimensional example. The exact solution is Poiseuille flow:

$$u^1 = x^2(1 - x^2), \quad u^2 = 0, \quad p = -2\mathrm{Re}^{-1}x^1, \quad \mathrm{Re} = 1.$$

The grid, shown in Fig. 13.8, is chosen rough deliberately. Figs. 13.9, 13.10 and 13.11 give streamlines and isobars for the staggered discretization, a finite element code using $Q_1 - P_0$ elements (quadrilaterals with bilinear basis functions for the velocity and constant basis functions for the pressure) and

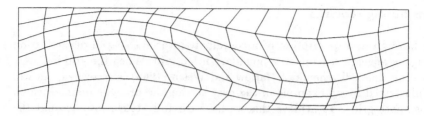

Fig. 13.8. Grid for Poiseuille flow.

a commercial code[1] using colocated discretization with pressure-weighted interpolation. Fig. 13.12 shows an even wilder grid, meant to investigate the

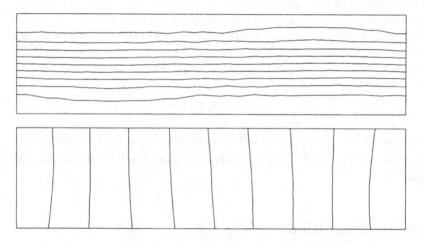

Fig. 13.9. Streamlines and isobars for staggered discretization.

effect of sudden refinement and derefinement. Results are shown in Figs. 13.13 and 13.14. We do not have results for the colocated code for this case. The streamlines should be straight and the isobars straight and equally spaced. Clearly, the staggered discretization is more accurate than the other two methods. This illustrates that staggered schemes are not inherently inaccurate on general grids. On the contrary, they can be quite accurate, provided the smoothness properties of the boundary-fitted coordinate mapping are carefully taken into account. This can be done in the way described above.

---

[1] Results kindly provided by Mr. R. Agtersloot (Delft Hydraulics).

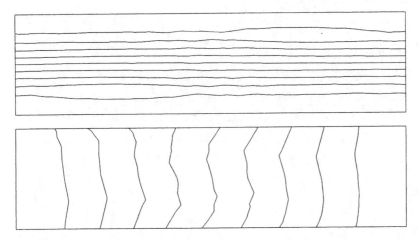

Fig. 13.10. Streamlines and isobars for finite element method.

13.7 Verification and validation

After results have been obtained with a computational fluid dynamics computer code, the credibility of these results needs to be assessed. Two types of errors may be distinguished: modeling errors and numerical errors. Modeling errors arise from not solving the right equations. Numerical errors arise from not solving the equations right. The assessment of modeling errors is called *validation*, whereas the assessment of numerical errors is called *verification*.

Validation

For simulation, a conceptual model of physical reality is built. Exactly what physical reality is turns out to be a surprisingly difficult question, that lacks a unanimous answer, and is at the core of the philosophical discipline of epistomology, and seems to transcend scientifc discourse. While justly wondering at what has been called the unreasonable effectiveness of mathematics in describing physical phenomena, we put our trust in the apparently immutable regularities and laws of nature, that exist independently from what science makes of this. In short, we accept physical reality as given. Conceptual models in fluid dynamics are mathematical models consisting of equations. The modeling error is the discrepancy between physical reality and the mathematical model. Modeling errors may arise from simplifications introduced in the Navier-Stokes equations, such as neglecting viscous terms, or making thin-layer approximations, or assuming two-dimensionality or some form of symmetry, or, most importantly of course, the use of a turbulence model. For validation, comparison with physical experiment is necessary, or comparison with computed results obtained with mathematical models involving fewer

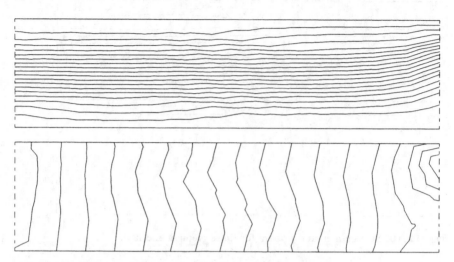

Fig. 13.11. Streamlines and isobars for colocated discretization.

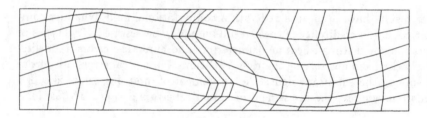

Fig. 13.12. Grid for Poiseuille flow.

assumptions, such as direct numerical simulation. Much experimental and numerical information for flows of varying degrees of complexity is available in databases that can be accessed via Internet. Good points of entry to this information are the following websites:

www.cfd-online.com

www.princeton.edu/~gasdyn/Fluid_DynLinkd/data.html

and the ERCOFTAC (European Research Community on Flow, Turbulence and Combustion) website:

    imhefwww.epfl.ch/ERCOFTAC

Of course, not only numerical but also experimental results are contaminated by errors. Validation makes sense only after verification, otherwise agreement between measured and computed results may well be fortuitous.

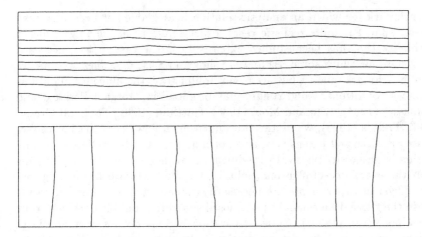

Fig. 13.13. Streamlines and isobars for staggered discretization.

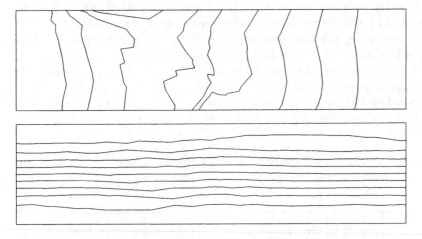

Fig. 13.14. Streamlines and isobars for finite element method.

Benchmark problems

For verification, comparison with results obtained with other computer codes for the same mathematical model is very valuable. Extensively studied flow cases using a well-defined mathematical model for which reliable numerical results are available are called benchmark problems. In benchmarking, results that are known to be correct should be reproduced. This helps to eliminate coding errors. Benchmarking is also useful for showing progress, such as enhanced efficiency, with respect to older methods. We will now give some pointers to the literature concerning widely used benchmark problems.

A benchmark for which an analytic solution is available has been discussed in Sect. 13.6. For more realistic cases, exact solutions are not available. A benchmark that has been studied extensively is the case of the backward facing step, for which some results were shown in Sect. 6.5. These results are not of benchmark quality, but are meant to illustrate the influence of outflow boundary conditions. Good results may be found in Morgan, Périaux, and Thomasset (1984), Kim and Moin (1985), Gartling (1990), Sani and Gresho (1994), Freitas (1995), where the two-dimensional flow over a backward facing step is discussed extensively, and much numerical information about the solution is gathered. Physical experiments provide only limited information about the exact two-dimensional solution for the backward facing step, because physical experiments are necessarily three-dimensional, and it seems that in the three-dimensional case a closed separation bubble does not occur (Wesseling *et al.* (1994)). Closed separation regions do not normally exist in three dimensions. In fact, separation is hard to define in three dimensions. In two-dimensional computations, the length of the bubble increases significantly with the Reynolds number. Consequently, if numerical diffusion generated by the discretization of the inertia terms is non-negligible (effectively decreasing the Reynolds number), the computed separation bubble will be too short. The first order upwind scheme is found to be wanting. This benchmark shows up the need for higher order discretization of the inertia terms, perhaps using a slope limited scheme (Sect. 4.8) to rule out spurious wiggles. This benchmark can be used to test codes on single block and multi-block Cartesian grids (Fig. 13.15), and on curvilinear boundary-fitted grids (Fig. 6.3).

Another frequently used benchmark is the flow in a closed cavity. This case can be handled already by simple codes, because only the simplest type of

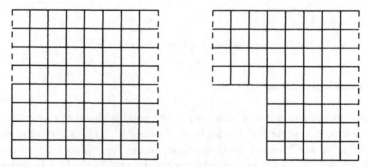

Fig. 13.15. Single block and multi-block Cartesian grids for flow over a backward facing step.

boundary condition occurs, namely the no-slip condition, and if the cavity is rectangular Cartesian grids suffice. For these reasons, this benchmark has been around a long time. Flow may be induced by having one of the walls move (lid-driven cavity) or by applying temperature differences to the walls (buoyancy driven cavity). In the latter case the Boussinesq equations are solved instead of the Navier-Stokes equations. For the lid-driven cavity, results of benchmark quality are given in Ghia, Ghia, and Shin (1982). Some publications in which this test case is discussed are Armaly, Durst, Pereira, and Schönung (1983), Burggraf (1966), Leschziner (1980), Kim and Moin (1985), Shyy, Thakur, and Wright (1992), Zang, Street, and Koseff (1994). Benchmark quality results for buoyancy-driven cavity flow are given in de Vahl Davis (1983), Hortman, Perić, and Scheuerer (1990).

If instead of being rectangular the shape of the cavity is skewed or otherwise distorted, a test case for codes employing non-orthogonal grids is obtained. Examples may be found in Demirdžić, Lilek, and Perić (1992), Oosterlee et al. (1993).

A third benchmark is the unconfined flow past a circular cylinder. At Reynolds numbers above about 40 the flow becomes unsteady, and periodic vortex shedding occurs, giving rise to a so-called Karman vortex street. So this test case is suitable to test both spatial and temporal discretization schemes. Results are given in Engelman and Jamnia (1990), Rosenfeld, Kwak, and Vinokur (1991), Marx (1994), Freitas (1995); further references may be found in these publications.

Verification

Verification is the process by which the degree of numerical accuracy is assessed. For benchmark problems checking numerical accuracy is relatively easy, because accurate results are available to compare with. But when a CFD code is used for the purpose for which it is designed, namely to predict flow properties that are a priori unknown, there is nothing to compare with. Nevertheless, numerical error assessment is possible to some extent, and we will briefly discuss this topic below.

Numerical errors may be divided into three kinds: the (global) discretization error, the iterative convergence error and the rounding error. The rounding error is due to the finite precision arithmetic of computers, and is in CFD applications usually negligible compared to the two other sources of numerical error. Therefore the rounding error will not be further discussed. The discretization error is due to the replacement of the differential equations by an algebraic system (discretization). The iterative convergence error arises from inexact solution of the algebraic system by some iteration method.

It was shown in Sect. 7.2 for the simple case of stationary iterative methods for linear problems, that the difference between two successive iterates does not give much indication about the size of the iterative convergence error. This holds a forteriori in more complicated situations. A better indication of the error can be got from looking at the residual. Of course, the relation between the residual and the error is complicated; for example, for linear systems the condition number, which is usually not known, enters into play. But often an interpretation of the residual can be given which provides a guideline as to how small it should be made. For instance, when iteration stops it is known that one has solved exactly a system in which the right-hand side is perturbed by the residual, or, in a time stepping method, the residual can be interpreted as a perturbation of the solution at the preceding time level.

If the discretization scheme is sound, the discretization error can be made smaller by decreasing the time step and by refining the spatial grid. The question is how fine the grid should be. In principle this question can be answered by comparing numerical solutions at different levels of refinement. If the difference between numerical solutions at two different levels of refinement is negligible from an engineering point of view, the numerical solution obtained may be said to be grid-converged, and the error is dominated by the modeling error. On modern computers the available memory is frequently large enough to allow sufficient refinement in two dimensions, but in three-dimensional applications it may easily happen that the grid is not sufficiently fine. The least one can do in this situation is to give an indication of the size of the error, by showing the discrepancy between numerical solutions at two refinement levels. Furthermore, one may try to estimate the zero mesh size limit of the solution (i.e. the grid-converged numerical solution) by *Richardson extrapolation*, which we will now briefly discuss.

Richardson extrapolation

The method is named after Richardson (1910), but the idea behind it is much older, and was already used by Archimedes (250 BC) in his method to approximate the value of π.

Let us consider a stationary problem discretized on a boundary-fitted grid. In $\boldsymbol{\xi}$-space the grid is uniform. Assume that the mesh size is $\Delta \xi^\alpha = c_\alpha h$ in the ξ^α-direction, with c_α constant, and h a parameter that governs the degree of refinement. Let $\boldsymbol{x}$ be a grid point belonging to all grids beyond a certain refinement level; many such points exist if the refinement parameter h is chosen as follows:

$$h \in H\{1, 1/2, 1/4, ...\}.$$

Let φ stand for one of the unknowns, for example a velocity component or the pressure. The global discretization error is defined as $\varepsilon_h \equiv \varphi_h - \varphi$, with φ_h the numerical solution (computed exactly), and φ the exact solution. It is assumed that ε_h has the following asymptotic expansion as $h \downarrow 0$:

$$\varepsilon_h(\boldsymbol{x}) = e(\boldsymbol{x})h^p + o(h^p) \tag{13.36}$$

for some known exponent p; how to find p will be discussed later. Although the function $e(\boldsymbol{x})$ is of course not known, we can use (13.36) to extrapolate the numerical solution to $h = 0$, by using the numerical solution at two refinement level in the following way. Writing

$$\varphi_0 = \lim_{h \downarrow 0} \varphi_h ,$$

we have

$$\varphi_h \cong \varphi_0 + eh^p ,$$
$$\varphi_{h/2} \cong \varphi_0 + e(h/2)^p .$$

Elimination of e results in

$$\varphi_0 \cong \frac{2^p \varphi_{h/2} - \varphi_h}{2^p - 1} .$$

It remains to determine p. The existence of the asymptotic expansion (13.36) has been rigorously proved and p is known a priori only for simple cases. If the exact solution is a sufficient number of times differentiable, and the numerical solution shows no trace of spurious wiggles, and the numerical boundary conditions have been implemented in a suitable way, then p equals the formal order of accuracy of the scheme. For example, for the convection-diffusion equation with a second order central scheme for the convection term we have $p = 2$. If the first order upwind scheme is used for the convection term then $p = 1$. But in realistic applications with complicated geometries, the exact solution may not be sufficiently differentiable near corners of the domain. Therefore it is safest to estimate p by introducing a third refinement level. Assuming that the $h/2$ level is the finest refinement that the available computer memory allows, we have to use a coarser level, namely $2h$. This gives

$$\varphi_{2h} \cong \varphi_0 + e(2h)^p .$$

We obtain

$$\frac{\varphi_{2h} - \varphi_h}{\varphi_h - \varphi_{h/2}} = 2^p , \tag{13.37}$$

from which p may be determined. If the left-hand side of (13.37) is not positive, h is too large to neglect the higher order terms in (13.36), or an asymptotic expansion of the form (13.36) does not exist.

Because in realistic applications usually the existence of the asymptotic expansion (13.36) has not been established theoretically, Richardson extrapolation is not a watertight procedure. But if a likely value of p, e.g. $p = 2$ for a formally second order scheme is confirmed by (13.37), then Richardson extrapolation may be said to work. If there is no a priori guess of p, but $|\varphi_h - \varphi_{h/2}|$ is significantly smaller that $|\varphi_{2h} - \varphi_h|$, and (13.37) gives approximately the same value of p in large parts of the domain (perhaps not in the vicinity of isolated points, which may indicate a local singularity), then again Richardson extrapolation may be said to work. If Richardson extrapolation works then φ_0 can be accepted as the grid-converged numerical solution. If $|\varphi_{h/2} - \varphi_h|$ is sufficiently small, then Richardson extrapolation is not necessary, and $\varphi_{h/2}$ can be accepted as the grid-converged numerical solution.

For time-dependent problems it is best to construct numerical solutions that are independent of the spatial grid in every time step, and then apply Richardson extrapolation in time.

A good source on uncertainty in numerical computing is Celik, Chen, and Roache (1993).

Concluding remarks

We emphasize again that when one computes flows with the purpose to make engineering predictions, or to increase physical understanding, or to contribute to improvements in flow modeling, the issue of numerical accuracy should at the very least be addressed, and preferably an estimate of the numerical error should be given, or grid-convergence of the numerical solution demonstrated.

Even grid-converged numerical results obtained with a well-tested code can be wrong if there is something amiss with the underlying mathematical model. To avoid this, the user of CFD codes needs to be familiar with classical fluid dynamics. Often this enables one to see that numerical results are not correct. For instance, symmetry conditions may be falsely imposed, resulting in a stationary symmetric solution, whereas it is known from fluid dynamics that in reality the flow is unsymmetric and oscillatory. Physical insight and knowledge of fluid dynamics are needed to choose the mathematical model and the boundary conditions correctly. Use of CFD codes without physical insight easily leads to results of unnecessarily poor quality or even to erroneous results.

A very good comprehensive discussion on verification and validation of computing methods is given in Roache (1998b). The topic is adressed by several authors in the fifth issue of Volume 36 (1998) of the AIAA Journal.

14. Unified methods for computing incompressible and compressible flow

14.1 The need for unified methods

If there are regions in the flow domain where the Mach number M is not small, the incompressible Navier-Stokes or Euler equations cannot be applied. In principle, the compressible equations of motion are uniformly valid as the Mach number ranges from zero to supersonic (until real gas effects set in). Therefore it suffices to forego the simplifications that incompressibility brings, and all one has to do is to employ the compressible equations of motion. However, as will be discussed below, the standard numerical methods that have been developed for compressible flows (discussed in Chap. 10 and 12) break down or do not function properly when M $\lesssim$ 0.2. Methods to remedy this will be discussed in the present chapter. It would be ideal if a unified method could be found that is accurate and efficient for both compressible and incompressible flows. Such unified methods with accuracy and efficiency more or less uniform in the Mach number have indeed been proposed, and will be discussed below.

Typical circumstances in which the Mach number is small but where compressibility cannot be neglected are the following. Due to the geometry of the flow domain, local high speed zones may be embedded in a low speed flow. For example, in an inlet of an internal combustion engine, the flow may be virtually incompressible (M $\lesssim$ 0.1) in the bulk of the inlet channel, but we may have M $\cong$ 0.5 near the valve. Another example of geometrically induced local high speed zones is the flow around the wing of a transport aircraft in landing or take-off configuration. For transport aircraft at minimum speed the freestream Mach number satisfies M $\cong$ 0.2. If this is the case uniformly in the flow domain, incompressible flow models give sufficiently accurate results. However, around aircraft in landing and take-off configurations, the local Mach number may be much larger than 0.2, and even regions of supersonic flow may occur around high-lift devices. This necessitates the use of the compressible Navier-Stokes equations for the prediction of maximum lift. At the opposite end of the velocity spectrum in aeronautical engineering, during the hypersonic flight phase of a re-entering space vehicle, thick boundary layers and separation zones develop, in which M $\lesssim$ 0.2, possibly resulting in inaccurate results and/or inefficient numerical processes for important quantities such as heat transfer at the body surface. Of course, low speed zones occur also in high speed flows near stagnation points, but in practice this does not seem to cause difficulties for standard compressible methods, presumably because these zones are small. Low speed flows may also be compressible because of density changes induced by strong heat addition, for example in flows with combustion.

First, we will explain why standard computing methods for compressible flow break down in the limit M $\downarrow$ 0. Then we will discuss the concept of preconditioning, by which methods for compressible flow may be extended to the low Mach number regime. Finally, we will generalize a method for incompressible flows to a method with approximately Mach-uniform accuracy and efficiency, for Mach numbers ranging from zero to supersonic.

14.2 Difficulties with the zero Mach number limit

We start with the one-dimensional Euler equations (10.1):

$$U_t + f(U)_x = 0, \quad U = \begin{pmatrix} \rho \\ m \\ \rho E \end{pmatrix}, \quad f = \begin{pmatrix} m \\ m^2/\rho + p(\rho, e) \\ mH \end{pmatrix}. \tag{14.1}$$

The following considerations show that numerical difficulties are to be expected when M $\downarrow$ 0, with M $= u/c$, $u = m/\rho$, and c the speed of sound.

Spectral number of Jacobian

According to (10.9), the eigenvalues of the Jacobian $f'(U)$ are given by

$$\lambda_1 = u - c, \quad \lambda_2 = u, \quad \lambda_3 = u + c.$$

The *spectral number* of the Jacobian is defined by

$$\kappa(f'(U)) = \rho(f')\rho((f')^{-1}),$$

with ρ the spectral radius. For simplicity we assume that $u \geq 0$. Then we find

$$\kappa(f') = (u + c) \max\{\frac{1}{u}, \frac{1}{|u - c|}\} = \max\{1 + \frac{1}{M}, \frac{M + 1}{|M - 1|}\}.$$

When $\kappa(f') \gg 1$, f' is ill-conditioned, and numerical difficulties may crop up. We see that f' is singular in the sonic limit M $\to$ 1 and in the incompressible limit M $\downarrow$ 0. As we saw in Chap. 10, in computing methods for compressible flows a carefully designed artificial viscosity term is incorporated, either implicitly as in approximate Riemann solvers and flux splitting methods, or explicitly as in the Jameson-Schmidt-Turkel scheme. This is found to take care of the singularity at $M = 1$. However, no special measures are taken for the limit $M \downarrow 0$, and in practice these methods are inapplicable when large regions with $M \lesssim 0.2$ are present. We will not delve deeply into the underlying causes for this. For more information, see for example Koren and van Leer (1995), Guillard and Viozat (1999). But we will give a few more indications of numerical troubles to be expected with standard methods for compressible flow (such as those presented in Chap. 10) as $M \downarrow 0$.

Stiffness of the equations

In Sect. 10.6 we found the following heuristic stability condition for the explicit Euler scheme:

$$\tau \leq h/(u+c) \,. \tag{14.2}$$

If acoustic effects are absent, in low-subsonic flow the physical time step τ_p that is in balance with the physical time scale L/u is given by

$$\tau_p = h/u \,.$$

We find

$$\tau/\tau_p \leq M/(1+M) \,,$$

so that the numerical time step needs to be much smaller than the physical time step if $M \ll 1$. When this happens a system of differential equations is called *stiff*. If no special measures are taken, numerical inefficiency results, caused by the need to resolve acoustic modes, which travel with speed $u \pm c$. However, if acoustic modes are not present in the flow, they need not be resolved, and the stringent stability condition (14.2) is a numerical artifact associated with standard computing methods for compressible flows.

The accuracy problem

Stiffness of the equations is not the only problem that besets standard compressible flow computing methods at low speed. There is also an accuracy problem. When the stiff system is solved by carrying out the required very large number of small time steps, often wrong results are obtained for low Mach numbers. This is shown and analyzed in Guillard and Viozat (1999), Turkel, Fiterman, and van Leer (1994).

Choice of units

Another indication of numerical trouble associated with the limit M ↓ 0 shows itself when the Euler equations are made dimensionless. The influence of the choice of units is more conveniently shown for the nonconservative formulation (10.7) than for the conservative formulation (14.1). This nonconservative formulation can be rewritten as follows, assuming a perfect gas:

$$\rho_t + m_x = 0 \,,$$
$$m_t + (um + p)_x = 0 \,, \qquad (14.3)$$
$$p_t + \gamma p u_x + u p_x = 0 \,.$$

We choose units to make these equations dimensionless. As noted in Sect. 1.5, there are four basic units to be chosen, for which we take the unit of length x_r, of velocity u_r, of mass $\rho_r x_r^3$. The unit of time is deduced from this as $t_r = x_r/u_r$. The unit of pressure follows as $p_r = \rho_r u_r^2$; but we may choose p_r differently, the consequence of which may be the introduction of some coefficient in the governing equations. For the moment the choice of p_r is left open. We let x_r be a characteristic dimension of the domain, and u_r a typical velocity magnitude, such as the velocity at infinity for the flow around a body. Our choice of the time unit reflects the fact that we are not interested in acoustics, for which the physical time scale is much smaller than t_r.

Dimensionless quantities are defined by $\rho' = \rho/\rho_r$, $p' = p/p_r$ etc. The dimensionless form of (14.3) is found to be, upon deleting primes for brevity:

$$\rho_t + m_x = 0 \,,$$
$$m_t + (um)_x + \frac{p_r}{\rho_r u_r^2} p_x = 0 \,,$$
$$p_t + \gamma p u_x + u p_x = 0 \,.$$

In compressible fluid dynamics it is customary and reasonable to choose for p_r a value that is representative for the magnitude of the pressure in the flow, for example the outlet or free stream pressure. Furthermore, naturally, ρ_r is chosen such that it is representative for the magnitude of the density of the flow. For example, ρ_r may be chosen equal to the density at the inflow boundary. As a consequence, $c_r = \sqrt{\gamma p_r/\rho_r}$ is an estimate for the magnitude of the sound speed in the flow. The dimensionless momentum equation can be written as

$$m_t + (um)_x + \frac{1}{\gamma M_r^2} p_x = 0 \,, \quad M_r = u_r/c_r \,,$$

and the reference Mach number M_r is representative for the magnitude of the Mach number that occurs in the flow. We see that this equation becomes singular as $M_r \downarrow 0$, which is again an indication of the difficulties to be expected for low subsonic flow with methods developed for compressible flow only.

14.3 Preconditioning

The principle

Standard methods for computing compressible flows, such as those described in Chap. 10, are numerically inefficient for weakly compressible flows, because of the stiffness of the governing equations, discussed in Sect. 14.2. One way to alleviate this problem is to change the governing equations. This looks like putting an end to the disease by killing the patient, but is not as bad as it seems. The procedure is called *preconditioning*, and is not to be confused with preconditioning in the field of numerical linear algebra. Preconditioning not only removes the stiffness problem, but solves the accuracy problem as well, as shown in Guillard and Viozat (1999), Turkel, Fiterman, and van Leer (1994). A practical advantage of the preconditioning approach is that existing codes for compressible flows can be easily modified to improve their performance for weakly compressible flows.

The principles of preconditioning will be explained for the one-dimensional inviscid case. The flow is governed by the Euler equations (14.1). Preconditioning consists of modifying the equations artificially by multiplication of the time-derivative by a matrix P^{-1}:

$$P^{-1}U_t + f(U)_x = 0 . \tag{14.4}$$

Stationary solutions are not affected by preconditioning, but time accuracy is lost. The preconditioning matrix $P(U)$ is to be chosen such that (14.4) is less stiff than the original system as M ↓ 0. The eigenvalues of the system are now given by $\lambda(Pf')$, and should remain closer together as M ↓ 0 than $\lambda(f')$.

An example of a preconditioner

The design of $P(U)$ has been and still is the subject of much research. Pointers to the literature will be given later. A generally accepted favorite preconditioner does not seem to have emerged yet.

We will discuss the preconditioner proposed in Weiss and Smith (1995). We switch to new primitive variables Q:

$$Q = \begin{pmatrix} p \\ u \\ T \end{pmatrix}$$

with T the temperature. The preconditioned system is

$$\Gamma Q_t + f(Q)_x = 0 \tag{14.5}$$

with

$$\Gamma = \begin{pmatrix} \theta & 0 & \rho_T \\ \theta u & \rho & u\rho_T \\ \theta H - 1 & \rho u & H\rho_T + \rho c_p \end{pmatrix} ,$$

where θ is a parameter to be chosen. H is the total enthalpy given by $H = h + \frac{1}{2}u^2$, $h = c_p T$, c_p is the specific heat at constant pressure, and $\rho_T = (\partial\rho/\partial T)_p$, i.e. the temperature derivative of ρ with p kept constant.

Because Γ is variable, (14.5) is not in conservation form, but the stationary part is in conservation form.

The eigenvalues

In order to see whether (14.5) is less stiff than (14.3) we have to inspect the eigenvalues $\lambda(\Gamma^{-1}f'(Q))$. For the parameter θ Weiss and Smith make the following choice:

$$\theta = 1/w^2 - \rho_T/(\rho c_p) , \tag{14.6}$$

where w is a velocity magnitude that will be chosen later. In Example 14.3.1 we find:

$$\lambda(\Gamma^{-1}f'(Q)) = \{u, \tilde{u} \pm \tilde{c}\} , \tag{14.7}$$
$$\tilde{u} = u(1 - \alpha) , \quad \tilde{c} = \sqrt{\alpha^2 u^2 + w^2} , \quad \alpha = (1 - \beta w^2)/2 ,$$
$$\beta = (\rho_p + \frac{\rho_T}{\rho c_p}) ,$$

where ρ_p is the pressure derivative with T constant. For w the following choice is made:

$$w = \begin{cases} \varepsilon c & \text{if } |u| \le \varepsilon c , \\ |u| & \text{if } \varepsilon c < |u| \le c , \\ c & \text{if } |u| > c , \end{cases}$$

with c the speed of sound. The small parameter ε ($\sim 10^{-5}$) is included to prevent singularities in points where $u = 0$. In the viscous case (not considered here), w is replaced by $\tilde{w}$, given by

$$\tilde{w} = \max(w, \nu/\tilde{h})$$

where $\tilde{h}$ is the intercell length scale over which diffusion occurs; ν is the kinematic viscosity coefficient.

We now inspect the behavior of the eigenvalues as M varies. For a perfect gas we have

$$\beta = \frac{1}{c^2} , \tag{14.8}$$

so that $\alpha = (1 - M_w^2)/2$, $M_w = w/c$. Hence, when $M = |u|/c \geq 1$ we have $\alpha = 0$, and the eigenvalues are equal to those of the unpreconditioned system:

$$\lambda(\Gamma^{-1}f'(Q)) = \{u, \ u \pm c\} .$$

At low speed ($|u| \ll c$) we have $\alpha \cong 1/2$, and $\tilde{u} \pm \tilde{c} \cong \frac{1}{2}u(1 \pm \sqrt{5})$, so that the magnitude of the three eigenvalues does not differ much. Hence, the stiffness has been removed from the system.

Example 14.3.1. Derivation of eigenvalues

In this example we derive (14.7). The eigenvalues $\lambda(\Gamma^{-1}f'(Q))$ of the preconditioned system (14.5) satisfy:

$$\det(f'(Q) - \lambda\Gamma) = 0 .$$

We find

$$|f' - \lambda\Gamma| = \begin{vmatrix} \rho_p u - \lambda\theta & \rho & \rho_T u - \lambda\rho_T \\ \rho_p u^2 + 1 - \lambda u\theta & 2\rho u - \lambda\rho & \rho_T u^2 - \lambda\rho_T u \\ \rho_p uH - \lambda H\theta + \lambda & \frac{3}{2}\rho u^2 + \rho h - \lambda\rho u & (\rho_T H + \rho c_p)(u - \lambda) \end{vmatrix} .$$

By subtraction of the first row multiplied by u from the second, and subtraction of the first row multiplied by H from the third we get:

$$|f' - \lambda\Gamma| = \rho \begin{vmatrix} \rho_p u - \lambda\theta & 1 & \rho_T(u - \lambda) \\ 1 & u - \lambda & 0 \\ \lambda & u(u - \lambda) & \rho c_p(u - \lambda) \end{vmatrix} = 0 .$$

This gives

$$(u - \lambda)\{\rho c_p - u^2(\rho_T + \rho\rho_p c_p) + \lambda u(2\rho_T + \rho\rho_p c_p + \theta\rho c_p) \\ -\lambda^2(\rho_T + \theta\rho c_p)\} = 0 .$$

Hence, we find one eigenvalue $\lambda = u$. We continue with the second factor. Substitution for θ from (14.6) gives

$$\lambda^2 - u(1 + \beta w^2)\lambda - w^2(1 - \beta u^2) = 0 .$$

The roots are given by

$$\lambda = \tilde{u} \pm \tilde{c} , \quad \tilde{u} = u(1 - \alpha) , \quad \tilde{c} = \sqrt{\alpha^2 u^2 + w^2} .$$

$\square$

A numerical scheme

We will now discretize (14.5) in space. For discretization in time the methods discussed in Chap. 10 can be followed. Applying finite volume integration (as in Chap. 10), we get

$$
h_j \Gamma_j \frac{dQ_j}{dt} + F(Q)|_{j-1/2}^{j+1/2} = 0 \,,
\tag{14.9}
$$

where h_j is the size of the cell. It remains to specify the cell face flux function $F_{j+1/2}$. Following Weiss and Smith (1995) we use a Roe type scheme for this. In Sect. 10.3 we saw that the Roe scheme gives

$$
F_{j+1/2} = \frac{1}{2}\{f(Q_j) + f(Q_{j+1})\} - \frac{1}{2}|A_{j+1/2}|(Q_{j+1} - Q_j) \,,
$$

where $A = \partial f/\partial Q$. This means that the spatial discretization is unaltered by the preconditioning. Practical experience and the analysis of Guillard and Viozat (1999), Turkel, Fiterman, and van Leer (1994) show that this gives unsatisfactory results at low Mach numbers. Much better results are obtained if the preconditioning is allowed to influence the artificial diffusion term. This can be done by choosing

$$
F_{j+1/2} = \frac{1}{2}\{f(Q_j) + f(Q_{j+1})\} - \frac{1}{2}\Gamma_{j+1/2}|\Gamma^{-1}A|_{j+1/2}(Q_{j+1} - Q_j) \,.
\tag{14.10}
$$

This does not change the consistency of the scheme, since we still have a central scheme with artificial diffusion. But it does change the numerical solution.

The next step is the evaluation of the artificial diffusion term. This can be done in the same way as for the Roe scheme. Let R be the matrix with the eigenvectors R_1, R_2, R_3 of $\Gamma^{-1}A$ as columns, and let R^1, R^2, R^3 be row vectors defined by

$$
R^q R_p = \delta_p^q \,.
$$

Define

$$
\alpha_p = R^p (Q_{j+1} - Q_j) \,.
$$

Then

$$
|\Gamma^{-1}A|_{j+1/2}(Q_{j+1} - Q_j) = \sum_{p=1}^{3} |\lambda_p| \alpha_p R_p \,,
$$

where, as we saw before, $\lambda_1 = u$, $\lambda_2 = \tilde{u} - \tilde{c}$, $\lambda_3 = \tilde{u} + \tilde{c}$. In Example 14.3.2 we show that

$$(R_1, R_2, R_3) = \begin{pmatrix} 0 & \alpha u + \tilde{c} & \alpha u - \tilde{c} \\ 0 & (\alpha u + \tilde{c})u - \frac{1}{\rho} & (\alpha u - \tilde{c})u - \frac{1}{\rho} \\ 1 & (\alpha u + \tilde{c})(H + \frac{1}{\rho c_p}) & (\alpha u - \tilde{c})(H + \frac{1}{\rho c_p}) \end{pmatrix},$$

and

$$\begin{pmatrix} R^1 \\ R^2 \\ R^3 \end{pmatrix} = \begin{pmatrix} -(H + \frac{1}{\rho c_p}) & 0 & 1 \\ \frac{1}{2\tilde{c}}(1 - \rho u(\alpha u - \tilde{c})) & \frac{\rho}{2\tilde{c}}(\alpha u - \tilde{c}) & 0 \\ \frac{1}{2\tilde{c}}(-1 + \rho u(\alpha \tilde{u} + \tilde{c})) & -\frac{\rho}{2\tilde{c}}(\alpha u + \tilde{c}) & 0 \end{pmatrix}.$$

The matrices $\Gamma_{j+1/2}$ and $A_{j+1/2}$ are not evaluated at the Roe average of Q_{j+1} and Q_j, because the nice properties of the Roe average are now lost. Instead, evaluation takes place at $(Q_j + Q_{j+1})/2$.

For extension of this method to the two-dimensional viscous case, see Weiss and Smith (1995), where successful applications to fully and weakly compressible flows are shown.

Example 14.3.2. Eigenvectors for artificial diffusion term

The eigenvectors of $\Gamma^{-1}A$ satisfy

$$(A - \lambda \Gamma)R =$$
$$\begin{pmatrix} \rho_p u - \lambda \theta & \rho & \rho_T u - \lambda \rho_T \\ \rho_p u^2 + 1 - \lambda u \theta & 2\rho u - \lambda \rho & \rho_T u^2 - \lambda \rho_T u \\ \rho_p uH - \lambda H \theta + \lambda & \frac{3}{2}\rho u^2 + \rho h - \lambda \rho u & (\rho_T H + \rho c_p)(u - \lambda) \end{pmatrix} \begin{pmatrix} r_1 \\ r_2 \\ r_3 \end{pmatrix}$$
$$= 0.$$

For $\lambda = \lambda_1 = u$ we see immediately that a solution is

$$R_1 = \begin{pmatrix} 0 \\ 0 \\ 1 \end{pmatrix}.$$

By subtraction of the first row multiplied by u from the second and subtraction of the first row multiplied by H from the third we get

$$\begin{pmatrix} \rho_p u - \lambda \theta & \rho & \rho_T(u - \lambda) \\ 1 & \rho(u - \lambda) & 0 \\ \lambda & \rho u(u - \lambda) & \rho c_p(u - \lambda) \end{pmatrix} \begin{pmatrix} r_1 \\ r_2 - u r_1 \\ r_3 - H r_1 \end{pmatrix} = 0.$$

Hence

$$r_2 = u r_1 + \frac{r_1}{\rho(\lambda - u)}, \qquad r_3 = H r_1 + \frac{r_1}{\rho c_p}.$$

For $\lambda = \lambda_2 = \tilde{u} - \tilde{c}$ this gives

$$r_2 = ur_1 - \frac{r_1}{\rho(\alpha u + \tilde{c})} \ , \quad r_3 = Hr_1 + \frac{r_1}{\rho c_p} \ .$$

We choose $r_1 = \alpha u + \tilde{c}$, and obtain:

$$R_2 = \begin{pmatrix} \alpha u + \tilde{c} \\ u(\alpha u + \tilde{c}) - \frac{1}{\rho} \\ (H + \frac{1}{\rho c_p})(\alpha u + \tilde{c}) \end{pmatrix} \ .$$

Similarly, for $\lambda = \lambda_3 = \tilde{u} + \tilde{c}$ we obtain the following eigenvector:

$$R_3 = \begin{pmatrix} \alpha u - \tilde{c} \\ u(\alpha u - \tilde{c}) - \frac{1}{\rho} \\ (H + \frac{1}{\rho c_p})(\alpha u - \tilde{c}) \end{pmatrix} \ .$$

The associated bi-orthonormal system of row vectors R^q satisfying $R^q \cdot R_p = \delta_p^q$ is given by

$$R^p = R_q \times R_r / R_1 \cdot (R_2 \times R_3) \ , \quad p, q, r \quad \text{cyclic} \ .$$

We find

$$\begin{pmatrix} R^1 \\ R^2 \\ R^3 \end{pmatrix} = \begin{pmatrix} -H - \frac{1}{\rho c_p} & 0 & 1 \\ \frac{1}{2\tilde{c}}(1 - \rho u(\alpha u - \tilde{c})) & \frac{\rho}{2\tilde{c}}(\alpha u - \tilde{c}) & 0 \\ \frac{1}{2\tilde{c}}(-1 + \rho u(\alpha u + \tilde{c})) & -\frac{\rho}{2\tilde{c}}(\alpha u + \tilde{c}) & 0 \end{pmatrix} \ .$$

$\square$

Unsteady flows

To compute unsteady flows with a preconditioned scheme, time accuracy has to be restored. One way to do this is to use *dual time stepping*. A pseudo-time s is introduced. The physical time is denoted by t. Equation (14.9) is replaced by

$$h_j \frac{dU_j}{dt} + h_j \Gamma_j \frac{dQ_j}{ds} + F(Q)|_{j-1/2}^{j+1/2} = 0 \ .$$

Time discretization methods in t and s are chosen. For simplicity, we choose the implicit Euler scheme in time and the explicit Euler scheme in pseudo-time, and obtain:

$$\frac{h_j}{\tau} \left(\frac{\partial U}{\partial Q} \right)_j^{m-1} (Q_j^m - Q_j^n) + \frac{h_j \Gamma_j}{\sigma}(Q_j^m - Q_j^{m-1}) + F(Q^{m-1})|_{j-1/2}^{j+1/2} = 0 \ ,$$

where τ is the time step and σ is the pseudo-time step. The superscript n counts physical time steps and the index m counts pseudo-time steps. For

$m = 0$ we put $Q^m = Q^n$. The next time step is executed by stepping forward with $m = 1, 2, ..., \bar{m}$, where $\bar{m}$ follows from the condition that steady state has been reached in pseudo-time:

$$\left| \frac{h_j \Gamma_j}{\sigma} (Q_j^{\bar{m}} - Q_j^{\bar{m}-1}) \right| < \text{tol} \ll 1 \, .$$

When this holds the pseudo-time difference can be neglected, so that we have completed a time step for the physical system

$$h_j \frac{dU_j}{dt} + F(Q)|_{j-1/2}^{j+1/2} = 0 \, .$$

We put $Q^{n+1} = Q^{\bar{m}}$, and the process can be repeated for the next physical time step.

Because the preconditioning influences the artificial diffusion term in $F(Q)$, different, and as it turns out, more accurate results are obtained for weakly compressible flows than with the original unpreconditioned equations.

In Weiss and Smith (1995) it is shown that dual time stepping is significantly more efficient than physical time stepping with the original system for low Mach numbers. Nevertheless, dual time stepping is expensive, because a large number of pseudo-time steps (30 in an example given in Weiss and Smith (1995)) is required for each physical time step. Hence, the efficiency lags behind that of incompressible flow solvers, so that dual time stepping methods cannot be called efficient.

In Guillard and Viozat (1999) it is shown that that pseudo-time stepping can be dispensed with. We can solve directly

$$\frac{dU_j}{dt} + F(Q)|_{j-1/2}^{j+1/2} = 0 \, ,$$

with $F_{j+1/2}$ given by (14.10). As remarked in Paillère et al. (1998), for explicit schemes stability requires a very small time step, so that an implicit time stepping scheme must be used. If this is necessary anyway, because of a large disparity in mesh sizes for example, this is no disadvantage. However, it turns out that the resulting nonlinear algebraic system is difficult, and iterative methods that have been used until now require much computing time. This still needs improvement.

Final remarks

For extensions to more dimensions, the viscous case, flows with combustion and chemistry and implementation of boundary conditions, the reader is referred to the literature. For a review of early literature, see Turkel (1993). The preconditioner of Weiss and Smith (1995) described above is an extension of the one introduced in Choi and Merkle (1985), Choi and Merkle (1993). Iterative methods for implicit preconditioned schemes are analyzed in Koren and van Leer (1995), Koren (1996), Darmofal (1998). Some recent publications are: Turkel, Radespiel, and Kroll (1997), Edwards and Roy (1998), Guillard and Viozat (1999), Mary, Sagaut, and Deville (2000).

Preconditioning is a way to extend the functionality of existing codes for fully compressible flows to almost incompressible flows. The efficiency of methods for incompressible flows is not matched. In the next sections the reverse route will be followed, namely extension of methods for incompressible flows to compressible flows. We will find that this results in methods that can match the efficiency of compressible flow methods for fully compressible flows, with superior efficiency for weakly compressible flows.

Exercise 14.3.1. Derive (14.8).

14.4 Mach-uniform dimensionless Euler equations

As we saw in Sect. 14.2, the Euler equations become singular as the Mach number tends to zero, if the equations are made dimensionless in the way that is customary for fully compressible flows. It is obvious that this difficulty also occurs for the Navier-Stokes equations. We will see presently that when the equations are not made dimensionless, the difficulty manifests itself in the fact that as $M \downarrow 0$, the pressure fluctuations that drive the flow become vanishingly small compared to the absolute pressure. To avoid numerical troubles, this has to be remedied. A different way to make the equations dimensionless has to be found. The unknowns have to remain bounded both for $M \downarrow 0$ and $M \gg 1$. To prepare ourselves, we first make an asymptotic analysis of the Euler equations as $M \downarrow 0$.

Asymptotic expansion

We consider the three-dimensional case. The nonconservative equations of motion (12.3)–(12.5) without heat addition and body force are made dimensionless in the same way as in Sect. 14.2, repeated here. The three space coordinates are made dimensionless by the same reference length x_r which is representative of the length scales occurring in the solution. The unit of velocity u_r is representative of the velocity magnitudes that occur, and T_r and ρ_r are of the order of the magnitudes of the temperature and the density of the flow. The reference pressure p_r is deduced from the equation of state. We obtain the following dimensionless equations:

$$\frac{D\rho}{Dt} + \rho \operatorname{div} \boldsymbol{u} = 0 \,, \tag{14.11}$$

$$\rho \frac{D\boldsymbol{u}}{Dt} + \frac{1}{\varepsilon} \operatorname{grad} p = 0 \,, \quad \varepsilon = \rho_r u_r^2 / p_r \,, \tag{14.12}$$

$$\frac{Dp}{Dt} + \gamma p \operatorname{div} \boldsymbol{u} = 0 \,. \tag{14.13}$$

For a perfect gas we have $p_r/\rho_r = c_r^2/\gamma$, with c_r the speed of sound belonging to the reference conditions, so that

$$\varepsilon = \gamma \mathrm{M}_r^2 \,, \quad \mathrm{M}_r = u_r/c_r \,.$$

In order to study the incompressible limit of (14.11)–(14.13), we postulate an asymptotic expansion of the following form:

$$\begin{aligned}
\boldsymbol{u}(t, \boldsymbol{x}) &= \boldsymbol{u}_0(t, \boldsymbol{x}) + \varepsilon \boldsymbol{u}_1(t, \boldsymbol{x}) + \mathcal{O}(\varepsilon^2) \,, \\
\rho(t, \boldsymbol{x}) &= \rho_0(t, \boldsymbol{x}) + \varepsilon \rho_1(t, \boldsymbol{x} + \mathcal{O}(\varepsilon^2) \,, \\
p(t, \boldsymbol{x}) &= p_0(t, \boldsymbol{x}) + \varepsilon p_1(t, \boldsymbol{x}) + \mathcal{O}(\varepsilon^2) \,.
\end{aligned} \tag{14.14}$$

This implies that $\boldsymbol{u}_0$, $\boldsymbol{u}_1$, ρ_0 etc. and their t- and x^α-derivatives are independent of ε. Since acoustic modes have ε-dependent time and length scales, these are excluded by the Ansatz (14.14). For asymptotic analysis including acoustics, see Klein (1995). If the terms $\boldsymbol{u}_0$, $\boldsymbol{u}_1$, ρ_0 etc. can be determined in a consistent way, the asymptotic expansion (14.14) may be safely assumed to be correct.

By substituting (14.14) in (14.11)–(14.13) and equating terms with like powers of ε we get the following results. To leading order, equation (14.12) gives $\operatorname{grad} p_0 = 0$, hence

$$p_0 = p_0(t) \,.$$

Because p_0 does not depend on $\boldsymbol{x}$, it has no dynamic effect. If there is no global compression or expansion, as for the flow around the body, $p_0(t) = $ constant, and p_0 represents the constant background pressure level. If there is net mass inflow or outflow in the flow domain, $p_0(t)$ represents the effect of global compression or expansion. This can be seen as follows. Equation (14.13) gives to leading order:

$$\operatorname{div} \boldsymbol{u}_0 = -\frac{1}{\gamma} \frac{d \ln p_0}{dt} \,. \tag{14.15}$$

Integration over the flow domain Ω gives

$$\frac{d \ln p_0}{dt} = -\frac{\gamma}{|\Omega|} \int_{\partial \Omega} \boldsymbol{u}_0 \cdot \boldsymbol{n} dS \,, \tag{14.16}$$

so that p_0 follows from the normal velocity component at the boundary. This makes it clear that $p_0(t)$ is due to global compression or expansion.

If the normal velocity is not prescribed along the whole boundary, $p_0(t)$ must be given, but if it is, $p_0(t)$ follows easily from (14.16). Hence, $p_0(t)$ may be assumed known. Equation (14.11) gives to leading order the equation that governs ρ_0:

$$(\frac{\partial}{\partial t} + \boldsymbol{u}_0 \cdot \nabla) \ln \rho_0 = \frac{1}{\gamma} \frac{d \ln p_0}{dt} . \tag{14.17}$$

We still need an equation to determine $\boldsymbol{u}_0$, since (14.15) is not sufficient. The equation for $\boldsymbol{u}_0$ is obtained from the next order terms of (14.12):

$$\rho_0 \frac{D\boldsymbol{u}_0}{Dt} + \mathrm{grad} p_1 = 0 . \tag{14.18}$$

Equations (14.15)–(14.18) determine $\boldsymbol{u}_0$, ρ_0 and p_1; p_1 acts as a Lagrange multiplier to allow $\boldsymbol{u}_0$ to satisfy the constraint (14.15). When ρ_0 and p_0 are constant we recover with (14.15) and (14.18) the familiar incompressible Euler equations. We may conclude that for the determination of $\boldsymbol{u}_0$, ρ_0, p_0, p_1 we have obtained a well-posed problem, which indicates that the asymptotic series (14.14) is correct.

Definition of dimensionless pressure

We have just seen that

$$p(t, \boldsymbol{x}) = p_0(t) + \varepsilon p_1(t, \boldsymbol{x}) + \mathcal{O}(\varepsilon^2) .$$

We see that for $\varepsilon \ll 1$ there is a danger of rounding errors in numerical approximations of $\mathrm{grad} p$. This can be avoided by working with $p - p_0$ instead of p. This is possible because, as argued above, $p_0(t)$ can be determined a priori. From (14.18) we see that p_1 is of the same size as $\rho_0 \boldsymbol{u}_0 \cdot \boldsymbol{u}_0$. We therefore choose the dimensionless pressure as follows:

$$p = \frac{\hat{p} - \hat{p}_0}{\hat{\rho}_r \hat{u}_r^2} , \tag{14.19}$$

where for clarity the hat symbol is used to denote dimensional quantities. The other quantities are made dimensionless as in Sect. 10.2.

Assuming a perfect gas, we have the following dimensional equation of state:

$$\hat{p} = R \hat{\rho} \hat{T} .$$

The dimensionless form of the equation of state becomes, noting that $\hat{c}^2 = \gamma R \hat{T}$,

$$p = \frac{R\hat{\rho}\hat{T} - \hat{p}_0}{\hat{\rho}_r \hat{u}_r^2} = \frac{1}{\gamma M_r^2}\rho T - \frac{\hat{p}_0}{\hat{\rho}_r \hat{u}_r^2} \, , \quad M_r = \hat{u}_r/\hat{c}_r \, , \tag{14.20}$$

which is singular as $M_r \downarrow 0$, no matter how we choose $\hat{p}_0$. This is a sign that in order to handle weakly compressible flows, we should use the pressure as a primitive variable and find the density from the equation of state, and not the other way around. Therefore the dimensionless equation of state is rewritten as $\rho = \rho(p, T)$:

$$\rho = \left(\frac{p}{T} + \frac{1}{T}\frac{\hat{p}_0}{\hat{\rho}_r \hat{U}^2}\right)\gamma M_r^2 \, . \tag{14.21}$$

We now restrict ourselves to the case where there is no global compression or expansion, so that $\hat{p}_0(t) = $ constant. This means that flows into or out of vessels are excluded, and only flows around bodies or through channels are considered. We relate $\hat{p}_0$ to $\hat{\rho}_r$ and $\hat{T}_r$ by the equation of state:

$$\hat{p}_0 = R\hat{\rho}_r\hat{T}_r \, . \tag{14.22}$$

Substitution in (14.21) gives the following dimensionless equation of state:

$$\rho = (1 + \gamma M_r^2 p)/T \, . \tag{14.23}$$

This is precisely what we want. When $M_r \to 0$ the dependence of ρ on p disappears, eliminating acoustics, but ρ still varies with temperature. Substitution of (14.22) in (14.19) gives our final definition for the dimensionless pressure:

$$p = \frac{\hat{p}}{\hat{\rho}_r \hat{u}_r^2} - \frac{1}{\gamma M_r^2} \, . \tag{14.24}$$

The dimensionless form of the nonconservative homogeneous equations of motion (12.3)–(12.5) is found to be

$$\rho_t + \text{div}\boldsymbol{m} = 0 \, ,$$
$$\boldsymbol{m}_t + \frac{\partial}{\partial x^\alpha}(u^\alpha \boldsymbol{m}) + \text{grad}p = 0 \, ,$$
$$M_r^2\{p_t + \text{div}(\boldsymbol{u}p) + (\gamma - 1)p\text{div}\boldsymbol{u}\} + \text{div}\boldsymbol{u} = 0 \, . \tag{14.25}$$

We see that as $M_r \downarrow 0$ the system does not become singular, and that the pressure equation reduces to the familiar solenoidality condition of incompressible flow.

Dimensionless conservative formulation

The dimensional form of the conservative energy equation is rewritten as

$$\left(c_v\rho T + \frac{1}{2}\rho u^2\right)_t + \left(\rho u c_p T + \frac{1}{2}\rho u^3\right)_x = 0 \, .$$

The dimensionless form becomes

$$\left(\rho T + \frac{\hat{u}_r^2}{c_v\hat{T}_r}\frac{1}{2}u^2\right)_t + \left(\gamma\rho u T + \frac{\hat{u}_r^2}{c_v\hat{T}_r}\frac{1}{2}\rho u^3\right)_x = 0 \, .$$

Noting that $c_v T = c^2/\gamma(\gamma - 1)$ we can rewrite this as

$$\left(\rho T + \gamma(\gamma - 1)\mathrm{M}_r^2\frac{1}{2}\rho u^2\right)_t + \left(\gamma\rho u T + \gamma(\gamma - 1)\mathrm{M}_r^2\frac{1}{2}\rho u^3\right)_x = 0 \, .$$

This can be put in a more familiar form if we define the dimensionless enthalpy and internal energy as

$$h = \gamma T \, , \quad e = T \, ,$$

and if, furthermore, we define the dimensionless total enthalpy and total energy as

$$\tilde{H} = h + \gamma(\gamma - 1)\mathrm{M}_r^2\frac{1}{2}u^2 \, , \quad \tilde{E} = e + \gamma(\gamma - 1)\mathrm{M}_r^2\frac{1}{2}u^2 \, ,$$

where the tilde serves to remind us that these definitions are not standard. Then the conservative energy equation takes on its familiar form:

$$(\rho\tilde{E})_t + (m\tilde{H})_x = 0 \, , \quad m = \rho u \, . \tag{14.26}$$

As $\mathrm{M}_r \downarrow 0$, solving p from the equation of state (14.23) is not possible. Therefore the conservative energy equation (14.26) should not be used for weakly compressible flow, but the nonconservative version (14.25).

Boundary conditions

For the units employed to make the equations dimensionless, appropriate values may be deduced from the boundary and/or the initial conditions. Boundary conditions for the Euler equations have been discussed in Sect. 10.2, 12.2 and 12.4. We assume that velocity and temperature are given at the inflow boundary, and the pressure at the outflow boundary if the normal outflow velocity is subsonic, and at the inflow boundary if the normal inflow velocity is supersonic. If the inflow velocity is zero, as in a shock tube problem, a suitable velocity unit $\hat{u}_r$ is deduced from the initial conditions. For instance, in a shock tube problem one may take

$$\hat{u}_r = 2\sqrt{|p_R - p_L|/(\rho_R + \rho_L)} \, ,$$

where R and L denote the initial right and left states. In this way we obtain $\hat{u}_r$, $\hat{p}_0$ and $\hat{T}_r$. Equation (14.22) gives $\hat{\rho}_r$. The reference speed of sound and Mach number are given by

$$\hat{c}_r = \sqrt{\gamma R \hat{T}_r} \; , \quad M_r = \hat{u}_r / \hat{c}_r \; .$$

The boundary conditions may be time-dependent, and can be summarized as follows, taking the one-dimensional case with subsonic outflow as an example:

$$u(t,0) = u_b(t) \equiv \hat{u}(t,0)/\hat{u}_r \; , \quad T(t,0) = T_b(t) \equiv \hat{T}(t,0)/\hat{T}_r \; ,$$

$$p(t,1) = p_b(t) \equiv \frac{\hat{p}(t,\hat{L})}{\hat{\rho}_r \hat{u}_r^2} - \frac{1}{\gamma M_r^2} \; .$$

Boundary conditions at solid walls and boundary conditions for the Navier-Stokes equations are the same as those discussed in Sect. 12.4 and 13.2.

14.5 A staggered scheme for fully compressible flow

In the next section the classical staggered scheme of Harlow and Welch (1965) for incompressible flow will be extended to compressible flows. But first, in the present section, we will develop a staggered scheme for the conservative fully compressible Euler equations. The dimensionless form of these equations was found in the preceding section to be given by

$$\rho_t + m_x = 0 \; ,$$
$$m_t + (um + p)_x = 0 \; ,$$
$$(\rho \tilde{E})_t + (m \tilde{H})_x = 0 \; .$$

The primary unknowns are ρ, m, $\tilde{E}$. As noted in the preceding section, this leads to trouble for weakly compressible flow, but the case of weak compressibility will be the topic of the next section. Here we want to investigate if a staggered scheme can be used for fully compressible flow.

Finite volume discretization

Fig. 14.1 shows a staggered grid with uniform cell-size h. The cell centers are located at $x_j \equiv (j - 1/2)h$, $j = 1, 2, ..., J$. The cell faces are located at $x_{j+1/2}$. The momentum unknowns are located at the cell faces, and the thermodynamic unknowns are located at the cell centers, quite similar to the

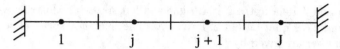

Fig. 14.1. A one-dimensional staggered grid.

staggered scheme of Sect. 6.4.

The dimensionless governing equations are discretized in space with the finite volume method. Integration of the mass conservation equation over a cell with center at x_j gives, with explicit Euler time stepping:

$$\rho_j^{n+1} - \rho_j^n + \lambda(\rho^n u^n)|_{j-1/2}^{j+1/2} = 0 , \quad \lambda = \tau/h .$$

The first order upwind scheme gives, if $u_{j+1/2} > 0$,

$$\rho_{j+1/2}^n = \rho_j^n ,$$

whereas the second order slope limited scheme of Sect. 4.8 gives

$$\rho_{j+1/2}^n = \rho_j^n + \frac{1}{2}\psi(r_j^n)(\rho_{j+1}^n - \rho_j^n) .$$

Similarly, the momentum equation is discretized as

$$m_{j+1/2}^{n+1} - m_{j+1/2}^n + \lambda(u^n m^n + p^n)|_j^{j+1} = 0 ,$$

where $(um)_j^n$ can be approximated with the first order upwind scheme or the slope limited scheme. Finally, the energy equation is discretized as follows:

$$(\rho\tilde{E})_j^{n+1} - (\rho\tilde{E})_j^n + (u\rho\tilde{H})^n|_{j-1/2}^{j+1/2} = 0 ,$$

where $(\rho\tilde{H})_{j+1/2}^n$ can be approximated with the first order upwind scheme or the slope limited scheme. After $(\rho\tilde{E})^{n+1}$ has been obtained, p^{n+1} follows from $p = (\rho e - 1)/\gamma M_r^2$, $(\rho e)_j = (\rho\tilde{E})_j - \gamma(\gamma - 1)M_r^2\frac{1}{4}\{(um)_{j+1/2} + (um)_{j-1/2}\}$, $u_{j+1/2} = 2m_{j+1/2}/(\rho_j + \rho_{j+1})$.

Boundary conditions

We assume that $u > 0$, and that inflow and outflow conditions are subsonic. The following boundary conditions are assumed:

$$u(t, 0) = u_b(t) , \quad \rho(t, 0) = \rho_b(t) , \quad p(t, 1) = p_b(t) .$$

We put $\rho_{J+1/2} = \rho_J$, corresponding to the use of the first order upwind scheme at the outflow boundary. For $m_{J+1/2}$, finite volume integration takes place over a half cell:

$$m_{J+1/2}^{n+1} - m_{J+1/2}^n + 2\lambda(u^n m^n)|_J^{J+1/2} + 2\lambda(p_b(t_n) - p_J^n) = 0 \ .$$

We write

$$(\rho\tilde{H})_{1/2} = \rho_b(t)\{h_1 + \gamma(\gamma-1)M_r^2\frac{1}{2}u_b(t)^2\} \ ,$$

and $(\rho\tilde{H})_{J+1/2} = (\rho\tilde{H})_J$, corresponding to the first order upwind scheme. The accuracy of numerical solutions for Riemann problems benefits if the initial discontinuous solutions are applied consistently. For example, if the discontinuity is at a velocity node $x_{j+1/2}$, we put

$$u_{j+1/2} = (u_L + u_R)/2 \ , \quad m_{j+1/2} = u_{j+1/2}(\rho_L + \rho_R)/2 \ .$$

and for all j

$$(\rho\tilde{E})_j = \gamma M_r^2(1 + p_j) + \gamma(\gamma-1)M_r^2\frac{1}{4}\{(um)_{j+1/2} + (um)_{j-1/2}\} \ .$$

Numerical results

Results for the Riemann problems of Chap. 10 are shown in Figs. 14.2–14.6. The slope limited scheme is used with the van Albada limiter of Sect. 10.8. The same time stepping scheme is used as in Sect. 10.8 for colocated schemes, namely the SHK Runge-Kutta scheme. The estimates for the stability limit λ_1 for λ are given in Table 10.4. In practice we found that sometimes λ has to be somewhat smaller than λ_1 for good results. Unless noted otherwise, $h = 1/48$. Comparison with the results of Sect. 10.8 shows that the staggered scheme has the same accuracy as the colocated schemes of Sect. 10.8. In the case of the supersonic shocktube there is a velocity overshoot at the beginning of the expansion fan. This is related to the smearing of the density, which is due to insufficient resolution. There are only a few cells in the narrow zones between the fan, the contact discontinuity and the shock. This hampers a staggered scheme more than a colocated scheme, because due to the staggered placement of the unknowns the stencil is effectively a bit wider.

The accuracy improves significantly if the mesh size is halved, as shown by Fig. 14.6. Fig. 14.7 shows results for a moving contact discontinuity. There are spurious wiggles of size less than 0.2% in velocity and pressure. These insignificant wiggles arise from imperfect representation of the initial discontinuity on the staggered grid, and spurious reflection from the outflow boundary.

The staggered scheme requires significantly less computing time than the colocated schemes of Sect. 10.8, because the numerical fluxes are much simpler. We conclude that the staggered scheme converges to genuine weak solutions that satisfy the entropy condition, and is quite suitable for compressible flows.

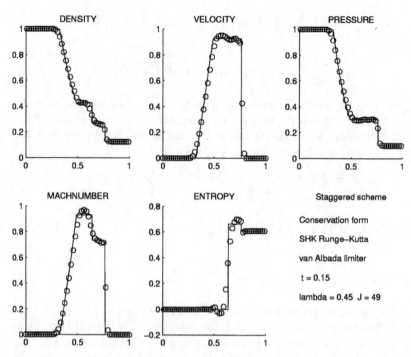

Fig. 14.2. Sod's test problem with staggered scheme.

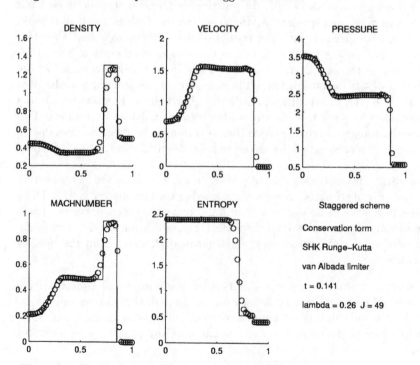

Fig. 14.3. Test case of Lax with staggered scheme.

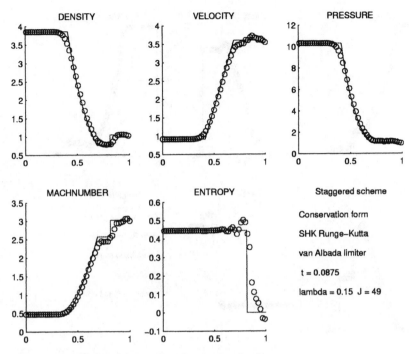

Fig. 14.4. Mach 3 test case with staggered scheme.

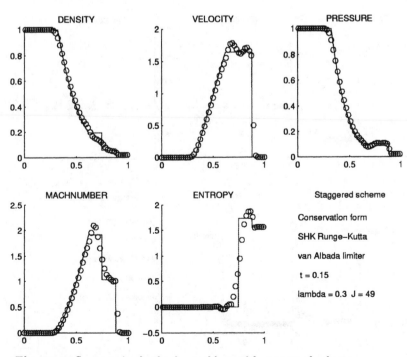

Fig. 14.5. Supersonic shocktube problem with staggered scheme.

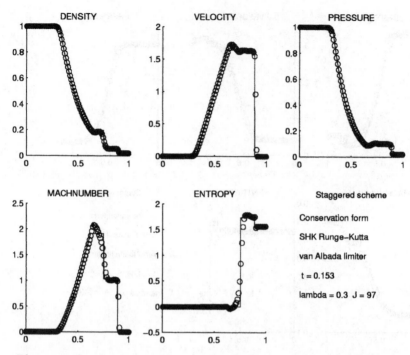

Fig. 14.6. Supersonic shocktube problem with $h = 1/96$.

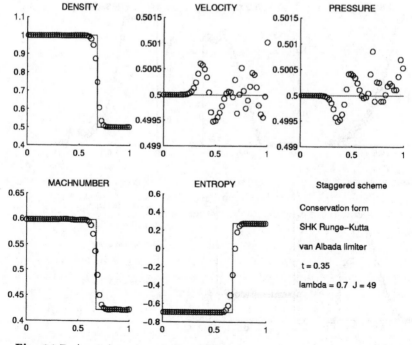

Fig. 14.7. A moving concact discontinuity.

14.6 Unified schemes for incompressible and compressible flow

The scheme of the preceding section breaks down when the flow is almost incompressible, because the pressure is not used as one of the primary unknowns. In this section a well-established discretization scheme for incompressible flows will be extended to compressible flows. Obviously, this may be expected to result in an efficient and accurate method for weakly compressible flows. It will come as a pleasant surprise that good performance is also obtained for fully compressible flows.

Point of departure is the classical staggered scheme for incompressible flow of Harlow and Welch (1965), discussed in Sect. 6.4. Generalization of this scheme to the compressible case has already been undertaken in Harlow and Amsden (1968), Harlow and Amsden (1971). There are mainly four differences between our approach and that of Harlow and Amsden. In the first place, we use the dimensionless pressure as defined by (14.19). This removes difficulties with rounding errors at very low Mach numbers. In the second place, we use a nonconservative form of the energy equation, for better efficiency. It turns out that this does not impair the approximation accuracy of genuine weak solutions. In the third place, we let the discretization of the mass conservation equation depend on the Mach number; this gives better accuracy for supersonic flow. Finally, we use the MUSCL approach and Runge-Kutta time stepping for second order accuracy.

Governing equations

The dimensionless equations are used. In order to handle weakly compressible and incompressible flows, we use the pressure instead of the density as primary unknown. The mass conservation equation is written as follows:

$$\rho_p p_t + \rho_T T_t + m_x = 0 . \tag{14.27}$$

The equation of state (14.23) gives

$$\rho_p = \frac{\gamma M_r^2}{T} , \quad \rho_T = -\frac{\rho}{T} .$$

The momentum equation is given by

$$m_t + (um + p)_x = 0 . \tag{14.28}$$

For the energy equation we start from the nonconservative dimensional version (14.13). By using the definition (14.24) of the dimensionless pressure and the equation of state (14.23) the energy equation can be cast in the following dimensionless form:

$$T_t + (uT)_x + (\gamma - 2)Tu_x = 0 . \tag{14.29}$$

We will also use a different set of equations. Instead of the temperature equation (14.29) the non-conservative pressure equation (14.13) is used. In that case, the dimensionless governing equations are:

$$\rho_t + m_x = 0 , \tag{14.30}$$

$$m_t + (um + p)_x = 0 , \tag{14.31}$$

$$\mathrm{M}_r^2\{p_t + (up)_x + (\gamma - 1)pu_x\} + u_x = 0 . \tag{14.32}$$

Finite volume scheme

The staggered grid of Fig. 14.1 is used. We propose the following finite volume scheme for (14.27)–(14.29):

$$\gamma \mathrm{M}_r^2(p_j^{n+1} - p_j^n) - \rho_j^n(T_j^{n+1} - T_j^n) +$$

$$\lambda T_j^n(s^n\rho^n u^n + (1 - s^n)m^{n+1})|_{j-1/2}^{j+1/2} = 0 , \tag{14.33}$$

$$m_{j+1/2}^{n+1} - m_{j+1/2}^n + \lambda(u^n m^n + p^{n+1/2})|_j^{j+1} = 0 , \tag{14.34}$$

$$T_j^{n+1} - T_j^n + \lambda(u^n T^n)|_{j-1/2}^{j+1/2} + \lambda(\gamma - 2)T_j^n u^n|_{j-1/2}^{j+1/2} = 0 , \tag{14.35}$$

where $\lambda = \tau/h$, $p^{n+1/2} = (p^n + p^{n+1})/2$, and $s^n = s(\mathrm{M}^n)$ is a Mach-dependent switch function, defined as

$$s(\mathrm{M}) = 0 , \quad \mathrm{M} \leq 1/2 ,$$
$$s(\mathrm{M}) = \mathrm{M} - 1/2 , \quad 1/2 < |\mathrm{M}| < 3/2 ,$$
$$s(\mathrm{M}) = 1 , \quad \mathrm{M} \geq 3/2 .$$

Here $\mathrm{M}_{j+1/2} = 2|u_{j+1/2}|/(c_j + c_{j+1})$ with c the speed of sound. It is found that the results are insensitive to the definition of $s(\mathrm{M})$, as long as $s(\mathrm{M})$ switches from 0 to 1 in some neighbourhood of $\mathrm{M} = 1$.

We see that in (14.33)–(14.35) some terms are evaluated at the new time level t^{n+1}, namely m in (14.33) and p in (14.34). This is done in order to recover the classical staggered scheme in the incompressible limit.

In (14.33), $\rho_{j+1/2}^n$ is evaluated with a slope limited scheme, as described in the preceding section. The same method is used to approximate $(u^n m^n)_j$ and $T_{j+1/2}^n$.

Pressure-correction method

In the one-dimensional case the nonlinear system that we have obtained for p^{n+1}, T^{n+1} and m^{n+1} is easily solved. But in more dimensions iterative methods are required with efficiency uniform or almost uniform in the Mach number M_r, especially as $M_r \downarrow 0$. We therefore generalize an iterative method that is efficient in the incompressible case to the compressible case. For nonstationary flows we take our point of departure in the pressure-correction method, described in Sect. 6.6. For stationary flows it is more efficient to generalize one of the iterative methods for incompressible flows described in Sect.7.6. But here we restrict ourselves to the nonstationary case.

First, T^{n+1} is obtained from (14.35). Making this possible is the reason for using a non-conservative form for the energy equation. Next, a momentum prediction m^* is obtained from (14.34), replacing $p^{n+1/2}$ by p^n:

$$m^*_{j+1/2} - m^n_{j+1/2} + \lambda(u^n m^n + p^n)|^{j+1}_j = 0 .$$

A momentum correction $\delta m = m^{n+1} - m^*$ is postulated as

$$\delta m_{j+1/2} = -\frac{1}{2}\lambda \delta p|^{j+1}_j , \quad \delta p = p^{n+1} - p^n .$$

Substitution of $m^{n+1} = m^* + \delta m$ in (14.33) gives the following pressure-correction equation for δp:

$$\gamma M_r^2 \delta p_j - \frac{1}{2}\lambda^2 T^n_j \{(1 - s^n_{j+1/2})\delta p|^{j+1}_j - (1 - s^n_{j-1/2})\delta p|^j_{j-1}\} =$$
$$\rho^n_j T_j|^{n+1}_n - \lambda T^n_j (s^n \rho^n u^n)|^{j+1/2}_{j-1/2} . \tag{14.36}$$

Boundary conditions

The boundary conditions are implemented as in the preceding section. For the pressure-correction equation (14.36) no additional boundary conditions are required, just as in the incompressible case discussed in Chapt. 6. We have, for $j = 1$,

$$\frac{1}{2}\lambda \delta p|^1_0 = -\lambda \delta m_{1/2} = -\lambda(\rho_b u_b)|^{t_{n+1}}_{t_n} .$$

For $j = J$ the momentum equation is integrated over a half cell:

$$m^*_{J+1/2} - m^n_{J+1/2} + 2\lambda(u^n m^n + p^n)|^{J+1/2}_J = 0 , \quad p^n_{J+1/2} = p_b(t^n) ,$$
$$\delta m_{J+1/2} = -\lambda(p_b|^{t_{n+1}}_{t_n} - \delta p_J) ,$$

which replaces in (14.36) for $j = J$ the term $\lambda \delta p|^{j+1}_j$.

Runge-Kutta method

Instead of the Euler method we can use other time stepping schemes, for better temporal accuracy, or for better efficiency, due to weaker stability restrictions on λ. For example, we may employ the SHK Runge-Kutta scheme that was used before. Each stage is an Euler time step, so that its implementation is obvious. In more dimensions, the most expensive part of the algorithm is the solution of the implicit system (14.36). For efficiency, one may freeze the implicit part at the old time level in the first three stages, and take it at the new time level only in the fourth stage, which is an Euler step with the full time step τ. This means that the $(m+1)^{\text{th}}$ stage becomes:

$$T_j^{(m+1)} - T_j^n + \alpha_{m+1}\lambda(u^n T^{(m)})|_{j-1/2}^{j+1/2} + \alpha_{m+1}\lambda(\gamma - 2)T_j^{(m)} u^n|_{j-1/2}^{j+1/2} = 0 \,,$$

$$m_{j+1/2}^{(m+1)} - m_{j+1/2}^n + \alpha_{m+1}\lambda(u^n m^{(m)} + p^n)|_j^{j+1} = 0 \,.$$

In the fourth stage the pressure correction is carried out:

$$\gamma M_r^2 \delta p_j - \frac{1}{2}\lambda^2 T_j^{(4)}\{(1 - s_{j+1/2}^{(4)})\delta p|_j^{j+1} - (1 - s_{j-1/2}^{(4)})\delta p|_{j-1}^j\} =$$
$$\rho_j^n(T_j^{(4)} - T_j^n) - \lambda T_j^{(4)}(s^{(4)}\rho^n u^n)|_{j-1/2}^{j+1/2} \,.$$

Numerical experiments

The above scheme has been designed to reduce to the well-established incompressible staggered scheme with pressure-correction, as the Mach number tends to zero. It remains to investigate its performance in the fully compressible case. We do this by carrying out numerical experiments for the same Riemann problems as in the preceding section. The SHK Runge-Kutta scheme is used, as in Sections 14.5 and 10.8. Pressure-correction is applied only in the final stage, as just described. The scheme is more stable than the conservative scheme of Sect. 14.5 (due to the implicitness of the momentum-pressure coupling), but in order to have a fair comparison of accuracy the same values of λ will be used as in Sect. 14.5. Unless noted otherwise, $h = 1/48$.

Numerical results for the test problems of Sect. 14.5 are shown in Figs. 14.8–14.13. Comparison with the results obtained with the conservative scheme in Sect. 14.5 shows that, except for the contact discontinuity, the accuracy obtained is about the same. With the pressure-correction method shock resolution is a bit less crisp, but shock speed and strength are correct. For the test case of Lax the accuracy is somewhat less than for the conservative scheme. Grid refinement gives the results shown in Fig. 14.13, which show a better approximation of the expansion fan, but still contain a density overshoot between the shock and the contact discontinuity. These are close together and both strong, which is a numerically difficult situation. For the contact discontinuity spurious wiggles of magnitude 4% are generated, which is rather more than for the conservative scheme.

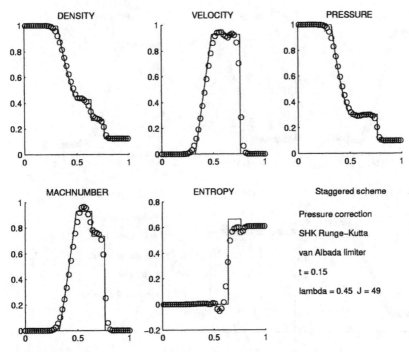

Fig. 14.8. Sod's test problem with pressure-correction scheme.

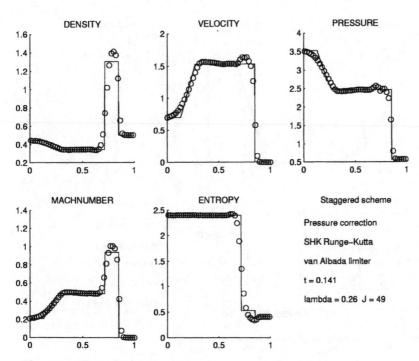

Fig. 14.9. Test case of Lax with pressure-correction scheme.

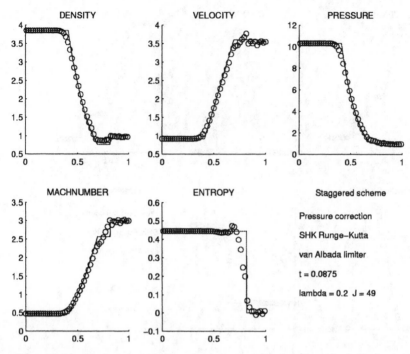

Fig. 14.10. Mach 3 test problem with pressure-correction scheme.

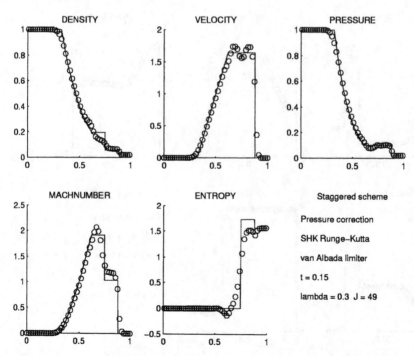

Fig. 14.11. Supersonic shocktube problem with pressure-correction scheme.

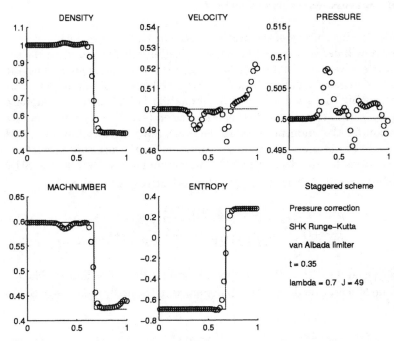

Fig. 14.12. A moving contact discontinuity with pressure-correction scheme.

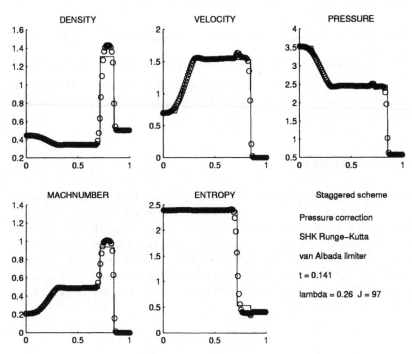

Fig. 14.13. Test case of Lax with $h = 1/96$.

Second pressure-correction method

Because of the inaccuracy of the preceding scheme for moving contact discontinuities we will derive a second pressure-correction scheme, based on the set of governing equations (14.30)–(14.32). Because $p = $ constant, $u = $ constant satisfies the pressure equation (14.32) exactly, we hope to obtain better accuracy for moving contact discontinuities. Equation (14.30) is used to update the density with a slope limited scheme, equation (14.31) is used to compute a prediction for the momentum using the pressure at the old time level, and equation (14.32) is used to obtain an equation for the pressure correction. The predictions for the density and the momentum are made with the SHK Runge-Kutta method. The $(m+1)^{\text{th}}$ Runge-Kutta stage is given by

$$\rho_j^{(m+1)} - \rho_j^n + \lambda(u^{(m)}\rho^{(m)})|_{j-1/2}^{j+1/2} = 0 \,,$$

$$m_{j+1/2}^{(m+1)} - m_{j+1/2}^n + \lambda(u^{(m)}m^{(m)} + p^n)|_j^{j+1} = 0 \,,$$

where $\rho_{j+1/2}^{(m)}$ and $(u^{(m)}m^{(m)})_j$ are approximated with a slope limited scheme. There is no Mach-dependent switch function, as in the first pressure-correction scheme. We put

$$\rho_j^{n+1} = \rho_j^{(4)} \,, \quad m_{j+1/2}^{n+1} = m_{j+1/2}^{(4)} - \frac{1}{2}\lambda\delta p|_j^{j+1} \,, \quad \delta p = p^{n+1} - p^n \,. \quad (14.37)$$

The pressure equation (14.32) is discretized as follows:

$$M_r^2\{\delta p_j + \lambda(u^{n+1}p^n)|_{j-1/2}^{j+1/2} + \lambda(\gamma-1)p_j^n u^{n+1}|_{j-1/2}^{j+1/2}\} + \lambda u^{n+1}|_{j-1/2}^{j+1/2} = 0 \,,$$

where $p_{j+1/2}^n$ is approximated with a slope limited scheme. Substitution of $u_{j+1/2}^{n+1} = (m/\rho)_{j+1/2}^{n+1}$, with m^{n+1} given by (14.37), gives a linear system for δp. The equivalence with the classical incompressible pressure-correction method as $M_r \downarrow 0$ is obvious.

Numerical experiments

The same test cases are computed as before, with the same values of h and λ. Results are shown in Figs. 14.14–14.19. Shock resolution is a bit more crisp than for the first pressure-correction method. The accuracy for the moving contact discontinuity is much better, which is the reason for introducing this pressure-correction method. The density overshoot in the density for the test problem of Lax is still present, and the approximation of the expansion fan is less accurate. Grid refinement gives the more satisfactory results of Fig. 14.19.

Further discussion

We have presented two pressure-correction methods, that reduce to the classical incompressible pressure-correction method of Harlow and Welch (1965) as the Mach number tends to zero. This makes it possible to design solution methods for which the computing work is almost uniform in the Mach number.

The two pressure-correction methods use different formulations of the energy equation. Although both formulations are non-conservative, weak solutions are found to be approximated correctly. But the first pressure-correction method may give rather large spurious oscillations for strong contact discontinuities, whereas the accuracy of the second method is much better for this case, and shock resolution is more crisp.

In Sect. 14.5 results have been presented for the conservative Euler equations discretized on a staggered grid. These results illustrate the kind of accuracy that may be obtained for fully compressible flows on a staggered grid. The accuracy is found to be about the same as for the commonly used colocated schemes discussed in Chap. 10. Because the staggered schemes use simple central and upwind-biased differences to compute the numerical flux, they require less computing time than the colocated schemes, which employ more computing-intensive flux splitting methods or second and fourth order artificial viscosity terms. Furthermore, for the staggered schemes the coupling between density, pressure and momentum is weaker, paving the way for more efficient solution methods for implicit schemes.

It was found both for the colocated and for the staggered schemes, that for crisp resolution of discontinuities it is necessary to use both a second order slope limited scheme and a time stepping scheme that is stable for the κ-scheme that corresponds to the limiter used. For example, the van Albada limiter corresponds to $\kappa = 1/2$, as shown in Sect. 4.8. As shown in Chap. 9, the explicit Euler scheme is TVD, hence stable. But it is unstable for the κ-scheme that corresponds to the limiter used, see Sect. 4.8. Hence, with the explicit Euler scheme the fruits of higher order spatial accuracy cannot be reaped, because the limiter will steer the scheme away from the κ-scheme, to maintain stability.

Extension of the staggered one-dimensional schemes for the Euler equations discussed in this chapter to more dimensions and to the Navier-Stokes equations is straightforward, and can be carried out along the lines laid out in Chap. 13. This results in a unified method to compute incompressible and compressible flows. Details and results for the first pressure-correction method presented above are given in Bijl and Wesseling (1998), where accuracy and computing work are found to be approximately uniform in the Mach number. An extension to hydrodynamic flow with cavitation, involving

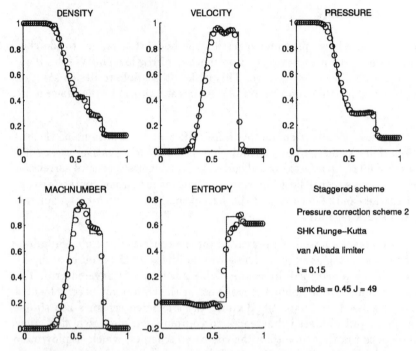

Fig. 14.14. Sod's test problem with second pressure-correction scheme.

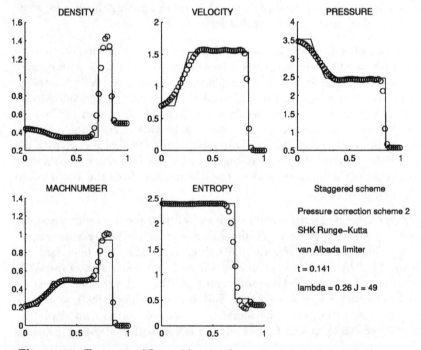

Fig. 14.15. Test case of Lax with second pressure-correction scheme.

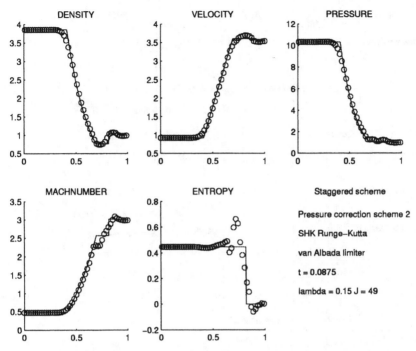

Fig. 14.16. Mach 3 test problem with second pressure-correction scheme.

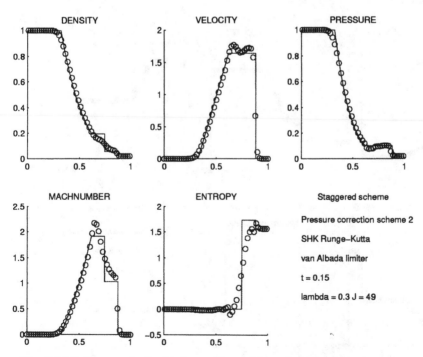

Fig. 14.17. Supersonic shocktube problem with second pressure-correction scheme.

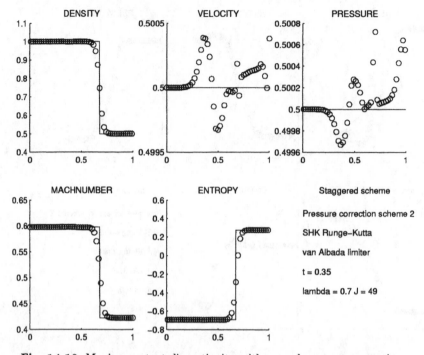

Fig. 14.18. Moving contact discontinuity with second pressure-correction scheme.

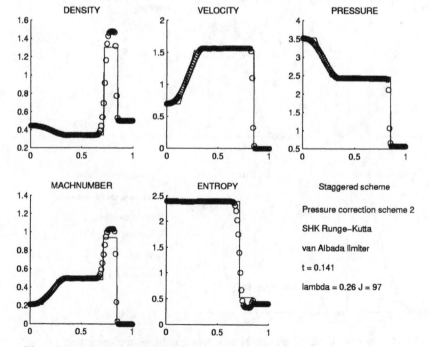

Fig. 14.19. Test case of Lax with $h = 1/96$.

a nonconvex equation of state, is given in van der Heul, Vuik, and Wesseling (1999). Here the Mach number ranges from 10^{-3} to 25, which makes a unified method for incompressible and compressible flows indispensable.

Extension of incompressible staggered schemes to the compressible case has also been undertaken in Harlow and Amsden (1968), Harlow and Amsden (1971), Issa, Gosman, and Watkins (1986), Karki and Patankar (1989), McGuirk and Page (1990), Shyy and Braaten (1988), Van Doormaal, Raithby, and McDonald (1987). In these methods the singularity at $M = 0$ is not removed, as we do by splitting the pressure according to (14.24). All generalize methods of SIMPLE or PISO type (Sect. 7.6) to the compressible case. It is shown in McGuirk and Page (1990) that the other methods give much shock smearing due to the fact that the velocity is updated in the pressure-correction step. This is improved in McGuirk and Page (1990) by updating the momentum instead of the velocity. In Shyy and Braaten (1988) better crispness is obtained by adaptive local grid refinement. All methods quoted derive the pressure-correction equation from the mass conservation equation, and are probably not very accurate for moving strong contact discontinuities. All use combinations of second order central and first order upwind schemes in space, and first order time stepping schemes. It is to be expected that the accuracy of these methods can be enhanced by use of second order slope limited schemes in space and higher order time stepping schemes.

Clearly, unified methods for incompressible and compressible flows are within reach and may be expected to come into widespread use soon.

References

Abbott, M.B. (1979). *Computational Hydraulics; Elements of the Theory of Free Surface Flows.* London: Pitman.

Acharya, S. and F.H. Moukalled (1989). Improvements to incompressible flow calculation on a nonstaggered curvilinear grid. *Num. Heat Transfer B* **15**, 131–152.

Anderson, W.K., J.L. Thomas, and B. van Leer (1986). A comparison of finite volume flux vector splittings for the Euler equations. *AIAA J.* **24**, 1453–1460.

Andersson, H.I., J.I. Billdal, P. Eliasson, and A. Rizzi (1990). Staggered and non-staggered finite-volume methods for nonsteady viscous flows: a comparative study. In K.W. Morton (Ed.), *Lecture Notes in Physics 371*, pp. 172–176. Twelfth International Conference on Numerical Methods in Fluid Dynamics, Berlin: Springer.

Arakawa, A. and V.R. Lamb (1977). Computational design of the basic dynamical processes of the UCLA general circulation model. In J. Chang (Ed.), *Methods in Computational Physics*, Volume 17, pp. 173–265. New York: Academic Press.

Aris, R. (1962). *Vectors, Tensors and the Basic Equations of Fluid Mechanics.* Englewood Cliffs, N.J.: Prentice-Hall, Inc. Reprinted, Dover, New York, 1989.

Armaly, B.F., F. Durst, J.C.F. Pereira, and B. Schönung (1983). Experimental and theoretical investigation of backward-facing step flow. *J. Fluid Mech.* **127**, 473–496.

Armfield, S.W. (1994). Ellipticity, accuracy and convergence of the discrete Navier-Stokes equations. *J. Comp. Phys.* **114**, 176–184.

Arnal, M. and R. Friedrich (1992). On the effects of spatial resolution and subgrid-scale modeling in the large eddy simulation of a recirculating flow. In J.B. Vos, A. Rizzi, and I.L. Rhyming (Eds.), *Proceedings of the Ninth GAMM-Conference on Numerical Methods in Fluid Mechanics*, pp. 3–13. Braunschweig: Vieweg.

Arora, M. and P.L Roe (1997). A well-behaved TVD limiter for high-resolution calculations of unsteady flow. *J. Comp. Phys.* **132**, 3–11.

Asselin, R. (1972). Frequency filter for time integrations. *Mon. Weather Rev.* **100**, 487–490.

Axelsson, O. (1994). *Iterative Solution Methods*. Cambridge, UK: Cambridge University Press.

Aziz, K. and A. Settari (1979). *Petroleum Reservoir Simulation*. London: Elsevier.

Balaras, E. and C. Benocci (1994). Subgrid scale models in finite difference simulations of complex wall bounded flows. In *Application of Direct and Large Eddy Simulation to Transition and Turbulence*, Conference Proceedings 551, pp. 2.1–2.6. Neuilly-sur-Seine: AGARD.

Barcus, M., M. Perić, and G. Scheuerer (1988). A control volume based full multigrid procedure for the prediction of two-dimensional, laminar, incompressible flow. In M. Deville (Ed.), *Proceedings of the Seventh GAMM-Conference on Numerical Methods in Fluid Mechanics*, pp. 9–16. Braunschweig: Vieweg. Notes on Numerical Fluid Mechanics 20.

Baron, A. and M. Quadrio (1994). A cheap DNS tool for turbulence models testing. In *Application of Direct and Large Eddy Simulation to Transition and Turbulence*, Conference Proceedings 551, pp. 11.1–11.10. Neuilly-sur-Seine: AGARD.

Barrett, R., M. Berry, T.F. Chan, J. Demmel, J. Donato, J. Dongarra, V. Eijkhout, R. Pozo, C. Romine, and H. van der Vorst (1994). *Templates for the Solution of Linear Systems: Building Blocks for Iterative Methods*. Philadelphia: SIAM.

Batchelor, G.K. (1967). *An Introduction to Fluid Dynamics*. Cambridge, UK: Cambridge University Press.

Bayliss, A. and E. Turkel (1980). Radiation-boundary conditions for wave-like equations. *Comm. Pure and Appl. Math.* **33**, 708–725.

Bell, J.B., P. Colella, and H.M. Glaz (1989). A second-order projection method for the incompressible Navier-Stokes equations. *J. Comp. Phys.* **85**, 257–283.

Belov, A.A., L. Martinelli, and A. Jameson (1994). A novel fully implicit multigrid driven algorithm for unsteady incompressible flow calculations. In S. Wagner, E.H. Hirschel, J. Périaux, and R. Piva (Eds.), *Computational Fluid Dynamics '94*, pp. 662–670. Chichester: Wiley.

Benocci, C. and A. Pinelli (1990). The role of the forcing term in the large eddy simulation of turbulent channel flow. In W. Rodi and E.N. Ganic (Eds.), *Engineering Turbulence*, pp. 287–296. New York: Elsevier.

Berger, M. and M. Aftosmis (1998). Aspects (and aspect ratios) of Cartesian mesh methods. In C.-H. Bruneau (Ed.), *Sixteenth International Conference on Numerical Methods in Fluid Dynamics*, pp. 1–12. Berlin: Springer.

Bijl, H. and P. Wesseling (1998). A unified method for computing incompressible and compressible flows in boundary-fitted coordinates. *J. Comp. Phys.* **141**, 153–173.

Biringen, S. and C. Cook (1988). On pressure boundary conditions for the incompressible Navier-Stokes equations using nonstaggered grids. *Num. Heat Transfer* **13**, 241–252.

Boersma, B.J., J.G.M. Eggels, M.J.B.M. Pourquié, and F.T.M. Nieuwstadt (1994). Large-eddy simulation applied to an electromagnetic flowmeter. In P.R. Voke, L. Kleiser, and J.-P. Chollet (Eds.), *Direct and Large-Eddy Simulation I*, pp. 325–333. Dordrecht: Kluwer.

Boersma, B.J., M.N. Kooper, F.T.M. Nieuwstadt, and P. Wesseling (1997). Local grid refinement in large-eddy simulations. *J. Eng. Math.* **32**, 161–175.

Boersma, B.J. and F.T.M. Nieuwstadt (1996). Large-eddy simulation of turbulent flow in a curved pipe. *Trans. ASME / J. Fluids Eng.* **118**, 248–254.

Botta, E.F.F., K. Dekker, Y. Notay, A. van der Ploeg, C. Vuik, F.W. Wubs, and P.M. de Zeeuw (1997). How fast the Laplace equation was solved in 1995. *Appl. Num. Math.* **24**, 439–455.

Bradshaw, P. (1997). Understanding and prediction of turbulent flow— 1996. *Int. J. Heat and Fluid Flow* **18**, 45–54.

Braess, D. (1986). On the combination of the multigrid method and conjugate gradients. In W. Hackbusch and U. Trottenberg (Eds.), *Multigrid Methods II*, pp. 52–64. Berlin: Springer.

Braess, D. and R. Sarazin (1996). An efficient smoother for the Stokes problem. *Appl. Num. Math.* **23**, 3–19.

Brandt, A. (1977). Multi-level adaptive solutions to boundary value problems. *Math. Comp.* **31**, 333–390.

Brandt, A. (1980). Multilevel adaptive computations in fluid dynamics. *AIAA J.* **18**, 1165–1172.

Brandt, A. and N. Dinar (1979). Multigrid solutions to flow problems. In S. Parter (Ed.), *Numerical Methods for Partial Differential Equations*, pp. 53–147. New York: Academic Press.

Braun, H., M. Fiebig, and N.K. Mitra (1994). Large-eddy simulation of separated flow in a ribbed duct. In *Application of Direct and Large Eddy Simulation to Transition and Turbulence*, Conference Proceedings 551, pp. 9.1–9.9. Neuilly-sur-Seine: AGARD.

Breuer, M. and D. Hänel (1990). Solution of the 3D incompressible Navier-Stokes equations for the simulation of vortex breakdown. In P. Wesseling (Ed.), *Proc. of the Eight GAMM Conf. on Num. Meth. in Fl. Mech.*, pp. 42–51. Vieweg, Braunschweig.

Breuer, M. and W. Rodi (1994). Large-eddy simulation of turbulent flow through a straight square duct and a 180° bend. In P.R. Voke, L. Kleiser, and J.-P. Chollet (Eds.), *Direct and Large-Eddy Simulation I*, pp. 273–285. Dordrecht: Kluwer.

Briggs, W.L. (1987). *A Multigrid Tutorial*. Philadelphia: SIAM.

Bristeau, M.O., R. Glowinski, and J. Périaux (1987). Numerical methods for the Navier-Stokes equations. Applications to the simulation of compressible flows. *Computer Physics Reports* **6**, 73–187.

Bruaset, A.M. (1995). *A Survey of Preconditioned Iterative Methods.* Pitman research notes in mathematics series 328. Harlow: Longman Scientific and Technical.

Burgers, J.M. (1948). A mathematical model illustrating the theory of turbulence. *Advances in Applied Mechanics* **1**, 171–199.

Burggraf, O.R. (1966). Analytical and numerical studies of the structure of steady separated flows. *J. Fluid Mech.* **24**, 113–151.

Burns, A.D., I.P. Jones, J.R. Kightley, and N.S. Wilkes (1987). The implementation of a finite difference method for predicting incompressible flows in complex geometries. In C. Taylor, W.G. Habashi, and M.M. Hafez (Eds.), *Num. Meth. in Lam. and Turb. Fl., vol.5*, pp. 339–350. Swansea: Pineridge Press.

Cantaloube, B. and T.-H. Lê (1992). Direct simulation of unsteady flow in a three-dimensional lid-driven cavity. In T.-H.Lê M. Deville and Y. Morchoisne (Eds.), *Numerical Simulation of 3-D Incompressible Unsteady Viscous Laminar Flows*, pp. 25–33. Braunschweig: Vieweg.

Cash, J.R. (1984). Two new finite difference schemes for parabolic equations. *SIAM J. Num. Anal.* **21**, 433–446.

Celik, I., C.J. Chen, and P.J. Roache (Eds.) (1993). *Quantification of Uncertainty in Computational Fluid Dynamics.* New York: American Society of Mechanical Engineers.

Chan, T.F. (1984). Stability analysis of finite difference schems for the advection-diffusion equation. *SIAM J. Num. Anal.* **21**, 272–284.

Chan, T.F. and T.P. Mathew (1994). Domain decomposition algorithms. In A. Iserles (Ed.), *Acta Numerica*, pp. 61–143. Cambridge, UK: Cambridge University Press.

Chang, J.L.C. and D. Kwak (1984). On the method of pseudo compressibility for numerically solving incompressible flows. AIAA Paper 84-0252.

Chapman, D.R. (1979). Computational aerodynamics development and outlook. *AIAA J.* **17**, 1293–1313.

Choi, D. and C.L. Merkle (1985). Application of time-iterative schemes to incompressible flow. *AIAA J.* **23**, 1518–1524.

Choi, H. and P. Moin (1994). Effects of the computational time step on numerical solutions of turbulent flow. *J. Comp. Phys.* **113**, 1–4.

Choi, H., P. Moin, and J. Kim (1993). Direct numerical simulation of turbulent flow over riblets. *J. Fluid Mech.* **255**, 503–539.

Choi, Y.-H. and C.L. Merkle (1993). The application of preconditioning in viscous flows. *J. Comp. Phys.* **105**, 207–223.

Chorin, A.J. (1967). A numerical method for solving incompressible viscous flow problems. *J. Comp. Phys.* **2**, 12–26.

Chorin, A.J. (1968). Numerical solution of the Navier-Stokes equations. *Math. Comp.* **22**, 745–762.

Chorin, A.J. (1969). On the convergence of discrete approximations to the Navier-Stokes equations. *Math. Comp.* **23**, 342–353.

Chorin, A.J. (1970). Numerical solution of incompressible flow problems. In J.M. Ortega and W.C. Rheinbold (Eds.), *Studies in Numerical Analysis 2*, pp. 64–71. Philadelphia: SIAM.

Chorin, A.J. and J.E. Marsden (1979). *A Mathematical Introduction to Fluid Mechanics.* Springer, New York.

Ciarlet, Ph.G. (1978). *The Finite Element Method for Elliptic Problems.* Amsterdam: North-Holland.

Coirier, W.J. and K.G. Powell (1995). An accuracy assessment of Cartesian mesh approaches for the Euler equations. *J. Comp. Phys.* **117**, 121–131.

Courant, R. and K.O. Friedrichs (1949). *Supersonic Flow and Shock Waves.* New York: Springer.

Courant, R., K.O. Friedrichs, and H. Lewy (1928). Über die partiellen Differenzgleichungen der Mathematischen Physik. *Mathematische Annalen* **100**, 32–74.

Courant, R., K. Friedrichs, and H. Lewy (1967). On the partial difference equations of mathematical physics. *IBM Journal* **11**, 215–234. English translation of Über die partiellen Differenzgleichungen der Mathematischen Physik, Mathematische Annalen 100:32-74, 1928.

Courant, R. and D. Hilbert (1989). *Methods of Mathematical Physics, Vol. 2. Partial Differential Equations.* New York: Interscience.

Courant, R., E. Isaacson, and M. Rees (1952). On the solution of nonlinear hyperbolic differential equations. *Comm. Pure and Appl. Math.* **5**, 243–255.

Crandall, M.G. and A. Majda (1980). Monotone difference approximations for scalar conservation laws. *Math. Comp.* **34**, 1–21.

Dahlquist, G. and Å. Björck (1974). *Numerical Methods.* Englewood Cliffs, N.J.: Prentice-Hall.

Darmofal, D.L. (1998). Towards a robust multigrid algorithm wiht Mach number and grid independent convergence. In K.D. Papailiou, D. Tsahalis, J. Périaux, and D. Knörzer (Eds.), *Computational Fluid Dynamics '98*, Volume 2, pp. 90–95. Chichester: Wiley.

Davies, D.E. and D.J. Salmond (1985). Calculation of the volume of a general hexahedron for flow predictions. *AIAA J.* **23**, 954–956.

de Saint-Venant, B. (1843). Mémoire sur la dynamique des fluides. *C. R. Acad. Sci. Paris* **17**, 1240–1242.

de Vahl Davis, G. (1983). Natural convection of air in a square cavity: a benchmark numerical solution. *Int. J. Num. Meth. in Fluids* **3**, 249–264.

Deardorff, J.W. (1970). A numerical study of three-dimensional turbulent channel flow at large Reynolds numbers. *J. Fluid Mech.* **41**, 453–480.

Deister, F., D. Rocher, E.H. Hirschel, and F. Monnoyer (1998). Three-dimensional adaptively refined Cartesian grid generation and Euler flow solutions for arbitrary geometries. In K.D. Papailiou, D. Tsahalis, J. Périaux, and C. Hirsch (Eds.), *Computational Fluid Dynamics '98*, Volume 1, pp. 96–101. Chichester: Wiley.

Demirdžić, I., Z. Lilek, and M. Perić (1992). Fluid flow and heat transfer test problems for non-orthogonal grids: bench-mark solutions. *Int. J. Num. Meth. in Fluids* **15**, 329–354.

Demirdžić, I. and M. Perić (1990). Finite volume method for prediction of fluid flow in arbitrary shaped domains with moving boundaries. *Int. J. Num. Meth. in Fluids* **10**, 771–790.

Deng, G.B. (1989). Numerical simulation of incompressible turbulent appendage-flat plate junction flows. In C. Taylor, W.G. Habashi, and M.M. Hafez (Eds.), *Numerical Methods in Laminar and Turbulent Flows*, Volume 6, Part 1, pp. 793–803. Swansea: Pineridge Press.

Dick, E. (1988). A flux-vector splitting method for steady Navier-Stokes equations. *Int. J. Num. Meth. in Fluids* **8**, 317–326.

Dick, E. (1989). A multigrid method for steady incompressible Navier-Stokes equations based on partial flux splitting. *Int. J. Num. Meth. in Fluids* **9**, 113–120.

Dick, E. and J. Linden (1992). A multigrid method for steady incompressible Navier-Stokes equations based on flux difference splitting. *Int. J. Num. Meth. in Fluids* **14**, 1311–1323.

Dol, H.S., K. Hanjalić, and S. Kenjereš (1997). A comparative assessment of the second-moment differential and algebraic models in turbulent natural convection. *Int. J. Heat and Fluid Flow* **18**, 4–14.

Drikakis, D. and M. Schäfer (1994). Comparison between a pressure correction and an artificial compressibility/characteristic based method in parallel incompressible fluid flow computations. In S. Wagner, E.H. Hirschel, J. Périaux, and R. Piva (Eds.), *Computational Fluid Dynamics '94*, pp. 619–626. Chichester: Wiley.

Dukowicz, J.K. and A.S. Dvinsky (1992). Approximate factorization as a high order splitting for the implicit incompressible flow equations. *J. Comp. Phys.* **102**, 336–347.

Dwyer, H.S., M. Soliman, and M. Hafez (1986). Time accurate solutions of the Navier-Stokes equations for reacting flows. In F.G. Zhuang and Y.L. Zhu (Eds.), *Tenth International Conference on Numerical Methods in Fluid Dynamics*, pp. 247–251. Berlin: Springer.

Eberle, A., A. Rizzi, and E.H. Hirschel (1992). *Numerical Solutions of the Euler Equations for Steady Flow Problems*. Braunschweig: Vieweg.

Eckhaus, W. (1973). *Matched Asymptotic Expansions and Singular Perturbations*. Amsterdam: North-Holland.

Edwards, J.R. and C.J. Roy (1998). Preconditioned multigrid methods for two-dimensional combustion calculations at all speeds. *AIAA J.* **36**, 185–192.

Eggels, J.G.M., F. Unger, M.H. Weiss, J. Westerweel, R.J. Adrian, R. Friedrich, and F.T.M. Nieuwstadt (1994). Fully developed pipe flow: a comparison between direct numerical simulation and experiment. *J. Fluid Mech.* **268**, 175–209.

Eisenstat, S.C., H.C. Elman, and M.H. Schultz (1983). Variable iterative methods for nonsymmetric systems of linear equations. *SIAM J. Num. Anal.* **20**, 345–357.

Ellison, J.H., C.A. Hall, and T.A. Porsching (1987). An unconditionally stable convergent finite difference method for Navier-Stokes problems on curved domains. *SIAM J. Num. Anal.* **24**, 1233–1248.

Engelman, M.S. and M.-A. Jamnia (1990). Transient flow past a circular cylinder: a benchmark solution. *Int. J. Num. Meth. in Fluids* **11**, 985–1000.

Engquist, B. and A. Majda (1979). Radiation boundary conditions for acoustic and elastic wave calculations. *Comm. Pure and Appl. Math.* **32**, 313–357.

Engquist, B. and S. Osher (1981). One-sided difference approximations for nonlinear conservation laws. *Math. Comp.* **36**, 321–351.

Esposito, P.G. (1992). Numerical simulation of a three-dimensional lid-driven cavity flow. In M. Deville, T.-H. Lê, and Y. Morchoisne (Eds.), *Numerical Simulation of 3-D Incompressible Unsteady Viscous Laminar Flows*, pp. 46–53. Braunschweig: Vieweg.

Faddeev, D.K. and V.N. Faddeeva (1963). *Computational Methods of Linear Algebra*. London: Freeman.

Farrell, P.A., P.W. Hemker, and G.I. Shishkin (1996). Discrete approximations for singularly perturbed boundary value problems with parabolic layers, I. *J. Comp. Math.* **14**, 71–97.

Farrell, P.A., J.J. Miller, E. O'Riordan, and G.I. Shishkin (1996). A uniformly convergent finite difference scheme for a singularly perturbed semilinear equation. *SIAM J. Num. Anal.* **33**, 1135–1149.

Favre, A. (1965). Equations des gaz turbulents compressibles. *Journal de Mécanique* **4**, 361–390.

Feistauer, M. (1993). *Mathematical Methods in Fluid Dynamics*. Harlow: Longman.

Ferziger, J.H. (1996). Recent advences in large-eddy simulation. In W. Rodi and G. Bergeles (Eds.), *Engineering Turbulence Modelling and Experiments 3*, pp. 163–175. Amsterdam: Elsevier.

Ferziger, J.H. and M. Perić (1996). *Computational Methods for Fluid Dynamics*. Berlin: Springer.

Fletcher, C.A.J. (1988). *Computational Techniques for Fluid Dynamics*, Volume 1,2. Berlin: Springer.

Forsyth, Jr., P.A. and P.H. Sammon (1988). Quadratic convergence for cell-centered grids. *Appl. Num. Math.* **4**, 377–394.

Forsythe, G.E. and W.R. Wasow (1960). *Finite Difference Methods for Partial Differential Equations.* New York: Wiley.

Freitas, C.J. (1995). Perspective: selected benchmarks from commercial CFD code. *Trans. ASME / J. Fluids Eng.* **117**, 208–218.

Friedrich, R. and F. Unger (1991). Large eddy simulation of boundary layers with a step change in pressure gradient. In O. Métais and M. Lesieur (Eds.), *Turbulence and Coherent Structures*, pp. 159–174. Dordrecht: Kluwer.

Fromm, J.E. (1968). A method for reducing dispersion in convective difference schemes. *J. Comp. Phys.* **3**, 176–189.

Fuchs, L. and H.S. Zhao (1984). Solution of three-dimensional viscous incompressible flows by a multigrid method. *Int. J. Num. Meth. in Fluids* **4**, 539–555.

Galperin, B. and S.A. Orszag (1993). *Large Eddy Simulation of Complex Engineering and Geophysical Flows.* Cambridge, UK: Cambridge University Press.

Gao, S. (1994). Numerical investigation of turbulent structures in thermal impinging jets. In P.R. Voke, L. Kleiser, and J.-P. Chollet (Eds.), *Direct and Large-Eddy Simulation I*, pp. 411–422. Dordrecht: Kluwer.

Garabedian, P.R. (1964). *Partial Differential Equations.* New York: Wiley.

Gartling, D.K. (1990). A test problem for outflow boundary conditions - flow over a backward facing step. *Int. J. Num. Meth. in Fluids* **11**, 953–967.

Gaskell, P.H. and K.C. Lau (1988). Curvature-compensated convective transport: SMART, a new boundedness-preserving transport algorithm. *Int. J. Num. Meth. in Fluids* **8**, 617–641.

Gavrilakis, S. (1993). Numerical simulation of low Reynolds number turbulent flow in a straight square duct. *J. Fluid Mech.* **244**, 101–129.

Gavrilakis, S., H.M. Tsai, P.R. Voke, and D.C. Leslie (1986). Large-eddy simulation of low Reynolds number channel flow by spectral and finite difference methods. In U. Schumann and R. Friedrichs (Eds.), *Direct and Large Eddy Simulation of Turbulence*, pp. 105–118. Braunschweig: Vieweg.

George, P.L. (1991). *Automatic Mesh Generation. Applications to Finite Element Methods.* New York: Wiley.

Gerz, T., U. Schumann, and S. Elgobashi (1989). Direct simulation of stably stratified homogeneous turbulent shear flows. *J. Fluid Mech.* **200**, 563–594.

Geveci, T. (1982). The significance of the stability of difference schemes in different l^p-spaces. *SIAM Review* **24**, 413–426.

Ghia, U., K.N. Ghia, and C.T. Shin (1982). High-Re solutions for incompressible flow using the Navier-Stokes equations and a multigrid method. *J. Comp. Phys.* **48**, 387–411.

Giles, M.B. (1990). Non-reflecting bounday conditions for Euler equations calculations. *AIAA J.* **28**, 2050–2058.

Glowinski, R. and J.F. Périaux (1987). Numerical methods for nonlinear problems in fluid dynamics. In A. Lichnewsky and C. Saguez (Eds.), *Supercomputing*, pp. 381–479. Amsterdam: North-Holland.

Godlewski, E. and P.-A. Raviart (1991). *Hyperbolic Systems of Conservation Laws*. Paris: Ellipses.

Godlewski, E. and P.-A. Raviart (1996). *Numerical Approximation of Hyperbolic Systems of Conservation Laws*. New York: Springer.

Godunov, S.K. (1959). Finite difference method for numerical computation of discontinuous solutions of the equations of fluid dynamics. *Mat. Sbornik* **47**, 271–306. (in Russian).

Goldstein, S. (Ed.) (1965). *Modern Developments in Fluid Dynamics Vol. 2*. New York: Dover.

Golub, G.H. and C.F. van Loan (1996). *Matrix Computations*. Baltimore: The Johns Hopkins University Press. Third edition.

Goodman, J.B. and R.J. Leveque (1985). On the accuracy of stable schemes for 2D scalar conservation laws. *Math. Comp.* **45**, 15–21.

Gourlay, A.R. and J.Ll. Morris (1980). The extrapolation of first order methods for parabolic partial differential equations. *SIAM J. Num. Anal.* **17**, 641–655.

Gourlay, A.R. and J.Ll. Morris (1981). Linear combinations of generalised Crank-Nicolson schemes. *IMA J. Num. Anal.* **1**, 347–357.

Greenbaum, A. (1997). *Iterative Methods for Solving Linear Systems*. Philadelphia: SIAM.

Greenbaum, A., V. Ptak, and Z. Strakos (1996). Any nonincreasing convergence curve is possible for GMRES. *SIAM J. Matrix Anal. Appl.* **17**, 465–469.

Gresho, M.P. and R.L. Sani (1987). On pressure boundary conditions for the incompressible Navier-Stokes equations. *Int. J. Num. Meth. in Fluids* **7**, 1111–1145.

Gresho, P.M. (1991). Some current CFD issues relevant to the incompressible Navier-Stokes equations. *Comp. Meth. Appl. Mech. Eng.* **87**, 201–252.

Gresho, P.M. (1992). Some interesting issues in incompressible fluid dynamics, both in the continuum and in numerical simulation. *Advances in Applied Mechanics* **28**, 45–140.

Gresho, P.M. and R.L. Lee (1981). Don't suppress the wiggles - they're telling you something! *Computers and Fluids* **9**, 223–255.

Grossmann, Ch. and H.-G. Roos (1994). *Numerik Partieller Differentialgleichungen*. Stuttgart: Teubner.

Grötzbach, G. (1982). Direct numerical simulation of laminar and turbulent Bénard convection. *J. Fluid Mech.* **119**, 27–53.

Grötzbach, G. and M. Wörner (1994). Flow mechanisms and heat transfer in Rayleigh-Bénard convection at small Prandtl numbers. In P.R. Voke, L. Kleiser, and J.-P. Chollet (Eds.), *Direct and Large-Eddy Simulation I*, pp. 387–397. Dordrecht: Kluwer.

Gu, C.-Y. (1991). Computations of flows with large body forces. In C. Taylor, J.H. Chin, and G.M. Homsky (Eds.), *Numerical Methods in Laminar and Turbulent Flow, Vol. 7, Part 2*, pp. 1568–1578. Swansea: Pineridge Press.

Guillard, H. and C. Viozat (1999). On the behavior of upwind schemes in the low Mach number limit. *Computers and Fluids* **28**, 63–86.

Gustafsson, B. (1975). The convergence rate for difference approximations to mixed initial boundary value problems. *Math. Comp.* **29**, 396–406.

Gustafsson, B., H.-O. Kreiss, and A. Sundström (1972). Stability theory of difference approximations for mixed initial boundary value problems. II. *Math. Comp.* **26**, 649–686.

Gustafsson, I.A. (1978). A class of first order factorization methods. *BIT* **18**, 142–156.

Hackbusch, W. (1985). *Multi-Grid Methods and Applications*. Berlin: Springer.

Hackbusch, W. (1986). *Theorie und Numerik Elliptischer Differentialgleichungen*. Stuttgart: Teubner.

Hackbusch, W. (1994). *Iterative Solution of Large Sparse Systems Equations*. New York: Springer.

Hageman, L.A. and D.M. Young (1981). *Applied Iterative Methods*. New York: Academic Press.

Hairer, E., S.P. Nørsett, and G. Wanner (1987). *Solving Ordinary Differential Eequations. Vol. 1. Nonstiff Problems*. Berlin: Springer.

Hairer, E. and G. Wanner (1991). *Solving Ordinary Differential Equations. Vol. 2. Stiff and Differential-Algebraic Problems*. Berlin: Springer.

Hall, C.A. and T.A. Porsching (1990). *Numerical Analysis of Partial Differential Equations*. Engewood Cliffs, NJ: Prentice Hall.

Hall, M.G. (1986). Cell-vertex multigrid schemes for solution of the Euler equations. In K.W. Morton and M.J. Baines (Eds.), *Numerical Methods for Fluid Dynamics II*, pp. 303–346. Oxford: Clarendon Press.

Hänel, D., R. Schwane, and G. Seider (1987). On the accuracy of upwind schemes for the solution of the Navier- Stokes equations. AIAA Paper 87-1105.

Hanjalić, K. (1994). Advanced turbulence closure models: a view of current status and future prospects. *Int. J. Heat and Fluid Flow* **15**, 178–203.

Hansen, W. (1956). Theorie zur Errechnung des Wasserstandes und der Strömungen in Randmeeren nebst Anwendungen. *Tellus* **8**, 289–300.

Harlow, F.H. and A.A. Amsden (1968). Numerical calculation of almost incompressible flows. *J. Comp. Phys.* **3**, 80–93.

Harlow, F.H. and A.A. Amsden (1971). A numerical fluid dynamics calculation method for all flow speeds. *J. Comp. Phys.* **8**, 197–213.

Harlow, F.H. and J.E. Welch (1965). Numerical calculation of time-dependent viscous incompressible flow of fluid with a free surface. *The Physics of Fluids* **8**, 2182–2189.

Harten, A. (1983). High resolution schemes for hyperbolic conservation laws. *J. Comp. Phys.* **49**, 357–393.

Harten, A. (1984). High resolution total-variation-stable finite-difference schemes. *SIAM J. Num. Anal.* **21**, 1–23.

Harten, A., B. Engquist, S. Osher, and S.R. Chakravarty (1987). Uniformly high order accurate essentially non-oscillatory schemes, III. *J. Comp. Phys.* **71**, 231–303.

Harten, A., J.M. Hyman, and P.D. Lax (1976). On finite difference approximations and entropy conditions for shocks. *Comm. Pure and Appl. Math.* **29**, 297–322. (with appendix by B. Keyfitz).

Harten, A. and P.D. Lax (1981). A random choice finite-difference scheme for hyperbolic conservation laws. *SIAM J. Num. Anal.* **18**, 289–315.

Hartwich, P.-M. and C.-H. Hsu (1988). High-resolution upwind schemes for the three-dimensional incompressible Navier-Stokes equations. *AIAA J.* **26**, 1321–1328.

He, P. and M. Salcudean (1994). A numerical method for 3d viscous incompressible flows using non-orthogonal grids. *Int. J. Num. Meth. in Fluids* **18**, 449–469.

Hemker, P.W. and G.I. Shishkin (1994). Discrete approximation of singularly perturbed parabolic PDEs with a discontinuous initial condition. *Comp. Fluid Dyn. J.* **2**, 375–392.

Hemker, P.W. and S.P. Spekreijse (1986). Multiple grid and Osher's scheme for the efficient solution of the steady Euler equations. *Appl. Num. Math.* **2**, 475–493.

Henshaw, W.B. (1996). Automatic grid generation. In A. Iserles (Ed.), *Acta Numerica 5*, pp. 121–148. Cambridge, UK: Cambridge University Press.

Hindmarsh, A.C., P.M. Gresho, and D.F. Griffiths (1984). The stability of explicit Euler time-integration for certain finite difference approximations of the multi-dimensional advection-diffusion equation. *Int. J. Num. Meth. in Fluids* **4**, 853–897.

Hinze, J.O. (1975). *Turbulence.* New York: McGraw-Hill.

Hirsch, C. (1988). *Numerical Computation of Internal and External Flows. Vol.1: Fundamentals of Numerical Discretization.* Chichester: Wiley.

Hirsch, C. (1990). *Numerical Computation of Internal and External Flows. Vol.2: Computational Methods for Inviscid and Viscous Flows.* Chichester: Wiley.

Hirt, C.W. (1968). Heuristic stability theory for finite difference equations. *J. Comp. Phys.* **2**, 339–355.

Ho, Y.-H. and B. Lakshminarayana (1993). Computation of unsteady viscous flow using a pressure-based algorithm. *AIAA J.* **31**, 2232–2240.

Hoffmann, G. and C. Benocci (1994). Numerical simulation of spatially-developing planar jets. In *Application of Direct and Large Eddy Simulation to Transition and Turbulence*, Conference Proceedings 551, pp. 26.1–26.6. Neuilly-sur-Seine: AGARD.

Hortman, M., M. Perić, and G. Scheuerer (1990). Finite volume multigrid prediction of laminar natural convection: bench-mark solutions. *Int. J. Num. Meth. in Fluids* **11**, 189–208.

Hou, T.Y. and P.G. Le Floch (1994). Why nonconservative schemes converge to wrong solutions: error analysis. *Math. Comp.* **206**, 497–530.

Hou, T.Y. and B.T.R. Wetton (1993). Second-order convergence of a projection scheme for the incompressible Navier-Stokes equations with boundaries. *SIAM J. Num. Anal.* **30**, 609–629.

Hugoniot, H. (1889). Sur la propagation du mouvement dans les corps et spécialement dans les gaz parfaits. *Journal de l'Ecole Polytechnique* **58**, 1–25.

Hundsdorfer, W., B. Koren, M. van Loon, and J.G. Verwer (1995). A positive finite-difference advection scheme. *J. Comp. Phys.* **117**, 35–46.

Huser, A. and S. Biringer (1992). Calculation of shear-driven cavity flows at high Reynolds numbers. *Int. J. Num. Meth. in Fluids* **14**, 1087–1109.

Huser, A. and S. Biringer (1993). Direct numerical simulation of turbulent flow in a square duct. *J. Fluid Mech.* **257**, 65–95.

Issa, R.I. (1986). Solution of the implicitly discretised fluid flow equations by operator-splitting. *J. Comp. Phys.* **62**, 40–65.

Issa, R. I., A. D. Gosman, and A. P. Watkins (1986). The computation of compressible and incompressible flows by a non-iterative implicit scheme. *J. Comp. Phys.* **62**, 66–82.

Jameson, A. (1985a). Numerical solution of the Euler equations for compressible inviscid fluids. In F. Angrand, A. Dervieux, J.A.Désidéri, and R. Glowinski (Eds.), *Numerical Methods for the Euler Equations of Fluid Dynamics*, pp. 199–245. Philadelphia: SIAM.

Jameson, A. (1985b). Transonic flow calculations for aircraft. In F. Brezzi (Ed.), *Numerical Methods in Fluid Mechanics*, pp. 156–242. Berlin: Springer. Lecture Notes in Mathematics 1127.

Jameson, A. (1988). Computational transonics. *Comm. Pure and Appl. Math.* **41**, 507–549.

Jameson, A. (1995a). Analysis and design of numerical schemes for gas dynamics, 1: artificial diffusion, upwind biasing, limiters and their effect on accuracy and multigrid convergence. *Int. J. Comp. Fluid Dyn.* **4**, 171–218.

Jameson, A. (1995b). Analysis and design of numerical schemes for gas dynamics, 2: artificial diffusion and discrete shock structure. *Int. J. Comp. Fluid Dyn.* **5**, 1–38.

Jameson, A. and T.J. Baker (1984). Multigrid solution of the Euler equations for aircraft configurations. AIAA-Paper 84-0093.

Jameson, A., W. Schmidt, and E. Turkel (1981). Numerical solution of the Euler equations by finite volume methods using Runge-Kutta time stepping schemes. AIAA Paper 81-1259.

Jin, G. and M. Braza (1993). A nonreflecting outlet boundary condition for incompressible unsteady Navier-Stokes equations. *J. Comp. Phys.* **107**, 239–253.

Johansson, B.C.V. (1993). Boundary conditions for open boundaries for the incompressible Navier-Stokes equation. *J. Comp. Phys.* **105**, 233–251.

Joslin, R.D., C.L. Streett, and C.-L. Chang (1993). Spatial simulation of boundary-layer transition mechanisms. In M. Napolitano and F. Sabetta (Eds.), *Proc. Thirteenth Internat. Conf. on Num. Methods in Fluid Dyn.*, pp. 160–164. Berlin: Springer.

Kajishima, T., Y. Miyake, and T. Nishimoto (1990). Large eddy simulation of turbulent flow in a duct of square cross section. In K.W. Morton (Ed.), *Proceedings of the Twelfth International Conference on Numerical Methods in Fluid Dynamics*, pp. 202–204. Berlin: Springer.

Karki, K.C. and S.V. Patankar (1988). Calculation procedure for viscous incompressible flows in complex geometries. *Num. Heat Transfer* **14**, 295–307.

Karki, K.C. and S.V. Patankar (1989). Pressure based calculation procedure for viscous flows at all speed in arbitrary configurations. *AIAA J.* **27**, 1167–1174.

Karni, S. (1992). Accelerated convergence to steady state by gradual far-field damping. *AIAA J.* **30**, 1220–1228.

Kenjereš, S. (1999). *Numerical modelling of complex buyoancy-driven flows.* Ph.D. thesis, Delft University of Technology, The Netherlands.

Kershaw, D. S. (1978). The incomplete Choleski-conjugate gradient method for the iterative solution of systems of linear equations. *J. Comp. Phys.* **26**, 43–65.

Kettler, R. (1982). Analysis and comparison of relaxation schemes in robust multigrid and conjugate gradient methods. In W. Hackbusch and U. Trottenberg (Eds.), *Multigrid Methods*, pp. 502–534. Berlin: Springer. Lecture Notes in Mathematics 960.

Kevorkian, J. and J.D. Cole (1981). *Perturbation Methods in Applied Mathematics.* New York: Springer.

Kim, C.A. and A. Jameson (1995). Flux limited dissipation schemes for high speed unsteady flows. AIAA Paper 95-1738-CP.

Kim, J. and P. Moin (1985). Application of a fractional-step method to incompressible Navier-Stokes equations. *J. Comp. Phys.* **59**, 308–323.

Kinmark, I.P.E. (1984). One step integration methods with large stability limits for hyperbolic partial differential equations. In V.R. Vichnevetsky and R.S. Stepleman (Eds.), *Advances in Computer Methods for Partial Differential Equations*, pp. 345–349. New Brunswick: IMACS.

Klein, R. (1995). Semi-implicit extension of a Godunov-type scheme based on low Mach number asymptotics 1: one-dimensional flow. *J. Comp. Phys.* **121**, 213–237.

Knupp, P. and S. Steinberg (1993). *Fundamentals of Grid Generation*. Boca Raton: CRC Press.

Kobayashi, M.H. and J.C.F. Pereira (1991). Numerical comparison of momentum interpolation methods and pressure-velocity algorithms using non-staggered grids. *Comm. Appl. Num. Meth.* **7**, 173–186.

Kobayashi, T. and M. Kano (1986). Numerical prediction of turbulent plane Couette flow by large eddy simulation. In U. Schumann and R. Friedrichs (Eds.), *Direct and Large Eddy Simulation of Turbulence*, pp. 135–146. Braunschweig: Vieweg.

Kobayashi, T., Y. Morinishi, and K. Oh (1992). Large eddy simulation of backward-facing step flow. *Comm. Appl. Num. Meth.* **8**, 431–441.

Koren, B. (1996). Improving Euler computations at low Mach numbers. *Int. J. Comp. Fluid Dyn.* **6**, 51–70.

Koren, B. and B. van Leer (1995). Analysis of preconditioning and multigrid for Euler flows with low-subsonic regions. *Advances in Comp. Meth.* **4**, 127–144.

Kost, A., L. Bai, N.K. Mitra, and M. Fiebig (1992). Calculation procedure for unsteady incompressible 3d flows in arbitrarily shaped domains. In J.B. Vos, A. Rizzi, and I.L. Rhyming (Eds.), *Proceedings of the Ninth GAMM-Conference on Numerical Methods in Fluid Mechanics*, pp. 269–278. Braunschweig: Vieweg.

Kost, A., N.K. Mitra, and M. Fiebig (1992). Numerical simulation of three-dimensional unsteady flow in a cavity. In M. Deville, T.-H. Lê, and Y. Morchoisne (Eds.), *Numerical Simulation of 3-D Incompressible Unsteady Viscous Laminar Flows*, pp. 79–90. Braunschweig: Vieweg.

Kowalik, Z. and T.S. Murty (1993). *Numerical Modeling of Ocean Dynamics*. Singapore: World Scientific.

Kreiss, H.-O. (1964). On difference approximations of the dissipative type for hyperbolic differential equations. *Comm. Pure and Appl. Math.* **17**, 335–353.

Kreiss, H.-O. and J. Lorenz (1989). *Initial-Boundary Value Problems and the Navier-Stokes Equations*. San Diego: Academic Press.

Kreiss, H.-O. and E. Lundqvist (1968). On difference approximations with wrong boundary values. *Math. Comp.* **22**, 1–12.

Krettenauer, K. and U. Schumann (1992). Numerical simulation of turbulent convection over wavy terrain. *J. Fluid Mech.* **237**, 261–299.

Kristoffersen, R. and H.I. Andersson (1993). Direct simulations of low-Reynolds-number turbulent flow in a rotating channel. *J. Fluid Mech.* **256**, 163–197.

Kröner, D. (1997). *Numerical Schemes for Conservation Laws.* Chichester/Stuttgart: Wiley and Teubner.

Kwak, D. and J.L.C. Chang (1984). A computational method for viscous incompressible flows. In R. Vichnevetsky and R.S. Stepleman (Eds.), *Advances in Computer Methods for Partial Differential Equations V*, pp. 277–288. New Brunswick: IMACS.

Kwak, D., J.I.C. Chang, S.P. Shanks, and S.R. Chakravarthy (1986). A three-dimensional incompressible Navier-Stokes flow solver using primitive variables. *AIAA J.* **24**, 390–396.

Lamb, H. (1945). *Hydrodynamics.* New York: Dover.

Landau, L.D. and E.M. Lifshitz (1959). *Fluid Mechanics.* London: Pergamon Press.

Laney, C.B. (1998). *Computational Gasdynamics.* Cambridge, UK: Cambridge University Press.

Lapidus, L. and G.F. Pinder (1982). *Numerical Solution of Partial Differential Equations in Science and Engineering.* New York: Wiley.

Lapworth, B.L. (1988). Examination of pressure oscillations arising in the computation of cascade flow using a boundary-fitted co-ordinate system. *Int. J. Num. Meth. in Fluids* **8**, 387–404.

Launder, B.E. (1990). Phenomenological medelling: present ... and future? In J.L. Lumley (Ed.), *Whither Turbulence? Turbulence at the Crossroads*, pp. 439–485. Berlin: Springer. Lecture Notes in Physics 357.

Launder, B.E. (1996). Turbulence modelling for flow in arbitrary complex domains. In J.-A. Désidéri, C. Hirsch, P. Le Tallec, E. O nate, M. Pandolfi, J. Périaux, and E. Stein (Eds.), *Computational Methods in Applied Sciences '96*, pp. 65–76. Chichester: Wiley.

Lawson, J.D. and J.Ll. Morris (1978). The extrapolation of first order methods for parabolic partial differential equations. *SIAM J. Num. Anal.* **15**, 1212–1224.

Lax, P.D. (1954). Weak solutions of nonlinear hyperbolic equations and their numerical approximation. *Comm. Pure and Appl. Math.* **7**, 159–193.

Lax, P.D. (1973). *Hyperbolic Systems of Conservation Laws and the Mathematical Theory of Shock Waves.* Philadelphia: SIAM.

Lax, P. and B. Wendroff (1960). Systems of conservation laws. *Comm. Pure and Appl. Math.* **13**, 217–237.

Le, H. and P. Moin (1991). An improvement of fractional step methods for the incompressible Navier-Stokes equations. *J. Comp. Phys.* **92**, 369–379.

Le, H., P. Moin, and J. Kim (1997). Direct numerical simulation of turbulent flow over a backward-facing step. *J. Fluid Mech.* **330**, 349–374.

Lê, T.H., J. Ryan, and K. Dang Tran (1992). Direct simulation of incompressible, viscous flow through a rotating channel. In J.B. Vos, A. Rizzi, and I.L. Rhyming (Eds.), *Proceedings of the Ninth GAMM-Conference on Numerical Methods in Fluid Mechanics*, pp. 533–541. Braunschweig: Vieweg.

Le Thanh, K.-C. (1992). Multidomain technique for 3-D incompressible unsteady viscous laminar flow around prolate spheroid. In M. Deville, T.-H. Lê, and Y. Morchoisne (Eds.), *Numerical Simulation of 3-D Incompressible Unsteady Viscous Laminar Flows*, pp. 131–139. Braunschweig: Vieweg.

Le Thanh, K.-C., B. Troff, and Ta Phuoc Loc (1991). Numerical study of unsteady incompressible separated viscous flows around an obstacle. *Recherche Aérospatiale* **1991-1**, 44–58.

Leendertse, J.J. (1967). *Aspects of a computational model for long period water-wave propagation.* Ph.D. thesis, Delft University of Technology. Also appeared as Rand Memorandum RM-5294-PR, Rand Corporation, Santa Monica, California, 1967.

Leonard, B.P. (1979). A stable and accurate convective modelling procedure based on quadratic upstream interpolation. *Comp. Meth. Appl. Mech. Eng.* **19**, 59–98.

Leonard, B.P. (1988). Simple high-accuracy resolution program for convective modelling of discontinuities. *Int. J. Num. Meth. in Fluids* **8**, 1291–1318.

Leonard, B.P. and J.E. Drummond (1995). Why you should not use 'hybrid', 'power-law' or related exponential schemes for convective modelling - there are much better alternatives. *Int. J. Num. Meth. in Fluids* **20**, 421–442.

Leone Jr., J.M. (1990). Open boundary condition symposium benchmark solution: stratified flow over a backward facing step. *Int. J. Num. Meth. in Fluids* **11**, 969–984.

Leschziner, M.A. (1980). Practical evaluation of three finite-difference schemes for the computation of steady-state recirculating flows. *Comp. Meth. Appl. Mech. Eng.* **23**, 293–312.

LeVeque, R.J. (1992). *Numerical Methods for Conservation Laws.* Basel: Birkhäuser.

Libby, P.A. (1996). *Introduction to Turbulence.* Washington DC: Taylor and Francis.

Lien, F.S. and M.A. Leschziner (1994). Multigrid acceleration for recirculating laminar and turbulent flows computed with a non-orthogonal, collocated finite-volume scheme. *Comp. Meth. Appl. Mech. Eng.* **118**, 351–371.

Liepmann, H.W. and A. Roshko (1957). *Elements of Gasdynamics*. New York: Wiley.

Lighthill, J. (1986). The recently recognized failure of predictability in Newtonian dynamics. *Proc. R. Soc. London* **A407**, 35–50.

Lions, P.-L. (1996). *Mathematical Topics in Fluid Mechanics. Vol. 1. Incompressible Models*. Oxford: Clarendon.

Liou, M.-S. and C.J. Steffen (1993). A new flux splitting scheme. *J. Comp. Phys.* **107**, 23–39.

Lorenz, E.N. (1993). *The Essence of Chaos*. London: University College of London Press.

MacCormack, R.W. (1969). The effect of viscosity in hyper-velocity impact cratering. AIAA Paper 69-354.

Majda, A. (1984). *Compressible Fluid Flow and Systems of Conservation Laws in Several Space Variables*, Volume 53 of *Applied Mathematical Sciences*. New York: Springer.

Manhart, M., G.B. Deng, T.J. Hüttl, F. Tremblay, A. Segal, R. Friedrich, J. Piquet, and P. Wesseling (1998). The minimal turbulent flow unit as test case for three different computer codes. In E.H. Hirschel (Ed.), *Numerical Flow Simulation I*, pp. 365–381. Braunschweig: Vieweg.

Manhart, M. and H. Wengle (1994). Large-eddy simulation of turbulent boundary layer flow over a hemisphere. In P.R. Voke, L. Kleiser, and J.-P. Chollet (Eds.), *Direct and Large-Eddy Simulation I*, pp. 299–310. Dordrecht: Kluwer.

Manteuffel, T.A. and A.B. White, Jr. (1986). The numerical solution of second-order boundary value problems on nonuniform meshes. *Math. Comp.* **47**, 511–535.

Marx, Y.P. (1994). Time integration schemes for the unsteady incompressible Navier-Stokes equations. *J. Comp. Phys.* **112**, 182–209.

Mary, I., P. Sagaut, and M. Deville (2000). An algorithm for low Mach number unsteady flows. *Computers and Fluids* **29**, 119–147.

Mason, P.J. (1989). Large-eddy simulation of the convective atmospheric boundary layer. *J. of the Atmos. Sci.* **46**, 1492–1516.

McGuirk, J.J. and G.J. Page (1990). Shock capturing using a pressure-correction method. *AIAA J.* **28**, 1751–1757.

Meijerink, J.A. and H.A. van der Vorst (1977). An iterative solution method for linear systems of which the coefficient matrix is a symmetric M-matrix. *Math. Comp.* **31**, 148–162.

Meinke, M. and D. Hänel (1990). Simulation of unsteady flows. In K.W. Morton (Ed.), *Proceedings of the Twelfth International Conference on Numerical Methods in Fluid Dynamics*, pp. 268–272. Berlin: Springer.

Melaaen, M.C. (1992). Calculation of fluid flows with staggered and non-staggered curvilinear nonorthogonal grids – Theory. *Num. Heat Transfer A* **21**, 1–19.

Merkle, C.L. and M. Athvale (1987). Time-accurate unsteady incompressible flow algorithms based on artificial compressibility. AIAA Paper 87-1137.

Merkle, C.L. and P.Y.L. Tsai (1986). Application of Runge-Kutta schemes to incompressible flow. AIAA Paper 86-0553.

Miller, J.J.H. (1971). On the location of zeros of certain classes of polynomials with applications to numerical analysis. *J. Inst. Math. Appl.* **8**, 397–406.

Miller, T.F. and F.W. Schmidt (1988). Use of a pressure-weighted interpolation method for the solution of the incompressible Navier-Stokes equations on a nonstaggered grid system. *Num. Heat Transfer* **14**, 213–233.

Mitchell, A.R. and D.F. Griffiths (1994). *The Finite Difference Method in Partial Differential Equations.* Chichester: Wiley.

Moeng, C.-H. (1984). A large-eddy simulation for the study of planetary boundary layer turbulence. *J. of the Atmos. Sci.* **41**, 2052–2062.

Mohammadi, B. and O. Pironneau (1994). *Analysis of the k-epsilon Turbulence Model.* Chichester: Wiley.

Moin, P. and J. Kim (1982). Numerical investigation of turbulent channel flow. *J. Fluid Mech.* **118**, 341–377.

Moin, P. and K. Madesh (1998). Direct numerical simulation: a tool in turbulence research. *Ann. Rev. Fluid Mech.* **30**, 539–578.

Morgan, K.J., J. Périaux, and F. Thomasset (Eds.) (1984). *Analysis of Laminar Flow over a Backward Facing Step.* GAMM Workshop held at Bièvres (Fr.), Braunschweig: Vieweg.

Morton, K.W. (1971). Stability and convergence in fluid flow problems. *Proc. Roy. Soc. London A* **323**, 237–253.

Morton, K.W. (1996). *Numerical Solution of Convection-Diffusion Problems.* London: Chapman and Hall.

Morton, K.W. and D.F. Mayers (1994). *Numerical Solution of Partial Differential Equations.* Cambridge, UK: Cambridge University Press.

Moukalled, F. and S. Acharya (1991). A local adaptive grid procedure for incompressible flows with multigridding and equidistribution concepts. *Int. J. Num. Meth. in Fluids* **13**, 1085–1111.

Murman, E.M. and J.D. Cole (1971). Calculation of plane steady transonic flows. *AIAA Journal* **9**, 114–121.

Na, Y. and P. Moin (1998). Direct numerical simulation of a separated turbulent boundary layer. *J. Fluid Mech.* **370**, 175–201.

Nakayama, Y. and W.A. Woods (Eds.) (1988). *Visualized Flow; Fluid Motion in Basic and Engineering Situations Revealed by Flow Visualization.* Oxford: Pergamon.

Navier, C.L.M.H. (1823). Mémoire sur les lois du mouvement des fluides. *Mém. Acad. R. Sci. Paris* **6**, 389–416.

Oleinik, O.A. (1957). Discontinuous solutions of nonlinear differential equations. *Uspekhi Mat. Nauk* **12**, 3–73. (Amer. Math. Soc. Transl. Ser. 2, 26, pp. 95–172).

Oosterlee, C.W. and P. Wesseling (1992a). A multigrid method for a discretization of the incompressible Navier-Stokes equations in general coordinates. In J.B. Vos, A. Rizzi, and I.L. Rhyming (Eds.), *Proceedings of the Ninth GAMM-Conference on Numerical Methods in Fluid Mechanics, Lausanne, sept. 1991*, pp. 99–106. Braunschweig: Vieweg.

Oosterlee, C.W. and P. Wesseling (1992b). A multigrid method for an invariant formulation of the incompressible Navier-Stokes equations in general co-ordinates. *Comm. Appl. Num. Meth.* **8**, 721–734.

Oosterlee, C.W. and P. Wesseling (1992c). A robust multigrid method for a discretization of the incompressible Navier-Stokes equations in general coordinates. In Ch. Hirsch, J. Périaux, and W. Kordulla (Eds.), *Computational Fluid Dynamics '92. Proc., First European Computational Fluid Dynamics Conf., Sept. 1992, Brussels*, pp. 101–108. Amsterdam: Elsevier.

Oosterlee, C.W. and P. Wesseling (1993). A robust multigrid method for a discretization of the incompressible Navier-Stokes equations in general coordinates. *Impact Comp. Sci. Eng.* **5**, 128–151.

Oosterlee, C.W. and P. Wesseling (1994). Steady incompressible flow around objects in general coordinates with a multigrid solution method. *Num. Meth. Part. Diff. Eqs.* **10**, 295–308.

Oosterlee, C.W., P. Wesseling, A. Segal, and E. Brakkee (1993). Benchmark solutions for the incompressible Navier-Stokes equations in general co-ordinates on staggered grids. *Int. J. Num. Meth. in Fluids* **17**, 301–321.

Oseen, C.W. (1910). Über die Stokessche Formel und über eine verwandte Aufgabe in der Hydrodynamik. *Ark. Mat. Astr. Fys.* **6**(No. 29).

Osher, S. (1981). Numerical solution of singular perturbation problems and hyperbolic systems of conservation laws. In O. Axelsson, L.S. Frank, and A. van der Sluis (Eds.), *Analytical and Numerical Approaches to Asymptotic Problems in Analysis*, pp. 179–204. Amsterdam: North-Holland.

Osher, S. and S. Chakravarthy (1983). Upwind schemes and boundary conditions with applications to Euler equations in general geometries. *J. Comp. Phys.* **50**, 447–481.

Osher, S. and F. Solomon (1982). Upwind difference schemes for hyperbolic systems of conservation laws. *Math. Comp.* **38**, 339–374.

Paillère, H., S. Clerc, C. Viozat, I. Toumi, and J.-P. Magnaud (1998). Numerical methods for low Mach number thermal-hydraulic flows. In K.D. Papailiou, D. Tsahalis, J. Périaux, and D. Knörzer (Eds.), *Computational Fluid Dynamics '98*, Volume 2, pp. 80–89. Chichester: Wiley.

Patankar, S.V. (1980). *Numerical Heat Transfer and Fluid Flow*. New York: McGraw-Hill.

Patankar, S.V. and D.B. Spalding (1972). A calculation procedure for heat and mass transfer in three-dimensional parabolic flows. *Int. J. Heat and Mass Transfer* **15**, 1787–1806.

Perić, M., R. Kessler, and G. Scheuerer (1988). Comparison of finite-volume numerical methods with staggered and collocated grids. *Computers and Fluids* **16**, 389–403.

Perot, J.B. (1993). An analysis of the fractional step method. *J. Comp. Phys.* **108**, 51–58.

Peyret, R. and T.D. Taylor (1985). *Computational Methods for Fluid Flow.* Berlin: Springer.

Platzman, G.W. (1959). A numerical computation of the surge of 26 June 1954 on Lake Michigan. *Geophysics* **6**, 407–438.

Poisson, S.D. (1831). Mémoire sur les équations générales de l'équilibre et du mouvement des corps solides élastiques et des fluides. *Journal de l'Ecole Polytechnique de Paris* **13**, 139–166.

Pourquié, M.J.B.M. (1994). *Large-eddy simulation of a turbulent jet.* Ph.D. thesis, Delft University of Technology.

Protter, M.H. and H.F. Weinberger (1967). *Maximum Principles in Differential Equations.* Englewood Cliffs: Prentice-Hall.

Quarteroni, A. and A. Valli (1994). *Numerical Approximation of Partial Differential Equations.* Berlin: Springer.

Rai, M.M. and P. Moin (1991). Direct simulations of turbulent flow using finite-difference schemes. *J. of Comp. Phys.* **96**, 15–53.

Randall, D.A. (1994). Geostrophic adjustment and the finite difference shallow-water equations. *Mon. Weather Rev.* **122**, 1371–1377.

Rankine, W.J.M. (1870). On the thermodynamic theory of waves of finite longitudinal disturbance. *Trans. Roy. Soc. London* **160**, 277–288.

Rannacher, R. (1989). Numerical analysis of nonstationary flow. In V. Boffi and H. Neunzert (Eds.), *Applications of Mathematics in Industry and Technology*, pp. 34–53. Stuttgart: Teubner.

Rannacher, R. (1993). On the numerical solution of the incompressible Navier-Stokes equations. *Zeitschrift für Angewandte Mathematik and Mechanik* **73**, 203–217.

Reynolds, O. (1895). On the dynamical theory of incompressible viscous fluids and the determination of the criterion. *Phil. Trans. A* **186**, 123–164.

Reynolds, W.C. (1990). The potential and limitations of direct and large eddy simulation. In J.L. Lumley (Ed.), *Whither Turbulence? Turbulence at the Crossroads*, pp. 313–343. Berlin: Springer. Lecture Notes in Physics 357.

Rhie, C.M. and W.L. Chow (1983). Numerical study of the turbulent flow past an airfoil with trailing edge separation. *AIAA J.* **21**, 1525–1532.

Richardson, L.F. (1910). The approximate arithmetical solution by finite differences of physical problems involving differential equations, with an application to the stress in a masonry dam. *Trans. Roy. Soc. London Ser. A* **210**, 307–357.

Richardson, L.F. (1922). *Weather Prediction by Numerical Process.* London: Cambridge University Press. Reprinted, Dover, New York, 1965.

Richtmyer, R.D. and K.W. Morton (1967). *Difference Methods for Initial Value Problems.* New York: Wiley.

Rizzi, A. (1978). Numerical implementation of solid-body boundary conditions for the Euler equations. *Zeitschrift für Angewandte Mathematik und Mechanik* **58**, 301–304.

Roache, P.J. (1972). *Computational Fluid Dynamics.* Albuquerque, NM: Hermosa.

Roache, P.J. (1998a). *Fundamentals of Computational Fluid Dynamics.* Albuquerque, NM: Hermosa.

Roache, P.J. (1998b). *Verification and Validation in Computational Science and Engineering.* Albuquerque, NM: Hermosa.

Rodi, W., S. Majumdar, and B. Schönung (1989). Finite volume methods for two-dimensional incompressible flows with complex boundaries. *Comp. Meth. Appl. Mech. Eng.* **75**, 369–392.

Roe, P.L. (1981). Approximate Riemann solvers, parameter vectors, and difference schemes. *J. Comp. Phys.* **43**, 357–372.

Roe, P.L. (1986). Characteristic-based schemes for the Euler equations. *Annual Review of Fluid Mechanics* **18**, 337–365.

Roe, P.L. (1989). Remote boundary conditions for unsteady multidimensional aerodynamic computations. *Computers and Fluids* **17**, 221–231.

Roe, P.L. and J. Pike (1984). Efficient construction and utilisation of approximate Riemann solutions. In R. Glowinski and J.L. Lions (Eds.), *Computing Methods in Applied Sciences and Engineering, VI*, pp. 499–518. Amsterdam: North-Holland.

Rogers, S.E. and D. Kwak (1990). Upwind differencing scheme for the time-accurate incompressible Navier-Stokes equations. *AIAA J.* **28**, 253–262.

Rogers, S.E., D. Kwak, and C. Kiris (1991). Steady and unsteady solutions of the incompressible Navier-Sokes equations. *AIAA J.* **29**, 603–610.

Roos, H.-G., M. Stynes, and L. Tobiska (1996). *Numerical Methods for Singularly Perturbed Differential Equations.* Berlin: Springer.

Rosenfeld, M., D. Kwak, and M. Vinokur (1991). A fractional step solution method for the unsteady incompressible Navier-Stokes equations in generalized coordinate systems. *J. Comp. Phys.* **94**, 102–137.

Roux, B. (Ed.) (1990). *Numerical Simulation of Oscillatory Convection in low-Pr Fluids.* Braunschweig: Vieweg. Notes on Numerical Fluid Mechanics 27.

Saad, Y. (1996). *Iterative Methods for Sparse Linear Systems.* Boston: PWS Publishing.

Saad, Y. and M.H. Schultz (1986). GMRES: a generalized minimal residual algorithm for solving non-symmetric linear systems. *SIAM J. Sci. Stat. Comp.* **7**, 856–869.

Sani, R.L. and P.M. Gresho (1994). Résumé and remarks on the open boundary condition minisymposium. *Int. J. Num. Meth. in Fluids* **18**, 983–1008.

Schmidt, H. and U. Schumann (1989). Coherent structures of the convective boundary layer derived from large-eddy simulations. *J. Fluid Mech.* **200**, 511–562.

Schmitt, L. and R. Friedrich (1988). Large-eddy simulation of turbulent backward facing step flow. In M. Deville (Ed.), *Proceedings of the Seventh GAMM-Conference on Numerical Methods in Fluid Mechanics*, pp. 355–362. Braunschweig: Vieweg.

Schumann, U. (1975). Linear stability of finite difference equations for three-dimensional flow problems. *J. Comp. Phys.* **18**, 465–470.

Sedov, L.I. (1971). *A Course in Continuum Mechanics, Vol. I. Basic Equations and Analytical Techniques*. Groningen, The Netherlands: Wolters-Noordhoff Publishing.

Shaw, G.J. and S. Sivaloganathan (1988). On the smoothing of the SIMPLE pressure correction algorithm. *Int. J. Num. Meth. in Fluids* **8**, 441–462.

Shen, J. (1992a). On error estimates of projection methods for Navier-Stokes equations: first-order schemes. *SIAM J. Num. Anal.* **29**, 57–77.

Shen, J. (1992b). On error estimates of some higher order projection and penalty-projection methods for Navier-Stokes equations. *Numer. Math.* **62**, 49–73.

Sheng, C., L.K. Taylor, and D.L. Whitfield (1995). Multigrid algorithm for three-dimensional high-Reynolds number turbulent flow. *AIAA J.* **33**, 2073–2079.

Shih, T.-H. (1997). Some developments in computational modeling of turbulent flows. *Fluid Dyn. Research* **20**, 67–96.

Shimomura, O. (1991). Large eddy simulation of MHD turbulent channel flow under a uniform magnetic field. In O. Métais and M. Lesieur (Eds.), *Tubulence and Coherent Structures*, pp. 553–567. Dordrecht: Kluwer.

Shishkin, G.I. (1990). Grid aproximations of singularly perturbed elliptic equations in domains with characteristic faces. *Sov. J. Numer. Anal. Math. Modelling* **5**, 327–343.

Shu, C.-W. (1988). Total-variation diminishing time discretizations. *SIAM J. Sci. Stat. Comp.* **9**, 1073–1084.

Shu, C.-W. and S. Osher (1988). Efficient implementation of essentially non-oscillatory shock-capturing schemes. *J. Comp. Phys.* **77**, 439–471.

Shyy, W. (1994). *Computational Modeling for Fluid Flow and Interfacial Transport*. Amsterdam: Elsevier.

Shyy, W. and M.E. Braaten (1988). Adaptive grid computation for inviscid compressible flows using a pressure correction method. AIAA Paper 88-3566-CP.

Shyy, W., S. Thakur, and J. Wright (1992). Second-order upwind and central difference schemes for recirculating flow computation. *AIAA J.* **4**, 923–932.

Sielecki, A. (1968). An energy-conserving difference scheme for the storm surge equations. *Mon. Weather Rev.* **96**, 150–156.

Sivaloganathan, S. (1991). The use of local mode analysis in the design and comparison of multigrid methods. *Computer Phys. Comm.* **65**, 246–252.

Sivaloganathan, S., G.J. Shaw, T.M. Shah, and D.F. Mayers (1988). A comparison of multigrid methods for the incompressible Navier-Stokes equations. In K.W. Morton and M.J. Baines (Eds.), *Numerical Methods for Fluid Dynamics III*, pp. 410–417. Oxford: Oxford Univ. Press.

Skote, M., D.S. Henningson, and R.A.W.M. Henkes (1998). Direct numerical simulation of self-similar turbulent boundary layers in adverse pressure gradients. *Flow, Turbulence and Combustion* **60**, 47–85.

Smith, B.F., P.E. Bjørstad, and W.D. Gropp (1996). *Domain Decomposition; Parallel Multilevel Methods for Elliptic Partial Differential Equations*. Cambridge, UK: Cambridge University Press.

Smoller, J. (1983). *Shock Waves and Reaction-Diffusion Equations*. New York: Springer.

Sod, G.A. (1978). A survey of several finite difference methods for systems of nonlinear conservation laws. *J. Comp. Phys.* **27**, 1–31.

Sod, G.A. (1985). *Numerical Methods in Fluid Dynamics: Initial and Initial Boundary-Value Problems*. Cambridge, UK: Cambridge University Press.

Soh, W.Y. and J.W. Goodrich (1988). Unsteady solution of incompressible Navier-Stokes equations. *J. Comp. Phys.* **79**, 113–134.

Sokolnikoff, I.S. (1964). *Tensor Analysis*. Englewood Cliffs, N.J.: Wiley.

Sommeijer, B.P., P.J. van der Houwen, and J. Kok (1994). Time integration of three-dimensional numerical transport models. *Appl. Num. Math.* **16**, 201–225.

Sonneveld, P. and B. van Leer (1985). A minimax problem along the imaginary axis. *Nieuw Archief voor Wiskunde* **3**, 19–22.

Spalart, P.R. (1988). Direct simulation of a turbulent boundary layer up to Rtheta = 1410. *J. Fluid Mech.* **187**, 61–98.

Spalart, P.R., R.D. Moser, and M.M. Rogers (1991). Spectral methods for the Navier-Stokes equations with one infinite and two periodic directions. *J. Comp. Phys.* **96**, 297–324.

Spalding, D.B. (1972). A novel finite difference formulation for differential expressions involving both first and second derivatives. *Int. J. Num. Meth. in Eng.* **4**, 551–559.

Spekreijse, S.P. (1995). Elliptic grid generation based on Laplace equations and algebraic transformations. *J. Comp. Phys.* **118**, 38–61.

Sperb, R.P. (1981). *Maximum Principles and Their Applications.* New York: Academic Press.

Spijker, M.N. (1971). On the structure of error estimates for finite-difference methods. *Numer. Math.* **18**, 73–100.

Steger, J.L. and R.F. Warming (1981). Flux-vector splitting of the inviscid gas-dynamic equations with applications to finite-difference methods. *J. Comp. Phys.* **40**, 263–293.

Stelling, G.S. (1983). *On the construction of computational methods for shallow water flow problems.* Ph.D. thesis, Delft University of Technology. Also appeared as Rijkswaterstaat Communications 35, 1984. Rijkswaterstaat, The Hague.

Stelling, G.S., A.K. Wiersma, and J.B.T.M. Willemse (1986). Practical aspects of accurate tidal computations. *J. Hydr. Eng.* **112**, 802–817.

Stoker, J.J. (1957). *Water Waves.* New York: Interscience.

Stokes, G.G. (1845). On the theories of the internal friction of fluids in motion, and of the equilibrium and motion of elastic solids. *Trans. Camb. Phil. Soc.* **8**, 287–305.

Stokes, G.G. (1851). On the effect of the internal friction of fluids on the motion of pendulums. *Trans. Camb. Phil. Soc.* **9, Pt. II**, 8–106.

Strikwerda, J.C. (1989). *Finite Difference Schemes and Partial Differential Equations.* Pacific Grove: Wadsworth and Brooks/Cole.

Swarztrauber, P.N. (1984). Fast Poisson solvers. In G.H. Golub (Ed.), *Studies in Numerical Analysis*, pp. 319–370. The Mathematical Association of America. Studies in Mathematics Vol. 24.

Sweby, P.K. (1984). High resolution schemes using flux-limiters for hyperbolic conservation laws. *SIAM J. Num. Anal.* **21**, 995–1011.

Takemoto, Y. and Y. Nakamura (1986). A three-dimensional incompressible flow solver. In F.G. Zhuang and Y.L. Zhu (Eds.), *Proceedings of the Tenth International Conference on Numerical Methods in Fluid Dynamics*, pp. 594–599. Berlin: Springer.

Takemoto, Y. and Y. Nakamura (1989). Numerical simulation of 2-D and 3-D channel flows using a third order accurate generalized QUICK scheme. In Y. Iwasa, N. Tamai, and A. Wada (Eds.), *Refined Flow Modeling and Turbulence Measurements.* Tokyo: Universal Academy Press.

Tamamidis, P., G. Zhang, and D.N. Assanis (1996). Comparison of pressure-based and artificial compressibility methods for solving 3D steady incompressible flows. *J. Comp. Phys.* **124**, 1–13.

Tannehill, J.C., D.A. Anderson, and R.H. Pletcher (1997). *Computational Fluid Dynamics and Heat Transfer.* London: Taylor and Francis.

Taylor, M.E. (1996). *Partial Differential Equations*, Volume 3. New York: Springer.

Temam, R. (1977). *Navier-Stokes Equations; Theory and Numerical Analysis.* Amsterdam: North-Holland.

Temam, R. (1985). *Navier-Stokes Equations.* Amsterdam: North-Holland.

Tennekes, H. and J.L. Lumley (1982). *A First Course in Turbulence.* Cambridge, Massachussets: MIT Press.

Thomas, J.L. and M.D. Salas (1986). Far-field boundary conditions for transonic lifting solutions to the Euler equations. *AIAA J.* **24**, 1074–1080.

Thomas, T.G. and J.J.R. Williams (1994). Large-eddy simulation of compound channel flow with one floodplain at Re 42000. In P.R. Voke, L. Kleiser, and J-P. Chollet (Eds.), *Direct and Large-Eddy Simulation I*, pp. 311–324. Dordrecht: Kluwer.

Thomeé, V. (1990). Finite difference methods for linear parabolic equations. In P.G. Ciarlet and J.L. Lions (Eds.), *Handbook of Numerical Analysis, Vol. I*, pp. 5–196. Amsterdam: North-Holland.

Thompson, J.F., Z.U.A. Warsi, and C.W. Mastin (1985). *Numerical Grid Generation, Foundations and Applications.* Amsterdam: North-Holland.

Thompson, M.C. and J.H. Ferziger (1989). An adaptive multigrid technique for the incompressible Navier-Stokes equations. *J. Comp. Phys.* **82**, 94–121.

Tikhonov, A.N. and A.A. Samarskii (1963). Homogeneous difference schemes on non-uniform nets. *USSR Comput. Math. and Math. Phys.* **2**, 927–953.

Toro, E.F. (1997). *Riemann Solvers and Numerical Methods for Fluid Dynamics.* Berlin: Springer.

Turek, S. (1994). Tools for simulating non-stationary incompressible flow via discretely divergence-free finite element models. *Int. J. Num. Meth. in Fluids* **18**, 71–105.

Turkel, E. (1993). Review of preconditioning techniques for fluid dynamics. *Appl. Num. Math.* **12**, 257–284.

Turkel, E., A. Fiterman, and B. van Leer (1994). Preconditioning and the limit of the compressible to the incompressible flow equations for finite difference schemes. In D.A. Caughey and M.M. Hafez (Eds.), *Frontiers of Computational Fluid Dynamics*, pp. 215–234. Chichester: Wiley.

Turkel, E., R. Radespiel, and H. Kroll (1997). Assessment of preconditioning methods for multidimensional aerodynamics. *Computers and Fluids* **26**, 613–634.

van Albada, G.D., B. van Leer, and W.W. Roberts (1982). A comparative study of computational methods in cosmic gas dynamics. *Astron. Astrophys.* **108**, 76–84.

van Beek, P., R.R.P. van Nooyen, and P. Wesseling (1995). Accurate discretization on non-uniform curvilinear staggered grids. *J. Comp. Phys.* **117**, 364–367.

van der Heul, D.R., C. Vuik, and P. Wesseling (1999). A staggered scheme for hyperbolic conservation laws applied to unsteady sheet cavitation. *Computing and Visualization in Science* **2**, 63–68.

van der Houwen, P.J. (1977). *Construction of Integration Formulas for Initial-Value Problems*. Amsterdam: North-Holland.

van der Vorst, H.A. (1992). Bi-CGSTAB: a fast and smoothly converging variant of Bi-CG for solution of non-symmetric linear systems. *SIAM J. Sci. Stat. Comp.* **13**, 631–644.

van der Vorst, H.A. and C. Vuik (1994). GMRESR: a family of nested GMRES methods. *Num. Lin. Alg. Appl.* **1**, 369–386.

Van Doormaal, J.P. and G.D. Raithby (1984). Enhancements of the SIMPLE method for predicting incompressible fluid flows. *Num. Heat Transfer* **7**, 147–163.

Van Doormaal, J.P., G.D. Raithby, and B.H. McDonald (1987). The segregated approach to predicting viscous compressible fluid flows. *Transactions of the ASME - J. of Turbomachinery* **109**, 268–277.

Van Dyke, M. (1975). *Perturbation Methods in Fluid Mechanics*. Stanford: The Parabolic Press.

Van Dyke, M. (1982). *An Album of Fluid Motion*. Stanford: The Parabolic Press.

van Kan, J.J.I.M. (1986). A second-order accurate pressure correction method for viscous incompressible flow. *SIAM J. Sci. Stat. Comp.* **7**, 870–891.

van Leer, B. (1977). Towards the ultimate conservative difference scheme III. Upstream-centered finite-difference schemes for ideal compressible flow. *J. Comp. Phys.* **23**, 263–275.

van Leer, B. (1979). Towards the ultimate conservative difference scheme. V. A second-order sequel to Godunov's method. *J. Comp. Phys.* **32**, 101–136.

van Leer, B. (1982). Flux-vector splitting for the Euler equations. In E. Krause (Ed.), *Eighth International Conference on Numerical Methods in Fluid Dynamics*, pp. 507–512. Berlin: Springer. Lecture Notes in Physics 170.

van Leer, B. (1984). On the relation between the upwind-differencing schemes of Godunov, Enquist-Osher and Roe. *SIAM J. Sci. Stat. Comp.* **5**, 1–20.

Vanka, S.P. (1986a). Block-implicit multigrid solution of Navier-Stokes equations in primitive variables. *J. Comp. Phys.* **65**, 138–158.

Vanka, S.P. (1986b). A calculation procedure for three-dimensional steady recirculating flows using multigrid methods. *Comput. Meths. Appl. Mech. Eng.* **59**, 321–338.

Varah, J.M. (1980). Stability restrictions on second order, three level finite difference schemes for parabolic equations. *SIAM J. Num. Anal.* **17**, 300–309.

Varga, R.S. (1962). *Matrix Iterative Analysis.* Englewood Cliffs, N.J.: Prentice-Hall.

Veldman, A.E.P. and K. Rinzema (1992). Playing with nonuniform grids. *J. Eng. Math.* **26**, 119–130.

Verboom, G.K. and A. Slob (1984). Weakly-reflective boundary conditions for two-dimensional shallow water flow problems. *Adv. Water Resources* **7**, 192–197.

Verstappen, R.W.C.P. and A.E.P. Veldman (1994). Direct numerical simulation of a 3D turbulent flow in a driven cavity at Re = 10,000. In S. Wagner, E.H. Hirschel, J. Périaux, and R. Piva (Eds.), *Computational Fluid Dynamics '94*, pp. 558–565. Chichester: Wiley.

Verstappen, R.W.C.P. and A.E.P. Veldman (1997). Direct numerical simulation of turbulence at lower costs. *J. Eng. Math.* **32**, 143–159.

Verstappen, R.C.W. and A.E.P. Veldman (1998). Spectro-consistent discretization of Navier-Stokes: a challenge to RANS and DNS. *J. of Engineering Math.* **34**, 162–179.

Versteeg, H.K. and W. Malalasekera (1995). *An Introduction to Computational Fluid Dynamics. The Finite Volume Method.* Harlow: Longman Scientific and Technical.

Vinokur, M. (1989). An analysis of finite difference and finite-volume formulations of conservation laws. *J. Comp. Phys.* **81**, 1–52.

Voke, P.R. and S. Gao (1995). Large-eddy simulations of plane impinging jets. *Int. J. Num. Meth. in Eng.* **38**, 489–507.

Voke, P.R., L. Kleiser, and J.-P. Chollet (Eds.) (1994). *Direct and Large-Eddy Simulation I*, pp. 325–333. Dordrecht: Kluwer.

Vreugdenhil, C.B. (1994). *Numerical Methods for Shallow-Water Flow.* Dordrecht: Kluwer.

Vuik, C. (1993). Solution of the discretized incompressible Navier-Stokes equations with the GMRES method. *Int. J. Num. Meth. in Fluids* **16**, 507–523.

Vuik, C. (1996). Fast iterative solvers for the discretized incompressible Navier-Stokes equations. *Int. J. Num. Meth. in Fluids* **22**, 195–210.

Wagner, C. and R. Friedrich (1994). Direct numerical simulation of turbulent flow in a sudden pipe expansion. In *Application of Direct and Large Eddy Simulation to Transition and Turbulence*, Conference Proceedings 551, pp. 6.1–6.11. Neuilly-sur-Seine: AGARD.

Warming, R.F. and R.M. Beam (1976). Upwind second-order difference schemes and applications in aerodynamic flows. *AIAA J.* **14**, 1241–1249.

Washio, T. and C.W. Oosterlee (1997). Krylov subspace acceleration of nonlinear multigrid schemes. *Electr. Trans. on Num. Anal.* **6**, 271–290.

Weiser, A. and M.F. Wheeler (1988). On convergence of block-centered finite differences for elliptic problems. *SIAM J. Num. Anal.* **25**, 351–375.

Weiss, J.M. and W.A. Smith (1995). Preconditioning applied to variable and constant density flows. *AIAA J.* **33**, 2050–2057.

Wendt, John F. (1996). *Computational Fluid Dynamics; An Introduction.* Berlin: Springer.

Werner, H. and H. Wengle (1989). Large-eddy simulation of turbulent flow over a square. In H.-H. Fernholz and H.E. Fiedler (Eds.), *Advances in Turbulence II*, pp. 418–423. Berlin: Springer.

Wesseling, P. (1973). On the construction of accurate difference schemes for hyperbolic partial differential equations. *J. Eng. Math.* **7**, 1–31.

Wesseling, P. (1992). *An Introduction to Multigrid Methods.* Chichester: Wiley.

Wesseling, P. (1995). A method to obtain von Neumann stability conditions for the convection-diffusion equation. In K.W. Morton and M.J. Baines (Eds.), *Numerical Methods for Fluid Dynamics V*, pp. 211–224. Oxford: Clarendon Press.

Wesseling, P. (Ed.) (1996a). *High Performance Computing in Fluid Dynamics.* Dordrecht: Kluwer. ERCOFTAC Series Vol. 3.

Wesseling, P. (1996b). Uniform convergence of discretization error for a singular perturbation problem. *Num. Meth. Part. Diff. Eqs.* **12**, 657–671.

Wesseling, P. (1996c). Von Neumann stability conditions for the convection-diffusion equation. *IMA J. Num. Anal.* **16**, 583–598.

Wesseling, P., C.G.M. Kassels, C.W. Oosterlee, A. Segal, C. Vuik, S. Zeng, and M. Zijlema (1994). Computing incompressible flows in general domains. In F.-K. Hebeker, R. Rannacher, and G. Wittum (Eds.), *Numerical Methods for the Navier-Stokes Equations*, pp. 298–314. Braunschweig: Vieweg.

Wesseling, P. and C.W. Oosterlee (2000). Geometric multigrid with applications to computational fluid dynamics. *J. Comp. Appl. Math.*, to appear.

Wesseling, P., A. Segal, and C.G.M. Kassels (1999). Computing flows on general three-dimensional nonsmooth staggered grids. *J. Comp. Phys.* **149**, 333–362.

Wesseling, P., A. Segal, C.G.M. Kassels, and H. Bijl (1998). Computing flows on general two-dimensional nonsmooth staggered grids. *J. Eng. Math.* **34**, 21–44.

Wesseling, P., M. Zijlema, A. Segal, and C.G.M Kassels (1997). Computation of turbulent flow in general domains. *Math. Comp. Sim.* **44**, 369–385.

Widlund, O.B. (1966). Stability of parabolic difference schemes in the maximum norm. *Numer. Math.* **8**, 186–202.

Wilcox, D.C. (1993). *Turbulence Modeling for CFD.* La Canada, California: DCW Industries Inc.

Wilders, P., Th.L. van Stijn, G.S. Stelling, and G.A. Fokkema (1988). A fully implicit splitting method for accurate tidal computations. *Int. J. Num. Meth. in Eng.* **26**, 2707–2721.

Wittum, G. (1989a). Linear iterations as smoothers in multigrid methods: Theory with applications to incomplete decompositions. *Impact Comp. Sci. Eng.* **1**, 180–215.

Wittum, G. (1989b). Multi-grid methods for Stokes and Navier-Stokes equations with transforming smoothers: Algorithms and numerical results. *Numer. Math.* **54**, 543–563.

Wittum, G. (1990a). On the convergence of multi-grid methods with transforming smoothers. *Numer. Math.* **57**, 15–38.

Wittum, G. (1990b). R-transforming smoothers for the incompressible Navier- Stokes equations. In W. Hackbusch and R. Rannacher (Eds.), *Numerical Treatment of the Navier-Stokes Equations*, pp. 153–162. Braunschweig: Vieweg. Notes on Numerical Fluid Mechanics 30.

Wittum, G. (1990c). The use of fast solvers in computational fluid dynamics. In P. Wesseling (Ed.), *Proceedings of the Eighth GAMM-Conference on Numerical Methods in Fluid Mechanics*, pp. 574–581. Braunschweig: Vieweg. Notes on Numerical Fluid Mechanics 29.

Wörner, M. and G. Grötzbach (1992). Analysis of semi-implicit time integration schemes for direct numerical simulation of turbulent convection in liquid metals. In J.B. Vos, A. Rizzi, and I.L. Rhyming (Eds.), *Proceedings of the Ninth GAMM-Conference on Numerical Methods in Fluid Mechanics*, pp. 542–551. Braunschweig: Vieweg.

Yang, K.-S. and J.H. Ferziger (1993). Large-eddy simulation of turbulent obstacle flow using a dynamic subgrid-scale model. *AIAA J.* **31**, 1406–1413.

Ye, T., R. Mittal, H.S. Udaykumar, and W. Shyy (1999). An accurate Cartesian grid method for viscous incompressible flows with complex immersed boundaries. *J. Comp. Phys.* **156**, 209–240.

Young, D.M. (1971). *Iterative Solution of Large Linear Systems.* New York: Academic Press.

Zang, Y., R.L. Street, and J.R. Koseff (1994). A non-staggered grid, fractional step method for time-dependent incompressible Navier-Stokes equations in curvilinear coordinates. *J. Comp. Phys.* **114**, 18–33.

Zeng, S., C. Vuik, and P. Wesseling (1995). Numerical solution of the incompressible Navier-Stokes equations by Krylov subspace and multigrid methods. *Advances in Comp. Math.* **4**, 27–50.

Zeng, S. and P. Wesseling (1994). Multigrid solution of the incompressible Navier-Stokes equations in general coordinates. *SIAM J. Num. Anal.* **31**, 1764–1784.

Zeng, S. and P. Wesseling (1995). An ILU smoother for the incompressible Navier-Stokes equations in general coordinates. *Int. J. Num. Meth. in Fluids* **20**, 59–74.

Zhang, Y. and Y.-K. Kwok (1997). Convergence analysis of a staggered pressure correction scheme for viscous incompressible flows. *Num. Meth. Part. Diff. Eqs.* **13**, 459–482.

Zijlema, M. (1996). On the construction of a third-order accurate monotone convection scheme with application to turbulent flow in general coordinates. *Int. J. Num. Meth. in Fluids* **22**, 619–641.

Zijlema, M., A. Segal, and P. Wesseling (1995). Finite volume computation of incompressible turbulent flows in general coordinates on staggered grids. *Int. J. Num. Meth. in Fluids* **20**, 621–640.

Zucrow, M.J. and J.D. Hoffman (1976). *Gas Dynamics. Vol. 1.* New York: Wiley.

Zucrow, M.J. and J.D. Hoffman (1977). *Gas Dynamics. Vol. 2. Multidimensional Flow.* New York: Wiley.

Index

Printing: Strauss GmbH, Mörlenbach
Binding: Schäffer, Grünstadt